ALLE ZEIT WACH
1842

W. Salomons B. L. Bayne
E. K. Duursma U. Förstner (Eds.)

# Pollution of the North Sea
## An Assessment

With 238 Figures

Springer-Verlag
Berlin Heidelberg New York
London Paris Tokyo

Dr. Wim Salomons
Institute for Soil Fertility
Delft Hydraulics Laboratory
P.O. Box 30003
9750 RA Haren (Gr), The Netherlands

Dr. Brian L. Bayne
Plymouth Marine Laboratory
Prospect Place, The Hoe
Plymouth PL 13 DH, United Kingdom

Professor Dr. Egbert Klaas Duursma
Netherland Institute for Sea Research
P.O. Box 59
1790 AB Den Burg/Texel, The Netherlands

Professor Dr. Ulrich Förstner
Arbeitsbereich Umweltschutztechnik
Technische Universität Hamburg-Harburg
Eißendorfer Str. 40
2100 Hamburg 90, Fed. Rep. of Germany

ISBN 3-540-19288-3 Springer-Verlag Berlin Heidelberg New York
ISBN 0-387-19288-3 Springer-Verlag New York Berlin Heidelberg

Library of Congress Cataloging-in-Publication Data. Pollution of the North Sea : an assessment / W. Salomons ... [et al.]. p. cm. Includes index. ISBN (invalid) 0-540-19288-3 (U.S.) 1. Marine pollution–North Sea. I. Salomons, W. (Willem), 1945-. GC1291.P65 1988 363.7'3942'09163–dc 19 88-20162



Typesetting: International Typesetters Inc., Makati, Philippines
Printing: Druckhaus Beltz, 6944 Hemsbach/Bergstraße;
binding: J. Schäffer, Grünstadt
2131/3130-543210 – Printed on acid-free paper

# Preface

This preface is being written at a time of exceptional public interest in the North Sea, following media headlines on toxic algal blooms, the mass mortality of common seals, and concern over pollution levels.

These headlines may suggest that pollution of the North Sea is a recent event. This is not the case. Although no data are available (methods simply did not exist), it is safe to assume that emission (both into air and water) of heavy metals already started to increase in the 19th century. The growth of cities and introduction of sewer systems led to the discharge of raw sewage and sewage sludge. The introduction of man-made (xenobiotic) organic chemicals and their subsequent emission into the North Sea commenced before the second world war.

The shallower and coastal areas of the North Sea receive the highest concentrations of these pollutants. Not unexpectedly, these areas – some Norwegian fjords, the Dutch coast, the German Bight – show signs of ecosystem deterioration and eutrophication. A certain percentage of the pollutants does not remain in the North Sea but is "exported" to the Atlantic. The North Sea therefore contributes to the global input of pollutants to the world's oceans. The major part of the pollutants accumulate in the North Sea and are incorporated in the bottom sediments. Although they are "out of sight", they should not be "out of mind".

Of course, opinions differ as regards the scale of disturbance to natural processes that may be due to these high levels of chemical input. Forming a balanced scientific judgment is rendered particularly awesome by the great difficulties in unraveling cause and effect relationships within the inherent complexity of the natural ecosystems. The theory of ecosystem structure is not yet advanced enough to allow detailed tracing of cause and effect, and it has been argued that in systems at this level of complexity predicting catastrophic events may be inherently impossible. It follows that to reduce the possibility of irretrievable damage to natural systems such as the

North Sea, caused by man's activities, will require a reduction in the inputs of potential pollutants, and a decline in other activities that contribute to the disturbance. In the meantime, we must accelerate the pace of scientific research, in order to identify the most sensitive areas and processes within the North Sea, coupled with careful monitoring to detect change, both as deterioration and recovery. The North Sea is one of the best-studied seas in the world, but much more additional research is required to allow the construction of comprehensive predictive management models.

This book has been written with the aim of presenting up-to-date scientific data and analysis on the status of the North Sea. The book does not attempt to offer immediate solutions to perceived problems. Rather it is written with two primary aims in mind. The first can be found in its multinational character, expressing remarkable consensus amongst the scientific community as to the vulnerability of the North Sea, and its finite capacity to assimilate waste. The second objective is even more important: the study of the North Sea is presented as a multidisciplinary problem, focusing the attention of a broad reach of scientific expertise which is committed to improving our understanding of this complex marine ecosystem.

We thank all the authors for their contributions and we hope that the result will prove stimulating to the public, to scientists in general, and to all those responsible for managing the North Sea system.

November 1988 W. Salomons, B. L. Bayne, E. K. Duursma, and U. Förstner

# Contents

## Part I The North Sea System: Physics, Chemistry, Biology

## Part II Input and Behavior of Pollutants

## Part III Impacts on Selected Areas and by Human Activities



## Part IV Biological Effects and Monitoring

# List of Contributors

You will find the addresses at the beginning of the respective contribution

# Part I
# The North Sea System: Physics, Chemistry, Biology

# The Hydrography and Hydrographic Balances of the North Sea

P.C. REID, A.H. TAYLOR, and J.A. STEPHENS[1]

## 1 Introduction

The North Sea is a shallow (30–200 m), rectangular-shaped basin with a shelving topography from south to north and a deep (up to 600 m) trough, the Norwegian Rinne, on its eastern margin. It encompasses a surface area of 575,000 $km^2$ with a volume of 40.3 $km^3$ (ICES 1983). The region is subject to oceanic influences in the north and to a minor extent via the Dover Straits; it is strongly influenced on the remaining three sides by terrestrial inputs, including a contribution from the Baltic through the Skagerrak. A knowledge of hydrography and water balances is needed to assess the impact of pollutant inputs from the major industrial nations which surround this partially enclosed sea. The physical oceanography of the North Sea has been reviewed by Hill (1973), Lee (1980) and Otto (1983); the present chapter seeks to summarise and update these descriptions, drawing attention to features of importance to pollution.

Various attempts have been made in the past to subdivide the water bodies of the North Sea according to their source or properties. For example Lee (1980) proposed a classification into six water masses (North Atlantic, Channel, Skagerrak, English Coastal, Scottish Coastal, and Continental Coastal). A subdivision into more rigid geographical boxes, which takes account of hydrographic and biological conditions, was chosen by the ICES study group established to consider the flushing times of the North Sea (ICES 1983). In the following sections, the oceanographic regime in the North Sea will be presented by means of these boxes (Fig. 1). Although this format has the disadvantage that each box contains more than one water mass, it illustrates the range of characteristics that may occur in each spatial area and allows comparisons with the fluxes and turnover times estimated by the ICES study group.

The text of this paper is divided into five sections, stratification and fronts (with a description of features seen from satellites), water masses, balances of water and heat, currents and circulation and a summary of research requirements.

[1]Natural Environment Research Council, Plymouth Marine Laboratory, Prospect Place, West Hoe, Plymouth PL1 3DH, Great Britain

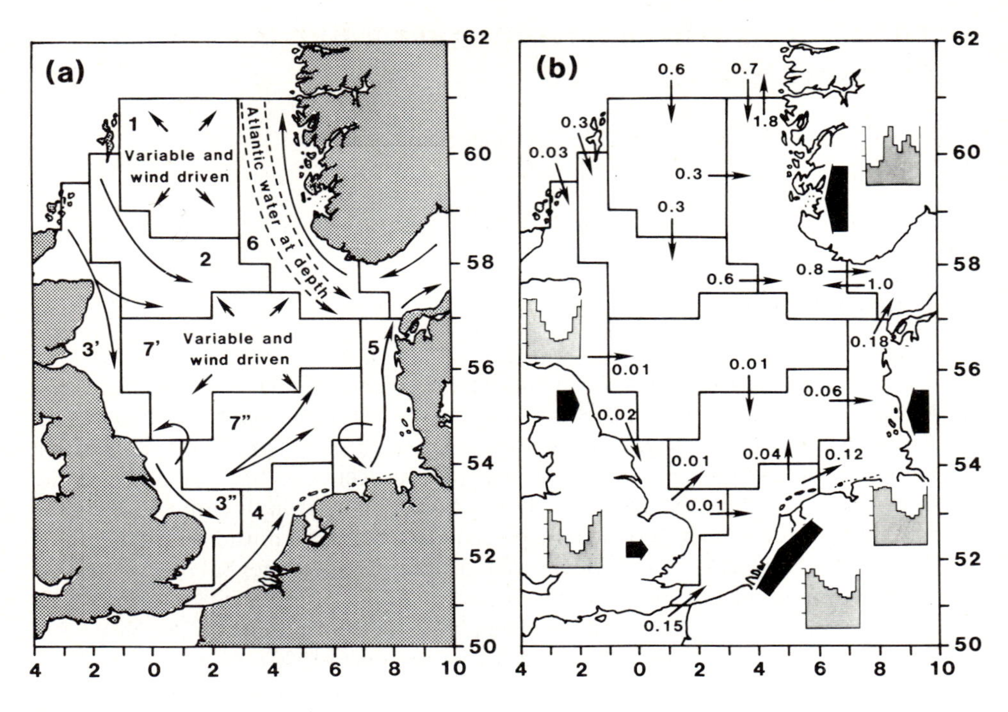


**Fig. 1. a** Subdivisions of the North Sea used by the ICES study group (ICES 1983). They are referred to in the text by the numbers *1, 2, 3′, 3″, 4, 5, 6, 7′* and *7″* shown. A schematic representation of the surface currents is superimposed. (Lee and Ramster 1981). **b** Net fluxes between the boxes (in $10^6$ $m^3$ $s^{-1}$) from ICES (1983). *Large arrows* indicate the sizes of the freshwater flows from the coasts. *Largest arrow* is 2889 $m^3$ $s^{-1}$. Seasonal cycles are shown for the runoff into each box; the divisions on each histogram represent 25% of the largest monthly runoff for that box. For areas 3″, 4 and 5 flows were calculated by the method of Taylor et al. (1981, 1983) using data from 1948 to 1981. Flows for area *6* were determined for the period 1948 to 1974 using data and hydrological maps provided by Dr J. Otnes, Norges Vassdrags - og Elektrisitetsvesen. Runoff in area *3′* was estimated from runoff minus actual evaporation by Mr M. Lees, Institute of Hydrology, using data from a variety of years

## 2 Stratification and Fronts

There are strong seasonal differences in the vertical structure of the water column in the North Sea. Most areas are vertically mixed throughout the winter. From the spring to autumn, however, the sea is divided into areas which remain mixed and others which become stratified (Fig. 2). Areas 3″ and 4 remain thermally mixed throughout the year with stratification developing to a minor extent in areas 3′ and 5. All other areas develop a strong seasonal thermocline by June which deepens throughout the summer, breaking down by November. The boundaries between these different regimes may be characterized by large gradients in temperature and/or salinity. Where such features are sharply defined at the surface they are termed "fronts". Such fronts may exhibit sharp contrasts in

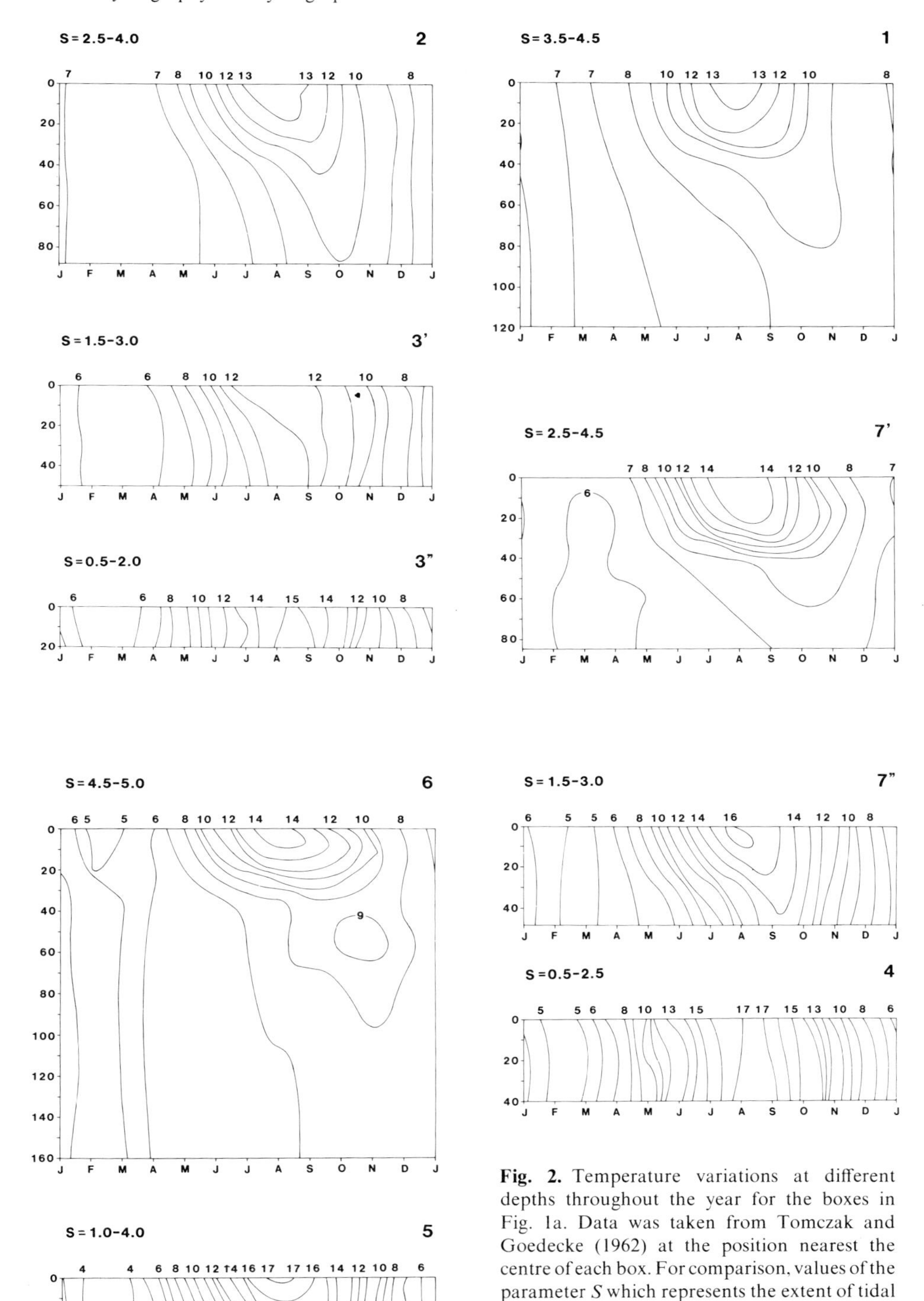


**Fig. 2.** Temperature variations at different depths throughout the year for the boxes in Fig. 1a. Data was taken from Tomczak and Goedecke (1962) at the position nearest the centre of each box. For comparison, values of the parameter $S$ which represents the extent of tidal mixing are also shown. Spatial variations within each box may be $\pm 1$–$1.5$°C

surface agitation, with convergence and alignment of debris, and are often accompanied by eddies. On cloud-free infrared satellite images, the temperature differences across fronts can be clearly distinguished.

Tidal stirring by the interaction of the sea bed with the tidal stream is the dominant process governing the downward mixing of solar heat input and the development of stratification. Simpson and Hunter (1974) and Pingree and Griffiths (1978) have quantified this process by the parameter

$$S = -\log_{10} \varepsilon = \log_{10} (H/U^3), \quad (1)$$

where H is the depth of the water and U is the maximum speed of the tidal stream at the surface. $\varepsilon$ ($m^2 s^{-3}$) is proportional to the tidal energy dissipation rate per unit mass. In general, areas where S is greater than 2 tend to be layered. The geographical spread of S in each of the boxes (Fig. 2) shows that the boxes without stratification have ranges that are almost entirely below 2. Soulsby (1983) has shown that the areas for which S is less than 2 are those in which the tidal bottom boundary layer stretches up to the surface. A critical S value of 1.5 determines the mean expected position of summer thermal fronts (Pingree and Griffiths 1978).

An indication of the time taken for any tracer to mix throughout the water column is provided by $H/U_m$, where $U_m$ is the depth mean velocity of the tidal stream averaged over half a tidal cycle. Talbot (1976) has calculated values of this ratio. For most of the northern regions, 1, 2, 6 and 7′ this mixing time is more than two tidal cycles. Close to the coasts and over much of the English Coast area 3″ the time is less than one fifth of a tidal cycle. Over the rest of the coastal and central areas 3′, 4, 5 and 7″ mixing times lie in the range between 0.2 and 2 tidal cycles. Figure 2 shows that the strongest stratification occurs in those boxes with the longest mixing times.

Over parts of the shelf other agents besides tides, such as winds, waves, eddy generation and river runoff may be important in governing the development of stratification. In particular, variations in salinity attributable to runoff may have a pronounced effect on stability in certain areas of the North Sea. A change of 0.2–0.3‰ salinity has approximately the same effect on buoyancy as a 1°C change in surface to bottom temperature. The impact of freshwater runoff on buoyancy is illustrated by the ranges of salinity at the surface and bottom of each representative box during September (Fig. 3). All of the boxes except 3″, 7′ and 7″ off the English Coast and in the Central North Sea show some degree of haline stratification, this being especially true in the areas 4 and 5 receiving runoff from Europe and area 6 into which Norwegian rivers and the Baltic outflow discharge. The summer thermocline tends to be deeper in the boxes 7′ and 7″ where reduced salinities do not contribute significantly to the vertical stability (Fig. 2). Haline stratification is present in boxes 4, 5 and 6 throughout the year, the salinity ranges in these boxes during March at the surface (bottom) being: 30–34.7 (33–34.8), 30–34.2 (31–34.6) and 32–35 (34.8–35.2), respectively.

Haline stratification also occurs throughout the year in the Skagerrak (Svansson 1975). In the central Skagerrak, possibly due to geostrophic effects, isotherms and isohalines dome producing a core of cold water within the centre of an anticyclonic circulation (Pingree et al. 1982).

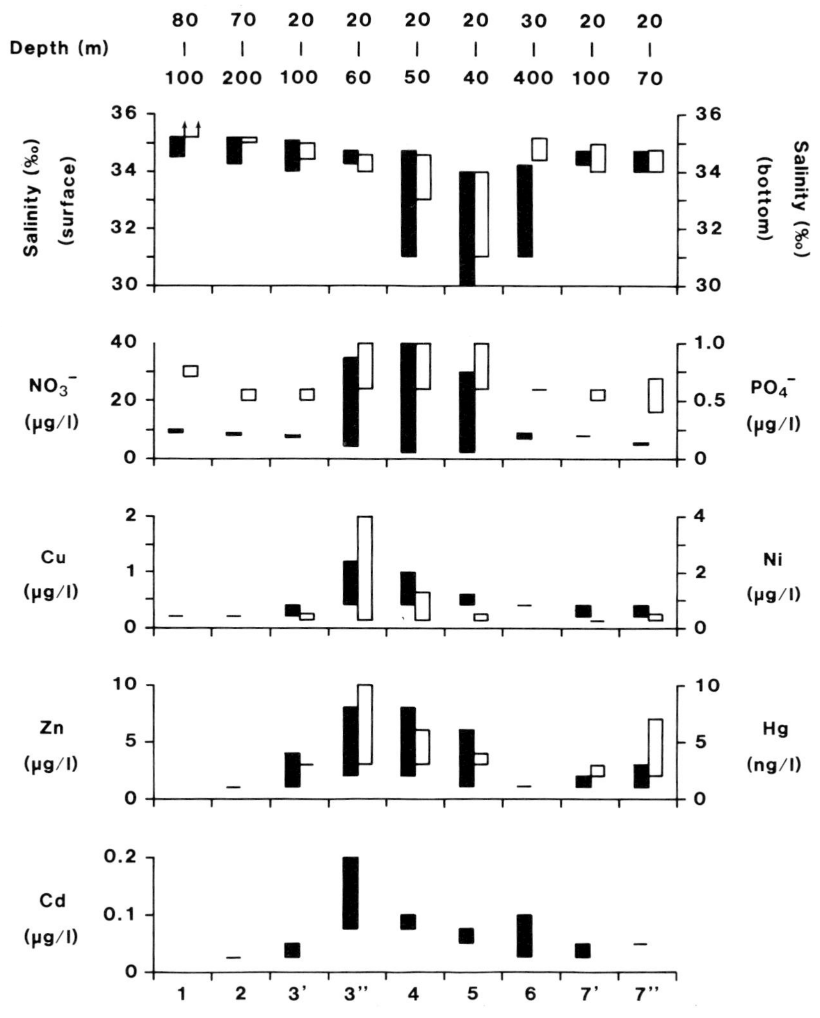


**Fig. 3.** Geographical ranges within the boxes of Fig. 1a for surface and bottom salinities in September (ICES 1962; Lee 1980), and dissolved concentrations of inorganic nitrate, inorganic phosphate, copper, nickel, zinc, cadmium and reactive mercury (all from Lee and Ramster 1981). In each case the *solid line* refers to the left axis, the *open line* to the right. The depth ranges of the boxes are included

A useful summary of the distribution of stratification, both haline and thermal, is provided by the potential energy per unit volume, $\bar{v}$ (Simpson et al. 1977).

$$\bar{v} = 1/H \int_{-H}^{0} (\rho - \bar{\rho})\, gz dz; \quad \bar{\rho} = (1/H) \int_{-H}^{0} \rho dz, \tag{2}$$

where z is the vertical coordinate (positive upwards), H is the depth, $\rho$ is the density with mean $\bar{\rho}$, and g is the acceleration due to gravity. The potential energy per unit volume is the work which would be done in redistributing the mass to bring about complete vertical mixing.

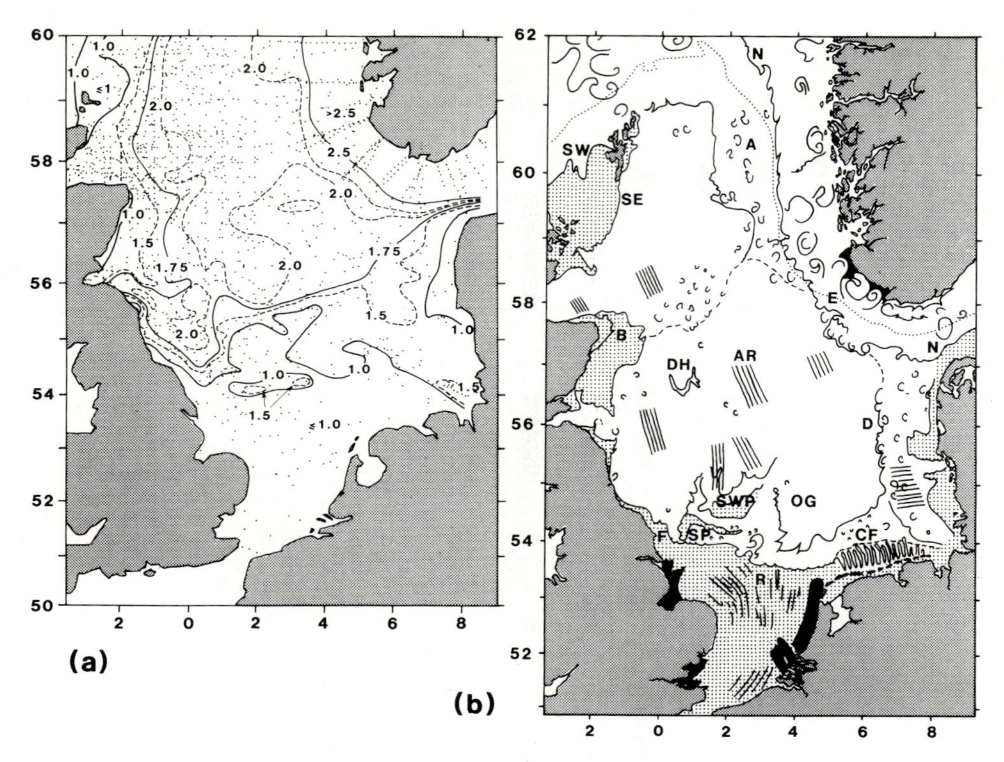


**Fig. 4. a** Contours of the mean potential energy per unit volume ($\log_{10} \bar{v}$) for July, calculated according to Eq. (2) (D.G. Hughes, unpublished data). *Dots* indicate station profiles. **b** A schematic representation of hydrographic features seen on infra-red satellite images. Examples of river plumes from the Humber, Thames and Rhine (two cases) are shown in *black*. Cold coastal water associated with eddies on the south Norwegian coast, which may be upwelling is also shown in *black*. Approximate sizes and shapes of typical eddies are indicated by the *swirls* in the diagram. The *dashed line* marks the approximate position of the 100-m contour in the central North Sea and the *dotted line* the 200-m contour. *A* mixed Atlantic and shelf water; *AR* aligned "ridges"; *B* Buchan fronts; *CF* Coastal "Flames"; *D* Danish Front; *DH* Devil's Hole; *E* Eigersundsbank; *F* Flamborough Front; *N* the western boundary of Norwegian coastal fronts; *OG* Oyster Ground; *R* striations marking cooler water; *SE* East Shetland Front; *SW* West Shetland Front; *SWP* South West Patch on the Dogger Bank; *SP* Silver Pit

Figure 4(a) shows contours of $\bar{v}$ for the whole North Sea in a representative summer month, employing data from several decades (Hughes 1976; Hughes pers. commun.). There is general agreement between areas of stratification predicted by S [Eq. (1)] and those shown by $\bar{v}$, although there are discrepancies along the southern and eastern margins of the stratified zone. A value of 2 for S in the tidal mixing model corresponds approximately to $\log_{10} \bar{v} = 1.0$, and so fronts are expected where $\bar{v}$ is less than this value.

The boundaries, as observed in satellite imagery on any occasion, show more structure than is revealed by the averaged distribution of $\bar{v}$ (or that of S). Care has to be taken in interpreting infrared (IR) imagery as low-lying fog may enhance or blur specific features and daytime warming of the surface may obliterate or confuse other details. In addition, the intensity and position of boundary features as seen from satellites varies to a considerable degree with season, state of the tide when the photograph was taken, and direction and strength of the wind. Imagery has now been produced by Tiros, Nimbus and other satellites for more than 10 years. While cloud-free conditions only occurred infrequently during this period, there are now sufficient scans available to produce a composite of frontal and other hydrographic features. Typical positions of fronts and some other features seen on IR images of the North Sea are given in Fig. 4(b).

Figure 4(b) shows that a continuous sequence of fronts extends from the Shetlands to the mainland of Scotland (SW, SE) with little evidence for a marked intrusion of Atlantic water at this site during the summer months. An area of mixed water with a frontal boundary (the Buchan Front) to the east extends from Dundee to Fraserburgh off the north-east of Scotland (B, two extreme positions are shown). A variable sequence of coastal fronts and upwelling (Lee 1980) develops south of Edinburgh to Flamborough Head. From here a clear boundary (the Flamborough Front, F) with the mixed waters of the southern North Sea passes eastwards to the Elbe and marks the northern boundary of boxes 3″ and 4. To the north of this "Flamborough Front", a complicated pattern of fronts and upwelling marks the position of the Dogger Bank and other shallow features in this area.

In the eastern North Sea a pronounced front (the Danish Front, D) between coastal lower salinity water and central North Sea water extends parallel to the continental coast approximately following the 40 m depth contour and the eastern edge of the Oyster Ground (OG). This front forms the western boundary of box 5. Nearshore fronts also occur off the Danish coast. In the early spring and late autumn, incipient fronts may be seen in the central North Sea running almost east to west.

In the mixed southern North Sea shallow banks and ridges may show as bands of cooler water (e.g. R). When the Rhine and other major rivers are in flood this may be reflected by plumes of warmer or cooler water, depending on the time of year. In nearshore waters off the coast of the Netherlands and Germany flame-like extensions of alternating cold and warm water may extend from the coast.

Off the Norwegian coast fronts delimit the outflow from the Skagerrak and coastal runoff from Norway; these approximate to the western boundary of box 6. The positions of these fronts is highly variable depending in particular on the seasonal meltwater from the Norwegian coast. In winter months when runoff over Scandinavia is frozen as snow and ice, the current along the Norwegian coast is narrow and confined to the Rinne. Melting of the snow cap over Norway in May and June is clearly evident and leads to a rapid eastward extension and shallowing of the low salinity water in the Rinne up to 150 nautical miles into the central North Sea (Carstens et al. 1984). At the northern entrance to the North Sea, adjacent to the "Norwegian Current", a southward directed tongue of mixed Atlantic and shelf water is evident for much of the year (box 1).

## 3 Water Masses

The different surface water bodies delineated by fronts exhibit other properties which help to distinguish them in satellite images. There is, for example, a marked difference in the size and frequency of eddies generated within each body of water (Fig. 4b). This variability may be due to varying water depth and density. Eddies were infrequently observed in the shallow, mixed, southern North Sea. A progressive decrease in the diameter of eddies is evident from oceanic Atlantic water, the Norwegian Rinne and other coastal zones to the central North Sea. For example, eddies range from 36–80 km in diameter in oceanic water to the west of Shetland, 24–36 km in the Norwegian Rinne, 12–20 km over the shelf adjacent to the Rinne, 12–16 km in the Atlantic inflow and 8–18 km in coastal frontal zones. Few eddies larger than 10 km in diameter are evident in the central North Sea south of the 100 m depth contour (the northern boundary of box 7′) except in the winter (Pingree 1978). This central region is characterised instead by alternating northwest to southeast (other orientations may occur) "ridges" (e.g. AR) of colder and warmer water 3.5 km apart. Patches of colder water are sometimes observed over the deep trenches of the Devil's Hole (DH).

Assignment of the waters in any of the boxes (Fig. 1a) to a mixture of distinct water masses from different sources can only be approximate because of the variations that occur across each box. In addition there may be significant changes seasonally or between years. Lee (1980, Table 14.3) based his classification into six water masses on temperature, salinity, and the concentrations of free phosphate, nitrate-nitrogen, soluble silicate, copper, cadmium, nickel and mercury. The levels of these observed in the boxes (Figs. 2 and 3) suggest that North Atlantic water occurs in the northern and central areas 1, 2, 3′, 6, 7′, and 7″. In the central boxes it may be mixed with Scottish coastal water, in 7″ with Channel water and in area 6 with Skagerrak water. Area 3′ is predominantly Scottish coastal water, and 3″ is predominantly English coastal water. Areas 4 and 5 contain mainly Continental coastal water with some Channel water in 4 or some Skagerrak water in 5.

The mixing of the water from Lee's separate sources, which the above scheme seeks to encompass, has been modelled (Delft Hydraulics Laboratory 1985; Klomp et al. 1985; van Pagee et al. 1985). Using the residual circulation in the North Sea from a two-dimensional model, steady-state concentrations arising from marine, terrestrial and atmospheric inputs were calculated. In most cases the predicted distributions are similar to those observed, but there are differences in levels (Fig. 5). Muller-Navarra and Mittlestaedt (1985) have also modelled the spreading of contaminants in the North Sea, but using a different approach.

The quantities of different waters reaching any area, and the time taken to do so, can also be estimated from modelling calculations (Fig. 6, redrawn from Delft Hydraulics Laboratory 1985). While these results are of limited reliability being based on a two-dimensional numerical model, they illustrate the main spatial variability. North Atlantic and Channel water reaching the central North Sea is relatively old as is all water in the German Bight. Water from the Rhine and Thames is seen to disperse in a northeasterly direction.

The North Sea is sufficiently shallow that the water masses can be considered to extend from the surface (or below the surface stratification) to the sea bed.

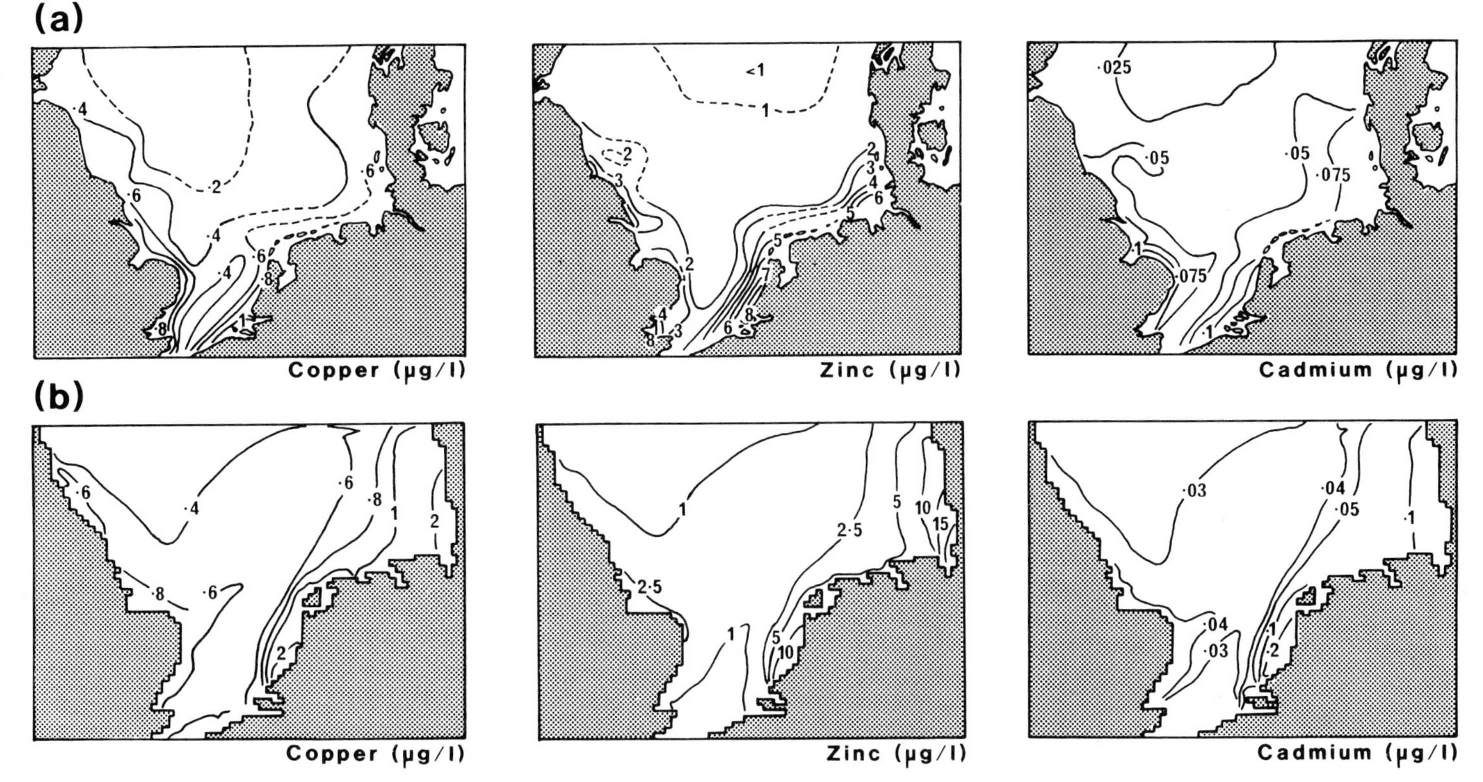


**Fig. 5a,b.** Distribution of copper, zinc and cadmium in the North Sea; **a** using data from surveys during 1971–8 (Lee and Ramster 1981), **b** simulated for winter 1980 conditions. (Delft Hydraulics Laboratory 1985)

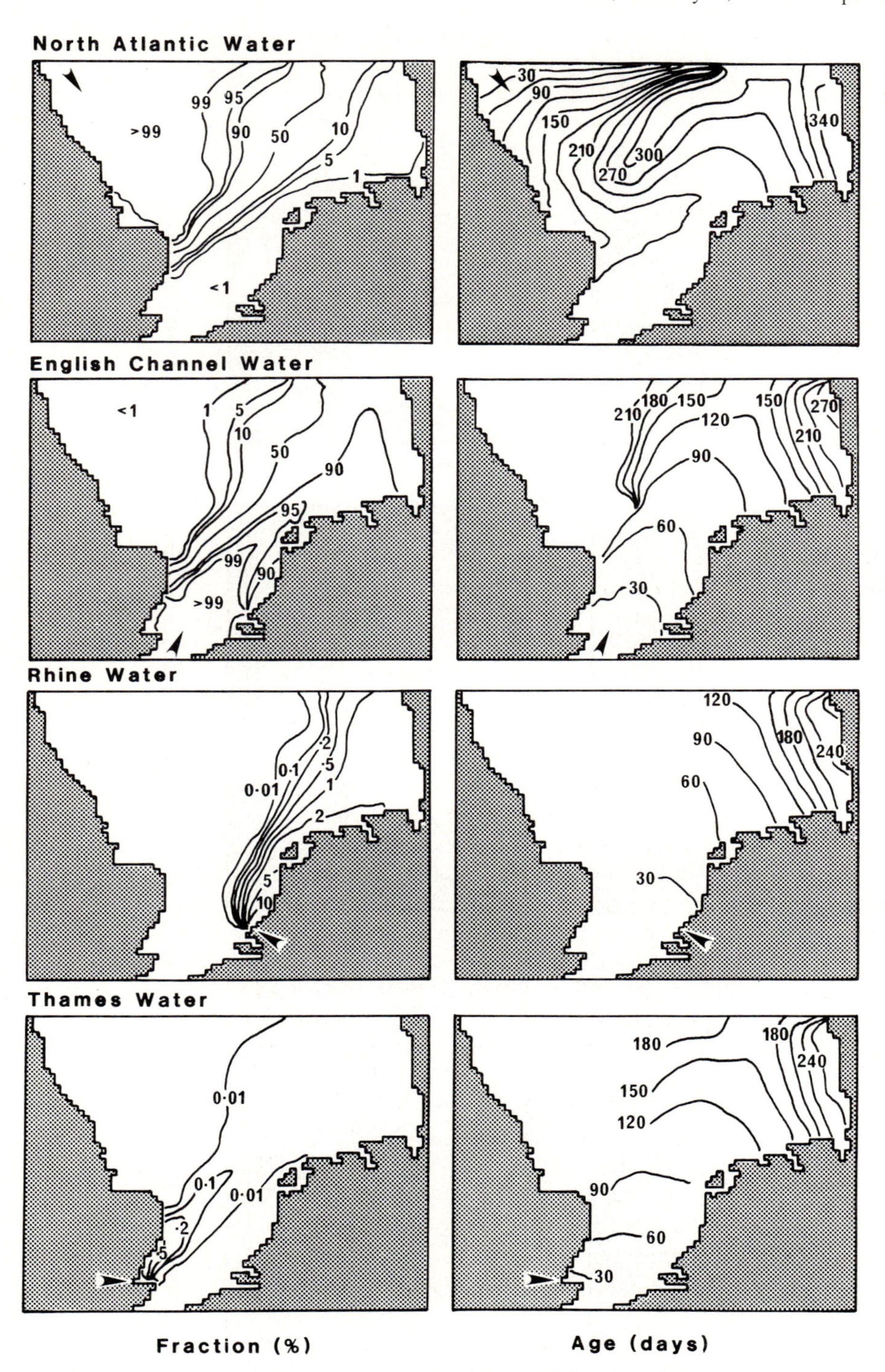


**Fig. 6.** Percentage and age of water from the North Atlantic, English Channel, River Rhine and River Thames reaching different points in the North Sea. *Arrows* indicate the sources

However, in the Skagerrak and Norwegian Rinne (box 6), where depths are much greater, the water column is occupied by two or more water masses. Along the western edge of the Norwegian Rinne, Atlantic water occurs as a subsurface flow (Fig. 1a) which corresponds approximately with the 200 m depth contour (Saetre 1978). In winter, warmer water from this inflow may extend to the surface, but by summer it is covered with warmer, low salinity water mainly of coastal origin. In winter the deep low salinity Norwegian coastal water blocks the shallow entrances to fiords causing stagnation by preventing renewal with Atlantic water (Carstens et al. 1984). During exceptionally cold winters dense water from the central North Sea cascades into the deep parts of the Norwegian Rinne and Skagerrak to form the Skagerrak Deep Water which has a temperature of about 4°C and a salinity below 35‰ (Ljoen and Svansson 1972). A substantial review of the Norwegian Current is given in Saetre and Mork (1981).

## 4 Balances of Water and Heat

The mean transport of water through the region is summarised in Fig. 1b, which also shows the seasonal patterns of river runoff. Freshwater inputs into the coastal boxes have been estimated by Otto (ICES 1983 p. 83) using an assemblage of published data, mainly prior to 1960. The values in Fig. 1b have been calculated by the method of Taylor et al. (1981, 1983) for the period 1948 to 1981. They are 3′: 1057 $m^3s^{-1}$, 3″: 541 $m^3s^{-1}$, 4: 2889 $m^3s^{-1}$, 5: 1473 $m^3s^{-1}$, and 6: 2215 $m^3s^{-1}$. The sum of all boxes, with the exception of box 6 (Otto's box 6 extends to 62°N), is 5360 $m^3s^{-1}$, which compares with Otto's estimate of 6088 $m^3s^{-1}$. Rivers in Britain peak in the winter with a minimum in the summer, whereas for those flowing into the German Bight the times of the maximum and minimum are spring and autumn respectively. The Rhine and adjacent rivers show an intermediate pattern. Runoff from the Norwegian coast exhibits two maxima in June and October with the lowest runoff in late winter, a pattern which results from the seasonal changes of storage in snow and ice. However, the pattern of changes in this area may have been distorted by the regulation of rivers for the production of hydro-electric power.

Becker (ICES 1983 p. 86) has derived mean distributions of precipitation-evaporation for individual quarters of the year by combining precipitation, estimated from occurrences observed on board merchant ships, with latent heat fluxes, calculated using bulk formulae (Becker 1979). Averaged throughout the year, the excess precipitation over evaporation supplied freshwater to the North Sea at a rate of 2700 $m^3s^{-1}$. This precipitation excess increases in a northward direction from 70 mm in the southern part of box 4 to 200–300 mm in the northern boxes 1, 2, and 6. In the western and southwestern North Sea evaporation dominates during the winter season, and Becker suggests that this may strengthen convective mixing. Throughout the summer, precipitation is greater than evaporation everywhere, which will encourage the formation of vertical temperature gradients.

Fluctuations in water balance will lead to changes in salinity; thus Schott (1966) has observed that salinity values in the North Sea tend to be correlated with river discharge. Dickson (1971) has attributed semi-regular periods of high salinities on the European shelf to a different aspect of the balance, namely the flow of Atlantic water on to the shelf. He suggested that occasional increases of this flow may be induced if an anomalously southerly wind occurs to the west of the British Isles, when the advected water will also have a higher salinity because of its more southerly origin. By constructing time-series of river flows, of evaporation and precipitation, and of transport through the Dover Strait (Prandle 1978), the interannual salinity budget of the eastern English Channel and the southern North Sea can be modelled (Taylor et al. 1981, 1983), whereupon the role of individual processes can be distinguished. The calculations (Fig. 7b) have shown that, for the period 1948 to 1974, more than 23% and as much as 56% of the variance of annual mean salinity in the southern North Sea may be the result of fluctuations in river flows, there being little evidence that advection from the Atlantic was important during this interval. Pingree (1980, p. 453) reached a similar conclusion about the high salinities in the English Channel during 1922.

Estimates of the heat content and the heat flux across the air-sea interface indicate (Becker 1981) that the North Sea is a source of heat energy for the atmosphere; the heat deficit being compensated for by deep water advection from the English Channel, North Atlantic and Norwegian Deep. Becker (1981) has discussed most aspects of the heat, mass and salt balances in the region. The seasonal heat budget of the English Channel and Southern Bight, considered as a single region, has been modelled by Maddock and Pingree (1982) by extracting an annual sine wave from each constituent of the heat budget. Seasonal and interannual changes in the heat contents of the two areas (Fig. 7a) have also been modelled separately (Taylor and Stephens 1983; Taylor 1983). These calculations show that the seasonal cycle of temperature is predominantly caused by changing surface heat fluxes which agrees with the conclusions of Maddock and Pingree. Figure 7a also indicates the influence of latent heat losses. The model demonstrates that, because of advection through the Dover Strait, the annual cycle of heating and cooling in the English Channel has a significant effect on the Southern Bight; in its absence the mean temperature range in the Southern Bight, 6°C to 17°C, is reduced by a third to 7.5°C to 14.5°C. Year-to-year changes in temperature (Fig. 7c) are also mainly the result of fluctuating surface heat flows, especially at the lowest frequencies; variations in the heat flux from the Atlantic are only of minor importance. Hill and Dickson (1978) have described long-term temperature changes, over the whole of the North Sea in terms of linear regressions of 5-year running means.

## 5 Currents and Circulation

Tidal motions in the North Sea have been reviewed by Huntley (1980). The $M_2$ tide rotates about an amphidromic point just off the Norwegian coast at about 58.5°N, a second point in the Southern Bight between boxes 3″ and 4 at about 52.5°N and a third point on the boundary between 7′ and 7″ at about 5°W. Maximum tidal streams have amplitudes of 0.1 m $s^{-1}$ in area 6, 0.25 to 0.5 m $s^{-1}$ in the northern boxes 1 and 2 and in the central regions 7′ and 7″, 0.4 to 1.0 m $s^{-1}$ in area 3′ off the Scottish

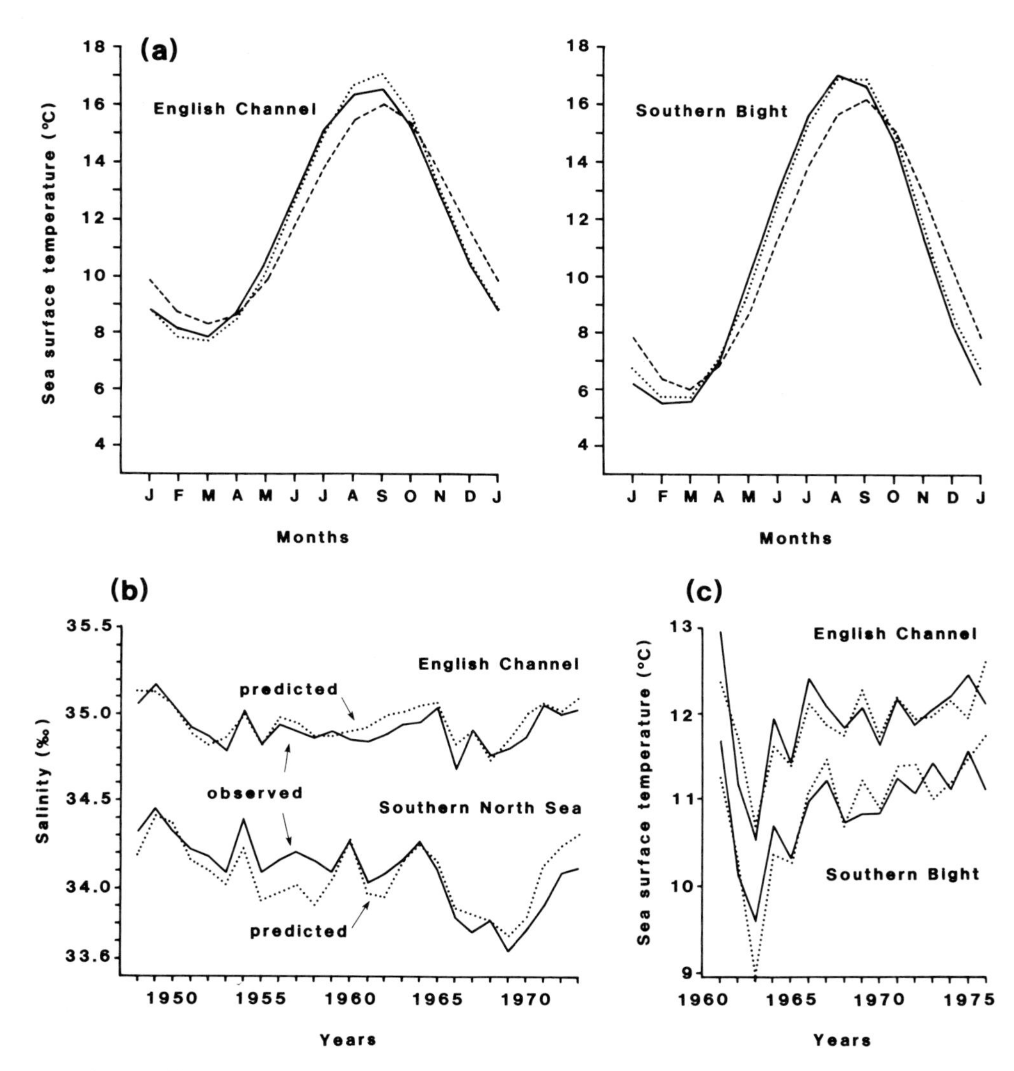


**Fig. 7.** **a** Observed seasonal cycle of temperature (solid) in the eastern English Channel and the North Sea south of 53°N (Southern Bight) compared with the prediction of a heat budget model (*dotted*), and of the model with constant latent heat fluxes from the two regions (*dashed*). **b** Predicted and observed time-series (1948–1973) of salinity for the North Sea south of 55°N and the eastern English Channel. **c** Observed time-series (1961–1976) of annual mean temperatures (*solid*) compared with the prediction of the heat budget model (*dashed*)

coast, and 0.75 to 1.5 m $s^{-1}$ in the southern boxes 3″, 4 and 5. The strongest currents are close to the coasts. These tidal currents can be reproduced by three-dimensional numerical models (Davies and Furnes 1980).

Figure 1 illustrates the residual circulation under typical wind conditions. The dynamics of residual currents under different wind forcing have been described by

Pingree and Griffiths (1980) by means of a two-dimensional model, and also by Nihoul and Runfola (1981) using a three-dimensional model. Davies (1980) has found that the depth mean residual currents from a wind-driven 2-D model were nearly identical with those from a 3-D model when there was no stratification even though the wind-induced currents varied considerably in magnitude and direction between surface and bottom. Although 2-D models can be used for storm surge predictions (Flather 1976), pollution studies will require 3-D models.

Nihoul (1980) has shown that non-linear interactions of tides and storm surges result in positive viscosity and negative viscosity eddy-like structures which may foster the mean flow in some regions and extract energy from it in other regions. Seasonal circulation patterns from a 3-D model have been described by Backhaus and Maier-Reimer (1983). The residual circulation of the North Sea is largely determined by the prevailing wind. Thus, the model results of Backhaus et al. (1985) show that the transport patterns in two individual winters may differ from the climatological mean winter value by as much as 100%. In the southern North Sea such circulation changes may lead to pronounced changes in turbidity (Dickson and Reid 1983). Off south Norway, intense transient currents may be generated by relaxation after west winds have piled up water in the Skagerrak (Carstens et al. 1984). Marked variations in the residual currents of the southern North Sea have been described by Van de Kamp (1983) during the period 1951 to 1977 from ebb and flood displacements collected by Dutch lightvessels.

An indication of the uncertainties in the residual circulation of the North Sea is provided by comparing the flushing times for the boxes (ICES 1983), estimated from observations and by means of models. These times (in years) are 1: 0.4 (0.6), 2: 0.3 (0.3), 3′ and 3″ combined: 3.9 (0.3), 4: 0.2 (0.1), 5: 0.2 (0.1), 6: 0.1 (0.4), 7′: 2.1 (0.3) and 7″: 1.5 (0.2), in each case the first number giving the value from observations and the value in brackets that of the model. Most boxes appear to have turnover times of well under 1 year. There are significant discrepancies between the two estimates along the British and Norwegian coasts and in the central North Sea. The higher values established on the basis of observations are the result of small net fluxes and it may be that variability at shorter periods (synoptic-meteorological time-scales) is dominating the exchange with surrounding boxes (Otto 1983).

Horizontal dispersion due to mixing processes with short time and space scales will vary across the region. In his model of the spreading of $^{137}$Cs around the shelf seas, Prandle (1984) expressed the dispersion coefficient as a function of the $M_2$ current amplitude vector. This function gives 1000 $m^2 s^{-1}$ in the English Channel; 50 $m^2 s^{-1}$ in the central and northern North Sea; and 300 $m^2 s^{-1}$ in the southern North Sea, German Bight and along the English east coast. These values indicate the geographical variation of the dispersion but may not be numerically reliable, for, although Pingree et al. (1975) obtained a similar value in the English Channel, Schott (1966) has estimated a lateral mixing coefficient of 100 $m^2 s^{-1}$ in the Southern Bight, and the RHENO experiment (ICES 1973) has given values of about 10 $m^2 s^{-1}$ in the northern North Sea.

## 6 Research Needs

Satellite imagery shows a number of features that should be investigated on a seasonal basis, for example, frontal boundaries and eddies. These, and results from the Coastal Zone Colour Scanner (CZCS), will require comparisons with sea truth data. Many unanalysed satellite data exist in archives.

The eddy field is of prime importance in the spreading of contaminants, and so the dynamics and statistics of these eddy motions need to be known before dispersion of materials can be reliably predicted.

Diffusion is frequently assumed to occur as an expanding cloud without significant structure; in reality the pattern may resemble spaghetti. Processes involved in dispersion need to be examined.

Transport of contaminants at larger scales will be dependent on the circulation. Three-dimensional numerical models incorporating an evolving density field will be needed to accurately simulate this transport. In addition there will be a need to determine fluxes of water at the open boundaries of the North Sea. Radar techniques offer the possibility of continuous real-time measurements of both surface waves and surface currents over large areas.

Because of its great depth the Norwegian Rinne could be a sink for contaminants entering the North Sea. The mechanisms by which such transfer may take place, e.g. winter cascades, are poorly understood.

The fluctuating balances of water, salt and heat provide a framework against which chemical changes may be quantitatively assessed. The determination of these balances are limited by the oceanographic and meteorological time series available. Vertical profiles of salinity are an important deficiency especially near the northern boundary. Three dimensional circulation models will allow the calculation of budgets with improved horizontal and vertical resolution.

*Acknowledgements.* We wish to thank Mr D.G. Hughes, Energy Conservation and Solar Centre, London for permission to publish Fig. 4a and Dr J. Otnes, Mr M. Lees, and Mr T. Marsh for providing river flow data. Satellite imagery was obtained from Mr. P. Baylis, University of Dundee and by access to a photographic archive courtesy of Dr. R. Pingree.

## References

Backhaus JO, Maier-Reimer E (1983) On seasonal circulation patterns in the North Sea. In: Sundermann J, Lenz W (eds) North Sea dynamics. Springer, Berlin Heidelberg New York, pp 63–84

Backhaus JO, Hainbucher D, Quadfesel D, Bartsch J (1985) North Sea circulation anomalies in response to varying atmospheric forcing. ICES CM 1985/C:29

Becker GA (1979) Mean net surface heat exchange of the North Sea. ICES CM 1979/C:36

Becker GA (1981) Contributions to the hydrography and heat budget of the North Sea. Dtsch Hydrogr Z 34:167–262

Carstens T, McClimans TA, Nilsen JH (1984) Satellite imagery of boundary currents. In: Nihoul JCJ (ed) Remote sensing of shelf sea hydrodynamics. Elsevier, Amsterdam Oxford New York Tokyo pp 235–256

Davies AM (1980) Application of numerical models to the computation of the wind-induced circulation of the North Sea during JONSDAP '76. Meteor Forsch Ergel A22:53–68

Davies AM, Furnes GK (1980) Observed and computed $M_2$ tidal currents in the North Sea. J Phys Oceanogr 10:237–257

Delft Hydraulics Laboratory (1985) Harmonisation of North Sea policies: North Sea water quality plan. Background document 4: framework for analysis. Rijkswaterstaat, Waterloopkundig Laboratorium, June 1985, The Netherlands

Dickson RR (1971) A recurrent and persistent pressure anomaly pattern as the principle cause of intermediate-scale hydrographic variation in the European shelf seas. Dtsch Hydrogr Z 24:97–119

Dickson RR, Reid PC (1983) Local effects of wind speed and direction on the phytoplankton of the Southern Bight. J Plank Res 5:441–455

Flather RA (1976) A tidal model of the North West European Continental Shelf. Mem Soc R Sci Liege 10:141–164

Hill HW (1973) Currents and water masses. In: Goldberg ED (ed) North Sea Science. MIT, Cambridge, pp 17–42

Hill HW, Dickson RR (1978) Long-term changes in North Sea hydrography. Rapp Cons Int Explor Mer 172:310–334

Hughes DG (1976) A simple method for predicting the occurrence of seasonal stratification and fronts in the North Sea and around the British Isles. ICES CM 1976/C:1

Huntley DA (1980) Tides on the North West European shelf. In: Banner FT, Collins MB, Massie KS (eds) The North West European shelf seas: the sea and the sea in motion. II Physical and chemical oceanography, and physical resources. Elsevier, Amsterdam pp 301–351

ICES (1962) Mean monthly temperature and salinity of the surface layer of the North Sea and adjacent waters from 1905 to 1954. Cons Int Explor Mer, Service Hydrographique, Charlottenlund Slot, Denmark

ICES (1973) The ICES Diffusion Experiment RHENO 1965. Weidermann H (ed) Rapp Cons Int Explor Mer 163, 111 pp

ICES (1983) Flushing times of the North Sea. Co-operative research report No 123. ICES, Charlottenlund, Denmark

Kamp G Van de (1983) Long-term variations in residual currents in the southern North Sea. ICES CM 1983/C:4

Klomp R, Pagee JA van, Glas PCG (1985) An integrated approach to analyse the North Sea ecosystem behaviour in relation to waste disposal. Waterloopkundig Laboratorium separate No. 85/05, The Netherlands

Lee AJ (1980) North Sea: physical oceanography. In: Banner FT, Collins MB, Massie KS (eds) The North West European shelf seas: the sea bed and the sea in motion. II Physical and chemical oceanography, and physical resources. Elsevier, Amsterdam, pp 467–493

Lee AJ, Ramster JW (eds) (1981) Atlas of the seas around the British Isles. MAFF, Lowestoft, UK

Ljoen R, Svansson A (1972) Long-term variations of sub-surface temperatures in the Skagerrak. Deep Sea Res 19:277–288

Maddock L, Pingree RD (1982) Mean heat and salt budgets for the eastern English Channel and the Southern Bight of the North Sea. J Mar Biol Assoc UK 62:559–575

Muller-Navarra S, Mittlestaedt E (1985) Schadstoffausbreitung und Schadstoffbelastung in der Nordsee. Dtsch Hydrogr Inst, Hamburg

Nihoul JCJ (1980) Residual circulation, long waves and mesoscale eddies in the North Sea. Oceanol Acta 3:309–316

Nihoul JCJ, Runfola Y (1981) The residual circulation in the North Sea. In: Nihoul JCJ (ed) Ecohydrodynamics. Elsevier Oceanogr Ser, vol 32, Elsevier, Amsterdam, pp 219–272

Otto L (1983) Currents and water balance in the North Sea. In: Sundermann J, Lenz W (eds) North Sea Dynamics, Springer, Berlin Heidelberg New York, pp 26–43

Pagee JA van, Gerritsen H, Ruijter WPM de (1985) Transport and water quality modelling in the southern North Sea in relation to coastal pollution research and control. Waterloopkundig Lab, November, 1985, The Netherlands

Pingree RD (1978) Cyclonic eddies and cross-frontal mixing. J Mar Biol Ass UK 58:955–963

Pingree RD (1980) Physical oceanography of the Celtic Sea and English Channel. In: Banner FT, Collins MB, Massie KS (eds) The North West European shelf seas: the sea and the sea in motion. II Physical and chemical oceanography, and physical resources. Elsevier, Amsterdam pp 415–462

Pingree RD, Griffiths DK (1978) Tidal fronts on the shelf seas around the British Isles. J Geophys Res 83:4615–4622

Pingree RD, Griffiths DK (1980) Currents driven by a steady uniform wind stress on the shelf areas around the British Isles. Oceanol Acta 3:227–236

Pingree RD, Pennycuick L, Battin GAW (1975) A time-varying temperature model of mixing in the English Channel. J Mar Biol Assoc UK 55:975–992

Pingree RD, Holligan PM, Mardell GT, Harris RP (1982) Vertical distribution of plankton in the Skagerrak in relation to doming of the seasonal thermocline. Cont Shelf Res 1:209–219

Prandle D (1978) Monthly-mean residual flows through the Dover Strait, 1949–1972. J Mar Biol Assoc UK 58:965–973

Prandle D (1984) A modelling study of the mixing of $^{137}Cs$ in the seas of the European Continental Shelf. Phil Trans Roy Soc Lond A310:407–436

Saetre R (1978) The Atlantic inflow to the North Sea and the Skagerrak indicated by surface observations. ICES CM 1978/C:17

Saetre R, Mork M (eds) (1981) The Norwegian Coastal Current. Univ Bergen, Bergen, Norway. Vols 1 and 2, 795 pp

Schott F (1966) The surface salinity of the North Sea. Dtsch Hydrogr Z (Reihe A) Erg H8:1–58

Simpson JH, Hunter JR (1974) Fronts in the Irish Sea. Nature (Lond) 250:404–406

Simpson JH, Hughes DG, Morris NCG (1977) The relation of seasonal stratification to tidal mixing on the continental shelf. In: Angel MV (ed) A voyage of discovery: George Deacon 70$^{th}$ anniversary volume. Pergammon, London, pp 327–340

Soulsby RL (1933) The bottom boundary layers of shelf seas. In: Johns B (ed) Physical oceanography of coastal and shelf seas. Elsevier, Amsterdam, pp 180–266

Svansson A (1975) Physical and chemical oceanography of the Skagerrak and Kattegat. 1. Open sea conditions. Fishery Board of Sweden, Inst Mar Res, Rep 1, 88 pp

Talbot JW (1976) Diffusion data. Fish Res Tech Rep MAFF Direct Fish Res, Lowestoft (28), 13 pp

Taylor AH (1983) Spectral response of a model of the English Channel and southern North Sea heat budgets 1961–1976. Cont Shelf Res 2:331–334

Taylor AH, Stephens JA (1983) Seasonal and year-to-year changes in the temperatures of the English Channel and southern North Sea, 1961–1976: a budget. Oceanol Acta 6:63–72

Taylor AH, Reid PC, Marsh TJ, Jonas TD, Stephens JA (1981) Year-to-year changes in the salinity of the eastern English Channel, 1948–1973: a budget. J Mar Biol Assoc UK 61:489–501

Taylor AH, Reid PC, Marsh TJ, Stephens JA, Jonas TD (1983) Year-to-year changes in the salinity of the southern North Sea, 1948–1973: a budget. In: Sundermann J, Lenz W (eds) North Sea Dynamics. Springer, Berlin Heidelberg New York, pp 200–219

Tomczak G, Goedecke E (1962) Monatskarten der Temperatur der Nordsee, dargestellt für verschiedene Tiefenhorizonte. Ergänzungsh Dtsch Hydrogr Z 7:16

# Suspended Matter and Sediment Transport

D. EISMA[1] and G. IRION[2]

## 1 Introduction

In the aquatic environment contaminants are transported in solution and in particulate form. In the latter case they are present as colloids or adsorbed onto particles (trace metals, trace organic compounds) or as single particles (e.g., fly ash, coal, cokes, colliery waste, organic waste). The contaminating particles range from very fine material with diameters in micron size to sand- and gravel-sized particles. For many contaminants transport in particulate form constitutes an important fraction (up to more than 70%) of the total transport.

The contaminating particles mix with the natural ones and follow the same transport paths. The fate of the contaminants involved depends very much on whether this transport is in suspension or along the bottom, like sands and gravel: depending on the transport mechanism they can be widely dispersed and mixed, or remain within a small area. Contaminated particles can also be exchanged against uncontaminated ones in bottom deposits when these are reworked: in this way older deposits can become contaminated to a certain extent without there being any net deposition. Generally fine-sized particles have a high content of adsorbed contaminants because of their large specific surface. Some contaminants may also adhere to sand grains, particularly when these are coated with organic matter or ironhydroxides. Also even relatively pure sands may contain a small admixture (less than 0.5%) of very fine particles that have settled in the pore space between the sand grains or are present as fecal pellets. By these means, also sand deposits may have a surprisingly high content of contaminants. The areas where suspended matter is being deposited, particularly those areas near to contaminant sources, are primarily the areas where adsorbed contaminants are concentrated.

## 2 Sources of Suspended Sediment

Suspended sediment ($<$ 125 $\mu$m in diameter) is supplied to the North Sea from a variety of sources: the North Atlantic Ocean, the Channel and the Baltic, rivers, coastal erosion, seafloor erosion, the atmosphere (dust) and primary production (Table 1). Most of the suspended matter comes from the North Atlantic and the Channel but in low concentrations (0.01–0.2 mg $l^{-1}$ and ca 3 mg $l^{-1}$ respectively).

[1]Netherlands Institute for Sea Research, P.O. Box 59, 1790 AB Den Burg, Texel, The Netherlands
[2]Forschungsinstitut Senckenberg, D-2940 Wilhelmshaven, FRG

**Table 1.** Suspended sediment in the North Sea

| Supply | $10^6$ t $a^{-1}$ |
|---|---|
| North Atlantic Ocean | 10.4 |
| Channel | 22–30 |
| Baltic | 0.5 |
| Rivers | 4.8 |
| Seafloor erosion | 9–13.5 (+?) |
| Coastal erosion | 2.2 |
| Atmosphere | 1.6 |
| Primary production | 1 |
| | 51.5–64.0 (+?) |
| **Outflow + Deposition** | |
| Outflow | 11.4 + < 3 |
| **Deposition** | |
| Estuaries | 1.8 |
| Waddensea + the Wash | 5 |
| Outer Silver Pit | 1–4 (?) |
| Elbe Rinne | ? |
| Oyster Grounds | 2(+?) |
| German Bight | 3–7.5 |
| Kattegat | 8 |
| Skagerrak + Norwegian Channel | 17(+?) |
| Dumped on land | 2.7 |
| | 51.9–62.4(+?) |

All water flowing into the North Sea flows out along the norwegian coast into the Norwegian Sea between Shetland and Norway (Fig. 1). The outflow is ca. 10% more than the inflow from the North Atlantic because the inflow through the Straits of Dover, from the Baltic, and from rivers is added to it. The suspended matter concentrations in the water flowing out are very similar to those in the water flowing in from the ocean. This indicates that approximately ca $45 \times 10^6$ t $a^{-1}$ of suspended matter (dry weight) is yearly deposited in the North Sea area, which includes the Waddensea, the Wash, rivermouths and the Norwegian Channel/Skagerrak/ northern Kattegat (Fig. 2). The balance of supply and outflow + deposition is shown in Table 1. It is based on a balance published by Eisma (1981a) with additional data (Eisma and Kalf 1987; Waterkwaliteitsplan Noordzee 1985; van Weering et al. 1987; van Alphen 1988). Uncertain are primarily the figures on seafloor erosion, the amounts of suspended matter deposited at present in the Outer Silver Pit, the Oyster Grounds and the Elbe Rinne, and the relation between deposition and local erosion in the Skagerrak. Therefore the almost perfect agreement between the average amounts of supply and outflow + deposition (both $57.5 \times 10^6$ t $a^{-1}$ $40 \times 10$ t $a^{-1}$) is misleading.

The suspended material consists of single mineral particles and aggregates of mineral particles and organic matter. The actual concentrations in the water are very much influenced by the seasonal and short-term variations in water movement (waves) and primary productivity. During storms and generally in winter, more fine particles are being eroded from the seafloor in the shallower parts of the North

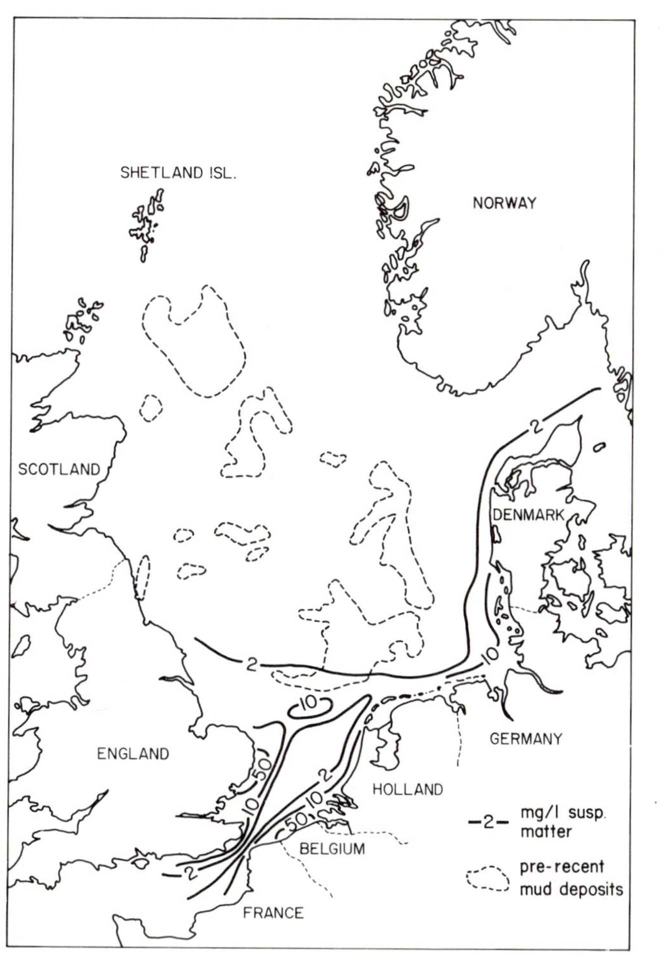


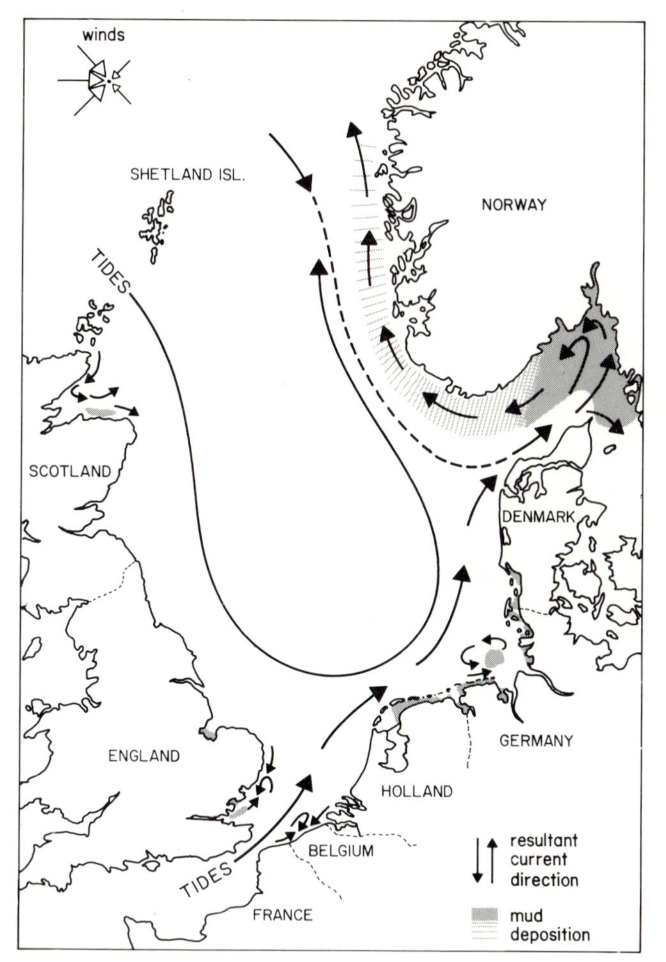


**Fig. 1.** Dispersal and deposition of suspended matter in the North Sea. (After Eisma 1981 and Eisma and Kalf 1987)

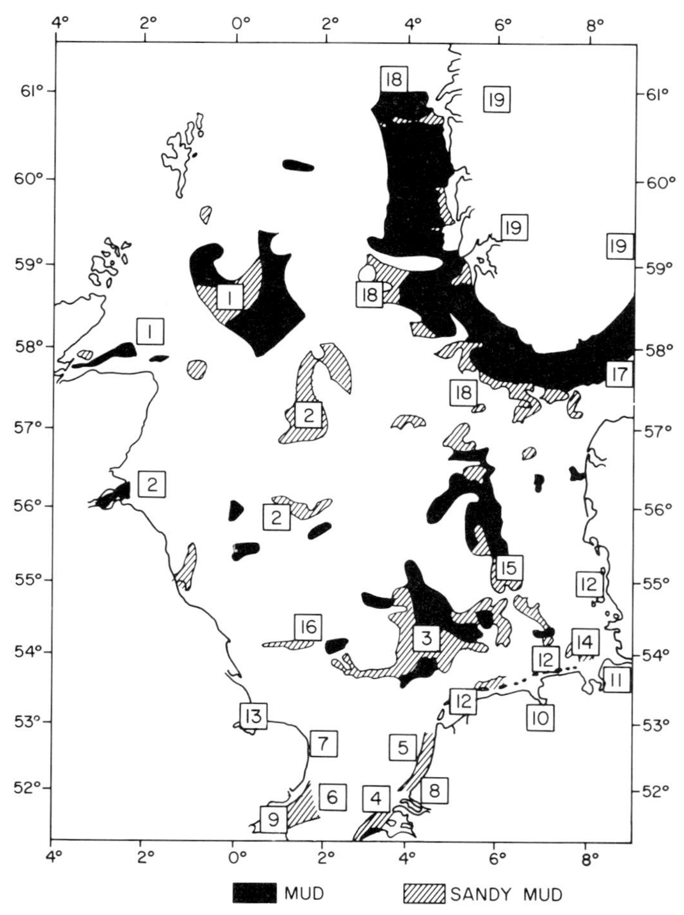


**Fig. 2.** Distribution of mud and sandy mud in the North Sea (from Eisma 1981). *1* Northern North Sea (chiefly Fladen Grounds); *2* Central North Sea (chiefly Devils Hole and NW of Doggerbank); *3* Oyster Grounds; *4* Flemish Banks; *5* Dutch coastal area; *6* East Anglia coast; *7* Yarmouth – Lowestoft; *8* Rhine-Meuse-Schelde estuaries; *9* Thames estuary; *10* Ems estuary; *11* Elbe-Weser estuary; *12* the Waddensea; *13* the Wash; *14* German Bight; *15* Elbe Rinne; *16* Outer Silver Pit area; *17* Skagerrak; *18* Norwegian Channel; *19* Norwegian and some Swedish fjords

Sea, so that the concentrations are several times higher than during the summer. Primary production produces organic particles and particles consisting of opal (diatom frustules) or carbonate (coccoliths etc.). During spring, summer, and autumn such particles temporarily form part of the suspended matter, sometimes in very large quantities, but almost all organic matter and opal produced during the year is mineralized, so that in winter little remains of it. Part of the organic matter comes from the ocean, where the suspended material predominantly consists of organogene particles. In the Northern North Sea during the winter, when there is virtually no primary production, the organic matter constitutes ca. 40% of the suspended material. In the shallow Southern North Sea, where uptake of mineral particles from the seafloor plays a large role, the organic content of the suspended material is less than 20%.

The mineralogy, elementary composition, and isotope composition of the inorganic particles in suspension have been used as tracers for the origin of the material, but mostly on a limited scale and largely related to the supply of suspended matter from rivers or other well-defined sources such as the cliffs at Calais, and dispersal of material supplied along the coast (Salomons et al. 1975; Irion et al. 1985; Bernard et al. 1986). Material coarser than suspended material – sand, gravel – is supplied to the North Sea only in very limited quantities. Sand reaches the sea from rivers only during exceptional conditions (high river discharge, spring ebb tide, strong offshore wind) or from erosion of cliffs, together with gravel. Virtually all sand and gravel on the North Sea floor dates from the Pleistocene and the Early to Middle Holocene.

## 3 Sediment Transport

### 3.1 River Mouths

The estuaries of the rivers entering the North Sea are all of the partially mixed type, which indicates that the circulation within the estuary is largely determined both by the tides and the river discharge. Usually there is a dominant inward flow along the bottom and a dominant seaward flow along the surface. In the tidal area, which usually includes an estuarine (brackish water) part and a freshwater part, sediment of fluvial and marine origin is mixed. Marine sediment can be dispersed by the tides far into the freshwater tidal area up to the point where the tidal influence on the waterflow is negligible. An example is the Ems estuary, where particulate matter of marine origin is dispersed inland as far as the weir at Herbrum (Eisma et al. 1985). Around the contact of freshwater and salt water, where the outward river flow and the inward estuarine flow meet, suspended matter is accumulated in a turbidity maximum. The turbidity maximum contains suspended matter supplied from the river and from the estuary or the adjacent sea and is not stationary: it moves with the tides and the seasons, pushed inward by the flood, outward by the ebb, sometimes being pushed out of the estuary during a period of high river discharge, or moving far inland when river discharge is very low. Local resuspension of (temporarily deposited) bottom sediment contributes considerably to the maximum. In most river estuaries around the North Sea the accumulating mud is being dredged away and the flow in the estuary is regulated by dikes and canalization. The Rhine-Meuse mouth has been changed by man to such an extent that the present mouth is a channel (it became operable in 1858) with an additional outlet through large sluices in the Haringvliet that are opened when river discharge rises above a certain limit. The regulation of the flow and the construction of numerous large harbour basins has resulted in an absence of a turbidity maximum.

## 3.2 The Waddensea and the Wash

In the Waddensea and the Wash, sand and suspended matter are transported inward through the tidal inlets by the tidal currents, which is enhanced by waves. Sand transport, concentrated along the bottom, is usually in the form of moving ripples. It is relatively slow, whereas the suspended matter moves with approximately the same speeds as the water. In the Waddensea and the Wash high concentrations of suspended matter are found along the inner edges and on the tidal watersheds. This is closely related to the deposition of suspended matter on the higher (inner) parts of the tidal flats. The mechanisms that contribute to this will be discussed in Section 4.2. Only a few percent of the total amount of sand and suspended matter moving through the Waddensea is retained in the Waddensea. Mainly from soundings, the yearly net-sediment transport into the Dutch Waddensea is estimated as 16–26 $\times$ 10 t of sand (Rijkswaterstaat 1979) and 1 $\times$ 10 t suspended matter (Eisma 1981a), so that in the entire Waddensea the net influx of sand will be in the order of one magnitude higher than the net influx of suspended matter.

## 3.3 Fjords

Sediment supply to the fjords largely comes from the surrounding land. The sediment is mostly locally transported and deposited, particularly where the rivers enter the fjord at the inner end. There is little evidence of suspended matter transport out of fjords into the North Sea, indicating that most or almost all of the supplied suspended material settles in the fjords (Eisma 1981a, unpublished data NIOZ). Sand or gravel are not transported out of the fjords because of the presence of a sill at the seaward end and the predominantly inward flow of the bottom water. The latter, if not actually measured, follows from the absence of anaerobic bottom waters in almost all fjords, indicating a regular renewal of the bottom waters. (An exception is Framvaren, which has very little water depth over the sill; Skei et al. 1981).

## 3.4 The North Sea

Sand transport in the North Sea occurs primarily along the coasts and in the shallower parts, where tides and waves have a strong effect on the bottom. Sand transport results in the formation of small ripples of several cm height, megaripples of 1–2 m height, flat sandy bottoms and in the formation of channels and banks along the coast and at the tidal inlets. The larger bedforms in the North Sea – the sand dunes and longitudinal ridges and the large banks like the Dogger Bank – are largely relict forms, i.e., they were formed during earlier periods of the Holocene. On these large structures sand is being transported, particularly during storms and part of the sand may then be removed by a combination of storm waves and tidal currents, but redeposition takes place during calmer periods so that there is hardly any resultant sand transport. Net sand transport is only of importance along the coast and particularly near to the inlets.

Transport of suspended matter supplied from the sources indicated in Table 1, largely follows the general movement of the water through the North Sea. The distribution of suspended matter in the North Sea (Fig. 3) shows the highest concentrations in the Southern Bight off East Anglia and the Belgian-Dutch coast and near to the bottom in the German Bight. High concentrations occur particularly where a large gyre is present nearshore or where resultant transport directions meet (Eisma and Kalf 1987). Concentration of suspended matter nearshore is partly related to the predominance of flow parallel to the coast with a relatively small component at right angles to the coast, so that suspended matter from nearshore sources will tend to stay nearshore. Further concentration of suspended matter in a shoreward direction is related to the nearshore water circulation. The nearshore water usually has a somewhat lower salinity than the water further offshore, so that the nearshore water has a lower density than the offshore water. The nearshore water therefore tends to flow seaward along the surface and the offshore tends to flow shoreward along the bottom (Dietrich 1955). This quasi-estuarine circulation is superposed on the tidal flow and also influenced by temperature effects and windstress. Suspended sediment flowing offshore along the surface and settling (particularly during periods of calm weather) will tend to be returned to the shore. In the German Bight the tides are also important in concentrating suspended matter in the inner part southeast of Helgoland (Eisma and Kalf 1987). In the bottom water the flood tide is much stronger, but of shorter duration, than tte ebb tide so that the net transport of suspended matter is inward into the German Bight. There is a resultant outward transport in the surface water, but at the surface the concentrations are much lower, so that suspended matter is accumulated at the bottom of the inner German Bight.

Although the suspended matter concentrations are highest along the coasts of the Southern Bight and further north along the eastern side of the Southern North Sea, the amounts of suspended matter transported through the remainder of the North Sea are large and at least equal in volume in spite of the much lower concentrations, because the volumes of water involved are much larger. In the Skagerrak the suspended matter concentrations decrease to less than 0.4 mg $l^{-1}$ and from there to less than 0.2 mg $l^{-1}$ in the northern part of the Norwegian Channel. There the concentrations are of the same order as in the adjacent ocean. The suspended matter settling out in the Norwegian Channel is at least partly transported back towards the Skagerrak by the inward flowing bottom water coming from the Atlantic.

## 4 Deposition

### 4.1 Estuaries

Deposition in estuaries is largely related to the presence of a turbidity maximum, to variations in water level and to the creation of artificial mud traps (harbor basins, navigation channels). In the turbidity maximum large amounts of mud are temporarily deposited and resuspended again depending on the variations in river discharge, tidal flow (spring tide, neap tide) and wind effects (blowing the water into or out of the estuary). Artificial deepening of channels and the excavation and dredging of harbor basins usually create areas where current velocities are low and mud accumulates. In some areas – like in the Rotterdam Waterway – a sufficient water flow can be maintained to prevent silting up. In rivermouths, and in nearshore areas where suspended matter concentrations are high, mud can be deposited in large quantities and fluid mud can be formed with suspended matter concentrations of more than 5 g $l^{-1}$. Fluid mud retains a certain fluidity because the particles settling out are in each other's way (hindered settling) and a mud-water mixture can be maintained for a long time (Kirby and Parker 1977). Fluid mud moves with the tides as a separate layer than can be recognized as such on an echo sounder. In this way large quantities of mud can be moved with the tides into a river mouth, which is probably at the base of the rapid mud deposition observed in channels and harbor basins. In time the fluid mud consolidates, expelling water, and a mud deposit is formed.

## 4.2 The Waddensea and the Wash

In the Waddensea and the Wash, both characterized by the presence of tidal flats, several mechanisms result in the accumulation of sediment. Suspended sediment settles more on the higher parts of the tidal flats because around high tide, when the water covers the flats, the water depth is much less than around low tide when the water is concentrated in the channels (Van Straaten and Kuenen 1957). Suspended matter, settling out, therefore reaches the bottom much earlier around high tide than around low tide. The suspended matter deposited on the higher parts of the tidal flats can become consolidated to a certain degree because the flats fall dry for many (up to 8) h. To erode this partly consolidated mud, higher velocities are needed than were present when the mud was deposited. Such higher velocities may occur, e.g., during spring tide as compared with neap tide, but occur predominantly because of surface waves. Mud that has been deposited during quiet weather is stirred up again when waves become larger. This effect is strongest on the tidal flats where water depth is relatively small and wind-induced surface waves can easily disturb the bottom sediment. Also there is a net transport of sediment over the tidal flats in the direction of the waves. Winds being mostly westerly and storm winds coming chiefly from the NW, waves will push the sediment over the tidal flats in the Waddensea towards the tidal watersheds and to the inner parts of the Waddensea. Mud deposits are therefore chiefly formed in the sheltered areas behind the islands or in the inner parts (the Jade, the Dollart particularly), where wave effects are minimal (Fig. 4).

In the inner parts of the Waddensea suspended matter deposition is enhanced by the change in the shape of the tidal wave, which in the tidal inlets is approximately symmetric but becomes asymmetric further inward with a slow change from flood to ebb around high tide and a rapid change from ebb to flood around low tide (Postma 1961). Therefore around high tide level the period with low current velocities favoring suspended matter deposition is much longer than around low tide. Organisms – pelagic as well as benthic – induce deposition of

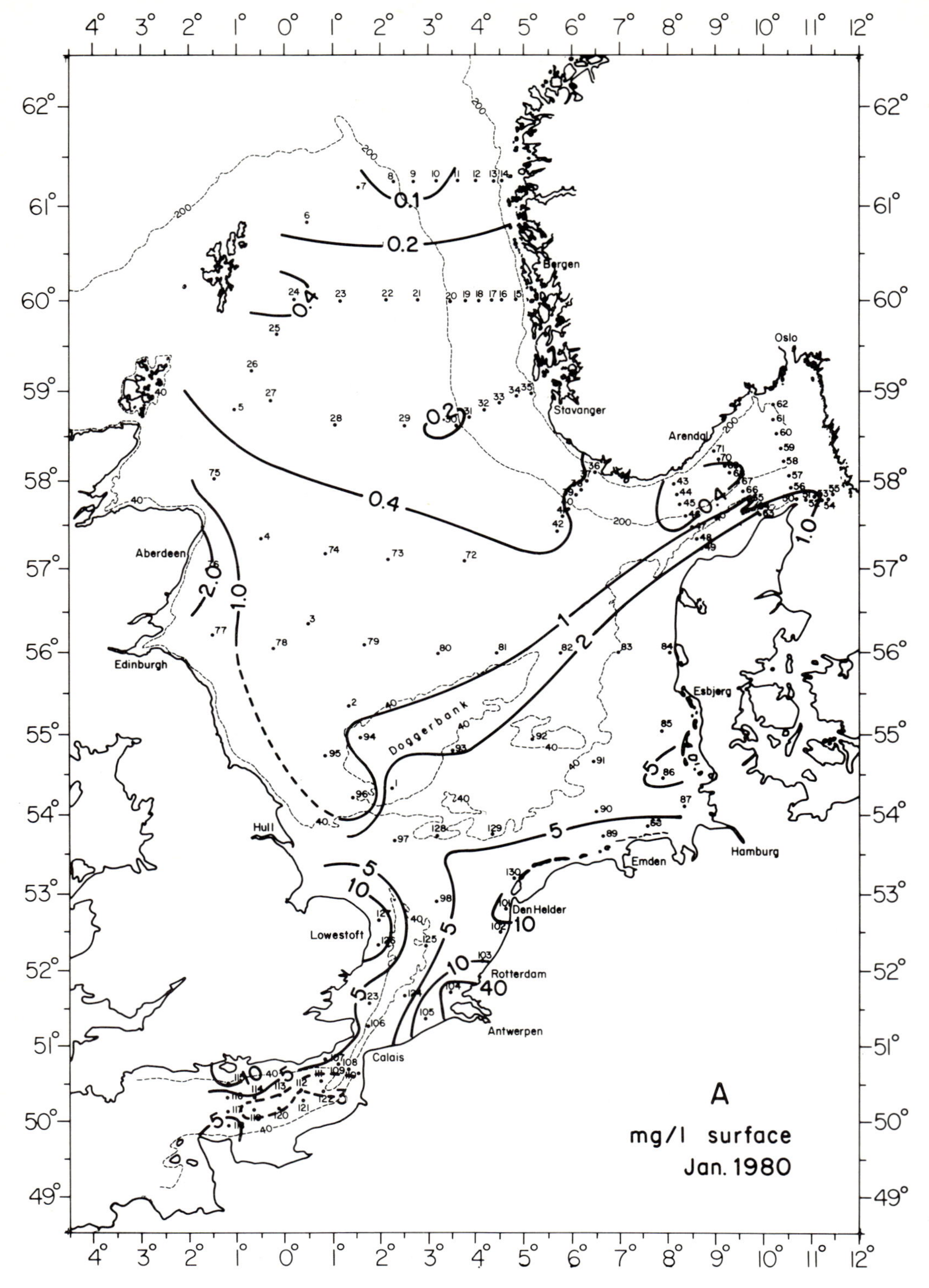


**Fig. 3A,B.** Concentration of suspended matter in the surface water (**A**) and in the bottom water (**B**) of the North Sea in January 1980 (in mg/l). (Eisma and Kalf 1987)

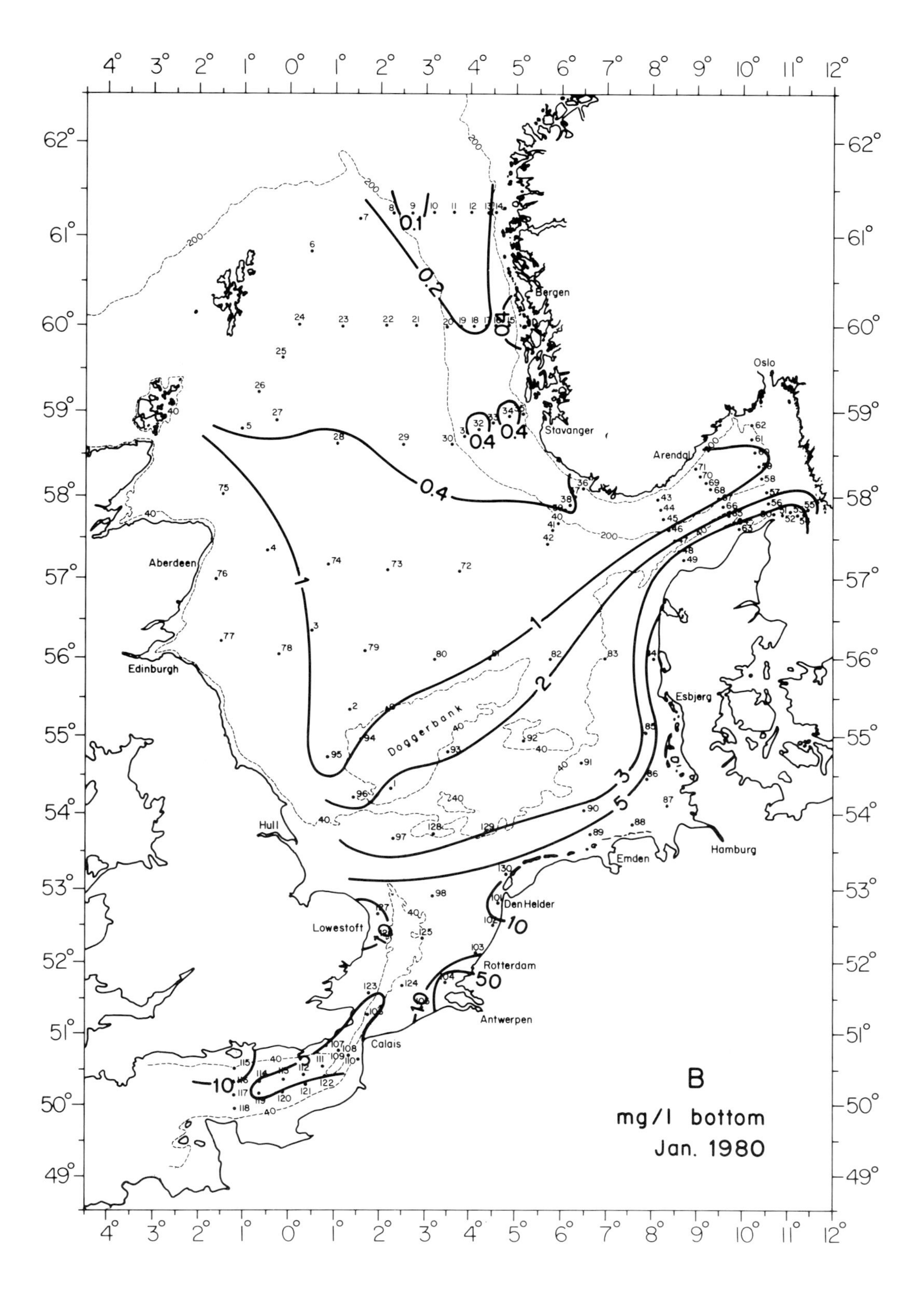
B
mg/l bottom
Jan. 1980
Bergen
Stavanger
Oslo
Arendal
Aberdeen
Edinburgh
Hull
Lowestoft
Calais
Antwerpen
Rotterdam
Den Helder
Emden
Hamburg
Esbjerg
Doggerbank
0.1
0.2
0.4
1
2
3
5
10
50

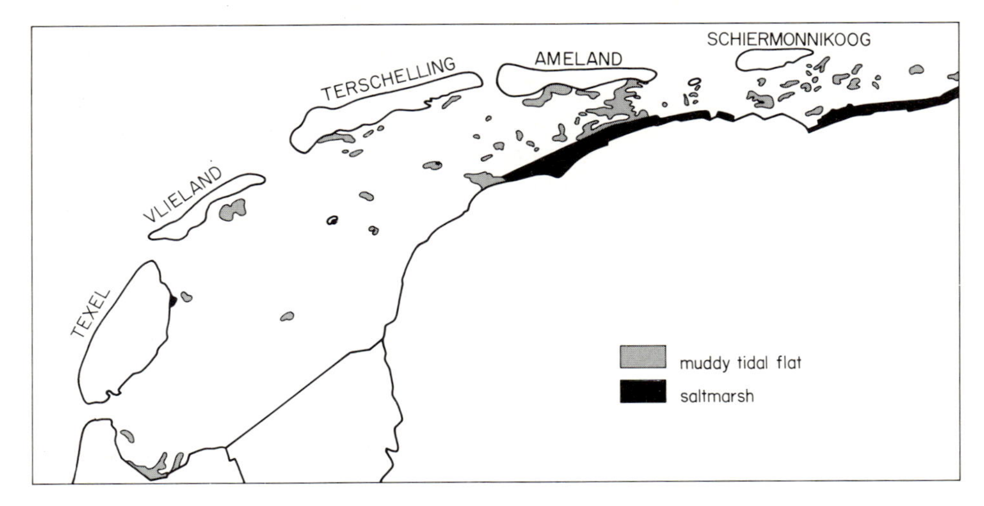


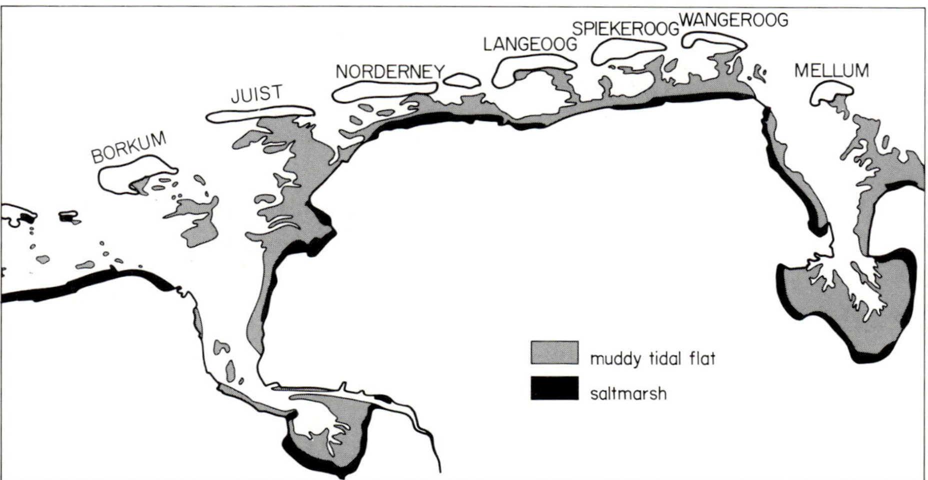


**Fig. 4.** Distribution of muddy tidal flats and salt marshes in the Waddensea. (After Abrahamse et al. 1976). Salt marshes also include present land reclamation areas

suspended matter by clogging particles together in feces and pseudofeces or generally by secreting sticky substances. Benthic organisms doing this (bottom fauna, benthic diatoms) are far more numerous on the flats than in the channels. The formation of pellets out of suspended material results in rapid deposition, but whether they are incorporated in a muddy deposit depends, again, largely on the effects of waves. They are easily rolled and after some time fall apart: the fine particles that are formed are easily resuspended. Salt marshes are rare in the Waddensea and the Wash: where they exist, suspended matter is retained by the vegetation when they are flooded during high tides.

Sand transport in the Waddensea and the Wash presents a very complicated pattern. In the inlet the flood-directed transport is mainly over the banks in the

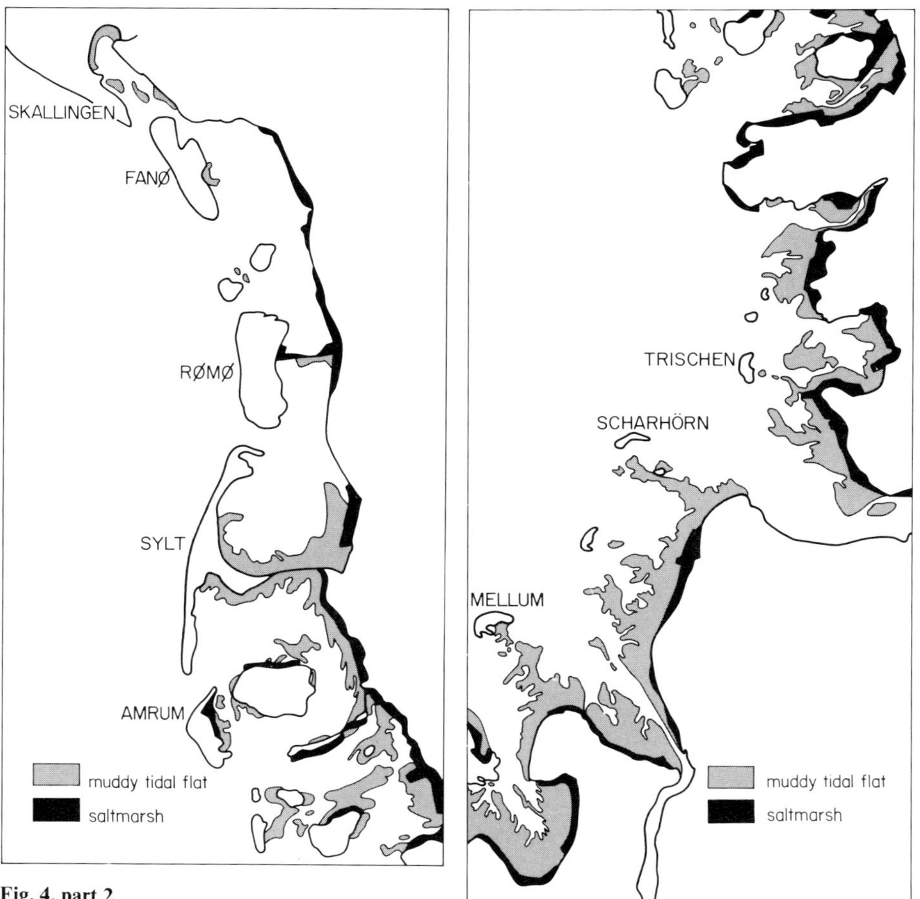


**Fig. 4, part 2**

outer delta, whereas the ebb transport mainly follows the channels. The complex movements of the sand are reflected in the changes in the configuration of the banks. Generally the banks move across the inlet in the general transport direction of the tidal wave, i.e., from west to east along the Dutch-German part of the Waddensea and from north to south along the Danish-German part. In the Dutch part, every 50–100 years a sandbank becomes attached to the tip of the next island lying to the east of the inlet. Part of the sand in the inlet, however, is transported into the Waddensea. That sand is accumulated there is evident from soundings (at least in the Dutch Waddensea) and from the fact that already for centuries the sedimentation in the Waddensea keeps pace with the relative rise of sea level of 10–20 cm/100 a. Also the construction of dikes has resulted in large sand accumulations caused by the filling in of old channels and the formation of new tidal flats. To give an idea of the quantities involved, data on the Dutch Waddensea are presented in Table 2. The sand is supplied by erosion of beaches and dunes along the North Sea coast and by erosion of the nearby seafloor. During the past years

**Table 2.** Dutch Wadden Sea

| | |
|---|---|
| Yearly sand supply through tidal inlets (based on soundings 1963/68–1975/78) | $8–13 \times 10^6$ t $a^{-1}$ |
| Average sand extraction (dredging) | $1.2 \times 10^6$ t $a^{-1}$ |
| Sand needed to compensate for subsidence | $1.3 \times 10^6$ t $a^{-1}$ |
| Exceptional sand extraction for public works: | 1977: $2.8 \times 10^6$t<br>1978/79: $2 \times 10^6$ t<br>1980: $2.4 \times 10^6$ t |
| Exceptional deposition because of dike construction:<br>Aflsuitdijk area 1932–1960: average $5 \times 10^6$ t $a^{-1}$<br>Lauwerszee area 1957–1963: average $3.5 \times 10^6$ t $a^{-1}$ | |

Data source: Rijkswaterstaat, Ministry of Public Works.

there has been a tendency for erosion of the channels in the Waddensea and the Wash and sedimentation on the tidal flats. The level of the tidal flats is probably mainly determined by waves, particularly storm waves. During the summer tidal flats are built up and during the winter they are eroded. In this way also the volume of the tidal basin behind each tidal inlet is maintained, so that the inlet retains its size. The difference between alternating sedimentation and erosion is a small average net deposition rate of 1–2 mm $a^{-1}$, which keeps the tidal flats at an approximately constant level with respect to relative sea level.

### 4.3 The German Bight and the Skagerrak/Kattegat/Norwegian Channel

In the German Bight a mud deposit is formed at 15–40 m water depth, although the tidal currents reach 1 m $s^{-1}$ in the surface water and 50 cm $s^{-1}$ at 1 m above the bottom (Reineckhet al. 1967; Fig. 5). The tidal mechanisms described above and the presence of a large gyre result in concentration of suspended matter, but the high concentrations at the bottom probably have their origin in settling of suspended matter during periods of calm weather when waves are absent or insignificant. Also deposition by fauna may be important. Because most of the area is too deep for settled mud to be resuspended by surface waves, mud accumulates at the bottom. Storm waves probably cause more erosion and transport of resuspended mud towards deeper water, but the tidal mechanism operating near to the bottom will return it to the inner German Bight. The formation of a mud deposit is at present probably enhanced by dredging in the Elbe river estuary during ebb tide, resulting in an increased mud supply to the inner German Bight, and by the dumping of dredge spoils and city sewage mud. The influence of these different sources of supply is not yet fully clear.

In the Skagerrak and the Kattegat, mud is concentrated in large gyres where it can settle out into deep water (maximum depth 700 m). The deposited mud comes from the Southern and Central North Sea (and probably a little from the North Atlantic) and from erosion of local fine-grained deposits, particularly on the Danish side of the Skagerrak (the shallow parts and probably also at the upper

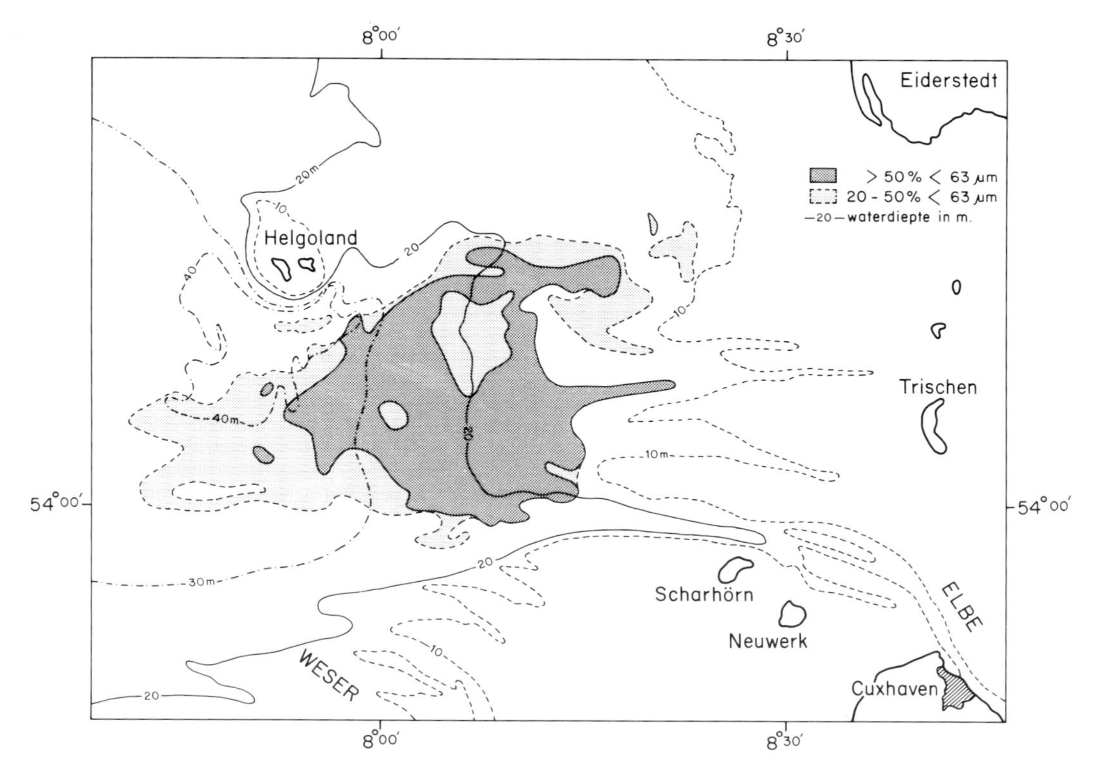


**Fig. 5.** Location of the mud deposit southeast of Helgoland in the German Bight. Waterdepths in meters. After the map published by the German Hydrographic Institute (1981) and the map of Hertweck (1983). (Eisma and Kalf 1987)

slope; Rodhe 1973; Van Weering pers. commun.). Also the sudden decrease in pollutant concentrations in the Skagerrak mud as compared with the Southern North Sea and the Kattegat indicates a strong admixture of older uncontaminated mud (Muller and Irion 1984). As indicated above: the water circulation tends to concentrate the mud in the Skagerrak (particularly the eastern side) and in the nearby Kattegat.

## 4.4 The Remaining North Sea

The formation of mud deposits in the North Sea occurs in a few isolated areas (mainly in rivermouths, the Waddensea and the Wash, the German Bight, and the Skagerrak/Kattegat/Norwegian Channel). The other muddy deposits in the North Sea are relict and at most reworked. They generally lie below waters with a very low suspended matter content and a very small supply of sediment from the coasts or from the shallower parts of the North Sea. Recent additions to these muddy sediments consist of organic matter (from pelagic organisms or from benthos living in the sediment) and some calcium carbonate particles from exoskeletons (mainly fragments). Some uncertainty whether recent mud deposition occurs exists with regard to the Outer Silver Pit, the Oyster Grounds, and the Elbe Rinne. In all three areas there may be some recent mud deposition in addition to reworking of the older sediment. Reworking, however, may and probably often does imply replacement of older uncontaminated particles by recent contaminated material, so that in this way particulate material contaminated with pollutants can become part of these deposits without any net deposition (Eisma 1981b). Fine material can also be trapped in the pore space between sand grains or gravel, but the quantities involved are very small.

## 5 Summary

Although in a general way the transport paths and depositional areas of sediments in the North Sea are reasonably well known, there are still important gaps, such as the contribution of sediment derived from seafloor erosion, the extent of reworking of older deposits and mixing of older material with recently supplied material, and the deposition rates in a number of areas. Sufficient quantitative knowledge on transport and mixing of material of different origin and on the fate of material supplied from a specific source is not available to predict the behavior and concentration of contaminants associated with these sediments. A second problem concerns the question to what extent contaminants associated with sediments behave conservatively : to what degree are contaminants released or adsorbed during transport and after deposition? In estuaries, as is well known by now, many contaminants are not conservative, but do such processes also play a role in the coastal North Sea where conditions are more uniform than in the estuaries? A third question concerns the degree of contamination of sandy deposits in the North Sea and what causes this contamination.

## References

Abrahamse J, Joenje W, Leeuwen-Seelt N van (1976) De Waddensee Harlingen

Alphen J van (1988) A mud balance for the Belgium-Dutch coastal waters between 1969 and 1986. Proc Int Symp Ecol North Sea 15–21 May

Bernard P, Grieken RE van, Eisma D (1986) Classification of estuarine particles using automated electron microprobe analysis and multivariate techniques. Environ Sci Technol 20(5):467–473

Dietrich G (1955) Ergebnisse synoptisch ozeanographischen Arbeiten in der Nordsee. Tag Ungs Ber Geographentag Hamburg, pp 376–383

Deutsches Hydrographisches Institut (1981) Karte der Sediment Verteilung in der Deutschen Bucht. Hamburg, mit Begleitheft (13 p)

Eisma D (1981a) Supply and deposition of suspended matter in the North Sea. Spec Publ Int Ass Sediment 5:415–428

Eisma D (1981b) The mass-balance of suspended matter and associated pollutants in the North Sea. Rapp P V Reun Cons Int Explor Mer 181:7–14

Eisma D, Kalf J (1987) Dispersal, concentration and deposition of suspended matter in the North Sea. Bull Geol Soc 144:161–178

Eisma D, Bernard P, Boon JJ, Grieken RE van, Kalf J, Mook WG (1985) Loss of particulate organic matter in estuaries as exemplified by the Ems and Gironde estuaries. Mitt Geol-Palaont Inst Univ Hamburg, SCOPE/UNEP Sonderb H 58:397–412
Hertweck G (1983) Das Schlickgebiet in der inneren Deutschen Bucht Senckenbergiana Maritima 15:219–249
Irion G, Wunderlich F, Schwedhelm E (1985) Deposition of finegrained sediments in the inner German Bight. Symp Modern and ancient clastic tidal deposits Utrecht aug 1985, Abstr, pp 181–184
Kirby R, Parker WR (1977) Physical characteristics and environmental significance of fine sediment suspensions in estuaries. In: Studies in geophysics. Estuaries: geophysics and the environment. Natl Acad Sci pp 110–120
Muller G, Irion G (1984) Chronology of heavy metal contamination in sediments from Skagerrak (North Sea). Mitt Geol-Palaont Inst Univ Hamburg 56:413–421
Postma H (1961) Transport and accumulation of suspended matter in the Dutch Waddensea. Neth J Sea Res 1:148–190
Reineck HE, Gutman WF, Hertweck G (1967) Das Schlickgebiet südlich Helgoland als Beispiel rezenter Schelfablagerungen. Senckenb Lethaea 48:219–275
Rijkswaterstaat (1979) Hydrografisch-sedimentologisch onderzoek met betrekking tot de winning van zand in de Waddenzee. Eindrapport I, mei 1979
Rodhe J (1973) Sediment transport and accumulation at the Skagerrak-Kattegat border. Rep 8 Oceanografiska Institutionen Goteborgs Univ
Salomons W, Hofman P, Boelens R, Mook WG (1975) The oxygen isotopic composition of the fraction less than 2 microns (clay fraction) in recent sediments from Western Europe. Mar Geol 18:23–28
Skei J, Knutzen J, Ormerod K, Rygg B, Sorensen K (1981) Framvaren ved Farsund. Et biogeokjemisk studium av en permanent anoksisk fjord. NIVA Rapport F 418, p 108
Straaten LMJU van, Kuenen PH (1957) Accumulation of fine-grained sediments in the Dutch Wadden Sea. Geol Mijnbouw 19:329–354
Waterkwaliteitsplan Noordzee (1985) Rijkswaterstaat State Printing Office, p 85 + 5 additional reports
Weering TCE van, Berger GW, Kalf J (1987) Recent sediment accumulation in the Skagerrak, northeastern North Sea. Neth Jour Sea Res 21(3), 177–189

# Geobiological Effects on the Mobility of Contaminants in Marine Sediments

M. KERSTEN[1]

## 1 Introduction

The marine environment of the North Sea is the major receptacle of waste products introduced by atmospheric and river input from the adjacent highly industrialized countries. However, this environment is too readily assumed to represent an efficient and inert sink. This assumption is based mainly on the concept that the sea acts as an infinite dilutor of the introduced pollutants, and on the concept that most of the inorganic and organic pollutants will ultimately be buried at sea bottom, at least within distinct sediment accumulation areas such as the tidal flats, the inner German Bight, and Skagerrak basin. However, contaminated sediments encountered in such areas within the North Sea may pose a serious problem in the near future through potential secondary pollution effects with yet quite unknown release potential for the next decades. This chapter will review the increasing alarming evidence that such secondary pollution effects may indeed arise from marine deposits.

First of all, we should gain an impression of the pollution inventory of such a sedimentation area from some simplified calculations based on more recent data. The so-called German Bight is a shallow sea located between 6°12′E and 55°15′N covering an area of ca. 31,150 $km^2$ with a mean depth of 24 m. It contains a water volume of about 736 $km^3$, which has a mean residence time of 2 months (Hainbucher et al. 1986; Schönfeld and Radach, pers. commun.). The dissolved cadmium content within this water column can be assessed using the Cd concentration values published recently by Mart and Nürnberg (1986). These researchers determined the dissolved and total Cd concentrations over several years at different seasons. Dissolved Cd concentrations found in the German Bight cluster around 30 ng $kg^{-1}$, with rather narrow ranges, representing 70%–80% of the total Cd contents within this water body. From this it follows that at any time ca. 22 t of Cd are dissolved in the whole German Bight, and ca. 132 t pass through this area annually.

For the comparison of this data with the Cd inventory of the sediment underlying this water column, we may take the data published for the confined mud accumulation site of the inner German Bight assuming a funnel-like coupling of both the sediment bottom and the overlying water body. Eisma (1981) calculated a recent deposition rate of 7,500,000 t $a^{-1}$ (dry weight), taking into account both

[1]Technische Universität Hamburg-Harburg, Arbeitsbereich Umweltschutztechnik, Postfach 90 14 03, D-2100 Hamburg 90, FRG

recent sedimentation rates and the extension of the site. The mean total Cd concentration within the upper 10 cm sediment layer of this area is 0.20 mg $kg^{-1}$ (dry weight) based on an average of 16 samples, which were recovered equidistantly on a south-north transect from 54°05′00″N to 54°12′28″N along 8°07′15″E (Kersten, unpublished data). These data result in a mean cadmium *net deposition* of "only" 1.5 t $a^{-1}$ within this mud accumulation area.

It is meanwhile well established that benthic regeneration in estuarine and shallow coastal environments is a potentially important source of nutrients to the overlying water column (Zeitschel 1980; Rutgers van der Loeff 1980). One may, however, for reasons discussed below assume also an "early diagenetic" remobilization of 50–80% of the sediment-bound cadmium subsequent to deposition (Kerner et al. 1986; Fernex et al. 1986). Taking into account all data presented, only a minor proportion of the dissolved Cd level in the complete water column passing the German Bight will be determined by sediment release. Could these pollutant fluxes to date represent significant harm to the North Sea ecosystem?

Studies on the rates and mechanisms controlling the flux of contaminants across the sediment-water interface have in fact received considerable attention in so far as recently Kremling (1983) and Nolting (1986) have succeeded in providing field evidence showing that remobilization from sea bottom controls elevated concentrations of dissolved Cu, Ni, Cd, and Zn levels near the southeastern English coast and in the Scottish shelf region, probably to an even greater extent than freshwater inputs. In both regions the fraction of these metals in solution was found to be consistently more important than the metal fraction in the suspended particulate phase. Similarly, mobilization from sediment is proposed as an explanation of the high dissolved Ni concentrations in N.E. Pacific continental slope water (Jones and Murray 1984). On the other hand, Kremling (1985) has suggested that in the water of the Southern Bight, river inputs are the main influence on the total concentrations of Cu, Cd, Ni, Mn, and Al. This is especially true in regions where the contributions of the particulate phases to the total concentrations of these elements cannot be neglected (Duinker and Nolting 1977). Similarly, Hunt and Smith (1983), using mesocosm experiments, showed that remobilization processes may operate in the Narragansett Bay system but are overwhelmed by the magnitude of freshwater pollutant inputs so that they are not detectable under field conditions. These authors suggested that if the remobilization rates observed in their mesocosm experiments are applicable to these bay and adjacent river systems, $< 5\%$ of the metals entering the water column would be released from sediments, which is in accord with the data calculated for the German Bight.

Such remobilization mechanisms, however, may become more important in two important cases. On the one hand, relaxed pollution pressure may not concur with an appropriate water quality improvement due to the potential long-term bleed of contaminants from the bottom sediments acting as a secondary pollution reservoir. On the other hand, there is increasing evidence that pollutant release may not only be steady and cyclic, but also that tremendous remobilization effects may arise in the course of unpredictably episodic "biological events". The latter effect may even be reinforced by diminishing in hydrographical fluxes, e.g., by stratification. Unfortunately, the relative importance of such effects cannot be determined from available data for the southern and especially for the central part of the North Sea with its important fish resources. In this chapter I will try to review briefly some more recent results derived from other shallow marine environments similar to those of the North Sea. The major emphasis in this paper is on recent interdisciplinary investigations conducted by both geochemists and biologists, which seem to be necessary for realistic secondary pollution assessments. This chapter should indeed be considered as an appeal for a rather more interdisciplinary approach to the study of this complex subject. Emphasis is placed here on biological alterations of sediments and their biogeochemistry rather than sediment and pollution effects on the biota. The latter topic is covered in other papers of this volume (cf. that of Dethlefsen et al.).

## 2 Geochemical Diagenesis Affecting Metal Mobility in Sediments

The importance of sediments to the geochemical cycling of pollutants such as trace metals (Salomons and Förstner 1984) or phosphorus (Krom and Berner 1981; Balzer 1984; and literature cited therein) is well known. Since release from the bottom of shallow seas can supply significant percentages of the nutrient requirements for the pelagic primary producers (Zeitschel 1980), it is of general interest to study the rates and extent of geochemical alterations of sediments, which is essential to an understanding of contaminant mobility and fluxes in marine environments. Once accumulated by sedimentation, significant release of these pollutants from sediments can originate from following processes (Förstner et al. 1986):

1. Post-depositional remobilization by oxidation and decomposition of organic detritus;
2. diffusion via the interstitial water subsequent to 'early diagenetic' effects such as reduction of ferromanganese oxides;
3. oxidation of reduced metallic sulfide solids, which are generally highly insoluble, to more soluble solid phases;
4. desorption from clay minerals and other substrates due to formation of soluble metal-organic complexes.

The intensity with which these processes occur in bottom sediments depends in turn on several sedimentological and physicochemical factors, which determine the specific form and reactivity of a pollutant-binding compound. A characteristic property of an anthropogenically derived pollutant is its introduction in a state far from any natural geochemical equilibrium. This results in its involvement in energy-consuming geochemical cycling within the aquatic environment. General experience from sediment studies on samples from the tributaries of the North Sea shows that the surplus of metal contaminants introduced into the aquatic system by man's activities is usually bound in relatively unstable chemical forms on surface sites of particles, and should, therefore, be more accessible for short- and middle-term geochemical processes – including biological uptake – than the detrital, predominantly naturally derived metal compounds (Calmano and Förstner 1983). Trace metals present in the sedimenting particulate matter of the southeastern North Sea, were found predominantly both in less stable oxidic or organic forms and more stable residual lithogenic associations (Kersten and Förstner 1985). Particulate Ni, Cr, As, Cd, and Cu are found to a large extent in organic forms in productive seasons, when terrigenous particles are diluted by biogenic seston, whereas particulate Pb, Zn, and Co have a great affinity for manganese and iron oxides at any time of the year.

Reactive particles settle upon the bottom and accumulate at the net rate of sedimentation, resulting in an unconsolidated sediment. Subsequent to sedimentation, the free exchange between particles and water is considerably reduced due to hindered exchange between pore water and overlying water column. Consequently, the settled compounds undergo a variety of early diagenetic processes (Stumm and Morgan 1970). One of the most important is the biogenic oxidation of decomposable organic matter by microorganisms. Sites of benthic decomposition are the depositional interface before burial and the sedimentary column from which net upward transport of released elements is provided by pore water diffusion and bioturbation. The aerobic degradation of freshly deposited organic detritus is greatest at the sediment surface, where the concurring oxygen depletion is counterbalanced by sufficient replenishment, as diffusion through the pore water. Thus this surfacial sediment layer contains free $O_2$, $CO_2$, $SO_4^{2-}$ $NO_3^-$ and ferric oxides, which impart a yellowish-brown color to the sediment (Lyle 1983). In sediments, however, oxygen is consumed not only by the degradation of organic matter but also by the oxidation of other reduced components which are produced in the reducing subsurface sediment. Examples of these are the manganous and ferrous ions, which upon oxidation precipitate as highly immobile hydrous oxides. In fine-grained organic-rich bottom sediments (so-called mud), the oxidized top layer is limited to the upper few millimeters (Revsbech et al. 1980). In deeper sediment strata, below which there is no more dissolved oxygen available, oxygen deficiency leads to development of reducing environmental conditions. Thus, the metabolic activities of microorganisms tend to create a vertical zonation of two main biogeochemical environments within the sediment. The marked boundary between these zones of aerobic and anaerobic microbial metabolism, respectively, is usually referred to as the redox potential discontinuity (RPD). This boundary layer is characterized by a rapid decrease in Eh with depth, which is easily mapped in the field using redox electrodes.

In the marine environment, sulfate is the most important alternative reactant for the oxydation of organic matter. Thus the concomitant production of reduced sulfur compounds largely controls the diagenetic milieu of the anoxic sediment zone. The production of $H_2S$ during sulfate reduction exerts a strong influence on the speciation of trace metals, which are precipitated as sulfides of very low solubility. Sediments of this environment are characterized by the grey or black color of precipitated iron sulfides. The mechanisms and consequences of sulfate reduction in shallow marine environments have been extensively studied by Jørgensen (1977), Goldhaber et al. (1977) and Hines and Jones (1985).

Assuming steady-state conditions, the microbial metabolic intermediates of organic matter decay (e.g., $HCO_3^-$, $HPO_4^{2-}$, carbohydrates and other low molecular organic ligands) and that of the coupled inorganic reduction processes [e.g., Fe(II), Mn(II), $S^{2-}$, $NH_4^+$] accumulate in the sediment until concentrations are limited by (physically or biologically intensified) diffusive transport, by subsequent microbial utilization, or by the authigenic formation of secondary ("authigenic") minerals such as metal sulfides or vivianite: $Fe_3(PO_4)_2\,8H_2O$ (Suess 1979; Berner 1981). Thus each of the biogeochemically different sediment strata reveals its own characteristic secondary mineral assemblages critical both in buffering interstitial water chemistry (Stumm and Morgan 1970) and in affecting transfer of dissolved trace metal compounds to the sediment-water interface (Emerson et al. 1984).

Clearly, it is the type of particle association (and the underlying elemental chemistry) of the contaminants that in fact controls their respective behavior during early diagenesis. Assuming one-dimensional vertical diffusion in a homogeneous sediment stratigraphy, the early diagenetic cycling of trace metals such as Co, Ni, and As predominantly associated with redox-sensitive manganese and iron oxyhydrates can be generally summarized as: (1) dissolution of the substrate upon burrowing in the reducing zone of the sediment column, (2) vertical migration of the released metals along the pore water gradient, which may be accelerated by organic ligand complexation reactions, and (3) readsorption at oxides (predominantly Fe oxyhydrate coatings on clay minerals) in the top oxidized sediment layer (e.g., Gendron et al. 1986; Peterson and Carpenter 1986). Similarly, interstitial water phosphate derives not only from microbial degradation of phosphorus-containing organic matter but also by release from adsorption sites on ferric oxyhydrate sediment coatings during anoxic reduction (Krom and Berner 1981). Up to 66% of the phosphorus input is returned to the water column from anoxic sediments, where sharply negative ferrous iron gradients are developed due to sulfide control with concomitant inhibition of Fe-P mineral formation (Balzer 1984). Mobilization of these pollutants to the overlying water column will usually proceed only as a result of severe disruption of this biogeochemical cycling, e.g., by physical release processes of turbulent mixing and resuspension of sediments in the course of storm events or dredging activities recovering part of the anoxic sediment zone, but more common when the entire sediment column becomes reducing.

In contrast, metals associated predominantly with organic matter are strongly depleted rather than enriched in the oxidized sediment strata (Gendron et al. 1986; Kerner et al. 1986). The biogeochemistry of these pollutants is dominated by the cycle of organic matter at the sediment-water interface, i.e., metals are liberated along with phosphate and other mineralization products in the top cm. In other words, considerable pollutant liberation occurs within a very short time after deposition from the aerobic degradation of the organic substrates. Metals released in dissolved form migrate along the pore water gradient downward into the reducing zone of the sediment where they can precipitate as a sulfide phase (or Fe-oxyhydrates in the case of phosphate). Both metals and phosphate may, however, also migrate upwards across the sediment-water interface into the overlying water column, occurring as bioavailable species with potential for benthic organism impact. Westerlund et al. (1986) have clearly demonstrated that the release of dissolved Cd, Cu, and Ni into the water column from the top sediment layer depends on the flux of oxygen into the sediment. The nonbiogenically mediated flux across the sediment/water interface of a component dissolved in sediment pore water arises mainly from diffusion. More important is, however, the flux controlled by interstitial water exchange with the water column, which may be enforced by wave and tidal action ("subtidal pumping": Riedl et al. 1972). While subtidal pumping is mainly restricted to estuarine and coastal flat environments, animal activities may represent another important effect on metal mobility in shallow marine environments which until now has often been neglected in assessments of secondary pollution potential of sediments.

## 3 Animal-Sediment Interactions

The importance of pelagic organisms in the cycling of elements in the marine environment has been recognized for some time (cf. Wangersky 1986). Investigations have focused mainly on pelagic remineralization and deposition effects and their role in element cycling and downward transport mechanism for solid material and sorbed pollutants to the sea bottom ("marine snow"). In shallow water environments, however, regeneration involves the sediment surface and the sediment, since settling times for organic detritus are relatively short and the supply is large (Suess 1980). The feeding, burrowing, respiratory activities, and adaptive strategies of benthic (bottom-dwelling) animals can, however, also have marked effects on the element cycling at the sediment-water interface (McCall and Tevesz 1982). Readily evident are the effects of benthic animals on the sedimentological properties of productive sediment areas. Perturbations of the sediment stratigraphy to 30–40 cm depth can occur within the North Sea as a consequence of the vertical tube-burrowing activities of benthic worms (Rachor and Bartel 1981). Macrobenthic animals (operationally defined as the adult stages of which remain in a 1-mm mesh sieve during the process of separating organisms from sediment) may strongly affect surfacial sediment properties. The effect of a benthic organism on sediment structure and chemistry depends on its mode of feeding and mobility in relation to the sediment substrate. Lee and Swartz (1980), for example, proposed a classification of benthic invertebrates based on three dichotomies (epifaunal/infaunal, mobile/stationary, and deposit/suspension feeding) and 12 subgroupings in relation to resulting bioturbation processes. Disrupting sediment structure by bioturbation may either alter its mass properties (e.g., by burrowing, excavation, tube construction, and biodeposition activities), or may alter particle properties (e.g., by ingestion/egestion and pelletization by voiding of fecal pellets).

The so-called conveyor-belt species commonly contribute an important "nonrandom bioadvective component" to sediment mixing in marine environments (Rhoads 1974). These animals are vertically oriented in the sediment with head end down and the posterior located at or near the sediment/water interface. They ingest thereby deeper sediments and deposit them as feces at the surface, resulting in the exposure of deep sediment particles and biodeposition of fecal pellet layers on the sediment surface. Conveyor-belt species appear to be stationary and are classified by Lee and Swartz (1980) as stationary subsurface deposit feeders. Feeding depths range from 1 cm to over 20 cm. Another important class are vagile and excavating subsurface deposit feeders representing such prominent groups as bivalves (e.g., *Yoldia, Nucula, Cyprina, Abra alba*), polychaetes (e.g., *Nephtys, Pectinaria)* and thalassinid crustaceans (e.g., *Diastylis rathkei*), which are dominant and productive sediment reworkers in many subtidal sandy to muddy habitats (Rhoads 1974). Stationary infaunal suspension feeders disrupt the sediment structure by U-shaped and mucus-coated tube burrows. Surface sediments are ingested by irrigation of overlying water into the tube holes, which slides material down the feeding funnel (funnel feeders *sensu* Lee and Swartz 1980). Prominent taxa of this class are lugworms (e.g., *Arenicola*) and spoonworms (e.g., *Echiurus*), which can be very abundant (i.e., over 100 individuals per $m^2$) in sandy intertidal and subtidal areas. They are in this case the dominant sediment reworkers in such areas (Lee and Swartz 1980). Epifaunal surface deposit feeders include such taxa as gastropods, echinoids, asteroids, and holothurians, but also demersal fish species. The burrowing activities of these species cause a uniform mixing of the surface layer. Schäfer (1972) reviewed extensively the types of sediment disruptions caused by epifaunal species of the North Sea. Vagile surface deposit-feeders include common bivalve molluscs (especially the Baltic tellin *Macoma balthica*) specific for benthic macrofauna assemblages in the German Bight (Salzwedel et al. 1985). All these species have moderate to high rates of sediment bioturbation. A compilation of ranges of individual particle reworking rates, annual reworking rates, and depth of reworking of these and other species not mentioned above has been given by Lee and Swartz (1980).

The local distribution and population density of benthic species vary considerably according to characteristics of the depositional environment such as temperature, grain size, organic detritus availability, and quality, resuspension and net deposition rates, and to some extent also to intra- and interspecific ecological interactions (Rhoads 1974). Macrobenthic invertebrates are commonly more abundant in fine-grained sediments characterized by high organic matter contents, which may also represent important pollutant accumulation areas. In an extensive survey of the macrobenthos of the sublittoral of the southeastern North Sea, five major bottom communities were identified, with a mean abundance of 2377 individuals per $m^2$ and mean biomass of 116 g total wet weight (Salzwedel et al. 1985). Generally, polychaetes, crustaceans, and bivalves combined comprise more than 80% of the total number of benthic species. The population densities found in autumn 1975 were relatively high as compared to some earlier surveys, probably due to more optimal trophic conditions for the bottom fauna as a consequence of some kind of eutrophication (Salzwedel et al. 1985). Overall pollution stress may modify the rate of bioturbation at least by altering community structures or even by reversal of ecological succession stages (Rhoads and Boyer 1982; Rubinstein et al. 1980; Gray 1982). Among the less sensitive and opportunistic organisms frequently bioturbating anoxic sediments is the priapulid *Halicryptus spinulosus* (Dicke 1986).

Populations of bottom-dwelling species may exert great influence on geotechnical and transport properties of the marine bottom due to alteration of shear strength, water content, surface relief, and generation of biogenic textures (e.g., Eckman 1985). Spatial and seasonal changes of sediment geotechnical properties and of erodibility may be related to considerable "patchiness" in benthic communities (Rhoads and Boyer 1982). Biogenic sediment reworking alters structure and texture (spatial arrangement of the particles) of a sediment, which in turn modifies the other physical properties. In cases where relatively fine material is preferentially ingested at depth and defecated at the sediment-water interface, a biogenic stratigraphy or graded bedding is formed, which generates a surface microtopography and increases bottom roughness and scour (Baumfalk 1979). Sediment surfaces consisting of fecal pellet deposits are easily eroded because of their high water content and low bulk density (Rhoads and Young 1970; Nowell et al. 1971). Lee and Swartz (1980) reviewed rates of resuspension over intensively reworked sediments based on near-bottom sediment trap measurements, which can be as high as 1–250 mg $cm^{-2}$ $d^{-1}$. The highest resuspension rates were found with protobranch bivalves (*Nucula, Yoldia* and *Tellina*) in the silty sands of Long Island Sound (McCall 1977) and Minas Basin (Risk and Moffat 1977). Where this is the case, fecal pellets may represent an important pathway for the lateral movements of co-entrained trace elements. Irion and Schwedhelm (1983) reported that the $< 20\,\mu m$ fraction of the sandy surface sediments common in the German Bight also consists of mainly fecal pellet biodeposits. It was not indicated, however, whether the pellets are benthic or pelagic in origin.

Sediment turnover rates – in this respect referred to as the amount of time required by a species population to completely rework all the sediment in a given area to the depth accessible to the individuals – are used to quantify the biological mixing by bioturbation. The annual rate of sediment turnover for polychaetes from continental slopes (250 m) was reported to be about 10 kg $m^{-2}$ dry sediment, which represents turnover rates of 4 to 5 years, assuming a medium sediment reworking depth of 5 cm (Nichols 1974). In shallow muddy areas, however, it is not uncommon to find that the sea bottom is recycled through the benthos at least once, and in some cases up to several times a year. Rachor and Bartel (1981) observed remarkable quantities of feces produced by *E. echiurus* in muddy areas of the German Bight. Their calculations show that this feces might amount to 1 $m^3$ during one season in densely populated areas (250 mean-sized individuals $m^{-2}$) with 1-weekly turnover of the uppermost cm of sediment layer. Sediment-reworking activities such as burrowing, feeding, ingestion, and egestion by infaunal species contribute to the subduction of nutritive material below the sediment-water interface (Aller 1982).

A characteristic feature of the North and Baltic Sea ecosystems are the spring blooms that are based on the nutrient standing stock of winter (Brockmann et al., this Vol.). Nutrient levels are highest in late winter, because uptake by phytoplankton is at its lowest due to low solar radiation and temperature, while mineralization of organic detritus from the bottom deposits continues to some extent during winter. In early spring when available light is high enough to allow development of phytoplankton, a rapid growth occurs up to the point where the accumulated nutrient stock is exhausted. From that moment, and during the whole summer, growth of plankton is controlled by the rate of nutrient remineralization or supply by the polluted rivers. Benthic release can contribute up to 100% to this pool in the case of phosphate, as observed in the Baltic Sea (Balzer 1984). Detritus of plankton bloom events reaching the seafloor by sedimentation, but also by active incorporation by benthic filter feeders, is of crucial importance for the maintenance of benthic life in both positive and negative manner. In the German Bight and the western coast of Denmark, when a temporarily stratified water column is developed along with deposition of high biomasses, severe oxygen depletion may result in bottom waters, leading to macrobenthos mortality (Rachor 1980). Such events may again recycle high release rates of nutrients, as shown by Balzer (1984). Recent studies on the kinetics of organic matter mineralization in the sediments indicate a rapid response (within a few days) to deposition of fresh organic detritus during or at the end of the blooms. In early spring bioturbation is enhanced after the settling of the bloom (in certain cases even 20-fold, Dicke 1986). However, part of it is more refractory and accumulates in the benthos, where it forms a very large stock. The slow biodegradation of this stock is responsible for a continuous input of nutrients back to the overlying water column. When the nutrient-rich bottom waters reach the photic zone in response to disruption of the stratification, e.g., during gales or in autumn, further phytoplankton blooms may occur (Brockmann and Eberlein 1986). Steady seasonal changing in benthic activities may be therefore superimposed by pelagic events, and vice versa, leading to a close coupling of the pelagic element cycle with the cycle in benthic activity (Graf 1987).

Another important active feedback mechanism altering the biogeochemical environment was found to exist between the abundance of deposit feeders and the abundance of microorganisms. The abundance of microorganisms is closely related to the surface area and fecal pellet abundance in the sediment, which in turn are increased by the macrobenthos feeding activities. Hylleberg (1975) called the stimulation of microbial growth by deposit feeders "gardening" if these sustained organisms are used as food source. Secretion of mucopolysaccharides as sediment-agglutinating agents for tube construction and of metabolites may create local "hot spots" of bacterial activity and early diagenetic reactivities (Aller 1982). Models of pore water distributions demonstrate that the production rates of $NH_4^+$ in sediments in Mud Bay, South Carolina, are increased by at least 20–30% in the presence of macrofauna compared to controls or anoxic incubations, regardless of the model used (Aller and Yingst 1985). This is attributed to the overall lowering of inhibitory metabolite concentrations as well as to stimulation of bacteria during grazing. An increase of both total bacteria numbers and ATP/bacteria ratios are found in the surface sediment in the presence of *Heteromastus filifornis* (Polychaeta) and *Tellina texana* (Bivalvia) relative to controls.

The advection of sediment particles from anoxic depth to the sediment-water interface by subsurface deposit feeders may link two distinctly different biogeochemical zones, and certainly enhances sediment oxygen consumption. McCall and Fisher (1980) showed that the presence of *T. tubifex* (105 individuals $m^{-2}$) enhanced sediment oxygen consumption by a factor of 2. Approximately 20% of the additional demand was found to be the result of tubificid respiration, 10–30% apparently were consumed by increased microbial activity, but the remaining 50–70% were attributed to the oxidation of FeS brought to the sediment-water interface by tubificid feeding activity.

Fluid bioturbation, resulting in pore water mixing and exchange, includes burrowing by vagile fauna and irrigation of stationary species. Burrowing mixes the pore water entrapped within the reworked sediment. The wall linings of inhabited burrows are always well oxygenated because the worms regularly pump fresh water through their holes with peristaltic movements, thus supplying the surrounding bottom substratum with oxygen. Individual rates of such bioirrigation of infaunal species range between 2 and over 6000 ml $h^{-1}$ (Lee and Swartz 1980). Investigations on individual species bred under controlled laboratory conditions ("microcosms"), however, may result in overestimations of the bioirrigation rates. Emerson et al. (1984) showed, by comparing literature-derived field data, that dissolved solute transport in nearshore and estuarine sediments caused by nondiffusive mechanisms varied by less than two orders of magnitude in five different environments located on the west and east coast of the United States. A portion of the overlying oxic water irrigated into the burrows may in turn diffuse into and through surrounding permeable sediment at a rate depending on tube lining permeability (Aller 1982). Rachor and Bartel (1981) calculated the inner surface area of mean-sized burrows and thus found that abundant populations of *E. echiurus* (100–150 individuals $m^{-2}$) can by their irrigation activities double the total oxygenated surface area in muddy sediments of the German Bight. Ott et al. (1976) observed an even higher increase of the oxygenated sediment areas by burrowing of crustaceans (six to ten fold). Certainly, such irrigation activities of marine macrobenthos affects the transport of diagenetic reactants and reaction products (e.g., $SO_4^{2-}$, $NH_4^+$, $HCO_3^-$) and enhances the flux of contaminants across the sediment-water interface. Consequently, in the presence of macrobenthos, surfacial sediments are not a homogeneous medium dominated by one-dimensional vertical diffusion of early diagenesis products. As indicated in Fig. 1, they are rather both horizontally and vertically heterogeneous, intensively mixed, and penetrated by a network of solute exchange conduits and reactive microniches. Aller and Yingst (1985) have shown that, despite the complexity of such an

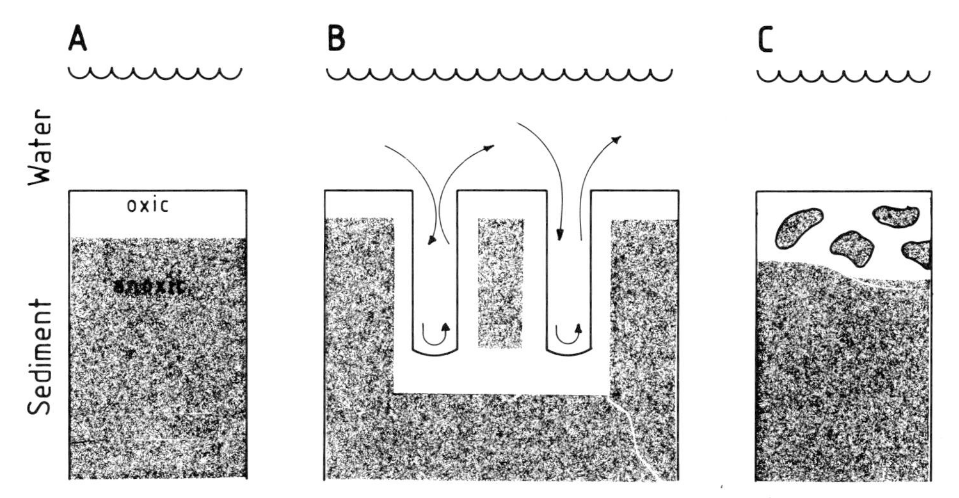


**Fig. 1A-C.** Major zonation of biogeochemical environments in sediments; **A** with no perturbation; **B** zonation around irrigated macrobenthos burrows; **C** development of reactive fecal pellet microniches. This concept of influence of macrofauna on solute transport in the sediment surface zone derives largely from the change in the geometry of diffusion brought about by burrow construction, defecation and irrigation. (Aller 1982)

**Table 1.** Qualitative effects of biogenic activities on sediment properties. (After Lee and Swartz 1980)

| Biogenic activity | Particle distribution | Erosion and resuspension | Sediment reworking | Sediment-water interface | Depth of aerobic layer | Flux of contaminants from sediment |
|---|---|---|---|---|---|---|
| Burrowing | Disaggregation of pellets and flocs | +(+ +) | +(+ +) | + | +(+ +) | +(+ +) |
| Excavation | Biogenic graded bedding, accumulation of fines | +(+ +) | + + | + + | +(+ +) | +(+ +) |
| Tubes | Agglutination and vertical aggregation | – | 0 | + + | + | + + |
| Fecal pellets | Pelletization and microtopography | + + | 0(+) | + | 0(–) | 0(+) |
| Digestion | Disaggregation | 0 | 0 | 0 | 0 | 0 |
| Irrigation | 0 | 0 | 0 | 0 | + + | + + |
| Microbe gardening | Disaggregation | – | 0 | 0 | – | – to + |

+ + = activity greatly increases sediment parameter.
0 = activity has no or minor effect on sediment parameter.
– = activity reduces sediment parameter.

environment, their transport-reaction models and two-dimensional or "nonlocal parametrization" models in particular provide a consistent basis for description of the effects of macroinfauna on bulk sediment properties, and allow for comparison of different species at similar population abundances.

Table 1 summarizes the primary effects of benthic species on sediment parameter in shallow marine environments, which can affect chemical sediment diagenesis in four ways:

1. Redistribution and addition of reactive materials in the form of organic metabolites and faecal pellets (filter feeding, biodeposition and bioentrainment), enhancing local element cycling, microbial activities ("hot spots") and rates at which diagenetic reactions proceed;
2. transfer of particles in the sediment column by bioadvection (burrowing, excavating and tube building) over vertical distances of up to several ten cm, moving of particles between oxidizing and reducing environments;
3. alteration of sediment fabric by intensive deposit feeding and burrowing. Production of a biotexture at and below the sediment surface by bioturbulent sediment reworking, biodeposition, and bioentrainment affecting the permeability and erodability of the sediments;
4. fluid bioirrigation ("biopumping"), which pumps water rapidly into and out of the bottom through vertically oriented tubes, increasing both the oxidized sediment-water interface and the surface area of the RPD, generally depressing the RPD at least by several cm in depth, depending on the organic matter inventory and sediment permeability, and resulting in a mosaic of biogeochemical sediment microenvironments rather than a vertically stratified zonation (see Fig. 1).

## 4 Mobility of Contaminants in Benthic Marine Ecosystems

Intensive sediment mixing and feeding activity of benthic animals is important in accelerating vertical diffusion and transport of trace element compounds adsorbed on particles (Schink and Guinasso 1978). Concentrations of trace elements are found to be higher in fecal materials than in surrounding sediments from which the animals select their food. Brown (1986) has shown that trace metal accumulation in feces is not caused by assimilation of organic material with its metal load for deposit feeders assimilating only a small percentage of material passing through their guts, but are likely to reflect the grading of materials prior to ingestion for fine particle size and for organic material. During feeding and burrow construction, animals are capable of selecting particles on the basis of particle position in the sediment, size, shape, surface texture, and density. Many conveyor-belt species or funnel feeders prefer ingesting and hence reworking fine-grained particles, which are known to carry the predominant fraction of contaminants. Thus any spatial segregation of particle types will result in a corresponding contaminant redistribution. Pollutant entrainment in fecal material further emphasizes the intimate link between benthic organisms and pollutant transport and fate in bottom sediments.

Since large numbers of infauna may occur in mud flats of the southeastern coast of the North Sea, sediment processing via fecal production of selective deposit feeders is likely to be an important pathway of pollutant cycling in these ecosystems. Little is known concerning the effects of residual digestive enzymes on metals and organic xenobiotics that are passed out with the feces, or the effects of decomposition on the fate of the pollutants accumulated in feces. Such studies would elucidate the role of benthic detritus feeders in the partitioning of metals between dissolved and particulate phases. Possible pathways are: leaching during the decomposition of organic matter concentrated in the feces (remobilization), or transfer into microorganisms such as bacteria and diatoms that colonize fecal pellets and uptake by coprophagous species, or by deposit feeders following pellet breakdown (Brown 1986). Uptake by microflora may render metals more bioavailable than sediment-bound metals to detritus-feeding organisms. Macrofaunal excretion can account for as much as 0.1 mmol $m^{-2}$ $day^{-1}$ to the bottom fluxes of phosphate found at summer temperatures, which represents a substantial proportion of the observed flux from mud sediments of Long Island Sound, and of the plankton requirement in this shallow marine environment (Aller 1980).

With regard to organic pollutants, Karickhoff and Morris (1985) have shown that tubifizid oligochaetes transported more than 90% of the xenobiotics hexachlorobenzene, pentachlorobenzene, and trifluralin to the sediment surface in laboratory microcosm experiments initiated with a uniform depth distribution of these toxic test chemicals in the sediment. Diffusive transport of the hydrophobic chemicals to the surface from depth via interstitial water organic colloids was orders of magnitude less than deposit-feeding macroinfauna-mediated transport via fecal pellets. Sorbed pollutants are transported by default in this process irrespective of the relative pollutant fugacities in the system, which indicates that secondary pollution assessments based on geochemical data alone may not imitate reality. Ultimately, however, the presence of benthic animals increased the release of the chemicals from the sediment to the overlying water by only four- to six-fold compared to the control sediment without benthos. This was less than the orders of magnitude expected based on the turnover rates. These researchers demonstrated by their experiments with fecal material that the strong entrapment of chemicals within the organic-rich fecal material in fact reduced the rate of release to the overlying water column. Similar studies have been reported on the recycling of organic pollutants by burrowing and feeding activities of experimental clams collected from the shore off Southamton, U.K. (Courtney and Denton 1977). Biogenic diffusion of contaminant-laden fine particles to the sediment-water interface may initiate horizontal dispersion and transport of these contaminants within the overlying water by resuspension, as discussed in the previous section. This process might explain the transport and cycling of many hydrophobic xenobiotics within productive marine environments of the German Bight and their occurrence in the Skagerrak and Norwegian trench basins.

Suspension-feeding animals anchored in the bottom may take advantage of vertically settling material but primarily feed on seston flowing horizontally past them. They prefer thereby ingesting fine-grained, organic-rich seston (Muschenheim 1987) carrying the predominant fraction of heavy metals. High metal concentrations may thereby accumulate in the small fine-grained fraction of sediments even when principally erosion is predominant in that area (Kersten and Klatt 1987). Selective feeding may also enhance organism pollutant concentrations. Gossett et al. (1983) found that the sediment and tissue concentrations of 27 selected xenobiotic organic compounds were positively correlated with each other and with the n-octanol/water partition coefficient but that these were negatively correlated with the water concentrations. Repetitive ingestion of fecal material or coprophagy is common in many invertebrate deposit feeders and demersal fish, which may significantly contribute to biomagnification of lipophilic contaminants by desorption during passage of material through the guts (Lee and Swartz 1980). Contaminated deposit-feeding epi- and infauna may significantly contribute to dietary uptake of toxic chemicals in demersal fish, resulting in food web transfer (e.g., Rubenstein et al. 1984). This class of effects, however, will not be considered here in more detail.

Vertical biogenic reworking of a pollutant, which proceeds at a rate faster than sedimentation, modifies its physicochemical and microbial environment, especially if particle excavation extends below the RPD layer. An important parameter controlling contaminant release in the course of early diagenesis is the particle residence time within the oxidized surfacial sediment layer. In order for a diagenetic process to alter the adsorptive characteristics and concentration of, for example, an organic substratum within the most reactive oxic sediment strata, its rate must be rapid with respect to the particle residence time. Due to the biological sediment mixing within the latter, the mean residence time of pollutants accumulating in the highly productive benthic ecosphere may far exceed that of the pollutants within underneath anoxic sediment layers. The pollutant residence time within the top oxidized cm of sediment is important in that aerobic microbial degradation is a major process by which many organic pollutants are removed from the aquatic environment, and many inorganic pollutants associated to degradable organic matter (e.g., $HPO_4^{2-}$ and Cd) are released to the overlying water column. While fecal elimination and degradation by benthic animals have been studied earlier (Ernst et al. 1977), little is known about the effect of bioturbation on microbial degradation rates of xenobiotics in marine sediments. An often cited first step is the study of Gordon et al. (1978), who demonstrated the more rapid decomposition and mineralisation of hydrocarbons in the burrows of lugworms *Arenicola marina*. Pollutant transformation and transfer is thus affected by sediment bioturbation through (1) increasing the oxidized sediment surface as schematized in Fig. 1, (2) enhancing particle residence time within this top sediment layer, and (3) increasing aerobic organic matter degradation by gardening effects on the microbiology.

Under oxidized conditions, the concentration of a metal in solution is determined by the equilibrium between the dissolved metal and the metal sorbed onto particle surfaces rather than that of precipitation-dissolution reaction found in anoxic zones. Salomons (1985) has stressed that this distinction is important in assessing the toxic effects of metals introduced into the aquatic ecosystem. If precipitation-dissolution processes control the trace metal availability, its dissolved concentrations will not depend on the solid metal content, and an increased input to the system will not affect its concentration in the interfacial water. On the other hand, when adsorption-desorption is the main process for binding metals to sediments, an increased input will cause an increase in soluble metal flux, as outlined in Fig. 2. This is especially true in environments subject to considerable pore water flushing by physical action of tides and waves (Morris et al. 1982; Kerner et al. 1986). It is worth noting that the reversible adsorption-bioturbation model explains much of the enigma of ferromanganese nodule growth in deep sea basin deposits, where element accumulations in the course of oxic sediment diagenesis are attributable neither to pore water transport nor to precipitation from the overlying water column (Lyle et al. 1984).

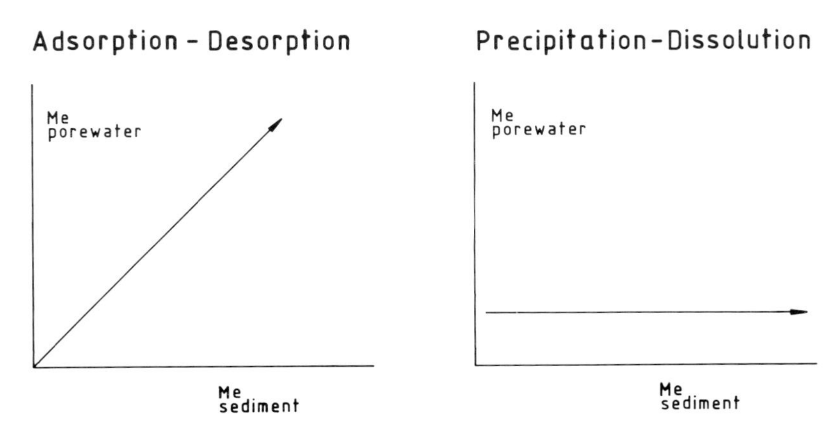


**Fig. 2.** Variation in pore water metal concentrations for adsorption-desorption (assuming Freundlich isotherms) versus precipitation-dissolution controlled particulate metal binding mechanisms. (Salomons 1985)

Enhanced biogenic ventilation of surfacial sediments by oxic overlying water affects the mobility of trace metals and phosphorus in two important ways: (1) by altering the chemistry of the redox-sensitive elements iron, manganese and sulfur, and (2) by providing an additional pore-water exchange mechanism at the sediment-water interface. The combined effect of both mechanisms is to enhance the role of the sediments as a source of dissolved metals released from oxidized sulfides, and a sink for phosphate precipitated with Fe/Mn-oxides. The importance of the latter mechanism to the $PO_4^{3-}$ flux is yet not clear. Dicke (1986), in considering her experimental results, arrived at the conclusion that bioturbation may counteract phosphate remobilization and eutrophication effects due to fixation in the increased oxic sediment layer. On the other hand, the results presented by Krom and Berner (1981) suggest that the flux of dissolved phosphate out of the sediment must arise largely from phosphate regeneration at the sediment-water interface at the time of burial, which means that the flux is closely related to pelagial biological events rather than to biologically controlled diffusive flux from beneath the sediment-water interface. Removal of sulfide in the sediment by oxidation with oxic water would facilitate the mobilization of Cd and Cu (and other "class B" metals) which form very insoluble sulfides in the course of anoxic diagenesis (Emerson et al. 1984). Kersten and Förstner (1986) have recently shown, using sequential chemical leaching experiments, that sulfides of these trace metals derived from recent anoxic muds are very sensitive to oxidation by irrigation with oxic water. These results are exemplified for cadmium in Fig. 3. In Narragansett Bay sediments Elderfield et al. (1981) found that in the absence of any bioturbation the concentration of Cd was below bottom water concentration in the surface pore waters and below detection limits a few centimeters beneath the sediment-water

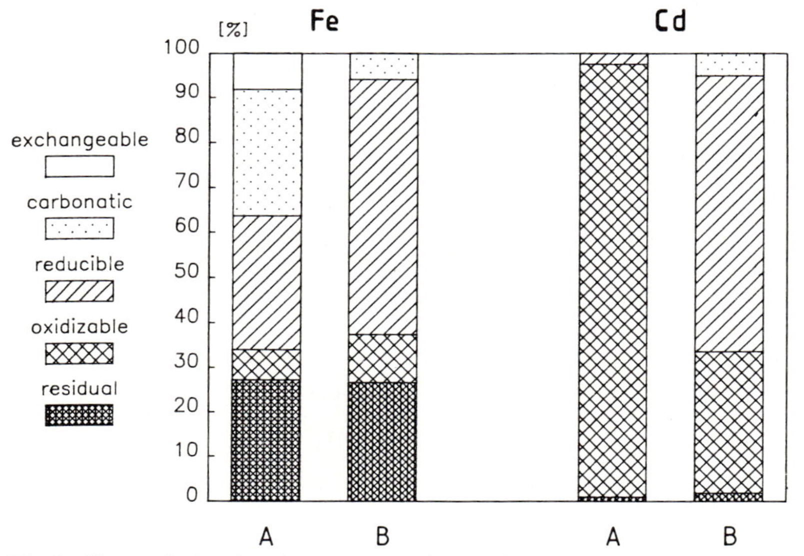


**Fig. 3.** Changes in fractionation patterns of iron and cadmium in anoxic sediment samples from Hamburg harbour basins, as analyzed by sequential chemical leaching (*A*) in situ and (*B*) subsequent to short-term (l-h) irrigation with oxic water. (Kersten and Förstner 1986)

interface. This they attributed to the strongly anoxic sediment environment beneath the sediment-water interface. The sediments of Narragansett Bay clearly act as a sink for this trace metal. On the other hand, Emerson et al. (1984) found that sediments in the Quartermaster Harbour (shallow embayment of the Puget Sound) were a continuing source for this metal due to the effective benthic bio-irrigation activity. In these sediments the pore waters were found to be highly enriched with Ni, Cu, and Cd in the surface few centimeters with strong vertical gradients for the latter two metals across the sediment-water interface. Jernelöv (1970) reported in an earlier study that the transfer of Hg from sediments was affected by tubificids (820,000 individuals per $m^2$). This caused methyl mercury release from a depth of 0–2.5 cm. In the absence of macrobenthos, Hg was released only from the top 1 cm sediment layer, probably due to the methyl mercury disproportionation to highly insoluble HgS in the anoxic environment arising below 1 cm depth.

Sensitivity of pollutants to biogenic pore water transport activities clearly depends on the chemical composition of the bioturbated zone of the marine sediments, which represents a balance between transport and reaction processes taking place within and around it. Various approaches have been used to evaluate the influence of benthos activity on the net flux of nutrients and trace metals. The magnitude and relative importance of this source in any given area can be determined in three independent ways (Aller 1980): (1) direct measurement by short-term incubation of bottom water and underlying sediment either in situ or in cores stored in laboratory; (2) calculation of fluxes by measurement of the appropriate pore water concentration gradient and employing a Fick's first law relation with an assumed diffusion coefficient; and/or (3) determination of the consumption/production rate term of the solute within the sediment body and modeling the resulting flux at steady state. Some or all of these approaches have been used to determine nutrient and trace metal fluxes in shallow marine environments (Goldhaber et al. 1977; Billen 1978; Aller 1980; Elderfield et al. 1981; Lyons et al. 1982; Balzer 1984; Dicke 1986; and references therein). Emerson et al. (1984) compared the contribution of biological advection and molecular diffusion to the transport of dissolved materials across the sediment-water interface using a combination of an in situ tritium tracer experiment and pore water silicate data. The results, as summarized in Table 2, indicated that biogenic transport dominates the flux in sediments only for compounds which do not have strong gradients at the sediment-water interface. This is true for alkalinity, silicate, ammonia, and Ni, but not for trace elements strongly associated to organic matter which are released within the top cm of sediment column. Silicate fluxes from the sediments in the Swedish Gullmarsfjorden were also found to be enhanced in summer and autumn two to ten times more than those expected on the basis of molecular diffusion. This was attributed to the activity of macrobenthos (Rutgers van der Loeff et al. 1984). Silicate fluxes are effectively quantifiable by theoretical models assuming eddy diffusive processes due to the fact that silicate regenerates at a sufficiently slow rate in comparison to phosphate regeneration, and is not strongly influenced by redox reactions taking place within the sediments. On the other hand, McCaffrey et al. (1980) found in their study of epibenthic fluxes in Narragansett Bay that even at a site of high benthic activity molecular diffusion is as important as the respiratory

**Table 2.** Flux between pore water and overlying water for the two main transport mechanisms: molecular diffusion (MD) and "nonlocal" biogenic (B) transport [average estimates made by Emerson et al. (1984) for Quartermaster Harbor sediments]. MD/B is the ratio of diffusion to biogenic flux calculated for the selected chemical species. The range in this value is caused by the range in the estimate of rate parameter $\alpha$ ($1-5 \times 10^{-7}$) used to evaluate the nonlocal source or sink term in the model upon that the calculation is based

| Chemical species | Fluxes Diffusive | Biogenic ($m\ cm^{-2} s^{-1} \times 10^{-5}$) | MD/B |
|---|---|---|---|
| $Si(OH)_4$ | 1071 ± 328 | 948– 4740 | 0.23–1.1 |
| $NH_4^+$ | 574 ± 155 | 575– 2875 | 0.20–1.0 |
| Alk | 1104 ± 562 | 5666–28333 | 0.04–0.2 |
| $Fe^{2+}$ | 200 ± 100 | 70– 350 | 0.57–2.9 |
| $Mn^{2+}$ | 60 | 29– 145 | 0.41–2.1 |
| Cu | 0.49 | 0.80– 0.42 | 1.20–5.8 |
| Ni | 0.12 | 0.10– 0.50 | 0.24–1.2 |
| Cd | 0.18 | 0.03– 0.14 | 1.30–6.0 |

activity of organisms in transporting metabolites to the overlying waters. They attribute this to the fact that most nutrient and metal release to pore waters by early diagenesis takes place in the upper centimeters of the sediment. At this depth, diffusion to overlying water column is rapid due to the very strong gradients, and is the dominant pathway of remobilization rather than biopumping. The flux data given in Table 2 suggest further that, in general, metal remobilization via pore water is about two orders of magnitude higher than that by biogenic particle reworking based on the resuspension rates given by Lee and Swartz (1980, cf. discussion in the preceding section). Supporting experimental evidence for the predominant influence of diffusion on the loss rate of metals was given by the tracer experiments conducted with $^{65}Zn$ in laboratory mesocosms (Renfro 1973).

Available evidence suggests that for trace metals in productive shallow marine environments, the main influence of macrobenthic activity is indirect, in that the enhanced sediment ventilation increases oxic diagenesis and metal recycling but inhibits the buildup of an anoxic environment, thus preventing metal sulfide formation and increasing the dissolved metal gradients in near-surface pore waters. In temperate environments, where temperature and benthic activities vary seasonally, iron and other trace metals in marine sediments may undergo seasonal dissolution and precipitation reactions as a result of changing redox regimes (Hines et al. 1984). Figure 4 reproduces a model of the seasonal variation in dissolved iron in Great Bay sediments, New Hampshire, as a result of a 3-year monitoring of pore water chemistry conducted by Hines and co-workers. This idealized annual trend of dissolved iron in bioturbated and nonbioturbated sites also presents the reactions which dominate the iron speciation throughout the year. In general, net iron mobilization was recorded twice a year; first during the spring when the first "marine snow" biodeposits arise, and second during the entire summer period of enhanced bioturbation activities. As pointed out by Hines et al. (1984) these trends are subject to large interannual variations and sensitive to weather events in many situations. Similar trends were found by the authors for manganese and copper.

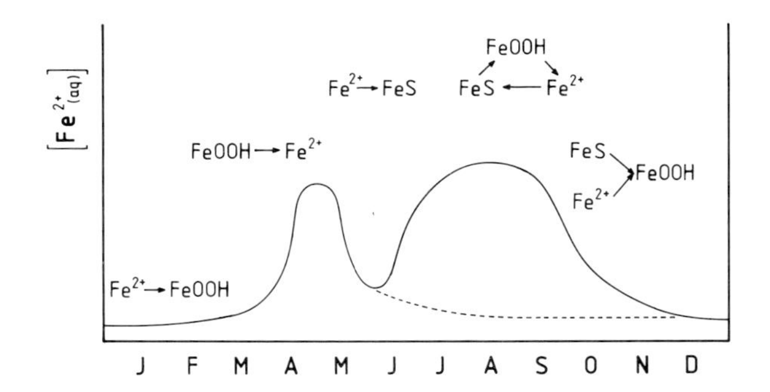


**Fig. 4.** Model of biologically induced seasonal variation in dissolved ferrous ions in Great Bay sediments. *Dashed line* indicates relative concentration in the absence of any bioturbation when the precipitation reaction $Fe^{2+} \rightarrow FeS$ continuous throughout the summer. (Hines et al. 1984)

They calculated a diffusional benthic copper flux into the overlying water during the pronounced spring event in 1978 of about $10^3$ g copper per day, which is approximately 20% of the daily river input into Great Bay Estuary at this time. An important consequence of this is that such biogenic events may contribute significantly to a flux of metals during a short period of time. This may be more characteristic for the geobiological influence on remobilization of pollutants than long-term remobilization at the lower mean levels as calculated in studies cited above.

# 5 Conclusions

Particle reworking and sediment irrigation activities by bioturbating macrofauna have a profound influence on sediment biogeochemistry and remobilization of pollutants in shallow marine environments. Sediments not only act as a sink for metals and organic xenobiotics but, under appropriate conditions, may function as a secondary pollution source of toxic contaminants affecting the overlying water quality. Hence we cannot consider North Sea sediments as permanent sinks, but rather as intermediate storage sites, or "Zwischenlager", to use a modern word created from nuclear technology. It has become apparent from the observations cited in this review that macrobenthos control sediment erodability and pore water solute concentration profiles, which in turn affect the efflux rates and redistribution of pollutants to the overlying water. Depending on the geochemical reactions governing the behaviour of a given pollutant, changes in transport processes brought about by bioturbation may influence the *net flux* of its dissolved form between sediment and overlying water. It has been shown that in at least the buildup of pore water solute concentrations and gradients of most pollutants is highly sensitive to the bioturbating activities of infaunal communities. The main process, in which enhanced trace metal remobilization from North Sea sediments might proceed, is suggested to function with the following mechanisms: (1) enhanced oxic chemical diagenesis of trace metal loaden organic matter by sediment bioturbating macrobenthos, (2) inhibition of released metal precipitation as highly insoluble sulfides by depressing of the oxic/anoxic interface to higher sediment depth through oxic water bioirrigating macrobenthos, (3) thereby generating increased dissolved trace metal pore water gradients at the sediment-water interface, and (4) efflux of dissolved trace metals from the sediments to the overlying water by molecular diffusion driven by the pore water gradients. Remobilization of organic xenobiotics might be enhanced by macroinfauna-mediated selective ingestion and reworking of contaminant-laden organic particles from the whole biologically active sediment layer to the sediment water interface, and subsequent physical resuspension and redistribution into the overlying water column. This can make contaminants available again which were thought to have been safely disposed of in the sedimentary Zwischenlager.

The mechanisms mentioned above are, of course, linked to trophic group/life habit type, which in turn is highly dependent on environmental parameters such as availability of organic matter and dissolved oxygen, and is thereby intimately coupled to pelagic biological activities. The most important finding is that contaminant remobilization may be episodic. Since the intensity of the above mechanisms will vary seasonally following biological events, assessments based on annually averaged data may clearly underestimate the significance of these mechanisms to the marine ecosphere. The significance of trace metal remobilization by epibenthic efflux have been indeed evidenced in several of the most intensely studied shallow marine environments of the world. However, appropriate studies for the coastal and shelf areas of the North Sea are as yet lacking. To simplify the monitoring task, inter-relationships tend to be ignored, and annually determinations of the elemental composition of isolated suspended matter and/or filtered water are commonly accepted as adequately defining the system. Basic line studies and empirical estimates of vertical distribution, resuspension, and epibenthic net fluxes on the trace metal distribution in the coastal waters of the Southern Bight were recently successfully accomplished (Baeyens et al. 1986). An interdisciplinary study of the mechanisms and process rates discussed above would prove a great challenge. Studies to quantify the importance of the detailed processes for modeling purposes still requires laboratory mesocosm experiments. The present literature review is part of such an interdisciplinary study involving biologists and geochemists of two German institutes funded by the German Ministry of Technology and Science. Detailed time series experiments will be conducted to examine the effects of bioirrigation of key species on the migration of metals during short- or long-term mobilization events. Radiotracer and luminophore tracer experiments are used to determine depth and rates of particle reworking. In particular, the effects of macrobenthos on trace metal speciation below and above the RPD will be analyzed. To verify the obtained results in field, development of new and far advanced in situ data-gathering techniques and appropriate fast analytical screening techniques is required, which warrants close international cooperation.

*Acknowledgments.* This chapter has greatly benefited from my collaboration with Dr. Gerhard Graf (University of Kiel). Dr. Monika Dicke (DHI, Hamburg) provided helpful material and comments on drafts of this paper. Ms. Karen Wiltshire (University of Dublin, on sabbatical at GKSS, Hamburg) kindly helped in improving the English version. Sincere appreciation is expressed to Prof. Dr. Ulrich Förstner for instructive discussions and encouragement, and to the German Ministry of Technology and Science for funding.

## References

Aller RC (1980) Diagenetic processes near the sediment-water interface of Long Island Sound. I. Decomposition and nutrient element geochemistry (S, N, P). Adv Geophys 22:237–350

Aller RC (1982) The effects of macrobenthos on chemical properties of marine sediment and overlying water. In: McCall PL, Tevesz MJS (eds) Animal-sediment relations. Plenum Press, New York, pp 53–102 (Topics in Geobiology, vol 2)

Aller RC, Yingst JY (1985) Effects of marine deposit-feeders *Heteromastus filiformis* (Polychaeta), *Macoma balthica* (Bivalvia), and *Tellina texana* (Bivalvia) on averaged sedimentary solute transport, reaction rates, and microbial distributions. J Mar Res 43:615–645

Baeyens W, Gillain G, Hoenig M, Dehairs F (1986) Mobilization of major and trace elements at the water-sediment interface in the Belgian coastal area and the Scheld estuary. In: Nihoul JCJ (ed) Dynamic biological processes at marine physical interfaces. Elsevier, Amsterdam, pp 453–466

Balzer W (1984) Organic matter degradation and biogenic element cycling in a nearshore sediment (Kiel Bight). Limnol Oceanogr 29:1231–1246

Baumfalk YA (1979) Heterogeneous grain size distribution in tidal flat sediment caused by bioturbation activity of *Arenicola marina* (Polychaeta). Neth J Sea Res 13:428–440

Berner RA (1981) Authigenic mineral formation resulting from organic matter decomposition in modern sediments. Fortschr Miner 59:117–133

Billen G (1978) A budget of nitrogen recycling in North Sea sediments off the Belgian coast. Estuar Coastal Mar Sci 7:127–146

Brockmann UH, Eberlein K (1986) River input of nutrients into the German Bight. In: Skreslet S (ed) The role of freshwater outflow in coastal marine ecosystems. NATO ASI Ser G7, Springer, Berlin Heidelberg New York Tokyo, pp 231–240

Brown SL (1986) Feces of intertidal benthic invertebrates: influence of particle selection in feeding on trace element concentration. Mar Ecol Prog Ser 28:219–231

Calmano W, Förstner U (1983) Chemical extraction of heavy metals in polluted river sediments in Central Europe. Sci Total Environ 28:77–90

Courtney WAM, Denton GRW (1977) Persistence of polychlorinated biphenyls in the hard-clam *Mercenaria mercenaria* and the effect upon the distribution of these pollutants in the estuarine environment. Environ Pollut 10:55–64

Dicke M (1986) Vertikale Austauschkoeffizienten und Porenwasserfluβ an der Sediment/Wasser-Grenzfläche. Thesis, University of Kiel (Berichte IfM Kiel, vol 155)

Dominik J, Förstner U, Mangini A, Reineck HE (1978) $^{210}$Pb and $^{137}$Cs chronology of heavy metal pollution in a sediment core from the German Bight (North Sea). Senckenberg Mar 10:213–227

Duinker JC, Nolting RF (1977) Dissolved and particulate trace metals in the Rhine estuary and the Southern Bight. Mar Pollut Bull 8:68–71

Eckman JE (1985) Flow disruption by an animal-tube mimic affects sediment bacterial colonization. J Mar Res 43:419–435

Eisma D (1981) Supply and deposition of suspended matter in the North Sea. Spec Publ Int Ass Sediment 5:415–428

Elderfield H, McCaffrey PJ, Luedtke N, Bender M, Truesdale VW (1981) Chemical diagenesis in Narragansett Bay sediments. Am J Sci 281:1021–1055

Emerson S, Jahnke R, Heggie D (1984) Sediment-water exchange in shallow water estuarine sediments. J Mar Res 42:709–730

Ernst W, Goerke H, Weber K (1977) Fate of $^{14}$C-labelled di-, tri- and pentachlorobiphenyl in the marine annelid *Nereis virens* II. Degradation and faecal elimination. Chemosphere 9:559–568

Fernex FE, Span D, Flatau GN, Renard D (1986) Behavior of some metals in surficial sediments of the northwest Mediterranean continental shelf. In: Sly PG (ed) Sediments and water interactions. Springer, Berlin Heidelberg New York Tokyo, pp 353–370

Förstner U, Ahlf W, Calmano W, Kersten M (1986) Mobility of pollutants in dredged materials – implications for selecting disposal options. In: Kullenberg D (ed) The role of the oceans as a waste disposal option. Reidel, Doordrecht, pp 597–615

Gendron A, Silverberg N, Sundby B, Lebel J (1986) Early diagenesis of cadmium and cobalt in sediments of the Laurentian Trough. Geochim Cosmochim Acta 50:741–747

Goldhaber MB, Aller RC, Cochran JK, Rosenfeld JK, Martens CS, Berner RA (1977) Sulfate reduction, diffusion, and bioturbation in Long Island Sound sediments: report of the FOAM Group. Am J Sci 277:193–237

Gordon D, Dale J, Keizer P (1978) Importance of sediment reworking by the deposit-feeding polychaete *Arenicola marina* on the weathering rate of sediment-bound oil. J Fish Res Bd Can 35:591–603

Gossett RW, Brown DA, Young DR (1983) Predicting the bioaccumulation of organic compounds in marine organisms using octanol/water partition coefficients. Mar Pollut Bull 14:387–392

Graf G (1987) Benthic energy flow during a simulated autumn bloom sedimentation. Mar Ecol Prog Ser 39:23–29

Gray JS (1982) Effects of pollutants on marine ecosystems. Neth J Sea Res 16:424–443

Hainbucher D, Backhaus JO, Pohlmann T (1986) Atlas of climatological and actual seasonal circulation patterns in the North Sea and adjacent shelf regions: 1969–1981. Tech Rep, Inst Mar Res Univ Hamburg

Hines ME, Jones GE (1985) Microbial biogeochemistry and bioturbation in the sediments of Great Bay, New Hampshire. Estuar Coastal Shelf Sci 20:729–742

Hines ME, Lyons WB, Armstrong PB, Orem WH, Spencer MJ, Gaudette HE, Jones GE (1984) Seasonal metal remobilization in the sediments of Great Bay, New Hampshire. Mar Chem 15:173–187

Hunt CD, Smith DL (1983) Remobilization of metals from polluted marine sediments. Can J Fish Aquat Sci 40 (Suppl 2):132–142

Hylleberg J (1975) Selective feeding by *Abarenicola pacifica* with notes on *Abarenicola vagabunda* and a concept of gardening in lugworms. Ophelia 14:113–137

Irion G, Schwedhelm E (1983) Heavy metals in surface sediments of the German Bight and adjoining areas. In: Müller G (ed) Intern Conf Heavy metals in the environment, Heidelberg, Sept 1983, vol 2. CEP Consultants, Edinburgh, pp 888–891

Jernelöv A (1970) Release of methyl mercury from sediments with layers containing inorganic mercury at different depth. Limnol Oceanogr 15:958–960

Jones CJ, Murray JW (1984) Nickel, cadmium and copper in the Northeast Pacific off the coast of Washington. Limnol Oceanogr 29:711–720

Jørgensen BB (1977) The sulfur cycle of a coastal marine sediment (Limfjorden, Denmark). Limnol Oceanogr 22:814–832

Karickhoff SW, Morris KR (1985) Impact of tubificid polychaetes on pollutant transport in bottom sediments. Environ Sci Technol 19:51–56

Kerner M, Kausch H, Kersten M (1986) Der Einfluß der Gezeiten auf die Verteilung von Nährstoffen und Schwermetallen in Wattsedimenten des Elbe-Aestuars. Arch Hydrobiol Suppl 75:118–131

Kersten M, Förstner U (1985) Trace metal partitioning in suspended matter with special reference to pollution in the southeastern North Sea. In: Degens ET, Kempe S, Herrera R (eds) Transport of carbon and minerals in major world rivers, Part 3. Mitt Geol-Paläontol Inst Univ Hamburg, SCOPE/UNEP Sonderbd 58:631–645

Kersten M, Förstner U (1986) Chemical fractionation of heavy metals in anoxic estuarine and coastal sediments. Wat Sci Tech 18:121–130

Kersten M, Klatt V (1988) Trace metal inventory and geochemistry of the North Sea shelf sediments. In: Degens ET, Kempe S, Liebezeit G (eds) TOSCH-Sonderband. Mitt Geol-Paläontol Inst Univ Hamburg (in press)

Kremling K (1983) Trace metal fronts in European shelf waters. Nature (Lond) 303:225–227

Kremling K (1985) The distribution of cadmium, copper, nickel, manganese, and aluminium in surface waters of the open Atlantic and European shelf area. Deep-sea Res 32:531–555

Krom MD, Berner RA (1981) The diagenesis of phosphorus in a nearshore marine sediment. Geochim Cosmochim Acta 45:207–216

Lee H, Swartz RC (1980) Biological processes affecting the distribution of pollutants in marine sediments. Part II: biodeposition and bioturbation. In: Baker RA (ed) Contaminants and sediments, vol 2. Ann Arbor Sci Publ, Ann Arbor, MI, pp 555–606

Lyle M (1983) The brown-green color transition in marine sediments: a marker of the Fe(III)-Fe(II) redox boundary. Limnol Oceanogr 28:1026–1033

Lyle M, Heath GR, Robbins JM (1984) Transport and release of transition elements during early diagenesis: sequential leaching of sediments from MANOP sites M and H. Part I: pH 5 acetic acid leach. Geochim Cosmochim Acta 48:1705–1715

Lyons WB, Loder TC, Murray SM (1982) Nutrient pore water chemistry, Great Bay, New Hampshire; benthic fluxes. Estuaries 5:230–233

Mart L, Nürnberg HW (1986) Cd, Pb, Cu, Ni and Co distribution in the German Bight. Mar Chem 18:197–213

McCaffrey RJ, Myers AC, Davey E, Morrison G, Bender M, Luedtke N, Cullen D, Froelich P, Klinkhammer G (1980) The relation between pore water chemistry and benthic fluxes of nutrients and manganese in Narragansett Bay, Rhode Island. Limnol Oceanogr 25:31–44

McCall PL (1977) Community patterns and adaptive strategies of the infaunal benthos of Long Island Sound. J Mar Res 35:221–266

McCall PL, Fisher JB (1980) Effects of tubificid oligochaetes on physical and chemical properties of Lake Erie sediments. In: Brinkhurst RO, Cook DG (eds) Aquatic oligochaete biology. Plenum Press, New York, pp 253–317

McCall PL, Tevesz MJS (eds) (1982) Animal-sediment relations. Plenum Press, New York (Topics in Geobiology, vol 2)

Morris AW, Bale AJ, Howland RJM (1982) The dynamics of estuarine manganese cycling. Estuarine Coastal Shelf Sci 13:175–192

Muschenheim DK (1987) The dynamics of near-bed seston flux and suspension-feeding animals. J Mar Res 45:473–496

Nichols F (1974) Sediment turnover by a deposit-feeding polychaete. Limnol Oceanogr 19:945–950

Nolting RF (1986) Copper, zinc, cadmium, nickel, iron and manganese in the Southern Bight of the North Sea. Mar Pollut Bull 17:113–117

Nowell AR, Jumars PA, Eckman JE (1981) Effects of biological activity on the entrainment of marine sediments. Mar Geol 42:133–153

Ott JA, Fuchs B, Fuchs R, Malasek A (1976) Observations on the biology of *Callianassa stebbingi* Borradaile and *Upogebia litoralis* Risso and their effect upon sediment. Senckenberg Marit 8:61–79

Peterson ML, Carpenter R (1986) Arsenic distributions in pore water and sediments of Puget Sound, Lake Washington, the Washington coast and Saanich Inlet, B.C. Geochim Cosmochim Acta 50:353–369

Rachor E (1980) The inner German Bight – an ecologically sensitive area, as indicated by the bottom fauna. Helgol Meeresunters 33:522–530

Rachor E, Bartel S (1981) Occurrence and ecological significance of the spoon-worm *Echiurus echiurus* in the German Bight. Veröff Inst Meeresforsch Bremerh 19:71–88

Renfro W (1973) Transfer of $^{65}$Zn from sediments by marine polychaete worms. Mar Biol 21:305–316

Revsbech NP, Jørgensen BB, Blackburn TH (1980) Oxygen distribution in sediments measured with microelectrodes. Limnol Oceanogr 25:403–411

Riedl RJ, Huang N, Machan R (1972) The subtidal pump: a mechanism of interstitial water exchange by wave action. Mar Biol 13:210–221

Rhoads DC (1974) Organism-sediment relations on the muddy sea floor. Oceanogr Mar Biol Annu Rev 12:263–300

Rhoads DC, Boyer LF (1982) The effects of marine benthos on physical properties of sediments – a successional perspective. In: McCall PL, Tevesz MJS (eds) Animal-sediment relations. Plenum Press, New York, pp 3–52

Rhoads DC, Young DK (1970) The influence of deposit-feeding organisms on sediment stability and community trophic structure. J Mar Res 28:150–178

Risk MJ, Moffat JS (1977) Sedimentological significance of faecal pellets of *Macoma balthica* in the Minas Basin, Bay of Fundy. J Sedim Petrol 47:1425–1436

Rubenstein N, Wilkes FG, D'Asaro CN, Sommers C (1980) The effects of contaminated sediments on representative estuarine species and developing benthic communities. In: Baker RA (ed) Contaminants and Sediments, vol 1. Ann Arbor, Ann Arbor, MI, pp 445–461

Rubenstein N, Gilliam WT, Gregory NR (1984) Dietary accumulations of PCBs from a contaminated sediment source by a demersal fish (*Leiostomus xanthurus*). Aquat Toxicol 5:331–342

Rutgers van der Loeff MM (1980) Nutrients in the interstitial water of the Southern Bight of the North Sea. Neth J Sea Res 14:144–171

Rutgers van der Loeff MM, Anderson LG, Hall POJ, Iverfeldt A, Josefson AB, Sundby B, Westerlund SFG (1984) The asphyxiation technique: an approach to distinguishing between molecular diffusion and biologically mediated transport at the sediment-water interface. Limnol Oceanogr 29:675–686

Salomons W (1985) Sediments and water quality. Env Technol Lett 6:315–326

Salomons W, Förstner U (1984) Metals in the hydrocycle. Springer, Berlin Heidelberg New York Tokyo

Salzwedel H, Rachor E, Gerdes D (1985) Benthic macrofauna communities in the German Bight. Veröff Inst Meeresforsch Bremerh 20:199–267

Schäfer W (1972) Ecology and paleoecology of marine environments. (Oertel I, Craig GY, translators), Univ Chicago Press, Chicago, Illinois

Schink DR, Guinasso NL Jr (1978) Redistribution of dissolved and adsorbed materials in abyssal marine sediments undergoing biological stirring. Am J Sci 278:687–702

Stumm W, Morgan JJ (1970) Aquatic chemistry. Wiley & Sons, New York

Suess E (1979) Mineral phases formed in anoxic sediments by microbial decomposition of organic matter. Geochim Cosmochim Acta 43:339–352

Suess E (1980) Particulate organic carbon flux in the oceans: Surface productivity and oxygen utilization. Nature (Lond) 288:260–263

Vanderborght JP, Wollast R, Billen G (1977) Kinetic models of diagenesis in disturbed sediments, I: mass transfer properties and silica diagenesis. Limnol Oceanogr 22:787–793

Wangersky PJ (1986) Biological control of trace metal residence time and speciation: A review and synthesis. Mar Chem 18:269–297

Westerlund SG, Anderson LG, Hall POJ, Iverfeldt A, Rutgers van der Loeff MM, Sundby B (1986) Benthic fluxes of cadmium, copper, nickel, zinc, and lead in the coastal environment. Geochim Cosmochim Acta 50:1289–1296

Zeitschel B (1980) Sediment-water interactions in nutrient dynamics. In: Tenore KR, Coull BC (eds) Marine benthic dynamics. Univ South Carolina Press, p 195

# The Nature and Functioning of Salt Marshes[1]

W.G. BEEFTINK[2] and J. ROZEMA[3]

## 1 Tidal Landforms of the North Sea Coast

Tidal landforms are defined as that part of coastal wetlands that includes all areas of marsh, mudflats, and stretches of water between (extreme) low- and high-water marks. They are mostly formed in coastal areas with a medium to large tidal range and shelter against the effects of wind-driven waves. Mudflats and salt marshes are separated from each other around the high-water line at neap tide (MHWN) where the pioneer vegetation closes. Salt marshes are in fact vegetated mudflats and other periodically inundated marine and estuarine grasslands.

Different types of salt marshes are found around the North Sea. The most common types are the foreland and barrier-connected salt marshes, followed by the estuarine salt marshes (Beeftink 1977; Dijkema et al. 1984). More local, around the Skagerrak, salt marshes with a limited sediment supply occur in drowned glacial erosion valleys (fjord heads) and on beach heads. Also land upheaval marshes are found there, developed on autochthonous material. Figure 1 shows the distribution of the salt-marsh areas along the North Sea coasts. About two third (502 km$^2$) of all North Sea salt marshes (654.5 km$^2$ in total) extend on the continental coasts, and 359 km$^2$ of that part is lying in the Dutch-German-Danish Wadden area (Dijkema et al. 1984). These figures must be enlarged to a multiple if mudflats proper are included.

## 2 Mudflat and Salt-Marsh Development

For the formation sand of mudflats and salt marshes the supply of, silt and clay is required, depositing in sheltered areas to above the mean high-water line. In fact there are four possible source areas for deposition of fine sediments in tidal wetlands (Pethick 1984):

1. Marine: derived from the sea bed. About 80% of the clay sediments deposited in the Dutch part of the Waddensea originate from the coastal North Sea zone.

---

[1]Communication Nr. 413 Delta Institute for Hydrobiological Research, Yerseke, The Netherlands

[2]Delta Institute for Hydrobiological Research KNAW, Vierstraat 28, 4401 EA Yerseke, The Netherlands

[3]Department of Ecology and Ecotoxicology, Free University, Post Box 7161, 1007 MC Amsterdam, The Netherlands

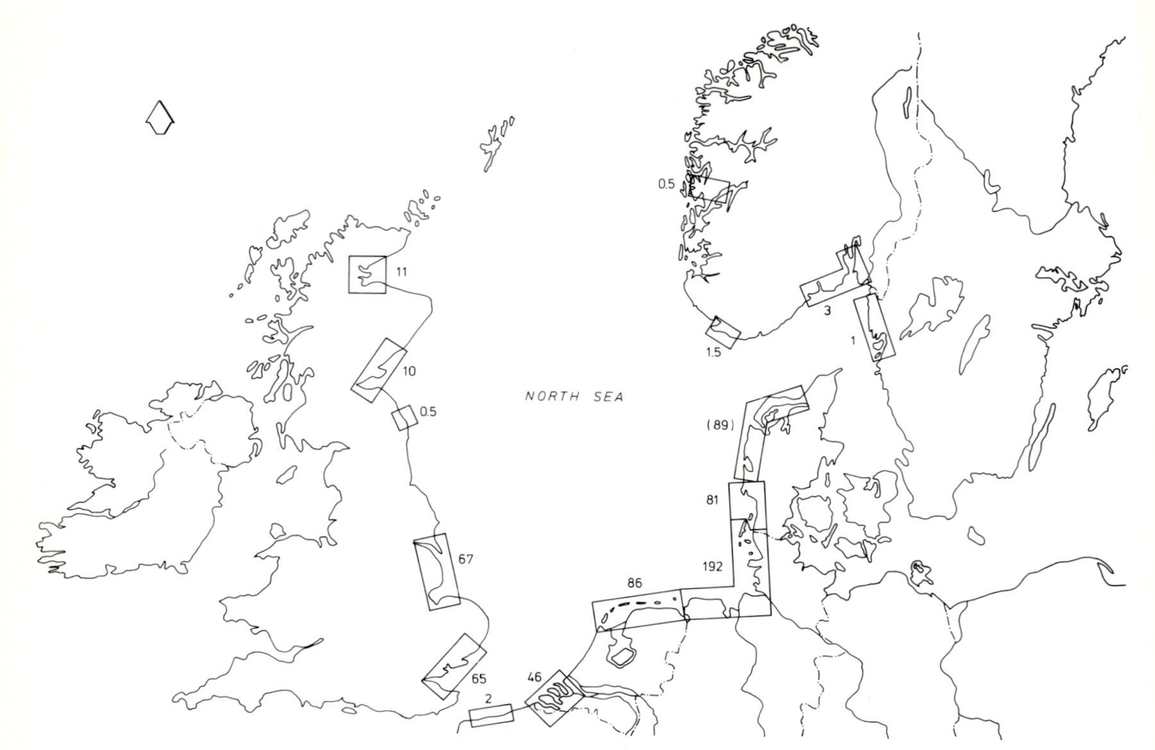


**Fig. 1.** Location of salt marsh areas (km² in surface) on the North-Sea coasts. (Data from Dijkema et al. 1984)

Most of the remaining 20% is contributed by the rivers Rhine and Meuse, and transported to the north by the tidal rest stream (Salomons et al. 1975). Likewise, most clay deposits in the Westerschelde come from the Vlaamse Banken off the Belgian coast (Bastin 1974).

2. Coaetal: derived from cliff ertsion. Sediments of the Wash are suggested to be mainly derived from the products of cliff erosion which were transported into the "giant mixing bowl" of the Wash, and deposited (Shaw 1973).

3. Fluvial: brought down by the rivers. In the Hollandsch Diep, part of the estuaries of the rivers Rhine and Meuse where the average chlorinity is less than 2‰, about 50% of the clay material is of fluviatile origin. In the Haringvliet (2–10‰ chlorinity) this percentage may drop to considerably lower levels (Salomons et al. 1975). This far-landward mixing process of fluviatile and marine clay materials is common in continental estuaries with a medium to large tidal range.

4. In situ reworking: derived from deposits within the estuary or bay. Guilcher and Berthois (1957) found that the clay material of mudflats and salt marshes in Breton river outlets have physical and chemical properties identical with the periglacial sediments bordering the estuaries. In the Oosterschelde, tidal-induced seaward transport of suspended matter prevails landward transport (Dronkers 1986).

Accretion of salt marshes in this sea-arm is therefore mainly obtained by material from eroding sediments and dredged spoils in the sea-arm.

On tidal coasts such as in the North Sea region, estuaries are considered sinks of materials both of marine and fluviatile origin as the result of the circulation of the overlying freshwater flowing outward, and the seawater wedging inward at depth. This mechanism traps suspended material within the brackish part of the estuary, which migrates back and forth in the estuary in response to the tides. The trapping and mixing efficiencies depend on the magnitudes of river flow, tides, waves at water surface, and wind-driven currents (Gordon 1981), and are therefore not a simple function of salinity (Fig. 2).

The fact that materials for the formation and accretion of mudflats and salt marshes may originate from different source areas means that input of pollutants associated with the finer fractions of these sediments may be the result of earlier contamination of the source areas. Spatial gradients in the relation in which sediments of different origin deposited are then usually reflected in a contaminant gradient. As an example, cadmium concentrations in salt marsh soils of the Westerschelde are shown in Fig. 3, demonstrating an upstream discharge of cadmium-containing effluents.

As a broad generalization, the salt-marsh development starts with an undulating tidal flat sparsely intersected with broad and shallow drainage channels (Fig. 4). At about MHWN the marsh flats have been accreted to more or less concave profiles, and the creeks are accordingly deeper. Vascular plant species (*Salicornia, Spartina, Puccinellia*) colonize the higher sites of these flats. Further accretion is aided by this vegetation by reducing wave energy and tidal currents. Isolated dense stands of *Spartina*, however, may promote scour locally, increasing geomorphological variation.

As the surface rises, the profile of the cross-section becomes more concave, with elevated and broadened creek levees and low-lying marsh flats or even salt pans in between. Other plant species establish themselves and the surface becomes fully vegetated, except for creeks and isolated depressions. At this stage the geomorphological variation and distribution of grain size of soil particles are optimal. Sand and silt are found in the creek levees and the clay particles are mainly deposited in the depressions.

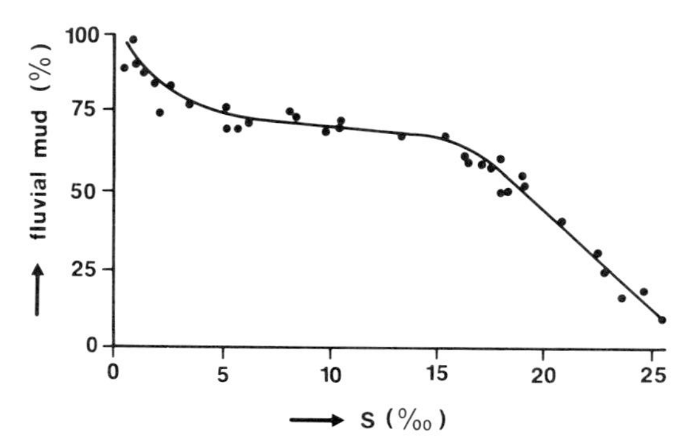


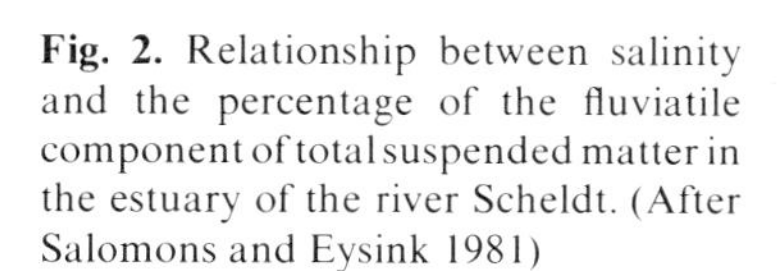
**Fig. 2.** Relationship between salinity and the percentage of the fluviatile component of total suspended matter in the estuary of the river Scheldt. (After Salomons and Eysink 1981)

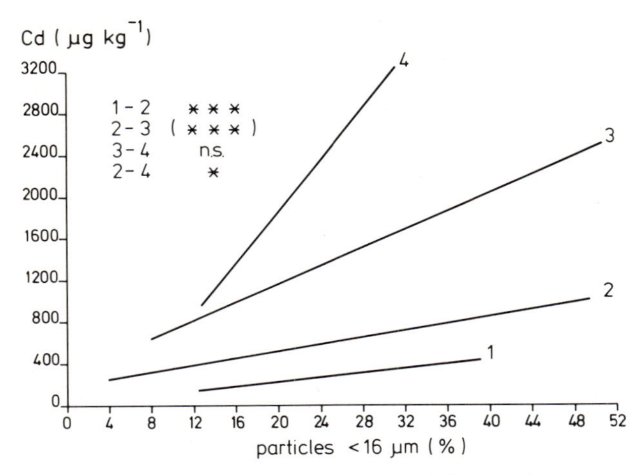


**Fig. 3.** Relationships between cadmium and clay ($< 16\ \mu m$) concentrations in the soil of four salt marshes. *2* Ellewoutsdijk; *3* Waarde; *4* Bath salt marsh, all Westerschelde in upstream sequence. *1* Krabbendijke salt marsh, Oosterschelde (reference area), symbols of significance in distance: 3 asterisks: $P < 0.001$; 2 asterisks: $P < .001$; 1 asterisk: $P < 0.05$; n.s. = not significant. Symbols within brackets indicate that significance found in the covariance tests for location is associated with significance in the tests for parallelism. (After Beeftink et al. 1982)

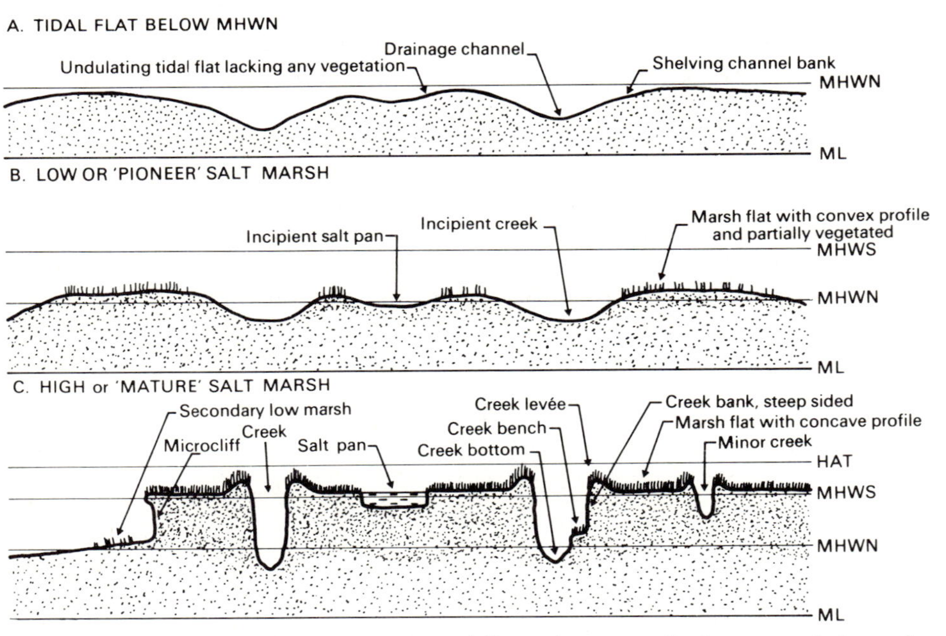


**Fig. 4A-C.** Cross-sections of tidal flat and salt marsh illustrating a generalized sequence of salt-marsh development with corresponding physiographic features, after Beeftink (1966) and Steers (1977), and following the nomenclature of Teal (1962). **A** High-level tidal flat. **B** Low-level marsh. **C** Higher marsh. (After Long and Mason 1983)

A final developmental stage may follow, leveling a great deal of this environmental variation. It starts if accretion diminishes the water storage capacity of the marsh at high water, so that the creeks are decreasing their drainage capacity by siltation and the accretion of the marsh flats is making up that of the creek levees. Considering the developmental process on the whole, large-scale environmental variation in the tidal flat and lower marsh stages is gradually transformed into smaller-scale variations in the mature stage.

Such processes are especially characteristic of situations where tidal currents predominate such as in most estuarine wetlands. In the foreland salt marshes taking up large stretches in the Wadden area, wave action is of equal importance with tidal currents. There terrace formation is more common as a consequence of local alternating erosion and sedimentation processes (Beeftink 1977). The barrier-connected salt marshes and those protected by sand or shingle laterals stand out by their sandy soil and usually show transitions to the adjacent sand and shingle formations. Accretion is here jointly due to the tides and the wind. Examples are found on the Wadden islands (Boschplaat, Terschelling) and on the north coast of Norfolk (Blakeney Point and Scolt Head Island).

## 3 Physical and Chemical Characteristics

The processes for wetland formation, either hydrologically based or wind-driven, result in a sorting-out of the available material into finer and coarser sediments. Not only the mineral components are dispersed according to their particle size, but also organic materials, such as carbonate particles and detritus. In a broad generalization, the dispersal of carbonate particles follows that of the mineral parts, but the distribution of detritus over the marsh depends more on its floating capacity and trapping with depositing silt.

Because metals tend to be concentrated in the finely grained sediment fractions, these substances are not uniformly distributed in the marshes. Therefore it is necessary to correct for these grain-size differences when comparing the metal concentrations of sediments from different localities, or comparing the metal concentration in polluted sediments with those in unpolluted deposits. Several authors have proposed standard fractions varying from $< 2$ μm to $< 63$ μm and more. For the discussion of the pros and cons of these proposals with respect to (ad)sorption and desorption of the different trace metals along with a number of practical reasons the reader is referred to Förstner and Salomons (1980), who recommended the fraction less than 63 μm as standard method for grain-size correction.

Organic matter, nutrients, and trace metals are known as pollutants which have natural background values in tidal wetlands. The character of estuaries as sinks of materials is the reason that both organic matter and nutrients usually accumulate in the circulation zone. These natural enrichments may be enhanced considerably by sewage discharged into the river from urban concentrations. Most pollution of organic waste dates back already from some centuries ago, and may be reflected by

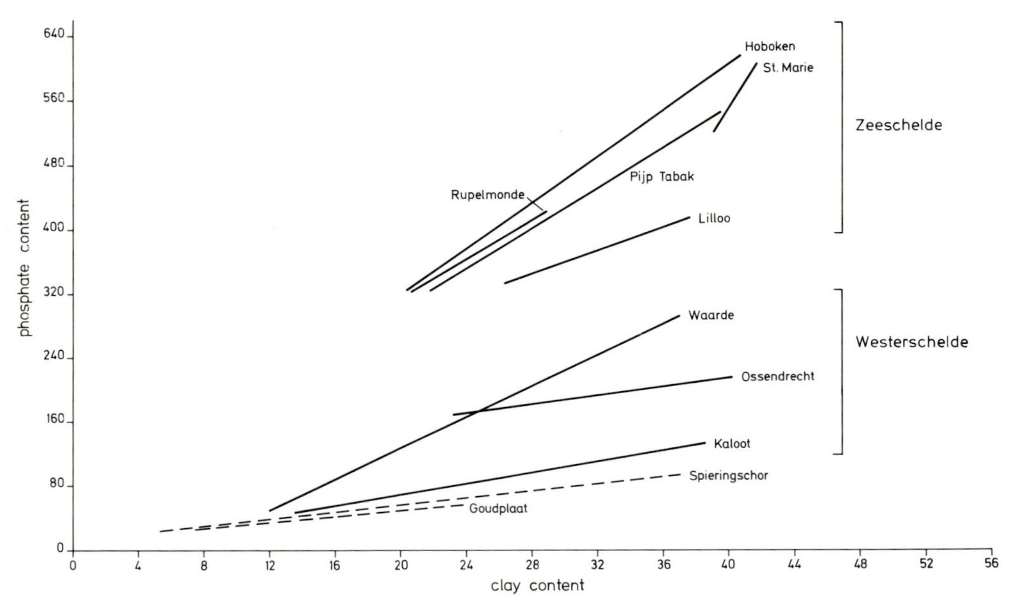


**Fig. 5.** Regression lines of phosphate content in the soils of individual salt marshes occurring in the Oosterschelde (*dashed lines*), the Westerschelde, and the Zeeschelde, the Belgian part of the Scheldt estuary

the birth and growth of the big trade cities. It is therefore unfeasible in most cases to trace back base-line data for this type of pollution. The less mobile phosphate may only serve as a rough indicator. Figure 5 shows that the brackish marshes in the Belgian part of the Scheldt estuary contain two to three times as much phosphate as their Dutch counterparts, calculated per 100 g clay, and 6–12 times that of unpolluted nonestuarine salt-marsh soils. Van der Kooÿ (1982) estimated the total amount of phosphates discharged by urban sewage, industrial effluents and agricultural drainage into the Scheldt estuary 8300 t $a^{-1}$. That is about ten times higher than the highest, and 100 times higher than the lowest phosphorus contents found in nonpolluted estuaries, according to De Pauw (1975).

Geochemical or lithogenical background values of trace metals, often referred to as "precivilizational", show that there is a relatively good correspondence for zinc, lead, mercury, and cadmium in the different sediments analyzed (Table 1). Metals such as chromium, nickel, and copper are more affected by local lithogenical influences, whereas strontium concentrations mainly depend on carbonate contents. These background values are necessary for establishing the enrichment factor after anthropogenic metal contamination. Müller (1979) introduced a qualitative measure for metal pollution in aquatic sediments:

$$I_{geo} = \log_2 \frac{C_n}{1.5 \times B_n} ,$$

where $I_{geo}$ is the "index of geoaccumulation", $C_n$ is the measured concentration of element n in the sediment fraction $< 2\ \mu m$, and $B_n$ is the geochemical background

**Table 1.** Base-line data for trace-metal concentrations ($\mu g\ g^{-1}$) in different types of sediments

| Sediments | Sr | Zn | Cr | Ni | Cu | Pb | Co | Hg | Cd |
|---|---|---|---|---|---|---|---|---|---|
| Shales (Turekian and Wedepohl 1961) | 300 | 95 | 90 | 68 | 45 | 20 | 19 | 0.4 | 0.3 |
| Near-shore sediments (Wedepohl 1960) | | 95 | 100 | 55 | 48 | 20 | | | |
| 16th Century Rhine deposits (De Groot et al. 1976) | | 93 | 63 | 33 | 21 | 31 | | 0.1 | 0.3 |
| Fossil Rhine sediments (Förstner and Wittman 1979) | 184 | 115 | 47 | 46 | 51 | 30 | 16 | 0.2 | 0.3 |
| Clays and shales (Vinogradov 1962) | | 80 | 100 | 95 | 57 | 20 | | 0.4 | 0.3 |

value of element n for fossil argillaceous sediments. Beeftink et al. (1982) used a simpler method in determining the enrichment:

$$A = \frac{C_n}{B_n} .$$

Where A is the concentration factor, $C_n$ and $B_n$ correspond to the symbols used by Müller (1979), provided that they are estimated from deposits with similar clay content.

Müller's index has the advantage that it allows a classification of pollution intensity of the sediments. Recently, Satsmadjis and Voutsinou-Taliadouri (1985) developed a sophisticated index of metal pollution where the amount of an element in a sediment is given as a function of the percentage of clay and silt. In general, however, metal enrichment has to be determined with the help of old data which do not really correspond to this method of calculation. For salt-marsh sediments in the Westerschelde, Beeftink et al. (1982) therefore estimated the concentration factor only, viz. 2–3 for chromium and copper, 4–5 for zinc, arsenic and lead, 10 for mercury, and 25 locally for cadmium. The Oosterschelde marshes showed a concentration factor of 2 to 3 only for zinc, arsenic, lead, and cadmium.

As compared with trace-metal data, only very few analyses exist on polycyclic aromatic hydrocarbons (PAH) concentrations and their speciation in pre-industrial times. PAH mixtures generate by combustion, both natural fires and burning of fossil fuels. Förstner and Müller (1981) summarized the data available and stated that with the present knowledge total PAH concentrations range from 17 to 4000 $\mu g\ g^{-1}$ fine-grained aquatic sediment (dry weight), perlene from 4 to 3900 $\mu g\ g^{-1}$, and benzo(a) pyrene from 1 to 47 $\mu g\ g^{-1}$ at most. The lower levels were found in lakes and rivers far away from urban settlements and industries, the higher levels in estuaries and bays. This large variation in background values makes it difficult to estimate enrichment factors. Förstner and Müller (1981), however, emphasized that PAH and trace-metal contamination (especially cadmium) often correlate positively because of a common source: coal burning, leading to an increase of both PAH and trace-metal levels in the total environment. According to this conclusion, PAH enrichment factors may be derived from analyzing both groups of pollutants in partially exposed substrates.

## 4 Hydrology of Salt Marshes

A major characteristic of the North Sea wetlands is that they are subjected to periodical inundations by the tides. The groundwater relations of these areas are therefore mainly governed by the so-called accumulation effect (Verhagen, unpubl.). This means that the inflow of water into the sediment during high tide is greater than the outflow of groundwater during low tide. This relation is mainly a function of the granular composition of the soil profile, the inclination and microrelief of the soil surface, tidal circulation, and the frequency of tidal immersion.

The accumulation effect results ultimately in a hydraulic equilibrium between input and output at a groundwater level of maximal about halfway the distance between mean tide level and mean high-tide level. Tidal circulations such as the spring-neap tide cycle, wind effects, rainfall, and evaporation give this equilibrium a dynamic character.

Mudflats and low marsh areas have therefore their groundwater level at or close to the soil surface generating oxidation/reduction processes in the habitat zone of benthic organisms and in the root zone of higher plants. The groundwater table of higher marshes reaches high levels after flooding of the soil surface, and may result in a total saturation with inundation water. This water soil oozes down slowly through the soil profile, if inundation is not repeated in the absence of rainfall. Only near the creeks where the soil is more sandy does the groundwater table more rapidly, in hours instead of weeks, as a result of lateral runoff.

## 5 Effects of Salinity and Flooding on Soil Chemistry and Plant Growth

In salt-marsh soils flooded by tides or rainspell, the gaseous diffusion of oxygen and carbon dioxide is no longer possible, and only diffusion in soil water determines the exchange between plant roots and the soil environment. Diffusion of $O_2$ and $CO_2$ in water is $10^4$ times slower compared with diffusion in the atmosphere. Therefore, the thickness of the oxygenated surface layer varies from a few millimeters (heavy clay) to some tens of centimeters (sandy soils) only. With tidal fluctuations, such as the spring-neap tide cycle, parallel fluctuations in the depth of the oxidation/reduction zone occur in the higher salt marshes (Armstrong et al. 1985), resulting in a process of alternating accumulation and drainage of the soil water. With this pumping mechanism, other substances, such as contaminants of various kind, may be carried along, and adsorbed on or incorporated into the soil. On the other hand, the limited gaseous diffusion in the soil water causes various gases to accumulate in the reduced zone, such as nitrogen, carbon dioxide, methane, hydrogen, hydrogen sulfide and dimethyl sulfide, or other plant toxins (Ponnamperuma 1984).

Several types of aerobic-anaerobic interfaces occur in waterlogged salt-marsh soils. First, the oxidized surface layer in contact with the anaerobic deeper soil layers. Second, the transition of the latter with the few-millimeters-thick oxidized rhizosphere of the plants with oxygen leaking from the root aerenchym (Armstrong et al. 1985). Third, on the very micro scale, the bacterial activity causing denitrification (reduction of $NO_3^-$ to $N_2$), reduction of manganese and iron, and of sulfate (e.g., to sulfide). At the same time, nitrogen fixation (resulting in $NH_4^+$ production) and methane formation may take place (for instance Zuberer and Silver 1978). Also microbial oxidation of the toxic sulfides has been reported (Ponnamperuma 1984).

Aerobic-anaerobic soil conditions have different consequences for the behavior of contaminants and trace metals. Organic matter in flooded soils is decomposed by mainly anaerobic bacteria and actinomycetes. Although the decomposition rate of organic matter in waterlogged soils is lower than under aerated conditions (Buth and De Wolf 1985), the breakdown of insecticides, herbicides, and fungicides seems to be enhanced in anaerobic soils (Ponnamperuma 1984). Already a low sulfide level, such as occurs under anaerobic conditions, may be sufficient to precipitate trace metals besides reduced $Fe^{2+}$ and $Mn^{2+}$ to insoluble sulfides. The latter chemical process is also the reason for the yellow-brown color of the aerated, oxidized upper soil layers turning to the grey and blue to black color of the reduced lower layers.

Salt-marsh species are considered to have developed secondarily in the evolutionary process of lower marine algae, occurring in shallow coastal waters to higher plants living on the continent. Some groups of these higher plants were the ancestors of relatives which returned to saline sites covering extensive parts of at least the Eurasian continent at certain geological times. In this secondary evolutionary process the plants had to cope with both an increased soil salinity and the effects of flooding. They succeeded in developing morphological and physiological features that escape or avoid salt stress, and, in muddy sites, conditions of a low oxygen tension (Rozema et al. 1985a).

Enzymes of halophytic higher plants are strongly inhibited by the presence of NaCl in seawater concentrations. This salt stress, however, is avoided through cellular compartmentation of the salts. The bulk of the NaCl is localized in the vacuole, while the cytoplasm is low in NaCl. Under salt stress the compatible solutes proline, glycinebetaine, sorbitol, or pinitol may accumulate in the cytoplasm, allowing enzymes to maintain their activity (Rozema et al. 1985a). Apart from this small-scale mechanism, salt-marsh plants have a more general adaptation for avoiding salt stress: besides the salt accumulators (accumulating NaCl in the vacuole or in plant tissues which may be shed), salt excluders and salt excretors can be distinguished (Albert 1982; Yeo and Flowers 1984; Rozema et al. 1985a). This distinction may also apply to the uptake and release of substances other than the sodium and chloride ions, such as trace metals (Ernst 1974).

Many higher plants inhabiting inland and coastal wetlands have a so-called aerenchyma, gas-filled channels in the cortex of the plant roots and directly connected with intercellular spaces of the shoot. In some halophyte species (*Spartina anglica*) this tissue is always present, also under nonflooded conditions, in other species (*Juncus gerardi, Triglochin maritima*) its development is induced in the anaerobic environment (Rozema et al. 1985b).

The primary function of the aerenchyma system is to allow aerobic respiration of the root tissue under conditions of low external oxygen tension (hypoxia). The oxidized rhizosphere, often shown as a red-brown coating of mainly iron-

(hydr)oxide around the root surface, protects the plant root against toxic sulfides, reduced iron and manganese, and the end products of microbial fermentation (Taylor et al. 1984). Trace metals, such as Cd, Pb, and Cu may be adsorbed to this Fe and Mn (hydr)oxide coating. Lion et al. (1982) concluded that this coating and also natural organic compounds (humics) substantially control the sorptive behaviour of these metals, and seem to play a crucial role in the inclusion or exclusion of these metals by the roots from the sediments (see Chap. 25 this Vol.).

## 6 Vegetation

The structure of the salt-marsh vegetation has a number of characteristic features. First, despite the amphibious nature of the habitat, by far most species are of terrestrial origin. Only *Zostera* species – and in brackish areas also *Ruppia* and some other Potamogetonaceae – are pure aquatic species, but originate evolutionarily from terrestrial higher plants. Second, most species are perennials, covering 70–95% of the area of different plant communities (Beeftink 1985). Moreover, most of these species have their hibernating buds at or beneath the soil surface. This enables them to overcome severe winter frost when great ice masses are covering the salt marsh, and uplift of the ice floes by the tides does not essentially damage the plants which are frozen in the ice. Only the sprouts of the grass *Spartina anglica* may be tough enough to allow the ice to uproot whole clumps together with the soil.

A third group of characteristic features is found in the distributional constraints of the species. Geographically, the North Sea region belongs to the Central Atlantic province, with the exception of SE England. There, some perennial species like the chenopods *Sarcocornia* (*Salicornia*) *perennis* and *Suaeda fruticosa*, and some *Limonium* species of Southern Atlantic or even West Mediterranean distribution reach their northern limit. In the NE of the North Sea region some species like *Halimione portulacoides* and *Elymus pycnanthus* are sensitive to climatic changes (Beeftink 1985).

Other zonation phenomena are found with respect to tidal immersion and salinity. In a single salt marsh including a number of plant communities, the dominant factor for their arrangement will be the tidal or altitudinal zones. On this basis, the vertical range of the marsh, from the seaward limit to the highest point of tidal influence, can be divided into four zones, i.e., the lower, middle, and upper marsh, and the marsh's upper edge.

The lower marsh or pioneer zone extends between MHWN and MHW. The middle marsh corresponds to the zone between MHW and MHWS, and the upper marsh to the zone between MHWS and the line above which tidal immersion occurs less than ten to five times a year.

The lower marsh includes only three to four species, one or two of which are far more abundant than the others, and there may be significant bare areas in between. In the middle marsh, more species are found, and the major lower marsh species have a reduced cover. The higher species' richness is mainly a reflection of differences in hydrology (creek levees and other ridges vs. depressions with more or less stagnant water), and clay content of the soil. Most species come to dominance or codominance depending on the strength of these environmental factors, and grazing. The upper marsh contains a mixture of other species, with species from the middle marsh in reduced vitality and cover. These species assemblages are mixed with nonhalophytic but salt-tolerant species towards the upper edge of tidal influence.

In Fig. 6 a simplified scheme is given of the succession in salt-marsh vegetation of euhaline and polyhaline zone (> 18‰ S). In the lower marsh, *Salicornia europaea* and *Spartina anglica* are the dominant species. The middle marsh is characterized by the grass *Puccinellia maritima* in combination with various other species with *Limonium vulgare*, *Halimione portulacoides*, and *Aster tripolium* along

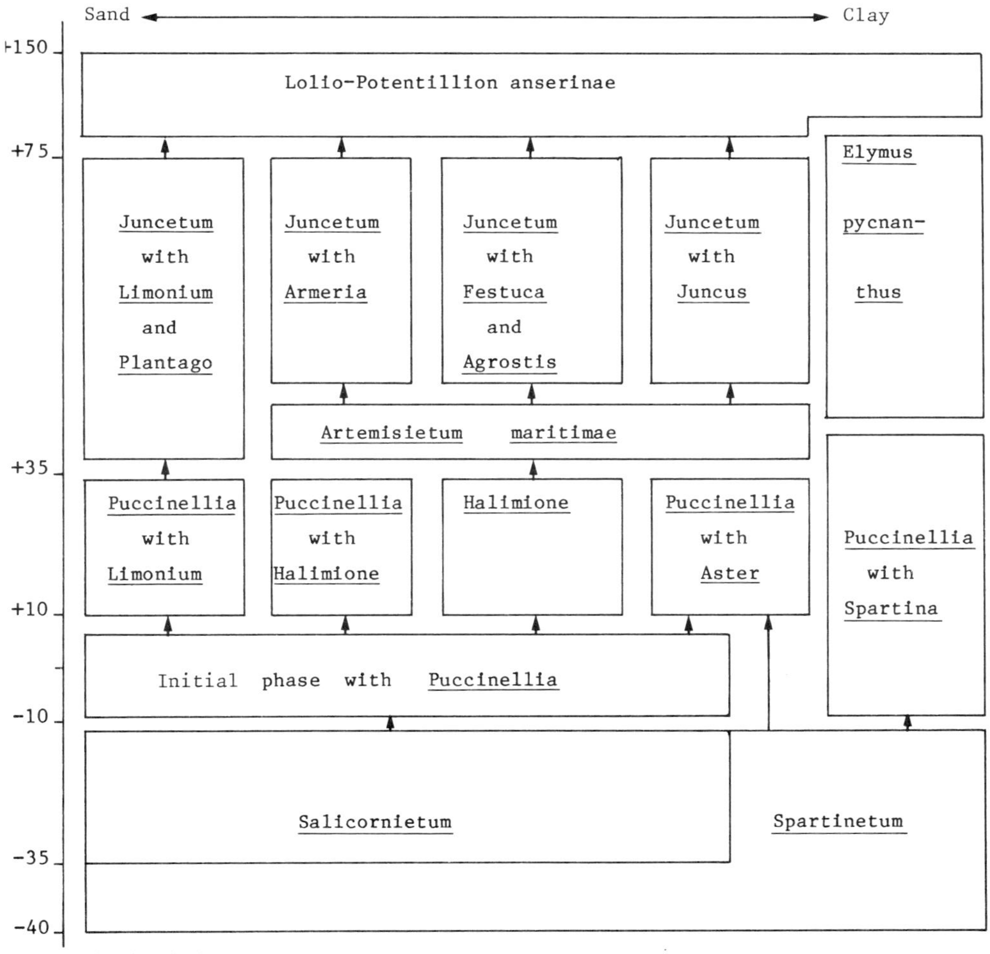


**Fig. 6.** Principal succession series in the vegetation of eu- and polyhaline salt marshes of the Waddensea. (After Dijkema 1983)

the sand-clay gradient. In the upper marsh the rush *Juncus gerardi* and the grass *Festuca rubra* are the major species, together with *Glaux maritima, Plantago maritima, Armeria maritima,* and *Agrostis stolonifera*. On the upper edge of the marsh species like *Carex distans, C. extensa, Juncus maritimus, Leontodon autumnalis* and some Leguminosae (*Trifolium fragiferum, T. repens, Lotus tenuis*) are added to the halophytes of the upper marsh, and known as the Lolio-Potentillion anserinae community group (Sykora 1982).

In estuaries the zonation in the higher salinity zones is usually more simple (Fig. 7). An overall deposition of finer sediments often includes a less hydrological and chemical variety than, for instance, barrier-connected salt marshes show. Moreover, the biomass productivity is higher under estuarine conditions, resulting in a more closed canopy of the vegetation in summer, and more plant debris washed ashore in the upper edge of the marsh in winter. The latter includes disturbance of the upper marsh vegetation, and a favoring of ruderal species, such as *Elymus pycnanthus* and *Atriplex* sp.

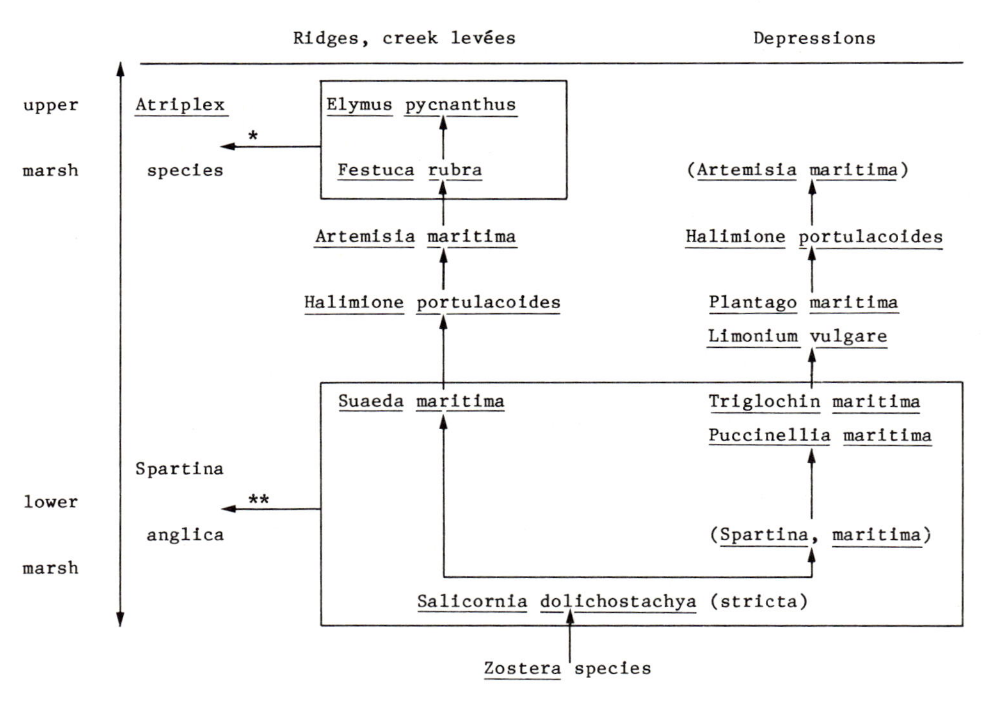


* On tidal drift washed ashore

** Amphipolyploid of Spartina alterniflora and S. maritima from about 1860 and since dispersed mainly by man nearly all around the North Sea coasts.

**Fig. 7.** Principal succession series of (co)dominant plant species in polyhaline estuarine salt marshes. Derived from Beeftink (1966)

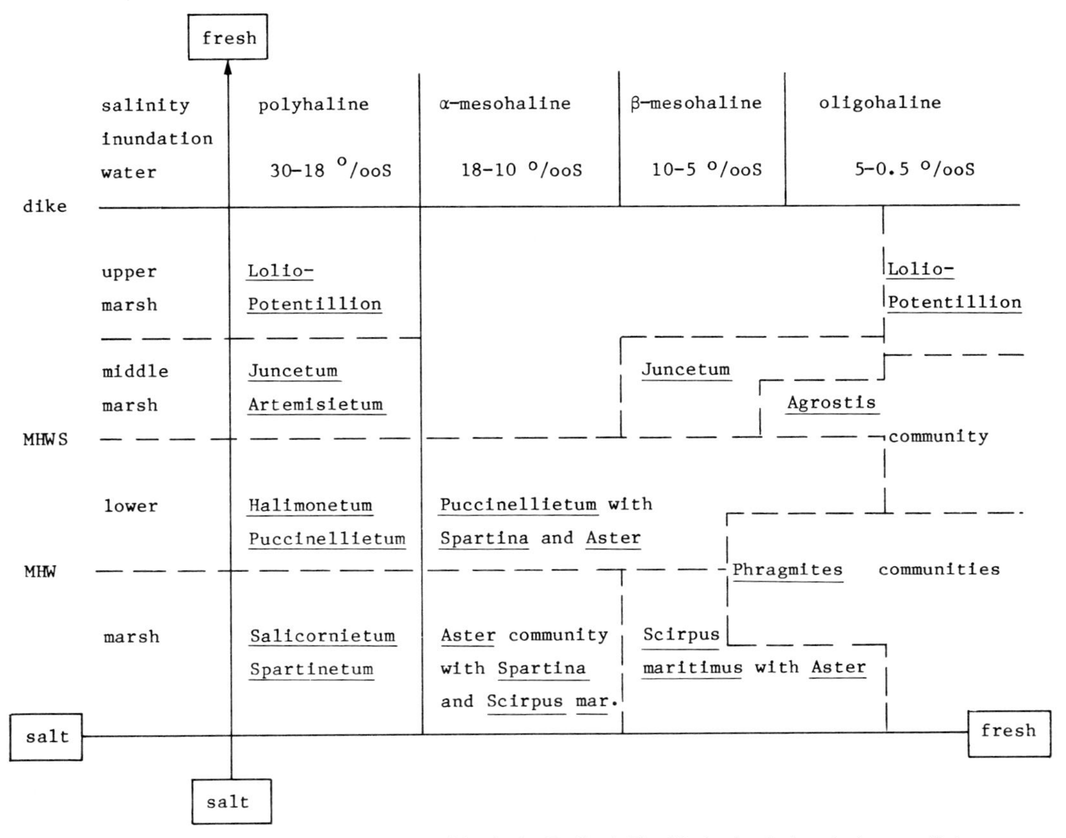


**Fig. 8.** Zonation pattern of plant communities in the Dollard, The Netherlands, in relation to salinity zones. (After Dijkema 1983)

Upstream, the estuaries salinity is the main governing factor determining vegetational conditions (Fig. 8). Lower salinities cause the pioneer vegetation to change into *Aster tripolium, Scirpus maritimus*, and *Phragmites australis* communities respectively. *Halimione* and *Artemisia* communities are more limited in the Dollard than in the Scheldt estuary, where they extend to into the α-mesohaline zone.

Background values of trace-metal concentrations in salt-marsh plants are scarce. The main reason for this is that these values are difficult to establish because soil-plant relationships may differ greatly with respect to:

1. The distribution and dispersal of the metal within the soil profile in relation to its chemistry within the soil.
2. The mechanism of metal uptake by the roots of a plant species in relation to root distribution, involvement of other ions, water relations, and microorganisms.
3. The mechanism of metal uptake by aboveground plant parts from tidal immersion with contaminated water and adhering clay particles.

4. Translocation of metals from roots to shoots and within aboveground parts of plants in relation to metabolism, growth, and development.
5. Loss of metals from plants via senescence, leaf-fall and leaching, exudation, etc.

As an example, Gambrell et al. (1977) found that cadmium uptake by the North American *Spartina alterniflora* was greatest under acid, oxidized conditions. A typically reduced soil of a lower salt marsh, which is subjected to oxidizing conditions such as severe drought by lack of tidal immersion, can thus be expected to release more of its cadmium to readily bioavailable forms.

The distribution and translocation of metals may also differ greatly (Fig. 9). Usually roots accumulate metals to highest levels indicating that especially these tissues take the metals in. Rozema et al. (1985c) demonstrated that a significant part of trace metals in the roots of *Spartina anglica* had adsorbed to the iron plaque *on* the root surface. *Aster tripolium*, however, stands out by highly elevated concentrations of cadmium, zinc, and mercury in the leaves as well, suggesting a second way of uptake, in this case from contaminated inundation water or clay particles brought by the water and adhering on the leaves.

A similar pattern of metal uptake was found in eelgrass (*Zostera marina*). Brix et al. (1983) reported that the concentrations of Cd, Cu, and Zn in aboveground parts were significantly higher than in belowground parts. Translocation of Zn was insignificant. The Zn content in aboveground parts was mainly derived from the ambient water, and that in belowground parts from interstitial water (Lyngby et al. 1982). However, Faraday and Churchill (1979) found that Cd was translocated from the leaves to the root rhizomes of eelgrass, while Brinkhuis et al. (1980) found a two-way transport. These results suggest that for some trace metals, parts of eelgrass will give an estimate of the bioavailability of these metals in the ambient and interstitial water (sediment). For other trace metals, however, translocation within the plants, age of the plant tissue, seasonality, and environmental salinity may interfere.

In Table 2 a rough estimate is given of maximal background concentrations of some trace metals in shoots of halophytes obtained from an Oosterschelde salt marsh. From these metals only Cd and Pb showed slightly elevated concentrations in the soil of this marsh (see Sect. 3). The figures suggest significant differences in metal uptake between plant species and metals.

## 7 Mudflat and Salt-Marsh Animals

Coastal wetland animals can be considered in five categories, viz. mammals, birds, arthropods, molluscs, and a number of benthic groups such as nematodes, crustaceans, flatworms and protozoans. Wild mammals usually visit the salt marsh for

→

**Fig. 9.** Distribution of Cu, Cd, Pb, Zn, As and Hg ($\mu g\ g^{-1}$ dry weight) in the tissues of four plant species from Waarde, Westerschelde (*black*), and Krabbendijke, Oosterschelde, salt marshes. Date of sampling 25 September 1978. Rosette leaves of *Aster tripolium* were sampled apart. *rts* roots; *st* stems (*b* at basis, *t* at top); *lvs* leaves; *fls* flowers; *ros* rosette leaves. (After Beeftink and Nieuwenhuize 1986)

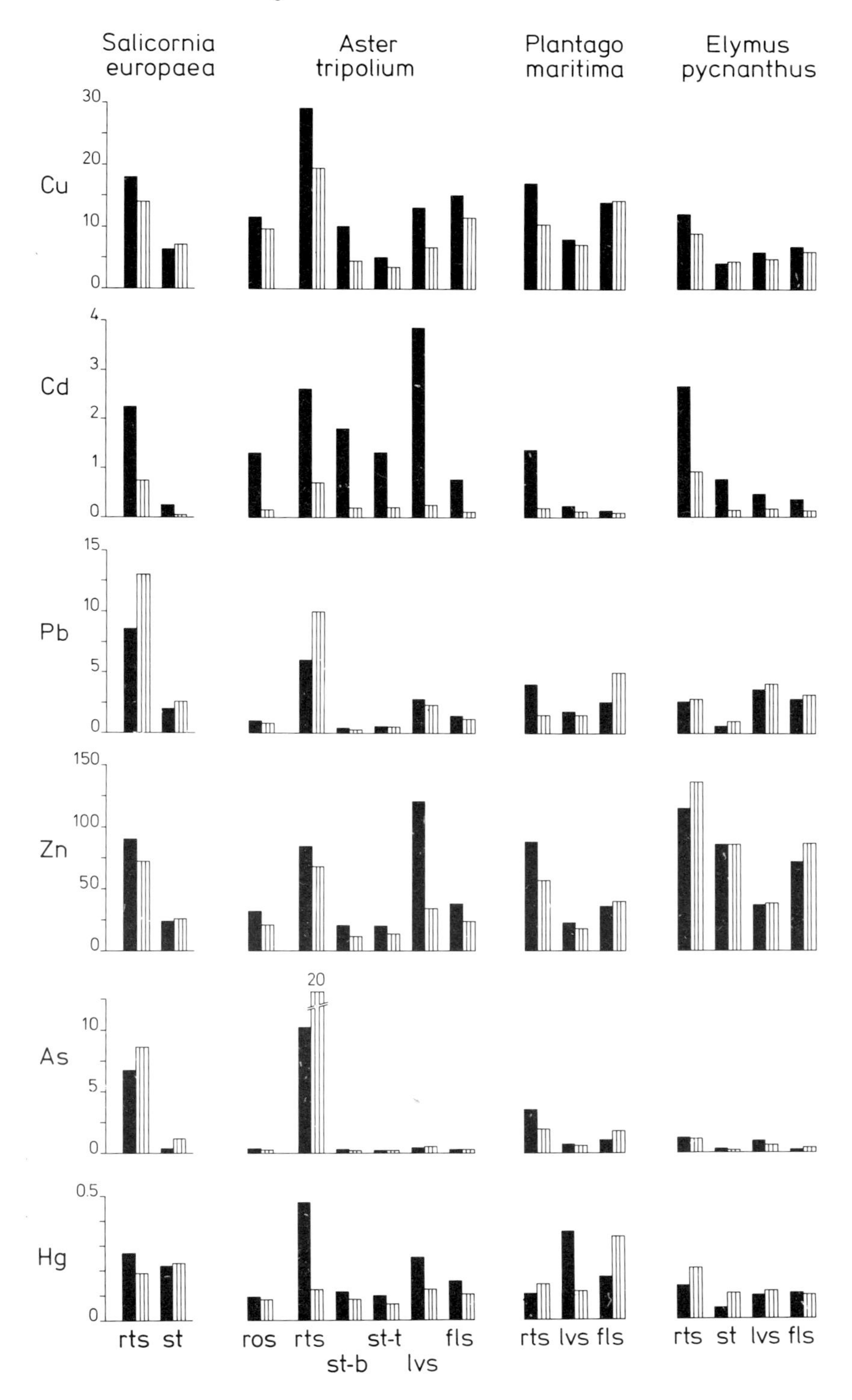
Salicornia europaea
Aster tripolium
Plantago maritima
Elymus pycnanthus
Cu
Cd
Pb
Zn
As
Hg
30
20
10
0
4
3
2
1
15
5
150
100
50
0.5
rts st
ros rts st-b st-t lvs fls
rts lvs fls
rts st lvs fls

**Table 2.** Maximal trace-metal concentrations ($\mu g\ g^{-1}$ dry weight) in plant shoots from a noncontaminated Oosterschelde salt marsh. n = number of plants analyzed. (After Beeftink et al. 1982)

| Enrichment factor in the sediment | Cd 2–3 | Hg — | Pb 2 | Cu — | Ni — | n |
|---|---|---|---|---|---|---|
| *Atriplex hastata* | 0.12 | 0.03 | 1.8 | 2.1 | 2.1 | 1 |
| *Salicornia europaea* | 0.19 | 0.07 | 10.0 | 8.3 | 4.2 | 3 |
| *Aster tripolium* | 0.26 | 0.03 | 3.5 | 7.6 | 1.1 | 3 |
| *Spartina anglica* | 0.07 | 0.04 | 4.3 | 6.0 | 1.8 | 3 |
| *Puccinellia maritima* | 0.44 | 0.10 | 14.2 | 8.1 | 5.0 | 3 |
| *Elymus pycnanthus* | 0.08 | 0.03 | 3.2 | 4.6 | 0.8 | 2 |
| *Triglochin maritima* | 0.11 | 0.07 | 11.2 | 7.9 | 3.2 | 3 |
| *Halimione portulacoides* | 0.25 | 0.06 | 8.9 | 5.4 | 3.3 | 3 |
| *Artemisia maritima* | 1.46 | 0.06 | 8.0 | 22.1 | 1.9 | 2 |

food only, or find habitats for living in the uppermost parts of the marsh where the tides hardly reach them. Grazing by cattle and sheep may utilize food sources more systematically, and can even be directed to selected food plants. Dependent on its intensity, grazing influences the species composition and structure of the vegetation considerably.

The North Sea coasts have prominant areas for wintering of ducks, geese, and waders in Europe (Fig. 10). The total number of estuarine birds in the Dutch part of the Wadden Sea varies from about 150,000 to nearly 1,000,000 per month (667,000 on an average) according to Smit (1980). In autumn especially dunlin (*Calidris alpina*), oystercatcher (*Haematopus ostralegus*), knot (*Calidris canutus*) and curlew (*Numenius arquata*) are numerous, in winter eider (*Somateria mollissima*), oystercatcher and dunlin, and in spring dunlin, knot and bar-tailed godwit (*Limosa lapponica*) (Table 3). These species mostly feed at low tide on the benthic fauna of the mudflats. Others, such as brent geese (*Branta bernicla*), feed on *Zostera, Puccinellia* and green algae. Salt-marsh plants, especially members of the Chenopodiaceae and Compositae, produce large crops of seed which attrack ducks and passerines. Grey-lag geese (*Anser anser*) like eating the root tubers of *Scirpus maritimus.*

Arthropods are represented in great numbers of both species and of individuals. The main groups are insects, arachnids, and crustaceans. Insects and arachnids are largely confined to the salt marshes, the crustaceans have their greatest species' richness in the mudflats.

Table 4 gives an impression of the most important macrobenthic species exemplified for the Dutch part of the Waddensea. About two-thirds of the biomass consists of molluscs, and one third of polychaetes. Other groups (crustaceans, echinoderms, coelenterates) are from this point of view relatively unimportant, with only a lower percentage of the total biomass.

As to the feeding type, suspension feeders rank first with about 55% of the total biomass, followed by deposit feeders with about 40%. Predators, scavengers, and grazers are minor groups. As part of the species shows mixed feeding types, the

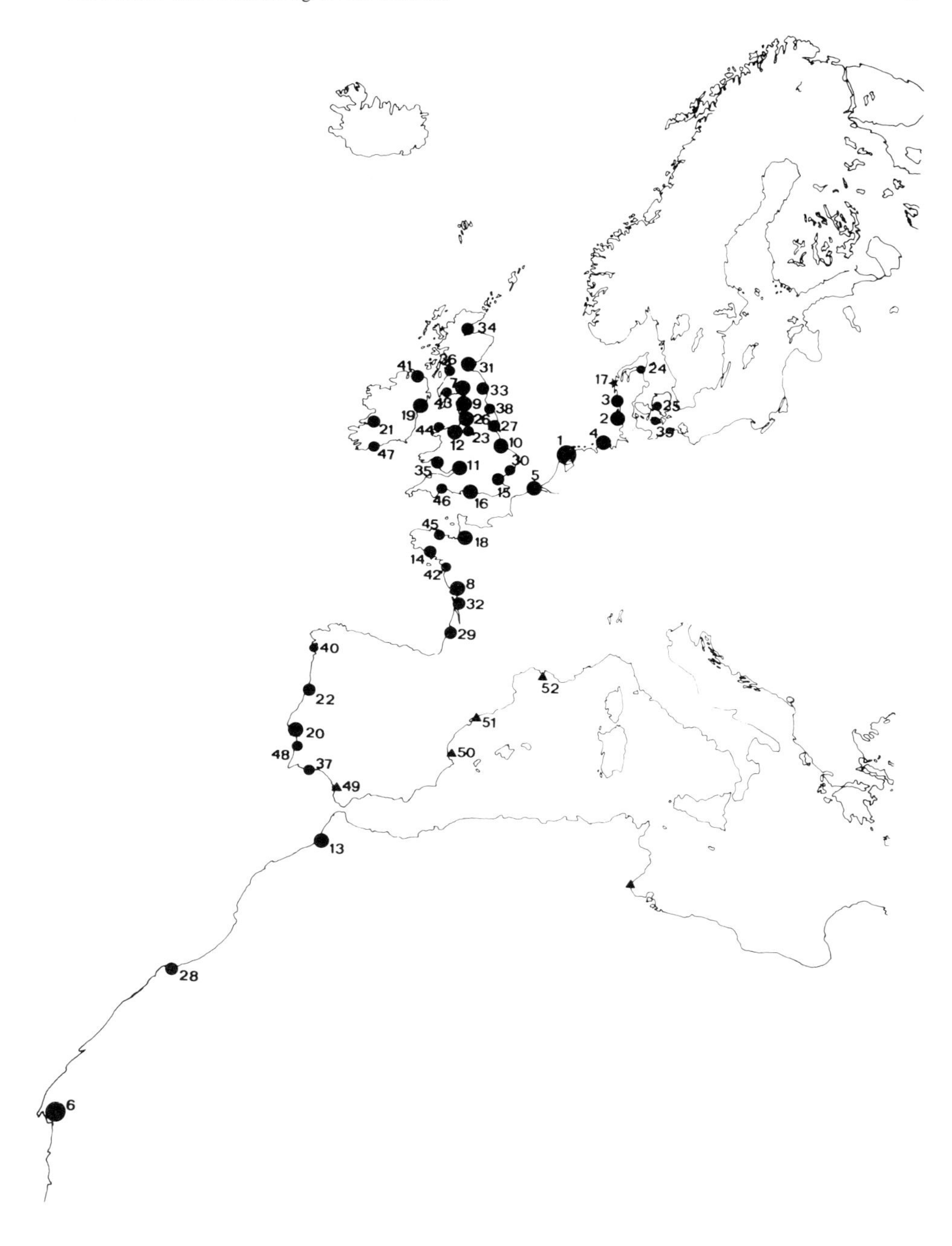


**Fig. 10.** Estuarine areas important for the wintering of ducks, geese, and waders in Europe and NW Africa. *Size of dots* refers proportionally to number of wintering birds: less than 10,000; 10,000–20,000; 20,000–50,000; 50,000–200,000; and more than 200,000 waters. * numbers unknown; ▲ birds not belonging to the W. European flyway population. Figures refer to number of locality. (After Smit 1980)

**Table 3.** Numbers of the main wader species in the Danish, German and Dutch Wadden Sea during three counts. (After Prater 1974, 1976)

| | January 1975 | April 1975 | September 1973 |
|---|---|---|---|
| Oystercatcher (*Haematopus ostralegus*) | 297,400 | 88,300 | 474,700 |
| Ringed plover (*Charadrius hiaticula*) | 4 | 1,150 | 17,500 |
| Grey plover (*Pluvialis squatarola*) | 4,000 | 6,500 | 33,600 |
| Turnstone (*Arenaria interpres*) | 1,700 | 3,400 | 3,760 |
| Curlew (*Numenius arquata*) | 84,300 | 45,400 | 155,500 |
| Black-tailed godwit (*Limosa limosa*) | 0 | 1,100 | 100 |
| Bar-tailed godwit (*Limosa lapponica*) | 21,500 | 104,900 | 78,100 |
| Redshank (*Tringa totanus*) | 15,600 | 18,000 | 55,400 |
| Spotted redshank (*Tringa erythropus*) | 3 | 1,640 | 5,500 |
| Greenshank (*Tringa nebularia*) | 9 | 390 | 9,500 |
| Knot (*Calidris canutus*) | 38,500 | 69,300 | 312,900 |
| Dunlin (*Calidris alpina*) | 220,100 | 458,000 | 728,800 |
| Sanderling (*Calidris alba*) | 1,950 | 1,800 | 6,200 |
| Ruff (*Philomachus pugnax*) | 150 | 350 | 3,400 |
| Avocet (*Recurvirostra avosetta*) | 2,500 | 6,340 | 45,300 |
| Total, including other species | 716,000 | 847,000 | 1,972,000 |

**Table 4.** Estimates of mean biomass and annual production in 1971–1972 of the most important macrobenthic animals living in the tidal flats of the Dutch Wadden Sea. All figures in percentage of total g ash-free dry weight (ADW) $m^{-2}$. (After Beukema 1981)

| Species | Biomass | Production |
|---|---|---|
| Total (g ADW $m^{-2}$) | 26.6 | 29.6 |
| Molluscs | 66 | 56 |
| *Mytilus edulis* | 23 | 24 |
| *Mya arenaria* | 17 | 7 |
| *Cerastoderma edule* | 16 | 17 |
| *Macoma balthica* | 8 | 5 |
| *Hydrobia ulvae* | 1 | 2 |
| Polychaetes | 30 | 33 |
| *Arenicola marina* | 19 | 12 |
| *Nereis diversicolor* | 5 | 9 |
| *Lanice conchilega* | 3 | 5 |
| Crustaceans | 3 | 8 |
| Other groups | 1 | 3 |

contribution of suspended food may be even higher than the above-mentioned 55% (Beukema 1981).

The salt marshes are inhabited by a host of insects and arachnids, most of them very specialized to the tidal environment. They can be grouped according to their feeding habit (Table 5). The molluscs are represented by eight species, of which *Hydrobia ulvae* is the most common. Apart from these macrofauna species, numerous species belonging to the micro- and meiofauna are found in the mudflat and salt-marsh soils: nematodes, turbellarians, oligochaetes, harpacticoid

**Table 5.** Species number of insects and arachnids requiring the salt-marsh environment of the North Sea Region to complete parts of their life cycle, specified according to their feeding habit. Data collected by Buth and coworkers (unpublished)

| Order | Total number | Herbivores | Carnivores | Detritivores |
|---|---|---|---|---|
| Diptera (flies, midges, etc.) | 90–110 | ? | ? | ? |
| Coleoptera (beetles) | 90–95 | 10–15 | 40–50 | 30–40 |
| Hemiptera (bugs, aphids, cicadas) | 50–60 | 50–55 | 5–10 | |
| Hymenoptera (bees, sawflies, wasps) | 30–40 | 1–5 | 30–35 | |
| Lepidoptera (butterflies) | 25 | 25 | | |
| Collembola (springtails) | ≈ 12 | | 1–4 | 4–10 |
| Thysanoptera (thrips) | 5–10 | 5–10 | | |
| Orthoptera (grasshoppers, etc.) | 2 | 2 | | |
| Araneae (spiders) | 10–20 | | 10–20 | |
| Acari (mites) | 10–20 | ? | ? | ? |
| Total, approximately | 325–375 | 93–112 | 86–120 | 34–50 |

copepods, halacarids, and foraminifers are the most important taxonomic groups. Most of these species show a zonal distribution, either in relation to the inundation frequency or to salinity regimes and food sources (Beeftink 1977).

Environmental pollution is mostly determined simply by analyzing the pollutant present in sediment and/or water. Concentrations measured in organisms may lie within the environmental range, but also above it, or even below it, according to the amount of pollutant taken in by the organism. Under certain conditions, natural background values can therefore also be considered in organisms for many pollutants. From that idea enrichment factors could be estimated for organisms with respect to their environment (sediment, water). However, the problem is that for instance the normal metal content of an organism is dependent on many variables, specific as well as environmental, and the intensity in which the organism (species) in question has been investigated, so that only a range of concentrations can be given (Table 6). Because unpolluted tidal areas are becoming very rare, especially in the North Sea Region, natural background values of metals have to be estimated around the minimal concentrations. Even then different environmental processes, such as chemical transformation of metals by microorganisms as a result of degradation and mineralization of organic matter, methylation of inorganic metal species, and bioturbation in the upper soil layers by benthic fauna influence the uptake of these pollutants.

The difference in trace-metal concentrations between species is also reflected in the concentration factor as an expression of the accumulating capacity of the species with respect to the concentration of the contaminant in the sediment in which it is living. Some examples illustrated in Table 7 suggest that there is a large difference in concentration factor between plants and animals, although for some metals exceptions may occur as in Sea Wormwood, *Artemisia maritima* (Table 2). Other pollutants, such as polycyclic aromatic hydrocarbons, polychlorobiphenyls (PCB's) and radionuclides should usually have extremely low natural background values, if any, in organisms (see also Sect. 3).

**Table 6.** Trace-metal concentrations in some intertidal organisms in less polluted coastal areas. Data in mg $kg^{-1}$ dry weight.

| Species | Cd | Cu | Pb | Hg | Zn | Reference |
|---|---|---|---|---|---|---|
| *Nereis diversicolor* | 0.08–3.4 | 22.0– 78.0 | | | 170–258 | Bryan and Hummerstone 1977 |
| *Arenicola marina* | | 6.7– 17.9 | | 0.1–0.4 | 105–175 | De Kock 1975 |
| *Mytilus edulis* | 0.84–2.64 | 3.9– 13.6 | | | 57–199 | Bryan and Hummerstone 1977 |
| *Cerastoderma edule* | 0.48–1.04 | 5.2– 27.2 | | | 46– 66 | Bryan and Hummerstone 1977 |
| *Macoma balthica* | 0.21–0.85 | 96 –615 | | | 510–1160 | Bryan and Hummerstone 1977 |
| *Scrobicularia plana* | 0.29–0.64 | 25 – 77 | 12 –21 | | 353–394 | Bryan and Hummerstone 1978 |
| *Zostera muelleri* | 0.03–1.69 | 3.1–10.6 | 1.6 –8.4 | | 21.0–67.0 | Harris et al. 1979 |
| *Zostera marina* | 0.09–2.92 | 1.86–16.6 | 0.47–37.5 | | 41.0–175.0 | Brix et al. 1983 |
| *Spartina alterniflora* | 0.1 –1.9 | 9.5–7.8 | | 0.07–0.37 | | Windom 1975 |

**Table 7.** Concentration factors (CF[a]) of some intertidal organisms for less polluted sediments

| Species | Cd | Cu | Pb | Hg | Zn | Reference |
|---|---|---|---|---|---|---|
| *Nereis diversicolor* | 10–20 | 1 | 0.025–0.25 | 2–3 | | Bryan et al. 1980 |
| *Arenicola marina* | 2–10 | 0.05–5 | 1 | | 1–10 | Packer et al. 1980 |
| *Macoma balthica* | 10–20 | 2–10 | 0.1 –0.5 | 1–5 | 4–10 | Bryan et al. 1980 |
| | | 8–30 | | 2–10 | 8–20 | De Kock 1975 |
| *Scrobicularia plana* | 60 | 8 | 2 | 2 | 10 | Bryan et al. 1980 |
| *Zostera marina* | 0.04–1.3 | 0.02–0.3 | 0.005–0.05 | | 0.06–0.38 | Brix and Lyngby 1983 |
| *Salicornia europaea* | 0.22–0.84 | 0.23–2.02 | 0.09 –0.56 | 0.15–0.88 | | Beeftink et al. 1982 |
| *Spartina anglica* | 0.14–0.27 | 0.33–1.44 | 0.04 –0.45 | 0.09–0.57 | | Beeftink et al. 1982 |
| *Aster tripolium* | 0.44–0.53 | 0.30–2.00 | 0.06 –0.24 | 0.06–0.43 | | Beeftink et al. 1982 |
| *Puccinellia maritima* | 0.16–1.96 | 0.26–1.98 | 0.15 –1.49 | 0.21–1.43 | | Beeftink et al. 1982 |
| *Halimione portulacoides* | 0.44–1.11 | 0.14–1.32 | 0.05 –0.94 | 0.15–0.86 | | Beeftink et al. 1982 |
| *Artemisia maritima* | 2.66–6.49 | 1.03–5.82 | 0.08 –0.45 | 0.18–0.36 | | Beeftink et al. 1982 |

[a] $CF = \frac{\text{concentration in organism (plants aboveground parts, animals whole soft parts)}}{\text{concentration in sediment (total)}}$ ($\mu g\ g^{-1}$ d. w.).

## 8 Functional Aspects of Coastal Wetland Ecosystems

In this section some aspects are briefly discussed of the internal functioning of the salt-marsh ecosystem and of its interrelationships with other coastal ecosystems.

As in other ecosystems, the salt-marsh biota all together have traits of a community living in close connection with the environment. This means that the salt-marsh ecosystem can be considered from two points of view:

1. The species' populations as compartments of the ecosystem functioning in different food chains and food webs,
2. materials, nutrients, and other substances, flowing through the ecosystem with different cycles, turnover rates, and input and output.

Figure 11 shows schematically the trophic levels and food-chain pathways present in a salt marsh. Besides these pathways at the production side, the ecosystem has equivalent pathways at the decomposition side, starting from dead organic material in each compartment, which serves as food for a host of lower organisms belonging to a great variety of taxonomic groups (bacteria, nematodes, fungi, etc.).

Primary production of salt marshes has been estimated by different methods. Some authors conceive the amount of organic matter which can be harvested at the end of the growing season. Others consider all organic material produced by the plants in the period under consideration (Ketner 1972; Groenendijk 1984). The latter estimate thus includes all parts of the plants washed away by the tides or mineralized on the spot, and the amount of organic matter consumed by animals. Although the latter method corresponds better to a description of the functional role of the salt marsh, the former is much easier to estimate. Usually, only the aerial parts of the plants are considered in production estimates. Belowground productivity is far more difficult to determine (Groenendijk and Vink-Lievaart 1986).

Production estimates of the salt marshes in the North-Sea Region are few. In the Waddensea area the values range from 300 to over 600 g dry weight $m^2$ $a^{-1}$ for aerial parts (Ketner 1972). These figures correspond to those found in Baltic salt marshes (Wallentinus 1973). Estuarine salt marshes show a much higher production. Wolff et al. (1980), using Smalley's method, found values ranging from 837 to 1030 g dry weight $m^{-2}$ $a^{-1}$ in an Oosterschelde salt marsh. Groenendijk (1984), comparing the production of four dominant plant species in the same salt marsh, came to estimates of the same order, but to much higher values if the more accurate method of Wiegert-Evans is used (Table 8). These estimates are of the same order of magnitude as the yield in agricultural systems.

The question now arises after the fate of the organic matter produced. In grazed salt marsh most organic matter will be consumed by sheep or cattle depending on the intensity of grazing. In ungrazed marshes, wild mammals use only a minor part of the food available. The same holds for the invertebrates (insects, mites, molluscs), although the number of species may be very large and some of them can reach densities of epidemic size. Birds, such as geese and passerines, may consume a great proportion of some species, but even then most

**Fig. 11.** Trophic levels and food-chain pathways at the production side of the salt-marsh ecosystem. (After Beefink 1977a)

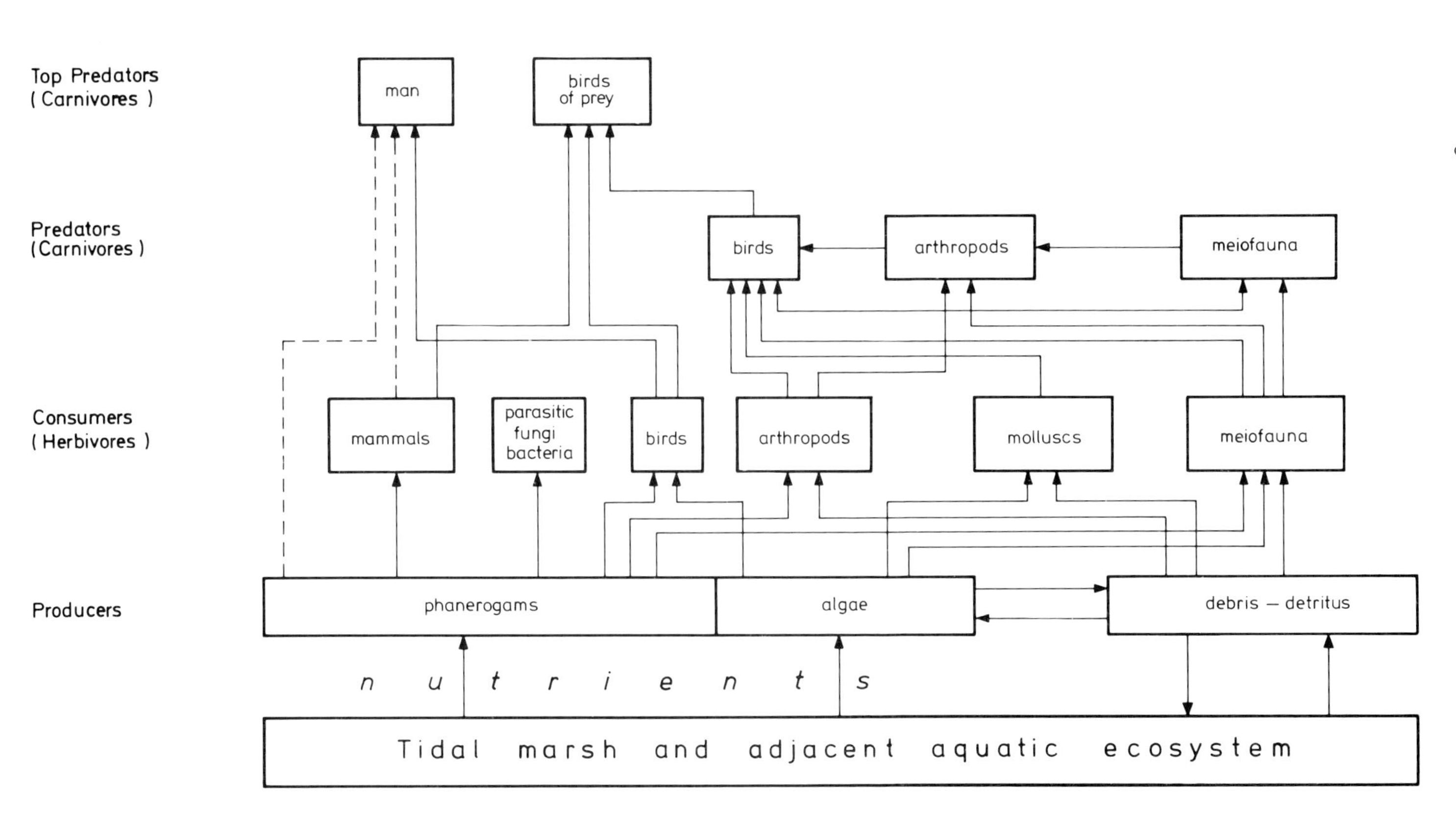

**Table 8.** Estimates of primary production (g dry weight $m^{-2}a^{-1}$) of aerial parts of four dominant plant species in an estuarine, polyhaline salt marsh. (After Groenendijk 1984)

| Species | Methods | |
|---|---|---|
| | Smalley | Wiegert-Evans |
| *Spartina anglica* | 1162–1649 | 2139–2659 |
| *Triglochin maritima* | 568– 783 | |
| *Halimione portulacoides* | 790–1434 | |
| *Elymus pycnanthus* | 474– 878 | 1416–1787 |

of the organic matter produced remains. As most plant species have their hibernating buds at the soil surface or beneath, or are annual species, the bulk of the aerial parts produced dies off at the end of the growing season.

The fate of this organic material is very difficult to establish. U.S. ecologists reported that a quarter to nearly half of the primary production of east coast salt marshes is exported by the tidal currents contributing to the food supply of the adjacent coastal water body (Odum 1971). Later research (e.g., Haines 1976) laid this statement open to doubt. In the North Sea region, the primary production of salt marshes seems to play a minor role in the nutrient supply of most estuarine and coastal waters, first because their surface is relatively lower than in east American estuaries, and second because it may be questioned whether the salt marshes really show a net export of particulate organic matter to the aquatic ecosystems.

Wolff et al. (1980) found for a salt marsh in the Oosterschelde a large net import of particulate organic matter in the size range 0.5–300 $\mu$m, and a small net export for particles over 300 $\mu$m. The total net import was very tentatively estimated 440 g dry weight $m^{-2}a^{-1}$. The problem was that by far the most organic matter is transported during storm floods, and measurements under those weather conditions must be abandoned in view of safety for the scientists. Assuming that part of the imported material will be buried and mineralized in the marsh soil, a net export of dissolved organic matter (including polycyclic aromatic hydrocarbons if present) and minerals (nitrate, ammonia, phosphate, trace metals) from the marsh to the estuary may be expected. Data, however, are lacking.

For a quantification of the distribution of a trace metal within an ecosystem and the estimation of transfers between the major components ecosystem audits may be helpful. Tyler (1971) made a first attempt for a Baltic seashore-meadow ecosystem (Table 9). Atmospheric and hydrologic inputs, retention by biomass, release by litter fall, leaching, and throughfall, as well as output in groundwater or coastal water were not estimated. Making a balance scheme for the elements, he found that Ca, Fe, Zn, Ni, and Pb were enriched during litter decomposition, at least in the early stages, contrary to K, Na, Mn, S, P, and Cu. He also found differences in metal uptake between aboveground litter of different plant species, as well as between retention in and release from litter of these species.

For the faunal elements of the ecosystem there is a variety of ways through which contaminants can be introduced into the organisms (Förstner and Wittmann (1979): (1) via respiration, (2) by adsorption onto the body surface, and (3) from foods. In general, contaminant uptake from food assumes a greater importance because its pollutant concentrations are prone to greater variations than those

**Table 9.** Ecosystem audits for trace metals in a south-Swedish sea-shore meadow. (Data from Tyler 1971)

| | Copper | Nickel | Zinc | Lead |
|---|---|---|---|---|
| Stocks (mmol $m^{-2}$) | | | | |
| Aboveground biomass (max.) | 1.5 | 0.5 | 5.9 | 1.2 |
| Aboveground litter | 5.7 | 4.7 | 18.3 | 14.3 |
| Belowground biomass and litter | 298.6 | 134.6 | 209.2 | 165.7 |
| Vegetation total | 305.8 | 139.8 | 233.4 | 181.2 |
| Soil[a] EDTA-extractable | 1715.6 | 514.8 | 784.4 | 1202.0 |
| Soil[a] total (x $10^3$) | 13.6 | 18.6 | 31.4 | ? |

[a] Depth 0–25 cm.

bio-available in water. However, the dietary practices are important, because food can be contaminated in a very different way.

The following feeding habits should be distinguished, (a) phytophagous (insects, gastropods, crustaceans, birds, mammals), (b) filter feeding (zooplankton, bivalves), (c) sediment feeding (poly- and oligochaetes), (d) detritus feeding (insects, gastropods, iso- and amphipods, nematods), (e) carnivorous (zooplankton, polychaetes, nematodes, crustaceans, insects, arachnids, fish).

Taking the tidal water body, phytoplankton, or salt-marsh plants as basic point of reference in unpolluted ecosystems, contaminant enrichment can take place at each next-highest trophic level (Förstner and Wittman 1979). So, for trace metals the natural background concentrations in the higher predatory organisms are even elevated in an unpolluted biotope, due to natural enrichment. A classic example is given by the xenobiotic organic pesticides (DDT, etc.) where the highest trophic levels coincides with the highest concentrations of the pollutant. In environments polluted with trace metals, the sediment has generally the highest concentrations, and consequently, sediment-feeding organisms and their predators will have higher metal concentrations than others.

## 9 Conclusions

Tidal landforms are present all around the North Sea, but the majority are found in the Dutch-German-Danish Wadden Sea area (Fig. 1). The sediment with which these tidal wetlands are built up by the tidal currents originate from different source areas. Hence contaminants, even in their natural background concentrations, come from different source areas as well, and may therefore vary from place to place in the North Sea region. Dissolved fractions such as in cadmium, however, may be much more mobile, covering larger areas than sorbed forms.

In the course of the development of mudflats and salt marshes large-scale environmental variations gradually turn into small-scale patterns in the mature stage of the salt marsh. It may be expected that similar processes are going on in the distribution of natural and polluted trace metals and hydrocarbons. Background values of trace-metal concentrations are reasonably well known, but polycyclic aromatic hydrocarbons may vary considerably in their concentration, owing to local differences in natural fires and burning of fossil fuels from earlier times.

Hydrologic relations are an important factor for the bioavailability of trace metals and hydrocarbons in mudflats and salt marshes. When comparing tidal salt marshes with nontidal (Baltic) marshes (Tyler 1971), metal concentrations in the aerial parts of plants of the former tend to be much higher than those in nontidal marshes. The inundation with metal-containing water (dissolved and/or sorbed to suspended matter) may therefore be a substantial factor in the uptake of metals even by terrestrial plant species. Background values for trace-metal concentrations in salt-marsh plants are scarce, and specific variations are found to be great.

Salt marshes are inhabited by a host of animals of taxonomically very diverse groups. Some of these groups have a crucial place in the functioning of the ecosystem as a whole, for instance bacteria, fungi, nematodes, and crustaceans in the decomposition processes. Others, such as the groups of large and small herbivores, active on the production side of the ecosystem, seem to have a place of minor importance from the viewpoint of functioning of the ecosystem. A decrease in the decomposition rate with accumulation of organic debris in the ecosystem, and perhaps a disturbance of nutrient cycles as a consequence, is one point. Loss of predators of animals in mudflats and salt marshes, decreasing species' richness and species diversity, is another point.

The variety in contaminant source, contaminant speciation, and pathways through the food web makes it impossible to conclude a generalized statement about enrichment processes. This seems valid even under unpolluted conditions, and makes it very difficult, if not impossible, to produce conclusions about the toxicology of pollutants for (groups of) organisms and their functioning in the ecosystem. Not only environmental parameters such as oxygen content, temperature, pH value, salinity, organic solution components, metal complexation, and the specific and mutuai (synergism, antagonism) chemical characteristics of the contaminants are determining. Also biological parameters, such as the physiology of uptake, translocation, storage and release of contaminants, many life-history characteristics, position in the system of trophic levels, and functioning in ecosystem processes are decisive as well.

In short, complexity within and between trophic levels is not only the strength of coastal wetland ecosystems against natural disturbances (Beeftink 1986), but it also prevents us from gaining a clear insight into the demolishing effects of pollution. Decreasing the degree of contamination and other human disturbances to safe levels is the only way to conserve the intrinsic and social values of coastal wetlands.

*Acknowledgments.* The authors wish to thank Mrs. M.J. van Leerdam-de Dreu and Messrs. A.A. Bolsius and J.A. van den Ende for preparing of text and figures.

## References

Albert R (1982) Halophyten. In: Kinzel H (ed) Pflanzenökologie und Mineralstoffwechsel. Ulmer, Stuttgart, pp 33–213

Armstrong W, Wright EJ, Lythe S, Gaynard TJ (1985) Plant zonation and the effects of the spring-neap tide-cycle on soil aeration in a Humber salt marsh. J Ecol 73:323–340

Bastin A (1974) Regionale sedimentologie en morfologie van de zuidelijke Noordzee en van het Schelde estuarium. Thesis Cath Univ Leuven, 91 pp

Beeftink WG (1966) Vegetation and habitat of the salt marshes and beach plains in the south-western part of the Netherlands. Wentia 15:83–108

Beeftink WG (1977) The coastal salt marshes of western and northern Europe: An ecological and phytosociological approach. In: Chapman VJ (ed) Wet coastal ecosystems. Elsevier, Amsterdam, pp 109–155

Beeftink WG (1977a) Salt marshes. In: Barnes RSK (ed) The coastline. A contribution to our understanding of its ecology and physiography in relation to land use and management and the pressures to which it is subject. Wiley, New York, pp 93–121

Beeftink WG (1985) Vegetation study as a generator for population biological and physiological research on salt marshes. Vegetatio 62:469–486

Beeftink WG (1986) Vegetation responses to changes in tidal inundation of salt marshes. In: Van Andel J, Bakker JP, Snaydon RW (eds) Disturbance in grasslands. Junk, Dordrecht pp 97–117

Beeftink WG, Daane MC, Liere JM van, Nieuwenhuize J (1977) Analysis of estuarine soil gradient in salt marshes of the south-western Netherlands, with special reference to the Scheldt estuary. Hydrobiologia 52:93–106

Beeftink WG, Nieuwenhuize J (1986) Monitoring trace metal contamination in salt marshes of the Westerschelde estuary. Environ Monit Ass 7:233–248

Beeftink WG, Nieuwenhuize J, Stoeppler M, Mohl C (1982) Heavy-metal accumulation in salt marshes from the Western and Eastern Scheldt. Sci Total Environ 25:199–223

Beukema JJ (1981) Quantitative data on the benthos of the Wadden Sea proper. In: Danker N, Kühl H, Wolff WJ (eds) Invertebrates of the Wadden Sea. Wadden Sea Working Group Report No 4, pp 134–142

Brinkhuis BH, Penello WF, Churchill AC (1980) Cadmium an manganese flux in eelgrass *Zostera marina*. II. Metal uptake by leaf and root-rhizome tissues. Mar Biol 58:187–196

Brix H, Lyngby JE (1983) The distribution of some metallic elements in eelgrass (*Zostera marina* L.) and sediment in the Limfjord, Denmark. Estuarine Coastal Shelf Sci 16:455–467

Brix H, Lyngby JE, Schierup HH (1983) Eelgrass (*Zostera marina* L.) as an indicator organism of trace metals in the Limfjord, Denmark. Mar Environ Res 8:165–181

Bryan GW, Hummerstone LG (1977) Indicators of heavy metal contamination in the Looe estuary (Cornwall) with particular regard to silver and lead. J Mar Biol Ass UK 57:75–92

Bryan GW, Hummerstone LG (1978) Heavy metals in the burrowing bivalve *Scrobicularia plana* from contaminated and un-contaminated estuaries. J Mar Biol Assoc UK 58:401–419

Bryan GW, Langston WJ, Hummerstone LG (1980) The use of biological indicators of heavy-metal contamination in estuaries. Mar Biol Assoc UK Occ Publ No 1, 73 pp

Buth GJC, Wolf L de (1985) Decomposition of *Spartina anglica, Elytrigia pungens* and *Halimione portulacoides* in a Dutch salt marsh in association with faunal and habitat differences. Vegetatio 62:337–355

De Groot AJ, Salomons W, Allersma E (1976) Processes affecting heavy metals in estuarine sediments. In: Burton JD, Liss PS (eds) Estuarine chemistry. Acad Press, London, pp 131–157

De Kock WC (1975) Milieutoxicologische waarnemingen over de verontreining van het Oosterschelde-areaal met zware metalen. Central Lab TNO Delft. Report CL 75/83, 53 pp

De Pauw N (1975) Bijdrage tot de kennis van milieu en plankton in het Westerschelde-estuarium. Thesis Ghent State Univ, 380 pp

Dijkema KS (1983) The salt-marsh vegetation of the mainland coast, estuaries and Halligen. In: Dijkema KS, Wolff WJ (eds) Flora and vegetation of the Wadden Sea islands and coastal areas. Wadden Sea Working Group, Report No 9, pp 185–220

Dijkema KS (ed), Beeftink WG, Doody JP, Géhu JM, Heydemann B, Rivas Martinez S (1984) Salt marshes in Europe. Europe Committee Conserv of Nature and Natural Resources, Strasbourg. Nature and Environment Series No 30, 178 pp

Dronkers J (1986) Tide-induced residual transport of fine sediment. In: Van de Kreeke J (ed) Physics of shallow estuarine embays, Miami, 1984:228–244

Ernst W (1974) Schwermetallvegetation der Erde. Fischer, Stuttgart, 194 pp

Faraday WE, Churchill AC (1979) Uptake of cadmium by the eelgrass *Zostera marina*. Mar Biol 53:293–298

Förstner U, Müller G (1981) Concentrations of heavy metals and polycyclic aromatic hydrocarbons in river sediments: geochemical background, man's influence and environmental impact. Geo Journal 5:417–432

Förstner U, Salomons W (1980) Trace metal analysis on polluted sediments. I. Assessment of sources and intensities. Environ Technol Lett 1:494–505

Förstner U, Wittman GTW (1979) Metal pollution in the aquatic environment. Springer, Berlin Heidelberg New York, 486 pp

Gambrell RP, Collard VR, Reddy CN, Patrick WH (1977) Trace and toxic metal uptake by marsh plants as affected by oxidation-reduction conditions, pH, and salinity. Laboratory for wetland soils and sediments. Louisiana State Univ Baton Rouge, 135 pp

Gordon RB (1981) Estuarine power and trapping efficiency. In: Martin JM, Burton JD, Eisma D (eds) River inputs to ocean systems. Proc workshop 1979, UNEP-UNESCO, pp 86–92

Groenendijk AM (1984) Primary production of four dominant salt-marsh angiosperms in the SW-Netherlands. Vegetatio 57:143–152

Groenendijk AM, Vink-Lievaart MA (1987) Primary Production and biomass on a Dutch salt marsh: emphasis on the below-ground component. Vegetatio 70:21–27

Guilcher A, Berthois L (1957) Cinq années d'observations sédimentologiques dans quatre estuaires-témoins de l'ouest de la Bretagne. Rev Geomorphol Dyn 5/6:67–86

Haines EB (1976) Stable carbon isotope ratios in the biota, soils and tidal water of a Georgia salt marsh. Estuarine Coastal Mar Sci 4:609–616

Harris JE, Fabris GJ, Statham PJ, Tawfik F (1979) Biogeochemistry of selected heavy metals in Western Port, Victoria, and use of invertebrates as indicators with emphasis on *Mytilus edulis planulatus*. Aust J Mar Freshwater Res 30:159–178

Ketner P (1972) Primary production of salt marsh communities on the island of Terschelling. Thesis Cath Univ Nijmegen, 180 pp

Lion LW, Altmann RS, Leekie JO (1982) Trace metal adsorption characteristics of estuarine particulate matter: evaluation of contributions of Fe/Mn oxide and organic surface coatings. Environ Sci Technol 16:660–666

Long SP, Mason CF (1983) Saltmarsh ecology. Blackie, Glasgow, 160 pp

Lyngby JE, Brix H, Schierup HH (1982) Absorption and translocation of zinc in eelgrass (*Zostera marina* L.). J Exp Mar Biol Ecol 58:259–270

Müller G (1979) Schwermetalle in den Sedimenten des Rheins. Veränderungen seit 1971. Umsch Wiss Tech 79:778–783

Odum EP (1971) Fundamentals of ecology. Saunders, Philadelphia, 574 pp

Packer DM, Ireland MP, Wootton RJ (1980) Cadmium, copper, lead, zinc and manganese in the polychaete *Arenicola marina* from sediments around the coast of Wales. Environ Pollut Ser A Ecol Biol 22:309–321

Pethick J (1984) An introduction to coastal geomorphology. Arnold, London, 260 pp

Ponnamperuma FN (1984) Effects of flooding on soils. In: Flooding and plant growth. Academic Press, London, pp 10–45

Prater AJ (1974) Wader research; coastal wader counts. Bull Int Waterfowl Res Bur 37:102–104

Prater AJ (1976) Wader research group. Bull Int Waterfowl Res Bur 41/42:60–62

Rozema J, Bijwaard P, Prast G, Broekman R (1985a) Ecophysiological adaptations of coastal plants from foredunes and salt marshes. Vegetatio 62:499–521

Rozema J, Luppes E, Broekman R (1985b) Differential response of salt-marsh species to variation of iron and manganese. Vegetatio 62:293–301

Rozema J, Arp W, Esbroek M van, Broekman R (1985c) Relaties tussen autotrofe en heterotrofe planten op kwelders. Vakbl Biol 65:465–468

Salomons W, Eysink WD (1981) Pathways of mud and particulate metals from rivers in the southern North Sea. In: Nio SD, Schüttenhelm RTE, Weering TCE (eds) Holocene Marine Sedimentation in the North Sea Basin. Spec Publ Int Assoc Sedimentol 5:429–450

Salomons W, Hofman P, Boelens R, Mook WG (1975) The oxygen isotopic composition of the fraction less than 2 microns (clay fraction) in recent sediments from Western Europe. Mar Geol 18:M23–M28

Satsmadjis J, Voutsinou-Taliadouri F (1985) An index of metal pollution in marine sediments. Oceanol Acta 8:277–284
Shaw HF (1973) Clay mineralogy of quaternary sediments in the Wash embayment, eastern England. Mar Geol 14:29–45
Smit CJ (1980) The importance of the Wadden Sea for estuarine birds. In: Smit CJ, Wolff WJ (eds) Birds of the Wadden Sea. Wadden Sea Working Group, Report No 6, pp 280–289
Steers JA (1977) Physiography. In: Chapman VJ (ed) Wet coastal ecosystems. Elsevier, Amsterdam, pp 31–60
Sykora KV (1982) Syntaxonomy and synecology of the Lolio-Potentillion anserinae Tüxen 1947 in the Netherlands. Acta Bot Neerl 31:65–95
Taylor GJ, Crowden AA, Rodden R (1984) Formation and morphology of an iron plaque on the roots of *Typha latifolia* L. grown in solution culture. Am J Bot 71:666–675
Teal JM (1962) Energy flow in the salt marsh ecosystem of Georgia. Ecology 43:614–624
Turekian KK, Wedepohl KH (1961) Distribution of the elements in some major units of the earth's crust. Geol Soc Am Bull 72:175–192
Tyler G (1971) Distribution and turnover of organic matter and minerals in a shore meadow system. Oikos 22:265–291
Van der Kooij LA (1982) De waterkwaliteit van de Westerschelde in de periode 1964–1981. Report No 82.063. State Institute for Purification of Sewage Lelystad,111 pp
Vinogradov AP (1962) Average contents of chemical elements in the principal igneous rocks of the earth's crust. Geokhimiya 7:555–571. Cited in: Parker RL (1967) Data of geochemistry D. Composition of the earth's crust. Prof Pap US Geol Surv 440–D
Wallentinus HG (1973) Above-ground primary production of a Juncetum gerardi on a Baltic seashore meadow. Oikos 24:200–219
Wedepohl KH (1960) Spurenanalytische Untersuchungen and Tiefseetonen aus dem Atlantik. Geochim Cosmochim Acta 18:200–231
Windom HL (1975) Heavy metal fluxes through salt-marsh estuaries. In: Cronin LE (ed) Estuarine research, vol 1. Chemistry, biology and the estuarine system. Academic Press, London, pp 137–152
Wolff WJ, Eeden MJ van, Lammens E (1980) Primary production and import of particulate organic matter on a salt marsh in the Netherlands. Neth J Sea Res 13:242–255
Yeo AR, Flowers TJ (1984) Mechanisms of salinity resistance in rice and their role as physiological criteria in plant breeding. In: Staples RC, Toenniessen GH (eds) Salinity tolerance in plants. Wiley, New York, pp 151–170
Zuberer DA, Silver WS (1978) Biological dinitrogenase (acetylene reduction) associated with Florida mangroves. Appl Environ Microbiol 35:567–575

# Estuaries

J.M. MARTIN and J.C. BRUN-COTTAN[1]

It is hardly possible, if not impossible, to devise an accurate definition of an estuary which would satisfy scientists of all disciplines concerned with such systems. Essential concepts have long been discussed (Lauff 1967) and the most broadly recognized definition is that of Pritchard. "An estuary is a semi-enclosed coastal body of water which has a free connection with the open sea and within which sea water is measurably diluted with freshwater derived from land drainage" (Pritchard 1967). However, study of estuarine sedimentation and estuarine chemistry generally requires a broader definition, including the tidal zone, where no salinity gradient is observed but where dynamical tide may transport sedimentary particles.

These observations lead Fairbridge (1980) to assess that the tidally affected freshwater region should be considered as an integral part of any estuary. Accordingly, he proposed the following definition "an estuary is an inlet of the sea reaching into a river valley as far as the upper limit of tidal rise, normally being divisible into three sectors: (a) a marine or lower estuary, in free connection with the open sea: (b) a middle estuary subject to strong salt and freshwater mixing and (c) an upper or fluvial estuary, characterized by freshwater but subject to daily tidal action".

Moreover, that no two estuaries are alike usually discourages generalizations, as one never knows whether general principles or unique details are being studied. However, there are some general concepts which are probably valid from one estuary to the other.

An estuary provides a particularly varied chemical environment which is characterized by strong physicochemical gradients (Fig. 1). It is there that seawater of ionic strength, approximately 0.7 M, mixes with river water of considerably lower salt content and often lower pH. In general, there is a slight pH minimum in the estuary itself, which is enhanced by biological processes: the classical carbonate equilibrium, i.e., the increase of apparent dissociation constants of carbonic acid with salinity is not always worth to compute the pH distribution in estuaries owing to the significant effect of biological production and respiration processes. Moreover, there is usually an oxygen minimum due to the abundance of primary producers related to the abundance of nutrients; this oxygen minimum may be considerably increased in polluted estuaries.

As stated by Morris (1985) "estuaries are classical example of complex thermodynamically open systems, subject to constantly changing input and output

[1] Institut de Biogéochimie Marine, 1, Rue Maurice Arnoux, 92120 Montrouge, France

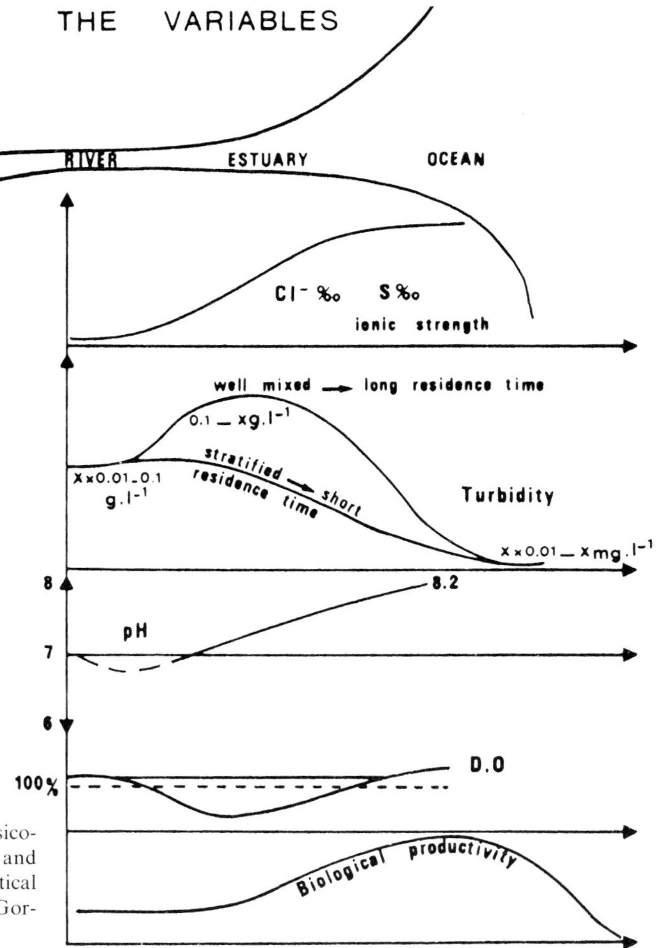


**Fig. 1.** The major physico-chemical, sedimentological and biological variables in theoretical estuaries. (After Martin and Gordeev 1986)

fluxes and to continuous internal chemical reactions", which do not usually reach a steady-state equilibrium. In recent years most of the reactions occurring in estuaries have been identified. However, we cannot usually predict the rates at which these various reactions should occur because of our almost total ignorance of the speciation of most trace metals, of the kinetics of many reactions as well as of the course of heterogeneous reactions involving both dissolved and solid phases. These different problems usually lead most estuarine scientists to characterize estuarine samples with respect to salinity. The actual distribution of one given element is usually compared with that predicted from the theoretical dilution curve of river water in seawater, the chlorinity being considered as a conservative index of mixing.

Curvilinear relationships are indicative of addition or removal processes (Fig. 2), depending upon the sense of curvature. However, the comparison is often misleading because of problems related to the definition of correct end-

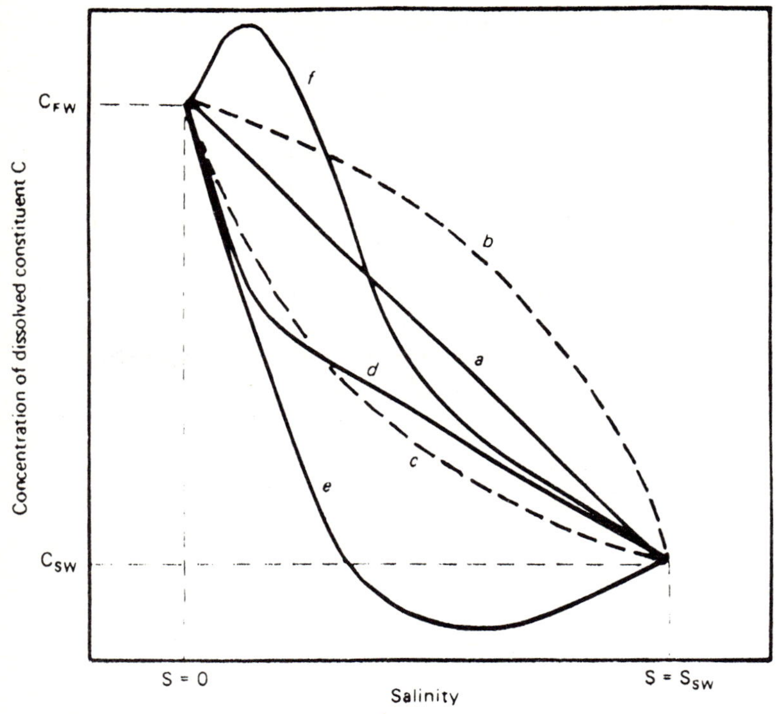


**Fig. 2.** Model dissolved constituent-salinity relationships in an estuary under steady-state conditions. $C_{FW}$ and $C_{SW}$ are the concentrations of constituent C in the freshwater and seawater mixing component, respectively. Line *a* defines the theoretical dilution line for a noninteractive constituent. Curves *b* and *c* indicate relatively widespread estuarine input and removal of C, respectively. Curve *d* is typical of removal occurring only in the upper estuary. Curve *e* is generated when the rate of removal of C in mid-estuary exceeds the riverine input. Curve *f* indicates net input of C to the upper estuary coupled with net removal further seaward. (After Morris 1985)

members, the role of the tributaries or bank flow in the saline part of the estuaries, and the occurrence of several water masses of different ages along the estuary.

The age and transit time of dissolved and particulate constituents are important parameters to understand estuarine processes. It is then necessary to define specific time concepts for any constituents, inside or leaving a given reservoir (Martin et al. 1986).

Let us consider a set of particles moving according to a given physical law and a virtual space volume, V (see Fig. 3). For each particle defined at time $t_x$ there are two associated times, $t_i$ and $t_o$. They are, respectively, the instant the particle enters and leaves the volume.

From these times, three different time parameters can be derived. We shall call $t_x - t_i$ the age of the particle, $t_o - t_x$ is life time. The age plus the life time is the transit time.

Due to the great number of particles in a natural aquatic system, it is not possible to apply these concepts to each of them, so that only average values can be used in practice.

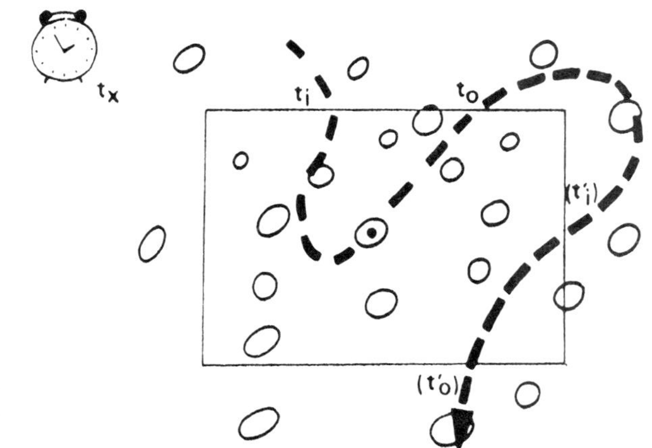


**Fig. 3.** Representation of the movement of particles in a virtual volume. (After Martin et al. 1986)

In a classic paper, Bolin and Rhode (1973) specified some of the time concept definitions and their inter-relationships. In order to avoid any confusion, their notation will be used here. They called $M_0$ the total number of particles inside the volume, and $F_0$ the number of particles entering or leaving the system during each unit of time. They defined two frequency functions $\psi(t)$ and $\phi(t)$, representing the age distribution of the particles inside the volume and the age (or transit time) of the particles leaving the volume. From the steady state hypothesis these parameters and functions do not vary with time. Bolin and Rhode (1973) showed three main relations between them:

$$\phi(t) = -M_0/F_0 \frac{\delta\psi}{\delta t}(t) \tag{1}$$

$\psi(t)$ is a non-increasing function and

$$\psi(O) = F_0/M_0 \tag{2}$$

$$\int_0^\infty t\phi(t)dt = M_0/F_0. \tag{3}$$

Note that, in Eq. (1), t is not the real time but is an age-scale for the particles in the system. Thus $\delta\psi/\delta t$ is not necessarily zero, despite our definition of steady state.

$M_0/F_0$, the mean age, or mean transit time, of the particles leaving the system, is a characteristic number of the reservoir. Bolin and Rhode (1973) called it residence time. It does not depend on the frequency functions of the system (i.e., on the movement of the particles inside the box).

Two other relations can be established involving the time notions specified by Bolin and Rhode (1973):

1. The mean transit time of the particles inside the volume is equal to twice their mean age (if it exists) so that their mean life time is equal to their mean age.
2. The mean age of the particles inside the volume is always greater than half of the residence time. The equality is valid only in the case of a plugflow system (i.e., a system in which all the particles have the same transit time).

This means that, at steady state, only two time parameters are relevant: the mean transit time of the particles leaving the system or residence time ($\tau$) and the mean age of the particles inside the volume usually called simply mean age ($\theta$). (see Martin et al. 1986).

Most estuaries can be considered as composed of several reservoirs. As pointed out by Gibilaro (1977), the residence time of the whole system does not depend on the flow rates between the reservoirs, but only on the total hold-up in the system and the overall flow rate; and the age distribution of the particles flowing out of a single reservoir can be obtained by convolution of the age distribution function of the particles flowing in with the specific frequency function of the reservoir.

Residence times and mean age can be computed for simple systems as exemplified Fig. 4.

As an example, let us consider system A in Fig. 4. We can compute that the mean age of the particles at the output is $(M_1 + M_2)/Q$ (which is the residence time), and the mean age of the particles flowing out of the lower reservoir is $P(\tau_1 + \tau_2)/Q$, where $\tau_1$ and $\tau_2$ are the residence times of the upper and the lower reservoirs.

It is possible to distinguish:

(1) Systems with dead space: a dead space is a region where fluid elements are retained for times of an order of magnitude greater than the mean residence time of total fluid.

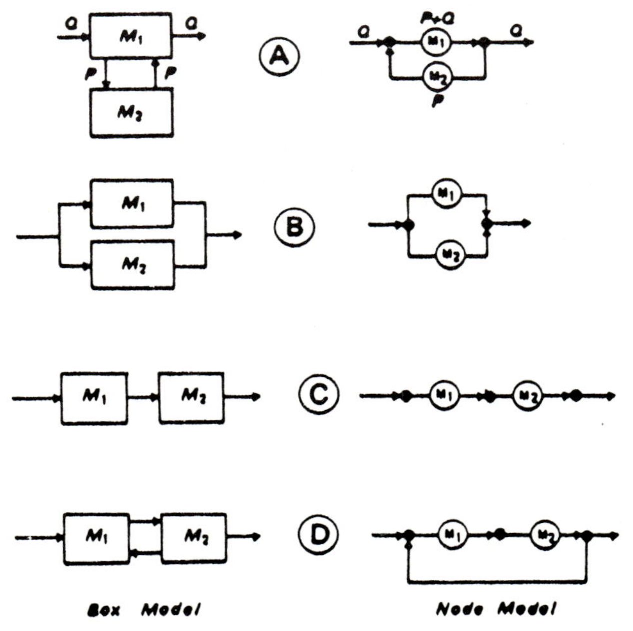


**Fig. 4 A-D.** Comparison of node models and box models for simple systems. **A** and **B** are systems with dead space or by-passing, **C** and **D** are flowing systems. **A** represents sedimentation/resuspension processes where the dead space $M_2$ is the sediment. **B** represents the flow transfer of particles, by the water ($M_1$) and by the bottom viscous layer or (bed load) of mud ($M_2$). **C** and **D** represent the mixing of marine and river waters. (After Martin et al. 1986)

(2) Systems with bypassing: fluid is bypassing if it passes through the vessel in a time an order of magnitude smaller than the residence time of the overall fluid.
(3) Plug or piston flows: as defined previously, a plug flow is a system in which all the particles have the same transit time. Velocities have to be uniform.

Another approach to the problem of determining the fate of dissolved constituents in estuaries has been proposed by Martin et al. (1976) and Elderfield (1978). They applied the concept of relative residence time (R), defined by Stumm and Morgan (1970) for lakes, i.e. the ratio between the residence time of a given element (tx) and the residence time of river water ($tH_2O$) so-called "flushing time". For conservative elements, R = 1. R < 1 means that the chemical element is removed from the solution, whilst R > 1 suggests an internal recycling. The residence time of water and material are probably important variables which determine the fate of the chemical elements in estuarine systems. It is quite obvious that long residence times occurring in well-mixed estuaries will enhance the influence of estuarine biogeochemical processes, as compared to the short residence times prevailing in stratified estuaries.

It is often difficult to compute the "flushing time" of the whole estuary because the actual volume of brackish water may extend in the coastal sea very far from the mouth.

Moreover, for a given section close to the mouth, an important fraction of freshwater can leave or reenter the estuarine zone, depending upon the currents induced by the tidal effects or by the forcing effect of wind over the upper water layer.

As a first approximation the "flushing time" computed for the western Scheldt, the Thames and the Rhine average 2 months, 1 month and 3–4 days, respectively (R. Wollast pers. comm.). It must be noted that the Rhine flushing time is surprisingly low as compared to the others. This is basically due to the important waterworks which have been implemented in the framework of the Deltaplan.

Some general comments can be made on the behavior of dissolved components in the estuarine zone.

As river water runs into the oceans, it enters a reservoir with a much longer holding time where an element can accumulate up to the age limit of the reservoir itself ($10^8$ years) unless its concentration is controlled at some intermediate level by interactions with particulate matter. It can be shown that the residence times of the elements in the ocean, i.e., the average time that a particular element spent in the oceans is known as its mean oceanic residence time (M.O.R.T.). (Whitfield 1979) are related to the ratio Ro = mean concentration in seawater) / (mean concentration in river water) (Whitfield and Turner 1979). The elements dissolved in seawater may be divided into "accumulated" ($R_o > 10$), "unchanged" ($10 > R_o > 0.1$) and "depleted" ($R_o < 0.1$) categories (Martin and Whitfield 1983). The "depleted" elements have MORT values which are less than the time required for a single circulation of the oceans (= 1500 years). These elements are therefore rapidly deposited and their estuarine chemistry is likely to be controlled by particulate removal. This is usually due to some adsorption or/and coagulation-flocculation processes involving humic acids or hydrous iron colloids with which they may be associated. The rate of removal will depend, among other things, on the affinity for organic ligands. The "accumulated" elements which are characterized by large MORT values, are able to accumulate in the world ocean; their concentration is remarkably constant over the world ocean. These elements will in general exhibit a conservative behavior during estuarine mixing, so that the mean concentrations of the "unchanged" elements are not altered on transfer from the river to the ocean. However, many elements belonging to this category are actively cycled by the biota and will consequently show marked spatial and temporal variations in concentration. Depending on the balance struck between the processes active in a particular estuary, these elements may exhibit conservative behavior (e.g., As, Si, Zn, Ni, N), removal (Zn, Si, Ni) or addition (Ba, As, P, Zn). The controlling reactions might be biological utilization (e.g. N, P), ion exchange (e.g., Ba), variations in redox chemistry (e.g., As), or adsorption (e.g., Al) (Martin and Whitfield 1983).

With regards to solid phase, most of the estuaries are characterized by a "turbidity maximum". The freshwater discharge generates a residual seaward flow in the upper layer, the sea-water which is entrained from the lower layer by this flow is compensated by a residual landward flow along the bottom (Fig. 5A). The

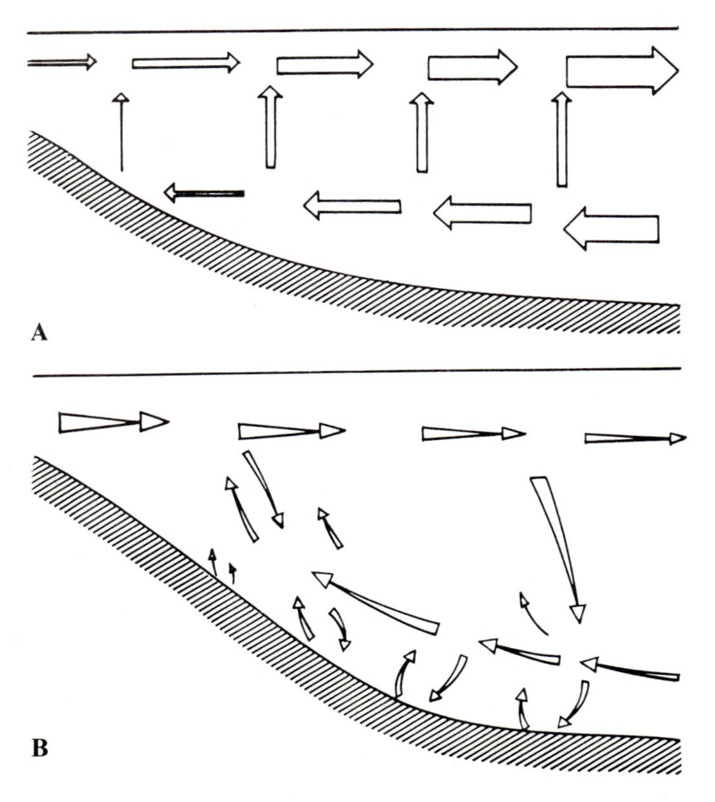


**Fig. 5 A,B.** Schematic representation of water (**A**) and particles (**B**) transport within a macrotidal estuary

river-borne particles which settle in this layer are taken up by the residual landward flow and trapped in the middle or upper part of the estuary through up-estuarine "tidal pumping" generated by the assymetry in the tidal ebb and flood velocities (Allen et al. 1980) (Fig. 5B). This process greatly enhances the residence time of particles, especially in well-mixed and partly mixed estuaries.

This process is well documented in the case of the Scheldt estuary. According to R. Wollast (1982) the salinity gradients influence the vertical distribution of the currents in order to modify the residual currents averaged over one tidal cycle.

The density currents slacken the ebb movement and accelerate the flood movement near the bottom. A reverse effect in a surface layer compensates this bottom movement. Consequently, the residual currents near the bottom are orientated upstreams in the lower zone. They are, however, orientated downstreams in the freshwater zone, and the two opposed movements cancel out in the area of the harbor of Antwerp, which is thus a highly favorable zone for the accumulation of sediments.

The existence of a vertical gradient of turbidity, associated with these water movements, will create a zone of maximum turbidity (Fig. 6).

Estuarine particulates provide a heterogeneous mixture of lithogenous, hydrogenous and biogenous origins.

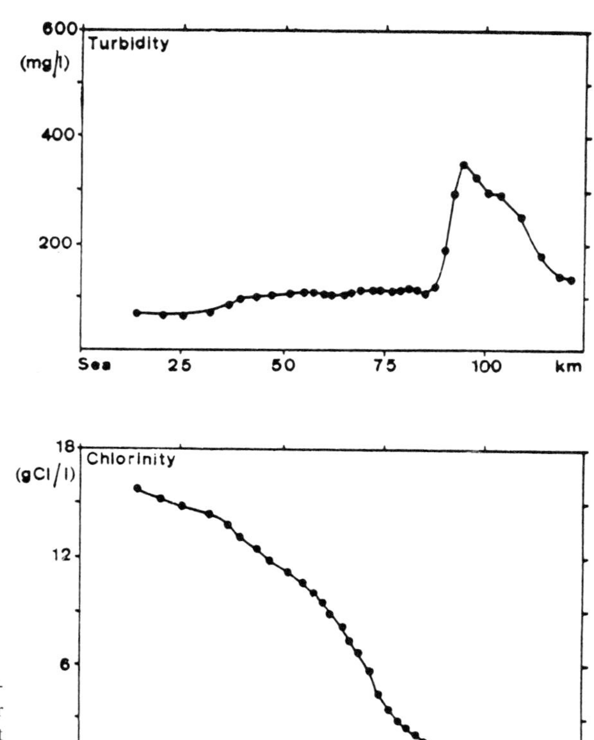


**Fig. 6.** Longitudinal profile of turbidity and salinity for mean river discharge conditions, in the Scheldt estuary (according to R. Wollast, 1982)

Conceptually, particles can be considered as a complex assemblage of different inorganic and organic components. The "matrix vehicle" or residual fraction (Jenne 1977) is associated with more labile and thermodynamically unstable components such as carbonates, amorphous alumino silicates; organic matter etc. Both these fractions are usually coated with Fe and Mn oxides and organics material (both living and nonliving).

It is obvious that the majority of interactive processes which determine the distribution of particle sizes and association of pollutants with solid particles are basically surface phenomena taking place at the solid-liquid interface.

The intensity of adsorption processes as well as that of any other processes occurring at the surface of particles should be directly proportional to the available surface of the solid. For all studied estuaries, SSA is lower in the riverine part, with a significant increase near the salt intrusion zone followed by a decrease near the sea end member (Fig. 7). This maximum could greatly enhance the availability of adsorption sites for dissolved microconstituents. It can be explained by a granulometric sorting due to the estuarine circulation and/or to flocculation or coagulation processes and by a change of the surface coating composition (Martin et al. 1985).

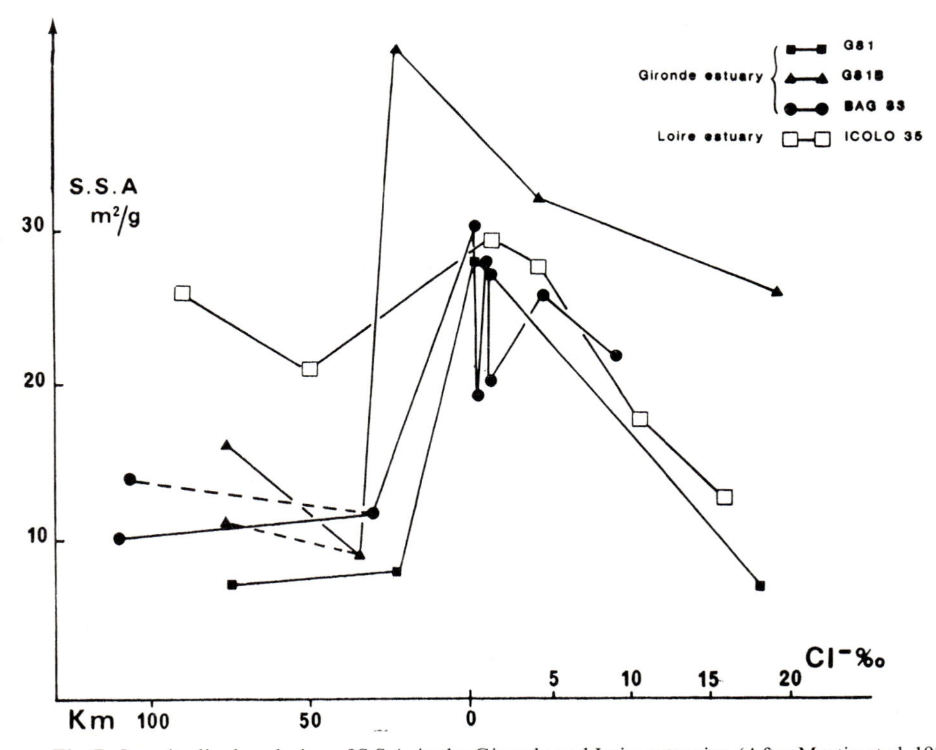


**Fig. 7.** Longitudinal evolution of S.S.A. in the Gironde and Loire estuaries. (After Martin et al. 1985)

The sedimentation through the water column and the resuspension from the sediment involve output and input fluxes of material of a given body of estuarine water.

The sedimentation flux depends on the settling velocity of particles, i.e., on the particle size distribution and the particle density versus particle size. The particle size distribution can be well described by one or more lognormal populations (Brun-Cottan 1986), then the mass concentration and the vertical mass flux can be calculated. The uncertainties on the particle density do not introduce an error in the estimates of the mass concentration larger than 50–100%, because in marine or estuarine waters, the wet density is usually comprised between 1 and 2. Unfortunately, when computing the vertical mass flux, the density uncertainties apply twice, on the mass determination and on the settling velocity. The effect on the settling velocity is proportional to p-1, then the uncertainties on the p values for the largest particles are critical for computing the total sedimentation flux. We can say that the error on the actual vertical mass flux value can exceed the theoretical estimation by two or three times.

It is possible to calculate the instantaneous suspended matter vertical flux at steady state, but in estuaries the particle size distribution can change very quickly and a large proportion of the suspended matter can coagulate or agglomerate into large flocs having very high settling velocities.

The major input process to an estuarine reservoir are advection/diffusion and the resuspension from bottom sediments. The residence time of the particles in the sediment depends on the intensity of resuspension flux and on the thickness of the sediment layer which is involved in this process. Because this thickness is a function of the flow velocity above the sediment and of the sediment structure, the resuspension is usually a nonsteady-state process over a short time scale ($<$ 1 h).

Depending on the time scale, the balance between sedimentation and resuspension fluxes can be regarded such as a steady or nonsteady state. With the use of a Lagrangian point of view regarding a wide body of water, O'Kane (1980) introduced the concept of quasi-steady state which is based on the use of the moving frame method; this method applies mainly for tidal effects such as the mixing between seawater and river water, but not when the system is submitted to stress, such as strong wind or flood discharges.

With the moving frame method, the estuary is conceived like an accordion. A reference frame taken on the bank is transformed to another reference frame which oscillates so as to maintain a constant upstream volume. This mathematical transformation removes almost all the tidal harmonics in the instantaneous mass balance of water and solutes in one-dimensional estuaries. Thus, the longitudinal coordinates are determined by the upstream volumes and not by upstream distances or salinities. Such modeling allows the use of dispersed data, since there is no need for tidal averages.

This last remark raises the tricky problem of the continuous variability of estuarine environment in response to the highly dynamic nature of the physical system (Morris 1985). As a matter of fact, most of the practical studies of estuaries have been carried out under the assumption of a steady-state condition, whilst the development of adequate methods of dealing with the variable state is one of the outstanding requirements of studying problems of estuarine circulation (Bowden 1967) and chemical activity.

Indeed the principal sources of power which drive the transfer of sedimentary particles in estuaries are primarily the tides and the river flow; in some instances wind-driven currents and waves, especially during storm periods and in shallow estuaries, may play a significant role. A first step toward the understanding of the fate of sedimentary particles and the exchange processes between these particles and the dissolved phase within an estuary is to determine the relative importance of these power sources. These problems have been discussed by Gordon (1981), who assessed the "trapping efficiency", i.e., the fraction of sediment discharged to a given estuary which is retained. Obviously the particles will be retained within the estuary when the specific power dissipation will be sufficiently low.

On a larger period of time, as stated by Fairbridge (1980) an estuary must be considered as an "ephemeral feature and be regarded as a dynamically evolving land-form that will go through a life cycle from valley creation and ending with the progressive infilling". The majority of estuaries were created by the post-glacial rise of sea level that stabilized at 6000 B.P., and almost all the world estuaries are dominated by this worldwide eustatic control, which overwhelms most of the local or regional tectonic modifications.

## References

Allen GP, Salomon JC, Bassoulet JR, Penhoat Y du, Grandpre CM de, Allen GP (1980) Effects of tides on mixing and suspended sediment transport in macrotidal estuaries, Sediment Geol 26:69–90

Bolin B, Rhode H (1973) A note on the concepts of age distribution and transit time in natural reservoirs. Tellus 25:58–62

Bowden KF (1967) Circulation and diffusion In: Lauff GH (ed) Estuaries. Washington DC Am Assoc Adv Sci, pp 15–36

Brun-Cottan JC (1986) Vertical transport of particles within the Ocean. NATO/ASI. The Role of Air-Sea Exchange in Geochemical Cycling, Reidel, 83–111

Elderfield H (1978) Chemical variability in estuaries. In: Biogeochemistry of estuarine sediments (1970) UNESCO, Paris, pp 111–178

Fairbridge RW (1980) The estuary: its definition and geodynamic cycle. In: Olausson E, Cato I (eds) Chemistry and biogeochemistry of estuaries. Wiley, New York, pp 1–35

Gibilaro LG (1977) Mean residence time in continuous flow systems. Nature 270:47–48

Gordon RB (1981) Estuarine power and trapping efficiency. In: Martin JM, Burton JD, Eisma D (eds) River inputs to ocean systems. UNEP/UNESCO

Jenne EA (1977) Trace element sorption by sediments and soils. Sites and processes. In: Petersen SK (eds) Proc. Symp. on Molybdenum in the environment, vol 2. In: Clapell, Dekker, New York, pp 425–453

Lauff GH (ed) (1967) Estuaries. Washington DC Am Assoc Adv Sci

Martin JM, Goordeev V (1986) River input to ocean system: a reassessment in estuarine processes: An application to the Tagus Estuary, UNESCO/IOC/CNA Workshop, Lisbon, pp 203–240

Martin JM, Whitfield M (1983) The significance of the river input of chemical elements to the ocean. In: Wong CS et al. (eds) Trace metals in sea water. Plenum, New York

Martin JM, Meybeck M, Salvadori F, Thomas AJ (1976) Pollution chimique des estuaires: Etat actuel des connaissances. Rapp Sci Tech CNEXO 22:286

Martin JM, Mouchel JM, Jednacak-Biscan J (1985) Surface properties of particles at the land-sea boundary. In: Lasserre P, Mailin JM (eds) Biogeochemical processes at the land-sea boundary. Elsevier, Amsterdam

Martin JM, Mouchel JM, Thomas AT (1986) Time concepts in hydrodynamic systems with an application to Be in the Gironde estuary. Mar Chem 18:369–392

Morris AW (1985) Estuarine chemistry and general survey strategy. In: Handbook A, Head PC (eds) Practical estuarine chemistry. Cambridge Univ Press, Cambridge

O'Kane JP (1980) Estuarine water quality management. Pitman, London, 156 pp

Pritchard DW (1967) Observations of circulation in coastal plain estuaries. In: Lauf GH (ed) Estuaries. Washington DC Am Assoc Adv Sci, pp 37–44

Stumm W, Morgan JJ (1970) Aquatic chemistry. Wiley, New York, p 583

Whitfield M (1979) The mean oceanic residence time (MORT) concept, a rationalisation Mar Chem 8:101–123

Whitfield M, Turner DR (1979) Water rock partition coefficients and the composition of sea water and river water, Nature 278

# Fjords

J.M. SKEI and J. MOLVAER[1]

## 1 Introduction

Fjords are found in high latitude coastlines as partly landlocked areas that overlie deep basins. According to Holtedahl (1975), a typical fjord is a relatively long and narrow, often curved or branched embayment with more or less steep sides and a considerable depth. The water depth is often greater than that on the adjacent continental shelf. As a class, fjords are deepest of all estuaries.

Considering the Norwegian coastline between the Swedish border and 62° 00′N (Fig. 1), there is a great variety in fjord morphology, varying from typical semi-enclosed polls with shallow sills to deep, large fjord systems with free connection with the coastal water. Fjords may be considered as the buffer zone between the land and the open sea (Syvitski et al. 1987), with respect to both exchange of water masses and transfer of sediments and pollutants.

In this chapter we intend to focus on the role which fjords play as a buffer zone for pollutant transfer between Norwegian land-based sources of pollution and the Skagerrak and the North Sea. The peculiarity of fjords and the difference from other estuaries with respect to hydrodynamics and pollutant transport, will be emphasized. This will be exemplified with three case histories (see pages 474–488, this Vol.) demonstrating how morphologically different fjords respond to varies types of pollutant loading. The overall objective is to make an assessment on the impact that polluted fjords have on the water quality of the Skagerrak and the North Sea.

## 2 The hydrodynamics of Fjords

For this discussion we will consider a typical fjord containing a deep basin with a shallow sill at its mouth. In most cases, one or several major rivers are entering the inner part of the fjord, i.e., the Dramsfjord and the Sognefjord. The sill depth and the runoff will determine the *stratification* of the fjord water, where three layers may often be identified (Fig. 2). In fjords without sills, it is usually sufficient to distinguish between a surface layer and the deep water below.

The water exchange between the fjords and the coastal water is normally dominated by four mechanisms.

[1]Norwegian Institute for Water Research, P.O. Box 33, Blindern, 0313 Oslo 3, Norway

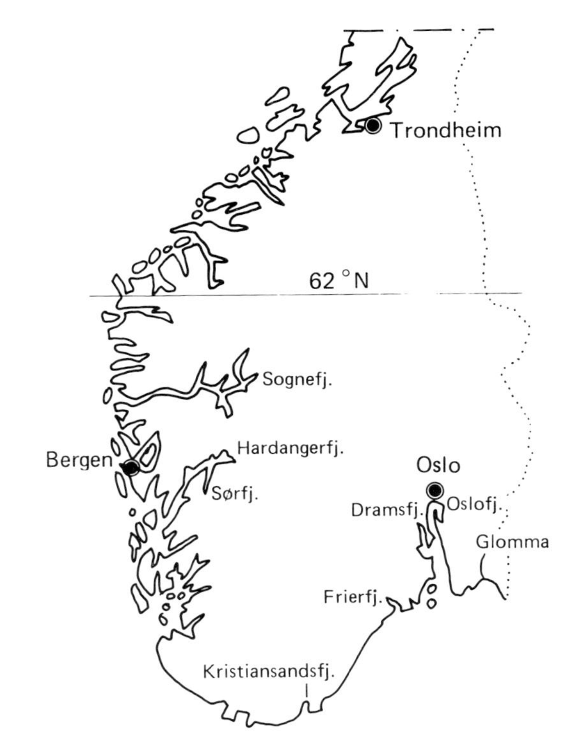


**Fig. 1.** Map of southern Norway

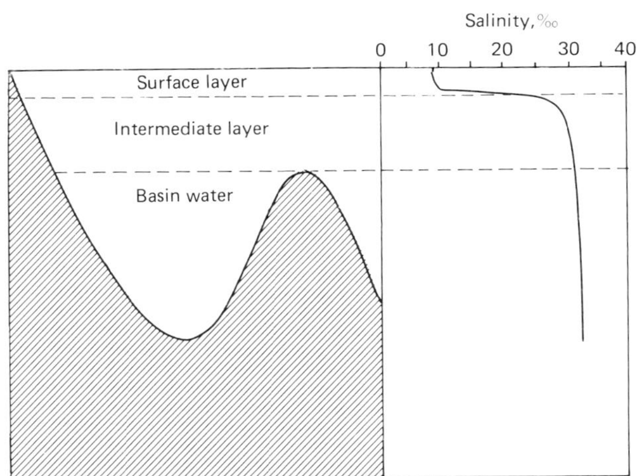


**Fig. 2.** General description of vertical salinity profile and water masses in a sill fjord

- estuarine circulation
- tidal currents
- effects of wind and air pressure in the fjord and in the coastal water
- diffusive processes.

A brief discussion of the water exchange processes of the three layers is presented in the following.

## 2.1 Surface Layer

The freshwater entering the fjord will rapidly be admixed with seawater from below, and establish a brackish layer flowing towards the sea. Along this distance the salinity may increase from nearly zero to nearly coastal water salinity (30–32‰). The seawater, which is thus removed from the fjord, is replaced by a return current (estuarine circulation). At the mouth of larger fjord systems, Saelen (1967) has estimated the volume flux of the brackish water to be two to six times the freshwater runoff. The freshwater discharge and the salinity of the surface layer may undergo large variations, implying considerable changes in the volume transport with time (Fig. 3).

Among fjords with well-established estuarine circulation in southern Norway are Dramsfjord, Frierfjord, Sørfjord, and Sognefjord (Fig. 1).

In fjords with relatively small freshwater supply (e.g., Inner Oslofjord) and in periods with low runoff, tidal currents, local wind stress in the fjord, and wind conditions offshore will dominate the circulation in the surface layer. On the Norwegian coast, the dominating tidal period is semi-diurnal. The mean difference between low and high water change from 0.1 m in the south to 2 m in the north. Considering fjords with surface areas in the order of 10–100 $km^2$, the tidal variations set very large water masses into motion, but to a large extent moving them back and forth with relatively small net exchange.

For topographical reasons, the local wind usually blows along the fjord axis, and the wind stress may strongly influence the circulation both in the surface layer and in the layers below. Periods of infjord wind may pile up water at the fjord mouth, while downfjord wind will normally increase the surface outflow. However, in periods when the water level along the coast is higher than in the fjord, an upfjord surface current may exist even with downfjord wind and considerable freshwater supply (Svendsen 1981).

The residence time for the surface layer in a fjord is typically in the order of days-weeks.

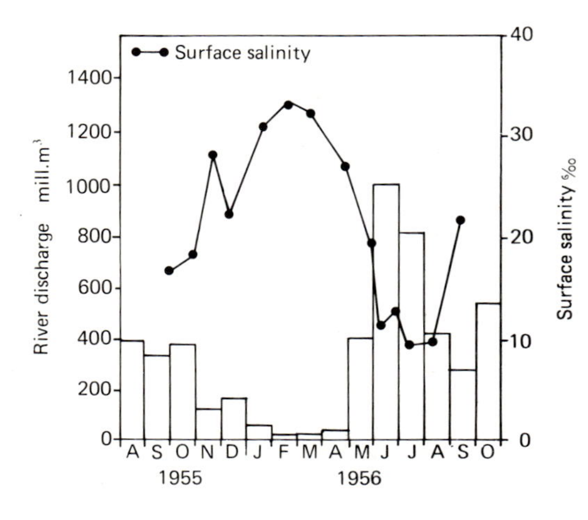


**Fig. 3.** Discharge of rivers Eio, Kinso, Opa, and Tysso into the innermost part of Hardangerfjord, and surface salinity at a station in the area. (Saelen 1967)

## 2.2 Intermediate Layer

This layer has an open connection to the coastal water, and the density distribution responds to the variations offshore. A more or less continuous replenishment of this water mass will therefore take place. This is also the case for the deep layer in fjords without sills.

Apart from the fjords where the estuarine return current is permanently or periodically important, the water exchange between the intermediate layer and the coastal water will be strongly influenced by the wind conditions offshore (Svendsen 1980). The mean circulation is often basically a two-layer flow. When the wind component along the coast has the shore to the left, an upwelling may take place, creating an inflow in the lower part of the intermediate layer with corresponding outflow in the upper part. Wind blowing in the opposite direction may reverse the circulation.

Typical residence time for the intermediate layer is in the order of weeks-months.

## 2.3 Basin Water

The combination of one or more sills and a stable density gradient create a blocking effect with periods of more or less stagnant basin water below sill level.

A thorough discussion of the hydrodynamics of basin water exchange processes is given by Gade and Edwards (1980). The basin water renewal is primarily connected to seasonal changes in the density of the coastal water. These changes are coupled to the "monsoonal" nature of the average wind field on the west coast, being predominantly northerly in the summer and southerly in winter. On the Skagerrak coast the corresponding directions are southwest and northwest.

The density variations are further coupled to fluctuations of air pressure above the Baltic and Skagerrak, the amount of freshwater in the coastal water due to Baltic outflow, and runoff from southern Norway. An example of the resulting density structure is shown in Fig. 4.

When the density of the coastal water above sill level is equal to or greater than the density of the basin water, a complete or partial basin water renewal will take place (Fig. 5).

For Inner Oslofjord, the annual renewal for the period 1973–84 varied from 20–100% (Magnusson 1985). For any given fjord, the residence time for the basin water will therefore often vary from 1 to 3 years. In fjord systems with several shallow sills or in very large fjords (i.e., the Sognefjord), the residence time may be in the order of 5–15 years.

Considering the relatively small volumes, the intermittent renewal of basin water in most cases is a small part of the water exchange between the fjord and the coastal water.

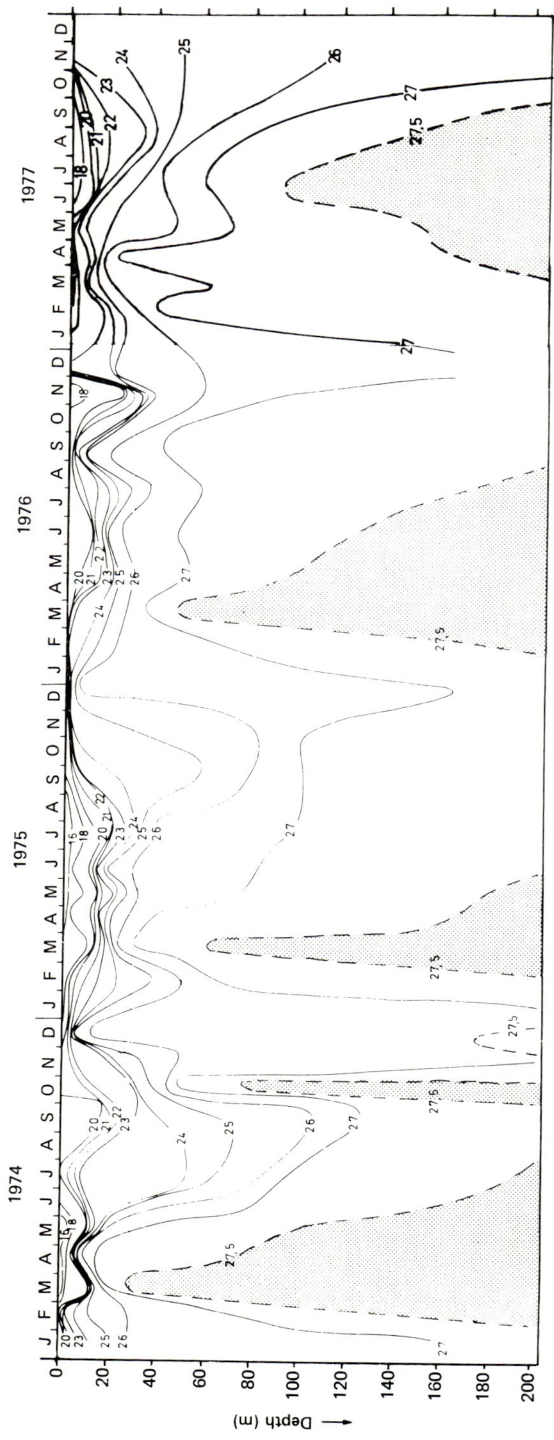


**Fig. 4.** Density variations in the coastal water outside the Frierfjord. *Vertical arrows* indicate renewals of the basin water. (After Molvaer 1980)

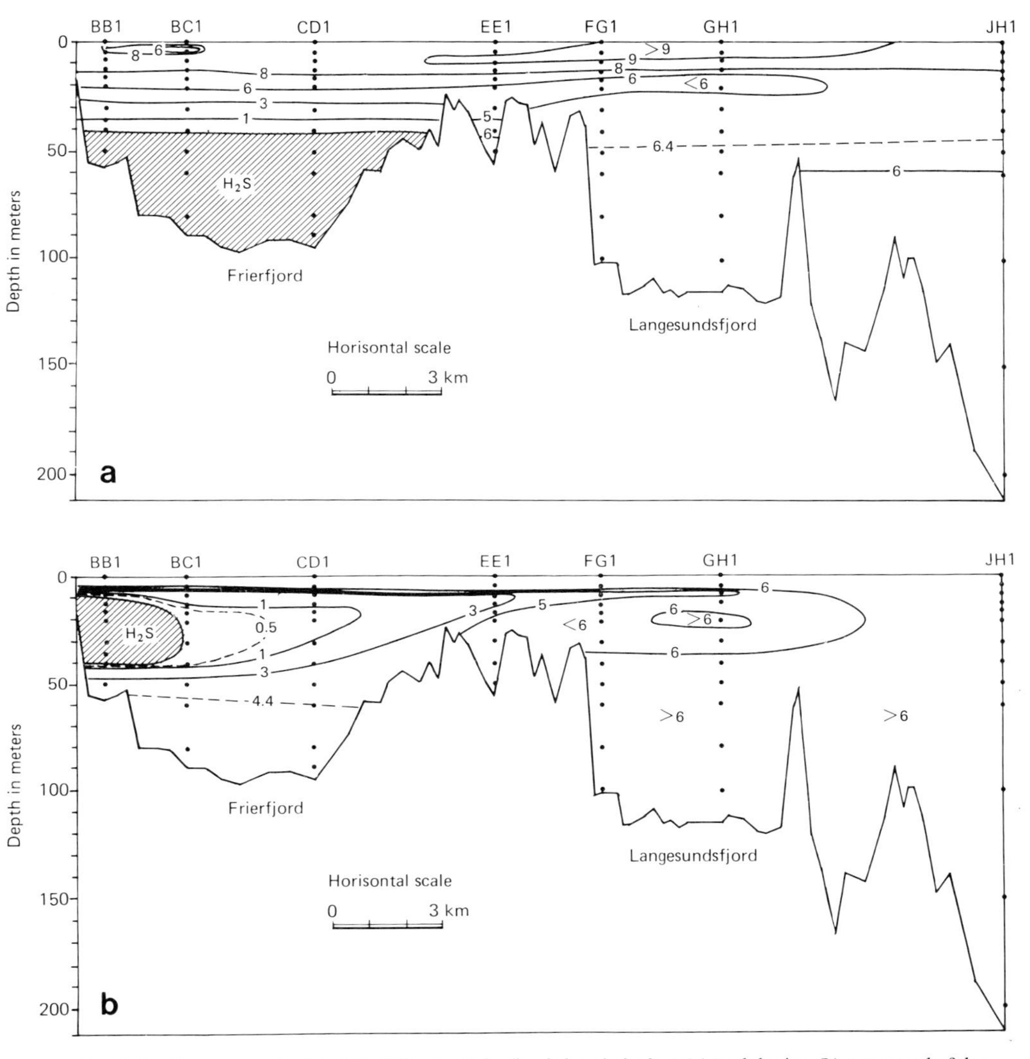


**Fig. 5a.b.** Oxygen content (ml $O_2$ / l) in the Frierfjord shortly before **(a)** and during **(b)** a renewal of the basin water

# 3 Transport and Sedimentation

The net export of freshwater, sediments, and pollutants from fjords to the coastal water will vary with time and local conditions. In most cases the export of sediments and pollutants follows that of freshwater, as the surface outflow acts as a conveyor belt. As the majority of pollutants show a strong particle affinity, the distribution of sediments also governs the fate of the pollutants (Skei 1981). Consequently, a major portion of the pollutant load, at least in terms of heavy metals and organochlorines, ends up in the bottom sediments. By looking at gradients along transects in the sediments from the fjords head to the mouth, information is obtained about the likelihood of fjords as sinks or sources of pollutants.

Pollutants introduced in the surface water of fjords will often follow the outflow above the halocline. Normally, this is the watermass with maximum particulate load in terms of river-borne sediments and organic matter of marine and terrestrial origin. Depending on the energy of the surface flow and the efficiency of the halocline in retarding particle settling, the sediment-associated pollutants are transported long or short distances. Experience from inestigations of ten polluted Norwegian fjords show that the concentrations of pollutants in the surface water decrease rapidly seawards. At the same time, substantial gradients of pollutants are seen in the bottom sediments, suggesting that fjords act to a large extent as pollutant sinks.

One exception may be nutrients (nitrogen and phosphorus compounds). These substances show little particle affinity and are only to a small extent accumulated in the fjord sediments. In fjords with large freshwater runoff, one may observe that only a small proportion of the nutrients released into the surface water of fjords will be utilized in the production of algae in the surface layer. This is due to two factors: (1) the short residence time of the surface water and (2) the inhibiting effect of the freshwater. As a result, the nutrients are transported into the coastal water to make a significant contribution to eutrophication. A typical example is the input of nutrients with the river Glomma (Fig. 1) into the fjord estuary, resulting in low primary production inside the islands but considerable productivity in the coastal zone outside.

Beside transport and sedimentation from the brackish surface layer, pollutants are frequently discharged below the halocline. This is generally a water mass with low levels of natural particulate matter. The fate of the pollutant will to a large extent depend on its state in the effluent and the density profile and dominating currents between the depth of discharge and the depth where the effluent plume will spread out after mixing with the ambient water.

If the pollutant is released into the estuarine return current, it may be transported towards the head of the fjord, either to become incorporated in the bottom sediments or to be mixed into the surface layer.

As multilayer flow may occur in fjords (Carstens 1970), a deep water discharge may imply a seaward transport. A long residence time will allow sedimentation to occur if the pollutant is particle-associated.

Discharge of particulate pollutants like mine tailings is less dependent on the general flow pattern in fjords. Due to the high density, such effluents act as a plume which may be recognizable long distances from the source. Occasionally, coarse mine tailings settle out near the discharge outlet, creating unstable underwater deltas and slope failure which sets up turbidity currents transporting tailing very long distances (Hay 1982).

## 4 Biological Characteristics

Beside chemical and physical gradients in water and sediments of fjords, there are also significant biological gradients (Syvitski et al., 1987). This is a consequence of the change in water quality from head to mouth of a fjord. Salinity is a critical environmental parameter and the organisms show variable ability to adapt to salinity changes (Fig. 6). Pollutant load is another stress factor, causing a selection of opportunistic and resistant species. Normally, the density of bottom fauna individuals is between 1000 and 2000 and the species between 60 and 90 per $m^2$ in fjords (Rygg and Skei 1984). Studies in a number of fjords in Norway showed negative correlation between the diversity of soft-bottom fauna and pollutant discharge amounts, proximity to discharge areas, and sediment pollutant load (Fig. 7).

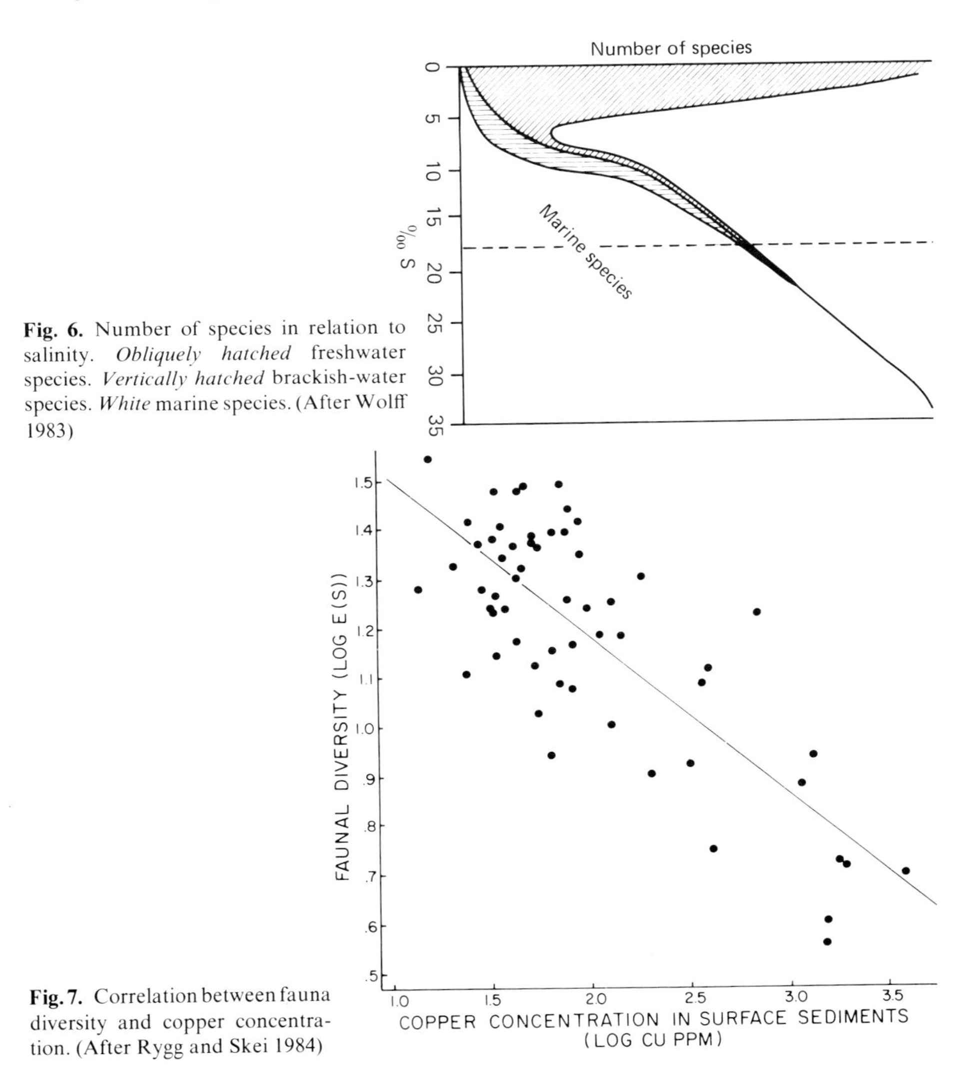


**Fig. 6.** Number of species in relation to salinity. *Obliquely hatched* freshwater species. *Vertically hatched* brackish-water species. *White* marine species. (After Wolff 1983)

**Fig. 7.** Correlation between fauna diversity and copper concentration. (After Rygg and Skei 1984)

In the buffer zone between the fjord and the coastal water there is an exchange of pelagic organisms. Phytoplankton may be passively transported in and out of the fjords, through the water exchange between the coastal water and the fjord upper and intermediate layers. This implies that a plankton bloom in the coastal water may trigger off a bloom in the fjords, and vice versa. The most conspicuous effect seems to be the increased occurrence of toxic dinoflagellates (*Prorocentrum minimum* and *Gyrodinium aureolum*), which may be transported by the Jutland current into the inner Skagerrak and cause mortalities in marine fishfarms along the coast of southern Sweden and Norway. The blooms may be accelerated by nutrient input from sources of pollution located in fjords.

In addition to the toxic effect of such blooms, they also represent an organic loading on the bottom water and the sediments. In recent years, oxygen deficiency has been frequently observed in the bottom water of the coastal zone of Skagerrak, resulting in anoxic sediments, devoid of bottom fauna (Rosenberg 1985).

## 5 Summary

Fjords, the buffer zone between the Scandinavian hinterland and the Skagerrak-North Sea, may act both as a sink and a source of pollutants. Industrial effluents consisting of metals and organochlorines are to a large extent trapped in the bottom sediments of the fjords, creating local problems. This is based on an approximate exponential increase of these pollutants in the sediments towards the source. Nutrients, on the other hand, if not utilized in the fjords due to low salinity water, may enhance the level of nutrients in the near coastal water and thus as a result contribute to possible eutrophication problems there.

With respect to the major water masses of the southern and northern North Sea, Norwegian fjords are not likely to influence the water quality.

## References

Carstens T (1970) Turbulent diffusion and entrainment in two-layer flow. Proc Am Soc Civil Eng WW1:97–104

Gade HG, Edwards A (1980) Deep water renewals in fjords. In: Freeland HJ, Farmer DM, Levings CD (eds) Fjord Oceanography. Plenum, New York, pp 453–489

Hay AE (1982) The effects of submarine channels of mine tailings disposal in Rupert Inlet, B.C. In: Ellis DU (ed) Marine Tailings Disposal. Ann Arbor Science, Ann Arbor, Michigan, pp 139–181

Holtedahl H (1975) The geology of the Hardangerfjord, west Norway. Nor Geol Unders Bull 36 (363):1–87

Magnusson J (1985) Overvåking av forurensningssituasjonen i Indre Oslofjord 1984, (in Norwegian). Norwegian Institute for Water Research Oslo, 58 pp

Molvaer J (1980) Deep water renewals in the Frierfjord. In: Freeland HJ, Farmer DM, Levings CD (eds) Fjord Oceanography. Plenum, New York, pp 531–537

Rosenberg R (1985) Eutrophication – the future marine coastal nuisance? Mar Pollut Bull 16:227–231

Rygg B, Skei JM (1984) Correlation between pollutant load and the diversity of marine softbottom fauna communities. In: Proc of the Int Workshop in Biol Testing of Effluents and Related Receiving Waters. OECD/USEPA/Environ Con, pp 153–183

Saelen OH (1967) Some features of the hydrography of Norwegian fjords. In: Lauff GH (ed) Estuaries. AAAS, Washington DC, pp 63–70
Skei JM (1981) Dispersal and retention of pollutants in Norwegian fjords. Rapp P V Reun Cons Int Explor Mer 181:78–86
Svendsen H (1980) Exchange processes above sill level between fjords and coastal water. In: Freeland HJ, Farmer DM, Levings CD (eds) Fjord Oceanography. Plenum, New York, pp 355–361
Svendsen J (1981) Wind-induced variations of circulation and water level in coupled fjord-coast systems. In: Saetre R, Mork M (eds) The Norwegian Coastal Current. University of Bergen, Bergen, pp 229–262
Syvitski JPM, Burrell DC, Skei JM (1987) Fjords: Processes and Products. Springer, Berlin Heidelberg New York, 379 p
Wolff EJ (1983) Estuarine benthos. In: Ketchum BH (ed) Ecosystems of the world 26. Estuaries and enclosed seas. Elsevier, Amsterdam, pp 151–182

# The Water-Air Interface

P.S. LISS[1], T.D. JICKELLS[1], and P. BUAT-MÉNARD[2]

## 1 Introduction

Transfer of matter across gas-liquid interfaces is governed by the same physical and chemical principles irrespective of whether the interface separates laboratory or environmental media. Similarly, in the environment the same general principles apply whether the atmosphere is exchanging with fresh or saline waters. Two important advantages stem from this: (1) ideas and results developed in simple or idealized laboratory set-ups (e.g., in chemical engineering) can often be used in field situations, which are generally less tractable; (2) field studies conducted on lakes are applicable to oceanic and coastal situations, and vice versa.

The total deposition of chemicals to a water surface will be the sum of the amounts transferred in the gas, liquid, and solid phases. Materials transferred in gas and solid phases are together referred to as dry deposition. Liquid, often referred to as "wet", deposition will comprise water and its dissolved gases and solutes, together with any insoluble particulate material contained therein. Net upward transfer is also possible and in some cases is of overriding importance. However, here the emphasis will be on deposition, since this tends to be of more interest in pollution situations, although many of the principles discussed are equally applicable to both net upwards and downwards transfer.

## 2 Estimation of Air-Water Mass Fluxes

### 2.1 Direct Measurement

A significant amount of the data in the literature for air-water fluxes of chemicals has been obtained by direct measurement. This is especially so for deposition in rain, since it is conceptually and practically so easy to collect the samples. However, there are very real problems with this simple approach. Unsophisticated funnel-in-bottle precipitation samplers collect dry deposition as well as rain. It is possible to arrange for the collector to be exposed only when rain is falling, but this increases the complexity of the apparatus. Another difficulty with direct measurements of

[1]School of Environmental Sciences, University of East Anglia, Norwich, NR4 7TJ, Great Britain
[2]Centre des Faibles Radioactivités, Laboratoire mixte CNRS-CEA, Domaine du CNRS, BP No. 1, 91190 Gif sur Yvette, France

**Table 1.** Range of concentrations measured in Bermuda rainwater, November 1981 to October 1982. (Jickells et al. 1984)

| | | Minimum concentration | Maximum concentration | Volume weighted average concentration |
|---|---|---|---|---|
| $H^+$ | μE/L | 0.5 | 50.1 | 14.6 |
| [a]$SO_4$ | μE/L | 4.0 | 48.1 | 12.7 |
| Na | mg/L | 0.67 | 17.9 | 2.9 |
| Mg | mg/L | 0.08 | 2.2 | 0.37 |
| Cd | μg/L | 0.02 | 0.11 | 0.06 |
| Cu | μg/L | 0.09 | 1.6 | 0.66 |
| Fe | μg/L | 1.9 | 13.0 | 4.8 |
| Mn | μg/L | 0.1 | 0.83 | 0.27 |
| Ni | μg/L | 0.04 | 0.8 | 0.21 |
| Pb | μg/L | 0.19 | 2.4 | 0.77 |
| Zn | μg/L | 0.2 | 2.75 | 1.15 |

Average of 18 samples. Rainfall range per sample: 0.5 to 7 cm.
[a] = non sea-salt sulfate.

wet deposition arises because of the natural variability of concentrations of trace substances in precipitation (Table 1), which means that a large number of rain events must be sampled and analyzed before meaningful average wet fluxes can be obtained. Great care also has to be taken to ensure that the samples do not become contaminated prior to analysis; this is particularly critical for substances present in trace amounts.

Despite these complexities, direct measurements have revealed several important features of the mechanisms controlling wet deposition, most of which are consistent with the chemistry of the constituents. It is clear that the concentration of a constituent in precipitation varies with its concentration in the atmosphere and is highest close to sources (Arimoto et al. 1985), although the relationship is not necessarily simple. For example, the presence of discrete layers of elevated particulate concentrations in the atmosphere makes direct comparisons of ground level aerosol concentrations and precipitation chemistry uncertain (Buat-Ménard and Duce 1986).

Soluble and reactive gaseous constituents (e.g., $HNO_3$ and $SO_2$) are readily removed from the atmosphere by wet deposition, while relatively insoluble gaseous constituents (e.g., organochlorines and mercury vapor) are removed very inefficiently by wet deposition (Slinn et al. 1978; Bidleman and Christensen 1979; Fitzgerald et al. 1983). The small amounts of these insoluble gaseous constituents found in rain probably reflect washout of the small proportion of these constituents associated with aerosol particles.

Particulate constituents are removed in wet deposition by collision with and capture by falling precipitation, or via the particles themselves acting as condensation nuclei. Some aerosol particles will readily dissolve in precipitation (e.g., sea salt, ammonium sulfate), while the extent of dissolution of less soluble particles such as fly ash or clays probably varies with the pH and pE (redox potential) of the precipitation. Both dissolved and particulate forms of many constituents can be expected to be present in precipitation, and the different environmental effects of these forms needs to be considered (Lindberg and Harriss 1983).

The efficiency of wet depositional air-sea transfer varies with the form of the precipitation (rain, snow etc.) and the meteorology of the storm (Buat-Ménard and Duce 1986). The importance of large-scale convective storms which penetrate the lower stratosphere as a removal mechanism for bomb and cosmic ray-produced radionuclides has been described (Burchfield et al. 1983).

Direct sampling of dry particle fluxes has been attempted using flat plate collectors having a variety of surfaces. Apart from the problems of contamination and natural inhomogeneity mentioned above, a fundamental difficulty is in knowing how well the collection surface mimics the water surface. This approach may be adequate only when gravitational settling of the large particles is responsible for a major fraction of the dry deposition (e.g., sea-salt particles and soil-dust particles).

Direct measurements of air-water gas fluxes are similarly fraught. In this case a major problem is that the water surface has to be covered by some sort of box arrangement which will clearly disturb the normal aero- and hydrodynamics of the interfacial transfer processes. For all of these reasons, resort is often made to indirect approaches for estimation of air-water fluxes.

## 2.2 Indirect Methods

Basically, in indirect approaches the air-water flux is calculated from the product of a concentration term (which drives the flux) and a kinetic parameter (which quantifies the rate of mass transfer). This concept is now briefly discussed for exchange of gases, "dry" particles, and rain.

*Gases.* In this case the flux (F) driving term is the concentration difference across the interface ($\triangle C$), and the rate expression is called a transfer velocity (K),

$$F = K \cdot \triangle C. \tag{1}$$

The $\triangle C$ term has to be obtained by direct field measurements of air and water concentrations. The transfer velocity may be dominated by near-surface processes in either the air or water phases, or both. Which phase(s) controls K depends on the Henry's Law constant for the gas concerned and its chemical reactivity in the water (Liss 1983). For gases of high water solubility/reactivity, the air phase processes will control K, and reasonably sophisticated estimates can be made of its value under a range of meteorological conditions (Hicks and Liss 1976). In the case of gases for which aqueous phase processes dominate transfer, it is apparently considerably more difficult to specify K (Liss 1983). More recently, measurements made by adding a purposeful gas tracer ($SF_6$) to lakes have helped to establish a relationship between transfer velocity and wind speed (Fig. 1), although the results are confined to the lower end of the environmental range of wind speeds. Liss and Merlivat (1986) have combined the results shown in Fig. 1 with the extensive data available from wind-tunnel studies in order to produce a relationship between K and the full spread of wind velocities operative over natural waters (Fig. 2). Application of Eq. (1) yields the *net* flux across the interface.

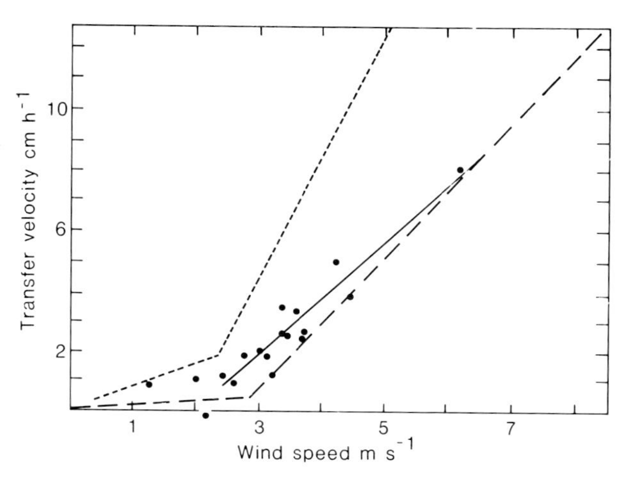


**Fig. 1.** Transfer velocity for $SF_6$ determined on Rockland Lake (New York State, U.S.A.) as a function of wind speed measured at 1 m above the lake. *Solid line* is the least-squares fit to the data above a wind speed of 2.4 m $s^{-1}$. *Short-dashed curve* is from the wind tunnel study of Broecker et al. (1978). *Long-dashed curve* is an estimate of the relation between the transfer velocity and wind speed if the wind was steady. (After Wanninkhof et al. 1985)

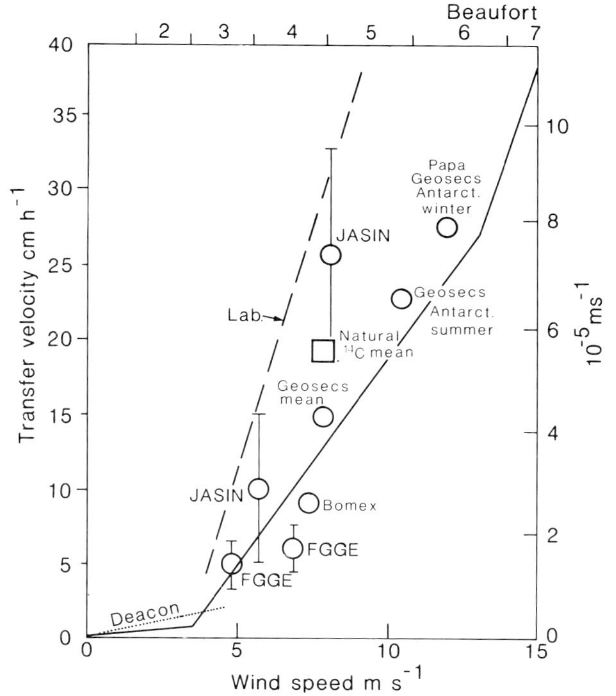


**Fig. 2.** Oceanic measurements of the liquid phase gas transfer velocity plotted as a function of wind speed measured at a height of 10 m and the corresponding Beaufort scale number. *Dotted line* represents predictions based on the Deacon (1977) smooth surface model. *Dashed line* represents the results for intermediate wind speeds from laboratory wind tunnel studies. All data are converted to Schmidt Number (Sc) = 600, corresponding to $CO_2$ at 20°C, by assuming that $K_w \propto Sc^{-1/2}$. (After Roether 1986). The predicted relationship between transfer velocity and wind speed of Liss and Merlivat (1986) is shown by the *three full lines*

*Particles.* In this case the flux is the product of the measured concentration of the particles in the air ($C_a$) and a deposition velocity ($V_d$),

$$F = V_d \cdot C_a. \tag{2}$$

The term $V_d$ is derived from theoretical considerations taking into account the processes involved, e.g., the settling of particles to the surface by gravitation, the impaction and the diffusion of particles to surfaces. The theory is very complex because each of these processes acts simultaneously, and because each of them is dependent on a number of variables (i.e., wind speed, particle size, relative humidity, air viscosity, sea-surface roughness, etc.). Even the most recent models (Slinn and Slinn 1980; Williams 1982) do not include all the relevant physics. In such models the atmosphere is separated into two layers. In the "constant flux layer", atmospheric turbulence and gravitational settling govern particle transport. In the second layer, the "deposition layer" which is just above the air-sea interface, particles grow in response to the higher relative humidity there. The overall resistance to air-to-sea particulate transfer is treated as several resistances in series. As an example, Fig. 3 gives the predictions of the model of Slinn and Slinn (1980), which clearly underline the influence of particle size, wind speed, and relative humidity on $V_d$. Because $V_d$ varies with the size of the particles, the $C_a$ term has to be obtained by accurate measurements on the size-fractionated aerosol, especially in the largest sized fraction. Acceptably accurate size distributions can be deduced from the use of cascade impactors, when proper corrections are made for particle loss in the largest size range. This correction is critical for sea-salt particles (McDonald et al. 1982) and in many cases for "soil-dust" particles. The use of such sampling devices (especially high-volume cascade impactors) may, however, cause serious problems for other types of aerosol particles, to which trace elements or substances may be attached. For example, it appears that particle "bounce" effects, especially for dry particles such as alumino-silicate minerals, can result in an apparent shift of the size distribution to smaller sizes (Buat-Ménard et al. 1983). The situation is even worse for some other materials, trace elements such as Pb, As,

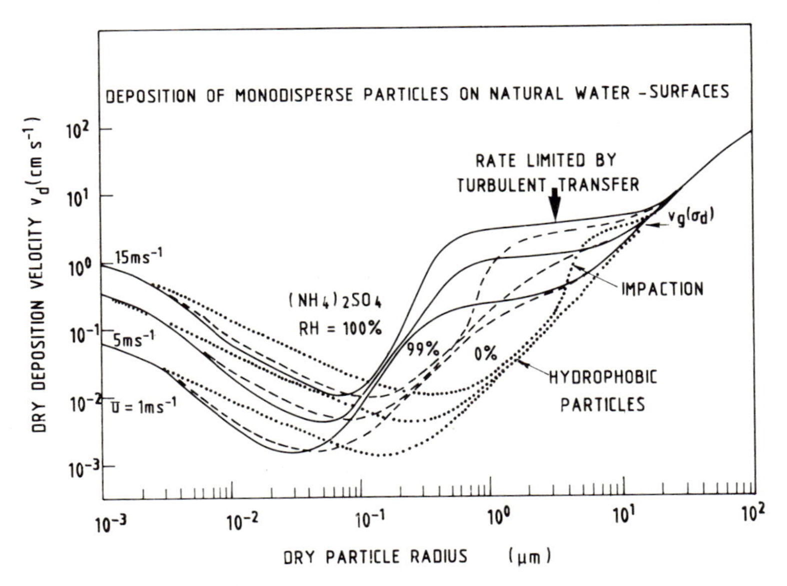


**Fig. 3.** Predictions of $V_d$ for three wind speeds and three types of particles: *dotted curves* hydrophobic (dry) particles; *dashed curves* particles that grow to equilibrium size for $(NH_4)_2SO_4$ particles exposed to a relative humidity of 99%; *solid curves* expected behaviour of $(NH_4)_2SO_4$ particles for the case of deposition to lakes with relative humidity near 100% in the deposition layer. (Slinn and Slinn 1980)

Cd, excess sulfate, and organic compounds, because a major fraction of their mass occurs on particles less than 0.25 $\mu$m in radius, and cascade impactors currently in use do not separate particles below that size. For such materials a further complication arises from the large differences in the deposition velocities than can be derived from the models for particles with dry radii of about 0.1 $\mu$m. As an example, for particles of this approximate size, the velocities calculated from a model by Williams (1982), which allows for differences in transfer velocities to smooth and broken water surfaces, are higher than those calculated from the Slinn and Slinn (1980) model, which only considers a smooth water surface, with a maximum difference of more than 30-fold, at 15 m $s^{-1}$ wind speed, 20% relative humidity, and assuming a broken surface transfer coefficient of 10 cm $s^{-1}$.

An alternative approach to estimating $V_d$ is to calculate it from measurements of atmospheric deposition fluxes to surrogate collector surfaces or from fluxes derived from radio-nuclide inventories. In both cases $V_d$ is obtained by dividing the flux by the concentration of aerosol in the atmosphere (Slinn et al. 1978; Uematsu et al. 1985).

*Rain.* Here the flux is formally expressed as the product of the precipitation rate (P) and the concentration of the substance of interest in the rain ($C_r$),

$$F = P \cdot C_r. \tag{3}$$

Equation (3) is often expressed not in terms of $C_r$, which effectively implies making a direct flux measurement, but through a washout factor (W), which is the ratio $C_r/C_a$, where

$$F = P \cdot W \cdot C_a. \tag{4}$$

By analogy between Eqs. (2) and (4), $P \cdot W$ is equivalent to a wet deposition velocity. The advantage of Eq. (4) over Eq. (3) is that provided W is known (see Buat-Ménard 1986), measurement of $C_a$ provides the concentration driving term for both equations and hence for estimation of deposition of "dry" particles and material in rain.

As with Eq. (2), Eqs. (3), and (4) give only gross fluxes. In the marine environment, the gross deposition of some materials (e.g., trace metals) to the ocean is composed of a net input as well as of a component associated with recycled sea spray. The importance of the atmosphere as a transport path for material from the continents to the ocean can only be assessed accurately if the relative contributions of the net and recycled components can be distinguished (Settle and Patterson 1982; Buat-Ménard 1983; Jickells et al. 1984; Arimoto et al. 1985). For example, there is strong evidence that atmospheric sea salt particles produced by bubbles bursting at the sea surface contain many metals in concentrations considerably higher than would be expected on the basis of metal-to-sodium ratios of near-surface water. It is apparent that some fraction of these metals is associated with particulate and surface active organic material and is scavenged by rising bubbles and concentrated on the sea salt particles produced both from the air-sea and bubble interface when the bubbles burst (Weisel et al. 1984). If this fractionation is not taken into account, the calculated net deposition to the ocean will be anomalously high.

As one approach to this problem, Arimoto et al. (1985) determined the metal/Na ratio on the largest atmospheric particles at Enewetak Atoll in the Marshall Islands, using only particles collected on the first stage of a cascade impactor. This represented, in general, particles with radii greater than about 3.5 $\mu$m. Most particles in this size range are sea salt, so analyses of material from this stage gave good estimates of metal/Na ratios in particles produced by bursting bubbles. These ratios also agreed well with direct measurements of the metal/Na ratios on sea salt particles produced and collected artificially using the Bubble Interfacial Microlayer Sampler in the North Atlantic by Weisel et al. (1984) (see also Buat-Ménard 1983 for an extensive discussion of this problem). The recycled component for any metal in the rain was calculated by

$$(M)_{recycled} = (M/Na)_{stage\ 1} \times (Na)_{rain}, \tag{5}$$

where $(M)_{recycled}$ is the concentration in rain of metal M recycled from the sea, in g kg$^{-1}$, $(M/Na)_{stage\ 1}$ is the metal/Na ratio on stage 1 of the impactor, and $(Na)_{rain}$ is the sodium concentration in rain in g kg$^{-1}$.

Recycled components in wet deposition that were calculated in this way for several metals ranged from 15% to about 50% and are comparable to the recycled fraction calculated using a similar approach by Jickells et al. (1984) at Bermuda. For Pb, about 30% was recycled in Enewetak rain according to Arimoto et al. (1985), compared with 10% at Enewetak calculated by Settle and Patterson (1982) and 17% at Bermuda determined by Jickells et al. (1984).

Recycled components calculated in the same way for dry deposition measurements to a surrogate surface were shown to range from 12 to 100%. For Pb, the percentage recycled varied considerably between samples. In two of the samples, approximately 50 to 60% of the gross Pb deposition was due to recycled material, but in a third sample, essentially all of the Pb could be attributed to sea spray. Recent data based on stable lead isotope measurements have provided unambiguous and direct evidence of the importance of this recycling process for pollution lead [C.C. Patterson (1986) pers. comm.].

These examples give an indication of the state of the art in this area. Clearly, more sophisticated techniques for accurately evaluating this recycled fraction must be developed. Future work should focus on two areas: (a) the use of adequate tracers (stable or radioactive) during field measurements, (b) an improvement of our knowledge of metal/Na ratios as a function of sea salt particle size through carefully designed in situ or laboratory experiments.

The "recycling component" may, however, be less important over semi-contaminated environments such as the North Sea or the Mediterranean, where the atmospheric concentrations of particulate pollutants are much higher than over remote marine areas. Furthermore the large-sized fraction of the aerosol may also be derived from land sources. Dulac et al. (1986) have shown that over the Western Mediterranean most of the Pb and Cd in this size range cannot be derived from material recycled from the sea surface.

## References

Arimoto R, Duce RA, Ray BJ, Unni CK (1985) Atmospheric trace elements at Enewetak Atoll: 2. Transport to the ocean by wet and dry deposition. J Geophys Res 90:2391–2408
Bidleman TF, Christensen EJ (1979) Atmospheric removal processes for high molecular weight organochlorines. J Geophys Res 84:7857–7862
Broecker HC, Petermann J, Siems W (1978) The influence of wind on $CO_2$-exchange in a wind-wave tunnel, including the effects of monolayers. J Mar Res 36:595–610
Buat-Ménard P (1983) Particle geochemistry in the atmosphere and oceans. In: Liss PS, Slinn WGN (eds) Air-sea exchange of gases and particles. Reidel, Dordrecht, pp 455–532
Buat-Ménard P (1986) The ocean as a sink for atmospheric particles. In: Buat-Ménard P (ed) The role of air-sea exchange in geochemical cycling. Reidel, Dordrecht, pp 165–183
Buat-Ménard P, Duce RA (1986) Precipitation scavenging of aerosol particles over remote marine regions. Nature (Lond) 321:508–510
Buat-Ménard P, Ezat U, Gaudichet A (1983) Size distribution and mineralogy of alumino-silicate dust particles in tropical Pacific air and rain. In: Pruppacher HR, Semonin RG, Slinn WGN (eds) Precipitation scavenging, dry deposition and resuspension, Vol 2. Elsevier, New York, pp 1259–1270
Burchfield LA, Akridge JD, Kuroda PK (1983) Temporal distributions of radiostrontium isotopes and radon daughters in rainwater during a thunderstorm. J Geophys Res 88:8579–8584
Deacon EL (1977) Gas transfer to and across an air-water interface. Tellus 29:363–374
Dulac F, Buat-Ménard P, Arnold M, Ezat U, Martin D (1986) Atmospheric input of trace metals to the western Mediterranean: 1. Factors controlling the variability of atmospheric concentration. J Geophys Res 92:8437–8453
Fitzgerald WJF, Gill GA, Hewitt AD (1983) Air-sea exchange of mercury. In: Wong CS, Boyle E, Bruland KW, Burton JD, Goldberg ED (eds) Trace metals in sea water. Plenum, New York, pp 297–315
Hicks BB, Liss PS (1976) Transfer of $SO_2$ and other reactive gases across the air-sea interface. Tellus 28:348–354
Jickells TD, Knap AH, Church TM (1984) Trace metals in Bermuda rainwater. J Geophys Res 89:1423–1428
Lindberg SE, Harriss RC (1983) Water and acid soluble trace metals in atmospheric particles. J Geophys Res 88:5091–5100
Liss PS (1983) Gas transfer: experiments and geochemical implications. In: Liss PS, Slinn WGN (eds) Air-sea exchange of gases and particles. Reidel, Dordrecht, pp 241–298
Liss PS, Merlivat L (1986) Air-sea exchange rates: introduction and synthesis. In: Buat-Ménard P (ed) The role of air-sea exchange in geochemical cycling. Reidel, Dordrecht, pp 113–127
McDonald RL, Unni CK, Duce RA (1982) Estimation of atmospheric sea salt dry deposition: wind speed and particle size dependence. J Geophys Res 87:1246–1250
Roether W (1986) Field measurements of gas exchange. In: Burton JD et al. (eds) Dynamic processes in the chemistry of the upper ocean. Plenum, New York pp 117–128
Settle DM, Patterson CC (1982) Magnitudes and sources of precipitation and dry deposition fluxes of industrial and natural leads to the North Pacific at Enewetak. J Geophys Res 87:8857–8869
Slinn SA, Slinn WGN (1980) Predictions for particle deposition on natural waters. Atmos Environ 14:1013–1016
Slinn WGN, Hasse L, Hicks BB, Hogan AW, Lal D, Liss PS, Munnich KO, Sehmel GA, Vittori O (1978) Some aspects of the transfer of atmospheric trace constituents past the air-sea interface. Atmos Environ 12:2055–2087
Uematsu M, Duce RA, Prospero JM (1985) Deposition of atmospheric mineral particles in the North Pacific Ocean. J Atmos Chem 3:123–138
Wanninkhof R, Ledwell JR, Broecker WS (1985) Gas exchange-wind speed relation measured with sulfur hexafluoride on a lake. Science 227:1224–1226
Weisel CP, Duce RA, Fasching JL, Heaton RW (1984) Estimates of the transport of trace metals from the ocean to the atmosphere. J Geophys Res 89:11607–11618
Williams RM (1982) A model for the dry deposition of particles to natural water surfaces. Atmos Environ 16:1933–1938

# The Ecosystem

P. DE WOLF[1] and J.J. ZIJLSTRA[2]

## 1 Introduction

A definition of an ecosystem is: a biological community and its abiotic environment (FAO 1978, Zijlstra 1988). Ecosystems are usually defined on a geographic basis (the North Sea ecosystem) and the biotic part is described in terms of structure (species, number of species, communities, species density, populations) and functions (primary and secondary production, food and energy transfer, mineralization, respiration, etc.). Here we will deal with both structural and functional aspects of the North Sea ecosystem.

The North Sea is a shelf sea, largely enclosed by highly industrialized land masses, with intensive fisheries which yield ±5% ($3 \times 10^6$ tons fresh weight per year) of the world fisheries, on a surface of 0.16% of the oceans. In addition, the southern part carries some of the world's busiest shipping lanes, and large oil- and gas resources have been exploited since the 1970's. As a result of this importance of the North Sea to man, it is relatively well known in many respects, due in no small part to the continuous efforts of the International Council for the Exploration of the Sea (ICES), dating back to 1902 (Went 1972). Originally ICES stimulated research on the physical aspects, and on plankton and fish stocks, more recently attention has been focused also on chemical (pollution) and geological studies.

### 1.1 The Environment

In this book the abiotic environment, which is part of the ecosystem is largely described in other chapters; e.g., watermasses and freshwater budgets on pages 3–19 (see also Laevastu 1963; Lee 1970; Hill 1973, and for residence times Maier-Reimer 1979; Otto 1983). Residence times (0.6–0.9 year) and watermasses result from and are influenced by tidal currents and their residual currents (for a review of North Sea tides see Thorade 1941). For tidal currents, residual currents and the bottom currents see Böhnecke (1922); Lee (1970) and Ramster (1965), respectively.

Shallow areas of the North Sea, with strong tidal currents, are permanently vertically mixed (mainly in the Southern Bight and along the British east coast)

[1]TNO, Division Technology for Society, post address NIOZ, P.O. Box 59, 1790 AB Den Burg, Texel, Netherlands
[2]Netherlands Institute for Sea Research, P.O. Box 59, 1790 AB Den Burg, Texel, Netherlands

(Dietrich 1953), while deeper areas with thermal stratifications in summer are mainly found in the homohaline central and northwestern North Sea (Dietrich 1954). In the northeastern part, seasonal and permanent haline stratification may be found (Dietrich 1950).

The temperature distribution of the North Sea in summer and winter is given in Fig. 1 for surface waters, and in Fig. 2 for bottom water (see also Engel 1983; Lamb 1973; Höhn 1973; Goedecke et al. 1967; Tomczak and Goedecke 1962; Becker et al. 1986). These figures indicate that the North Sea contains at least two systems, as far as abiotic factors are concerned: a shallow, well-mixed southern part and a deeper, central, and northern part with summer stratification. The southern part has a strong river influence and is hence liable to eutrophication, and the northern part has a much smaller human influence in terms of eutrophication and pollution.

Although nutrients are nowadays regarded almost as a nuisance, and are described only in terms of eutrophication (see also van Bennekom et al. 1975) and "impact" on the ecosystem (see pages 348–389, this Vol.), it must be kept in mind that also before the second industrial revolution nutrients were a part of seawater and the ecosystem. A review of the changes in nutrient contents in recent years has been given by Postma (1973, 1978), mainly for the Southern Bight (see also Johnston and Jones 1965; Johnston 1973; Folkard and Jones 1974). It must be noted that although an ICES meeting in 1974 recommended regular monitoring of nutrients in the North Sea (see also Postma 1978; Hempel 1978b), and accordingly at least some parts are monitored, it is remarkable that after 1975 no primary papers on North Sea nutrients were published. (With the advance of automated analyses it seems that thinking came to a halt!)

The composition of the North Sea bottom (deposited sediment) is treated on pages 36–58 (see also Lee and Ramster 1979), and suspended matter and sediment transport on pages 20–35. Papers by McCave (1973) and Postma (1978) are on the transport of mud and organic material respectively, while Stride (1973) also covered sand transport (see also Eisma 1986). Although these authors give, in general, ideas on the whole North Sea, locally large deviations from general distributions may exist, e.g., along tidal fronts. During the last 10 years the scientific interest in fronts or frontal zones has been on the rise, although their existence had been common knowledge among fishermen for a long time (Michitaka Uda, in Bowman et al. 1978). There are a number of fronts in the North Sea, such as the Flamborough Head Front (Pingree and Griffiths 1978), Frisian Front (Creutzberg and Postma 1979; Creutzberg 1985), German Bight (Krause et al. 1985; Czitrom et al. 1985), and others. Fronts probably have important biological consequences: frontal zones usually have a high productivity for all members of the food chain from phytoplankton to marine mammals (Bowman et al. 1978; see also Holligan 1981; Creutzberg 1985; Krause et al. 1985; Fogg 1985) while they also exert an influence on the horizontal and vertical transport of pollutants (Bowman et al. 1978). Hence, frontal zones enhance the variation in the ecosystem. In more general terms: the relative uniformity of the sea is probably an illusion (see also Steele 1974, p. 3).

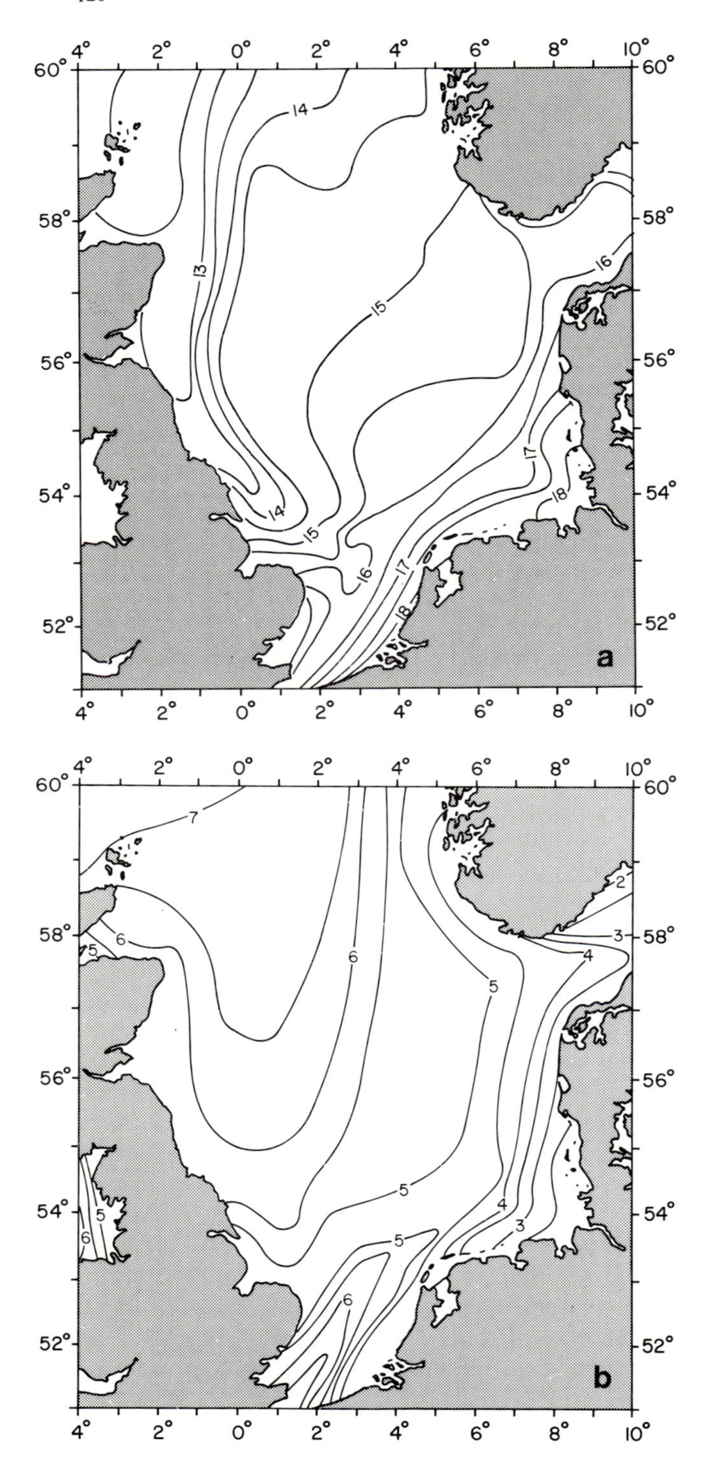


**Fig. 1a,b.** Surface temperatures. **a** in summer (August) and **b** in winter (February)

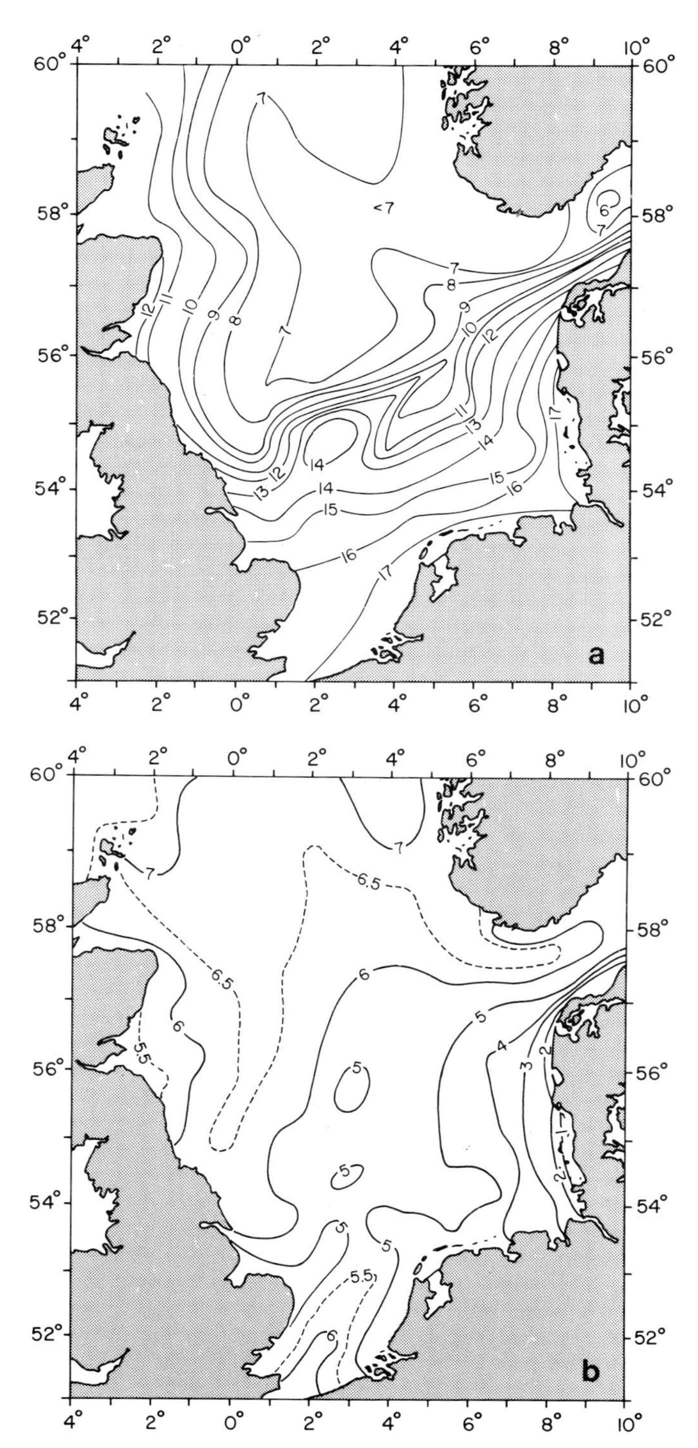


**Fig. 2a,b.** Bottom temperatures. **a** in summer (August) and **b** in winter (February)

## 1.2 The Living System

The biotic part of the ecosystem is usually described in a simplified way in the form of a food chain, built from a number of functional units such as phytoplankton, zooplankton (pelagic herbivores, invertebrate carnivores) etc. (Fig. 3). Phytoplankton (small algae such as diatoms, dinoflagellates, microflagellates) provides by photosynthesis the chemical energy for all biota; see Section 2. Higher stages of the food chain or food web are the zooplankton (Sect. 3) (in itself containing a food chain), consisting of pelagic hervibores (e.g., copepods), preyed upon by invertebrate carnivores (arrowworms and jellyfish); zooplankton itself is preyed upon by pelagic fish (Sect. 5), which in its turn is eaten by larger fish (Sect. 5), (small) cetaceans (Sect. 7.2) and birds (Sect. 6) and caught in the fisheries (pages 152–163 and pages 538–550). Part of the phytoplankton and zooplankton (and their remains: feces) sink to the bottom of the sea and enter there the benthic food chain. This consists of a bacterial degradation, and bacteria, and their unicellular predators as protozoans and the remains mentioned above are eaten by macrobenthos (shellfish, worms) and meiobenthos (Sect. 4) which serve both as food for benthic invertebrate carnivores (crabs, starfishes) (Sect. 4) and demersal fish (Sect. 5). The latter are eaten by larger fish and other carnivores, such as cetaceans and man.

The food chain as mentioned here is a highly simplified version of a much more complicated reality, as can be shown by the food web of a pelagic fish, the herring (Fig. 4); also this food web, for one species, is not complete! A recent review on food chains has been given by Wyatt (1976b). The high degree of complication is, of course, due to the presence of very many species of plants and animals, each of which fits into its own niche; they form the basic units of biology.

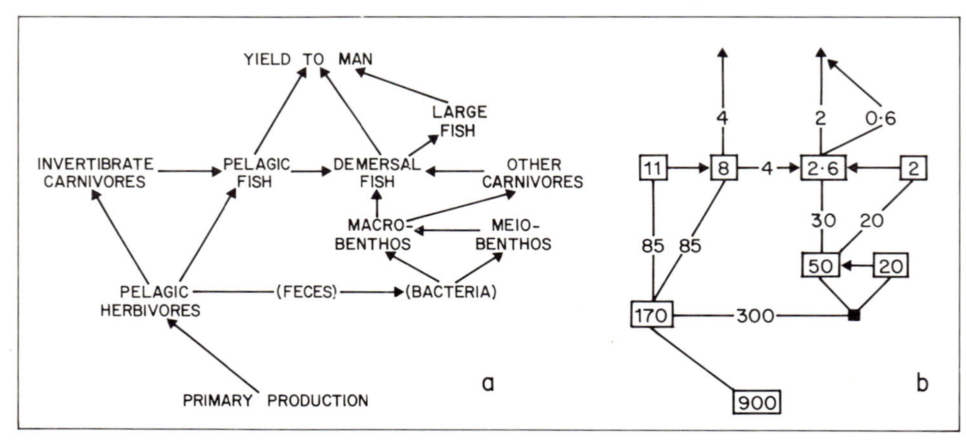


**Fig. 3. a** A North Sea food web based on main functional groups. (After Steele 1974). **b** Values for yearly production and transport between functional groups (Steele 1974)

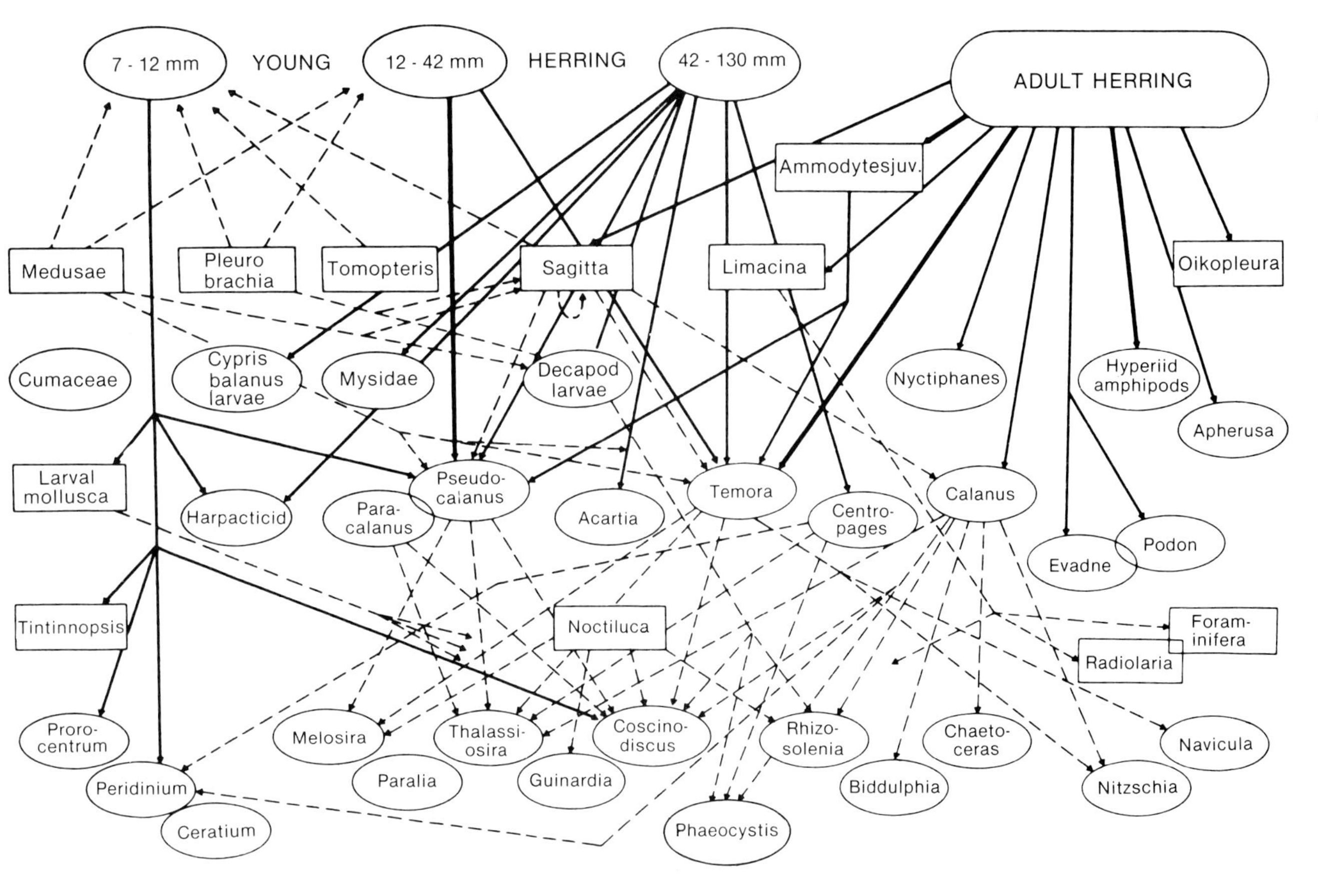


**Fig. 4.** The food chains for herring of different size classes. (Hardy 1924)

## 1.3 Sampling Problems

A large part of the structure and function of the biotic part of the ecosystem can at best be described in terms of abundant species (see e.g. Sect. 5, for fishes), but adequate sampling of rare species, or those with an intrinsic low density, is usually impossible, or demands a very large effort. Moreover, sampling of bottom fauna is in most cases methodically inadequate and/or extremely time-consuming, so that a large part of the North Sea bottom is virtually a terra incognita, at least in terms of its biology. Sampling of most biota is literally groping in the dark; unlike in land ecosystems, selective sampling for a certain species is hardly possible. Such a situation does not only hold for a hypothetical small plankton species, but also for the low density ctenophore *Beroe gracilis* (Greve et al. 1976), for the huge basking shark (*Cetorhinus maximus*) (Parker and Boeseman 1954; Wheeler 1969; Parker and Stott 1965; Whitehead et al. 1984), and for quite a few of the small cetaceans (FAO 1978). For all of these, and many other species, even life cycles are imperfectly known.

A review of the sampling problem in the sea has been given by Kelley (1976). The fisheries are an example of an adequate sampling mechanism for a limited number of species, although Hempel (1978a, p. 448) noted that for fish statistics there is a need for more fishery-independent data. Relatively well-sampled are the large species of phytoplankton and zooplankton, through the Continuous Plankton Recorder Surveys (CPRS), in which a small instrument, containing a continuously moving band of silk gauze (270 $\mu$) samples at a constant depth of 10 m, towed behind a merchant ship on a constant route (Colebrook et al. 1961).

The sampling problem, of course, is due to the occurrence of small- and large-scale variations, in space and time, and to our inability to distinguish these in the "noise" of a series of samples. A review of short-term variation in space, the patchiness problem (Steele 1976), contains a beautiful example of the variability of the dry weight of zooplankton (in itself already a highly "lumped" parameter) over a period of 4 days in an area of $40 \times 60$ km$^2$ (Fig. 5).

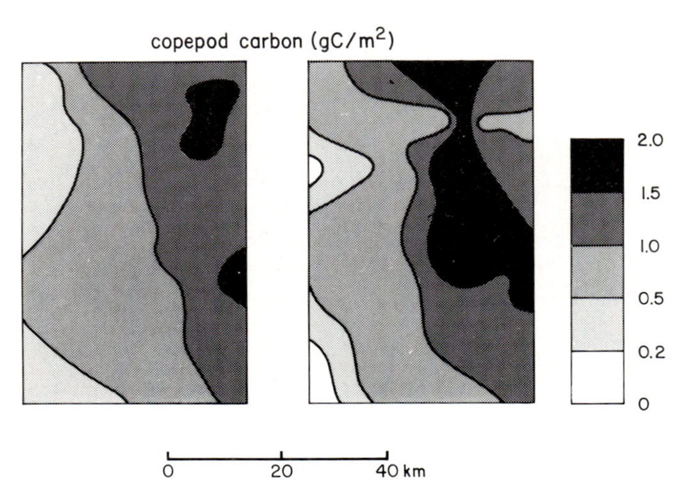


**Fig. 5.** Distribution of copepod carbon on two surveys in the northern North Sea, 2 days apart. (Steele 1974)

At the other end of the time scale, Cushing (1984) discussed sources of variability over a period of many years in the North Sea. An excellent series of variations in the penetration of Atlantic water (as labeled by zooplankton species) into the northern North Sea is to be found in Fraser (1973), indicating at the same time that at least some fronts may be found at varying places from year to year.

## 1.4 Models

During the last 10 years, a number of attempts have been made to describe the North Sea ecosystem, or parts thereof, for smaller or larger areas, in mathematical models. Strictly, the description of a *Calanus* "patch" off the north east coast of England over a period of 3 months by Cushing et al. (1963) can be regarded as a first attempt to integrate a part of the ecosystem from nutrients via algae to pelagic herbivores. Steele (1974) published a "closed" model of the food web of the North Sea, using the large functional groups referred to above (Fig. 3). This model is closed, as the beginning (primary production in g C $m^{-2}$) for the North Sea is known (70–90 g C $m^{-2}$), as well as the fish production (8.0 g fresh weight $m^{-2}$ for pelagic fish, and 2.6 g for demersal). The food web in between is given in Fig. 3. Estimates of food consumption by fish suggested that the secondary production was only just sufficient to feed the fish with no food available for tertiary feeding or for invertebrate carnivores, based on transfer efficiencies between steps in the food chain of 10% (based on Slobodkin 1961). Jones (1984) argued, using Steele's model, that the above shortcomings could only be fulfilled by assuming transfer efficiencies greater than 10%. The verification of this assumption asks for determination of the transfer efficiencies in quite a number of food chains! Ursin and Andersen (1978) made a model based on data by Hagel and van Alkemade (1973) on the effects of eutrophication (by the Rhine in the Southern Bight) on fisheries, indicating that the doubling of the yield of the North Sea fisheries in the 1960's was not due to eutrophication. This model started on light and phosphate, and contained phytoplankton and four species of animals: Copepods, Euphausids, Herring and Cod, and they note that for this simplification their model showed some oddities. Andersen and Ursin (1978) made a multispecies fishery model showing that the observed doubling of the fisheries yield can be reproduced by manipulating mainly the fishing mortality coefficients.

Dynamic interactions in lower trophic levels of the web on the Fladen Ground (Radach 1982) during the spring plankton bloom showed that validation could only be successful if in addition to standing stocks and concentrations also flux estimates or measurements had been performed.

Joiris et al. (1982) showed that, at least in Belgian coastal waters, the "classical" structure of the marine food web, wherein phytoplankton production flows via intermediate stages to fish only, does not hold. Here, cycling of primary production through microheterotrophs in water and sediments dominates over the longer trophic chains to pelagic fish. The question arises whether this holds for the entire Southern Bight, and whether this, together with a relatively low fish production, is indication for an unbalance in the system there. The results of Joiris et al. (1982) can be taken as an indication of serious eutrophication, but it is just as possible that this system exists in other parts of the North Sea as well, although it has not been measured (cf. Radach 1982). Recently Fransz and Verhagen (1985) have modeled the phytoplankton cycle in relation to river-borne nutrient loads, but found a too high primary production in their model, when compared to field data.

From the examples given, it follows that there is a considerable amount of uncertainty; some models have only local application, others are open-ended for lack of data, either to build the model or for verification.

In the following paragraphs the different functional groups will be described; coastal zone and open sea will be treated separately, when applicable.

Lastly, models in terms of functional groups, carbon, and/or calories are useful, but the driving functions behind the behavior of individuals or species are eating and mating, functions which are infinitely small in energy demand. In other words, the solution of our difficulties in making a closed C budget model for the North Sea is to be found at the level of the population or even the individual.

## 2 Phytoplankton

From the title of this section, it follows that we consider the larger algae and higher plants, growing on solid substrates in coastal water, to be of relatively minor importance. Higher algae have been described for the Waddensea (Den Hartog 1979; van den Hoek et al. 1979) and for the Dutch Delta area (Den Hartog 1959). The number of species in the Waddensea is approximately 60, for the Dutch coast larger. The number of phytoplankton species for the Wadden Sea is approximately 500 (including a number of benthic microalgae (van den Hoek et al. 1979), for the North Sea undoubtedly much larger.

Phytoplankton of the open North Sea has been studied by two different methods. The first is the data collection on the structure of the ecosystem by the Continuous Plankton Recorder Surveys (CPRS), recording spatial and temporal occurrence and variation since 1958 (Glover 1967; Glover et al. 1974). The second method describes the functional properties of the phytoplankton and studies phytoplankton dynamics in relation to nutrients, light availability, hydrographic properties such as salinity, temperature, and mixing, grazing of zooplankton etc. These studies are usually intensive over a relatively small area, in the Southern Bight (Mommaerts 1973a,b; Gieskes and Kraay 1975, 1977a,b; Joiris et al. 1982), in the German Bight (Hagmeier 1978), in the central North Sea (Cushing et al. 1963; Gieskes and Kraay 1983) and in the northern North Sea (Steele 1956, 1958, 1962; Radach 1982). A general introduction is given by Wyatt (1976a).

### 2.1 The Phytoplankton Community

The CPRS recovered 68 species or genera of phytoplankton; most of these have a pan-North Sea distribution, but some were more numerous in the northern, others in the southern part (Fig. 6) (Anonymous 1973). However, the CPRS collects only the larger, armored cells. There is an increasing amount of information on smaller

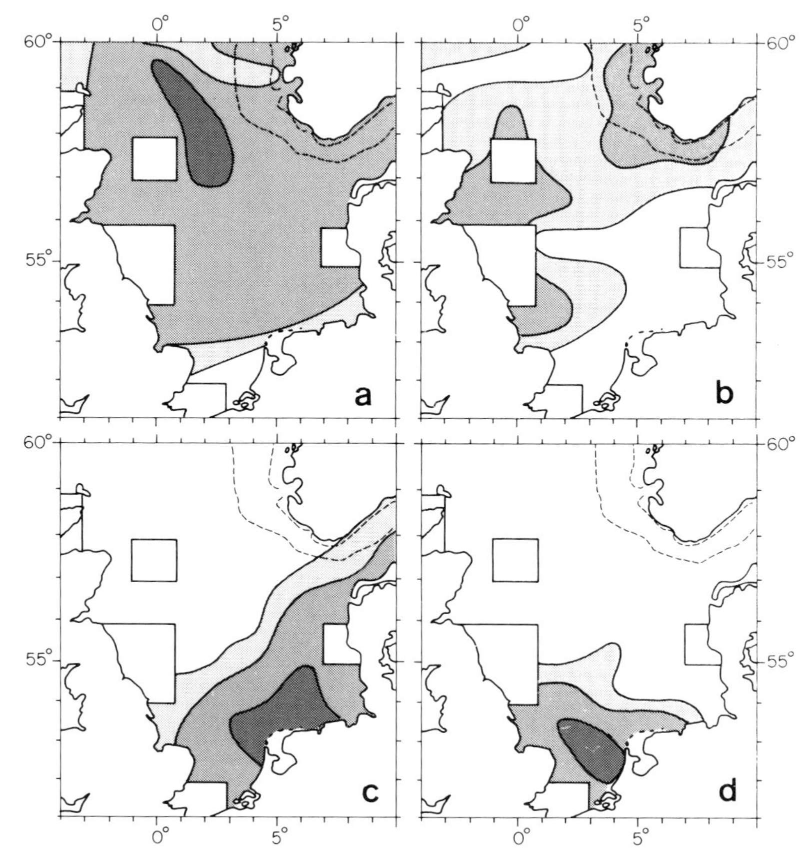


**Fig. 6a-d.** Distribution of phytoplankton species; **a,b** species with a northerly distribution; **c,d** species with a southerly distribution. **a** *Ceratium longipes;* **b** *Thalassionema nitzschioides;* **c** *Biddulphia sinensis;* **d** *Phaeocystis spp.* (After Robinson 1961)

cells (nanoplankton and microplankton), but data are too scarce to form a coherent picture of spatial and temporal distribution; they might possibly be of greater importance than the larger species, especially in terms of primary production, particularly in the Southern Bight (Mommaerts 1973a,b; Gieskes and Kraay 1975, 1983; Reid 1978). For instance, Leewis (1985) mentions the presence of over 270 smaller species in the Southern Bight area. The succession in time starts, in net-phytoplankton, with diatoms; they are succeeded in most areas by flagellates, blooming during summer (in the south often *Phaeocystis* spp.) (Gieskes and Kraay 1975; Hagmeier 1978; Wandschneider 1980). Usually, but not always, diatoms bloom again in autumn, but are then less numerous (Reid 1978). The diatom succession is related to the time scale of silicate mineralization and other factors; the role of microplankton is not yet clear.

## 2.2 Seasonal Variation in Biomass

In the CPRS, clear seasonal variations in color were observed on the silk of the recorder, which have been taken as a relative expression of variations in biomass of phytoplankton (Fig. 7) (Colebrook and Robinson 1965; Robinson 1970; Colebrook 1979). These authors suggested that the long duration of the season in the south may be attributed to the shallowness and the absence of stratification, allowing a continuous exchange of nutrients between bottom and water surface. In this area the spring outburst occurs early (March) and is mainly affected by light (Gieskes and Kraay 1977a,b). In all other areas, the start of the spring bloom is thought to depend on the onset of stratification (Colebrook 1979), and the early start of the spring bloom in the Baltic outflow ($B_1$ in Fig. 7) is possibly related to the permanent halocline there.

Algal biomass can also be recorded by chlorophyl a measurements, or by counting and measuring cells; such observations are scarce for the North Sea, and are virtually limited to the southern part. Joiris et al. (1982) and Fransz and Gieskes (1984) indicate along the Belgian and Dutch coasts a clear spring maximum of 15–25 mg $m^{-3}$ chlorophyll a in April-May. In the German Bight, Hagmeier (1978) found a late maximum in August (mean of 13 years). Less complete series for the northern North Sea indicate a single maximum in April-May of up to 10 mg chlorophyll $m^{-3}$ (Steele and Baird 1965; Steele and Henderson 1977; Mommaerts 1980; Radach 1983).

Of these observations only those of the Southern Bight do not conform to the CPRS data, for this area; however, the Southern Bight observations resemble the CPRS data for the upstream area of the neighboring English Channel, which might suggest that the eastern part of the Southern Bight is an extension of the English Channel system.

## 2.3 Primary Production

Series of primary production data are available for restricted areas only, mainly in the Southern Bight and the northern North Sea; a complete coverage of this important parameter is lacking. Steele (1956, 1974) estimated primary production on basis of phosphate utilization, in the north, as 70–90 g C $m^{-2}a^{-1}$. During the Flex experiment, covering the spring bloom only, comparable values were found (Joiris et al. 1982; Radach 1983). However, other authors (Jones 1984; Fransz and Gieskes 1984) expect higher values for the area, but recent, complete measuring series are lacking. For the Southern Bight, Joiris et al. (1982) estimate total net primary production at 320 g C $m^{-2}a^{-1}$, which agree with estimates of Gieskes and Kraay (1975), based on a less complete data set. No data are available for variability from year to year in the North Sea although Boalch et al. (1978) mentioned such a variation in the English Channel from $< 100$ to nearly 300 g C $m^{-2}a^{-1}$. The high production in the Southern Bight is supposed to be a result of nutrient enrichment; it is remarkable, however, that primary production was highest offshore, where the influence of freshwater runoff is small, as shown by salinity measurements (Gieskes and Kraay 1975).

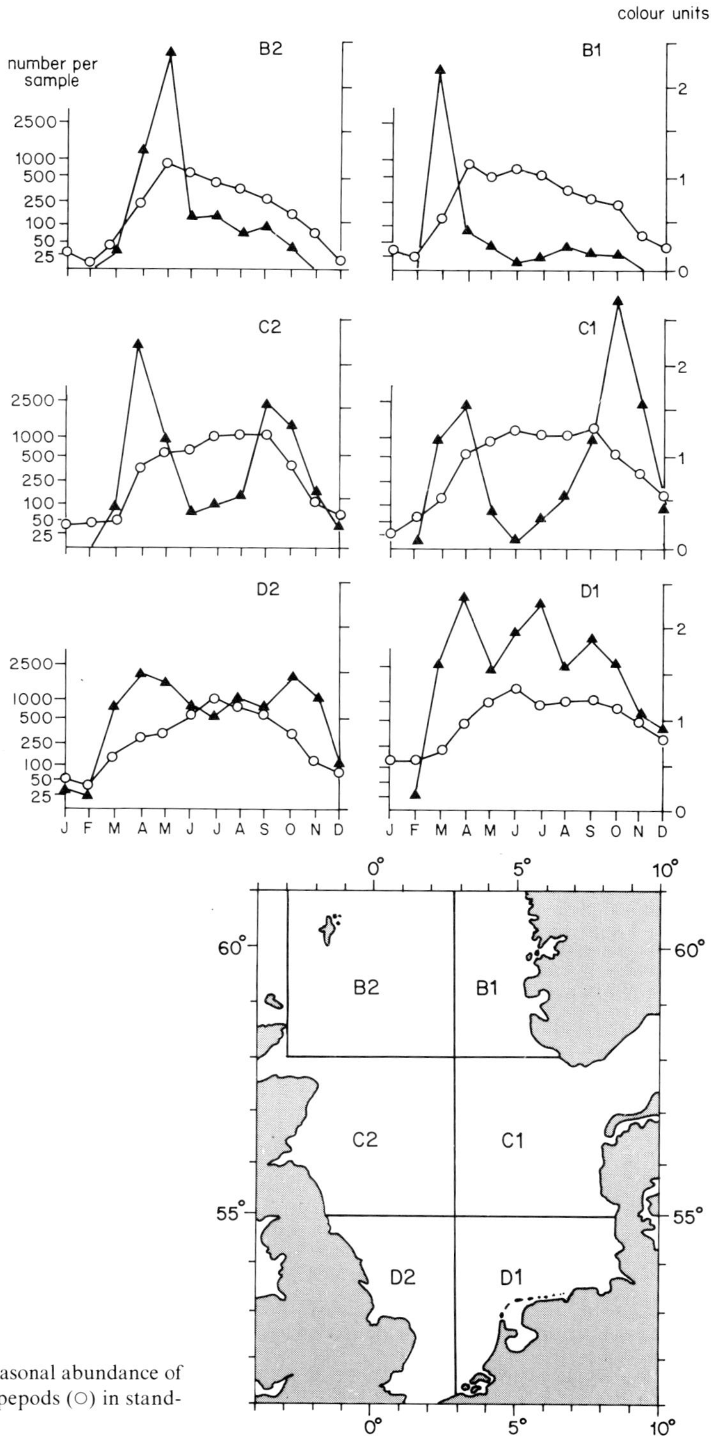


**Fig. 7.** Graphs of average seasonal abundance of phytoplankton (▲) and copepods (○) in standard areas of the North Sea

Opinions on the use of the primary product vary between authors. Steele (1974) and Jones (1984) assumed that most of the consumption was due to copepod grazing. The Flex experiments showed that there could indeed be a balance (Radach 1983). Joiris et al. (1982) agree with this opinion for the northern North Sea, but think that in the Southern Bight a major part of the primary production is consumed by small heterotrophs. Lastly, Fransz and Gieskes (1984) argue that for the whole North Sea a large part of the important spring bloom is *not* consumed by copepods and that copepod grazing is not responsible for the decline of the spring bloom; they think that, analogous to the situation in the Baltic Sea (Elmgren 1984), a large part of the bloom sinks to the bottom, or is consumed by microbial plankton. Cushing and Vucetic (1963), however, suggested that during the spring bloom copepods destroyed algae without eating them.

### 2.4 Long-Term Trends

Data from CPRS between 1958 and 1973 indicate significant long-term changes (Reid 1978; Reid and Budd 1979). The most conspicuous change took place in diatom abundance since 1968, leading to a disappearance of the autumn bloom. However, total phytoplankton tended to increase, in particular in the west and south, which is tentatively attributed to microflagellates. Concurrently with these changes was a delay in the spring bloom, which affected the whole north-east Atlantic (Glover et al. 1974), which makes a relation with eutrophicaton unlikely.

These changes in phytoplankton were accompanied by a general decline in zooplankton (see Sect. 4.4) and are thought to affect large areas of the north Atlantic (Glover et al. 1984; Radach 1982; Cushing 1982); the changes may be related to the position of the Gulf Stream, and to climatic changes (Reid and Budd 1979; Garrod and Colebrook 1978; Radach 1982).

## 3 Zooplankton

Data availability for North Sea zooplankton is similar to that for phytoplankton: a major source is the CPRS, wherein the value of the zooplankton data exceeds that for phytoplankton, as the mesh size of the instrument is better adapted. Small forms (protozoa), small developmental stages (larvae) and fragile planktonic animals (e.g., medusae) are incompletely known. As in phytoplankton, the other source is provided by ecosystem studies, restricted in time and space.

### 3.1 Zooplankton

The zooplankton community of the North Sea as caught by CPRS contains five species of Branchiopoda, 22 species of Calanoid copepods, three genera of Cyclopoid copepods, seven families of Malacostraca, four species of planktonic Gastropods, four species of Tunicates; a few species of Chaetognatha, and the pelagic Polychaeta genus *Tomopteris* (Anon 1973) have been regularly observed. Of these, only a few species are really dominant. In addition, larval stages of benthic animals are commonly encountered.

Some of the species are more abundant or only present in the northern part; others, partly more neritic species, are more abundant in the south. In general, the larger species, such as *Calanus* and the Euphausids, dominate in the north, while in the south smaller species are more abundant; thus copepods are more numerous in the south (Colebrook et al. 1961; Colebrook and Robinson 1965; Colebrook 1979). Knowledge of plankton not recorded by CPRS is much more fragmentary. In most coastal areas the ctenophore *Pleurobrachia pileus* occurs in large numbers (Greve 1981) as do Scyphomedusae (Möller 1980). Hydromedusae, occurring in many species, are generally poorly studied. Microplankton is virtually unknown. The total number of species of zooplankton in the North Sea is $> 300$. Zooplankton of coastal areas and estuaries is generally poorer in species; for the Wadden Sea see Dankers et al. (1981).

### 3.2 Seasonal Variations

Seasonal variations in total copepod numbers are given in Fig. 7 for the period 1948–1960 (Colebrook and Robinson 1965; see also Colebrook 1979). It appears that the start of the copepod season is closely related to the beginning of the phytoplankton bloom [but see Fransz and Gieskes (1984), who recently found that this fit does not always exist nowadays]. The zooplankton peak usually occurs up to 3 months later than the phytoplankton maximum; however, variation in zooplankton numbers does not follow the large variations in phytoplankton stock; zooplankton density appears to be much more stable. For grazing problems the reader is referred to Colebrook (1979).

### 3.3 Biomass

There are few reliable estimates of functional properties, such as biomass, production, and respiration of zooplankton. Robertson (1968) attempted to convert copepod numbers (from CPRS) into biomass; he found biomass to vary between areas from 12–19 mg $m^{-3}$, averaged over the year. More incidental information is the following: in the northern North Sea a biomass of 2.5 g C $m^{-2}$ (yearly mean) (Steele 1974), 0.4 g C $m^{-2}$ in March, and 2.0 g C $m^{-2}$ in April-May and November (Adams and Baird 1970); on the Fladen Grounds 3.5–4 g C $m^{-2}$ from mid April-June (Krause and Radach 1980; Fransz and Gieskes 1984) and 0.9 g C $m^{-2}$ in mid-April for *Calanus* alone (Fransz and van Arkel 1980). Further south, in the central North Sea values are lower: southwest of Doggerbank 0.8–2.0 g C $m^{-2}$ in summer 1947–52 (Wimpenny 1953); and southeast of Doggerbank 0.2 g C $m^{-2}$ in April-May, 0.8 g C $m^{-2}$ in July and 0.3 g C in September (Fransz and Gieskes 1984). The southern Bight has still lower values, along the Belgian coast a yearly mean of 0.3 g C, but farther offshore 0.6 g C $m^{-2}$ (Fransz and Gieskes 1984).

There are very few *production* estimates for the zooplankton of the North Sea. Cushing (1973) estimated 20 g C $m^{-2}$ $a^{-1}$, but in view of CPRS data this is likely an overestimate; Steele (1974) offered tentatively 18 g C $m^{-2}$ $a^{-1}$ for the northern North Sea, Fransz and Gieskes (1984) mention a crude estimate of 20 g C $m^{-2}$ $a^{-1}$ for the northern North Sea, and 10 g C $m^{-2}$ $a^{-1}$ for the Southern Bight. Fransz et al. (1984) give a more precise estimate, for copepod production on the Oyster Ground: 9 g C $m^{-2}$ $a^{-1}$. Further data will be necessary to sustain the idea that biomass and production of zooplankton are lower in the south that in the north, as opposite to phytoplankton. Joiris et al. (1982) observed this discrepancy and suggested that in the south more energy is dissipated via a microbial foodchain, while in the north the larger part would be channelled via the classical foodchain of larger zooplankton (Steele 1974). That oxygen-demanding mineralization is of the same magnitude and closely follows the development of primary production in the Southern Bight (Gieskes and Kraay 1977) fits with Joiris's idea. The almost instantaneous reaction of small heterotrophs to production of organic matter and the observation of Fransz (1976) that the development of copepods is retarded by low temperatures make it likely that small heterotrophs are responsible. This hypothesis would also offer an explanation for the lower fish production in the southern North Sea (Hummel and Zijlstra 1978).

### 3.4 Long-Term Trends

As in phytoplankton, an analysis of CPRS data reveals important changes in the abundance of zooplankton. For many of the species, there appears to be a linear downward trend in abundance over 30 years (Colebrook et al. 1978). The decline in zooplankton biomass was accompanied by a reduction in the zooplankton season, by a delayed onset of the production cycle, parallel to the delay in the spring phytoplankton bloom. The production season of the zooplankton was gradually reduced from 7–8 to 5–6 months, which means that perhaps one or two generations of copepods have been lost (Cushing 1982). This decline was not restricted to the North Sea, but affected the whole NE Atlantic area, and at least in the North Sea not only the open sea species but also coastal species. It is thought that the decline is due to a general climatic deterioration in the arctic and subarctic of the north Atlantic Ocean, together with a more southward displacement of the Gulfstream. It is not fully understood by which processes the climate interacts with the plankton (Cushing 1982; Radach 1982).

## 4 Zoobenthos

Early research on North Sea benthos was restricted to easily accessible coastal areas, usually intertidal. In offshore parts studies were confined to restricted areas: Southern Bight, German Bight, Doggerbank, Northumberland coast, and some areas in the north. A few studies, however, covered large areas: Ursin (1960) on echinoderms, and Kirkegaard (1969) on polychaetes, studied areas larger than the central North Sea. Up to 1950, studies were strongly influenced by the ideas of Petersen (1911, 1914), who distinguished benthic communities characterized by dominant and persistent species. Later, this concept was questioned on various grounds (Stephen 1933; Eisma 1966; Stephensen et al. 1972; Creutzberg 1986).

In addition to the regional character there was lack in standardization of methods for sampling and sorting. Various types of grabs have been used for the infauna (Petersen's, van Veen's, boxcorers) which yield different results (Beukema 1974; Heip et al. 1978), and mesh sizes of sieves varied between authors. For epifauna, trawls of various makes have been used (Creutzberg et al. 1979–1985). As a result it is almost impossible to reach a comprehensive picture of North Sea benthos (McIntyre 1978).

In the last 30 years, the interest has shifted from a faunistic to a more functional approach. There have been many attempts to relate benthos to sediment type, temperature, and depth; Glémarec (1973) proposed a system of three "étages": an *infralittoral étage* with depth < 40 m and temperature variations > 10°C; a *coastal etage* with depth 40–100 m, with temperature < 12°C and temperature variations < 5°C, and an *open sea etage*, with depth > 100 m, temperature < 10°C and little temperature variation. Glémarec's system has been adopted by McIntyre (1978) for his review of North Sea benthos. Waddensea benthos has been reviewed by Dankers et al. (1981). Benthos is usually divided into three size groups: the smallest microbenthos passes through a sieve with 50 $\mu$ mesh: bacteria and protozoans. Meiobenthos passes through a 1 mm mesh, but is larger than 50 $\mu$: nematodes, harpacticoid copepods, small polychaetes and oligochaetes, ostracods, kynorhynchids, and juveniles of larger animals. The larger animals form the macrobenthos: bivalves, gastropods, polychaetes and oligochaetes, echinoderms and crustaceans. In addition, benthos is divided into an infauna, living in, and epibenthos, living on the bottom.

It is tacitly assumed that the North Sea bottom does not have a bottom film of diatoms, such as intertidal areas have (Admiraal 1984). Recently, however, Cadée (1984) mentioned the occurrence of living bottom-dwelling diatoms on the Oyster Ground at 20–30 m, while Lindeboom (pers. commun.) mentioned a benthic diatom bloom at 24 m depth. It thus appears that these may play a role in the Southern Bight.

## 4.1 Faunal Composition

A large number of species have been recorded in the North Sea macrobenthos for various taxonomic groups (Table 1).

The numbers of species vary considerably, even in the same area. Ursin (1960) indicates that the number of species of echinoderms increases with the number of samples studied. Mulder et al. (1987) note that up to 25 grabs per station the number of species found still increases. It follows that attempts to characterize areas by numbers of species are abortive. Very few studies have been made in the deeper area (Stephen 1923, 1933, 1934; McIntyre 1978).

Information on meiofauna is limited to the Southern Bight (Heip et al. 1982, 1983, 1984) and the German Bight (Stripp 1969a).

Virtually no information is available on the microbenthic fauna.

**Table 1.** Numbers of species in infauna and epifaunal macrobenthos

| *Infauna* Area | Polychaetes | Molluscs | Crustaceans | Echinoderms | Reference |
|---|---|---|---|---|---|
| German Bight | 90 | 54 | 52 | 12 | Salzwedel et al. 1985 |
| German Bight | 68 | 38 | 55 | 9 | Ziegelmeier 1978 |
| Central North Sea | | | | 57 | Ursin 1960 |
| Doggerbank | 143 | | | | Kirkegaard 1969 |
| Oyster Ground | | 63 | | | Cadée 1984 |
| West of Scheveningen | 62 | 16 | 49 | 5 | Mulder and de Wolf 1987 |
| *Epifauna* | | | | | |
| 57°20′–54°40′N and 2°20′– 5° E | 2 | 75 | 56 | 20 | Creutzberg et al. 1979–1985 |

### 4.2 Numerical Abundance Macrofauna

The total number of animals per m$^2$ varies from 750–2700, these figures, however, are based on averages of a large number of samples, so that in fact the variation is much larger. There is no clear trend of abundance with depth. The dominant groups are usually polychaetes and bivalves, together 70–90% of the numbers; only in deeper water (open sea étage) are crustaceans of equal importance. In the German Bight Stripp (1969a) has shown that the number of animals per m$^2$ increases when the grain size becomes smaller. However, McIntyre (1978) could not find this in a shallow area along the Scottish coast.

*Meiofauna*. Data on meiofauna abundance in the North Sea are few (Stripp 1969b; Warwick and Buchanan 1970; Gerlach 1978; McIntyre 1978; Heip et al. 1983, 1984); the material suggests a large within-area variance, related to sediment type, with high numbers (up to $1.3 \times 10^6$ m$^{-2}$) in sandy mud and low numbers ($8 \times 10^4$m$^{-2}$) in coarse sand. As usual in meiofauna there is always a dominance of nematodes.

### 4.3 Biomass and Production

There are slightly more estimates of the biomass of both macro- and meiofauna than for their abundance, and generally for the same regions. The macrofauna weight seems to range between 2 and 15 g dry weight m$^{-2}$, and biomass is related to sediment type (Stripp 1969a; Creutzberg et al. 1983) or to depth (Rachor 1982). However, de Wilde et al. (1984) and de Wilde et al. (1986), sampling the southeastern and northwestern North Sea respectively with a boxcorer, found rather higher values for macrofauna, ranging from on average 21 g AFD m$^{-2}$ in the southeastern North Sea to 11 g AFD m$^{-2}$ in the northwest.

Data on meiofauna do not allow for any firm conclusions; within the German Bight weight varies between 0.2–2.0 g m$^{-2}$, with the lowest biomass in coarse sand and the highest in mud bottoms.

Production estimates are extremely scarce. Buchanan and Warwick (1974) estimated off the Northumberland coast: 1.74 g dry weight m$^{-2}$a$^{-1}$, at an average biomass of 3.98 g m$^{-2}$. Buchanan et al. (1974) found in later years a slightly higher production (1.85 g $m^{-2}a^{-1}$) at a slightly higher biomass. Gerlach (1978) in the German Bight estimated for the macro-infauna a production of 12 g $m^{-2}a^{-1}$, at a biomass of 6 g $m^{-2}$; for the meiofauna he assumes P = 10 g $m^{-2}a^{-1}$ at B = 1 g $m^{-2}$. De Wilde et al. (1984) estimated the benthic fauna production of the Oyster Ground at least 36 g $m^{-2}a^{-1}$, in which the macrofauna accounted for 21 g $m^{-2}a^{-1}$ and the meiofauna for only 6 g $m^{-2}a^{-1}$.

The P/B ratio of Gerlach for macrofauna is high, but that of Buchanan and Warwick (1974) (0.44) low. Ankar and Elmgren (1976) found P/B = 1 for the Baltic. Assuming the last value for the North Sea and using Gerlach's P/B ratio for meiofauna production (10), and excluding microfauna, the North Sea benthic production could be tentatively estimated at 16–35 g dry weight $m^{-2}a^{-1}$.

### 4.4 Long-Term Trends

Time series of observations of benthos are few and cover short periods. Mostly comparisons are made between distinct periods, with no observations in between, in the way Birkett (1953) noted that a dense *Spisula* population on the Doggerbank present earlier (Davies 1923) had disappeared; and Stripp (1969b), who observed no major changes in the German Bight since Hagmeier (1925).

Ziegelmeier (1978) followed many elements of the macrofauna of the German Bight from 1950 to 1974, as did Rachor and Salzwedel (1976) during a shorter period, for bivalves, and Rachor and Gerlach (1978) from 1967–1975; all found severe winter influences (see Chap. 3.4.2) but no obvious trends, although total bivalve numbers increased from 1969–1975.

Gray and Christie (1983) suggest cycles of 6–7 and 10–11 years in some macrobenthos species; however, other species, mostly deep-living in the bottom, may show constancy through decades, which may be due to relative longevity of bivalves.

## 5 Fishes

As noted earlier, fishes are the best known of all foodchain groups, due to the very intensive fisheries in the North Sea, in which most species occur either as catch or as by-catch (Tiews 1978a). On the other hand, the ecosystem is changed by this fishing effort to a considerable extent, with the fisheries adapting rapidly to the changes at a rapid rate in their turn.

### 5.1 The Fish Fauna

The North Sea contains at least 160–170 species (Wheeler 1969, 1978). Most of these species are fairly common, so that if all rare species and occasional visitors were included, the number would increase considerably. On the other hand, Sahrhage (1964, 1967) recorded only 71 species during two winter surveys and 75 species in two summer surveys, fishing all offshore parts in the North Sea with a herring trawl; relatively low numbers due to mesh selection and absence of samples from coastal water. Therefore one could conclude that only 25–30 species are really numerous. Half of the species are found all over the North Sea; 51 species are restricted to deeper, in summer colder, northern parts, 37 species are almost limited to the shallow southern area. Some 33 species occur mainly in coastal waters. Of the 11 species, making up 95% of the fishing yield (Cole and Holden 1973), five exploit the coastal zone as a nursery (Zijlstra 1972); these five (herring, sprat, plaice, coal fish, sole) made up 40% of the North Sea catch in the late 1960's, which shows the importance of the coastal zone for the fisheries. Redeke (1941) suggested that about one third of North Sea fishes are regular or occasional immigrants from the south (Bay of Biscay, Channel) or from the North (subarctic Atlantic); 80% of the immigrants being of southern origin, during the summer half year. As the annual temperature differences are far larger in the southern than in the northern part, this seasonal immigration is more spectacular in the south.

The abundance of immigrants can vary greatly between years, and may show trends superimposed on variations (Michaelis 1978, for *Mugil* spp.). Such variations are recorded in the fisheries catch statistics for commercially important species such as pilchard, horsemackerel and sea bream. Postuma (1978) concluded from catch statistics that variations in the catch reflect abundance in the main distribution area, south of the North Sea (e.g., in the pilchard), but this may be obscured by variations in fishing effort.

### 5.2 Species of Commercial Importance (Table 2)

The commercial importance of a species is largely determined by its abundance and its sustained production, but also by its size and market value. However, in the late 1960's over half of the landings concerned were industrial catch for fish meal production (and in 1974 two thirds'), in which quality is less important. Therefore, the landings as shown in Table 2 reflect roughly the relative abundance and production of the species. Thus, herring and mackerel are the most numerous species in the 1960's, but no longer so in 1974 (Table 2) (see also Chaps. 1.9 and 3.4.2). In recent years, after a fishing stop between 1977–1983, the herring stock show a strong recovery.

In general, pelagic fish [including Norway pout, see Andersen and Ursin (1977)], depending on plankton for food, would contribute 65% of the North Sea fish fauna. Also (Table 2) species with a relatively long timespan determine to a large degree the abundance and production of the North Sea fish fauna (two thirds of the landings) while short-lived species are of minor importance (13%). The significance of this observation could be that in an unstable (unpredictable) environment such as the North Sea, long-living species, capable of dealing with longer periods of poor reproductive success, have an advantage. This is demonstrated by the large year-to-year variation in biomass of the Norway pout (a short-lived species) as compared to herring and cod (Andersen and Ursin 1977). For the years 1965–1969, these authors estimated the biomass of the 11 most important exploited fish species at least 5 million metric tons (fresh weight), that is about 10 g fresh weight or 2 g dry weight per $m^2$. However, Yang (1982) considering

**Table 2.** Data on 11 important fish species in the North Sea fisheries

| Species | Average landing[a] 1965–1969 in thousands of metric tons, fresh weight | % of total landings | Industrial catch 1974[d] | Habitat | Distribution | Longevity[c] | biomass[b] |
|---|---|---|---|---|---|---|---|
| Herring | 933 | 30.7 | 88 | Pelagic | All North Sea | Long | 1133 |
| Mackerel | 618 | 20.4 | 230 | Pelagic | All North Sea | Long | 1599 |
| Haddock | 287 | 9.5 | 44 | Demersal | North, central | Medium | 518 |
| Cod | 227 | 7.5 | 5 | Demersal | All North Sea | Long | 304 |
| Norway pout | 179 | 5.9 | 753 | Pelagic | North, central | Short | 239 |
| Sandeel | 157 | 5.2 | 512 | Pelagic | Central, south | Short | 251 |
| Whiting | 139 | 4.6 | 117 | Demersal | All North Sea | Medium | 173 |
| Plaice | 106 | 3.5 | – | Demersal | Central, south | Long | 372 |
| Coalfish | 86 | 2.8 | – | Demersal | North | Long | 334 |
| Sprat | 77 | 2.5 | 243 | Pelagic | All North Sea | Short | ? |
| Sole | 25 | 0.8 | – | Demersal | Central, south | Long | ? |
| Other species | | | 116 | | | | |
| Total | 3036 | 100 | | | | | |

[a] After Cole and Holden (1973).
[b] Estimate, calculated from a computer simulation by Andersen and Ursin (1977).
[c] Long = 10 years or more, medium = 5–10 years, short = 2–5 years.
[d] Data from Popp Madsen (1978).

the total fishing fauna of the North Sea estimated its biomass at 10 million tons in 1977–1978 (approx. 4 g dry weight per $m^2$), with a biomass of almost 6 million tons for the ten most important commercial species.

When studying the distribution of fishes over the North Sea on the basis of fisheries statistics, it must be kept in mind that information from individual fishermen is generally unreliable, as they are not inclined to disclose their fishing ground; also various fish species (herring!) are migratory, and may be caught in one area after growing in another. Nevertheless, flatfish have the highest landings per $m^2$ in the shallow Southern Bight, but most flatfish are caught in the central North Sea. Cod and cod-like fish, and pelagic fish (herring, mackerel, and Norway pout) have the highest yield from the northern North Sea, in total landings and in landings per $m^2$. Landings of all species together are highest from the north and lowest from the southern part (Hummel and Zijlstra 1978).

## 5.3 Some Life Cycles

The life cycles of a few species that have been studied relatively well because of their commercial importance are given here. The life cycles of herring, cod, and plaice are certainly not representative for the less well-known large number of small and short-lived species.

The *herring* is an example of species which form local, mainly self-contained stocks, is migratory, and has separate nursery grounds. Adult North Sea herring have a preference for cool waters (6°–9°C) and are therefore found north and west of the Doggerbank during summer; in winter they penetrate into the shallow southern part. In the north they have a strong diurnal, vertical migration, moving to the warmer surface waters during the night.

There are mainly three separate stocks: the Scottish (Buchan) stock spawns in July-September in the northwestern North Sea, the Bank stock mainly in the central western part in August-October, and the Downs stock spawns just north and south of Dover Straits in November-December. The eggs, small and numerous in the north, and larger and fewer in the south, are attached to stones and pebbles on the bottom. After hatching, the planktonic larvae drift during winter to nursery areas in the east and southeast, in coastal waters; possibly the immigration of coastal waters is an active process. Larval development is slow (6–7 months for the Buchan stock, 4–5 months for Downs herring). Up to a length of 10 cm the young remain in coastal waters (first summer); larger juveniles move to offshore nurseries in the shallow eastern part of the North Sea (second summer). At the end of their second year they move to feeding grounds in the central and northern North Sea.

*Herring* is a typical zooplankton feeder with a preference for copepods (e.g., *Calanus*) but other food items such as fish larvae are also taken. Nowadays most herring mature at the age of 3 years, but formerly (in the 1920's) maturation occurred up to an age of 5 years. The main feeding grounds of the adults are found in the central western and northwestern North Sea. The Downs fish overwinter mainly in the Southern Bight, the Buchan and Bank fish overwinter in the Skagerrak and the northeastern North Sea; the migration routes are largely anticlockwise. North Sea herring live to an age of about 15 years, reaching a length of 30 cm and a weight of 200–250 g.

*Cod*, in contrast with herring, does not show a clear separation in stocks, and lacks separate nursery grounds. Separation between very young and older fish is present as juveniles up to 1 year old live pelagically and adults are demersal.

Adults spawn over the whole North Sea, in the south mainly in February, in the north in March (Daan 1978), but not in coastal waters; there are, however, areas where egg concentrations are higher than average: Southern Bight, German Bight, around Doggerbank and in the northwest. Young, under 1 year-old cod, live pelagically in large parts of the North Sea; at the end of the first year they tend to concentrate in coastal areas in the east. During the second year they start demersal life; they become mature at an age of 2–5 years at a length of 50–60 cm. Cod is potentially long-living (> 10 years) and may reach a length > 1 m. Smallest, pelagic, cod feeds on plankton, the intermediate sizes eat benthic animals, and the older and larger cod eat progressively more fish, including young cod (Daan 1973). Cod does not separate in well-defined stocks, like the herring, although it follows from tagging experiments that there is little exchange between the south and north. Possibly the cod from the North Sea, which shows genetic differences from those of the Baltic and the Norwegian Seas (de Ligny 1969), is best described as a system of clines (Daan 1978). Adult cod has a preference for cool water (6–10°C) and largely leaves the shallow parts in summer.

*Plaice*, as adult is mainly found in the shallower, permanently mixed waters in the southern and southeastern North Sea, has only a weak tendency to form separate stocks, is hardly migratory, and has a spatial separation of juveniles and adults by the presence of a coastal nursery. Spawning takes place in the eastern part of the Channel in December, and moves up the Southern Bight, to the German Bight and off Flamborough Head (March). There are indications for separate spawning concentrations, while they return to their previous spawning ground (de Veen 1961); there is, however, no clear migration but dispersal with the spawning ground as central point (de Veen 1962, 1978b). The eggs are pelagic, and eggs and larvae drift with the residual current usually to the northeast, and immigrate coastal waters at the end of their larval life, in April-June. Here they metamorphose to a bottom mode of life, in particular in estuarine areas, where they remain for 2–3 years, exploiting the tidal areas (Kuipers 1973; Rijnsdorp et al. 1985). Maturation occurs at an age of 3–5 years, at a length around 35 cm (Bannister 1978).

Potentially plaice is long-living, reaching a length of 40–45 cm in males and 50–60 cm in females. Plaice is apparently adapted to large seasonal temperature variations, as its life cycle is completed in the shallower southern part of the North Sea.

These three life cycles are more or less representative for the other large fish stocks. The life cycle of sprat and sandeels might show some similarity to that of herring, although sandeels probably lack a coastal nursery. The life cycle of sole resembles that of plaice (de Veen 1978a). Life cycles of haddock, whiting, coalfish, and possibly also Norway pout, are similar to that of cod, although haddock, coal fish, and Norway pout are mainly deeper-water fish, and coalfish may have a coastal nursery (for haddock see Jones and Hislop 1978; Sahrhage and Wagner 1978). Mackerel in the North Sea is a single stock (Hamre 1978).

In general, North Sea fish life histories are well known as far as eggs and larvae are concerned from special programs or as a byproduct of plankton research and in the adult stage from the fisheries. However, the knowledge of young fish, too small to be retained by the meshes of fisheries nets, is limited.

## 5.4 Long-Term Trends

Landings of fish, as published by ICES in Bulletins Statistiques since 1909, form the longest series of observations on the fish fauna on earth (Holden 1978). In the last few decades considerable changes have occurred in the commercial fish of the North Sea; whether similar changes took place in unexploited fish species is virtually unknown (Tiews 1978b, c). These changes affected both size and composition of the landings, and hence the biomass, recruitment, growth-rate and distribution (Hempel 1978; Cushing 1982). Here we will restrict ourselves to some major events. In the period 1909–1960, the landings remained at a level of 1–1.5 million tons, notwithstanding a probable increase in fishing intensity; from 1960 the landings increased to around 3 million tons in 1968–1975. Landings of herring and mackerel rose to about 1967, and declined afterwards. Total landings of pelagic fish followed the trend of herring and mackerel to 1970, but then rose again due to the increase in catch of Norway pout, sandeels and sprat, the fishery on these species being relatively new (Table 2). In the 1960's there was an enormous upsurge in the landings of demersal fish (cod, haddock, whiting, coalfish), followed by a restricted decline in the early 1970's. The changes in size and composition of the landings between 1960 and 1974 are thought to be a result of stock changes: herring and mackerel stocks declined from 2.5 million tons (1950–1960) to 250,000 tons (early 1970's). The stock size of the other species seems to have increased, in particular the cod, haddock, whiting and coalfish, including Norway pout, by a factor 3. Again, there is little doubt that herring and mackerel were overfished (Burd 1978; Hamre 1978), but this does not explain the simultaneous increase in the landings of other species.

There are two lines of reasoning to explain these changes: one considers predation and food competition as the major factors responsible, and starts at the reduction of the herring and mackerel stocks. These large stocks would have preyed heavily on larvae of other stocks; and in addition would have consumed large amounts of food, which in these active pelagic fish was mainly used for maintenance and swimming, resulting in slow growth (production). Removal of these large stocks of herring and mackerel would lessen the predation pressure on eggs and larvae of other species, resulting in a better reproductive success of these species, while large amounts of pelagic food would become available to opportunistic species with a high growth rate such as Norway pout, sandeels and sprat.

This line of reasoning has been incorporated in a mathematical simulation model by Andersen and Ursin (1977, 1978), giving an adequate explanation of the events in the 1960's.

The other line of thinking considers the changing plankton population in the North Sea as fundamental (Cushing 1982) for the changing in the fish stocks. In particular the increase of cod and codlike fishes in the 1960's would be connected with a period of cooling, which started in the 1950's and induced a delay in the plankton bloom in spring. These fishes' larvae would have profited from this delay by better matching the bloom of their planktonic food. Because of the increased reproductive success of the demersal species they would have increased against a strong fishing pressure. The herring and mackerel stocks, in this line of thinking, are thought to be decimated to a point where egg production limits reproductive success (recruitment overfishing).

## 6 Birds

As all birds breed on land, the birds of the North Sea can be divided into the following groups:
– coastal birds, breeding along the coasts, and collecting their food either at sea, on intertidal flats or, inland; they usually migrate over sea at the end of the breeding period;
– open sea birds, breeding along the coasts, and collecting their food on the open sea; during the winter half year they usually stay in open sea.

The total number of species in the North Sea area is well known. Bellamy et al. (1973) lists 71 species for both groups together. Evans (1973) mentions for the open sea birds a breeding population of 19 species in the North Sea area and a winter population of 21 species. More species migrate along the coast of the Netherlands: 68 species (Camphuysen and van Dijk 1983). Coastal birds for the Waddensea area are described in Smit and Wolff (1980). Total population sizes of all species are generally well known from censusses during breeding periods, and for coastal birds also at other periods by direct observation; however, there are only very limited data available on numbers of open sea birds present in the North Sea area (Nelson 1980). Swennen (1985) has shown for the Waddensea and IJssel Lake the existence of a very clear zonation, at a distance from the coasts for a number of species; it seems reasonable to assume that the same situation may hold for the North Sea. Evans (1973) based a calculation of food used by an open sea bird population on numbers that were *thought* to be present! Serious efforts to determine bird abundance on the open sea by means of ships, aircraft, and from oil-platforms, have been made by Blake et al. (1984), and are presently being made by others (Swennen, pers. commun.). However, in view of the high demand on aircraft- and shiptime for this kind of observations, it must be expected to take a long time before an adequate data base will become available.

## 7 Mammals

Mammals in the North Sea belong to two different groups: seals (Pinnipedia) and whales and dolphins (Cetacea); of these, the seals usually occur along sea boards, as they live part time on land or on dry tidal flats (breeding seasons). As a result of this difference, seals are much better known than whales, due to accessibility and possibilities of observation.

### 7.1 Seals

Six species of seals are known from the North Sea; of these four are rare or occasional stragglers, and two are common. The grey seal lives mainly in the west and northwest of the North Sea (Summers et al. 1978) and strays only occasionally to the east. The North Sea population size in the U.K. is estimated at 29,000–32,000 individuals. Formerly the grey seal used to be common in the German Bight (Reijnders 1978). The common or harbor seal lives along the seaboard of the German Bight (the Waddensea) and the Wash, and the English and Scottish east coast. The population size is estimated at 4500–5000 individuals in the German Bight, 6000 individuals in the Wash, and 7000–10,000 individuals along the British east coast.

The life cycle of the grey seal has been described by Bonner (1972, 1975); Bonner and Hickling (1971), van Haaften (1982b), and of the common seal by van Haaften (1982a); the population dynamics of the common seal by Reijnders et al. (1982). There are clear indications that the population of the common seal suffers in the Waddensea from poisoning by chlorinated hydrocarbons (PCB's!) (Reijnders 1982, see also pages 596–603; Smit and van Wijngaarden 1981). However, over the whole of its extended range the population appears to be stable (FAO 1978) despite widespread habitat disturbance and casual hunting.

Total food consumption of the seal populations in the North Sea is estimated at 100,000 t $a^{-1}$. (common seal 5 kg fish $d^{-1}$, grey seal 5–8 kg $d^{-1}$); as the grey seal feeds on salmon and cod, this leads to competition with owners of salmon rights.

### 7.2 Cetaceans (Whales and Dolphins)

Nearly 30 species are known to occur in the North Sea; most of these (21 species) are rare or occasional stragglers, and are known only because of accidental strandings. About eight species do occur regularly and in larger numbers, all dolphins. All of these are top-predators, but the biology of most species is poorly known.

Only two species, with a tendency to a coastal distribution, are better known, again because of strandings: the harbor porpoise and the bottlenose dolphin (Verweij and Wolff 1982). Both species were formerly much more numerous than at present: it is thought that pollution and a reduction of their prey is causal in the decline (see Andersen 1984). The decrease of the population is still continuing, mainly due to accidental catch by faster fishing ships (FAO 1978). The other more or less common species, such as the common dolphin, white beaked dolphin, white-sided dolphin, and in the north the beluga and the pilot whale, are hardly known. In the period prior to 1960 the minke whale (*Balaenoptera acutorostrata*) used to be observed rather frequently in the western part of the North Sea (Zijlstra, pers. commun.), but it seems to have declined strongly due to whaling.

For all species it holds that nothing is known about population size, now or formerly, about migration routes, food or whatever. We expect, however, that most, if not all, populations are diminishing in size and are subject to the same stresses as harbor porpoise and bottlenosed dolphin.

## 8 Conclusions

This chapter presents an attempt to review the large amount of knowledge available on the North Sea, in detail as well as at a high level of integration. However, we are aware that, especially at the high integration level, much of our knowledge is apparent only, based as it often is on an insufficient data set. This inadequacy is, without doubt, partly due to the size of the North Sea ($\pm$500,000 km$^2$), and partly to sampling problems. In the following the main defects in our knowledge are summarized.

First of all it must be realized that the uniformity of the sea is apparent only; this can be illustrated by the different watermasses that can be distinguished. These have been known for a long time, but although the boundaries between them (fronts) are just as old, the scientific interest in them, as areas of potentially high biological activity, is of recent date. Further it must be realized that although residual currents are in general following the courses as indicated on the maps, they may make about-turn at times, with remarkable effects (van de Kamp 1983; Essink 1986). Therefore, much of our insight is based on a concept of a more or less uniform sea area with a constant and predictable abiotic environment, although there are exceptions (Cushing 1982). Considering the inadequacy of data sets, it is remarkable that hardly any primary paper on nutrients has been published since 1975, although the amount of data in the grey-paper literature is undoubtedly larger than ever. In view of the possible eutrophication of the Southern Bight this is a precarious situation.

In the field of biology the problems of unsufficient data and inadequate insight are due, at least partly, to sampling problems; the two fields where adequate sampling has been realized (plankton and fish) are much ahead of others such as microplankton, small fish, and all bottom fauna. However, attempts to fill in these gaps are being made at present, for instance for bottom fauna (Rachor 1982; de Wilde et al. 1984, 1986).

Knowledge of the open-sea birds is far below par as far as their distribution, feeding behavior, and abundance is concerned outside the breeding season. In view of the effects of enhanced mortalities in these birds due, for instance, to oil pollution, this is a subject that deserves attention in the years to come, for it would be a pity to exterminate another species unknowingly, after the killing of the last great auk (*Alca impennis*) in 1834.

Our knowledge of whales and dolphins in the North Sea is virtually nonexistent.

Quite a few models of (parts of) the North Sea have been constructed in recent years; nearly all contain far-reaching extrapolations and are usually based on far to small data sets; this holds for construction as well as verification.

## References


Adams JA, Baird IE (1970) Chlorophyll and zooplankton standing stock in the North Sea 1969. Ann Biol 26:113–114

Admiraal W (1984) The ecology of estuarine sediment-inhabiting diatoms. Prog Phycol Res 3:269–322

Andersen SH (1984) Bycatches of the harbour porpoise (*Phocaena phocaena*) in Danish fisheries (1980–1981) and evidence for overexploitation. Rep Int Whal Comm 34:745

Andersen KP, Ursin E (1977) A multispecies extension to the Beverton and Holt theory of fishing with accounts of phosphorus circulation and primary production. Medd Dan Fisk Havunders 7:319–435

Andersen KP, Ursin E (1978) A multispecies analysis of the effect of variations of (fishing) effort upon stock composition of eleven North Sea fish species. Rapp P V Réun Cons Int Explor Mer 172:286–291

Ankar S, Elmgren R (1976) The benthic macro- and meiofauna of the Askö-Landsort area – a stratified random sampling survey. Contr Askö Lab, Univ Stockholm 11:1–115

Anonymous (1973) Continuous plankton records: a plankton atlas of the North Atlantic and the North Sea. In: Lucas CE, Glover RS (eds). Bull Mar Ecol 7:1–174

Bannister RCA (1978) Changes in plaice stocks and plaice fisheries in the North Sea. Rapp P V Réun Cons Int Explor Mer 172:86–101

Becker GA, Frey H, Wegner G (1986) Atlas der Temperatur an der Oberfläche der Nordsee. Wöchentliche und monatliche Mittelwerte für den Zeitraum 1971–1980. Dtsch Hydrogr Z, Ergänzungsheft B 17, pp 1–127

Bellamy DJ, Edwards P, Hirons MJD, Jones DJ, Evans PR (1973) Resources of the North Sea and some interactions. In: Goldberg D (ed) North Sea Science. MIT Press, Cambridge Mass, pp 383–399

Bennekom AJ van, Gieskes WWC, Tijssen SB (1975) Eutrophication in Dutch coastal waters. Proc R Soc Lond B Biol Sci 189:359–374

Beukema JJ (1974) The efficiency of the van Veen grab compared with the Reinecke box sampler. J Cons Perm Int Explor Mer 35:319–327

Birkett L (1953) Changes in the composition of the bottom fauna of the Dogger Bank area. Nature 171:265

Blake BF, Tasker ML, Jones PH, Dixon TJ, Mitchell R, Langslow DR (1984) Sea bird distribution in the North Sea. Huntingdon, Nat Conserv Counc, pp 1–432

Boalch GT, Harbour DS, Butler EI (1978) Seasonal phytoplankton production in the western English Channel. 1964–1974. J Mar Biol Assoc UK 58:943–954

Böhnecke G (1922) Salzgehalt und Strömungen der Nordsee. Veröff Inst Meereskd Univ Berl, Neue Folge A, Geol-Naturwiss Reihe 10:1–34

Bonner WN (1972) The grey seal and common seal in European waters. Oceanogr Mar Biol Annu Rev 10:461–507

Bonner WN (1975) Population increase of grey seals at the Farne Islands. Rapp P V Réun Cons Int Explor Mer 169:366–370

Bonner WN, Hickling G (1971) The grey seal of the Farne Islands. Trans Nat Hist Soc Northumbria 17:141–162

Bowman MJ et al. (1978) Oceanic fronts in coastal processes. Springer, Berlin Heidelberg New York

Buchanan JB, Warwick RM (1974) An estimate of benthic macrofaunal production in the offshore mud of the Northumberland coast. J Mar Biol Assoc UK 54:197–222

Buchanan JB, Kingston PF, Sheader M (1974) Long-term population trends of the benthic macrofauna in the offshore mud of the Northamberland coast. J Mar Biol Assoc UK 54:785–795

Burd AC (1978) Long-term changes in North Sea herring stocks. Rapp P V Réun Cons Int Explor Mer 172:137–153
Cadée GC (1984) Macrobenthos and macrobenthic remains on the Oyster Ground, North Sea. Neth J Sea Res 18:160–178
Camphuysen CJ, Dijk J van (1983) Zee- en kustvogels langs de Nederlandse kust, 1974–79 (Engl summary). Limosa 57:81–230
Cole HA, Holden MJ (1973) History of the North Sea Fisheries, 1950–1969. In: Goldberg ED (ed) North sea science. MIT Press, Cambridge Mass, pp 337–360
Colebrook JM (1978) Changes in the zooplankton of the North Sea, 1948–1973, Rapp P V Réun Cons Int Explor Mer, 172:390–396 pp
Colebrook JM (1979) Continuous plankton records: Seasonal cycles of phytoplankton and copepods in the North Atlantic Ocean and the North Sea. Mar Biol 51:23–32
Colebrook JM, Robinson GA (1965) Continuous plankton records: seasonal cycles of phytoplankton and copepods in the north-eastern Atlantic and the North Sea. Bull Mar Ecol 6:123–139
Colebrook JM, John DE, Brown WW (1961) Continuous plankton records: contribution towards a plankton atlas of the north-eastern Atlantic and the North Sea; part II: Copepoda. Bull Mar Ecol 5:90–97
Creutzberg F (1985) A persistent chlorophyll a maximum coinciding with an enriched benthic zone. In: Gibbs PE et al. (eds) Proceedings of the 19th European Marine Biol Symp (1984). Cambridge Univ Press, pp 97–108
Creutzberg F (1986) Distribution patterns of two bivalve species (*Nucula turgida, Tellina fabula*) along a frontal system in the southern North Sea. Neth J Sea Res 20:305–311
Creutzberg F, Postma H (1979) An experimental approach to the distribution of mud in the southern North Sea. Neth J Sea Res 13:99–116
Creutzberg F et al. (1979–1985) Aurelia cruise reports on the benthic fauna of the southern North Sea, rep. 1–10. Neth Inst Sea Res, Aurelia Cruise reports 1–10, 825 p
Creutzberg F, Wapenaar P, Duineveld G, Lopez Lopez N (1983) Distribution and density of the benthic fauna in the southern North Sea in relation to bottom characteristics and hydrographic conditions. Rapp P V Réun Cons Int Explor Mer 183:101–110
Cushing DH (1973) Productivity of the North Sea. In: Goldberg ED (ed) North sea science. MIT Press, Cambridge Mass, pp 249–266
Cushing DH (1982) Climate and fisheries. Academic Press, London, pp 1–373
Cushing DH (1984) Sources of variability in the North Sea ecosystem. In: Sündermann J, Lenz W (eds) North Sea Dynamics. Springer, Berlin Heidelberg New York, pp 498–516
Cushing DH, Vucetic T (1963) Studies on a *Calanus* patch III. The quantity of food eaten by *Calanus finmarchicus*. J Mar Biol Assoc UK 43:349–371
Cushing DH, Tungate DS, Nicholson HF, Vucetic T (1963) Studies on a *Calanus* patch I-V (a series of papers). J Mar Biol Assoc UK 43:327–389
Czitrom SPR et al. (1985) (manuscript) A heat-induced front in an area influenced by land runoff
Daan N (1973) A quantitative analysis of the food intake of North Sea cod, *Gadus morhua*. Neth J Sea Res 6:479–517
Daan N (1978) Changes in the cod stocks and cod fisheries in the North Sea. Rapp P V Réun Cons Int Explor Mer 172:39–57
Dankers N, Kühl H, Wolff WJ (eds) (1981) Invertebrates of the Wadden Sea, rapport 4. Balkema, Rotterdam, pp 1–221
Davies FM (1923) Quantitative studies on the fauna of the sea bottom. No 1. Preliminary investigation of the Dogger Bank. Min Agric Fish, Fishery Invest, Ser 2, 6 no 2
Dietrich G (1950) Die natürlichen Regionen von Nord- und Ostsee auf hydrografischer Grundlage. Kiel Meeresforsch 7:35–69
Dietrich G (1953) Verteilung, Ausbreitung und Vermischung der Wasserkörper in der südwestlichen Nordsee auf Grund der Ergebnisse der "Gauss"-Fahrt im Februar/März 1952. Ber Dtsch Wiss Komm Meeresforsch 13:104–129
Dietrich G (1954) Einfluss der Gezeitenstromturbulenz auf die hydrografische Schichtung der Nordsee. Arch Meteorol Geophys Bioclimatol (ser A) 7:391–405
Eisma D (1966) The distribution of benthic marine molluscs off the main Dutch coast. Neth J Sea Res 3:107–163

Eisma D (1987) The North Sea, an overview. Trans R Soc Lond B Biol Sci, B 316, 461–485
Elmgren R (1984) Trophic dynamics in the enclosed brackish Baltic Sea. Rapp P V Réun Cons Int Explor Mer 183:152–169
Engel M (1983) Evaluation of North Sea hydrocasts for modelling purposes. In: Sündermann J, Lenz W (eds) North Sea Dynamics. Springer, Berlin Heidelberg New York, pp 429–435
Essink K (1986) De opmars van de Amerikaanse Zwaardschede, Ensis directus. Rijkswaterstaat, dienst Getijdewateren, AOBB-86. 151, pp 1–5
Evans PR (1973) Avian resources of the North Sea. In: Goldberg ED (ed) North sea science. MIT Press, Cambridge Mass, pp 400–412
FAO (1978) Mammals in the sea, vol 1. FAO Fisheries Series no 5, pp 1–264
Fogg GE (1985) Biological activities at a front in the western Irish Sea. In: Gibbs PE et al. Proc 19th EMBS (1984). Cambridge Univ Press, pp 87–95
Folkard AR, Jones PGW (1974) Distribution of nutrient salts in the southern North Sea during early 1974. Mar Pollut Bull 5:181–185
Fransz HG (1976) The spring development of calanoid copepod populations in Dutch coastal waters as related to primary production. Proc 10th EMBS, Ostende Belgium, 2, pp 247–269
Fransz HG, Arkel WG van (1980) Zooplankton activity during and after the phytoplankton spring bloom at the central station in the Flex box, northern North Sea, with special reference to the calanoid copepod *Calanus finmarchicus* (Gunn). Meteor Forschungs ergeb Reihe A 22:113–121
Fransz HG, Gieskes WWC (1984) The unbalance of phytoplankton and copepods in the North Sea. Rapp P V Réun Cons Int Explor Mer 183:218–225
Fransz HG, Verhagen JHG (1985) Modelling research on the production cycle of phytoplankton in the Southern Bight of the North Sea in relation to riverborne nutrient loads. Neth J Sea Res 19:241–250
Fransz HG, Miquel JC, Gonzales SR (1984) Mesoplankton composition, biomass and vertical distribution, and copepod production in the stratified central North Sea. Neth J Sea Res 18:92–96
Fraser JH (1973) Zooplankton of the North Sea. In: Goldberg ED (ed) North Sea science. MIT Press, Cambridge Mass, pp 267–289
Garrod DJ, Colebrook JM (1978) Biological effects of variability in the North Atlantic Ocean. Rapp P V Réun Cons Int Explor Mer 172:128–144
Gerlach SA (1978) Food chain relationships in subtidal silty sand marine sediments and the role of meiofauna in stimulating productivity. Oecologia (Berl) 33:55–69
Gieskes WWC, Kraay GW (1975) The phytoplankton spring bloom in Dutch coastal waters of the North Sea. Neth J Sea Res 9:166–196
Gieskes WWC, Kraay GW (1977a) Primary production and consumption of organic matter in the southern North Sea during the spring bloom of 1975. Neth J Sea Res 11:146–167
Gieskes WWC, Kraay GW (1977b) Continuous plankton records: changes in the plankton of the North Sea and its eutrophic Southern Bight from 1948–1975. Neth J Sea Res 11:334–364
Gieskes WWC, Kraay GW (1983) Dominance of Cryptophyceae during the phytoplankton spring bloom in the Central North Sea by HPLC analysis of pigments. Mar Biol 75:146–167
Glémarec M (1973) The benthic communities of the European North Atlantic continental shelf. Oceanogr Mar Biol Annu Rev 11:263–289
Glover RS (1967) The continuous plankton recorder survey of the North Atlantic. Symp Zool Soc Lond 19:189–210
Glover RS, Robinson A, Colebrook JM (1974) Marine biological surveyance. Environ Change 2:395–402
Goedecke E, Smed J and Tomsczak G, Monatskarten des Salzgehalts der Nordsee Ergänzungsheft zur D Hydrogr Z, Reihe B (4°) no. 9, 100 pp
Gray JS, Christie H (1983) Predicting long-term changes in marine benthic communities. Mar Ecol Prog Ser 13:87–90
Greve W (1981) Invertebrate predator control in a coastal marine ecosystem: the significance of *Beroe gracilis* (Ctenophora). Kiel Meeresforsch, Sonderheft 5:211–217
Greve W, Stockner J, Fulton J (1976) Towards a theory of speciation in Beroe. In: Mackie GO (ed) Coelenterate ecology and behavior. Plenum, New York, pp 251–258
Haaften JL van (1982a) The life history of the harbour seal in the Wadden Sea. In: Reijnders PJH, Wolff WJ (eds) Marine mammals of the Wadden Sea. Balkema, Rotterdam, pp 15–19
Haaften JL van (1982b) The grey seal. In: Reijnders PJH, Wolff WJ (eds) Marine mammals of the Wadden Sea. Balkema, Rotterdam, pp 48–50

Hagel P, Alkemade van Rijn JWA (1973) Eutrophication of the North Sea. ICES, CM 1973/L:22, 17 p mimeo
Hagmeier A (1925) Vorläufiger Bericht über die vorbereitenden Untersuchungen der Bodenfauna der Deutschen Bucht mit dem Petersen-Bodengreifer. Ber Dtsch Wiss Komm Meeresforsch NF 1:247–272
Hagmeier E (1978) Variations in phytoplankton near Helgoland. Rapp P V Réun Cons Int Explor Mer 172:361–363
Hamre J (1978) The effect of recent changes in the North Sea mackerel fishery on stock and yield. Rapp P V Réun Cons Int Explor Mer 172:197–210
Hardy AC (1924) The herring in relation to its animate environment, Part I The food and feeding habits of the herring. Fish Invest Lond, Ser II, 7, 3:1–53 pp
Hartog C den (1959) The epilithic algal communities occurring along the coasts of the Netherlands. Thesis, Univ Amsterdam, pp 1–241
Hartog C den (1979) The epilithic algae and lichens of the Wadden Sea. In: Wolff WJ (ed) The flora and vegetation of the Wadden Sea. Balkema, Rotterdam, pp 119–123
Heip C, Willems KA, Goossens A (1978) Vertical distribution of meiofauna and the efficiency of the van Veen grab on sandy bottoms in Lake Grevelingen. Hydrobiol Bull 11:35–45
Heip C, Vincx M, Smol N, Vranken G (1982) The systematics and ecology of free-living marine Nematodes. Helminthological Abstr B 51:1–31
Heip C, Herman R, Vincx M (1983) Subtidal meiofauna of the North Sea, a review. Biol Jb Dodonea 51:116–170
Heip C, Herman R, Vincx M (1984) Variability and productivity of meiobenthos in the Southern Bight of the North Sea. Rapp P V Réun Cons Int Explor Mer 183:51–56
Hempel G (1978) Synopsis of the symposium on North Sea fish stocks – recent changes and their causes. Rapp P V Réun Cons Int Explor Mer 172:445–449
Hempel G (ed) (1978) North Sea fish stocks – recent changes and their causes. Rapp P V Réun Cons Int Explor Mer 172:1–449
Hill HW (1973) Currents and water masses. In: Goldberg ED (ed) North Sea science, MIT Press, Cambridge Mass, pp 17–42
Hoek C van den, Admiraal W, Colijn F, Jonge VN de (1979) The role of algae and seagrasses in the ecosystem of the Wadden Sea: a review. In: Wolff WJ (ed) Flora and vegetation of the Wadden Sea, report 3, Balkema, Rotterdam, pp 9–118
Höhn R (1973) On the climatology of the North Sea. In: Goldberg ED (ed) North Sea science, MIT Press, Cambridge Mass, pp 183–236
Holden MJ (1978) Long-term changes in landings of fish from the North Sea. Rapp P V Rëun Cons Int Explor Mer 172: 11–26
Holligan PM (1981) Biological implications of fronts on the north west European continental shelf. In: Swallow JC et al. (eds) Circulation and fronts in continental shelf seas. R Soc London, pp 547–562
Hummel H, Zijlstra JJ (1978) Some notes on the production of the North Sea. ICES, CM 1978/Gen: 11 pp 1–10 mimeo
Johnston R (1973) Nutrients and metals in the North Sea. In: Goldberg ED (ed) North Sea Science. MIT Press, Cambridge Mass, pp 293–307
Johnston R, Jones PGW (1965) Inorganic nutrients in the North Sea. Serial atlas of the marine environment, Folio 11, Am Geogr Soc, New York
Joiris C, Billen G, Lancelot C, Daro MH, Mommaerts JP, Bertels A, Bossicart M, Nys J, Hecq JH (1982) A budget of carbon cycling in the Belgian coastal zone; relative roles of zooplankton, bacterioplankton and benthos in the utilization of primary production. Neth J Sea Res 16:260–275
Jones R (1984) Some observations on energy transfer through the North Sea and George Bank food webs. Rapp P V Réun Cons Int Explor Mer 183:204–217
Jones R, Hislop JRG (1978) Changes in North Sea haddock and whiting. Rapp P V Réun Cons Int Explor Mer 172:58–71
Kamp G van de (1983) Long-term variations in residual currents in the southern North Sea. ICES, CM 1983 C4
Kelley JC (1976) Sampling in the sea. In: Cushing DH, Walsh JJ (eds) The ecology of the seas. Blackwell, Oxford, pp 361–387
Kirkegaard JB (1969) A quantitative investigation of the central North Sea Polychaeta. Spolia Zool Mus Havn 29:1–285

Krause M, Radach G (1980) On the succesion of developmental stages of herviborous zooplankton in the northern North Sea during Flex 76. I. First statements about the main groups of the zooplankton community.Meteor Forschungsergeb Reihe A 22:133-149

Krause G et al. (1985) Frontal systems in the German Bight and their physical and biological effects. Proc 17th Liege Symp (in press) Elsevier, Amsterdam

Kuipers B (1973) On the tidal migration of young plaice (*Pleuronectes platessa*) in the Wadden Sea. Neth J Sea Res 6:376-388

Laevastu T (1963) Serial atlas of the marine environment. Folio 4, Publ Am Geogr Soc, New York

Lamb HH (1973) The effect of climatic anomalies in the oceans on long term atmospheric circulation behaviour and currents in the North Sea and surrounding regions. In: Goldberg ED (ed) North Sea science. MIT Press, Cambridge Mass, pp 153-182

Lee AJ (1970) The currents and water masses of the North Sea. Oceangr Mar Biol Annu Rev 8:33-71

Lee AJ, Ramster JW (1979) Atlas of the sea around the British Isles. MAFF Lowestoft

Leewis RJ (1985) Phytoplankton off the Dutch coast. Thesis, Nijmegen, pp 1-147

Ligny W de (1969) Serological and biochemical studies on fish populations. Oceanogr Mar Biol Annu Rev 7:411-513

Maier-Reimer E (1979) Some effects of the Atlantic circulation and of river discharges on the residence circulation of the North Sea. Dtsch Hydrogr Z 32:126-130

McCave IN (1973) Mud in the North Sea. In: Goldberg ED (ed) North Sea Science. MIT Press, Cambridge Mass, pp 75-100

McIntyre AD (1978) The benthos of the western North Sea. Rapp P V Réun Cons Int Explor Mer 172:405-417

Michaelis H (1978) Recent biological phenomena in the German Wadden Sea. Rapp P V Réun Cons Int Explor Mer 172:276-277

Möller H (1980) A summer survey of large zooplankton, particularly Scyphomedusae, in North Sea and Baltic. Meeresforschung 28:61-68

Mommaerts JP (1973a) The relative importance of nannoplankton in the North Sea primary production. Br Phycol J 8:13-20

Mommaerts JP (1973b) On primary production in the South Bight of the North Sea. Br Phycol J 8:217-231

Mommaerts JP (1980) Seasonal variations of the parameters of the photosynthesis-light relationship during the Fladen Ground Experiment 1976. Proc ICES/Jonsis workshop on JONSDAP '76, ICES, CM 1980/C3:31-48

Mulder M, Lewis WE, van Arkel MA (1987) Effecten von oliehoudend boorgruis op de benthische fauna, Neth Inst Sea Res, report, 3:1-60 pp

Nelson B (1980) Sea birds, their biology and ecology. Hamlyn, London, pp 1-224

Otto L (1983) Currents and water balance in the North Sea. In: Sündermann J, Lenz W (eds) North Sea dynamics. Springer, Berlin Heidelberg New York, pp 26-43

Parker HW, Boeseman M (1954) The basking shark, *Cetorhinus maximus*, in winter. Proc Zool Soc Lond 124:185-194

Parker HW, Stott FC (1965) Age, size and vertebral calcification in the basking shark, *Cetorhinus maximus* (Gunnerus). Zool Meded (Leiden) 40:305-319

Petersen CGJ (1911) Valuation of the sea. I. Animal life of the sea bottom, its food and quantity. Rep Dan Biol Stn 20:1-81

Petersen CGJ (1914) Valuation of the sea. II. The animal communities of the sea bottom and their importance for marine zoogeography. Rep Dan Biol Stn 21:1-68

Pingree RD, Griffiths DK (1978) Tidal fronts on the shelf seas around the British Isles. J Geophys Res 83 (C9):4615-4622

Popp Madsen K (1978) The industrial fisheries in the North Sea. Rapp P V Réun Cons Int Explor Mer 172:27-30

Postma H (1973) Transport and budget of organic matter in the North Sea. In: Goldberg ED (ed) North Sea science. MIT Press, Cambridge Mass, pp 326-334

Postma H (1978) The nutrient contents of the North Sea water: changes in recent years, particularly in the Southern Bight. Rapp P V Réun Cons Int Explor Mer 172:350-357

Postuma KH (1978) Immigration of southern fishes into the North Sea. Rapp P V Réun Cons Int Explor Mer 172:225-229

Rachor E (1982) Biomass distribution and production estimates of macro-endofauna in the North Sea. ICES, CM 1982/L2
Rachor E, Gerlach SA (1978) Changes of macrobenthos in a sublitoral sand area of the German Bight, 1967–1975. Rapp P V Réun Cons Int Explor Mer 172:418–431
Rachor E, Salzwedel H (1976) Studies on the population dynamics and productivity of some bivalves in the German Bight. Proc 10th Eur Mar Biol symp 2. Ostend. Universa, Belgium, pp 575–588
Radach G (1982) Dynamic interactions between the lower trophic levels of the marine food web in relation to the phsyical environment during the Fladen Ground experiment. Neth J Sea Res 16:231–246
Radach G (1983) Simulations of phytoplankton dynamics and their interactions with other system components. In:Sündermann J, Lenz W (eds) North Sea dynamics. Springer, Berlin Heidelberg New York, pp 584–610
Ramster JW (1965) Studies with the Woodhead sea-bed drifter in the southern North Sea. Lab Leafl Fish Lab Lowestoft (N.S.) 6:1–4
Redeke HC (1941) De Visschen van Nederland. Sijthoff, Leiden, p 331
Reid PC (1978) Continuous plankton records: large-scale changes in the abundance of phytoplankton in the North Sea from 1958–1973. Rapp P V Réun Cons Int Explor Mer 172:384–389
Reid PC, Budd TD (1979) Plankton and environment in the North Sea in the period 1948–1977. ICES, CM 1979; L26
Reijnders PJH (1978) De grijze zeehond in het Waddengebied. Waddenbulletin 13:500–502
Reijnders PJH (1982) Threats to the harbour seal population in the Wadden Sea. In: Reijnders PJH, Wolff WJ (eds) Marine mammals of the Wadden Sea. Balkema, Rotterdam, pp 38–47
Reijnders PJH, Drescher HE, Haaften JL van, Bogebjerg Hansen E, Tongaard S (1982) Population dynamics of the harbour seal in the Wadden Sea. In: Reijnders PJH, Wolff WJ (eds) Marine mammals of the Wadden Sea. Balkema, Rotterdam, pp 19–32
Rijnsdorp AD, Stralen M van, Veer HW van der (1985) Selective tidal transport of North Sea plaice larvae *Pleuronectes platessa* in coastal nursery areas. Trans Am Fish Soc 114:461–470
Robertson A (1968) The continuous plankton recorder: a method for studying the biomass of calanoid copepods. Bull Mar Ecol 6:185–223
Robinson GA (1961) Contribution towards an plankton atlas of the North-Eastern Atlantic and the North Sea, part I Phytoplankton. Bull Mar Ecol 5: 81–89
Robinson GA (1970) Continuous plankton records: variations in the seasonal cycles of phytoplankton. Bull Mar Ecol 6: 333–345
Sahrhage D (1964) Über die Verbreitung der Fischarten in der Nordsee I. Juni-Juli 1959 und Juli 1960. Ber Dtsch Wiss Komm Meeresforsch 17:165–278
Sahrhage D (1967) Über die Verbreitung der Fischarten in der Nordsee, Teil II, Januar 1962 und 1963. Ber Dtsch Wiss Komm Meeresforsch 19:66–179
Sahrhage D, Wagner G (1978) On fluctuations in the haddock population of the North Sea. Rapp P V Réun Cons Int Explor Mer 172:72–85
Salzwedel H, Rachor E, Gerdes D (1985) Benthic macrofauna communities in the German Bight. Veröff Inst Meeresforsch Bremerhaven 20:199–267
Slobodkin LB (1961) Growth and regulation of animal populations. Rinehart & Winston, New York
Smit CJ, Wijngaarden A van (1981) Threatened mammals in Europe. Akademische Verlagsges, Wiesbaden, pp 1–259
Smit CJ, Wolff WJ (eds) (1980) Birds of the Wadden Sea. Balkema, Rotterdam, pp 1–308
Steele JH (1956) Plant production of the Fladen Ground. J Mar Biol Assoc UK 35:1–33
Steele JH (1958) Plant production in the northern North Sea. Scott Home Dep Mar Res 7:1–36
Steele JH (1962) Environmental control of photosynthesis in the sea. Limnol Oceanogr 7:137–150
Steele JH (1974) The structure of marine ecosystems. Harvard Univ Press, Cambridge Mass, London
Steele JH (1976) Patchiness. In: Cushing DH, Walsh JJ (eds) The ecology of the seas. Blackwell, Oxford, pp 98–115
Steele JH, Baird IE (1965) The chlorophyll a content of particulate organic matter in the northern North Sea. Limnol Oceanogr 10:261–267
Steele JH, Henderson EW (1977) Plankton patches in the northern North Sea. In: Steele JH (ed) Fisheries mathematics. Academic Press, London, pp 1–19
Stephen AC (1923) Preliminary surveys of the Scottish waters of the North Sea by the Petersen grab. Sci Invest Fish Scott 1922, 3:1–21

Stephen AC (1933) Studies on the Scottish marine fauna: the natural faunistic divisions of the North Sea as shown by the quantitative distribution of the Molluscs. Trans R Soc Edinb 57:601–616

Stephen AC (1934) Studies on the Scottish marine fauna: Quantitative distribution of the echinoderms and the natural faunistic divisions of the North Sea. Trans R Soc Edinb 57:777–787

Stephensen W, Williams WT, Cook SA (1972) Computer analyses of Petersen's original data on bottom communities. Ecol Monogr 42:387–415

Stride AH (1973) Sediment transport in the North Sea. In: Goldberg ED (ed) North Sea science. MIT Press, Cambridge Mass, pp 101–130

Stripp K (1969a) Die Associationen des Benthos in den Helgoländer Bucht. Veröff Inst Meeresforsch Bremerhaven 12:95–142

Stripp K (1969b) Das Verhältnis von Makrofauna und Meiofauna in den Sedimenten der Helgoländer Bucht. Veröff Inst Meeresforsch Bremerhaven 12:143–148

Summers CF, Bonner WN, Haaften J van (1978) Changes in the seal populations in the North Sea. Rapp P V Réun Cons Int Explor Mer 172:278–285

Swennen C (1985) Iets over de vogels van het open water van IJsselmeer, Waddenzee en Noordzee. Vogeljaar 33:208–214

Thorade H (1941) Ebbe und Flut, ihre Entstehung und ihre Wandlungen. Springer, Berlin Heidelberg New York, pp 1–114

Tiews K (1978a) On the disappearance of blue fin tuna in the North Sea and its ecological consequences for herring and mackerel. Rapp P V Réun Cons Int Explor Mer 172:301–309

Tiews K (1978b) The German industrial fisheries in the North Sea and their by-catches. Rapp P V Réun Cons Int Explor Mer 172:230–238

Tiews K (1978c) Non-commercial fish species in the German Bight: Records of by-catches of the brown-shrimp fishery. Rapp P V Réun Cons Int Explor Mer 172:259–265

Tomczak G, Goedecke E (1962) Monatskarten der Temperatur der Nordsee, dargestellt für verschiedene tiefe Horizonte. Dtsch Hydrogr Z, Erg Reihe, B8

Ursin E (1960) A quantitative investigation of the echinoderm fauna of the central North Sea. Medd Dan Fisk Havunders NS 2:1–204

Ursin E, Andersen KP (1978) A model of the biological effects of eutrophication in the North Sea. Rapp P V Réun Cons Int Explor Mer 172:366–377

Veen J de (1961) The 1960 tagging experiments on mature plaice in different spawning areas in the southern North Sea. ICES, CM 1961, 44

Veen J de (1962) On the subpopulations of plaice in the southern North Sea. ICES, CM 1962, Near Northern Seas Comm 94

Veen J de (1978a) The changes in North Sea sole stocks (*Solea solea* L). Rapp P V Réun Cons Int Explor Mer 172:124–136

Veen J de (1978b) On selective tidal transport in the migration of North Sea plaice (*Pleuronectes platessa*) and other flatfish species. Neth J Sea Res 12:115–147

Verweij J, Wolff WJ (1982) The common or harbour porpoise (*Phocaena phocaena*). The bottlenose dolphin (*Tursiops truncatus*). In: Reijnders PJH, Wolff WJ (eds) Marine mammals of the Wadden Sea. Balkema, Rotterdam, pp 51–64

Wandschneider K (1980) Variations in the composition of phytoplankton populations during Flex 1976. Proc ICES/Jonsis workshop on Jonsdap '76. ICES, CM 1980/C3:131–139

Warwick RM, Buchanan JB (1970) The meiofauna off the coast of Northumberland. I. The structure of the nematode population. J Mar Biol Assoc UK 50:129–146

Went AEJ (1972) Seventy years agrowing. A history of the international council for the exploration of the sea, 1902–1972. Rapp P V Réun Cons Int Explor Mer 165:1–252

Wheeler A (1969) The fishes of the British Isles and Northwest Europe. McMillan, London, pp 1–613

Wheeler A (1978) Key to the fishes of Northern Europe. Warne, Frederick, London, pp 1–380

Whitehead PJ, Bauchot ML, Hureau J-C, Nielsen J, Tortonese E (1984) Fishes of the north-eastern Atlantic and the Mediterranean, vol 1. Unesco, Paris 1984, pp 1–510

Wilde PAWJ de, Berghuis EM, Kok A (1984) Structure and energy demand of the benthic community of the Oyster Ground, central North Sea. Neth J Sea Res 18:143–159

Wilde PAWJ de, Berghuis EM, Kok A (1986) Biomass and activity of benthic fauna on the Fladen Ground (northern North Sea). Neth J Sea Res 20:313–324

Wimpenny RS (1953) The dry weight and fat content of plankton with estimates of flagellate counts. Ann Biol 9:119–122

Wyatt T (1976a) Plants and animals of the sea. In: Cushing DH, Walsh JJ (eds) The ecology of the seas. Blackwell, Oxford, pp 81–97
Wyatt T (1976b) Food chains in the sea. In: Cushing DH, Walsh JJ (eds) The ecology of the seas. Blackwell, Oxford, pp 341–358
Yang I (1982) An estimate of the fish biomass in the North Sea. J Cons Int Explor Mer 40:161–172
Ziegelmeier E (1978) Macrobenthos investigations in the eastern part of the German Bight from 1950 to 1974. Rapp P V Cons Int Explor Mer 172:432–444
Zijlstra JJ (1972) On the importance of the Wadden Sea as a nursery area in relation to the conservation of the southern North Sea fishery resources. Proc Symp Zool Soc Lond 29:233–258
Zijlstra JJ (1988) The North Sea Ecosystem. In: Goodall DW (ed) Ecosystems of the world part 27. Postma H, Zijlstra JJ (eds) Ecosystems of the continental shelves. Elsevier, Amsterdam, p 231–278

# Fishery Resources

A.D. McINTYRE[1]

## 1 Introduction

It is frequently said that the North Sea is one of the world's richest fishing grounds, and even a cursory examination of its physical and biological characteristics indicates the basis of such a claim. It is a relatively shallow temperate shelf area, flushed by nutrient-rich oceanic water and enhanced by terrestrial inputs from several large rivers. Its water column is subject to total vertical overturn at least annually, so nutrients are well distributed, encouraging high biological production. The North Sea has a great diversity of habitats on rock, gravel, sand and mud ranging from estuarine and coastal situations to shallow banks and deeper offshore areas. It is not unexpected, therefore, that the region, defined (ICES 1985) as ICES sub-area IV, (i.e. excluding the Skagerrak) is capable of supporting substantial fishery resources and these in recent years have been exploited mainly by the six countries bordering the North Sea – Belgium, Denmark, Federal Republic of Germany, the Netherlands, Norway and the UK, but some other countries participate to a lesser extent – the Faroe Islands, France, Poland and Sweden. In the widest sense, such resources may be defined as those species of animals and plants for which a market exists or can be developed. They include predominantly demersal and pelagic fish species and a variety of shellfish, mainly molluscs and crustaceans. In 1984, the most recent year for which full statistics are available, the total landings of fish and shellfish from the North Sea amounted to almost 2.6 million tonnes, and details of these are given in Table 1 broken down by country and species. This shows that Denmark, Norway, the UK, the Netherlands and the Federal German Republic accounted for 93% of the total landings, and that only ten species made up 92% of the fish, but this does not reflect the relative values, since shellfish are mostly much higher priced. It should be noted that the 1984 statistics are merely a snapshot of a dynamic situation which is discussed in more detail below. In addition to the catch of wild species, there has in recent years been an increasing yield from marine fish farming, which although small in comparison with the wild catch, does indicate an expanding and diversifying industry. The wild species are under pressure from man's exploitation, and all the living resources are exposed to the effects of human uses of the environment and are subject to natural variations which can produce major fluctuations in abundance of the stocks. Against this background, sound management of the North Sea's living resources is

[1]Department of Agriculture and Fisheries for Scotland, Marine Laboratory, P.O. Box No. 101, Victoria Road, Aberdeen AB9 8D8, Scotland

**Table 1.** Nominal catch ('000s metric tons) of various fish species and of invertebrates (Inv) from the North Sea, IV, by each member country of ICES in 1984

| | | | | | Norway Sand | | | | | | | |
|---|---|---|---|---|---|---|---|---|---|---|---|---|
| | Cod | Had | Pla | Sai | Whi | Pout | Eel | Her | Mac | Spr | Total | Inv |
| Belgium | 6 | + | 10 | + | 3 | — | — | 3 | + | — | 32 | 2 |
| Denmark | 47 | 16 | 23 | 5 | 20 | 205 | 620 | 40 | 9 | 121 | 1183 | 79 |
| Faroe | — | — | — | — | — | 19 | 11 | — | — | — | 30 | — |
| France | 8 | 8 | 1 | 44 | 19 | — | — | 6 | — | — | 99 | + |
| German F Rep | 13 | 3 | 2 | 25 | + | — | — | 12 | + | 1 | 64 | 78 |
| Netherlands | 25 | 1 | 61 | + | 9 | + | — | 44 | — | + | 157 | — |
| Norway | 7 | 4 | + | 90 | + | 174 | 29 | 99 | 23 | 10 | 468 | 3 |
| Poland | + | + | — | + | + | — | — | — | — | — | + | — |
| Sweden | 1 | 2 | + | + | + | — | — | 1 | + | — | 4 | + |
| U.K. (E & W) | 36 | 12 | 13 | 8 | 5 | — | — | 2 | + | 1 | 86 | 14 |
| U.K. (Scot) | 54 | 87 | 4 | 7 | 43 | + | 33 | 31 | + | + | 277 | 10 |

No catches were taken from the North Sea by the following member countries: — Finland, German Democratic Republic, Greenland, Iceland, Ireland, Portugal, Spain and U.S.S.R.

essential for their continued availability. While it cannot be argued that food from the sea represents a major part of the riches of countries round the North Sea (fish and shellfish landings make up less, in some cases very much less, than 1% of the Gross Domestic Product) it is relevant that fishery resources have substantial added value, they tend to support distinct and significant communities, and the importance of fish for a healthy human diet is increasingly appreciated.

## 2 Finfish

The pattern of fisheries exploitation in the North Sea shows progressive development as new techniques are introduced and new equipment becomes available. Steam began to replace sail in the last quarter of the 19th century and the beam trawl was eventually superseded by the otter trawl. Motor followed steam and new improved trawls, such as the more efficient Vigneron-Dahl gear appeared in the 1920's. In the 1930's, various arrangements of kited headlines were introduced on trawls towed fast to catch spawning herring, and after 1946 a wide variety of new trawls, midwater as well as demersal, were developed. The pelagic fisheries tended to retain the older methods for longer, and drift nets were still widely used in the North Sea in the 1950's, but the invention of the power block towards the end of the 1950's allowed purse-seines, which had long been shot and hauled by hand in sheltered waters, to be operated hydraulically, so that these huge nets came into widespread and highly effective use for shoaling fish in the open sea. While the bulk of the landings from the North Sea are now made by trawl or purse-seine, a variety of gears are in use and Table 2 shows the distribution of effort among them in 1976, the most recent year for which a complete breakdown in terms of gear is available from ICES.

Traditionally the bulk of the fish landed from the North Sea was for direct human consumption, and while more than 50 species or species groups are listed in the statistical returns (Table 3), the bulk of the landings were made up of the pelagic herring and mackerel, together with the demersal gadoids (cod, haddock, saithe, and whiting) and flatfish (particularly plaice and sole). However, in recent years the so-called "industrial" species used for the production of meal and oil and animal food have assumed importance. The main species are Norway pout (*Trisopterus esmarkii*), various sandeels (*Ammodytidae*) and sprat, but young age groups of species which support fisheries for human consumption, such as herring, haddock and whiting are also landed, sometimes in considerable quantities, as a by-catch in those fisheries.

As a background to a discussion of North Sea fisheries, the landings of the main species, and the total landings, are set out in Table 4 for the two decades up to 1984. In addition, to illustrate the pattern of change, data are provided at 5-year intervals back to 1950. Before that, in the three decades prior to the second world war, the total fish landings from the North Sea were relatively stable (except for 1914–1918) ranging between 1.0 and 1.5 million tonnes annually. When fishing was resumed after 1945 the landings for the next 15 years were generally higher, between 1.3 and 2.0 million tonnes. They rose rapidly, however, after 1961 and have exceeded 2

**Table 2.** Landings (thousands of tonnes) of all species by individual gears in the North Sea taken by ICES member countries in 1976

| | |
|---|---|
| Purse seine | 362.0 |
| Boat or vessel seines | 128.0 |
| Beach seines | 0.2 |
| Bottom trawls (general) | 194.0 |
| Bottom otter trawls | 333.0 |
| Bottom pair trawls | 5.0 |
| Bottom beam trawls | 89.0 |
| Midwater trawls (general) | 37.0 |
| Midwater otter trawls | 39.0 |
| Midwater pair trawls | 43.0 |
| Otter trawls (not specified) | 239.0 |
| Pair trawls (not specified) | 58.0 |
| Other trawls | 19.0 |
| Traps (not specified) | 75.0 |
| Hand-lines and pole-lines (mechanised) | 0.9 |
| Hooks and lines (not specified) | 3.0 |
| Drift lines (longlines drift) | 0.2 |
| Longlines (not specified) | 0.1 |
| Set gillnets | 0.1 |
| Drift gillnets | 4.0 |
| Gillnets (not specified) | 0.3 |
| Gear not known or specified | 1247.0 |
| Miscellaneous gears | 0.2 |
| Total | 2877 |

million tonnes every year since then (Holden 1978). A maximum of 3.44 million tonnes was reached in 1974, but there have been significant changes in the relative importance of the different species.

## 3 Pelagic Fisheries

In the 25 years after the war, herring dominated the North Sea landings, reaching maxima of 1.4 million tonnes in 1955 and 1.2 million tonnes in 1965. In the early part of this period the traditional drift net was still the main gear used, but it was later joined by the bottom trawl and then by the more efficient pair trawl, a midwater trawl towed by two vessels. Pressure was building up on the pelagic stocks, but the major blow was struck when the Atlanto-Scandian herring further north collapsed and the powerful Norwegian purse-seining fleet was deployed in the North Sea in 1964. Landings immediately rose to a second peak in the following year, but thereafter showed a steady decline. The spawning stock of North Sea herring, estimated at nearly 2 million tonnes in the early 1960's, fell to around 0.13 million tonnes in 1975, and in 1977 a ban on fishing herring was imposed.

Mackerel, the other main pelagic species in the North Sea, suffered a similar fate. It had never been as important as the herring, but its annual landings exceeded 100,000 tonnes for the first time in 1965 and reached a maximum of over 0.9 million

**Table 3.** Landings (thousands of tonnes) in the North Sea of the various species of fish taken by member countries of ICES in 1982

| | | | |
|---|---|---|---|
| Cod | 255.8 | Monk | 3.2 |
| Haddock | 174.5 | Redfishes | 0.6 |
| Hake | 1.5 | Sandeels | 627.4 |
| Ling | 16.7 | Sea breams | + |
| Norway pout | 468.3 | Various demersal percomorphs | 0.9 |
| Pollack | 2.1 | Garfish | + |
| Poutassou (blue whiting) | 51.2 | Horse mackerel | 170.2 |
| Saithe | 150.3 | Various pelagic percomorphs | + |
| Tusk | 6.3 | Herring | 47.7 |
| Whiting | 100.0 | Pilchard | 0.1 |
| Various gadiforms | 2.9 | Sprat | 298.0 |
| Brill | 1.3 | Various clupeoids | 0.2 |
| Dab (common) | 9.6 | Bluefin tuna | + |
| Flounder (European) | 2.7 | Mackerel | 27.3 |
| Halibut | 0.1 | Picked dogfish | 9.9 |
| Greenland halibut | + | Dogfishes and hounds | 0.2 |
| Lemon sole | 7.2 | Skates and rays | 4.2 |
| Megrim | 0.6 | River eel | 1.2 |
| Plaice (European) | 112.9 | Atlantic salmon | 0.2 |
| Sole (common) | 21.5 | Smelt (European) | 0.6 |
| Turbot | 4.5 | Trouts, chars | + |
| Witches | 1.7 | Various diadromous fishes | + |
| Various pleuronectiforms | 3.7 | Shads | + |
| Catfishes | 2.1 | Various non-teleost fishes | + |
| Conger eel | + | Fishes unassorted, unidentified | 276.0 |
| Gurnards | 2.2 | | |

tonnes in 1967 when, with the reduction of the herring, the full effort of the purse seine fleet was transferred to mackerel. At that time the bulk of the Norwegian purse seine mackerel were used for meal and oil (Hamre 1978). Unfortunately, recruitment to the North Sea mackerel stock has been consistently poor since the 1969 year class which recruited to the spawning stock in 1972. This, together with the greatly increased fishing effort, has resulted in the decline of the mackerel stock, which being highly migratory, is in addition fished significantly outside the North Sea itself.

## 4 Demersal Fisheries

In the demersal fisheries the dominant gear was the various modifications and improvements of the otter trawl as first shot on twin warps over the side but later replaced by stern trawling. The use of two vessels (pair trawling) added some advantages, while the Danish seine provided an alternative which was particularly effective on some grounds. Demersal landings in general tended to be less variable than the pelagic. By the 1930's most of the main demersal species in the North Sea were so heavily exploited that some reduction in fishing mortality was desirable and the second world war provided a respite. Up to 1950 demersal landings ranged,

**Table 4.** Nominal catch ('000s metric tons) of various fish species and of invertebrates (Inv) from the North Sea, area IV, by member countries of ICES since 1950

| | Cod | Had | Pla | Sai | Whi | Norway Sand Pout | Eel | Her | Mac | Spr | Total | Inv |
|---|---|---|---|---|---|---|---|---|---|---|---|---|
| 1950 | 67 | 56 | 67 | 20 | 45 | n/a | n/a | 1107 | 32 | 10 | 1530 | n/a |
| 1955 | 83 | 88 | 63 | 41 | 72 | n/a | n/a | 1411 | 53 | 24 | 2018 | 178 |
| 1960 | 104 | 66 | 86 | 29 | 53 | n/a | n/a | 787 | 73 | 16 | 1553 | 176 |
| 1965 | 179 | 222 | 97 | 69 | 107 | 59 | 131 | 1230 | 152 | 76 | 2597 | 227 |
| 1966 | 220 | 269 | 100 | 87 | 155 | 53 | 161 | 1039 | 505 | 107 | 2920 | 217 |
| 1967 | 250 | 167 | 101 | 73 | 91 | 180 | 189 | 819 | 910 | 69 | 3065 | 189 |
| 1968 | 285 | 139 | 109 | 97 | 145 | 469 | 194 | 850 | 809 | 65 | 3361 | 221 |
| 1969 | 199 | 639 | 122 | 106 | 199 | 135 | 113 | 725 | 714 | 65 | 3237 | 209 |
| 1970 | 225 | 672 | 130 | 170 | 182 | 274 | 191 | 749 | 290 | 51 | 3173 | 202 |
| 1971 | 320 | 258 | 114 | 206 | 112 | 359 | 382 | 644 | 228 | 89 | 2959 | 198 |
| 1972 | 346 | 213 | 123 | 199 | 109 | 493 | 358 | 605 | 182 | 92 | 2907 | 217 |
| 1973 | 236 | 196 | 130 | 182 | 143 | 437 | 297 | 599 | 318 | 228 | 2990 | 228 |
| 1974 | 211 | 193 | 113 | 253 | 188 | 823 | 524 | 327 | 292 | 326 | 3440 | 230 |
| 1975 | 186 | 174 | 109 | 250 | 140 | 642 | 428 | 295 | 252 | 652 | 3311 | 230 |
| 1976 | 213 | 205 | 108 | 283 | 191 | 532 | 496 | 163 | 297 | 610 | 3312 | 216 |
| 1977 | 185 | 151 | 107 | 176 | 120 | 434 | 786 | 44 | 252 | 311 | 2722 | 237 |
| 1978 | 261 | 89 | 93 | 129 | 103 | 322 | 787 | 6 | 143 | 402 | 2560 | 241 |
| 1979 | 231 | 87 | 108 | 106 | 141 | 365 | 603 | 6 | 146 | 396 | 2382 | 214 |
| 1980 | 249 | 104 | 101 | 115 | 109 | 523 | 728 | 12 | 79 | 408 | 2576 | 217 |
| 1981 | 287 | 133 | 95 | 120 | 96 | 304 | 588 | 40 | 56 | 314 | 2205 | 219 |
| 1982 | 256 | 174 | 113 | 150 | 100 | 468 | 627 | 48 | 27 | 298 | 2456 | 134 |
| 1983 | 237 | 165 | 103 | 151 | 99 | 499 | 548 | 190 | 32 | 198 | 2419 | 160 |
| 1984 | 197 | 133 | 116 | 179 | 99 | 399 | 693 | 238 | 33 | 133 | 2401 | 187 |

for the most part, between 0.4 and 0.5 million tonnes per year, but after that there was a steady substantial increase to over 2 million tonnes by the 1970's. As discussed below, it should be noted that this demersal bonanza is roughly coincident with the rise of industrial and the decline of the pelagic fisheries. The demersal species for direct human consumption are dominated by the four gadoids, cod, haddock, saithe and whiting, with cod particularly important in the southern part of the area, and haddock in the north. The big rise in demersal landings was evident in all the main gadoid species. It began in the early 1960's and for cod reached a maximum of 0.35 million tonnes in 1972, and of 0.67 million tonnes for haddock in 1970. For saithe, the maximum was 0.28 million tonnes in 1976 and whiting showed very wide fluctuations, reaching 0.20 million tonnes in 1969 and 0.19 million in 1976. Of the various flatfish species caught commercially in the North Sea, the plaice is dominant and it also showed a marked upward trend in landings, with high landings continuing into the 1980,s.

For all these demersal human consumption species it can be said that landings increased to relatively high levels at the end of the 1960's or during the 1970's. A study of the data by ICES working groups shows that these landings were based on significantly increased stock size. During the 1960's cod and whiting stocks in the North Sea roughly doubled and the haddock and saithe stocks increased by a factor of about 4 (Jones 1983). For all these species the increase in stocks and landings was preceded by the appearance of strong year classes which later recruited to the fishery. The dates of these good classes are tabulated on the next page:

| Cod | Haddock | Whiting | Saithe |
|---|---|---|---|
| 1969 | 1962 | 1962 | 1966 |
| 1970 | 1967 | 1967 | 1967 |
| 1976 | 1971 | 1972 | 1968 |
| 1979 | 1973 | 1974 | 1973 |
| | 1974 | | |
| | 1979 | | |

## 5 Industrial Fisheries

As already indicated, fishing in the North Sea has traditionally been for species intended for human consumption. The so-called "industrial" fishery, however, lands fish mainly for reduction to fish meal and fish oil, or in some cases (about 5% of the total) for food for pets or fish or mink farms. The early history of the North Sea industrial fishery from 1950 to 1974 is reviewed by Popp Madsen (1978), who shows that initially it made up less than 2% of the total landings and in the early 1950's was due mainly to German, U.K. and Danish herring catches. However, the percentage of total landings reached 46% in 1965 and 61% in 1974, and this high level of the industrial fishery is now a regular feature. More recently the catches have diversified, and over the period 1970 to 1974, Popp Madsen's list of industrial landings includes ten species and one "other species" category. Of these, herring, sprat, sandeel, mackerel, Norway pout, whiting and haddock all exceed 100,000 tonnes in at least 1 year. Some industrial landings, on occasions very large, are made from gears such as the purse-seine and midwater trawl, and involve herring, mackerel, and sprat, but it is often difficult to categorise such landings accurately, since the catches may become "industrial" only if they turn out to be surplus to human consumption requirements. On the other hand, there are some fisheries that are clearly directed to the industrial category, using small meshed nets and focusing mainly on Norway pout, sandeels, sprats, whiting, immat7re herring and immature blue whiting (Gauld et al. 1986). Landings of this type in recent years are shown in Table 5, and it is clear that the most important target species are Norway pout, sandeels and sprats. The by-catch presents no problem in sandeel fisheries, but the other industrial species usually produce a significant by-catch of more valuable species (e.g. immature herring along with sprats, and immature haddock and whiting with pout). International regulations limit the by-catch of such protected species to 10% of the total amount of fish retained on board. In addition, except for sandeels, the mesh used is restricted to a minimum of 16 mm.

The pout fishery, which peaks usually from August to November, is concentrated in the northern North Sea, and since for the most part it takes fish less than 2 years old, it is highly dependent on good year broods, fluctuating according to the strength of the broods. Of the five species of sandeels found round western Europe, *Ammodytes marinus* dominates the North Sea catches. It is an important part of the food chain, being eaten by many other species of fish and by sea birds. In recent years it has even been identified as a significant component of the diet of seals. The

**Table 5.** Total annual landings (thousand tonnes) in the small-mesh trawl fisheries in the North Sea (ICES Divisions IVa-c) 1974–85. (After Gauld et al. 1986)

| Year | Norway pout | Sandeel | Blue whiting | Sprat[a] | Herring | Bycatch[b] | Total |
|---|---|---|---|---|---|---|---|
| 1974 | 736 | 525 | 62 | 314 | – | 220 | 1,857 |
| 1975 | 560 | 428 | 42 | 641 | – | 128 | 1,799 |
| 1976 | 435 | 488 | 36 | 622 | 12 | 198 | 1,791 |
| 1977 | 390 | 786 | 38 | 304 | 10 | 147 | 1,675 |
| 1978 | 270 | 787 | 100 | 378 | 8 | 68 | 1,611 |
| 1979 | 320 | 578 | 64 | 380 | 15 | 77 | 1,434 |
| 1980 | 471 | 729 | 76 | 323 | 7 | 69 | 1,675 |
| 1981 | 236 | 569 | 62 | 209 | 84 | 85 | 1,245 |
| 1982 | 360 | 620 | 118 | 153 | 153 | 57 | 1,461 |
| 1983 | 423 | 537 | 118 | 91 | 155 | 38 | 1,362 |
| 1984 | 355 | 669 | 79 | 80 | 35 | 34 | 1,252 |
| 1985 | 197 | 621 | 73 | 50 | 63 | 29 | 1,033 |

[a] Includes sprat landings for reduction made by purse-seine and pair-trawl.
[b] Protected (Annex V) species defined in Council Regulation (EEC) No. 171/83.

main landings have been made from the southern and more especially the central North Sea, but in recent years landings from the northern North Sea (north of 59°) have increased in importance. The bulk of the landings have been made by Danish and Norwegian vessels but U.K. effort has been increasing. While sandeels may live for up to 8 years, the fishery depends largely on young fish, in some cases (e.g. off the Shetlands) on fish in their first 2 years of life. The fishery may thus be expected to fluctuate annually and there is a good case for protecting the youngest stages until they have spawned.

The North Sea sprat fishery began to increase in the mid 1960's (Johnson 1985) when catches were mainly taken in winter from the U.K. coast for human consumption and in summer in the eastern half of the central North Sea by Danish and German industrial fishing vessels. In the early 1970's, encouraged by two very good brood years, the landings increased, with peaks in 1975 and 1976. Thereafter stocks in most areas have been declining and the heavy exploitation has changed the age structure so that significant numbers of fish over 3 years old are no longer found in the catches, the bulk being immature fish between 1 and 2 years old. At present the stock remains low, and this must be a cause for serious concern until a new strong year class appears. It is difficult to know when this will occur, since for sprats, as for other species, there is no obvious relationship between spawning stock and recruitment, and there is no clear explanation of what caused the advent of earlier good year classes, although one may speculate on the linkage of those with the decline in the North Sea herring populations in the 1970's, since the two species may compete on common feeding grounds in the early stages of their lives.

**Table 6.** Landings (tonnes) in the North Sea of invertebrates taken by member countries in ICES in 1982

| | |
|---|---|
| Edible crab | 4431 |
| Lobster | 410 |
| Norway lobster | 5974 |
| Pandalid shrimps | 3060 |
| Crangonid shrimps | 32532 |
| Various crustaceans | 717 |
| Whelk | 1888 |
| Periwinkles | 813 |
| Oyster (flat) | 260 |
| Oyster (*Crassostrea*) | 18 |
| Mussel (blue) | 73676 |
| Escallop | 1162+ |
| Queen scallop | 364+ |
| Cockle | 6184 |
| Cuttlefishes | 56 |
| Squids | 1075 |
| Various molluscs | 1448 |

## 6 Shellfish

The main shellfish species of commercial importance in the North Sea (Table 6) are crustaceans (particularly lobsters, Norway lobsters, crabs, and shrimps) and molluscs (particularly squid, oysters, mussels, scallops and cockles). The more mobile species are caught in demersal trawls, while the less active are taken in fixed gears such as creels and traps. Although some of the species (especially squids and prawns) are taken in the open sea, the majority are exploited in coastal waters often by small boats close to the shore. Population dynamics are difficult to determine and the individuals may be distributed in "stocklets", the interactions of which are virtually unknown. For these reasons there is less possibility of achieving effective international regulation of shellfisheries and the management of stocks tends to be done on a national basis using a more ad hoc approach than is adopted for finfish stocks. The high demand and low supply helps to keep prices high.

## 7 Aquaculture

With the levelling off of world catches of wild fish in the 1970's, there has been an increasing interest in the development of aquaculture, an activity with a long history. A wide range of species, both fish and invertebrates, is available for aquaculture, but in the North Sea the focus must be on marine organisms and both climatic conditions and topography limit the possibilities, with a few species of bivalve mollusc and more recently salmon as the main contenders.

Of the cultured molluscs in the North Sea, the mussel (*Mytilus edulis*) is probably the dominant species, with the Netherlands, producing 104,000 tonnes in 1985, as the most important country. Oysters, although generally favourite organisms for culture, are not of great importance in the North Sea, where both temperature and topography are against them. They are mainly grown in trays in tidal waters, with the Netherlands, England and Norway as the main producers.

Turning to fish, while there are efforts to develop the culture of turbot, plaice, sea bream, cod and halibut, the major success in the North Sea has been with salmon. The development of sea cage cultivation of North Atlantic salmon represents the greatest advance in European aquaculture, so that farmed salmon in 1984 outweighed the wild product by a factor of more than 3. The conditions required are clean, sheltered, relatively deep sea water and ideal situations in the North Sea are provided by the fjords of Norway which produced 23,300 tonnes of farmed salmon in 1984.

There are significant environmental, ecological and legal constraints on marine aquaculture, but this is nevertheless an expanding prospect in the North Sea.

## 8 Management

Given the modern highly efficient methods of detecting and catching fish, and the mobility of most species between the zones of jurisdiction of several countries, the need for international action in managing the stocks is widely recognised. Management in the North Sea was first significantly applied by the so-called Overfishing Convention of 1946 (fully ratified in 1953), which covered the NE

Atlantic and confined its provisions to mesh sizes and size limits of fish landed. It suffered from being voluntary and having no legal powers of enforcement, and an apparently more effective measure was the NE Atlantic Fisheries Convention, ratified in 1963. This aimed to control fishing effort by limiting its amount, place or duration, and also by catch limits, but the agreement with its Commission, which was required for enforcement, was never reached. The International Council for the Exploration of the Sea (ICES) was the body which provided scientific advice to the Commission of the NEAFC, assessing the stocks and the consequences of any action. ICES provides a similar service to the EEC, and when UK, Ireland and Denmark acceded to the EEC in 1973 and when a 200-mile exclusive fisheries zone was declared by countries round the North Sea in 1977, the EEC became a powerful balancing force in the area. More recently, when it established a Common Fisheries Policy, the scene was set for adequate control, and management of fish stocks in the North Sea is carried out jointly by the EEC and Norway, through a series of annual negotiations. For most species, particularly demersal fish, a percentage allocation of the annual total allowable catch (TAC) between Norway and the EEC has been agreed (termed zonal attachment), but for some pelagic species (notably herring) there is as yet no long-term agreement on allocation.

## 9 Conclusions

Exploited fish stocks in the North Sea are currently all under pressure. The pelagic species have declined and for herring, which is now recovering, it has been shown by the 4-year closure that management measures can be effective. After the unusually high landings of the 1970's, the demersal stocks are much reduced. Even the industrial landings in 1985 were lower than for many years. Earlier in this chapter reference has been made to the effects of intensive fishing, the importance of good brood years to sustain the catches and the possible effects of food chain effects such as predation. However the causes of the fluctuations are not clear. A recent study (Corten 1986) speculates that the changes in the North Sea herring stocks in the 1970's may be explained by changes in water movements which diverted the young stages from their preferred nursery areas, but this is by no means generally accepted, and perhaps all that can be said is that the changes are almost certainly related to as yet incompletely understood environmental variations. It is possible, too, that fluctuations in recruitment include a cyclical component, and that conditions in the North Sea are now in the process of returning to a pre-1960 situation, but it is too early to be certain about this. One thing is clear: in setting the facts about fisheries in context with the information on water quality presented in later chapters of the volume, it will be important to recognise the massive effects that fishing and natural events can have on the stocks.

# References

Corten A (1986) On the causes of the recruitment failure of herring in the central and northern North Sea in the years 1972–1978. J Cons Int Explor Mer 42:281–294

Gauld JA, McKay DW, Bailey RS (1986) The current state of the industrial fisheries. Fishing Prospects, pp 49–53

Hamre J (1978) The effect of recent changes in the North Sea mackerel fishery on stock and yield. Rapp P V Réun Cons Int Explor Mer 172:197–210

Holden MJ (1978) Long-term changes in landings of fish from the North Sea. Rapp P V Réun Cons Int Explor Mer 172:11–26

I.C.E.S. (1985) Bulletin Statistique des Pèches Maritimes. Vol 67. Cons Int Explor Mer, Copenhagen

Johnson PO (1985) North Sea sprat. Fishing Prospects 19–23

Jones R (1983) The decline in herring and mackerel and the associated increase in other species in the North Sea. In: Sharp GD, Csirke J (eds) Proceedings of the Expert Consultation to Examine Changes in Abundance and Species Composition of Neritic Fish Resources. FAO Fish Rep 291(2):507–519

Popp Madsen K (1978) The industrial fisheries of the North Sea. Rapp P V Réun Cons Int Explor Mer 172:27–30

# Natural Events

J.J. ZIJLSTRA and P. DE WOLF[1]

## 1 Introduction

The impact of pollution and other kinds of human interference on the marine environment may take either of two extreme forms, i.e., mass mortalities or a slow and gradual change in the size of populations. However, such mortalities or changes may also occur under the influence of natural events and the effects may be similar, causing a sudden, often conspicuous catastrophic mortality in part of the marine biota or resulting in a slow change in the abundance of a population. Such gradual changes may either be the result of an alteration in mortality rates, experienced in the subadult and adult part of a population or, more often, be due to changes in the reproductive success, affecting recruitment.

In contrast to mass mortalities, a gradual decline or increase in the size of a population is often difficult to detect, as natural populations are usually all but stable in size. They commonly show large year-to-year fluctuations, as demonstrated in Fig. 1 for a representative of the bottom fauna of the Wadden Sea, a North Sea zooplankter and the annual recruitment of a fish species in the North Sea. These large between-year variations tend to obscure consistent changes and therefore demand observations over an extended period to detect a significant trend – upwards or downwards – in abundance of a population. In only few cases the factors, underlying the between-year variations in natural populations, have been analyzed to some extent. Most often a relationship has been suggested with some simple environmental factors as temperature, salinity or windfield (Fig. 2). It seems doubtful, however, whether such factors are always directly or even mainly responsible for the fluctuations observed in reproductive success or survival, as they may for instance also affect biotic factors as predation rate by influencing the metabolic rate of predators (temperature). Alternatively, fluctuations in factors like temperature and salinity may reflect variations in the circulation pattern of the North Sea or even larger ocean areas which could be of major importance for the survival of plankton or planktonic stages. Therefore, the situation is probably much more complex, as suggested by some simple relations observed with a single environmental factor.

However, it is outside the scope of this contribution to consider the factors responsible for the short-term variations observed in an abundance of North Sea species, although it should be mentioned that a better knowledge on this subject

[1]Netherlands Institute for Sea Research, P.O. Box 59, 1790 AB Den Burg, Texel, The Netherlands

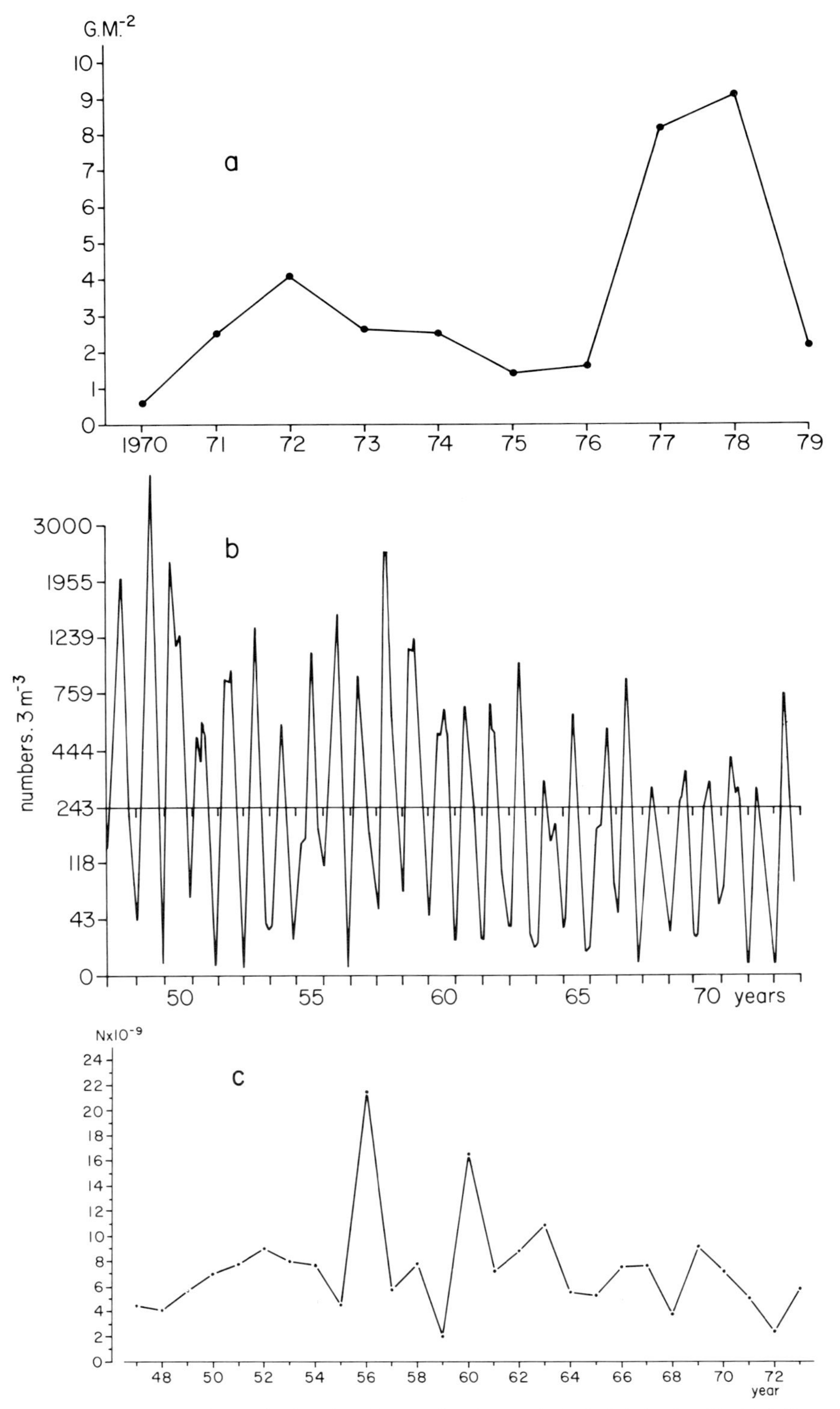


**Fig. 1a-c.** Between-year variation in the abundance of **a** *Cerastoderma edule* in the Wadden Sea (Beukema 1979); **b** *Pseudocalanus* (Colebrook 1978); **c** recruitment of sole (de Veen 1978)

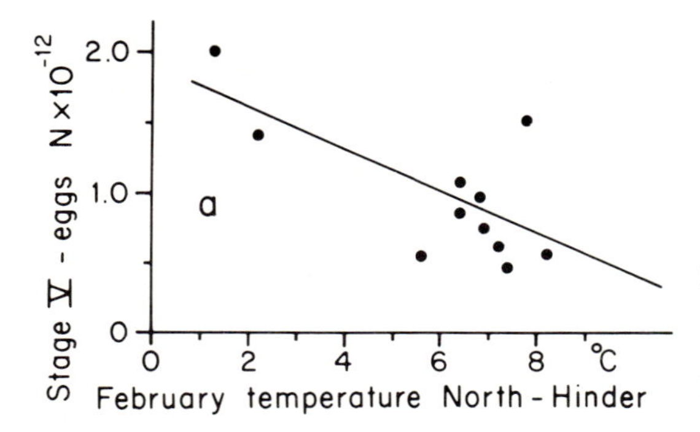


**Fig. 2.** Relation between sea-water temperature in the spawning period of the North Sea plaice (*Pleuronectes platessa*) and subsequent recruitment. (Zijlstra and Witte 1985)

would be of considerable help in analyzing the effects of, e.g., pollution. In this chapter some natural events will be described which have been shown or are suspected to cause occasional (mass)mortalities among North Sea animals. In addition, some long-term changes recorded in parts of the North Sea biota will be mentioned, which probably resulted from natural (climatic) events.

## 2 (Mass)Mortalities

Most reports on increased mortalities, including mass mortalities observed in the North Sea and related to natural events, concern the effect of cold winters. For reasons of perceptibility, (mass)mortalities were in particular recorded in intertidal and coastal areas and among commercial fish species or in species captured by fishing gear.

In coastal and in particular intertidal areas, enhanced mortalities among the benthic fauna were recorded after the cold winters of 1929, 1946–1947, 1962–1963, and 1978–1979, with the effects depending on the severity of the winter. Besides, the various species present were not affected to the same extent. For instance, around the British Isles 26 out of 83 coastal species studied, that is 31%, suffered a heavy mortality in the severe winter of 1962–1963, whereas 37% were completely unaffected (Crisp et al. 1964). In the Wadden Sea, 9 out of 24 more or less numerous species of the macrobenthic fauna of a tidal flat area (37%) were significantly affected by the cold winter of 1978–1979, but the reduction in macrobenthic biomass was only 25% (Beukema 1979, 1985). The largest mortality was encountered on the lower parts of the flats where species occurred with a mainly subtidal distribution. With the exception of the cockle (*Cerastoderma edule*), the typical intertidal fauna was found to be most resistant. As could be expected, mortality was most severe in species with a southern distribution. In particular, the cockle (*C. edule*) showed a high mortality, about 90–95% in the Wadden Sea in 1978–1979 (Beukema 1979; Dörjes 1980) and 40–92% in the Thames estuary in 1963 (Hancock and Urquhart 1964). A heavy mortality of 70–95% was also observed among oysters along the English coast (Waugh 1964) and in the Dutch delta in 1962–1963 (Korringa 1963). In the mussel (*Mytilus edulis*) high mortalities were sometimes encountered on intertidal and shallow banks (0–1 m), but seldom in subtidal areas (Blegvad 1929; Dörjes 1980).

Also in the subtidal offshore area of the shallow southern and southeastern North Sea, where temperatures may decline to or even below 0°C (Fig. 3), heavy mortalities have been recorded in the benthic fauna, in particular in molluscs and echinoderms (Crisp et al. 1964; Ziegelmeier 1964). For instance, mass mortalities of the molluscs *Ensis ensis* and *Mactra coralina* have been observed in the southern and southeastern North Sea (Woodhead 1964a; Ziegelmeier 1964; Dörjes 1980) with a subsequent washing up on the beaches.

The mortality inflicted by the cold winters to the benthic fauna has been attributed to various causes. Some animals die probably as a direct effect of the low temperature, as possibly *Cerastoderma*, whereas others become too torpid to protect themselves from attack, as for instance mussels in the intertidal area. Some animals have been reported to die of asphyxia, because of a lowered ciliary activity which render them incapable of removing silt from the mantle space, as in *Ostrea*. Others again may suffer from scour of ice-floes or die of lack of oxygen, sulfide poisoning or an increased salinity under an ice cover (Blegvad 1929; Crisp et al. 1964; Beukema 1979).

There are various reports on fish mortality in relation to severe winters (Johansen 1929; Lumby and Atkinson 1929; Simpson 1953; Crisp et al. 1964; Woodhead 1964a). Fishermen reported the presence of dead fish in the offshore

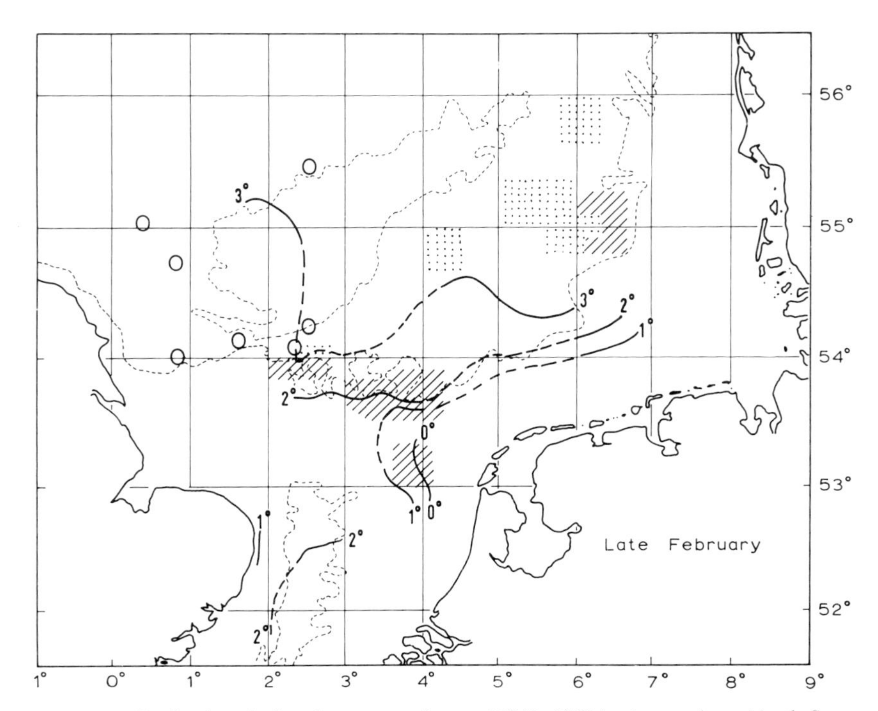


**Fig. 3.** Temperature distribution during the severe winter of 1962–1963 in the southern North Sea. (Woodhead 1964a)

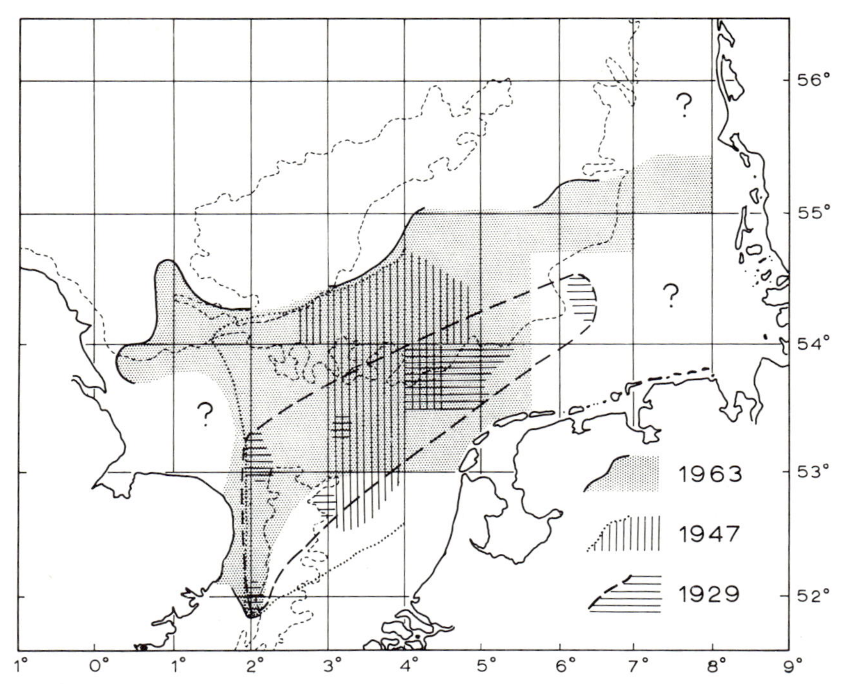


**Fig. 4.** Area where dead fish were reported by fishermen in the winters of 1929, 1947, and 1963. (Woodhead 1964a)

area of the southern North Sea in the winters of 1929, 1947, and 1963 (Fig. 4). Some ten species of fish were found dead and decaying, usually in March or early April, of which sole, cod, whiting, dab and plaice were most consistently present. Most of the area where the dead fish were collected by the fishermen had a temperature between 0°–2°C (Figs. 3, 4). The fish most commonly reported dead was the sole (*Solea solea*), a species reaching its northern limit in the North Sea (Woodhead 1964a).

According to Woodhead (1964a), the sole population of the North Sea suffered heavy losses during severe winters not only because of a direct mortality, but also due to skin diseases. Skin diseases were also reported in some of the other species. Indications were obtained both in sole and in cod, that mortality could be in part related to a breakdown of the osmotic system in the animals at low temperatures (Woodhead 1964a). De Veen (1969) found evidence that sole mortality became effective at temperatures below 3.5°C after a period of 30–40 days. A further reduction of the sole population during severe winters was caused by enhanced catches (Table 1), which are attributed to a concentration of the soles in a smaller area (on the western verge of the low temperature region) in combination with a behavioral change, in which the fish is no longer burying in the sand (Woodhead 1964b). The extent of the mortality occurring in the sole population as a result of cold winters is difficult to estimate. Woodhead (1964a) mentions that in the 1962–1963 winter, in the coldest water, 50–100% of the soles present died. However,

**Table 1.** Trawl landings of soles, total for January -April from southern North Sea by English and Dutch vessels

| Year | Total English landing (tons) | kg per 100 hrs fishing by vessels landing soles | Total Dutch landing (tons) |
|---|---|---|---|
| 1959 | 561.4 | 320.0 | 2,051.7 |
| 1960 | 669.1 | 335.3 | 2,808.7 |
| 1961 | 579.3 | 309.9 | 3,451.3 |
| 1962 | 814.6 | 426.7 | 4,778.9 |
| *Mean* | | | |
| *1956–62* | *656.1* | *350.5* | *3,272.7* |
| 1963 | 2,486.8 | 1,214.1 | 7,350.3 |

many, in particular the larger and older fish, escaped westward, to be decimated there by an increased fishing pressure. From a comparison of the catch-per-effort in 1963 to 1964 (de Veen 1978) one might tentatively deduce that mortality connected to the severe winter of 1962–1963 was as high as about 50%. This estimate includes mortality from both natural death and removal by the more effective fishing. In addition to benthic fauna and fish, also harbor porpoises (*Phocaena phocaena*) have been found dead, possibly as a result of the low temperatures during cold winters (Johansen 1929; Woodhead 1964c). Cold winters in the North Sea are known not only because of their effect on mortality in several species, but also in relation to a high reproductive success in the summer following the winter. For instance, in the Waddensea, a strong recruitment was observed for *Cerastoderma* after the severe winters of 1929, 1947 and 1963; for *Mya* and *Mytilus* (Beukema 1982) after the winter of 1947. In British estuaries an exceptional recruitment was observed by, e.g., Hancock (1973) for *Cerastoderma* after the winter of 1963 and by Savage (1956) for *Mytilus* after the winter of 1940. In the tidal area various benthic species appeared to reproduce successfully after the winter of 1978–1979, which could hardly be attributed to a resettling on open spaces, as the fauna had only been reduced by 25% (Beukema 1982, 1985). Therefore, it has been suggested that predation on spat was reduced after a severe winter (Beukema, pers. commun.). Thus, in the offshore German Bight a mass settlement of a polychaete worm *Sphiophanes bombyx* after the winter 1962–1963 was attributed to the temporary absence of the mollusc *Angulus fabula*, a predator on settling and newly settled larvae (Ziegelmeier 1970). Also in various fish species, reproductive success is above normal after a severe winter, for instance in the sole and possibly the plaice.

There are far fewer reports on catastrophic mortalities due to high summer temperatures, but they seem to occur in coastal, in particular intertidal, areas. There are, for instance, indications that early post-larval plaice, settling on the tidal flats, may occasionally experience a mass mortality due to high temperatures (Berghahn 1983).

A special case of temperature-dependent mortality in the North Sea may occur in late autumn, when species of southern origin, which have extended their area of distribution northward during summer, are caught in their southward migration in the North Sea during autumn. Moving south, they enter in late autumn an area of low temperatures in the shallow southern North Sea, where the temperature is declining much faster than in the northern area. It seems likely that such species, of which *Brama brama* is a well-known example, become torpid and thus strand occasionally on the beaches (e.g., von Brandes 1952; Nijssen and de Groot 1976). However, this stranding never attains large dimensions and is usually, even in invasion years, restricted to some tens of animals.

A completely different form of mass mortality, also occurring mainly in the shallow southern and southeastern North Sea (German Bight) has been described by Rachor and Gerlach (1978). They suggest that in areas with a water depth of less than 30 m or less, part of the bottom fauna would occasionally be destroyed through the effect of heavy winter gales, by which the sediment is eroded away by wave action to be redeposited later in the same or other areas. In this way they explained the decline in both numerical abundance and species number during each winter, in particular after two extremely stormy seasons. That erosion and redeposition of sediment does occur under severe storm conditions in the area has been shown by Hickel (1969) and Gadow and Reineck (1969). Mortality is thought to occur mainly in shallow-buried species and in animals unable to withstand coverage by sediment. There are indications that the benthic fauna in other shallow areas along the unsheltered continental North Sea coasts of Denmark, Germany, The Netherlands, and Belgium will suffers occasionally from the same circumstances.

Catastrophic mortalities have also been reported for the same general area off the continental coast during summer, probably due to a strongly reduced oxygen content of the bottom water. In the summers of 1981 and 1982, during and following prolonged periods of still weather with weak offshore winds, large areas in the German Bight, along the coast of Germany and Denmark (Fig. 5) showed reduced oxygen concentrations in the bottom water in areas, where some stratification occurred. Oxygen values as low as 1–2 mg $l^{-1}$ and 18–20% saturation were measured. These circumstances are held responsible for an enhanced mortality observed in the local benthic fauna, in particular in some echinoderms and molluscs and in demersal fish species as flatfishes (Dethlefsen and Westerhagen 1983; Rachor and Albrecht 1983). The oxygen depletion in the area was, at least in 1982, preceded by a heavy bloom of the dinoflagellate *Ceratium tripos* (e.g., Gerlach 1985). To the west and northwest of the area concerned, where stratification normally occurs in summer, lower oxygen concentrations (appr. 50% saturation) have already been observed at the beginning of the century (Gehrke 1916).

Similar circumstances, with low oxygen concentrations and unusual mortalities in benthic fauna and demersal fish, may have occurred off the northwest coast of Holland, in the summer of 1972, following a strong bloom of the diatom *Coscinodiscus concinnus* (Kat 1972; Gieskes 1973), but the case was less well-documented and concerned a much smaller area. Low oxygen concentrations in summer have also been reported for the Wadden Sea (Broekhuysen 1935; Tijssen and van Bennekom 1976), with sometimes uncontrolled rumours of mass mortalities in the local benthic fauna.

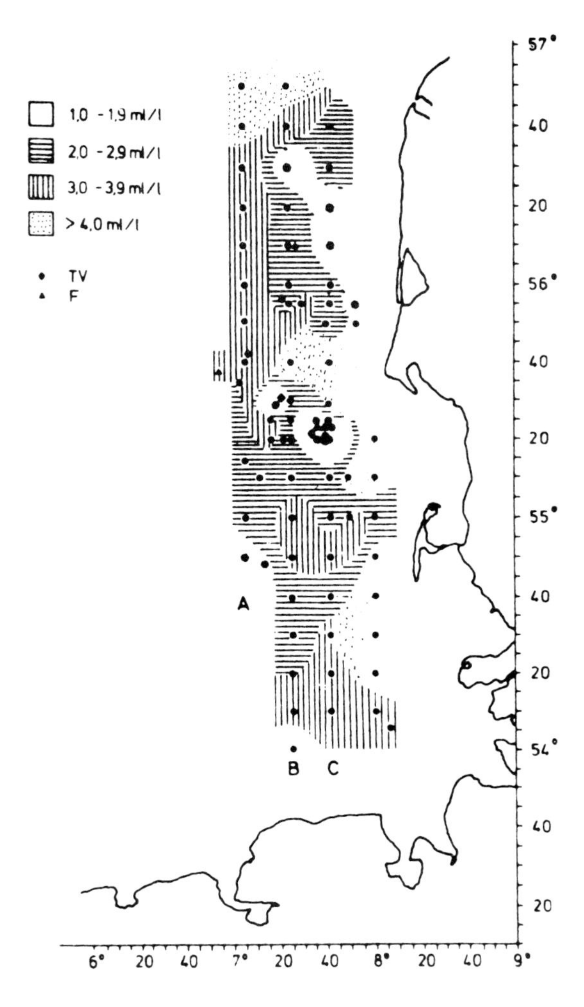


**Fig. 5.** Map showing area with reduced oxygen concentration in the bottom water in 1982. (Dethlefsen and Westerhagen 1983)

Whether these situations of locally occurring mass mortalities in benthic fauna and fish by lack of oxygen observed along the shallow coasts of Germany, Denmark, and the Netherlands can be attributed completely to natural events remains to be seen. There are indications that eutrophication of the coastal part of the continental shelf of the North Sea, in particular along the continental coast, might be in part responsible, in combination with unusual weather conditions (Gerlach 1985).

Finally, mass mortalities are a well-known phenomenon among coastal and sea birds. Part of such mortalities, in particular in coastal birds, is probably connected with cold winters, but usually the stranding of large numbers of birds are no doubt related to oil pollution. However, studies on alcid sea-birds such as auks, razorbills, and guillemots in captivity suggested that occasional mass mortalities among such pelagic birds may occur as a result of a phytoplankton bloom in combination with still weather. Under such circumstances, exudates or substances of decaying cells (lipids) rise to the surface, reducing the surface tension to values below 60 dyne $cm^{-1}$. The causes of death in the birds would be similar to that in oil pollution, being related primarily to a reduction of the water-repellent properties of the coat of feathers (Swennen 1977).

## 3 Long-Term Changes

Natural events such as climatic changes or large-scale shifts in the ocean circulation pattern are likely to affect the abundance of marine populations and to change the structure of the marine communities. Because of the high year-to-year variability in marine populations and communities (Fig. 1), such trends can only be recognized in data-sets, collected over a large number of years with a sufficient constancy in methods. Such data sets are rare in any part of the world ocean and also in the North Sea, but the plankton data collected by the Continuous Plankton Recorder Surveys (C.P.R.S.) in the period 1948 to the present offer the best example (Glover 1967). In addition, there are long series of observations on commercial fish stocks (1909 to the present) (e.g., Holden 1978), on the benthic fauna of the western Wadden Sea (1969 to the present) (Beukema, 1986) and on the phytoplankton in the German Bight (1962–1984) (Hagmeier 1978; Berg and Radach 1985). The commercial fish stocks are considered on pages 538–550, and are certainly not only affected by natural events, whereas both the western Wadden Sea and the German Bight could be and probably are influenced by eutrophication as well as pollution. Nevertheless, some attention will be given to these time series.

The C.P.R.S. data are collected by a plankton recorder (Glover 1967) towed at a standard depth of 10 m behind merchant ships and Ocean Weather Ships along a number of more or less fixed routes at, as far as possible, monthly intervals. The plankton recorder contains a continuously moving band of silk gauze, mesh size 270 $\mu$m, which samples some 3 $m^3$ of water per 10 nautical miles traveled. Because of the mesh size, the instrument is primarily aimed at collecting data on distribution and abundance of zooplankton, recording their spatial and temporal variation (Glover et al. 1974).

However, some phytoplankton cells are retained by the gear, in particular the larger or chain-forming diatoms and larger dinoflagellates, which can be determined up to species level (Colebrook 1960). In addition, different degrees of green color ("phytoplankton color") can be recognized on the silk, which have been classified into color categories (Robinson 1970). From an analysis of data on phytoplankton, collected over various parts of the North Sea in the period 1958–1973, Reid (1978) concluded that major changes had occurred (Fig. 6). In all areas the abundance of diatoms, which up to 1965 showed a spring and an autumn bloom, had drastically declined, in particular during the autumn bloom, which was practically absent in the later years. By contrast, phytoplankton color had increased, with a longer season in most areas. The dinoflagellates *Ceratium* sp. showed no consistent trend, remaining constant or declining slightly. Reid (1977,

1978) concludes that changes in phytoplankton color are not related to changes in any known component of the phytoplankton and suggests them to be associated to microflagellates. These would disintegrate on the silk, but their chloroplasts would survive to add to the coloration. He interprets the changes observed in the sense that an unidentified component of the phytoplankton was increasing in the period when diatoms were declining.

Similar large-scale changes have been recorded in the zooplankton of the North Sea. During the period of observation (1948–1982), many of the zooplankton species showed an almost linear downward trend in abundance in most areas of the North Sea, as shown for *Calanus* in the western North Sea in the period 1948–1973 (Colebrook 1978; Fig. 7). The decline was not restricted to only some of the species, but affected the total zooplankton biomass as well as the length of the zooplankton season (Fig. 8), in which the biomass appears to have declined to a level of approximately a third of its value of 30 years ago (Reid 1984).

The changes in both phytoplankton and zooplankton seem to be part of a pattern affecting large areas of the north Atlantic, where similar changes have been reported (e.g., Reid 1977; Colebrook 1978; Radach 1982; Cushing 1982), as shown in Fig. 9 for zooplankton. Because of the extent of the area involved, it would seem unlikely that the changes are related to local factors such as eutrophication or pollution, which could be effective in an area as the southern North Sea with a large river outflow (e.g., Postma 1973; Gerlach 1985). Also other local factors, such as the fisheries affecting the abundance and composition of fish stocks in the North Sea, as suggested as a possible cause by Steele and Frost (1977), are unlikely to have caused the changes observed. It would seem more plausible that the trends are related to environmental changes operating on large time and space scales such as, for instance, the general climate deterioration in the postwar period (Wallén 1984). In fact, relations between plankton abundance and climate- and ocean circulation-related environmental factors have been observed (e.g., Colebrook 1978; Taylor and Stephens 1980). However, although there is little doubt that the changes recorded in the plankton are "related to climatic changes on an ocean-wide scale" (Colebrook 1978), the processes by which the climate interacts with the plankton are not fully understood (Cushing 1982; Radach 1982).

A second series of long-term observations concerns the areas along the north coast of Germany and the Netherlands. Around Helgoland (German Bight), Hagmeier (1978) reported a general increase of phytoplankton biomass, which concerned both diatoms and dinoflagellates, in the period 1962–1974, which was confirmed by Gillbricht (1983) for a partly later period (1971–1981). Berg and Radach (1985) reanalyzed the data set for the years 1962–1984 and found a strong upward trend in phytoplankton-carbon, together with similar trends in phosphate and nitrate. Most of the four fold phytoplankton-C increase between 1962–1984 appears to be made up by flagellates, whereas diatoms tended to decline or remain constant.

In the western part of the Wadden Sea, Cadée (1984) observed during spring an increase in chlorophyll level in the water in the period 1974–1983. In the same area the chlorophyll level in the top layer of the sediment at a tidal flat and the rate of primary production of microphytobenthos increased during the period

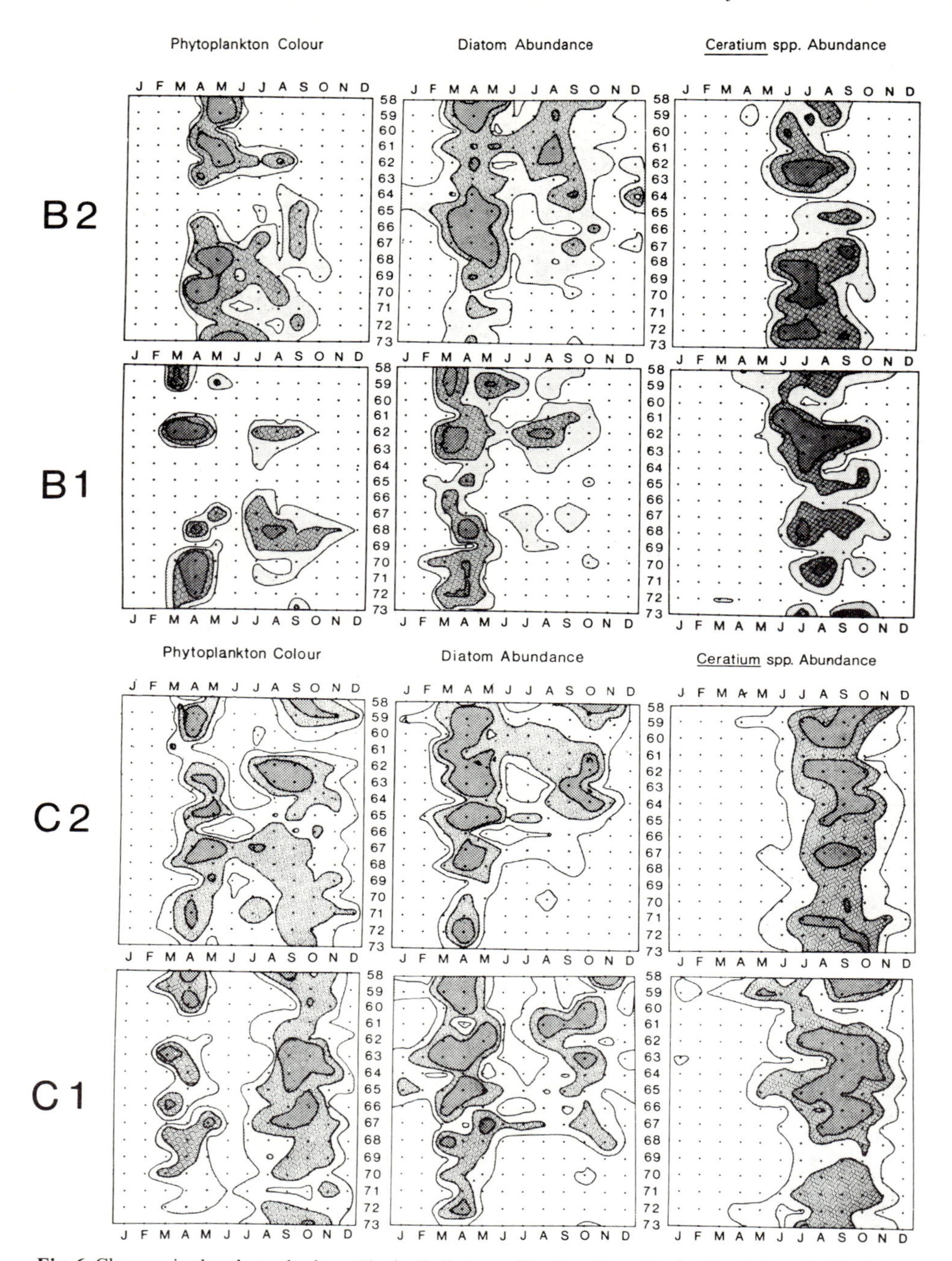


**Fig. 6.** Changes in the phytoplankton ("color", diatoms, dinoflagellates) in the North Sea. (Reid 1978)

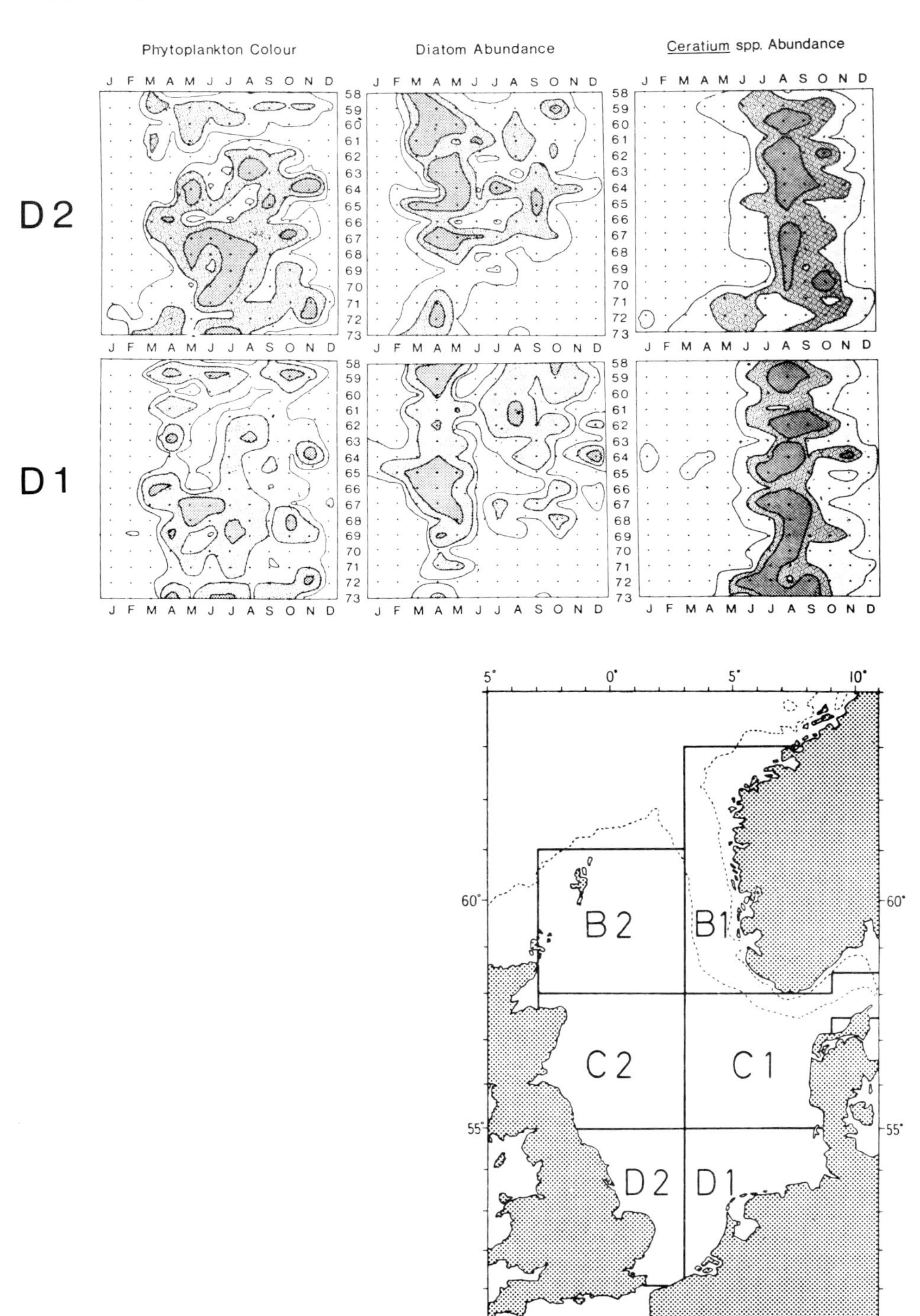
Phytoplankton Colour
Diatom Abundance
Ceratium spp. Abundance
J F M A M J J A S O N D
58 59 60 61 62 63 64 65 66 67 68 69 70 71 72 73
D2
D1
5° 0° 5° 10°
60°
55°
50°
B2
B1
C2
C1
D2
D1

**Fig. 6, part 2**

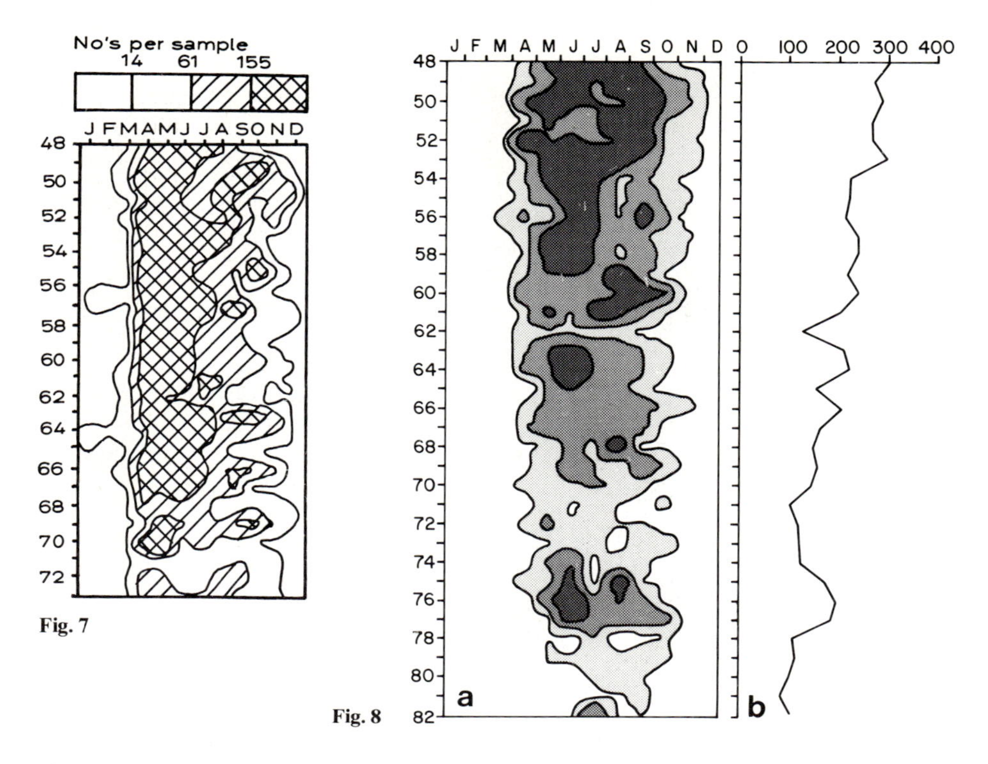


**Fig. 7.** Abundance of *Calanus* sp. in the western part of the North Sea in the years 1948–1973. (Colebrook 1978)

**Fig. 8a,b.** Zooplankton biomass in the North Sea, showing the distribution in the course of each season between 1948–1982 **(a)** and the average level per year **(b)**. (Reid 1984)

1968–1981 (Cadée 1984). In about the same period (1970–1984), both biomass and annual production of the macrozoobenthos living in the tidal flats of the western Wadden Sea are reported to have doubled (Beukema and Cadée 1986). The increase was strongest in short-living species and virtually absent in the long-living *Mya arenaria* (Fig. 10).

The trends observed in phytoplankton, primary production, and benthic fauna along the coasts of Germany and the Netherlands are thought to be related to eutrophication, due to an increased nutrient outflow from the rivers in the area. If that tentative conclusion is correct, the observed changes are man-made and not dependent on natural events. Moreover, they would be complementary to the other possible effect of eutrophication mentioned earlier, resulting in the occasional occurrence of low-oxygen areas.

In this section we have shown that natural events, such as cold winters, warm summers, strong gales and long-term changes in climate or ocean circulation, may influence the biota of the North Sea. In some cases, the effect of such natural events was obvious and conspicuous, being demonstrated by mass mortalities among

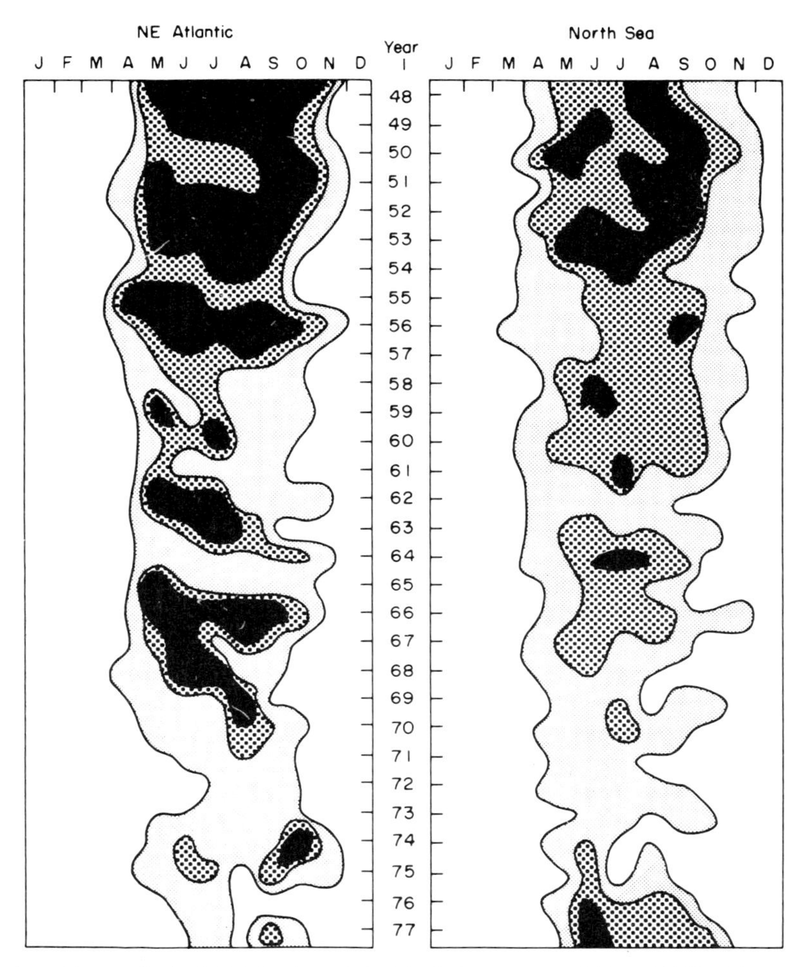


**Fig. 9.** Zooplankton biomass in the NE Atlantic and the North Sea between 1948 and 1979. (Glover 1979)

certain species, but in other instances gradual changes occurred, which were masked by strong between-year fluctuations. To evaluate properly the effect of man on an ecosystem such as the North Sea one should be aware of the existence of the short- and long-term events described here.

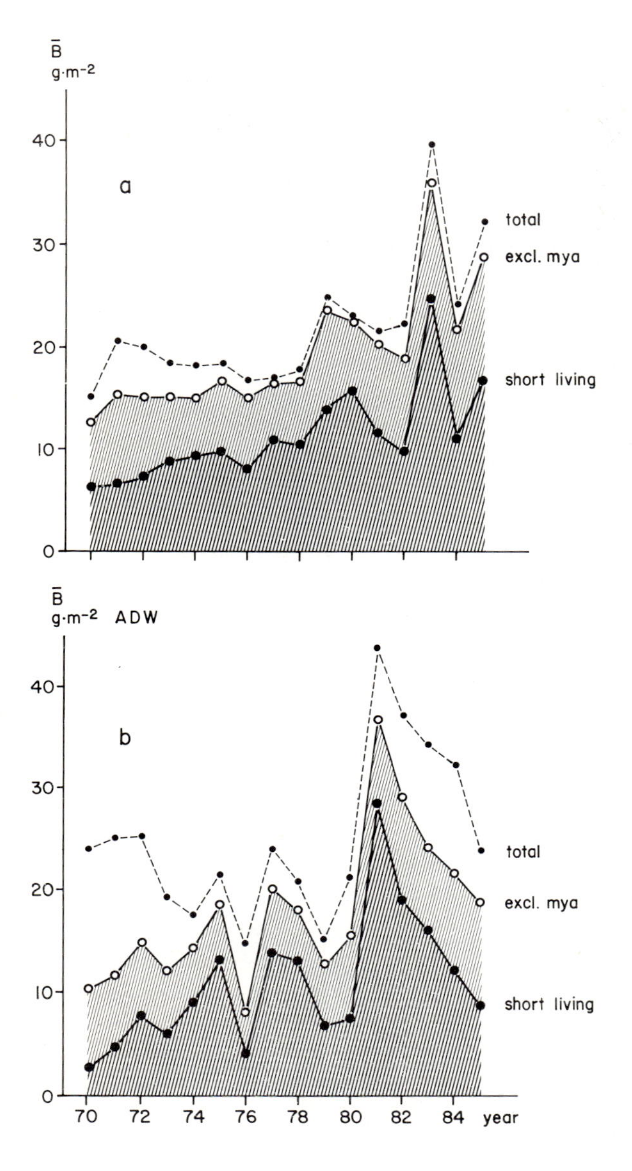


**Fig. 10 a,b.** Changes during a 16-year period in the macrozoobenthic biomass in an intertidal area in the western part of the Wadden Sea. **a** Annual means of three sampling stations. **b** means of spring values at 12 transects. (Beukema and Cadée, 1986)

## References


Berg J, Radach G (1985) Trends in nutrients and phytoplankton concentrations at Helgoland Reede (German Bight) since 1962. ICES, CM 1985, L2 (mimeo)

Berghahn R (1983) Untersuchungen an Plattfischen und Norseegarnelen (*Crangon crangon*) im Eulitoral des Wattenmeeres nach dem Ubergang zum Bodemleben. Helgol Wiss Meeresunters 36:136–181

Beukema JJ (1979) Biomass and species richness of the macrobenthic animals living in a tidal flat area in the Dutch Wadden Sea: Effects of a severe winter. Neth J Sea Res 13:203–223

Beukema JJ (1982) Annual variation in reproductive success and biomass of the major macrozoobenthic species living in a tidal flat area of the Wadden Sea. Neth J Sea Res 16:37–45

Beukema JJ (1985) Zoobenthos survival during severe winters on high and low tidal flats in the Dutch Wadden Sea. In: Gray JS, Christiansen ME (eds) Marine biology of polar regions and effects of stress on marine organisms. Wiley, New York, pp 351–361
Beukema JJ, Cadée GC (1986) Zoobenthos response to eutrophication of the Dutch Wadden Sea. Ophelia, 55–64 pp
Blegvad H (1929) Mortality among animals of the littoral region in ice-winters. Rep Dan Biol Stn 35:49–62
Brandes CH von (1952) Uber das Auftreten der Brachsenmakrele, *Brama rayi* Bl., in den nordeuropäischen Gewässern . Veröffent Inst Meeresforsch Bremerhaven 1:37–46
Broekhuysen GJ (1935) The extremes in percentages of dissolved oxygen to which the fauna of a *Zostera* field in the tide zone at Nieuwediep can be exposed. Arch Néerl Zool 1:339–346
Cadée GC (1984) Has input of organic matter into the western part of the Dutch Wadden Sea increased during the last decades? Neth Inst Sea Res Publ Ser 10:71–82
Colebrook JM (1960) Continuous plankton records: Methods of analysis 1950–1959. Bull Mar Ecol 5:51–64
Colebrook JM (1978) Changes in the zooplankton of the North Sea, 1948 to 1973. Rapp P V Réun Cons Int Explor Mer 172:390–396
Crisp DJ, Moyse J, Nelson-Smith A (1964) The effects of the severe winter of 1962–63 on marine life in Britain. J Anim Ecol 33:165–210
Cushing DH (1982) Climate and fisheries. Academic Press, London, pp 1–373
Dethlefsen V, Westerhagen H von (1983) Oxygen deficiency and effects on bottom fauna in the eastern German Bight 1982. Meeresforschung 30:42–53
Dörjes J (1980) Auswirkung des kalten Winters 1978/1979 auf das marine Makrobenthos. Natur Museum 110:109–115
Gadow S, Reineck HE (1969) Ablandiger Sandtransport bei Sturmfluten. Senckenb Marit (1) 50:63–78
Gehrke J (1916) Uber die Sauerstoffverhältnisse in der Nordsee. Ann Hydrogr Marit Meteorol 44:177–193
Gerlach SA (1985) Wurde der 1981 in der Deutschen Bucht beobachten Sauerstoffmangel durch antropogene Nährstoff-Frachten begünstigt? Wasser Berlin 1985, AML Berlin, Wissenschaftsverlag von Spiess, pp 430–451
Gieskes WWC (1973) De massale planktonsterfte op de kust bij Callantsoog juli 1972. Waddenbull 1:36–37
Gillbricht M (1983) Eine "red-tide" in der südlichen Nordsee und Beziehungen zur Umwelt. Helgol Wiss Meeresunters 36:393–426
Glover RS (1967) The continuous plankton recorder survey of the North Atlantic. Symp Zool Soc Lond 19:189–210
Glover RS (1979) Natural fluctuations of populations. Ecotoxicol Environ Saf 3:190–203
Glover RS, Robinson GA, Colebrook JM (1974) Marine biological surveillance. Environ Change 2:395–402
Hagmeier E (1978) Variations in phytoplankton near Helgoland. Rapp P V Réun Cons Int Explor Mer 172:361–363
Hancock DA (1973) The relationship between stock and recruitment in exploited invertebrates. Rapp P V Réun Cons perm Int Explor Mer 164:113–131
Hancock DA, Urquhart AE (1964) Mortality of edible cockles (*Cardium edule* L.) during the severe winter of 1962–63. J Anim Ecol 33:176–178
Hickel W (1969) Sedimentsbeschaffenheit und Bakteriengehalt im Sediment eines künftigen Verklappungsgebied von Industrieabwässern nordwestlich Helgolands. Helgol Wiss Meeresunters 19:1–20
Holden MJ (1978) Long-term changes in the landings of fish from the North Sea. Rapp P V Réun Cons Int Explor Mer 172:11–26
Johansen AC (1929) Mortality among Porpoises, Fish and larger Crustaceans in the waters around Denmark in severe winters. Rep Dan Biol Stn 35:63–97
Kat M (1972) Stank aan de Noordhollandse kust, natuurlijk of onnatuurlijk. Visserij 25:545–551
Korringa P (1963) Winter inflicts a deadly blow to the Dutch oyster industry. Int Counc Explor Sea, CM, Shellfish Comm, Document 81, p 3
Lumby JR, Atkinson GT (1929) On the unusual mortality among fish during March and April 1929, in the North Sea. J Cons Int Explor 4:309–332

Nijssen H, Groot SJ de (1976) The occurrence of *Brama brama* (Bonnaterre 1788) along the coast of the Netherlands in 1974 and 1975 (Pisces, Perciformes, Bramidae). Bull Zool Mus Univ Amst 5:131–137
Postma H (1973) Transport and budget of organic matter in the North Sea. North Sea Science. MIT, Cambridge, pp 326–333
Radach G (1982) Variations in the plankton in relation to climate. ICES, CM 1985, Gen 5
Rachor E, Albrecht H (1983) Sauerstoffmangel im Bodenwasser der Deutschen Bucht. Veroff Inst Meeresforsch Bremerhaven 19:209–227
Rachor E, Gerlach SA (1978) Changes of macrobenthos in a sublittoral sand area of the German Bight, 1967 to 1975. Rapp P V Réun Cons Int Explor Mer 172:418–431
Reid RC (1977) Continuous plankton records: Changes in the composition and abundance of the phytoplankton of the North-Eastern Atlantic Ocean and the North Sea, 1958–1974. Mar Biol 40:337–339
Reid RC (1978) Continuous plankton records: large-scale changes in the abundance of phytoplankton in the North Sea from 1958 to 1973. Rapp P V Réun Cons Int Explor Mer 172:384–389
Reid RC (1984) Year-to-year changes in zooplankton biomass, fish yield and fish stock in the North Sea. ICES, CM 1984, L39:1–7
Robinson GA (1970) Continuous plankton records: variations in the seasonal cycles of phytoplankton in the North Atlantic. Bull Mar Ecol 6:333–345
Savage RE (1956) The great spat fall of mussels (*Mytilis edulis* L.) in the river Conway estuary in spring 1940. Fish Invest Lond 20:1–22
Simpson AC (1953) Some observations on the mortality of fish and the distribution of plankton in the southern North Sea during the cold winter, 1946–1947. J Cons Int Explor 19:150–177
Steele J, Frost BW (1977) The structure of plankton communities. Philos Trans R Soc Lond B Biol Sci 280:485–534
Swennen C (1977) Laboratory research on sea-birds. Publ Neth Inst Sea Res, Texel, pp 44
Taylor DH, Stephens JA (1980) Latitudinal displacements of the Gulf Stream (1966 to 1977) and their relation to changes in temperature and zooplankton abundance in the NE Atlantic. Oceanol Acta 3:145–149
Tijssen SB, Bennekom AJ van (1976) Lage zuurstofgehaltes in het water op het Balgzand. $H_2$ O9:28–31
Veen JF de (1969) Vlekzieke en dode tong in de Duitse Bocht in het voorjaar van 1969. Visserij 22:482–487
Veen JF de (1978) Changes in North Sea sole stocks (*Solea solea* L.) Rapp P V Réun Cons Int Explor Mer 172:124–136
Wállen CC (1984) Present century climate fluctuations in the northern hemisphere and examples of their impact. World Climate Programme, Publ WPC-87
Waugh GD (1964) The effects of the severe winter of 1962–1963 on oysters and the associated fauna of oyster grounds of southern England. J Anim Ecol 33:173–175
Woodhead PMJ (1964a) The death of North Sea fish during the winter of 1962/63, particularly with reference to the sole, *Solea vulgaris*. Helgol Wiss Meeresunters 10:283–300
Woodhead PMJ (1964b) Changes in the behaviour of the sole, *Solea vulgaris*, during cold winters, and the relation between the winter catch and sea temperatures. Helgol Wiss Meeresunters 10:328–342
Woodhead PMJ (1964c) The death of fish and sublittoral fauna in the North Sea and the English Channel during the winter of 1963. J Anim Ecol 33:169–173
Ziegelmeier E (1964) Einwirkung des kalten Winters 1962/63 auf das Makrobenthos im Ostteil der Deutschen Bucht. Helgol Wiss Meeresunters 10:276–282
Ziegelmeier E (1970) Uber Massenvorkomen verschiedener makrobenthaler Wirbelloser während der Wiederbesiedlungsphase nach Schädigungen durch "katastrophale" Umwelteinflüsse. Helgol Wiss Meeresunters 21:9–20
Zijlstra JJ, Witte JIJ (1985) On the recruitment of O-group plaice in the North Sea. Neth J Zool 35:360–376

# Part II
# Input and Behavior of Pollutants

# The Scheldt Estuary

R. WOLLAST[1]

## 1 Hydrodynamical Characteristics

The Scheldt estuary constitutes the southern branch of the "Golden Delta" formed by the rivers Rhine, Meuse, and Scheldt. The Scheldt river (Fig. 1) and its tributaries drain 21580 square kilometers in the northwest France, west Belgium, and southwest Netherlands. The hydrographic basin covers one of the most heavily populated regions of Europe, where a highly diversified industrial activity has developed. Most of the discharges are uncontrolled and as a consequence large amounts of domestic and industrial wastes are carried by the river.

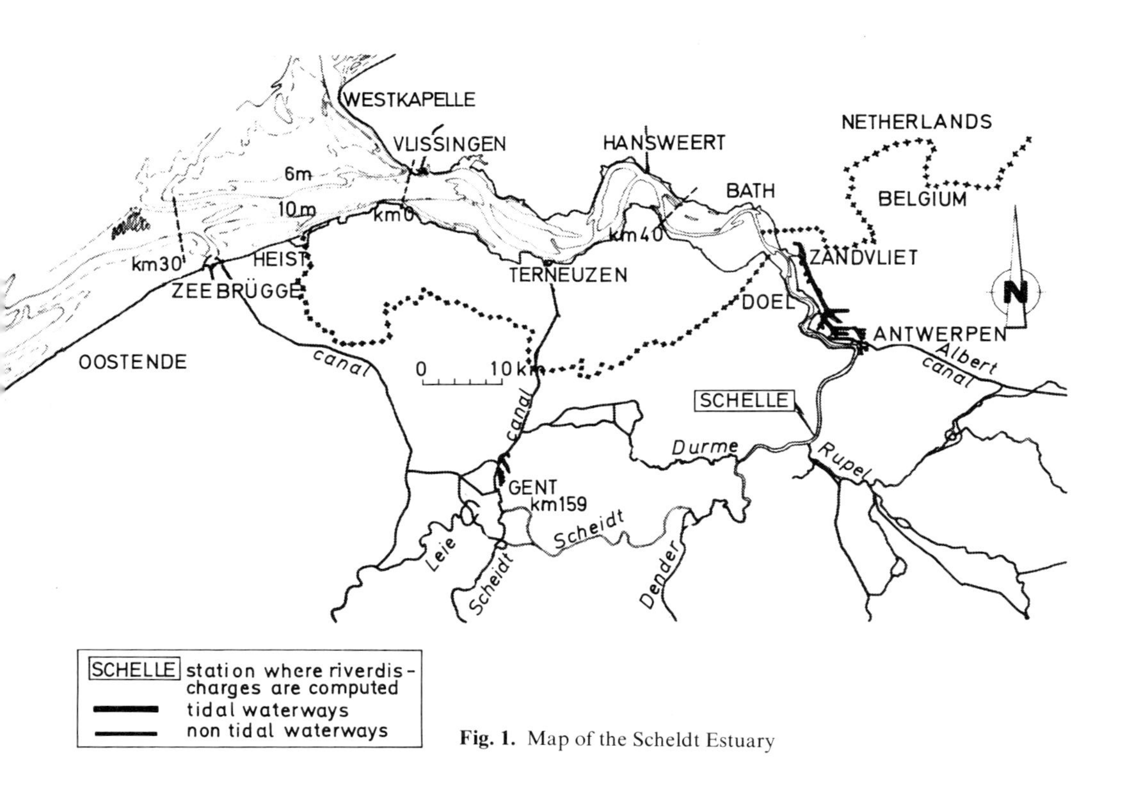


**Fig. 1.** Map of the Scheldt Estuary

[1]Oceanography Laboratory, University of Brussels, Campus de la Plaine, C.P. 208, 1050 Brussels, Belgium

The mean river discharge amounts to 120 $m^3 s^{-1}$ at the mouth, or 5 million $m^3$ during one tidal period, while the volume of the seawater flowing up the estuary during the flood tide is about 1 billion $m^3$. The Scheldt may be considered as a well-mixed estuary with only a small and local vertical salinity gradient. The mixing zone of fresh and salt water extends over a distance of 70 km to 100 km.

One of the most important characteristics for the transport of pollutants in an estuarine system is the residence time of the freshwater masses in the mixing zone. In the case of the Scheldt estuary, the residual currents averaged over one complete tidal cycle in a cross-section drop from 0.08 m $s^{-1}$ at km 100 to 0.02 m $s^{-1}$ at km 50. The total residence time in the brackish water zone, which extends over 100 km, comprised between 1 and 3 months. From an environmental point of view, this implies a high accumulation of the persistent pollutants and intense modifications of the chemically or biologically active substances in the estuarine region. The composition of the water at the mouth of the estuary reveals that important physical, chemical, and biological processes occurring in the mixing zone modify strikingly the transport of pollutants to the sea.

## 2 Transport and Accumulation of Sediments

Mixing of fresh and salt water induces complicated water movements and influences the physicochemical behavior of both suspended and dissolved species. The measurements of vertical profiles of salinity, temperature, and currents permit one to distinguish two zones with different hydrodynamical characteristics. The lower one, extending from the sea to km 50, is constituted by well-defined flood and ebb channels which contribute to the intense mixing. The upper zone extending from km 50 to the freshwater zone (km 100) is characterized by a single and narrower channel where vertical salinity gradients of about 0.2‰ S $m^{-1}$ are observed. Despite their relatively low values, these salinity gradients markedly influence the vertical distribution of the currents (Fig. 2). The residual currents near the bottom averaged over one complete tidal cycle are orientated upstream in the lower zone. They are, however, orientated downstream in the freshwater zone, and the two opposed movements cancel out in the area of the harbor of Antwerp, which is thus a highly favorable zone for the accumulation of sediments. This region corresponds roughly also to the transition from fresh to brackish water. The suspended matter transported by the river water is mainly composed of colloidal particles which flocculate as the salinity increases. Laboratory experiments carried out with suspended matter of the river Scheldt show that an intense flocculation occurs as soon as the chlorinity reaches 1‰ Cl and is completed for a chlorinity of 2.5‰ Cl. The optimum values of salinity for flocculation occur in the zone propitious to sedimentation and accumulation, leading to intense shoaling of mud in a restricted area.

As in many other river systems, anthropogenic activities have considerably increased the particulate load transported by the Scheldt. Although this parameter is not often discussed in terms of pollutant, it deserves some attention in the case of the Scheldt.

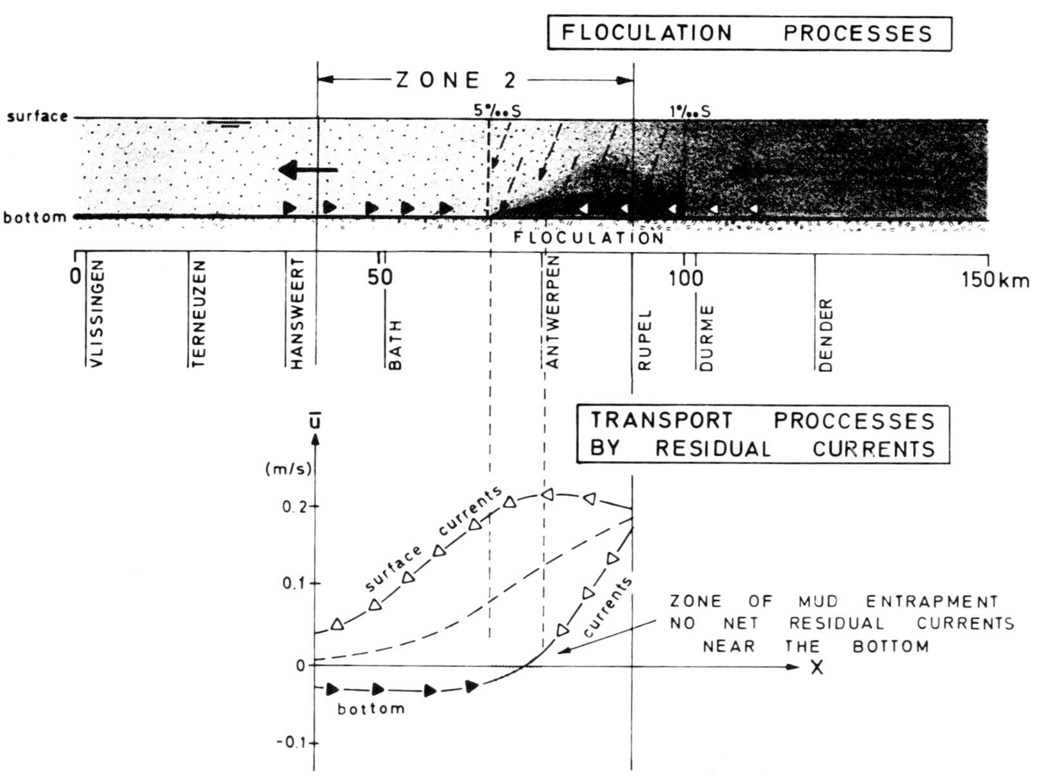


**Fig. 2.** Influence of density currents on the residual transport processes of particulates and accumulation of mud in the Scheldt. (After Wollast and Peters 1983)

Suspended matter by itself may be considered as a pollutant, since it reduces the light penetration and thus the productivity. Enhanced deposition of suspended material related to an increase of particulate load has deleterious effects on the benthic organisms. Intense shoaling in a harbor such as Antwerp requires continuous removal of the sediments by costly dredgings. Finally, these deposited sediments maintain in the estuary a large fraction of inorganic or persistent organic pollutants introduced into the river or estuarine system.

We have tried to estimate separately various sources of particulate material in the Scheldt estuary by considering domestic and industrial activities on one hand and the input of solids due to erosion in the various hydrographic sub-basins of the river system on the other.

These estimations are rather rough, but give some insight into the influence of man's activities on the flux of particulate matter in the Scheldt (Table 1).

Supposing that erosion has not been significantly affected by agricultural activities in this typically coastal plain region, Table 1 shows nevertheless that at least two thirds of the suspended load is directly related to human activities. The suspended matter due to domestic and agricultural activities introduced into the river is characterized by a high organic matter content (40–60% by weight), whereas

**Table 1.** Sources of particulate load to the estuary of the Scheldt ($10^3$ t $a^{-1}$)

| Source | Load | % of total |
|---|---|---|
| Domestic | 190 | 25 |
| Industrial | 290 | 39 |
| Land erosion | 270 | 36 |
| Total | 750 | 100 |

the industrial suspended load is often strongly contaminated by heavy metals and contains large amounts of unusual minerals for the Scheldt, like gypsum, slag dust, and metallic iron spherules.

We estimate that during low river discharge two thirds of the suspended load carried by freshwater is deposited in a restricted zone, 30 km long (km 55–85). The accumulation areas of this highly polluted solid material can be easily identified by investigating the distribution of organic matter and of some typical trace metals like Cd, Cu, Pb, and Zn in the sediments of the estuary. These results of a recent survey are given in Figs. 3 and 4.

The deposited sediments are, however, not necessarily permanently removed from the water column and are resuspended during short periods of the flood events. Transport of particulate matter to the sea during these short periods may be considerably more important than that during extended periods of normal river discharge.

Furthermore, the freshly deposited sediments of the Scheldt estuary are intensively dredged and thus are largely removed from the river system.

Nevertheless, it is obvious from the record of the evolution of the morphology of the Scheldt estuary and of the adjacent coastal zone that over the last 2000 years a slow but continuous accumulation of sediments has occurred in the estuary itself or in the adjacent coastal zone (Van Veen 1950). Significant amounts of suspended matter and of pollutants are thus definitely entrapped in these zones.

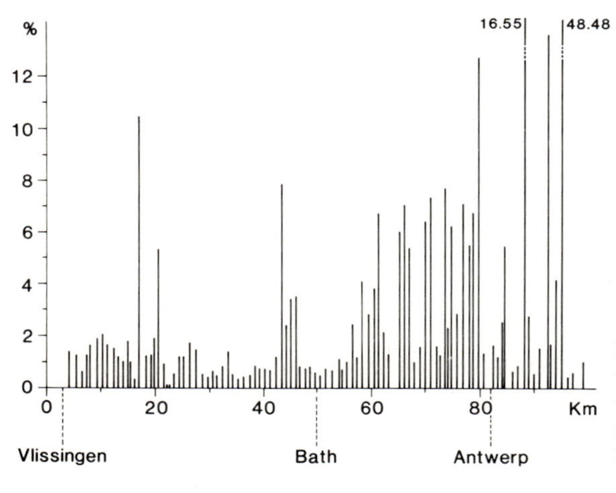


**Fig. 3.** Longitudinal distribution of the organic matter content (% by weight) in sediments in the estuarine zone. The mouth of the estuary is at 0 km. (After Wollast et al. 1985)

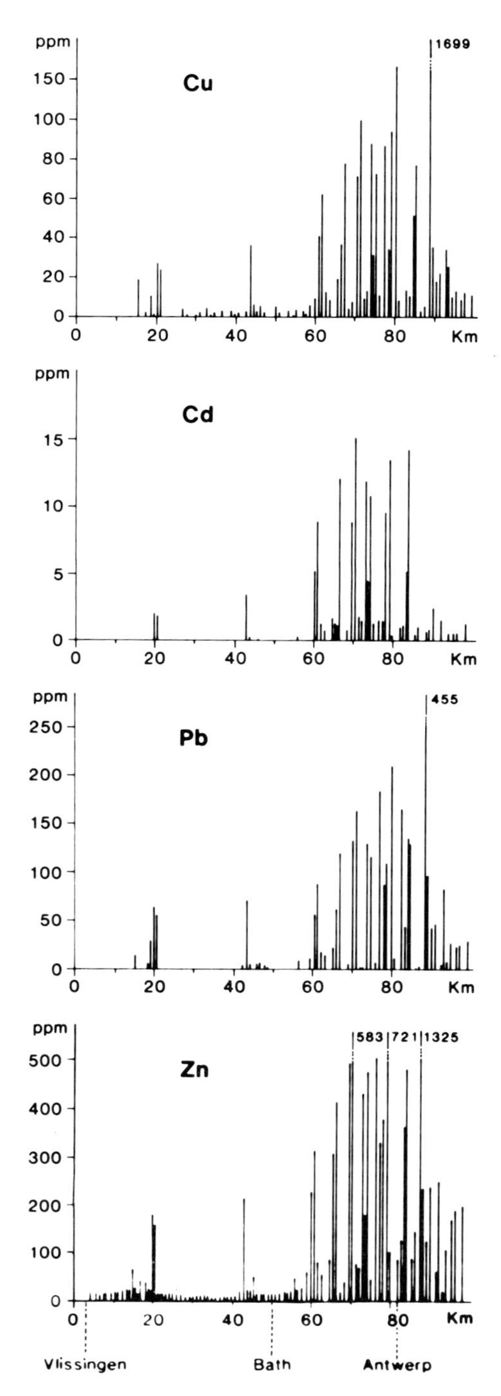


**Fig. 4.** Longitudinal distribution of Cu, Cd, Pb, and Zn in the sediments of the Scheldt estuary obtained after attack by $HNO_3/HCl$. The mouth of the estuary is at 0 km. (After Wollast et al. 1985)

## 3 Organic Matter and Nutrients

The Scheldt has to support an unusually high organic and nutrient load owing to the large number of inhabitants in the drainage basin and to the development of agricultural and industrial activities. For comparison, the number of inhabitants per unit runoff is equal to $73 \times 10^3$ inh $m^{-3}$ $s^{-1}$ for the Scheldt, to $15 \times 10^3$ inh $m^{-3}$ $s^{-1}$ for the Rhine and to a mean of $6.5 \times 10^3$ inh $m^{-3}$ $s^{-1}$ for all European rivers.

A consequence of the high organic load is the existence of a permanent anoxic zone in the estuary over a distance of 30 km, which may exceed 70 km during the summer. The presence of these anaerobic waters in the upper part of the estuary plays an important role in the transfer of various elements through the estuary and will be discussed later.

In the zone characterized also by long residence times of the water masses, the nonrefractory organic matter is almost completely mineralized by the heterotrophic bacteria either in the water column or in the first few centimeters of the freshly deposited sediments. Mass balance calculation and measurements of the organotrophic activity by $H^{14}CO_3$ incorporation or by biological oxygen demand give very similar results and show that the amount of organic carbon degraded in this zone reaches between 100 and $150 \times 10^3$ t C $a^{-1}$ (Somville and Wollast 1981). As a consequence, one may estimate that no more than 10% of the terrestrial organic carbon input in the Scheldt estuary is transferred to the coastal zone.

As in many other estuaries, the decrease of turbidity with increasing salinity and the large supply of nutrients by the freshwaters produce in the Scheldt phytoplankton blooms during spring, summer, and early autumn. During these periods the organic matter produced by photosynthesis in the lower part of the estuary equals almost the amount of terrestrial organic carbon removed by respiration and sedimentation in the upper part (Wollast and Peters 1978). Most of this fresh organic matter is transferred to the coastal zone, where it is often difficult to distinguish from the coastal primary production.

The influence of man's activities is still more pronounced when we consider the concentrations and fluxes of the nitrogen and phosphorus species. A detailed nitrogen budget for the Scheldt hydrographical basin has recently been established by Billen et al. (1985) and is summarized in Table 2. This table shows that only a very small fraction of the nitrogen discharged to the river systems reaches the sea and approximately only one half of the nitrogen entering the estuary is transferred to the coastal zone.

The three main processes which modify the speciation of nitrogen in aquatic systems (nitrification, denitrification, and biological uptake), are commonly very active in estuarine systems and may significantly affect the transfer of forms of nitrogen to the adjacent coastal waters and to the atmosphere in the case of denitrification. When occurring, denitrification has a pronounced effect on the transfer of nitrogen because nitrate is released as $N_2$ or $N_2O$ to the atmosphere. However, denitrification occurs only in sections of estuaries that exhibit oxygen depletion and where nitrate is used by heterotrophic bacteria as an oxidant. This is the case with heavily polluted rivers, with long residence times, or stratified estuaries where organic matter, even of natural origin, accumulates. In the Scheldt,

**Table 2.** Nitrogen budget in the Scheldt hydrographical basin. (After Billen et al. 1985); in $10^3$ t $a^{-1}$

| Inputs | River system | Estuarine system | Total |
|---|---|---|---|
| Agricultural | 30 | 11 | 41 |
| Domestic | 23 | 3 | 26 |
| Industrial | 38 | 10 | 48 |
| Total | +101 | +24 | +115 |
| Transfer from river to estuary | − 28 | +28 | |
| Total | | +52 | |
| Transfer from estuary to sea | | −27 | − 27 |

for instance, denitrification is observed during the winter when the bacterial activity is at its lowest.

A typical longitudinal profile of oxygen and nitrate concentration corresponding to a winter situation is presented in Fig. 5. It shows the important reduction in nitrate concentration occurring in the anoxic part of the estuary. Rapid nitrification in the downstream aerobic part of the estuary explains the subsequent rise of nitrate concentration (Billen 1975). According to Billen et al. (1985), withdrawal of particulate N by sedimentation in the estuarine zone is of much less importance than that due to denitrification.

As one may expect from the intensive human activities in the Scheldt drainage basin, this river exhibits also unusually high concentrations of phosphorus. A tentative mass balance for this element in the Scheldt estuary is given in Table 3 (Wollast 1983).

The speciation of inorganic phosphate is much simpler than that of nitrogen as it occurs mainly as orthophosphate.

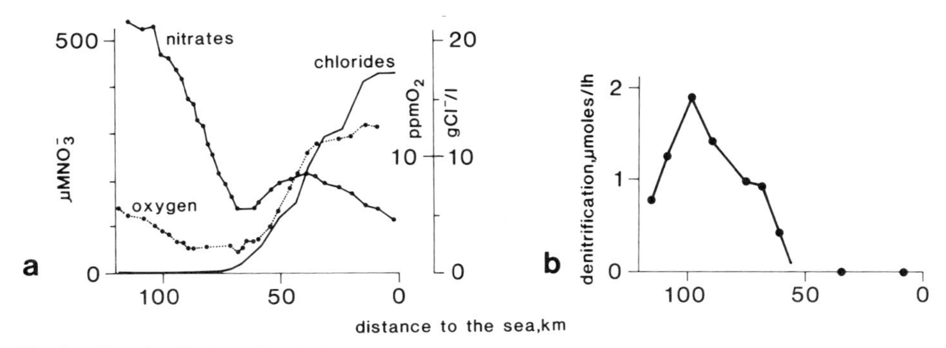


**Fig. 5. a** Longitudinal profile of nitrate, oxygen and chloride concentration in the Scheldt estuary at low tide in February. **b** Direct measurement of denitrification rate in the water column. (After Billen et al. 1985)

**Table 3.** Tentative mass balance for phosphorus in the Scheldt (in $10^3$ t P $a^{-1}$)

| | Output | Input | Chemical precipitation | Plankton uptake | Sedimentation |
|---|---|---|---|---|---|
| Dissolved P | 5.6 | −3.3 | −0.75 | 0 | 1.5 |
| Particulate P | 1.5 | +3.3 | +0.75 | −4.9 | 0.7 |
| Total P | 7.1 | – | – | −4.9 | 2.2 |

Processes affecting the behavior of phosphate in estuaries are, however, very complex and probably not entirely identified and certainly not sufficiently understood. First, orthophosphate is a chemically active compound which may be involved in various reactions of dissolution-precipitation or adsorption-desorption. The solid phases resulting from these reactions are so complex that basic properties like their solubilities or exchange equilibria are poorly known. It seems, however, that these chemical reactions are reversible and rather fast. They act as a buffering mechanism which tends to maintain dissolved phosphate in a narrow range of concentration (around 1 $\mu mol^{-1}$) during the mixing of river and seawater (Liss 1976).

The behavior of phosphorus in the Scheldt estuary is further complicated by the existence of an extended anaerobic zone and the long residence time of the water masses. Low redox potential conditions lead to the reduction of the most reactive iron hydroxides which are dissolved as $Fe^{2+}$. The phosphate earlier adsorbed onto this particulate phase is then released to the dissolved phase. When the dissolved oxygen is restored by reaeration and by mixing with seawater, $Fe^{2+}$ is re-oxidized and precipitates again as iron hydroxide sequestering large amounts of dissolved phosphate by coprecipitation and adsorption. This particulate material accumulates mainly by sedimentation and is not transported to the lower part of the estuary. In the aerobic zone, phytoplankton growth is responsible for a supplementary uptake of dissolved phosphate.

## 4 Trace Metals

As already shown in Figs. 3 and 4, the Scheldt is also heavily polluted by trace metals. The behavior of these elements during estuarine mixing is strongly affected by the large changes in the chemical properties of the water masses. The distribution of trace metals between solution and particulates may be strongly affected by salinity, pH and redox conditions occurring in the Scheldt (see Fig. 6). As an example, the distribution of Mn, Cd, Zn and Ni in the particulate phase and in solution as a function of salinity have been carefully examined by Duinker et al. (1982). Their results are given in Fig. 7. The behavior of each metal differs considerably.

The concentration of Ni in the particulate phase remains constant over the entire range of salinity. The concentration of dissolved Ni follows a perfectly

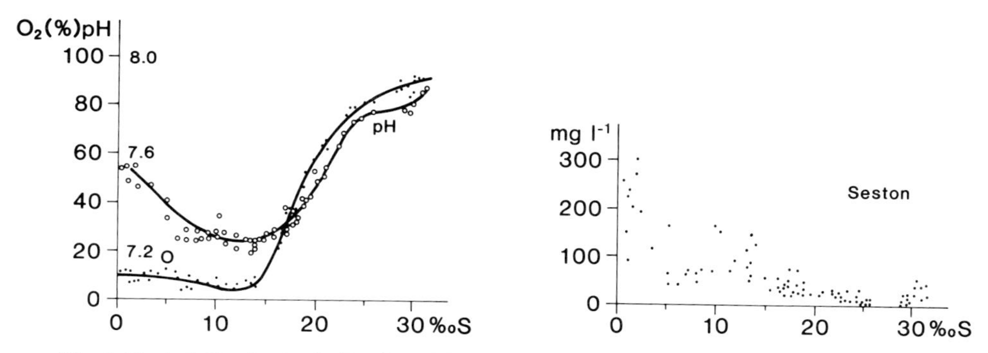


**Fig. 6.** Typical Distribution in October of dissolved oxygen, pH, and turbidity as a function of salinity in the Scheldt. (After Duinker et al. 1982)

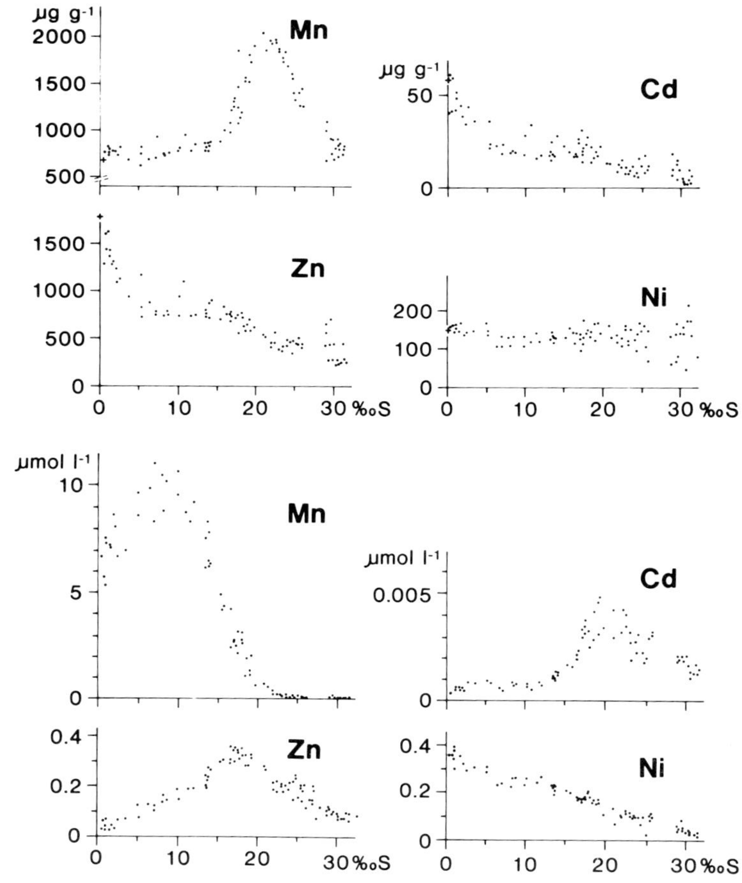


**Fig. 7.** Distribution of the concentration of Mn, Cd, Zn, and Ni in suspended matter (*upper part*) and dissolved (*lower part*) as a function of salinity in the Scheldt ). (After Duinker et al. 1982). Same conditions as in Fig. 6

straight line as a function of salinity suggesting a conservative behavior of this element.

The behavior of Mn can easily be explained by the effect of redox and pH changes in the estuary. This element is very soluble under reducing conditions and precipitates as soon as the concentration of oxygen increases. The distribution of Mn between the dissolved and particulate phases in the Scheldt can be described by a simple thermodynamical model assuming chemical equilibrium for this element in the water column (Duinker et al. 1979; Wollast et al. 1979).

Concentrations of both Zn and Cd in the particulate phase decrease with increasing salinity. This is clearly associated with the upstream region of high turbidity, where particles of small size dominate and the resuspension-deposition process is effective. The continuous decrease of the Zn and Cd content in particles in the downstream region of the turbidity maximum probably indicates mainly mixing of riverine material highly contaminated with suspended material of marine origin or with resuspended bottom material.

Both dissolved Cd and Zn present a maximum at a salinity close to 20‰.

A possible explanation for the low concentrations of dissolved Cd in the anaerobic part of the estuary is the removal of this compound as the highly insoluble sulfide. It is rather exceptional that $H_2S$ is produced in the water column during the summer, but even a small leak by diffusion of this component from the sediments, where sulfate reduction is abundant, is sufficient to explain the low values of dissolved Cd in the anaerobic zone. The oxidation of cadmium sulfide in the oxic zone and the formation of chloride complexes may well explain the sudden increase of dissolved Cd.

There is no clear explanation concerning the linear increase of Zn concentration with salinity in the 0–20‰ range. One may speculate that the intensive mineralization of detrital organic matter in that region is responsible for the release of Zn. Finally, the plots of Cd and Zn suggest a conservative behaviour during mixing of brackish water with sea-water.

## 5 Organic Micropollutants

There is a considerable lack of data concerning the presence and distribution of organic micropollutants like the harmful chlorinated hydrocarbons (PCB's) or polyaromatic hydrocarbons. A recent survey of the distribution of dissolved and particulate PCB's in the Scheldt estuary (van Zoest and van Eck, in press) indicates that the concentration of dissolved PCB's is maximum in fresh water where it reaches 1 mg $l^{-1}$, a value similar to that found in other Western European rivers. The dissolved PCB's behave conservatively during mixing with seawater. The concentrations of particulate PCB's, normalized with respect to organic carbone, show the highest values in the turbidity maximum due to the resuspension of bottom sediments. The annual total PCB load to the Scheldt estuary is estimated by these authors to be 50 kg $a^{-1}$.

## 6 Conclusions

The existence of a permanent anaerobic zone in the Scheldt estuary has generated many interesting and fundamental studies concerning the origin of the organic load, the contribution of various oxido-reduction processes to the anaerobic degradation of organic matter and their influence on the behavior of some nutrients and heavy metals. The distribution and behavior of organic microcontaminants like the PCB's or polyaromatic hydrocarbons are not well documented in the case of the Scheldt estuary, and it is recommended that a special effort be devoted to the occurrence and fate of these harmful organic contaminants.

Due to the long residence time of water in the estuarine zone and the accumulation by sedimentation of organic and inorganic pollutants, the Scheldt estuary has been dramatically affected by man's activities. Restoration of the water quality in this estuary will also require, for these particular reasons, an especially large effort in the control of discharge of pollutants in the waters of the watershed.

## References

Billen G (1975) Nitrification in the Scheldt estuary. Est Coast Mar Sci 3:79–89

Billen G, Somville M, De Becker E, Servais P (1985) Nitrogen budget of the Scheldt hydrographical basin. Neth J Sea Res 19(3/4):223–230

Duinker JC, Nolting RF, Michel D (1982) Effects of salinity, pH and redox conditions on the behaviour of Cd, Zn, Ni and Mn in the Scheldt Estuary. Thalassia Jugosl 18(1–4):191–202

Duinker JC, Wollast R, Billen G (1979) Behaviour of manganese in the Rhine and Scheldt Estuaries. II. Geochemical cycling. Est Coast Mar Sci 9:727–738

Liss PS (1976) Conservative and non-conservative behaviour of dissolved constituents during estuarine mixing. In: Burton JD, Liss PS (eds) Estuarine chemistry. Academic Press, London, pp 93–100

Somville M, Wollast R (1981) Modélisation de la qualité des eaux de l'estuaire de l'Escaut. Studiedag Kustwater en Estuariumverontreiniging Brugge, Koninklijke Vlaamse Ingenieursvereniging

Veen J van (1950) Eb en vloed schaarsystemen in de Nederlandse getijwateren. Tijdsch Koninklijke Nederl Aardrijkskundig Genootschap 67:303–325

Wollast R (1983) Interactions in estuaries and coastal waters. In: Bolin B, Cook RB (eds) The major biogeochemical cycles and their interactions. Wiley, New York, Scope 21:385–407

Wollast R, Peters JJ (1978) Biogeochemical properties of an estuarine system: the river Scheldt. In: Goldberg ED (ed) Biogeochemistry of estuarine sediments. UNESCO, Paris, pp 279–293

Wollast R, Peters DJ (1983) Transfer of materials in estuarine zones. In: Pearce JB (ed) Review of water quality and transport of materials in coastal and estuarine waters. International Council for the Exploration of the Sea, Copenhagen

Wollast R, Billen G, Duinker JC (1979) Behaviour of manganese in the Rhine and Scheldt Estuaries. I. Physico-chemical aspects. Est Coast Mar Sci 9:161–169

Wollast R, Devos G, Hoenig M (1985) Distribution of heavy metals in the sediments of the Scheldt estuary. In: Grieken R van, Wollast R (eds) Proceedings Progress in Belgian Oceanographic Research. Royal Academy of Sciences, Brussels

Zoest van, Eck GTM van (1986) The Behaviour of PCBs in the Scheldt estuary. Internal report, Rijkswaterstaat, Middelburg, Netherlands

# The Rhine/Meuse Estuary

K.J.M. KRAMER[1] and J.C. DUINKER[2]

## 1 Hydrodynamical Characteristics of the Rhine/Meuse System

The Rhine is a major river (length 1328 km, catchment area 224,000 $km^2$) flowing through several densely populated and industrialized areas in Switzerland, the Federal Republic of Germany, and the Netherlands, draining parts of France and Luxembourg as well. Important contributions of contaminants originate from the rivers Aare, Neckar, Main, Lahn, Wupper and Ruhr. The connections of the Rhine with the North Sea are the (open) Nieuwe Waterweg and the sluices in the enclosure dams of Ysselmeer and Haringvliet (Fig. 1). The long-term annual mean discharge (1901–1975) is 2200 $m^3s^{-1}$ at Lobith (Fig. 2) (RIZA 1982; RIWA 1984). Seasonal variations are between 980 and 9400 $m^3s^{-1}$, mainly due to irregular precipitation (Fig. 3a represents monthly averages for 1982 and 1983). About 10% of the Rhine discharge is directed through the river Yssel that discharges into the Ysselmeer (with a surface area of 1230 $km^2$ and a mean depth of 4.5 m). The discharge through the Yssel is kept above 285 $m^3s^{-1}$ for shipping purposes. The residence time of water in Ysselmeer is about 6 months before it is sluiced into the Waddensea. Because of the strong decrease in water velocity at the entrance of the lake, the bulk of suspended matter is deposited in the Ketelmeer area. Turbidity in the Ysselmeer is lower, and biological activity, pH, and dissolved oxygen concentrations are higher than in the Rhine and Yssel rivers. This affects the behaviour of several contaminants.

Other branches of the Rhine are the Waal and the Lek (the latter is canalized). The Waal joins with the river Meuse (length 925 km, catchment area 33,000 $km^2$), draining parts of France and Belgium and the southern part of the Netherlands. The water discharge of the Meuse is about 10% of that of the Rhine. Rain being the sole water source, seasonal variations are considerable; extreme values are 1800 $m^3s^{-1}$ in winter and 10 $m^3s^{-1}$ in summer (annual mean value 230 $m^3s^{-1}$). In discussions of the "Rhine" estuary, the contribution of the Meuse is usually tacitly included. This is also adopted here.

The sluices in the Haringvliet dam (construction was completed in 1971) are kept closed, unless the Rhine discharge rate (at Lobith) exceeds 2200 $m^3s^{-1}$. The residence time of water in the Haringvliet is in the order of 1.5–3 months. Only at river discharge rates above 9400 $m^3s^{-1}$ (occurring on the average only once per year) the sluices are fully opened. The Haringvliet then behaves like an estuary with a

[1]Marine Research Laboratory MT-TNO, PO Box 57, 1780 AB Den Helder, The Netherlands
[2]Institut fur Meereskunde a/d Univ. Kiel, Düsternbrookerweg 20, D-2300 Kiel, FRG

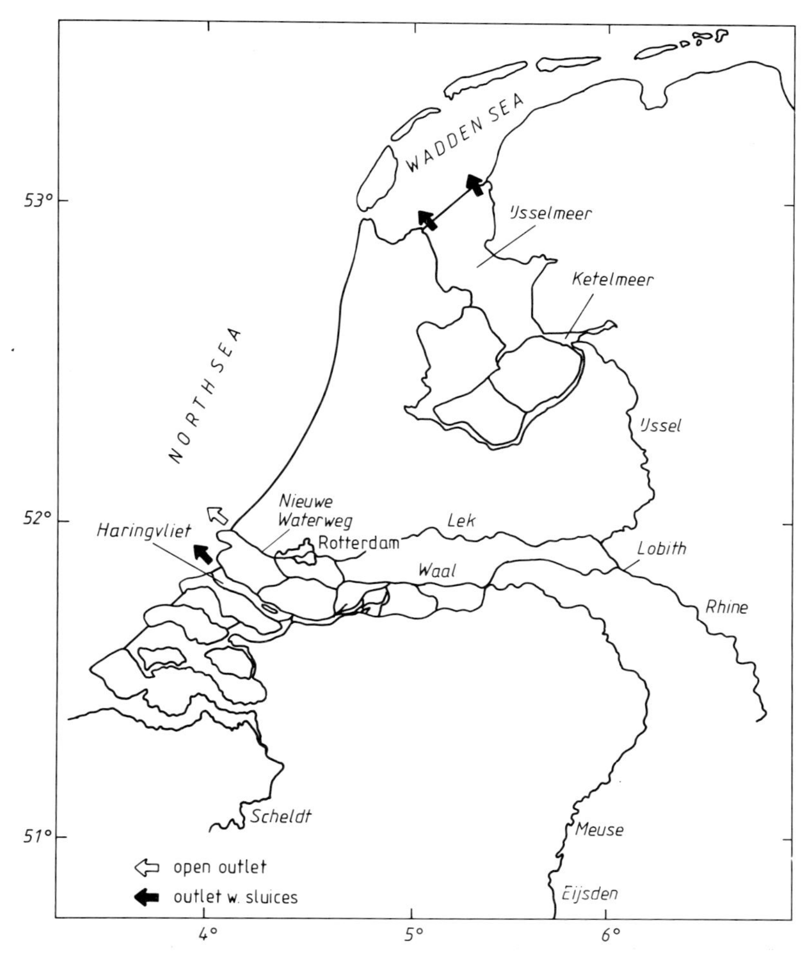


**Fig. 1.** Map of the Rhine – Meuse delta, the river Ijssel and the input locations into the North Sea. *Open arrow* open outlet; *closed arrow* sluices

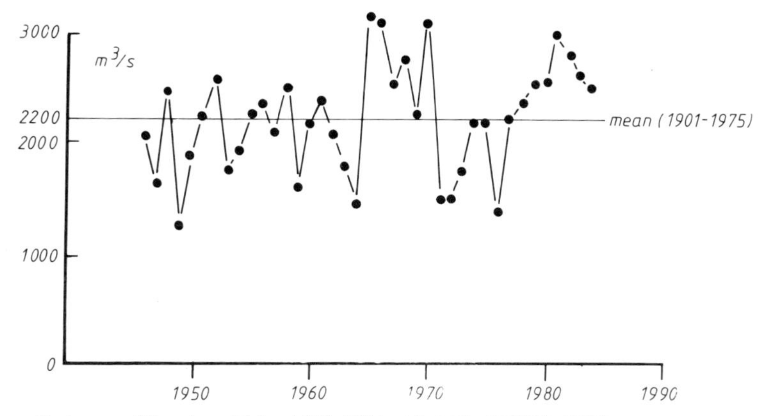


**Fig. 2.** Annual mean discharge of the river Rhine 1946–1984 at Lobith, (RIWA 1985)

corresponding decrease of water residence time. Under normal conditions, it acts as a freshwater sink for suspended particulate matter. The annual mean discharge rate into the North Sea is 1145 $m^3s^{-1}$ (RIZA 1983).

The major part of Rhine water is discharged into the North Sea through Nieuwe Waterweg (Rotterdam Waterways). The upper part of the Rhine estuary has characteristics of a salt wedge type estuary with partial mixing characteristics in the lower part of the estuary (Bowden 1980). A sill at about 37 km from the mouth marks the maximum intrusion of the salt wedge. The residence time of freshwater in the estuary is of the order of a few days. The estuary is dredged practically continuously. Dissolved oxygen concentrations in the river and in the estuary are typically in the 60–100% range, and seldom below 50%. Suspended matter concentrations in the river vary between 5–100 mg $dm^{-3}$ (Fig. 3b). This is to some extent connected with variations in discharge (Fig. 3a). A typical SPM concentration value is 30 mg $dm^{-3}$.

## 2 Transport of Contaminants

### 2.1 Introduction

Natural and man-made inorganic and organic chemicals in rivers are transported in solution and/or in association with particulate matter. Not all the material can be said to represent an influx to the sea as a whole. A certain proportion of the constituents is retained within freshwater, estuarine, or near-shore sediments, and does not succeed in reaching the truly off shore marine environment. The fresh-

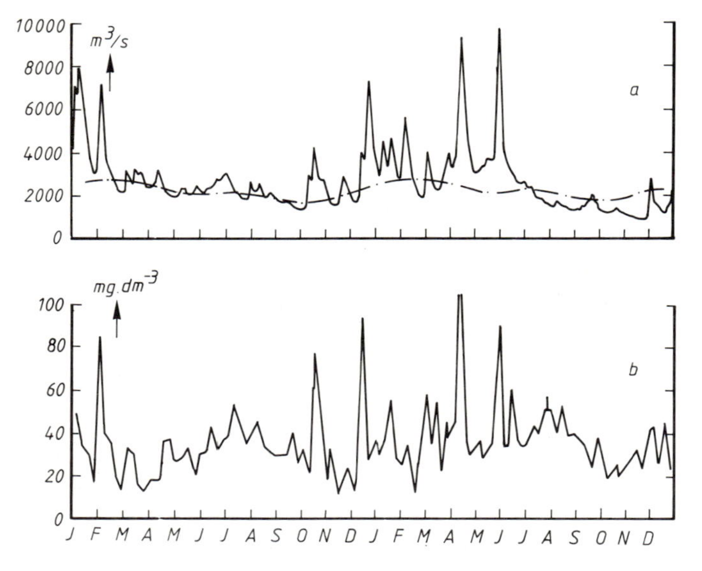


**Fig. 3. a** Water discharge of the Rhine at Lobith ($m^3$/s) in 1982 and 1983 (—) and long-term mean values (—·—). (RIWA 1984). **b** Concentrations of suspended particulate matter (in mg $dm^{-3}$) in 1982 and 1983 at Lobith (RIWA 1984)

water regions, where significant amounts of Rhine suspended particulate matter (SPM) are deposited, are Haringvliet, Ketelmeer (and Ysselmeer) and in some riverine sections upstream of the maxium salt intrusion region. Most estuaries have their origin in the rising level after the ice age about 20,000 years ago. The length of time has been sufficient to reach practically equilibrium between sediment supply and removal in the shallow estuaries around the North Sea. We are dealing with dynamical equilibria: sedimentation and erosion are intermittent phenomena. Net sedimentation may occur locally. The situation is complicated by redistribution of sediments within estuaries by dredging operations as well as estuarine circulation processes. It is therefore of considerable interest to the marine chemist to have an estimate of the gross riverine fluxes of the chemicals in which he is interested. Of equal or even more interest from a marine chemistry and contamination-pollution perspective are the net river fluxes, which represent the proportions of the gross fluxes that survive estuarine and near-shore removal and sedimentation processes and can escape to off-shore regions. The determination of net fluxes is, however, considerably more complicated than the measurement of gross fluxes. We shall consider these two aspects for the Rhine/Meuse system using available data on trace elements and organic compounds.

## 2.2 Gross River Fluxes of Inorganic and Organic Contaminants by the Rhine and Meuse

Information on gross river fluxes of contaminants requires data on river water discharge as well as on concentrations of contaminants, measured simultaneously at appropriate time intervals (e.g., once a week, fortnight, or month), including extremely dry and wet periods. Such data for the Rhine and Meuse are being collected by various governmental organizations in the bordering countries.

It should be realized that in several of these monitoring programs total samples are analyzed, without distinction of contributions from solution and particulates to the total concentration. Alternatively, data may be given of total and dissolved concentrations, which originate from different subsamples. This does not limit their use for estimating gross fluxes but such data are less useful for estimating the fate of the contaminants within the estuary and their effects on the adjacent marine environment (Wollast and Duinker 1982).

Table 1 summarizes annual flux data in terms of average (total) concentrations of some trace elements and organochlorines in the rivers Rhine and Meuse and their calculated annual gross fluxes for 1984. The evaluation of data collected monthly over a period of several years has revealed that the average concentration of cadmium has been decreasing significantly since 1977 (Fig. 4). This decrease in average concentration was also observed for lead, copper, zinc, chromium, and, to a lesser extent, for mercury (RIZA 1982; RIWA 1984). However, as the mean water discharge rate in these years has been above the average value (see also Fig. 2), the decrease in gross flux is much less evident. Only for cadmium, chromium, and mercury can a decrease be observed; for the other three elements an almost constant gross flux is observed in these years.

The (total) concentrations in the Ysselmeer are lower, due to a decrease in SPM concentration. The trend of the concentration of Cd in the Ysselmeer is opposite

**Table 1.** Mean (total) concentrations (in $\mu g\ dm^{-3}$) and gross annual flux (in $t\ a^{-1}$) of selected trace elements and chlorinated hydrocarbons in river Rhine water (at Lobith) and river Meuse water (at Eysden) in 1984. (calculated from RIZA 1984)

| | Rhine | | Meuse | |
|---|---|---|---|---|
| | $\mu g\ dm^{-3}$ | $t\ a^{-1}$ | $\mu g\ dm^{-3}$ | $t\ a^{-1}$ |
| Zinc | 55.2 | 4284 | 160.5 | 1670 |
| Cadmium | 0.2 | 16 | 1.6 | 16 |
| Lead | 6.5 | 504 | 23.0 | 240 |
| Copper | 8.6 | 670 | 15.2 | 158 |
| Mercury | 0.07 | 5.6 | 0.13 | 1.3 |
| Chromium | 8.7 | 672 | 12.7 | 132 |
| Nickel | 4.8 | 377 | 7.3 | 76 |
| Cobalt | 1.0 | 82 | 2.2 | 23 |
| HCB | 0.0034 | 0.266 | – | – |
| α-HCH | 0.0037 | 0.286 | – | – |
| γ-HCH | 0.0093 | 0.720 | 0.0322 | 0.335 |

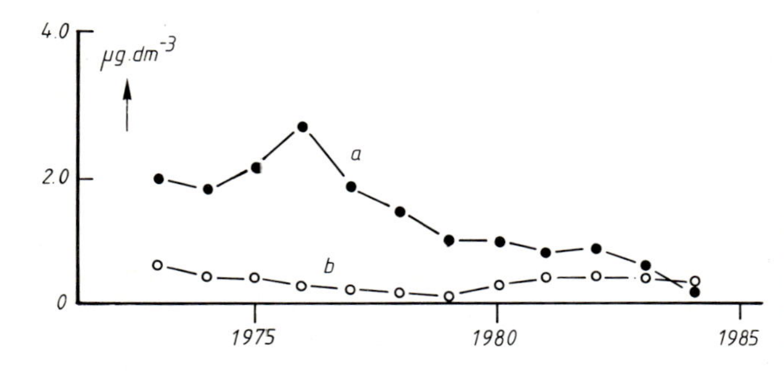


**Fig. 4a,b.** Annual mean concentration of total cadmium in river Rhine (**a**) and Ijsselmeer (**b**) water (1973–1984). (RIWA 1984)

to the situation in the Rhine. The increase in Cd levels observed in the western part in recent years (since 1979) results from erosion of and remobilization from particulate material, originally deposited in the eastern part (Ketelmeer area) and successive transport through the Ysselmeer. This increase, in combination with the decrease in levels of the Rhine (since 1977), results in presently higher Cd concentration levels in the western Ysselmeer than in the Rhine river. A similar behaviour [i.e., different trends of concentration levels in Rhine and Ysselmeer are also observed for Zn and Cr (RIWA 1985)].

As we have seen, the river water discharge rate may affect the concentrations of chemicals significantly. In the monitoring programs carried out in the Rhine (e.g., RIWA 1984; RIZA 1984; ARW 1985), the quantitative relations between discharge and concentrations have been evaluated in detail for several classes of contaminants. These relations, involving a normalization procedure, allow an evaluation of the river quality trend with time. For instance, comparison should involve data obtained at the same (standard) discharge rate. An example of such relations is represented in Fig. 5a for organically bound chlorine, based on data collected at different water discharge rates over a 3-year period. Trend analyses using such plots have shown that the contamination of the Rhine by total organically bound chlorine has decreased significantly during recent years (ARW 1985). Quality improvement has not been observed for total dissolved organic carbon (DOC). In a plot of total and dissolved copper concentrations against discharge in the Rhine (at Lobith), no general trend can be observed (Fig. 5b). One would expect an increase in total concentration with increasing discharge (higher SPM) and a lower dissolved concentration due to dilution. Still, when plotting both gross flux of copper and discharge a similarity is visible (Fig. 5c). An explanation for this discrepancy is that only a limited part of the sediment is available for resuspension.

Organically bound chlorine and DOC are both descriptions of important aspects of water quality. They describe bulk properties, however, and are thus non-specific. In both a quantitative and qualitative sense, they are less accurate than data for well-defined compounds. Such information is being collected within the monitoring programs; e.g., the research has resulted in the identification of many well-defined organic compounds of which many are potentially mutagenic or carcinogenic. The reader is referred to the annual reports of the various organizations (ARW, RIWA, RIZA) for full details. Not enough information is yet available for estimations of quantitative reliable gross river fluxes of individual compounds, but such information is becoming available.

## 2.3 Estuarine Processes and Their Effects on Net Fluxes to the Sea

The pathway of a contaminant through the estuary depends strongly on its partition over dissolved and particulate forms, as water and particles are subjected to different transport mechanisms in the estuary. In general, partition in an estuary differs greatly from the situation in the river. Various factors contribute to this fact. The first factor relates to the concentration and composition of SPM: concentrations in the estuary are usually different from the river and also considerably more variable; in addition, the grain size distribution (affecting contaminant contents in total SPM to a large extent) may be quite different because of selective sedimentation effects, starting already in the upper (freshwater) section of the estuary (Duinker and Nolting 1978). Secondly, transitions between solution and particulate forms may occur (involving chemical changes), in response to the large gradients of SPM and of other important parameters such as salinity, major-ion composition, pH, and dissolved $O_2$ concentration. These may affect the chemical behaviour of any contaminant. The extent to which this occurs depends on reaction tendency and kinetic factors. A complicating factor is that pH, dissolved $O_2$, and SPM concentration often do not vary linearly with the parameter describing the extent of mixing of fresh and marine water. Depending on the situation, minimum or maximum values occur in the longitudinal profile of the estuary, for the various parameters usually at different locations and salinities. During residence in the estuary, a river-borne component is exposed to these gradients. Finally, the distribution of its concentration in solution and suspension within the estuary, representing the net effect of all mechanisms and processes involved, may, in

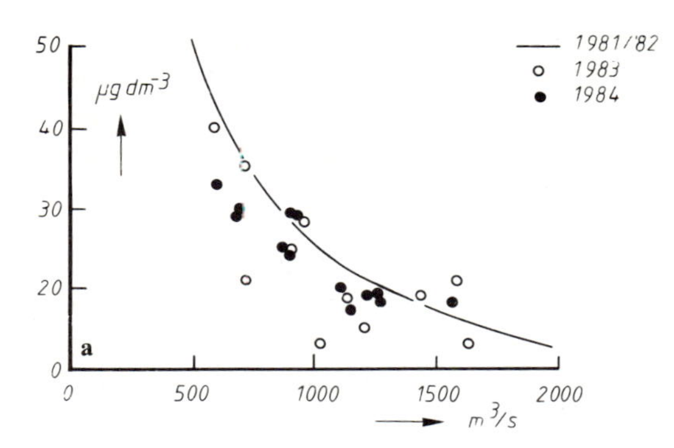


**Fig. 5. a** Adsorbable organic halogens in the river Rhine at Basel, 1981–1984 in relation to water discharge rates. (ARW 1985) **b** Concentrations of total (•) and dissolved (o) copper in the river Rhine at Lobith at different water discharge rates in 1984–1985 (RIZA 1984). **c** Gross flux of (total) copper (a, in $g.s^{-1}$) and water discharge (b as in Fig. 3) at Lobith in 1982–1983 (RIWA 1984)

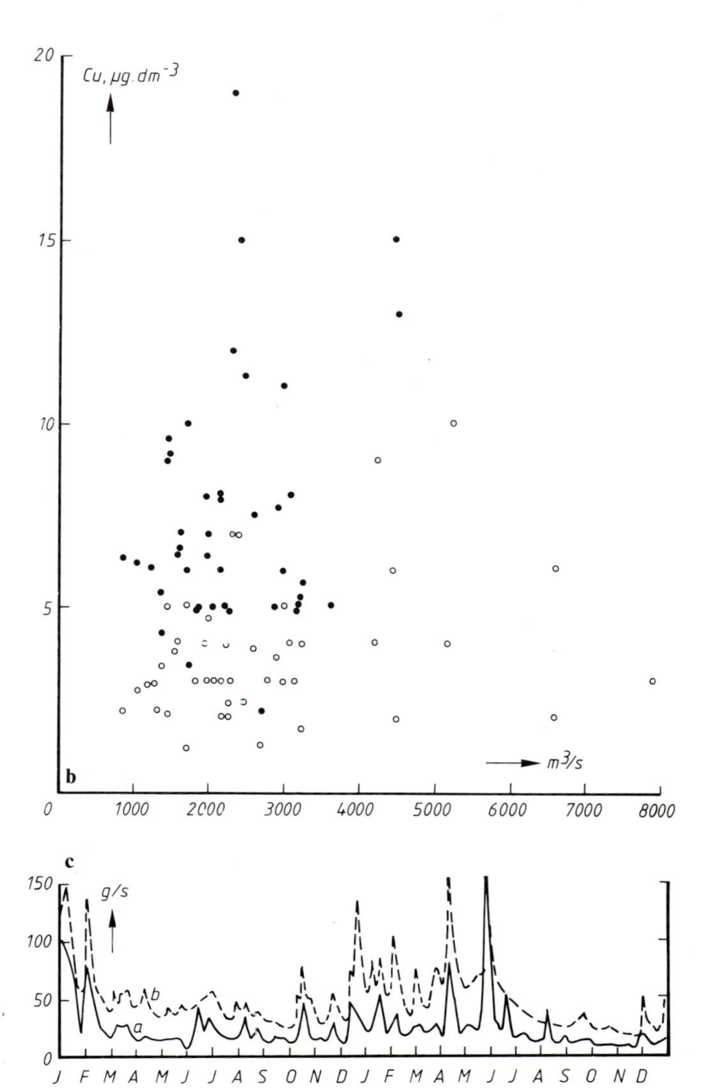

addition, depend on river discharge rate, meteorological conditions, the phase in the neap-spring tide cycle, etc., causing variations in sediment deposition-erosion and transport. Under extreme meteorological conditions, net transport of water and sediment may be in an upstream direction. It is not surprising, therefore, that contaminants in estuaries often show quite unexpected distributions (Wollast and Duinker 1982).

A river-borne constituent in dissolved form will have a residence time in the estuary that is characterized by the residence time of the freshwater. Particulate forms usually have a considerably longer residence time, but small particles may behave like dissolved constituents in this respect. It is important, therefore, to investigate how estuarine processes affect the distribution of a contaminant over dissolved and particulate forms.

## 2.4 Trace Elements

### 2.4.1 Trace Elements in Solution

Removal of river-borne Fe, Cu, and Cd from solution during estuarine mixing was suggested from the data obtained during several cruises in the Rhine estuary (Duinker and Nolting 1976, 1977). The results for Zn varied between conservative behavior and removal from solution. Manganese is the most reactive of the elements investigated: it is cycled between reduced forms (with higher solubility) and oxidized (particulate) forms with lower solubility during transport within the estuary. This results in a positive deviation from linearity in the dissolved concentration vs. salinity plot (at low salinities, Fig. 6) and a corresponding maximum in the manganese content of SPM ($\mu g\ g^{-1}$, not represented here) in the lower estuary. The formation of Mn- and Fe(hydr-)oxides in the early stages of estuarine mixing (as interpreted by a thermodynamical model for Mn species (Duinker et al. 1979; Wollast et al. 1979) may well contribute to the effective estuarine removal of trace elements like Cu, Zn, and Cd from solution.

Calculation of the net input into the sea of an element like Zn requires an estimate of the amounts that survive estuarine trapping mechanisms. The dissolved concentration to be considered for net input of an element which behaves conservatively during estuarine mixing is essentially the river water concentration. The value for an element which is removed from solution can, in principle, be obtained by extrapolation of the linear parts of the concentration vs. salinity plots (in the case of Fig. 6 at salinities above $5 \times 10^{-3}$) to $S = 0.4 \times 10^{-3}$ (characterizing the freshwater). For example, the value at $S = 0.4 \times 10^{-3}$ for dissolved Zn obtained by extrapolation is 7 $\mu g\ dm^{-3}$. This is an order of magnitude smaller than the river water concentrations often found. Similar calculations result for copper in 1.5 $\mu g\ dm^{-3}$ and 0.05 $\mu g\ dm^{-3}$ for cadmium. The field data obtained in estuaries often result in plots which look more complicated and confusing than the situation represented in Fig. 6, due to natural variability in the processes involved. This requires considerable efforts to obtain sufficient series of measurements for reliable net input data.

Various oxy-anions in solution (i.e., of V, As, Se and Sb) were found to behave conservatively during estuarine mixing in the Rhine estuary, in contrast to the

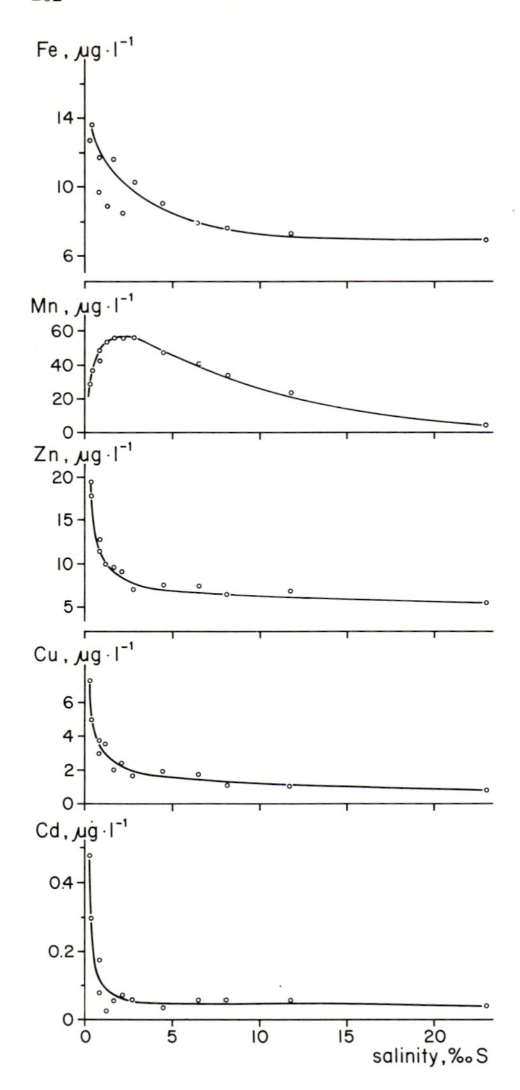


**Fig. 6.** Dissolved metal-salinity plots for a longitudinal section in the Rhine estuary. (Duinker and Nolting 1978)

situation found in the Scheldt estuary (Van der Sloot et al. 1985). This may be caused by the smaller driving force and smaller time available for reactions resulting from the considerably higher dissolved oxygen concentration and smaller residence time of water and dissolved components in the Rhine estuary compared with the Scheldt estuary (Duinker et al. 1982; Wollast, this Vol.). River water concentrations of these elements therefore form an appropriate basis for the calculation of dissolved concentrations in coastal and off-shore waters (Van der Sloot et al. 1985).

No chemical speciation study for trace elements has been reported for the Rhine estuary, as it has been for the river Scheldt (Kramer and Duinker 1984). Unpublished results from the Rhine estuary indicate, however, a stronger complexation (log K′ = 8.1) and a higher complexing capacity for copper ($CC_{Cu}$ = 430 nM Cu). This may affect the geochemical processes and the biological availability differently; these aspects need to be studied in more detail.

### 2.4.2 Trace Elements in Suspension and Dissolved/Particulate Partition

Particulate suspended matter is an important (not necessarily the dominant) carrier for many trace elements in river water. Significant amounts of SPM can be removed from the water column in the early stages of mixing in the Rhine estuary. The concentrations of SPM in the mixing zone can also be above the values in the source waters, depending on hydrodynamical and meteorological conditions (Fig. 7). Mass balance considerations of particulate matter and of particulate contaminants are, therefore, not justified without series of data obtained at regular intervals covering longer time periods, under variable conditions.

The calculation of the net or gross flux of particulate contaminants, based on weekly or monthly sampling intervals, can be problematic. The highly variable transport of SPM and the relatively few occasions at which high transport fluxes

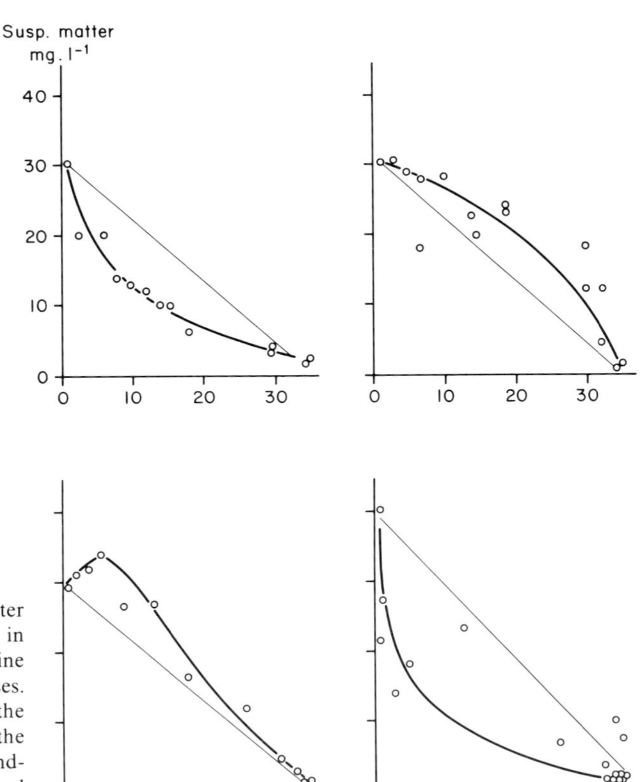


**Fig. 7.** Suspended matter concentrations vs. salinity in the surface layer of the Rhine estuary during four cruises. The straight lines connect the typical concentrations of the fresh- and marine end-members. (Duinker and Nolting 1976)

occur may result in an underestimation of the particle-associated flux of pollutants (Walling and Webb 1985).

The various estuarine processes have different effects on the concentrations of an element (per unit volume) in solution and in suspension. An element may be present in river water mainly in particulate form, but in seawater mainly in solution. For another element, the dominant form in the river may be dissolved, with particulate and dissolved forms being similar in seawater. The observations made on the concentrations of Cd, Mn, Cu, and Zn in solution and in suspension ($\mu g\ dm^{-3}$) during a cruise in the longitudinal section of the Rhine estuary show that large differences may exist between elements (Fig. 8). A complicating factor in the interpretation of distribution patterns is the observation that SPM consists of

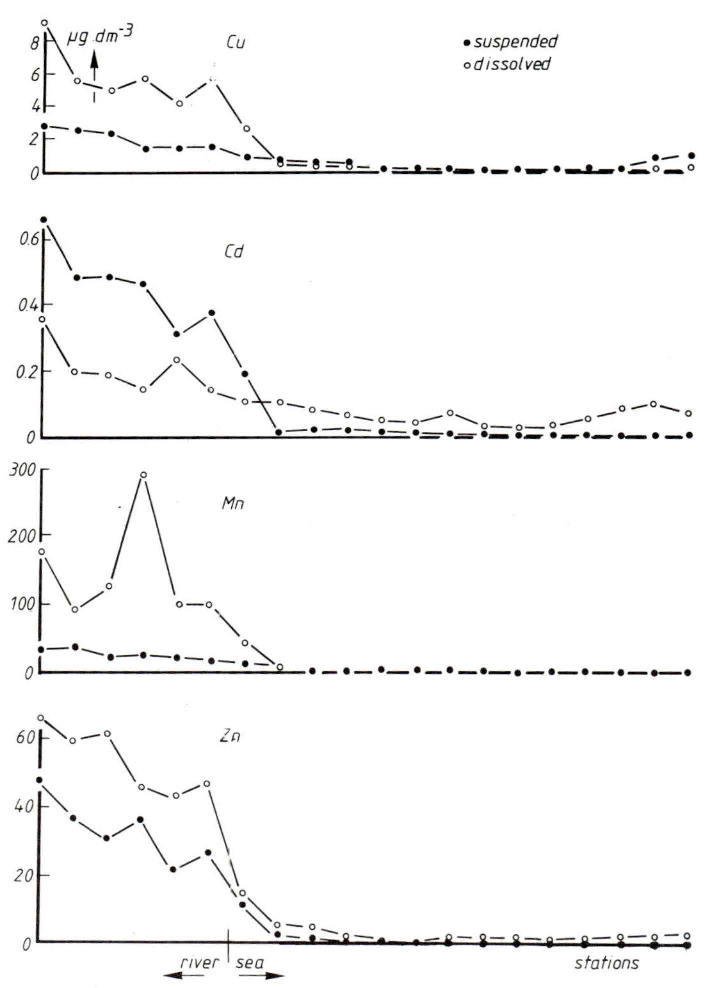


**Fig. 8.** Particulate (•) and dissolved (o) concentrations of Cu, Cd, Mn and Zn from a longitudinal transect of the Rhine estuary

different more or less continuously changing size/density fractions, with different element contents. Different fractions can be collected with a continuous centrifugation technique. With appropriate precautions, each fraction can then be analyzed for its trace element contents (Van der Sloot and Duinker 1981). The presence of different fractions in total suspended matter can also be detected and interpreted without physical separation techniques. This can be accomplished conveniently by plotting element content ($\mu g\ g^{-1}$ SPM) versus SPM concentration (mg $dm^{-3}$) (Fig. 9, Duinker 1981). Small, low-density particles, dominating at low SPM concentrations, have higher contents of Cu, Cd, Zn, Pb, and organic C (as well as lower contents of K, Al and Fe) than larger/denser particles which dominate at higher SPM concentrations (typically above 5 mg $dm^{-3}$). The smaller/less dense particles can escape from the estuary, giving rise to increased contents of Cu, Zn etc. in off-shore suspended matter. The larger/denser fractions are subjected to repeated sedimentation and resuspension processes in different phases of a tidal cycle. Their varying relative contributions give rise to variations in trace element contents in suspended matter sampled during a tidal cycle. The variations in trace element contents of SPM in coastal regions can also be readily explained in terms of mixing of such fractions (Duinker 1983). A strong covariance is observed between the contents of organic C and of Cu and some other elements in SPM. Whether these elements are associated with organic matter, or alternatively, with mineral fragments occurring in association with organic matter, is an open question. From Fig. 9 it will be clear that differences are observed in typical concentrations of copper, zinc, and organic C between the estuaries of Rhine and Scheldt. This stresses the importance of (geo)chemical speciation studies, especially concerning complexation and sorption processes in estuarine regions.

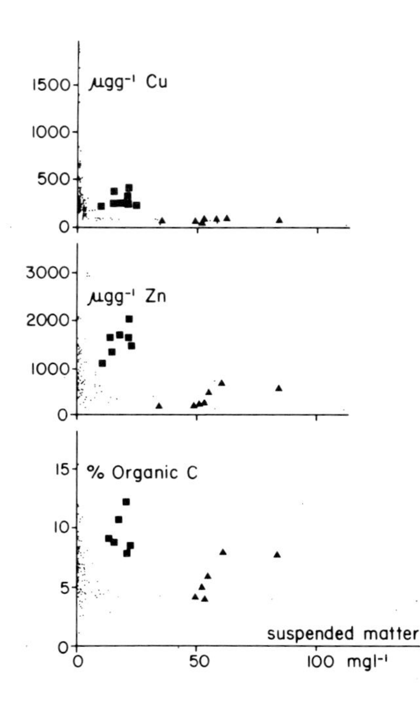


**Fig. 9** Elemental contents of Cu and Zn (in $\mu g/g$) and organic C (in %) in relation to total suspended matter concentration (in mg $dm^{-3}$) for the Rhine (■) and Scheldt (▲) estuaries and coastal waters. (Duinker 1981)

**Table 2.** Mean flux into the North Sea through the Nieuwe Waterweg (I), Haringvliet (II) and Ijsselmeer (III) and mean input into the North Sea by dredgings from the Rotterdam Harbour/ Nieuwe Waterweg (IV). Averages for the period 1979–81 in t $a^{-1}$, – = no data available. (RIZA 1983)

| | I | II | III | IV |
|---|---|---|---|---|
| Discharge ($m^3/s$) | 1463 | 1145 | 504 | – |
| Seston | 860400 | 464900 | 318100 | – |
| Total N | 250800 | 178700 | 66600 | – |
| Total P | 28780 | 11700 | 4200 | 14350 |
| Si | 102800 | 100200 | 14200 | – |
| Copper | 482 | 230 | 75 | 416 |
| Zinc | 3030 | 2040 | 421 | 3125 |
| Lead | 392 | 278 | 67 | 1077 |
| Cadmium | 37 | 22 | 67 | 49 |
| Mercury | 4.4 | 4.5 | 1.3 | 7.3 |
| Nickel | 432 | 214 | – | 281 |
| Chromium | 637 | 239 | – | 871 |
| HCB | 0.6 | 0.5 | – | 0.3 |
| α-HCH | 0.03 | 0.15 | – | – |
| γ-HCH | 0.74 | 0.69 | 0.07 | – |
| Aldrin | 0.08 | n.m. | – | – |
| Dieldrin | 0.32 | 0.06 | – | – |
| Endrin | 0.01 | 0.03 | – | – |

The latter section shows that an accurate estimation of the net input of contaminants into the North Sea is difficult. For the outlets of the river Rhine (Nieuwe Waterweg, Haringvliet and Ysselmeer), the average annual net discharge into the North Sea has been calculated for the period 1979–1981 (Table 2; RIZA 1983). As the data are derived from calculation of the fractions of river water that are discharged by the different outlets, and not by measurements in, for example, the estuary, they can only be considered as a crude estimate.

### 2.4.3 Trace Elements in Bottom Sediments

Figure 10 shows the contents of several elements in the < 16 μm fraction of bottom sediments in the Rhine estuary. These decrease from a position near the maximum salt intrusion in a seawater direction (de Groot et al. 1982). This is most likely due to mixing of fluviatile and marine-derived particles with different contaminant levels (Müller and Förstner 1975; Duinker and Nolting 1976). Stable C and O isotope data demonstrate the importance of the mixing mechanism (Salomons 1975; Salomons and Mook 1977).

It should be mentioned that trace element contents (e.g., Cu, Cd, Zn) in total SPM – even without correction for grain size – may exceed those of the < 16 μm fraction of bottom sediments. As discussed earlier, this is caused by the presence of

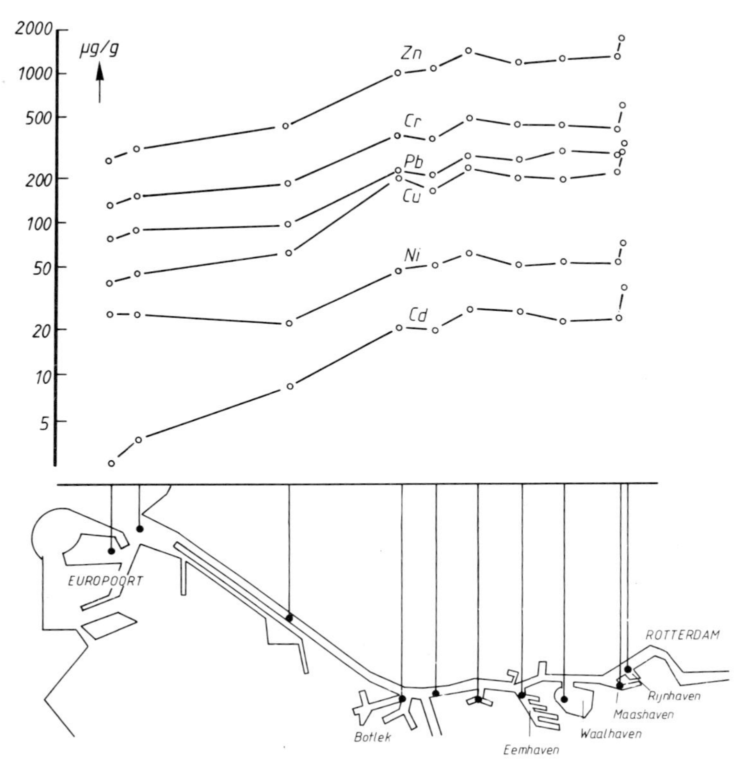


**Fig. 10.** Contents of trace metals in the fraction < 16 $\mu$m of river Rhine estuarine sediments (in $\mu g\ g^{-1}$) for a transect from Rotterdam Harbors to the mouth of the estuary. (de Groot, et al., 1982)

suspended particles with high contents of these elements and which resist settling in the estuarine and coastal environment (Duinker et al. 1974).

The contribution to the concentration of an element in particulate matter (suspended or in bottom sediments) from originally dissolved forms, which are removed during estuarine mixing, may result in an increase of the content of that element in particulate matter. For the Rhine estuary this has been observed for Mn. For other elements, the contribution from dissolution processes may be so small with respect to the concentration in suspension, that it is obscured by the effect of all other mechanisms involved. This seems to be the case for Cd and Cu.

Bottom sediments are dredged continuously from the harbors and main shipping channel. About 80% of the relatively uncontaminated dredgings is dumped into the North Sea, about 5 km north of the mouth. Table 2 summarizes the amount introduced in this way per year. The most heavily contaminated sediments are stored on land under controlled conditions.

## 2.5 Organochlorine Compounds

Information on organochlorine compounds in the Rhine and Meuse and their estuary is limited, especially as far as transport through the estuary is concerned.

### 2.5.1 Dissolved Organochlorines

Studies carried out in the Rhine/Meuse rivers and their estuary in the period 1974–1981 have shown that several organochlorine compounds behave conservatively during estuarine mixing (Duinker and Hillebrand 1979). This was found for penta- and hexachlorobenzene, $\alpha$-, $\gamma$-, and $\beta$-hexachlorocyclo-hexane (HCH) and polychlorinated biphenyls (PCB's), as determined by packed column GLC-ECD. Table 3 lists the range of concentrations determined in a series of samples of two branches of the Rhine (i.e., Waal and Lek) and in the Meuse. Concentrations in both systems varied considerably; the values in the Meuse were at least an order of magnitude smaller than in the Rhine. Taking the freshwater values in the Nieuwe Maas as characteristic for the freshwater endmember and central off-shore water with $S = 35 \times 10^{-3}$ for the marine endmember, calculated and observed concentrations in Dutch coastal waters and Marsdiep water agreed rather well, as can be expected for the case of conservative mixing.

### 2.5.2 Organochlorines in Suspension

The contents in suspension measured in the freshwater of the Waal, Lek, and Meuse were also quite variable (Table 3). Again, values in the Meuse were considerably lower (typically by one order of magnitude). Although particle size distribution determines to a large extent the content of several organochlorines in particulates, the differences between the two rivers are mainly caused by different levels of contamination.

### 2.5.3 Partition of Organochlorines over Particulates and Solution

The amount of organochlorines in suspension per unit volume depends on SPM concentrations. Thus, in the estuary large variations with time and place can be expected. This is confirmed by observations (Duinker and Hillebrand 1979; Gemeentewerken Rotterdam 1984). In the estuary, concentrations in solution were usually smaller than in suspension (ng $dm^{-3}$) for penta- and hexachlorobenzene and PCB as Aroclor 1254. The dominance of particulate matter as carrier of PCB was also found on the basis of individual congeners (Duinker 1986).

The HCH isomers have considerable larger concentrations in solution, both in freshwater, the estuary, and in seawater. In the rivers, PCB's are roughly equally distributed over solution and suspension. Penta- and hexachlorobenzene concentrations in solution usually exceed those in suspension.

**Table 3.** Range of concentrations of organochlorines in solution (ng $dm^{-3}$), in suspension (ng $dm^{-3}$ and ng $g^{-1}$ dry weight) and in bottom sediments (ng $g^{-1}$ dry weight) in samples of the Rhine (Waal and Lek) in June 1974 and in the Meuse (August 1976). (Duinker and Hillebrand 1979). Concentrations of some individual PCB components, identified by numbers suggested in the literature on the basis of IUPAC rules (Ballschmiter and Zell 1980) and by the position of the chlorine atoms in the biphenyl framework, in water (ng $dm^{-3}$) and in suspension (ng $g^{-1}$) in Waal water, sampled in the freshwater section of the Rhine just upstream of the salt intrusion maximum (Duinker and Hillebrand 1983; Duinker 1986). The ranges of concentrations for these components in the estuary are from Gemeentenwerken Rotterdam 1984

| Organochlorine | Solution (ng $dm^{-3}$) | | | Suspension: (ng $dm^{-3}$) | | Suspension: (ng $g^{-1}$) | | | Sediment (ng $g^{-1}$) | | |
|---|---|---|---|---|---|---|---|---|---|---|---|
| | Waal | Meuse | Estuary | Waal | Meuse | Waal | Meuse | Estuary | Waal and Lek | | Estuary |
| | | | | | | | | | Fine | Coarse | Total |
| Pentachlorobenzene | 1– 5 | 2– 4 | 6–18 | 1– 8 | 0.3– 0.5 | 6–130 | 13– 25 | 25–120 | 50– 80 | 0.3–0.5 | 1–20 |
| Hexachlorobenzene | 3– 8 | 1– 2 | 3–10 | 2– 19 | 0.2– 1 | 40– 300 | 13– 45 | 20– 80 | 100– 400 | 1 –8 | < 10–50 |
| α-HCH | 100–1200 | 2– 9 | 3– 4 | < 1– 10 | 0.2– 1 | < 10– 200 | 13– 50 | 13– 50 | 22– 50 | 0.1–1 | < 1 |
| γ-HCH | 80– 170 | 6–12 | 6– 8 | 5– 19 | 1 – 3 | 75– 300 | 75– 100 | 55– 80 | 17– 60 | 0.2–9 | < 2 |
| β-HCH | 25– 50 | 3– 5 | 1– 3 | 3– 11 | < 1 | 60– 200 | < 5 | 60–135 | 9– 60 | 0.1–0.9 | < 1– 3 |
| Dieldrin | < 1 | 9–16 | 1– 5 | < 1 | < 1 | < 20 | < 20 | < 20 | 70– 140 | 0.2–1.5 | 2–20 |
| Endrin | < 1 | < 1 | < 1 | 20– 35 | < 1 | 350– 500 | < 20 | < 20 | 20– 350 | 0.2–0.3 | 1–10 |
| PCB (Aroclor 1254) | 40– 100 | 13–26 | – | 50–230 | 25–40 | 1200–3700 | 1500–1800 | – | 2400–3400 | < 40 | – |
| 1.0 PCB no. 18 | – | – | – | – | – | 25 | – | – | – | – | – |
| 0.4 PCB no. 28 | – | – | – | – | – | 29 | – | – | – | – | < 10–60 |
| 0.7 PCB no. 52 | – | – | – | – | – | 50 | – | – | – | – | < 10–50 |
| 0.2 PCB no. 101 | – | – | – | – | – | 22 | – | – | – | – | < 10–50 |
| 0.2 PCB no. 138 | – | – | – | – | – | 26 | – | – | – | – | < 10–50 |
| 0.2 PCB no. 153 | – | – | – | – | – | 23 | – | – | – | – | < 10–50 |
| 0.1 PCB no. 180 | – | – | – | – | – | 149 | – | – | – | – | < 10–50 |

The introduction of improved techniques for separating the many organochlorine compounds in environmental samples by using capillary rather than packed GC columns has led to the identification and quantitation of a number of well-defined individual PCB congeners in water and suspended matter of the Rhine and its estuary (Duinker and Hillebrand 1983). The presence in river- and seawater of PCB components other than the ones present in the most commonly used reference solution when working with packed columns (Aroclor 1254 or 1260) had already been established in earlier work (Duinker and Hillebrand 1979), but identification and quantitation was not feasible at that time. Table 3 specifies the typical concentrations of some PCB congeners in water and SPM of the Waal (Duinker and Hillebrand 1983). The sum of concentrations of all individual congeners is equivalent to what is reported in Table 3 as indicated by the Aroclor data. The present data allow a more detailed analysis of partitioning between solution and particulates. The partition coefficient $K_d$ defined as

$$K_d = \frac{C_p}{C_w},$$

where $C_p$ = mass of compound/g of particulates and $C_w$ = mass of compound/g of solution generally increases with the number of chlorine atoms; using the IUPAC identification numbers (Ballschmitter and Zell 1980) the following $K_d$ values were calculated: 18: $2 \times 10^4$; 28: $7 \times 10^4$; 52: $8 \times 10^4$; 101, 138 and 153: $1 \times 10^5$ and 180: $3 \times 10^6$ (Duinker 1986; see also Duursma et al., 1986).

The amounts per unit volume (A) in each of the compartments are related by the simple expression

$$\frac{A_{susp}}{A_{sol}} = 10^{-3} \cdot K_d \cdot [SPM] \qquad \text{where [SPM] in g dm}^{-3}.$$

Thus, compounds with $K_d < 3 \times 10^4$ are found mainly in solution at [SPM] $\approx$ 30 mg dm$^{-3}$ (typical value for Rhine water). Twenty five % of the total concentration (per volume water) of a component with $K_d = 1 \times 10^5$ is expected to be in solution at this SPM concentration. The dissolved phase would be the dominant carrier phase at lower SPM concentrations such as in off-shore waters, as has been confirmed experimentally (Duinker 1986).

### 2.5.4 Organochlorines in Bottom Sediments

The grain size distribution in bottom sediments has an important effect on the content of organochlorines in bulk (unfractionated) sediments. Table 3 summarizes the range of contents in fine and coarse sediments of Waal and Lek. Differences for a particular compound between fine and coarse sediment are up to two orders of magnitude. Similar to the situation for some trace elements, contents in suspended matter are considerably higher than in the sediments with even a high percentage of fine material. This is caused by the presence of fine particles in SPM, which have high contents of organochlorines and resist settling in the estuarine and coastal environment (Duinker 1986). The concentration levels measured in total

SPM can be explained in terms of mixing of different size/density fractions, with different characteristic specific organochlorine contents (ng $g^{-1}$), in different mixing ratios.

## 3 Conclusion

The estuary of the rivers Rhine and Meuse is one of the most polluted estuaries in the world. One would expect a thorough understanding of its processes and fluxes. However, net and/or gross fluxes into the North Sea are usually derived from calculations from the river flux data, which often do not discriminate between dissolved and particulate forms. Also, estuarine processes seem too complex to be incorporated into these calculations yet.

The database from which the conclusions mentioned above have been derived has to be improved significantly before accurate mass balances can be constructed. However, they may be useful as a first guess and for understanding processes and mechanisms.

## References

ARW (1985) Jahresbericht 1984. Arbeitsgemeinschaft Rhein-Wasserwerke, Karlsruhe, pp 333

Ballschmitter K, Zell M (1980) Analysis of Polychlorinated Biphenyls (PCB) by glass capillary gas chromatography. Composition of technical Arochlor and Clophen – PCB mixtures. Z Anal Chem 302:20–31

Bowden KF (1980) Physical factors: salinity, temperature, circulation, and mixing processes. In: Olausson E, Cato I (eds) Chemistry and biochemistry of estuaries. Wiley, New York, pp 37–70

Duinker JC (1981) Partitioning of Fe, Mn, Al, K, Mg, Cu and Zn between particulate organic matter and minerals, and its dependence on total concentrations of suspended matter. Spec Publs Int Assoc Sediment 5:451–459

Duinker JC (1983) Effects of particle size and density on the transport of metals to the oceans. In: Wong CS, Boyle E, Bruland KW, Burton JD, Goldberg ED (eds) Trace metals in seawater. Plenum, New York, pp 209–226

Duinker JC (1986) The role of small, low density particles and the partition of selected PCB congeners between water and suspended matter (North Sea area). Neth J Sea Res 20:229–238

Duinker JC, Hillebrand MTJ (1979) Behaviour of PCB, Pentachlorobenzene, $\alpha$ HCH, $\gamma$ HCH, $\beta$ HCH, Dieldrin, Endrin and p,p′DDD in the Rhine – Meuse estuary and the adjacent coastal area. Neth J Sea Res 13:256–281

Duinker JC, Hillebrand MTJ (1983) Analyses van gechloreerde koolwaterstoffen in zeewater oplossing en suspensie. Report to Rijks Instituut voor Zuivering Afvalwater, Lelystad, Mimeo, pp 17

Duinker JC, Nolting RF (1976) Distribution model for particulate trace metals in the Rhine estuary, Southern Bight and Dutch Wadden Sea. Neth J Sea Res 10:71–102

Duinker JC, Nolting RF (1977) Dissolved and particulate trace metals in the Rhine estuary and the Southern Bight. Mar Pollut Bull 8:65–71

Duinker JC, Nolting RF (1978) Mixing, removal and mobilisation of trace metals in the Rhine estuary. Neth J Sea Res 12:205–223

Duinker JC, Eck GTM van, Nolting RF (1974) On the behaviour of copper, zinc, iron and manganese, and evidence for mobilization processes in the Dutch Wadden Sea. Neth J Sea Res 8:214–239

Duinker JC, Wollast R, Billen G (1979) Manganese in the Rhine and Scheldt estuaries. Part 2, Geochemical cycling. Est Coast Mar Sci 9:727–738

Duinker JC, Nolting RF, Michel D (1982) Effects of salinity, pH and redox conditions on the behaviour of Cd, Zn, Ni and Mn in the Scheldt estuary. Thalassia Jugosl. 18:191–202

Duursma EK, Nieuwenhuize J, Liere JM van, Hillebrand MTJ (1986) Partitioning of organochlorines between water, particulate matter and some organisms in estuarine and marine systems of the Netherlands. Neth J Sea Res 20:239–251

Gemeentewerken Rotterdam (1984) Milieuaspecten onderhoudsbaggerspecie. Granulaire samenstelling baggerspecie en gehalten aan organische en anorganische microverontreinigingen. Analyseresultaten monstercampagne 1984, B Report 110.11 – R 8424, pp 35

Groot AJ de, Driel W van, Salomons W, Kerdijk H (1982) Disposal of river sludge in the Netherlands. In: Thomee-Kozmiensky KJ (ed) Recycling International. Freitag, Berlin pp 438–444

Kramer CJM, Duinker JC (1984) Complexation capacity and conditional stability constants for copper of sea- and estuarine waters, sediment extracts and colloids. In: Kramer CJM, Duinker JC (eds) Complexation of trace metals in natural waters. Nijhoff/Junk, The Hague, pp 217–228

Müller G, Förstner U (1975) Heavy metals in the Rhine and Elb estuaries: mobilisation or mixing effects. Environ Geol 1:33–39

RIWA (1984) De samenstelling van het Rijnwater in 1982 en 1983. Rijks Instituut voor Waterleiding Artikelen, Amsterdam, p 150

RIWA (1985) Jaarverslag '84 – deel A: de Rijn. Rijks Instituut voor Waterleiding Artikelen, Amsterdam, p 96

RIZA (1982) De waterkwaliteit van de Rijn in Nederland in de periode 1970–1981. Nota 82-061, Rijks Instituut voor Zuivering Afvalwater, Lelystad, p 112

RIZA (1983) De waterkwaliteit van de Noordzee 1975–1982. Nota 83.084, Rijks Instituut voor Zuivering Afvalwater, Lelystad, p 95

RIZA (1984) Kwaliteitsonderzoek in de rijkswateren, 4 issues. Rijks Instituut voor Zuivering Afvalwater, Lelystad

Salomons W (1975) Chemical and isotopic composition of carbonates in recent sediments and soils from Western Europe. J Sediment Petrol 45:440–449

Salomons W, Mook WG (1977) Trace metal concentrations in estuarine sediments: mobilisation, mixing or precipitation. Neth J Sea Res 11:119–129

Sloot HA van der, Duinker JC (1981) Isolation of different suspended matter fractions and their trace metal contents. Environ Technol Lett 2:511–520

Sloot HA van der, Hoede D, Wijkstra J, Duinker JC, Nolting RF (1985) Anionic species of V, As, Se, Mo, Sb, Te and W in the Scheldt and Rhine estuaries and the Southern Bight (North Sea). Est Coast Shelf Sci 21:633–651

Walling DE, Webb BW (1985) Estimating the discharge of contaminants to coastal waters: some cautionary comments. Mar Pollut Bull 16:448–492

Wollast R, Duinker JC (1982) General methodology and sampling strategy for studies on the behaviour of chemicals in estuaries. Thalassia Jugosl 18:471–491

Wollast R, Billen G, Duinker JC (1979) Manganese in the Rhine and Scheldt estuaries. Part 1, Physico-chemical aspects. Est Coast Mar Sci 9:161–169

# The Estuaries of the Humber and Thames

A.W. MORRIS[1]

## 1 Historical Perspectives

The Humber and the Thames are the two largest English estuaries discharging into the southern North Sea and are highly important centres of population, supporting a wide variety of industrial, commercial and recreational activities. For much of their history, these estuaries and their riverine sources have served as convenient local recipients of domestic and industrial wastes without due regard for the environmental impact of this assault.

Historically, attention to pollution problems has been focussed mainly on the dissolved oxygen content in these estuaries and its dependence on inputs of BOD and reduced nitrogen species. The situation in the Humber to 1949 was documented briefly in a report by the Ministry of Agriculture and Fisheries (1951). It appears that up to 1904, the Humber Estuary was well oxygenated, with little sign of organic pollution other than in the immediate vicinity of untreated sewage outfalls off Hull and Grimsby, although serious pollution of contributary rivers was noted. Subsequently, oxygenation of the upper estuary declined progressively. In summer 1949 the upper estuary (salinities up to 10‰) was found to be at less than 20% saturation. A major proportion of the oxygen demand was introduced by the influent rivers.

Post-1964 survey results have been collated by Gameson (1982) in an assessment of recent trends in the oxygen content of the estuary and its contributary rivers. There has been a steady improvement in the quality of the Trent, and consequently of the Humber Estuary, up to 1981 when the average summer dissolved oxygen content of this river was over 80% saturated, whereas the Ouse had not improved significantly through this period.

Pollution of the Thames accompanied by gross oxygen depletion has long been a major environmental problem. In contrast to the Humber, the oxygen demand originates predominantly from local discharges rather than from the riverine input. Historically, remedial measures have been continuously overtaken by increasing demands for water usage and effluent discharge by an increasing industrial and domestic population. A succinct historical account can be found in the report prepared by The Ministry of Housing and Local Government (1961) and a more detailed graphic account has been prepared by Wood (1982).

[1]Natural Environment Research Council, Institute for Marine Environmental Research, Prospect Place, The Hoe, Plymouth PL1 3DH, Great Britain

The most recent severe deterioration of oxygenation in the Thames developed progressively through the first half of the present century till by mid-century stretches of the estuary seaward of London Bridge were at times totally devoid of oxygen. This led to a concerted study of the sources of organic pollution and their effects, supported for the first time by systematic scientific investigations of the estuary, including modelling studies (Department of Scientific and Industrial Research 1964). Actions based on these investigations have led to progressive improvement in the oxygenation and ecological status of the Thames since the late 1950's; this has been documented in a number of reports including Wheeler (1979), Andrews and Rickard (1980), Wood (1980, 1982) and Andrews (1984).

Much less attention was paid until quite recently to contaminants other than oxygen consumers introduced into these estuaries from a diversity of local industrial and domestic activities. However, there has been a considerable amount of legislation in the recent past aimed at improving the condition of British estuaries with respect to a widening range of potential pollutants.

Amelioration of estuarine pollution was helped initially by legislation aimed primarily at reducing the pollution of inland waters [The Rivers (Prevention of Pollution) Acts 1951 and 1961]. Additionally, the Clean Rivers (Estuaries and Tidal Waters) Act of 1960 gave responsible authorities the power to control new discharges to rivers and estuaries. Existing discharges were not covered and could be regulated only by obtaining a Ministerial Order. However, implementation of Part II of the Control of Pollution Act 1974, which is imminent, will rationalise this situation by bringing all discharges under the control of the enforcing agencies, the Regional Water Authorities. Legislative requirements for estuarine and river water quality control are now being further strengthened by European Economic Community actions, especially through Directives covering pollution by specified dangerous substances.

In response to these measures, the Regional Water Authorities, instituted in 1974, have made considerable progress in assessing the present pollution status of estuaries, in formulating quality objectives for these systems and, to a lesser extent, in advancing knowledge of the complexity of processes controlling the distribution and dispersion of polluting substances in estuaries so that equitable control measures may be applied. Woodward (1984) has outlined present pollution control measures for the Humber Estuary. Present policy for pollution control management in the Thames Estuary has been outlined in Cockburn et al. (1980), Andrews et al. (1983) and Lloyd and Cockburn (1983).

## 2 Basic Characteristics of the Estuaries of the Humber and Thames

### 2.1 The Humber

The catchment area of the Humber system (ca. 25,000 km$^2$) is the largest in the U.K. (Fig. 1). From its mouth at Spurn Head the estuary extends 62 km to Trent Falls, where it divides into two river systems, the Ouse (tidal length, 62 km) and the Trent (85 km). The Ouse has three principal tidal tributaries – Wharfe, Aire and Don. A

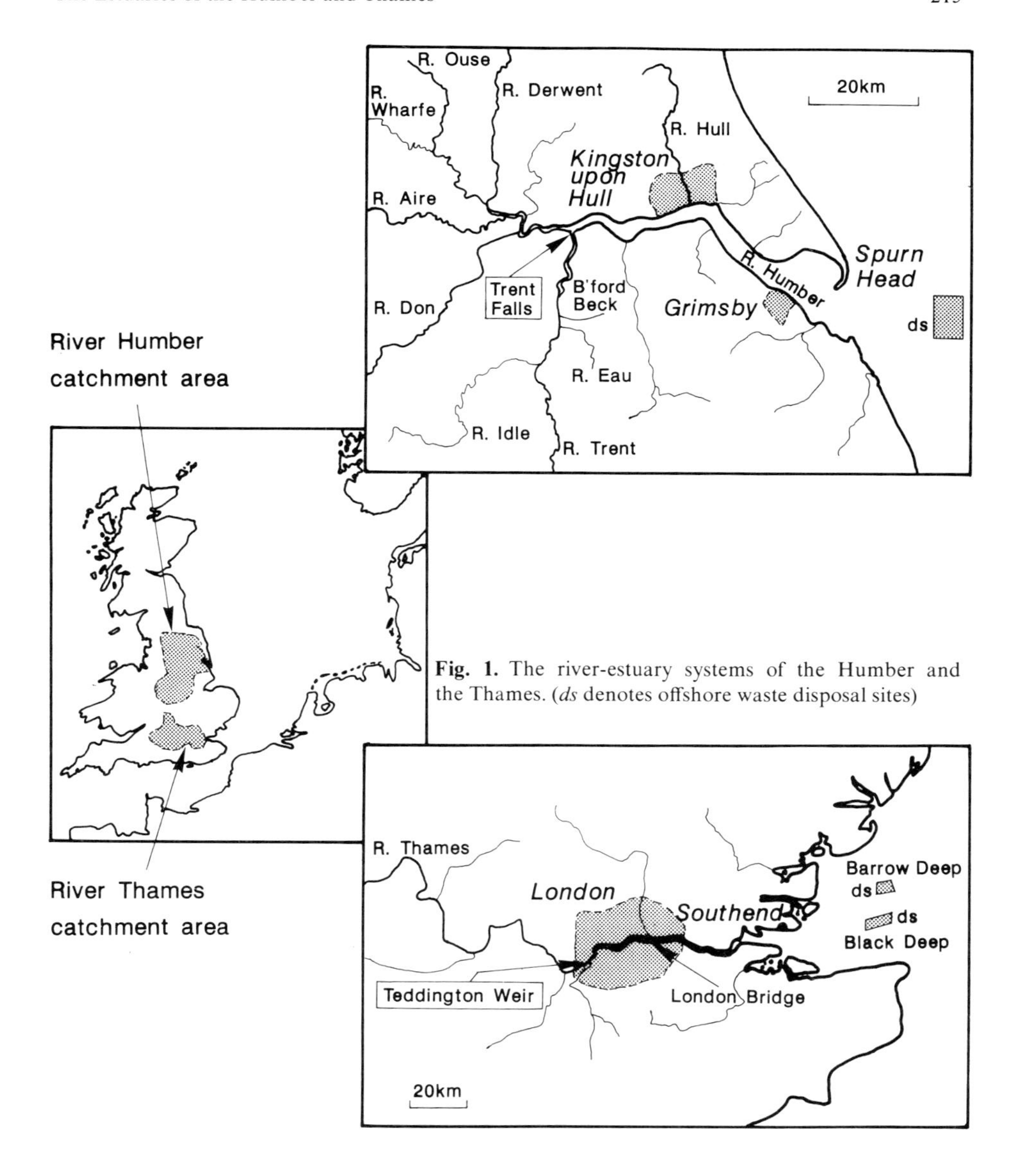


**Fig. 1.** The river-estuary systems of the Humber and the Thames. (*ds* denotes offshore waste disposal sites)

further tributary, the Derwent, has been barriered. The Trent has three major tributaries – Bottesford Beck, Eau and Idle – all non-tidal through engineered flow control. A number of small rivers discharge directly into the Humber Estuary, the most important is the Hull, which enters at 34 km from Spurn Head and is tidal for 32 km. Altogether the Humber system comprises over 300 km of tidal water.

The Humber Estuary approaches the description 'macrotidal' with a tidal range at its mouth of about 3.5 m at neaps and more than 6 m at springs. Large areas of estuary bed are exposed at low water. The tidal range is locally amplified in mid-estuary, where boundary convergence outweighs frictional tidal energy dissipation. At Trent Falls, the tidal range is similar to or even slightly larger than at the mouth, depending on the tidal phase. The estuary is essentially well-mixed, although minor vertical gradients in salinity have been recorded (Gameson 1976, 1982).

Accurate quantification of river discharges to the Humber Estuary is restricted by a shortage of suitable gauged data. Gameson (1976) reported an annual mean input of 246 $m^3 s^{-1}$. Regression analysis of flow data acquired between 1972 and 1981 produced a range, excluding the upper and lower five percentiles, of 60 to 450 $m^3 s^{-1}$ around a fifty percentile flow of about 160 $m^3 s^{-1}$ (Gameson 1982). During floods, the freshwater flow may temporarily exceed 1500 $m^3 s^{-1}$ (Denman 1979).

Penetration of sea salt into the system varies with river discharge and tidal state. At Trent Falls, median salinities vary from near zero at low water to 6–7‰ at high water; at the mouth, median salinities range from 27 to 30‰. With extremely high river flow, freshwater may extend some 20 km below Trent Falls; on the other hand, salt water is detectable to 40 km upriver of Trent Falls under drought conditions (Gameson 1976, 1982). A mean tidal excursion of around 15 km prevails throughout the estuary seaward of Trent Falls, but reduces sharply riverward of this point (Gameson 1976, 1982).

Strong tidal currents support a highly dynamic sediment system accompanied by migrations of sediment shoals (Denman 1979; Gameson 1982). The distribution of sediment types in the estuary has been described by Denman (1979). The waters of the estuary generally contain high suspended particulate loads which increase with depth and vary systematically in response to the tidal energy cycles. A turbidity maximum is located in the low salinity region, usually in the vicinity of Trent Falls. Here, suspended particulate loads had a geometric mean close to 400 $mg l^{-1}$ for 70 survey observations between 1977 and 1981, but ranged from around 50 to greater than 5000 $mg l^{-1}$ (Gameson 1982). Well developed turbidity maxima of this type generally indicate a net riverward transport of tidally resuspendable sediment particles within the estuary.

## 2.2 The Thames

The Thames Estuary (Fig. 1) extends almost 100 km from its mouth at Southend to Teddington Weir, which marks the limit of tidal energy propagation; the main river extends a further 236 km. The river above Teddington Weir drains a catchment area of 9900 $km^2$. Below the weir, there are a number of subsidiary riverine inputs. Many of these in the metropolitan region have been enclosed to form parts of the waste discharge/storm overflow system.

Fresh water entering the estuary across the weir at Teddington averages 67 $m^3 s^{-1}$. This is less than the natural supply (82 $m^3 s^{-1}$) due to abstractions a short distance upstream (Department of Scientific and Industrial Research 1964). Gauged flow at Teddington for the period 1920 to 1954, excluding the upper and lower five percentiles, ranged approximately from 9 $m^3 s^{-1}$ to 210 $m^3 s^{-1}$ around a median of 47 $m^3 s^{-1}$. With normal winter river flow, freshwater prevails roughly as far as London Bridge, situated 31.5 km seaward of Teddington; in late summer, average salinities of 1 to 5‰ occur at London Bridge and salt water penetrates a further 10 to 20 km (Department of Scientific and Industrial Research 1964).

The mean tidal range at Southend is 5.1 m at springs and 3.3 m at neaps (Inglis and Allen 1957), so that the estuary may be considered 'macrotidal'. The tidal amplitude increases riverward, maximising at about 5.8 m in the vicinity of London Bridge. The average tidal excursion is fairly uniform at between 12 and 16 km for the stretch of estuary from 60 km below to 10 km above London Bridge but then drops off sharply through the next 20 km. Minor vertical stratification has been recorded in the middle reaches of the estuary, but strong tidal currents ensure that for most purposes the estuary can be considered as a well mixed system. Like the Humber, the Thames has a well-developed turbidity maximum; an associated net riverward transport of tidally resuspendable sediment in the estuary has been demonstrated (Inglis and Allen 1957). Inglis and Allen (1957) have described the distribution of sediment types in the estuary.

## 3 Chemical Inputs and Distributions

### 3.1 The Humber

A shortage of reliable information on natural and pollutant chemical inputs and distributions in the Humber Estuary prior to the present decade is evident from an assessment of knowledge of the Humber Estuary prepared in 1979 (Natural Environment Research Council 1979). Woodward (1982) has detailed earlier and subsequent monitoring activities.

Urquhart (1979) calculated from data for the period 1964–1973 that the average daily flow into the Humber from the Ouse (127 $m^3 s^{-1}$) contains 12 $m^3 s^{-1}$ of sewage effluent and 8.3 $m^3 s^{-1}$ of trade effluent. The Trent supplies 119 $m^3 s^{-1}$ containing 2.5 $m^3 s^{-1}$ of sewage effluent and 3.9 $m^3 s^{-1}$ of trade discharge. In addition, 2.5 $m^3 s^{-1}$ of sewage and 3.1 $m^3 s^{-1}$ of trade effluent, excluding cooling waters, are discharged directly into the Humber. In total, there are 45 sewage discharges, 41 direct industrial outfalls and 17 cooling water returns to the tidal waters of the Humber system (Urquhart 1979). The Humber Estuary Committee of Water Authorities responsible for the quality of the system have attempted a direct assessment of polluting loads entering the estuary via these discharges. However, it is presently illegal to disclose information regarding individual effluent quality and only a small amount of summary data has been made available (Woodward 1984).

Routine shore-line sampling of the Humber Estuary was started in 1962 for a few constituents including oxygen, organic and inorganic nitrogen species, BOD and suspended solids load. Continuously recording oxygen meters have subsequently been installed at a number of sites. Between 1976 and 1981, a series of 31 surveys along the main channel were carried out by helicopter. The routine shore-line observations were supplemented by the addition of orthophosphate and chlorophyll determinations and bacterial examination. Attempts to determine dissolved metal concentrations were abandoned after a few earlier helicopter surveys because of contamination problems.

Nutrient data for the principal river inputs to the Humber Estuary are given in Table 1. The ammonia data show that the chemistry of the Humber system has changed appreciably through the recent past in response to changes in anthropogenic inputs to these rivers. Significant changes in dissolved oxygen have been noted earlier. Ammonia concentrations in the Humber Estuary fall quite sharply (and non-conservatively) with distance seaward from the vicinity of Trent Falls where significant nitrite is detectable, indicating nitrification within this zone (Gameson 1982).

**Table 1.** Average nutrient concentrations in the lower reaches of principal river inputs to the Humber Estuary. (Data from Gameson 1982 and Woodward 1984)

| River | Period | Constituent | Concentration (mg $l^{-1}$) |
|---|---|---|---|
| Trent | 1964–1966 | Ammonia-N | 2.82 |
| | 1979–1981 | Ammonia-N | 0.26 |
| | 1976–1981 | Phosphate-P | 0.77 |
| | 1981 | Nitrate + nitrite-N | 7.75 |
| | 1977–1980 | Organic-N | 1.55 |
| Ouse | 1964–1966 | Ammonia-N | 1.70 |
| | 1979–1981 | Ammonia-N | 0.63 |
| | 1976–1981 | Phosphate-P | 0.34 |
| | 1981 | Nitrate + nitrite-N | 5.35 |
| | 1977–1980 | Organic-N | 2.20 |

It can be estimated from data in Murray et al. (1980) that the total inputs of nitrogen and phosphorus to the Humber Estuary from rivers, sewage and industrial discharges are 155,000 and 37,000 kg $d^{-1}$, respectively. Above salinities of a few ‰, both phosphate and nitrate + nitrite are close to conservatively distributed through the Humber Estuary (Gameson 1982). Some minor deviations from linear mixing curves are evident, presumably arising from subsidiary discharges.

Estimates by Murray et al. (1980) of metal inputs to the Humber Estuary system via river (dissolved component), sewage and industrial discharges are compared in Table 2 with estimates based on metal concentration gradients recorded in the outer estuary. The direct estimates are necessarily imprecise, being based partly on averaged values for the composition of British rivers and effluents. An updated assessment of riverine supplies based on more recent data is included Table 2.

Available information on the distributions of dissolved metals in the Humber Estuary has been discussed by Gardiner (1982) and some additional data are given in Woodward (1984). Jones and Jefferies (1983) reported dissolved metal concentrations at the mouth of the estuary. It appears from this limited amount of information that river inputs have a major influence on distributions of dissolved cadmium, copper, lead, nickel and zinc in the estuary, although little can be

**Table 2.** Direct and indirect estimates of metal inputs (kg $d^{-1}$) to the Humber Estuary

| Source | | Cd | Cu | Cr | Ni | Pb | Zn | Hg | As |
|---|---|---|---|---|---|---|---|---|---|
| a. Rivers[a] | | 15 | 74 | 15 | 4 | 44 | 147 | 1 | – |
| b. Sewage[a] | | 2 | 111 | 4 | 12 | 23 | 128 | 0.5 | – |
| c. Industrial[a] | | 8 | 44 | 94 | 35 | 9 | 9025 | – | – |
| Sum a + b + c | | 25 | 229 | 113 | 51 | 76 | 9300 | 1.5 | – |
| Total inputs estimated from dissolved efflux at estuary mouth[a] | | 38 | 186 | – | 595 | – | 1098 | – | – |
| Dissolved river inputs estimated | (a) | 89 | 825 | – | – | 795 | 1639 | – | – |
| from direct measurements[b] | (b) | 22 | 440 | 260 | 411 | 352 | 1284 | 4 | 26 |

[a] from Murray et al. (1980).
[b] (a) from Gardiner (1982): (b) from Hill et al. (1984).

deduced at present about the behavior of these metals within the system. However, it appears either that the high industrial zinc input recorded in Table 2 has been overestimated, or that it is discharged and maintained in a particulate form. Only arsenic appears to be strongly influenced by direct discharges to the estuary. Metal contents of sediments and organisms in the system have been reported by Jaffe and Walters (1977), Jones (1979), Gardiner (1982) and Woodward (1984). The general conclusion from this present evidence is that although metal contamination of the Humber system is clearly evident, only arsenic, copper and possibly mercury present potential environmental quality problems from a legislative viewpoint (Gardiner 1982).

### 3.2 The Thames

In contrast to the abundance of reports on oxygen consuming constituents (BOD loading, reduced nitrogen species) and their control of dissolved oxygen in the Thames Estuary, published information on inputs and distributions of persistent trace chemicals is scarce. Nevertheless, a wide range of constituents in effluents and in the estuary is included in the monitoring programmes of the responsible authorities (Wood 1982) although, as for the Humber, detailed information on individual discharges is not disclosed.

The most recent estimates of riverine dissolved metal inputs to the Thames Estuary are given in Table 3. Table 4 lists the average concentrations of some dissolved metals and organohalogen compounds in unfiltered samples from the Thames Estuary collected during the period 1975–79. Since the highest recorded concentrations tend to coincide with the region of highest suspended load, direct pollutant effects are not readily distinguishable. Generally, concentrations in-

**Table 3.** Riverine dissolved metal inputs (kg $d^{-1}$) to the Thames Estuary. (Hill et al. 1984)

| Cd | Cu | Cr | Ni | Pb | Zn | Hg | As |
|---|---|---|---|---|---|---|---|
| 5 | 22 | 11 | 34 | 16 | 198 | 0.2 | 7 |

**Table 4.** Average concentrations of metals (1976–1979; $\mu g\ l^{-1}$) and organohalogen compounds (1975–1978; ng $l^{-1}$) in the Thames Estuary. (Data from Wood 1982)

| Distance below London Bridge (km) | Zn | Cu | Ni | Pb | Cd | Hg | αBHC | γBHC | Aldrin | Dieldrin | DDT | PCB[a] |
|---|---|---|---|---|---|---|---|---|---|---|---|---|
| -25.8 | 52 | 23 | 20 | 15 | 2.3 | 0.36 | 2.2 | 4.5 | 1.0 | 2.0 | 1.3 | 3.2 |
| -24.1 | 62 | 19 | 26 | 18 | 2.3 | 0.25 | | | | | | |
| -22.0 | 76 | 25 | 30 | 18 | 2.3 | 0.67 | | | | | | |
| -20.9 | 86 | 31 | 35 | 28 | 2.4 | 0.29 | | | | | | |
| -17.7 | 65 | 23 | 28 | 27 | 2.3 | 0.32 | 1.6 | 3.7 | 1.6 | 2.0 | 2.0 | 4.8 |
| -11.9 | 91 | 29 | 31 | 27 | 2.7 | 0.22 | | | | | | |
| -7.9 | 84 | 45 | 32 | 29 | 2.7 | 0.44 | 1.4 | 2.8 | 1.3 | 2.0 | 2.4 | 8.6 |
| -4.5 | 106 | 43 | 31 | 32 | 3.0 | 0.85 | | | | | | |
| -2.4 | 117 | 40 | 32 | 38 | 3.0 | 0.42 | 1.1 | 1.4 | — | 2.0 | 2.7 | 8.2 |
| 0 | 96 | 37 | 47 | 39 | 2.7 | 0.36 | | | | | | |
| 1.9 | 84 | 31 | 42 | 31 | 2.3 | 0.63 | | | | | | |
| 4.7 | 112 | 39 | 43 | 32 | 2.7 | 0.60 | | | | | | |
| 7.7 | 118 | 37 | 43 | 31 | 2.9 | 0.66 | 1.1 | 3.0 | — | 2.1 | 3.7 | 10.2 |
| 11.4 | 80 | 29 | 41 | 28 | 2.8 | 0.90 | | | | | | |
| 14.7 | 85 | 34 | 48 | 34 | 3.2 | 0.89 | | | | | | |
| 18.4 | 92 | 28 | 37 | 23 | 2.5 | 0.39 | 1.2 | 4.0 | — | 1.8 | 3.6 | 7.2 |
| 21.9 | 79 | 28 | 41 | 20 | 2.7 | 0.51 | — | 2.7 | — | 1.4 | 1.6 | 7.6 |
| 26.6 | 70 | 24 | 37 | 15 | 2.0 | 0.60 | | | | | | |
| 34.8 | 81 | 21 | 38 | 15 | 2.0 | 0.59 | — | 1.7 | — | 1.4 | 2.4 | 5.6 |
| 42.5 | 99 | 31 | 34 | 13 | 2.0 | 0.35 | | | | | | |
| 47.7 | 67 | 30 | 29 | 11 | 2.1 | 0.81 | | | | | | |
| 53.2 | 57 | 19 | 19 | 25 | 2.0 | 0.30 | | | | | | |
| 62.5 | 59 | 16 | 22 | 17 | 2.0 | 0.25 | — | — | — | — | 1.6 | 3.1 |
| 69.7 | 40 | 22 | 16 | 18 | 2.4 | 0.29 | | | | | | |

— denote $<$ 1.0 ng/l of organohalogen compounds.
[a] As Arochlor 1254.

crease seaward through the freshwater reaches above the estuarine mixing zone towards broad maxima centred in the low salinity, high turbidity region of the estuary around London Bridge. Concentrations then decrease towards the estuary mouth. Irregularities about this basic pattern probably reflect the distribution of major inputs.

Mercury distributions differ in showing a number of distinct peaks along the waterway. Smith et al. (1971) reported that dissolved and particulate mercury were concentrated in the estuary relative to levels in the influent river and offshore seawater and attributed this to mercury in sewage effluents. Most (82–97%) of the mercury in the waters was in particulate form.

Nelson (1979) has reported the concentrations of a few trace elements in sediments and suspended particulate material of the outer Thames Estuary, but there is little other available information on these components. Concentrations of metals and chlorinated hydrocarbons in fish from the tidal Thames have been reported by Rickard and Dulley (1983), who concluded that levels of persistent chemicals in the resident fish were low in comparison with other industrialised estuaries.

## 4 Offshore Waste Disposal

Marine disposal of wastes emanating from both the Humberside and Thameside regions is effected by boat transport to dumping sites just offshore of their respective estuary mouths (Fig. 1). Disposal of sewage sludge to the outer reaches of the Thames Estuary was introduced as early as 1887 as a remedial action to counter gross pollution of the inner estuary and has continued to the present. Since 1967, sewage sludge alone has been dumped in the Barrow Deep. Previously, both sewage sludge and dredged spoils were disposed into the Black Deep. Sewage sludge and industrial waste from the Humberside region has been dumped at the Spurn Head site, located 20 km east of the mouth of the Humber, since 1971.

Table 5 shows estimates of quantities and average elemental compositions of materials dumped off the two estuaries. For the Spurn Head site, Murray et al. (1980) estimated that dumped quantities represented less than 20% of all sources to the site, except for copper and mercury. Only 0.7% of the total nitrogen and 0.8% of the total phosphorus supplied to the dumping site were attributed to dumping, the remainder to local river inputs, coastal discharges and atmospheric deposition. Murray et al. (1980) reported enrichments of organic carbon and mercury in sediments at the disposal site and attributed them to dumping activities. However, they concluded that rapid local dispersion meant that there was little probability of long-term accumulations of dumped material.

**Table 5.** Offshore dumping of waste (in tonnes) from the Humberside and Thameside regions in 1976

| Total mass | Solid content | N | P | Cd | Cr | Cu | Ni | Pb | Zn | Hg |
|---|---|---|---|---|---|---|---|---|---|---|
| Humber Estuary, Spurn Head site: sewage sludge[a] | | | | | | | | | | |
| $1.7 \times 10^5$ | $6.9 \times 10^3$ | 300 | 104 | 0.1 | 10.1 | 7.8 | 3.3 | 5.1 | 31.3 | 0.2 |
| Humber Estuary, Spurn Head site: industrial waste[a] | | | | | | | | | | |
| $1.8 \times 10^4$ | 485 | 10 | nd | <0.1 | 0.5 | 32.5 | <0.1 | 0.1 | 18.5 | <0.1 |
| Thames Estuary, Barrow Deep: sewage sludge[b] | | | | | | | | | | |
| $4.4 \times 10^6$ | $1.1 \times 10^5$ | 8800 | 1900 | 6.9 | 54.4 | 93.7 | 28.9 | 118.0 | 395.7 | 1.4 |

[a] Data from Murray et al. (1980).
[b] Data from Norton et al. (1981).

Much larger quantities of waste material are dumped in the outer Thames (Table 5) and these represent larger proportions of the total supply to the dumping site. Dumping has been estimated to contribute 9% of the total nitrogen and 12% of the total phosphorus delivered to the outer Thames Estuary; estimated metal contributions ranged from 16% for cadmium to 34% for nickel (Department of the Environment/National Water Council 1979).

Studies of the dispersion and local chemical and biological effects of sewage dumping in the outer Thames Estuary have been reported by Norton et al. (1981) and Talbot et al. (1982). Areas of seabed showing accumulations of organic matter and metals were identified.

## 5 Summary

Very little information is available on natural and contaminant constituents in both the Humber and Thames estuaries, despite their being principal centres of domestic and industrial activity. Only aspects of deoxygenation in these estuaries and its interrelationships with BOD and nitrogen species appear to have been examined sufficiently to enable predictive modelling based on sound scientific principles to be applied to pollutant budgeting and control decisions.

Accurate assessments of discharges of natural and pollutant chemicals from the Humber and Thames estuaries to the North Sea require firstly, quantitative data on inputs to the estuaries from all sources and secondly, an understanding of the behaviour of those inputs within the estuarine systems. Monitoring of discharges is presently carried out in these estuaries by their responsible Regional Water Authorities. However, even though details of discharges have not been made available, it is probable that the acquired information is insufficient for accurate estuarine chemical budgeting purposes. The practical problems in quantifying discharges arising from their frequent wide fluctuations in both rate of flow and composition have often been stressed. Nevertheless, whatever the demand on resources, an appropriately detailed input sampling programme should be carried out as a necessary requirement for budgeting, process evaluation and modelling purposes. This information will provide a firm scientific basis for effective management decisions in the future.

Similarly, although pollutant concentrations in the water column and sediments are recorded for monitoring purposes, more specific sampling schemes will be required for quantifying chemical behaviour within these estuaries and for assessing their potential for internally cycling and accumulating persistent pollutants. Ideally, synoptic surveys of the whole estuary, carried out simultaneously with quantification of inputs are needed. These should cover the ranges of natural cyclic events, e.g. spring-neap tidal oscillations and seasonal and river run-off variations, and sporadic events, e.g. heavy precipitation and storms. Because these are macro-tidal systems, with pronounced internal movements of sediment and with well-developed turbidity maxima, quantitative investigations of the dynamics of resuspendable particles will also be necessary to assess fully the effects of particle-water exchange processes on chemical accumulations and fluxes.

*Acknowledgements.* The assistance of Robin Howland in preparing this contribution is most gratefully acknowledged. This study was carried out within the Estuarine Chemistry Programme of the Institute for Marine Environmental Research, a component of the Natural Environment Research Council, and was supported, in part, by the Department of the Environment under Contract No. PECD 7/7/076.

# References

Andrews MJ (1984) Thames Estuary: pollution and recovery. In: Sheehan PJ, Miller DR, Butler GC, Bourdeau P (eds) Effects of Pollutants at the Ecosystem Level. SCOPE 22, Wiley and Sons, Chichester, pp 195–227

Andrews MJ, Rickard DG, (1980) Rehabilitation of the inner Thames Estuary. Mar Pollut Bull 11:327–332

Andrews MJ, Steel JEC, Cockburn AG (1983) Biological considerations in the setting of quality standards for the tidal Thames. Water Pollut Res 82:52–60

Cockburn AG, Griggs RW, Lloyd PJ (1980) The equitable approach to pollution control management in the Thames Estuary. Water Res 14:1119–1124

Denman NE (1979) Physical characters of the Humber. In: The Humber Estuary. Nat Environ Res Counc Publ, Ser C, 20:5–8

Department of the Environment/National Water Council (1979) Report of the Sub-Committee on the Disposal of Sewage Sludge to Sea 1975–8. Nat Water Counc, Lond, 66 pp

Department of Scientific and Industrial Research (1964). Effects of polluting discharges on the Thames Estuary. Water Pollut Res, Tech Paper, HMSO Lond, 11:609

Gameson ALH (1976) Routine surveys of the tidal waters of the Humber basin. 1 — Physical parameters. Water Res Centre, Tech Rep TR25:51

Gameson ALH (ed) (1982) The Quality of the Humber Estuary, 1961–1981. Yorkshire Water Authority, Leeds, 88 pp

Gardiner J (1982) Nutrients and persistent contaminants. In: Gameson ALH (ed) The Quality of the Humber Estuary, 1961–1981. Yorkshire Water Authority, Leeds, pp 27–33

Hill JM, Mance G, O'Donnell AR (1984) The quantities of some heavy metals entering the North Sea. Water Res Cent Tech Rep TR 205:21

Inglis CC, Allen FH (1957) The regimen of the Thames as affected by currents, salinities, and river flow. Proc Inst Civil Eng 7:827–878

Jaffe D, Walters JK (1977) Intertidal trace metal concentrations in some sediments from the Humber Estuary. Sci Total Environ 7:1–15

Jones LH (1979) Heavy metals in the Humber Estuary and its organisms. In: The Humber Estuary. Nat Environ Res Counc Publ, Ser C, 20:13–16

Jones PCW, Jefferies DF (1983) The distribution of selected trace metals in United Kingdom shelf waters and the North Atlantic. Can J Fish Aquat Sci 40:111–123

Lloyd PJ, Cockburn AG (1983) Pollution management and the tidal Thames. Water Pollut Control 82:392–401

Ministry of Agriculture and Fisheries (1951) Pollution of the Humber. Fish Invest Ser 1, Vol V, HMSO Lond 4:20

Ministry of Housing and Local Government (1961) Pollution of the tidal Thames. HMSO Lond, 68 pp

Murray LA, Norton MG, Nunny RS, Rolfe MS (1980) The field assessment of effects of dumping wastes at sea: 6. The disposal of sewage sludge and industrial waste off the River Humber. Fish Res Tech Rep MAFF, Lowestoft, 55:35

Natural Environment Research Council (1979) The Humber Estuary. Nat Environ Res Counc Publ Ser C, 20:36

Nelson LA (1979) Minor elements in the sediments of the Thames Estuary. Estuarine Coastal Mar Sci 9:623–629

Norton MG, Eagle RA, Nunny RS, Rolfe MS, Hardiman PA, Hampson BL (1981) The field assessment of effects of dumping wastes at sea: 8. Sewage sludge dumping in the outer Thames Estuary. Fish Res Tech Rep MAFF, Lowestoff 62:62

Rickard DG, Dulley MER (1983) The levels of some heavy metals and chlorinated hydrocarbons in fish from the tidal Thames. Environ Pollut Ser B, 5:101–119

Smith JD, Nicholson RA, Moore PJ (1971) Mercury in waters of the tidal Thames. Nature (Lond), 232:393–394

Talbot JW, Harvey BR, Eagle RA, Rolfe MS (1982) The field assessment of the effects of dumping wastes at sea: 9. Dispersal and effects on benthos of sewage sludge dumped in the Thames Estuary. Fish Res Tech Rep MAFF, Lowestoft 63:42

Urquhart C (1979) Water quality investigations in the Humber Estuary. In: The Humber Estuary. Nat Environ Res Counc Publ Ser C 20:9–12

Wheeler A (1979) The Tidal Thames: a History of a River and Its Fishes. Paul, Lond 228 pp

Wood LB (1980) The rehabilitation of the tidal River Thames. Public Health Eng 8:112–120

Wood LB (1982) The Restoration of the Tidal Thames. Hilger Bristol, 202 pp

Woodward GM (1984) Pollution controi in the Humber Estuary. Water Pollut Control 83:82–90

Woodward GM (1982), Appendix B. Monitoring detail. In: The Quality of the Humber Estuary, 1961–1981 Grameson ALH, (ed), Yorkshire Water Authority, Leeds pp 70–74

# Dredged Materials

U. FÖRSTNER[1] and W. SALOMONS[2]

## 1 Introduction

Sediments are increasingly recognized as both a carrier and a possible source of contaminants in aquatic systems. Pollutants are not necessarily fixed permanently by sediment, but may be recycled via biological and chemical agents, both within the sedimentary compartment and the water column (see Kersten, this Vol.). This is especially valid for "dredged materials". As shipping demands minimal water action or current within the harbor basins, this means that optimal conditions have been created for the sedimentation of river or sea-borne material. In order to keep these ports and channels accessible to (marine) shipping, this material has to be removed regularly by dredging (Van Driel et al. 1984).

The International Association of Ports and Harbors (IAPH 1981) received 108 responses from 37 countries on a questionnaire of the year 1979: There were 350 million tonnes of maintenance dredging and 230 million tonnes average annual new construction dredging; this survey found that about one-fourth of all dredged material is ocean-dumped and another two-thirds is deposited in wetlands and nearshore. In the river mouths to the southern coast of the North Sea, approx. 20 million $m^3$, have to be dredged from Rhine/Meuse (Rotterdam Harbor) and approx. 10 million $m^3$ from the rivers Scheldt (Antwerp), Weser (Bremerhaven), and Elbe (Hamburg) (d'Angremond et al. 1978). The total quantity of sediment which is dredged in The Netherlands, Belgium, and West Germany amount to about 12 times the total suspended matter supply from the Rhine (Van Driel et al. 1984). Table 1 summarizes the data for the coast of the southern North Sea (Salomons and Eysink 1981).

The possibilities of disposal of these enormous quantities of material are severely limited because of the pollutants present in the dredged material. An example is given in Fig. 1: In the Rotterdam harbor the amount of material which has to be dredged annually increased from 0.4 million $m^3$ in 1920 to more than 20 million $m^3$ at present. In the last 80 years the cadmium concentrations increased by a factor of approximately 100 (note logarithmic scale in Fig. 1).

Economic and environmentally safe disposal options have to be found for these materials. In this review two major disposal strategies – disposal on land and intertidal sites; marine dispersion and containment – will be discussed.

[1]University of Technology Hamburg-Harburg, P.O. Box 90 14 03, D-2100 Hamburg 90, FRG
[2]Institute for Soil Fertility, P.O. Box 30003 9750 RA Haren (Gr), The Netherlands

**Table 1.** Mean annual maintenance dredging of harbors along the coast of the Southern North Sea. (Salomons and Eysink, 1981)

| Harbour | Mean annual maintenance dredging approx. values in million $m^3$ per year |
|---|---|
| Dutch harbors at the Western Scheldt | 2.5 |
| Antwerp | 10 |
| Rotterdam | 21 |
| Scheveningen | 0.2 |
| IJmuiden | 2.5 |
| Emden/Delfzijl | 15 |
| Bremen/Bremerhaven | 10 |
| Hamburg/Cuxhaven | 11 |
| London | 0.7 |
| Hull | 4.8 |

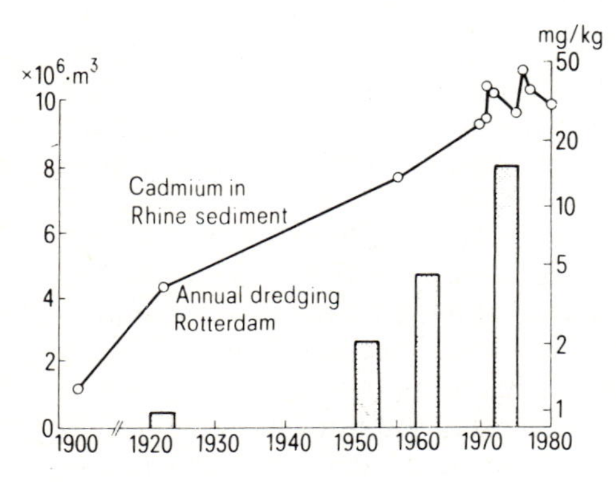


**Fig. 1.** Increase in cadmium concentrations in Rhine sediments due to increased industrial use and the increase in annual dredging in the Rotterdam harbor due to harbor extension. (Salomons and Förstner 1980)

## 2 Origin and Composition of Dredged Material

Suspended matter, originating from the weathering and erosion of soils and rocks as well as from anthropogenic sources, consists of a variety of components including clay minerals, cajbonates, quartz, feldspars, and organic solids. These components are usually coated with hydrous manganese and iron oxides or by organic substances, which to a large extent affect the interaction processes between solids and dissolved components. Depending on internal conditions (the composition and adsorption characteristics of the suspended matter) and external conditions (for example, pH, Eh, the presence of natural and man-derived organic and inorganic ligands), a redistribution of trace metals, nutrients and organic components takes place during transport and deposition of suspended matter, particularly during diagenesis of the resulting sediment (see Kersten, this Vol.). The recycling of mineralized organic matter and the pore-fluid transfer processes are essential components in the nutrient and pollutant dynamics of aquatic systems.

Sediments and dredged materials are classified physically on the basis of grain size. Grain size of sediment is important because it determines the conditions under which sediment will be resuspended or deposited, it determines the basic habitat available for benthic organisms, and it also determines the surface area to volume ratio of the solid phase, which is important in chemical exchange processes with the aqueous phase (Kester et al. 1983).

Water content, which significantly affects mechanical properties of the sediment, is increased greatly during hydraulic dredging. Table 2 (after Christiansen et al. 1982, Tent 1982, Municipality of Rotterdam 1986) lists the range of chemical constituents which are expected to be important for dredged materials. The variability of dredged material must be recognized in considering its disposal in the environment (Kester et al. 1983).

**Table 2.** Composition of dredged sludge from Rotterdam and Hamburg harbors (Christiansen et al. 1982, Tent 1982, Rotterdam Municipality 1986). All numbers for dry matter

| General characteristics | Hamburg harbor | Rotterdam harbor |
|---|---|---|
| Water content | 70 to 85 wt % | |
| Grain size fraction < 63 m | 70 to 90% d.m. | 40% d.m. (< 16 m) |
| Loss of ignition | approx. 15% d.m. | 6–8% (organic m.) |
| Carbonate content | approx. 2% d.m. | 16.5–30% |
| pH | original pH > 7 | |

| Main Constituents | | Metals | Rotterdam[a] | Hamburg[b] | (Natural[c] background) |
|---|---|---|---|---|---|
| $SiO_2$ | 50–60% | Arsenic | ca. 40 | 28– 95 mg kg$^{-1}$ | ( 13 mg kg$^{-1}$) |
| $Al_2O_3$ | 10% | Lead | 80 –230 | 112– 438 mg kg$^{-1}$ | ( 20 mg kg$^{-1}$) |
| $Fe_2O_3$ | 5% | Cadmium | 3 – 16 | 6– 20 mg kg$^{-1}$ | ( 0.3 mg kg$^{-1}$) |
| CaO | 2.5% | Chrom | 97 –187 | 94– 244 mg kg$^{-1}$ | ( 60 mg kg$^{-1}$) |
| $K_2O$ | 1.5% | Kupfer | 39 –142 | 170– 897 mg kg$^{-1}$ | ( 50 mg kg$^{-1}$) |
| MgO | 1% | Nickel | 24 – 44 | 50– 100 mg kg$^{-1}$ | ( 35 mg kg$^{-1}$) |
| $SO_3$ | 1% | Mercury | 0.8– 7 | 3– 12 mg kg$^{-1}$ | ( 0.4 mg kg$^{-1}$) |
| $P_2O_5$ | 1% | Zinc | 256 –1016 | 1020–2450 mg kg$^{-1}$ | (100 mg kg$^{-1}$) |
| $Na_2O$ | 0.7% | Oil | 520 –2320 mg kg$^{-1}$ | | |
| $TiO_2$ | 0.5% | PCB's[d] | 0.32–1.09 mg kg$^{-1}$ | | |

[a] Europoort (smaller values)/Waalhaven (higher values) for 1984.
[b] 80%-range of samples with more than 10% loss of ignition.
[c] Fine-grained shallow water sediment (Salomons and Förstner 1984).
[d] Sum of Aroclor 1242, 1248, 1254 and 1260.

The data in Table 2 indicate characteristic differences of the dredged materials from Rotterdam and Hamburg harbors in the content of carbonate; there is a particular high enrichment of cadmium (factor 50), mercury (factor 20), zinc (factor 17.5), lead (factor 14), and copper (factor 10) in the average composition of dredged material from Hamburg harbor, compared to natural (geogenic) concentrations of these metals. These characteristics are valid for approximately 1–2 million cbm of sediments to be dredged annually in the harbor area.

In the Rotterdam harbor area, both fluvial and marine sediments are deposited. In the most easterly harbors the percentage of marine mud is low, whereas in the westerly harbors (Europoort) up to 90% of the mud may be derived from the

North Sea. This decrease in the amount of fluvial, contaminated material is reflected in decreasing heavy metals in the seaward direction. From these compositional and geographical differences, four classes of dredged materials are distinguished, for which different disposal options are being applied (Sect. 5).

## 3 Treatment of Strongly Contaminated Sludges

During the last few years, several unit operations have been developed for the treatment of contaminated residues such as dredged materials. A detailed discussion of the available technologies and future perspectives for the management of dredged materials is presented in several contributions to the book of Salomons and Förstner (1988). In a review article of Van Gemert et al. from TNO the following techniques can be applied for the treatment of contaminated dredged materials:

**A** – Large-scale concentration techniques. These techniques are characterized by large-scale applicabilities, low costs per unit of residue to be treated and a low sensitivity to variations in circumstances. It is advantageous that these techniques may be constructed in mobile or transportable plants. "A"-techniques include methods such as hydrocyclonage, flotation, and high gradient magnetic separation; highly effective combination of hydrocyclonage and elutriator has been designed by Werther (1988) for the processing of harbor sludge from Hamburg.

**B** – Decontamination or concentration techniques which are especially designed for relatively small-scale operation. These techniques are generally suited for the treatment of residues which contain higher concentrations of contaminants; they involve higher operating costs per unit of residue to be treated; furthermore, they are more complicated, require specific experience of operators and are suitably constructed in stationary or semi-mobile plants. "B"-techniques include biological treatment, acid leaching of inorganic compounds, ion exchange methods, and solvent extraction of organic compounds. A combined scenario for the treatment of contaminated dredged sediment is shown in Fig. 2 (Van Gemert et al. 1988).

Stabilization techniques proposed for solid waste materials and contaminated soils (Wiedemann 1982; Rulkens et al. 1985) may also be considered for polluted sediments (Calmano 1988).

## 4 Disposal Alternatives for Dredged Material

The major disposal alternatives for dredged sediments and their modifications are listed in Table 3 (Gambrell et al. 1978):

These categories differ primarily in the biological population exposed to the contaminated sediments, oxidation-reduction conditions, and transport processes potentially capable of removing contaminants from dredged materials at the

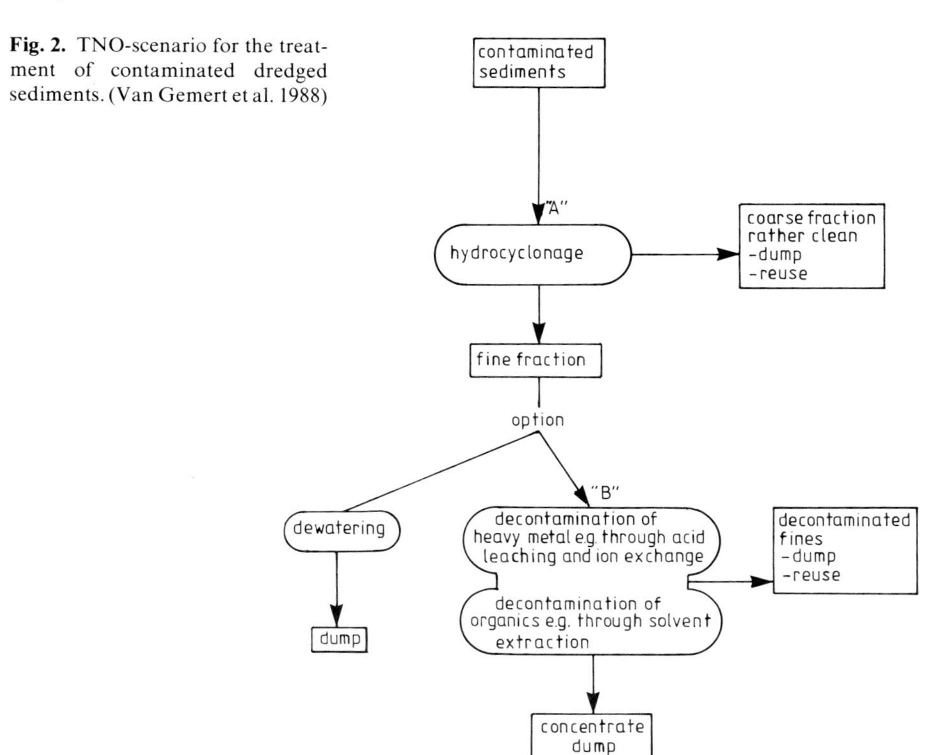


**Fig. 2.** TNO-scenario for the treatment of contaminated dredged sediments. (Van Gemert et al. 1988)

**Table 3.** Disposal alternatives for dredged materials. (After Gambrell et al. 1978)

| |
|---|
| Upland disposal |
| – Long-term confinement (ponded/no drainage; nonponded/dewatering) |
| – Interim confinement (dewatering and consolidation prior to transport) |
| – Unconfined upland[a] |
| – Habitat development[a] |
| – Agricultural soil amendment[a] |
| Intertidal Sites |
| – Unconfined (mudflats, marsh) |
| – Confined by boundary structures, e.g., for habitats |
| Subaqueous Disposal |
| – Unconfined disposal |
| – Confined deposition (mounded deposits or capped borrow pits) |

[a] For less contaminated materials.

disposal site (Gambrell et al. 1978). A review on the decisionmaking framework for the various disposal options of dredged material has been given by Peddicord et al. (1986); implications for marine disposal have been discussed by Kester et al. (1983). Biological aspects of dredged material disposal in the sea are discussed by Marquenie and Tent in this Volume.

## 4.1 Upland Disposal

With respect to upland disposal, it has been exemplified by Salomons et al. (1982) from studies in the Rhine river flood plain that contaminated sediments affect both groundwater quality and agricultural products:

Calculations by Kerdijk (1981) of dispersion processes indicated that chloride, showing conservative behavior, will appear in the adjacent polders in the year 2100 approximately, the heavy metals one to three centuries, and pesticides several thousand years later.

The cattle grazing in the river flood plains is exposed to three sources of heavy metals in the diet; the drinking (river) water, the herbage and the ingested soil particles. Contribution of drinking water is less than 1% of the total heavy metal intake. The contribution of the contaminated grass and soil particles, however, is quite significant for the daily intake of As, Cd, Hg, and Pb by a dairy cow on the river flood plain as compared with an animal in an uncontaminated situation.

Pot experiments and field studies on dredged materials from Rotterdam harbor that in particular cadmium concentrations approximate or exceed the tolerable level in several crops. In green leafy vegetables, wheat grain, and other cereals and in most green fodder crops, the cadmium concentrations must be considered too high for human and animal consumption.

While the levels of chlorinated hydrocarbons in Rhine sediments are relatively low and do not induce intolerable high concentrations in the consumable products, sometimes pesticides and byproducts of the synthesis thereof have been dumped in certain harbor areas; dredged materials originating from these areas and in use for agriculture gave rise to intolerable content of chlorinated hydrocarbons in potatoes and carrots.

In the Hamburg harbor area, where contaminated mud is still pumped into large polders for sedimentation, these materials are no longer permitted for agricultural use. Due to the low carbonate content (see Table 2), metals are easily transferred to crops during lowering of pH (see below) and permissible limits of cadmium have been exceeded in as much as 50% of wheat crops grown on these materials (Herms and Tent 1982). In a recent study, high concentrations of metals have been measured in oxidized pore waters from sedimentation polders in the Hamburg harbor area (Maaβ et al. 1985). From these findings, two major implications during land disposal of dredged material can be demonstrated.

### 4.1.1 Metal Mobilization in Acid Solutions

There is a significant decrease of pH values in less buffered, low-carbonate dredged sludges, such as those from Hamburg harbor, after some time interval (months to several years) subsequent to land spreading. This can be explained by the ability of certain bacteria (*Thiobacillus thiooxidans* and *T. ferrooxidans*) to oxidize sulfur and ferrous iron; while decreasing the pH from 4–5 to about 2.0, the process of metal dissolution from dredged sludge is enhanced. In a laboratory system acidification with sulfurous acid to pH 4.0 and subsequent bacterial leaching solubilized the following metal percentages from the dredged sediments of Hamburg harbor (Calmano et al. 1983): Cd and Co, 98%; Mn, 91%; Cu, 84%; Ni, 66%; Cr, 45%; Fe, 27%; and Pb, 17%. Despite this, the original objective to "detoxify" the material to the quality standard required for agricultural application (e.g., Müller and Riethmayer 1982) could not be reached, even by the combined method of acid/bacterial leaching.

### 4.1.2 Pollutant Mobilization from Porewaters

The composition of interstitial water in sediments is perhaps the most sensitive indicator of the types and the extent of reactions that take place between trace metal-laden sediment particles and the aqueous phase which contacts them. Significant enrichment of trace metals in porewaters has been found in anoxic sediment samples and has been explained by effects of complexation by organic substances. On the other hand, iron in pore solutions shows a typical temporal evolution, which seems to be controlled by precipitation reactions rather than by complexation: In a large-scale experiment on $80 \times 30 \times 6$ m pits in the Rhine estuary, filled with dredged sediments, iron concentrations in the pore water reached a maximum after 40–50 days and then decreased as a result of precipitation of iron sulfide (marine conditions) or iron carbonate (Förstner and Salomons 1983).

Early sediment changes and element mobilization from porewater in a man-made estuarine marsh has been investigated by Darby et al. (1986). This study exemplifies both the mechanisms of release of metals via porewater extraction and subsequent changes by the effect of oxidation (Table 4).

Compared to the river water concentration, the channel sediment porewater is enriched by a factor of 200 for Fe and Mn, 30–50 for Ni and Pb, approximately 10 for Cd and Hg, and 2–3 for Cu and Zn. When the expected concentration of metals following hydraulic dredging, which were calculated from a rate of

**Table 4.** Mobilization of metals and nutrients during dredging. (After Darby et al. 1986). All concentrations are in mg $l^{-1}$ except Hg

| Metal | Channel sediment porewater (a) | River water concn. (b) | a/b | Effluent at man-made marsh | | | |
|---|---|---|---|---|---|---|---|
| | | | | Expected concn. | Measured concn. | % | Change |
| Mn | 6.94 | 0.03 | 230 | 1.34 | 1.19 | – | 11 |
| Fe | 57.3 | 0.26 | 220 | 11.12 | 6.01 | – | 46 |
| Ni | 0.054 | 0.001 | 54 | 0.011 | 0.035 | + | 218 |
| Pb | 0.077 | 0.002 | 38 | 0.016 | 0.142 | + | 788 |
| Hg (g/l) | 3.2 | 0.26 | 12 | 0.82 | 2.0 | + | 144 |
| Cd | 0.009 | 0.001 | 9 | 0.0025 | 0.019 | + | 660 |
| Cu | 0.012 | 0.004 | 3 | 0.0055 | 0.051 | + | 827 |
| Zn | 0.12 | 0.052 | 2 | 0.065 | 5.30 | + | 8069 |

porewater to river water of about 1:4, were compared with the actual measurements at the pipe exiting the dredging device, characteristic differences were observed (Table 4). When the actual concentration at the pipe's exit was less than expected, that element was removed from solution, presumably by scavenging or precipitation; this was valid for iron (approx. half of the expected concentration) and to a lesser extent for manganese. When the exiting solution was greater than the expected concentration in Table 4, mobilization from sediments was assumed; in this respect, highest rates of release were found for zinc, followed by copper, lead, and cadmium.

According to Darby et al. (1986), the levels of heavy metal mobilization were higher at this time than at any time in the subsequent 2 years of marsh maturation. The amount of metal mobilization detected at the effluent pipe of the disposal area during dredging could be accounted for by the release of relatively small amounts of those elements bound to labile sediment phases; while only 3–5% of the labile phase Cu and Ni would account for the measured increases reported in Table 4, up to 36% of the labile phase Pb and Zn was required to account for the higher than expected concentrations of these elements (see also Adams and Darby 1980).

Hoeppel et al. (1978) were among the first to observe changes in element forms during dredged spoil disposal on land containment areas. They compared five influent and effluent samples of suspended matter in such ponding systems (Table 5). From their findings it is obvious that cadmium concentrations increased significantly in the carbonate phase of the effluent samples, presumably as a direct result of the transfer from organic/sulfidic associations present in the influent slurries. This findings are confirmed by our data from sequential leaching procedures on Hamburg harbor sediments (Kersten and Förstner 1987; see Kersten, this Vol.): Following the application of the elutriate test, the oxidizable sulfidic/organic portion of Cd decreases drastically and is now found in the easily reducible fraction. Coprecipitation and adsorption of Cd with the precipitated oxyhydrates may have removed the liberated metal from solution. After freeze- and oven-drying of the initially anoxic samples, cadmium proportions were found even in the most mobile operationally defined carbonatic and exchangeable fractions. The high concentration of cadmium present in these fractions may have a hazardous impact on water quality during dredging and disposal operations as well as upland disposal of these sediments (Gambrell et al. 1978; Khalid 1980).

**Table 5.** Cadmium forms in solids from confined land disposal. (Hoeppel et al. 1978)

| Chemical fraction | Percent of total cadmium content | |
|---|---|---|
| | Influent | Effluent |
| Exchangeable fraction[a] | 21.0% | 18.0% |
| Carbonatic fraction[b] | 21.4% | 56.7% |
| Easily reducible fraction[c] | 9.2% | 11.8% |
| Remaining phases | 49.3% | 13.5% |

[a] Ammonium acetate extractable.
[b] 1 M acetic acid extractable.
[c] 0.1 M hydroxylamine hydrochloride in 0.01 M nitric acid.

## 4.2 Disposal on Intertidal Sites

Riverborne metals entering an estuarine environment can be affected by a change in pH, chlorinity, turbidity maximum, and formation of new particulate matter (Salomons 1980). Significant changes occur even in the low-salinity region of an estuary; the removal of riverborne iron and, in some estuaries, of manganese at low salinities is well established (Duinker 1980). However, apart from the flux of riverine material into an estuarine environment, the deposited particulates may provide a source of dissolved and newly formed particulate components. As a result of biological or biochemical pumping, the intertidal flats in many estuaries act as a source of dissolved metals (Morris et al. 1982).

### 4.2.1 Effect of Salinity

Field investigations by Ahlf (1983) and Calmano et al. (1985) on longitudinal sections of the Elbe and Weser estuaries indicate characteristic mobilization of cadmium, different from other trace metals studied, at the salinity gradient. Similar effects have been reported from other estuarine examples, e.g., from the Scheldt estuary and the Gironde estuary, and has been interpreted by oxidation processes and by intensive breakdown of organic matter whereafter the released metals become complexed with chloride and/or ligands from the decomposing organic matter in the water (Salomons and Förstner 1984). In this way the uptake by or precipitation on the suspended matter may be inhibited; in addition, it has been suggested by Millward and Moore (1982) that the major cations, magnesium and calcium, are probably co-adsorbed; competition from these species for adsorption sites increases with increasing salinity.

Results of experiments on anoxic Rhine sediments have been performed by Salomons et al. (1982) (Table 6). Whereas in freshwater no remobilization was observed for all metals studied (in some cases even an additional adsorption from the water phase), treatment with seawater affects approximately 50% of the cadmium concentration to be released into the water. Similar effects were observed by Rohatgi and Chen (1976) at investigations on the release of trace elements from waste effluents on mixing with seawater, where up to 95% of cadmium was released from the suspended solids to the oxygenated seawater.

Recent data by Prause et al. (1985) suggest that the reaction kinetics, mainly with respect to the initial mobilization of metals from solids, are controlled by

**Table 6.** Metal mobilization from anoxic Rhine sediments by treatment with freshwater and seawater. (Salomons et al. 1982)

| | Zn | Cu | Ni | Cd | Pb |
|---|---|---|---|---|---|
| River water | −0.8%[a] | 0.9% | −2.0%[a] | 1% | 0% |
| Seawater | 2.2% | 2.0% | 2.5% | 49% | 0.1% |

[a] Negative values indicate adsorption of metals from solution.

microbial activity. This would explain the findings of Salomons et al. (1982), that maximum mobilization of cadmium from anoxic sediments occurs only after 6 weeks of suspended interaction with seawater. Subsequent differentations, however, are mainly influenced by thermodynamic factors, e.g., stability of chlorocomplexes, hydrolysis, or readsorption to suspended solids (Salomons 1980; Förstner 1984).

#### 4.2.2 Effect of Oxidation

An example of oxidative remobilization of cadmium and other heavy metals has been studied in a tidal freshwater flat in the upper Elbe estuary near Hamburg (Kersten et al. 1986). This mudflat – diurnal tidal water fluctuations in the range of 3 m affect this productive site – is colonized by dense monodominant reed stands providing an effective trap for heavy metal-laden suspended matter from upstream. Examination of sediment cores taken at this site showed a distinct pattern of redox potential and heavy metal fractionation profiles (Fig. 3). The particulate cadmium binding forms reveal a behavior inverse to that of iron. While in the anoxic zone approximately 60% to 80% of Cd is found in the oxidizable fraction, high percentages of Na-acetate-extractable forms are found in the oxic and post-oxic zones of the sediment cores. The higher amounts of labile cadmium forms are accompanied by a marked depletion in the total content of the toxic metal compared to that in the anoxic sediment zone. Comparison of the fractionation patterns and total contents of other diagenetically less mobile metal examples indicates that a significant proportion of cadmium is leached from the surface sediment by a process of "oxidative pumping" by tidal water drainage in this high-energetic environment. This could result in migration of the remobilized metal into either the deeper anoxic zone, where it can precipitate again to contribute to the enhanced oxidizable sulfidic/organic fraction, or to the surface water, from where it can be exported into the outer estuary. It could, however, also contribute to bioavailable cadmium portions such as indicated by the enhanced macrophyte cadmium concentrations.

These data demonstrate the problematic effect of dispersing anoxic waste materials in ecologically productive, high-energy nearshore, estuarine, and inlet zones (Khalid 1980). This may also pertain to procedures such as "sludge-harrowing" as is occasionally performed in the cold season in some sections of Hamburg harbor (Kausch and Förstner 1986). By application of these techniques, highly contaminated sediments are transferred into the zone of lower pollution; oxygen-consuming substances, such as ammonia, are released from the porewater; increased turbidity affects "light climate" and thus the ecosystem in the lower reaches of the estuary.

### 4.3 Marine Disposal Options

The Definitive Environmental Impact Report (D-EIR) Disposal of Dredged Material of the Netherlands Ministry of Public Health from September 1979 compares the nature of environmental effects produced by the disposal of dredged

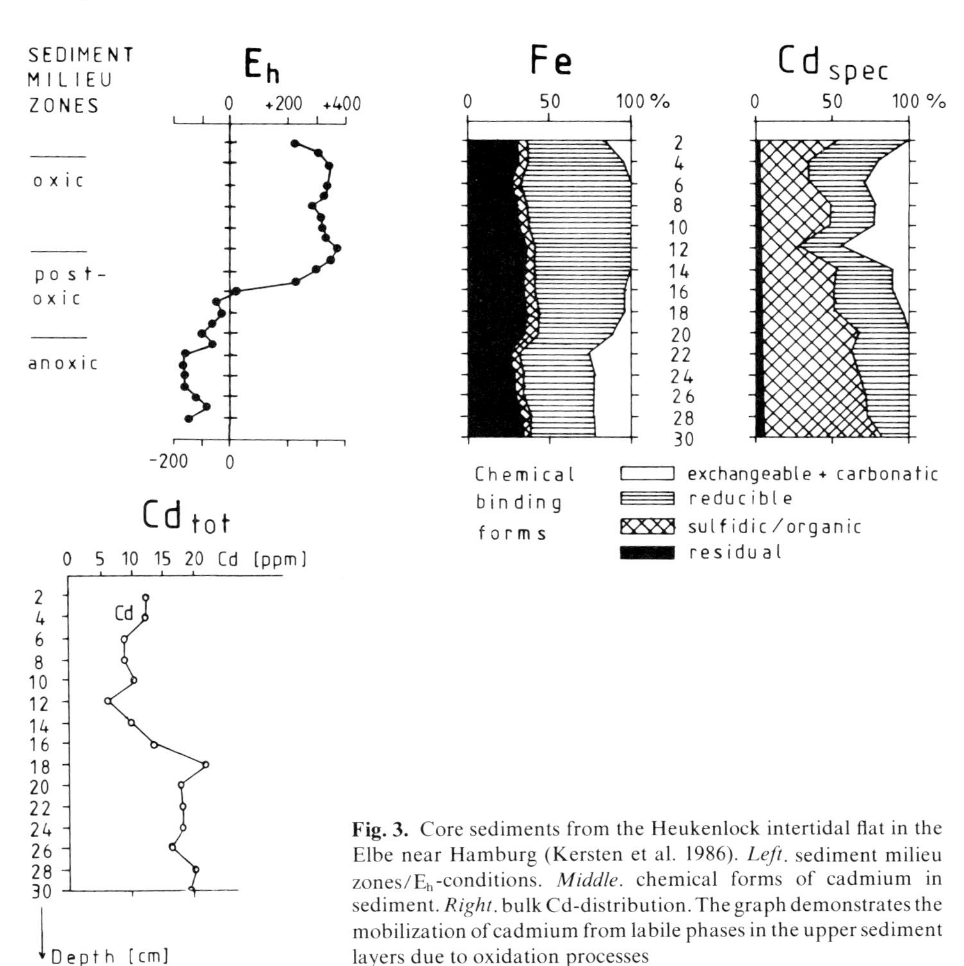


**Fig. 3.** Core sediments from the Heukenlock intertidal flat in the Elbe near Hamburg (Kersten et al. 1986). *Left*. sediment milieu zones/$E_h$-conditions. *Middle*. chemical forms of cadmium in sediment. *Right*. bulk Cd-distribution. The graph demonstrates the mobilization of cadmium from labile phases in the upper sediment layers due to oxidation processes

material at sea, on land and in lakes; Table 7 lists several criteria which are particularly important for the marine disposal option.

With regard to the environmental impact of dredged material in the sea, the EIR concluded that in view of the open character of the food chain, the inability to control the dispersion of contaminants is one of the most important disadvantages of this disposal option. In the EIR the term "open character" is understood as the large uninterrupted geographical dispersion and as the significant dependence/interdependence of the organisms in the marine environment. While at first glance, disposal in the sea seems to have its merits due to considerable dilution, it has to be stated that with such dispersive processes the effects are mostly unpredictable.

**Table 7.** The nature of the environmental impact produced by the disposal of dredged material at sea. (Ministerie van Volksgezondheid en Milieuhygiene 1979)

| | |
|---|---|
| Abiotic environment | |
| Soil | Silt formation through sedimentation, contamination of the soil in the sedimentation area |
| Groundwater | Not applicable |
| Surface Water | Turbidity caused by suspended silt particles; dispersal of contaminants by sea currents |
| Air | No important effects |
| Biotic environment | |
| Disposal Site | Interference with existing ecosystems; changes in the soil structure resulting in changes in composition and species of flora and fauna; accumulation of contaminants in benthic organisms |
| Environs of Disposal Site | Dispersion of contaminants with suspended silt particles and plankton as a result of currents; accumulation in the food chain of fishes and bottom-living organisms via plankton and bottom sediment |
| Human environment | |
| Foodstuffs | Possibility of disappearance of commercially important species of fish; contamination of fish or crustaceans |
| Recreation | Possible negative effects on beach formations |

### 4.3.1 Pollutant Release from Bottom Sediments

Transfer of pollutants from marine sediments into the ecosphere is discussed in the contribution by Kersten in this Volume; it is shown that one of the major mechanisms of pollutant release is oxidation of organic and sulfidic particulate matter, subsequent to mechanical or biological turbation of bottom sediments. In this respect, enclosure studies in the Marine Ecosystem Research Laboratory (MERL) suggest that rates of release of potentially toxic metals are higher from contaminated sediments than from less polluted ones (Hunt and Smith 1983); extrapolation from data on spiked sediments indicates that background levels of Cd will be reached only after 3 years, whereas Cu and Pb are mobilized to attain natural levels after 44 and 400 years, respectively. Other findings with respect to metal mobilization include that after oxidation of the surface sediment the ecosystem is rapidly recovering and that an oxidized, bioturbated surface layer constitutes an effective barrier against the transfer of most trace metals from below into the overlying water (Salomons 1985). These data demonstrate the problematic effect of dispersing polluted sediments, and suggest that containment is generally the more appropriate option for disposal of waste materials (Förstner et al. 1986).

### 4.3.2 Pollutant Behaviour under Anoxic Conditions

Incorporation in naturally formed minerals, which remain stable over geological times, constitutes favorable conditions for the immobilization of potentially toxic metals in large-volume waste materials both under environmental safety and economic considerations. There is a particular low solubility of metal sulfides,

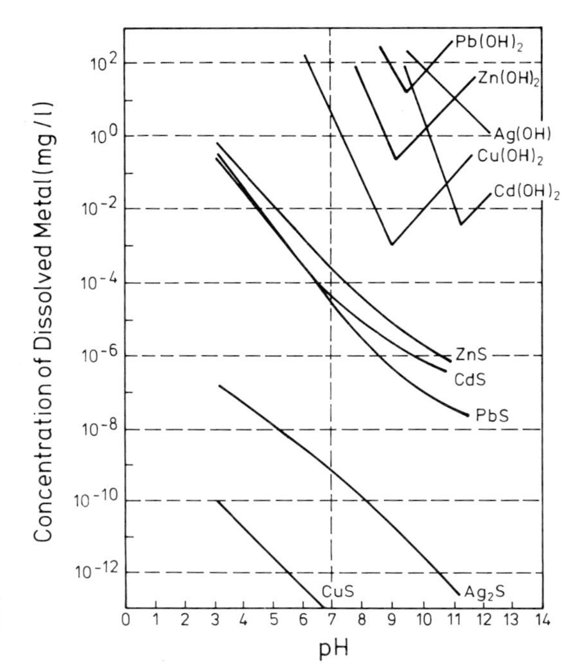


**Fig. 4.** Graph showing the higher solubility of metal hydroxides compared to metal sulfides

compared to the respective carbonate, phosphate, and oxide compounds (Fig. 4). One major prerequisite is the microbial reduction of sulfate; thus, this process is particularly important in the marine environment, whereas in the anoxic freshwater milieu there is a tendency for enhancing metal mobility due to the formation of stable complexes with ligands from decomposing organic matter. Marine sulfidic conditions, in addition, seems to repress the formation of mono-methyl mercury, one of the most toxic substances in the aquatic environment, by a process of disproportionation into volatile dimethyl-mercury and highly insoluble mercury-sulfide (Compeau and Bartha 1983; Craig and Moreton 1984). There are indications that degradation of highly toxic chlorinated hydrocarbons is enhanced under in the sulfidic environment relative to oxic conditions (Sahm et al. 1986; Kersten 1988). A summary of the positive and negative effects of anoxic conditions on the mobility of heavy metals, arsenic, methyl mercury and organochlorine compounds in dredged sludges is given in Table 8.

### 4.3.3 Sub-Sediment Deposition

In a review of various marine disposal options, Kester et al. (1983) suggested that the best strategy for disposing contaminated sediments is to isolate them in a permanently reducing environment. Disposal in capped mound deposits above the prevailing sea-floor, disposal in subaqueous depressions, and capping deposits in depressions provide procedures for contaminated sediment (Bokuniewicz 1983; Morton 1983); in some instances it may be worthwhile to excavate a depression for the disposal site of contaminated sediment than can be capped with clean sediment.

**Table 8.** Summary of positive and negative effects of anoxic (sulfidic conditions on the mobility of heavy metals, metalloids, methyl mercury, and organochlorine compounds in sludges. (Kersten 1988)

| Element or compound | Advantageous effects | Disadvantageous effects |
|---|---|---|
| Heavy Metals (e.g. cadmium) | Sulfide precipitation | Formation of mobile polysulfide and organic complexes under certain conditions with low Fe-oxide concentrations; strong increase of mobility under post-oxic and acidic conditions |
| Metalloids (e.g. arsenic) | Capture by sulfides | Highly mobile under postoxic and neutral to slightly alkaline conditions |
| Methyl Mercury | Degradation and inhibition of $CH_3Hg^+$ formation by HgS precipitate formation | Formation of mobile polysulfide complexes, especially at low Fe concentrations |
| Organochlorine compounds | Initiation of biodegradation by reductive dechlorination (methanic environment is more favorable) | Formation of harmful terminal residues with certain compounds, especially in sulfidic environments; remobilization through colloidal matter suspended in porewater |

This type of waste deposition under stable anoxic conditions, where large masses of polluted materials are covered with inert sediment became known as "subsediment-deposit"; the first example was planned for highly contaminated sludges from Stamford Harbor in the Central Long Island Sound following intensive discussions in the U.S. Congress (Morton 1980). From studies performed by Brannon et al. (1984) is seems that even a 50-cm sand layer is an effective barrier against the transfer of PCB-compounds into the surface water.

There are as yet only few data on the material exchange processes during the input phase. Initial results on the behavior of sludge from Hamburg harbor indicate redistribution of cadmium and zinc between sulfidic and more easily exchangeable phases; in test algae copper was increasingly accumulated under more saline conditions, and it can be expected that photosynthesis of primary producers is inhibited by the introduction of dredged material to the (contained) disposal site (Förstner et al. 1986). With respect to the mechanical processes during deposition of highly polluted sediments, a diffusor technique is applied in Rotterdam harbor, which takes the dredged material to the bottom as a cohesive flow, thus limiting turbidity around the discharge point (Municipality of Rotterdam 1986).

Another type of difficulty associated with the deposition of organic-rich sediments is not yet solved: It is necessary to de-gas the dredged material. Methane and other gaseous components formed under these conditions tend to escape, thus disturbing the structures and causing an increase in turbidity in the surface water. Such effects may also inhibit solidification processes within the deposit, which is particularly important on coastal zones.

## 5 Dredged Material in the North Sea: Dispersion vs. Containment

The Oslo Commission Guidelines for the disposal of dredged material into the sea regulate (13th Meeting of the Standing Advisory Committee for Scientific Advice, Amsterdam, 10–14 March 1986), among others, that

I. In accordance with Article 5 of the Oslo Convention, Contracting Parties shall prohibit the dumping of dredged material containing substances listed in Annex I unless the dredged material can be exempted under Article 8 (2) (trace contaminants) or, in the case of organohalogen compounds, "rapidly converted in the sea into substances which are biologically harmless" (Annex I, paragraph 1). The "black list" contains substances such as organosilicone compounds, mercury, and cadmium compounds as well as cancerogenic substances, e.g., polycyclic aromatic hydrocarbons.

II. Furthermore, in accordance with Article 6 of the Convention, Contracting Parties shall issue specific permits for the dumping of dredged material containing significant quantities of the substances listed in Annex II and, in accordance with paragraph 1 of Annex II, shall ensure that special care is taken in dumping such dredged material. The following interpretation of "significant quantities" have been agreed by the Oslo Commission: Pesticides and their by-products not covered by Annex I and lead and lead compounds – 0.05% or more by weight; all other substances in Annex II, paragraph 1(a) – 0.1% or more by weight.

III. For the dredged material to be disposed of at sea the following information should be obtained:

a) Amount and composition
b) Amount of substances and materials to be deposited per day (per week, per month)
c) Form in which it is presented for dumping, i.e. whether as a solid, sludge or liquid
d) Physical (especially solubility and specific gravity), chemical, biochemical (oxygen demand, nutrient production) and biological properties (presence of viruses, bacteria, yeasts, parasites, etc.)
e) Toxicity
f) Persistence
g) Accumulation in biological materials or sediments
h) Chemical and physical changes of the waste after release, including possible formation of new compounds
i) Probability of production of taints reducing marketability of resources (fish, shellfish, etc.)

These guidelines include advice on dredged material sampling and analysis, e.g., as to suitable numbers of separate stations for a certain amount of dredged material, frequency of sampling, etc. Characteristics of dumping site and method of deposit have to be evaluated, e.g., location in relation to living resources and to amenity areas, initial dilution achieved, dispersal, horizontal transport and vertical mixing characteristics, etc.

All dredged materials, whether contaminated or not, have a significant physical impact at the point of disposal, which includes covering of the seabed (and smothering of benthic organisms) and local enhancement of suspended solids levels. In certain circumstances disposal may interfere with migration of fish (e.g., the impact of high turbidity on salmonids) or of crustacea (e.g., if deposition occurred in the coastal migration path of crabs).

It has been stressed by the Standing Advisory Committee for Scientific Advice of the Oslo Commission (1986) that monitoring is an essential component of management action. Sediment transport studies may be relevant if redistribution, especially of fines, is likely. Where re-colonization studies are appropriate, these can include activities such as remote camera/TV surveys, trial fishing activity or benthic faunal studies.

## 5.1 Dumping of Dredged Material

Dumping of dredged material in the North Sea is practised by a number of countries. In cases like the Scheldt estuary and the Ems estuary, huge amounts of sediments are removed but dumped in the estuary itself; this type of dumping, which does not directly affect the North Sea, will not be considered here. The amounts of dredged material dumped in the North Sea are shown in Table 9, which is based on data from ICES.

**Table 9.** Dredged material dumped in the North Sea by different countries. The data are given in percentage of the total amount ($111 \times 10^6$ tonnes)

| | |
|---|---|
| Belgium | 47.3% |
| Denmark | 0.7% |
| France | 6.1% |
| Ireland | 0.1% |
| Netherlands | 33.5% |
| Portugal | 0.7% |
| United Kingdom | 11.5% |

One important fact which should be taken into account is that even without dredging contaminated fluvial sediments would enter the North Sea to a similar extent as it takes place by dredging activities. The major effect of man is to transport contaminated sediments to certain (dumping) areas, whereas the natural pathways would lead to a more diffuse dispersion.

With respect to the more problematic dumping activities it was decided by the Definitive Policy Plan of the Netherlands (1982) that of the dredged sludge from Rotterdam harbor only class 1 material from the western harbor area, which is primarily of marine origin, is permitted for disposal in the sea at the Loswal Noord site (approximately 13 million m$^3$ per year). Since beginning of 1985, the Municipality of Rotterdam has not been given an exemption under the Sea Water Act to discharge dredged material from the class 2 area (Botlek area) into the sea.

That the disposal of class 1 dredged material at sea is considered acceptable is based on present knowledge. However, in the Policy Plan it is recommended that further research into this is necessary and it is also established that the formulation of target values for the quality of sea water and bottom should be developed, in relation to the disposal of dredged material. In this context, is seems particularly important to know the routes of fine-grained and more polluted components even of class 1 material which may be washed out from bulk sediment during disposal and may enter the circulation patterns of the southern North Sea (see contribution by Taylor, this Vol.).

## 5.2 Marine Near-Shore Containment of Dredged Material

For the disposal of approximately 10 million $m^3$ dredged sludge of the classes 2 and 3 from the harbor area, the Port of Rotterdam and the Netherlands Waterways Administration, after several years of intensive and costly planning, has now started to construct a "sludge island" in the form of a peninsula as a containment for approximately 150 million $m^3$ of sediment. The deposit will consist of a 20 m deep hole; the excavated material will form an 18-m-high, high-tide resistant ring wall, containing approximately 30 million $m^3$ sand. With an area of 300 ha, the net capacity is about 90 million $m^3$. Owing to consolidation during filling, a considerable volume reduction will occur; thus, approximately 150 $m^3$ of wet sludge will eventually be deposited in this structure (Göhren et al. 1986).

The sludge is transported via pipeline over a distance of approximately 2 km from "Mississippi harbor", to which the circulation water is pumped back after passing a purification unit. While the supernatant water resulting from the consolidation will be cleaned as well in this plant, part of the aqueous solutions from the deposit will enter the bottom sediment together with mobilized pollutants. Model calculations, however, suggest that the concentration of most contaminants will not affect groundwater composition significantly; it is expected that pollutants discharged to the seafloor will have only minor effects on the surrounding ecosystems (Municipality of Rotterdam/Rijkswaterstaat 1984).

It seems, that due to the short distance between the source and deposition areas in the case of Rotterdam harbor the large-scale "island solution" (5–6 D.Fl per $m^3$ original sediment) is economically competitive with the old inland sites (5–10 D.Fl.) and to the sea disposal of dredged materials class 1 (3–7 D.Fl) (Municipality of Rotterdam 1986). However, it has definitively been stated that the Municipality of Rotterdam has no intention to create further large-scale sites after 2002 (the official date by which time the present site is expected to be filled), and measures have to be undertaken to improve the quality of the sediments, particularly from municipal and industrial dischargers in the Rhine River catchment area, to be acceptable for other uses, such as civil engineering construction work, in the ceramics industry or for agricultural purposes (Municipality of Rotterdam 1986).

For the Hamburg harbor dredged sludge similar considerations are now undertaken with respect to a marine near-coast subsediment deposit (Göhren et al. 1986). Initial discussions indicate (Führböter 1985): (1) The optimal location would be an area with water depth between 2 and 5 m and a ground consisting of sand at least of 25 m thickness; (2) the site should be located in an area which is morphologically stable with a low erosion potential for current and wave forces; (3) the technological concept will involve a pre-dug hole which is sealed with un-

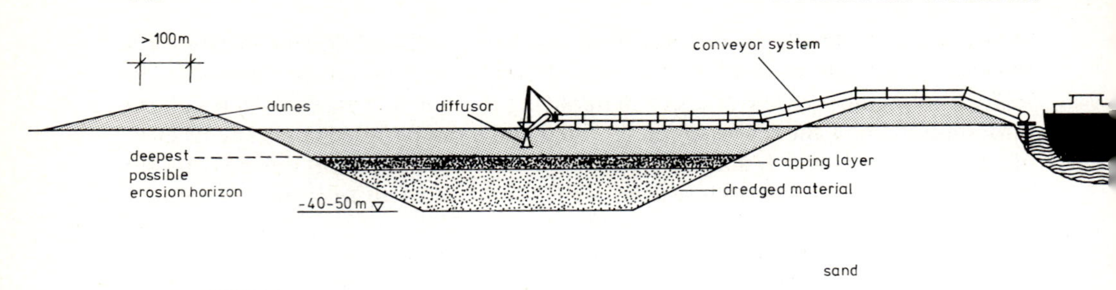


**Fig. 5.** Cross-section of a near-shore, high-tide resistant subsediment deposit for dredged material. (After Göhren et al. 1986)

contaminated material after completion of disposal; (4) the site should be enclosed by an annular dam to prevent discharge of contaminants; (5) the dredged material can be transferred by spoiling or in partially dewatered form (Fig. 5). The disposal site would be operated for a period of 20 years. With an annual volume of dredged sediment of 0.6 million m$^3$ this would be are relative small structure compared to Rotterdam large-scale deposit; however, due to the long distances to acceptable, near-shore marine sites, the price of such a project would be near to 50% of the Rotterdam deposit, much higher than the costs for the actual dumping in polder areas, and probably as expensive as the middle-term project of heap-like deposits in the harbor area.

In this context the question should be raised about the responsibilities for these cost-intensive measures in the future. Since extraction and safe storage of highly contaminated materials, which would otherwise be dispersed, constitutes beneficial action for the whole North Sea system, it would be adequate for refunding at least part of these expenses by the North Sea communities.

## 6 Conclusions

Dredging as such probably does not result in a significantly increased pollutant transport to the North Sea, since without dredging and harbor construction the pollutant load would reach the sea anyhow. The impact of dredged material dumped into the North Sea is difficult to assess, since most of the more problematic constituents do not stay at the dump site. If the dredge spoil is not dumped at a sheltered area, the material (especially the fine-grained proportions) is removed, dispersed, and the dumping area acts as a point source for polluted suspended matter (and, under certain conditions, also for dissolved contaminants).

Generally it is not advisable to disperse polluted materials in the environment, and containment is the appropriate option for the disposal of contaminated sediments. With respect to the various containment strategies it has been argued that upland containment, e.g., on heap-like deposits, could provide a more controlled management than containment in the marine environment. However, contaminants released either gradually from an imperfect impermeable barrier (also to groundwater) or catastrophically from failure of the barrier could produce substantial damage (Kester et al. 1983). On the other hand, near-shore marine containment, e.g., in capped mound deposits, offers several advantages, particularly with respect to the protection of groundwater resources, since the underlying water is saline and chemical processes are favorable for the immobilisation or degradation of priority pollutants.

Research is needed for the recognition of critical pathways of pollutants dumped into the North Sea together with the dredged material; one primary issue concerns the transport patterns from the "classical" dumping grounds into the North Sea circulation system. Research should be continued on the different aspects associated with the near-shore containment technology, in particular with respect to the long-term mechanical stability of these constructions.

## References

Adams DD, Darby DA (1980) A dilution-mixing model for dredged sediments in freshwater systems. In: Baker RA (ed) Contaminants and sediments, vol 1. Ann Arbor Sci Publ, pp 373–392

Ahlf W (1983) The River Elbe: behaviour of Cd and Zn during estuarine mixing. Environ Technol Lett 4:405–410

Bokuniewicz HJ, Kester DR, Ketchum BH, Duedall IW, Park PK (eds) (1983) Submarine borrow pits as containment sites for dredged sediments. In: Kester DR et al. (eds) Dredged-material disposal in the ocean, vol 2. Waste in the ocean. Wiley & Sons, New York, pp 215–227

Brannon JM, Hoeppel RE, Gunnison D (1984) Efficiency of capping contaminated dredged material. In: Dredging and dredged material disposal, vol 2. Qroc Conf Dredging '84, Clearwater Beach, Fla, pp 664–673

Calmano W (1988) Stabilization of dredged mud. In: Salomons W, Förstner U (eds) Environmental management of solid waste. Springer, Berlin Heidelberg New York Tokyo

Calmano W, Ahlf W, Förstner U (1983) Heavy metal removal from contaminated sludges with dissolved sulfur dioxide in combination with bacterial leaching. Proc Int Conf Heavy metals in the environment, Heidelberg, Sept 6–9, 1983. CEP Konsultants, Edinburgh, pp 952–955

Calmano W, Wellershaus S, Liebsch H (1985) The Weser estuary: a study on the heavy metal behaviour under hydrographic and water quality conditions. Veröff Inst Meeresforsch, Bremerhaven

Christiansen H, Ohlmann G, Tent L (1982) Probleme im Zusammenhang mit dem Anfall von Baggergut im Hamburger Hafen. Wasserwirtschaft 72:385–389

Compeau G, Bartha R (1983) Effects of sea salt anions on the formation and stability of methylmercury. Bull Environ Contamin Toxicol 31:486–493

Craig PJ, Moreton PA (1984) The role of sulphide in the formation of dimethylmercury in river and estuary sediments. Mar Pollut Bull 15:406–408

d'Angremond K, Brakel J, Hoekstra AJ, Kleinbloesem WCH, Nederlof L, De Nekker J (1978) Assessment of certain European dredging practices and dredged material containment and reclamation methods. US Army Eng Waterw Exp Stn, Vicksburg, Miss, Tech Rep D–78–58

Darby DA, Adams DD, Nivens WT (1986) Early sediment changes and element mobilization in a man-made estuarine marsh. In: Sly PG (ed) Sediment and water interactions. Springer, Berlin Heidelberg New York Tokyo, pp 343–351

Driel W van, Kerdijk HN, Salomons W (1984) Use and disposal of contaminated dredged mateial. Land Water Intern 53:13–18

Duinker JC (1980) Suspended matter in estuaries: adsorption and desorption processes. In: Olausson E, Cato I (eds) Chemistry and biogeochemistry of estuaries. Wiley, Chichester, pp 121–153

Förstner U (1984) Effects of salinity on the metal sorption onto organic particulate matter. In: Laane RWPM, Wolff WJ (eds) The role of organic matter in the Wadden Sea. Netherlands Inst Sea Res, Publ Ser 10–1984, pp 195–209

Förstner U, Salomons W (1983) Trace element speciation in surface waters: interactions with particulate matter. In: Leppard GG (ed) Trace element speciation in surface waters and its ecological implications. Proc NATO Adv Res Worksh, Nervi/Italy. Plenum, New York, pp 245–273

Förstner U, Ahlf W, Calmano W, Kersten M (1986) Mobility of pollutants in dredged materials – implications for selecting disposal options. In: Kullenberg G (ed) Role of the ocean as a waste disposal option. Reidel, Dordrecht, pp 597–615

Führböter A (1985) Optimierungsbetrachtungen für eine Tiefdeponie auf einer künstlichen Sandinsel in der Nordsee (Einbau von kontaminiertem Hafenschlamm im Klappverfahren). Internal study on behalf, Amt für Strom- und Hafenbau, Hamburg

Gambrell RP, Khalid RA, Patrick WH, Jr (1978) Disposal alternatives for contaminated dredged material as a management tool to minimize environmental effects. Tech Rep DS-78-8, US Army Eng Waterw Exp Stn, Vicksburg, Miss

Gemert WJT van, Quakernaat J, Veen HJ van (1988) Methods for the treatment of contaminated dredged sediments. In: Salomons W, Förstner U (eds) Environmental management of solid waste. Springer, Berlin Heidelberg New York Tokyo

Göhren H, Tamminga P-G, Duchrow H (1986) Baggergutablagerung im Küstenmeer. Schiff Hafen Kommandobrücke 9:71–73

Herms U, Tent L (1982) Schwermetallgehalte im Hafenschlick sowie in landwirtschaftlich genutzten Hafenschlick-Spülfeldern im Raum Hamburg. Geol Jahrb F12:3–11

Hoeppel RE, Meyers TE, Engler RM (1978) Physical and chemical characterization of dredged material influents and effluents in confined land disposal areas. US Army Eng Waterw Exp Stn, Vicksburg, Miss, Tech Rep D-78-24

Hunt CD, Smith DL (1983) Remobilization of metals from polluted marine sediments. Can J Fish Aquat Sci 40:132–142

International Association of Ports and Harbors (ed) (1981) A survey of world port practices in the ocean disposal of dredged material as related to the London Dumping Convention. Report of the Ad Hoc Dredging Committee. AJ Tozzoli, Chairman. Port Authority of New York and New Jersey, One World Trade Center, Room 64W, New York, 38 pp

Kausch H, Förstner U (1986) Gutachterliche Stellungnahme zum Effekt des Schlickeggens in der Tideelbe. Internal study on behalf, Amt für Strom- und Hafenbau, Hamburg

Kerdijk HN (1981) Groundwater pollution by heavy metals and pesticides from a dredge spoil dump. In: Fuyvenboden W van, Glasbergen P, Lelyveld H van (eds) Quality of groundwater. Elsevier, Amsterdam, pp 279–286

Kersten M (1988) Geochemistry of priority pollutants in anoxic sludges: cadmium, arsenic, methyl mercury, and chlorinated organics. In: Salomons W, Förstner U (eds) Environmental management of solid waste. Springer, Berlin Heidelberg New York Tokyo

Kersten M, Förstner U (1987) Effect of sample pretreatment on the reliability of solid speciation data of heavy metals – implications for the study of diagenetic processes. Mar Chem 22:299–312

Kersten M, Förstner U, Kerner M, Kausch H (1986) Remobilisation of Cd and N from periodically inundated soil. In: Trans 13th Congr Int Soc Soil Sci, Hamburg, vol 2, pp 348–349

Kester DR, Ketchum BH, Duedall IW, Park PK (eds) (1983) Wastes in the ocean, vol 2. Dredged-material disposal in the ocean. Wiley, New York, 299 pp

Khalid RA (1980) Chemical mobility of cadmium in sediment-water systems. In: Nriagu JO (ed) Cadmium in the environment, vol 1. Ecological cycling. Wiley, New York, pp 257–304

Maaβ B, Miehlich G, Gröngröft A (1985) Untersuchungen zur Grundwassergefährdung durch Hafenschlick-Spülfelder. II. Inhaltsstoffe in Spülfeldsedimenten und Porenwässern. Mitt Dtsch Bodenkundl Ges 43/I:253–258

Millward GE, Moore RM (1982) The adsorption of Cu, Mn and Zn by iron oxyhydrate in model estuarine solution. Water Res 16:981–985

Ministerie van Volksgezondheid en Milieuhygiene/The Netherlands (ed) (1979) Definitief milieueffectrapport berging baggerspecie

Morris AW, Bale AJ, Howland RJM (1982) The dynamics of estuarine manganese cycling. Estuar Coast Shelf Sci 13:175–192

Morton RW (1980) "Capping" procedures as an alternative technique to isolate contaminated dredged material in the marine environment. In: Dredge spoil disposal and PCB contamination: Hearings before the Committee on Merchant Marine and Fisheries. House of Representatives, 96th Congr, 2nd Sess. Exploring the various aspects of dumping of dredged spoil material in the ocean and the

PCB contamination issue, March 14, May 21, 1980. USGPO Ser No 96–43, pp 623–652, Washington DC
Morton RW (1983) Precision bathymetric study of dredged-material capping experiment in Long Island Sound. In: Kester DR (ed) Wastes in the ocean, vol 2. Dredged-material disposal in the ocean. Wiley, New York, pp 99–121
Müller G, Riethmayer S (1982) Chemische Entgiftung: das alternative Konzept zur problemlosen und endgültigen Entsorgung schwermetallbelasteter Baggerschlämme. Chem Z 106:289–292
Municipality of Rotterdam/Rijkswaterstaat (ed) (1984) Grootschalige locatie voor de berging van baggerspecie uit het benedenrivierengebied. Projectrep/Environ Compat Stud, Oct 1984, 334 pp
Municipality of Rotterdam (ed) (1986) Rotterdam dredged material: Approach to handling, 40 pp
Peddicord RK, Lee CR, Palermo MR, Francingues NR, Jr (1986) General decisionmaking framework for management of dredged material – example application to Commencement Bay, Washington. US Army Eng Waterw Exp Stn, Vicksburg, Miss, Misc Pap D-86-16
Prause B, Rehm E, Schulz-Baldes M (1985) The remobilization of Pb and Cd from contaminated dredge spoil after dumping in the marine environment. Environ Technol Lett 6:261–266
Rohatgi NK, Chen KY (1976) Fate of metals in wastewater discharge to ocean. J Environ Eng Div ASCE 102:675–685
Rulkens WH, Assink JW, Gemert WJT van (1985) On-site processing of contaminated soil. In: Smith MA (ed) Contaminated land – reclamation and treatment. Plenum, New York, pp 37–90
Sahm H, Brunner M, Schobert SM (1986) Anaerobic degradation of halogenated aromatic compounds. Microbial Ecol 12:147–153
Salomons W (1980) Adsorption processes and hydrodynamic conditions in estuaries. Environ Technol Lett 1:356–365
Salomons W (1985) Sediments and water quality. Environ Technol Lett 6:315–368
Salomons W, Eysink W (1981) Pathways of mud and particulate trace metals from rivers to the southeastern North Sea. In: Nio SD, Schuettenhelm RTE, Weering TCE van (eds) Holocene marine sedimentation in the North Sea basin. Spec Publ Int Assoc Sedimentol 5:429–450
Salomons W, Förstner U (1980) Trace metal analysis on polluted sediments. II Evaluation of environmental impact. Environ Technol Lett 1:506–517
Salomons W, Förstner U (1984) Metals in the hydrocycle. Springer, Berlin Heidelberg New York 349 pp
Salomons W, Förstner U (eds) (1988) Environmental management of solid waste. Springer, Berlin Heidelberg New York Tokyo
Salomons W, Driel W van, Kerdijk H, Boxma R (1982) Help! Holland is plated by the Rhine (environmental problems associated with contaminated sediments). Effects of waste disposal on groundwater. Proc Exeter Symp IHAS 139:255–269
Standing Advisory Committee for Scientific Advice for the Oslo Commission (ed) (1986) Oslo Commission guidelines for the disposal of dredged material. 13th Meet, Amsterdam, March 10–14, 1986
Tent L (1982) Auswirkungen der Schwermetallbelastung von Tidegewässern am Beispiel der Elbe. Wasserwirtschaft 72:60–62
Tent L (1987) Contaminated sediments in the Elbe estuary – ecological and economic problems for the port of Hamburg. In: Thomas RL (ed) The ecological effects of in-situ sediment contaminants. Proc Worksh Aberystwyth/Wales. Hydrobiologia 149:189–199
Werther J (1988) Classification and dewatering of sludges. In: Salomons W, Förstner U (eds) Environmental management of solid waste. Springer, Berlin Heidelberg New York Tokyo
Wiedemann HU (1982) Verfahren zur Verfestigung von Sonderabfällen und Stabilisierung von verunreinigten Böden. Rep Fed Environ Prot Ag 1/82. Schmidt, Berlin

# Sewage Sludge Disposal in the North Sea

M. PARKER[1]

## 1 Introduction

The treatment of sewage leads to final products which cannot be reduced or eliminated by process changes at source. Throughout the ages, sewage has caused injury to health and aesthetic offence but in the modern world these problems are managed in three ways. Firstly by excluding sewage from the human environment (e.g. by development of sanitary engineering and sewerage), secondly by treatment to reduce its deleterious properties and thirdly by disposal of the raw or treated materials so as to avoid public health and aesthetic impact and, preferably to make use of natural processes to further "treat" the materials, which is largely effected by processes of decay. Disposal of sludge to land may also permit a benefit to be gained from the fertilising content of the material. The problems of sewage and sludge disposal have been exacerbated since the nineteenth century by diverting industrial aqueous wastes to the treatment facilities used for human sewage, which although permitting effective treatment of their degradable components, increases the contamination of sewage by persistent and/or toxic materials. This paper examines the options available for disposal of sludge from sewage treatment in North Sea bordering countries, the use made of marine disposal, and assessment of the factors involved in controlling the use of this option without unacceptable environment detriment.

To avoid repetition of citation, it should be noted that an important source of information on sea disposal of sludge has been the Annual Report of the Oslo Commission, and papers presented to its Standing Advisory Committee for Scientific Advice. Other sources are cited in the text in the normal fashion.

## 2 Sources, Quantities and Disposal Routes of Sewage Sludge in North Sea Countries

Sewage sludge production depends upon the size of the population of a country, the degree to which sewerage is provided and the level of treatment applied to the sewage sludge. Sludge production per head is an index of the development of sanitary systems (Table 1).

[1] Ministry of Agriculture, Fisheries and Food, Directorate of Fisheries Research, Fisheries Laboratory, Burnham-on-crouch, Essex CMO 8HA, Great Britain
Present address: Great Westminster House, Horseferry Road, London, SW1P2AE

**Table 1.** Sewage sludge production in North Sea bordering countries and some other Oslo Convention signatory states

| Country | Population (M) | % Pop'n sewered | % Pop'n served by treatment | Total waste water load (mpe)[a] | % Discharged to sewer | % Sewage served by treatment | Current sludge production ($kt\ a^{-1}$) | Sludge production per head of population ($kg\ a^{-1}$) |
|---|---|---|---|---|---|---|---|---|
| North Sea-Bordering Countries | | | | | | | | |
| UK | 55.9 | 95 | 84 | 102 | 75 | 88 | 1500 | 27 |
| FRG | 61.7 | 86 | 66 | 182 | 62 | 78 | 78 | 36 |
| Netherlands | 13.9 | 40 | 79 | 31 | 68 | 86 | 230 | 17 |
| Denmark | 5.1 | 92 | 57 | 15 | 31 | 61 | 130 | 26 |
| Belgium | 9.9 | 55 | 22 | 31 | 52 | 24 | 70 | 7 |
| Norway | 4.1 | | 25[b] | 4.1[c] | | | 55 | 13 |
| Other OSCOM Countries | | | | | | | | |
| France | 53.6 | 56 | 45 | 146 | 38 | 74 | 840 | 16 |
| Ireland | 3.4 | 58 | 26 | 6 | 48 | 25 | 20 | 6 |
| Luxembourg | 0.4 | 76 | 55 | 1 | 83 | 80 | 11 | 26 |
| Spain | 37.0 | | | 37[c] | | | 45 | 1 |
| Sweden | 8.3 | 86 | | 8[c] | | | 210 | 25 |

Notes: Source – recalculated from Vincent and Critchley (1982)
[a] Including industrial waste water.
[b] A further 41% is treated by septic tank.
[c] Domestic load only.

The choice of disposal route for the sludge is related to the availability of each outlet and its environmental, social and economic costs (Table 2). The major proportion of sludge in all European countries is disposed of to land, either to sanitary landfill or for use in agricultural, forestry or park lands as a fertiliser. The primary constraints on the beneficial use of sludge on land is one of land availability, for which population density, combined with a total land usage index, provide an approximate guide in Table 2; this ignores the important constraint also applied by the proportion of land which is built-up, mountainous, heathy or wetland, and therefore unsuitable for spreading or tipping. Constraints also apply in relation to pathogen contamination and to contamination by persistent toxic substances (Commission of the European Communities 1986). Thus, in densely populated areas, tipping, incineration or disposal to sea is often used. The availability of land tipping areas is also constrained, not least by the large and growing requirements for tipping domestic and industrial wastes. Marine disposal is also limited in most continental countries by lack of access to relatively short coastlines and, particularly in the Waddensea areas of the Netherlands, the Federal Republic of Germany (FRG) and Denmark, by the sensitivity of the local coastal ecology. Limitations on the incineration of sludge arise because of substantial energy costs and lack of social acceptance; incinerators are particularly affected by the 'not in my back yard' (NIMBY) syndrome. This latter political constraint applies in varying degrees to all outlets for sewage sludge. In many countries, sea disposal is politically unacceptable; even controlled land spreading, the most obviously beneficial re-use of sludge, is not looked on favourably because of the association in the public mind of sewage and disease.

Thus, of the North Sea bordering countries, only Britain and the Netherlands (both with a high coastal population density) dispose of significant amounts of sludge to sea; elsewhere, Ireland and Spain, both with long coastlines and extensive mountainous or wetland areas also use the marine route. The coastal states of the USA also use this outlet; their constraints on land disposal include high coastal population density.

## 3 Trends in Marine Disposal

Trends in marine disposal since the signing of the Oslo Convention have varied between countries as a result of basic differences in philosophy concerning sea disposal. Until 1980, the FRG dumped ca. 15–20,000 dry tonnes per year of sludge from Hamburg into the German Bight. From 1980, however, this load was diverted from the North Sea to a point beyond the shelf break the south western approaches to the Celtic Sea, but dumping eventually ceased in 1983. Heavy metal contamination prevented use of this sludge in agriculture and it is now exported to the German Democratic Republic for disposal.

In the Netherlands, primary sludge from the Hague conurbation is discharged through a 10-km-long pipeline. About 21,000 t dry weight (1.9 mt wet weight) is discharged by these means, but this figure has remained stable, while sludge production generally in the Netherlands has risen from 160,000 t dry weight in 1975

**Table 2.** Sewage sludge disposal in North Sea bordering countries and some other Oslo Convention signatory states

| Country | Sludge production (kt $a^{-1}$) | Disposal routes (% of total) | | | | | | Population density nos $ha^{-1}$ [a] | Land usage index[b] | Length of coastline (km) | Coastline per unit land area ($\times 10^3$) |
|---|---|---|---|---|---|---|---|---|---|---|---|
| | | Agriculture | Sanitary landfill | Incineration | Pipeline to sea | Dumped at sea | Unspecified | | | | |
| North Sea-Bordering Countries | | | | | | | | | | | |
| UK | 1500 | 41 | 26 | 4 | 2 | 27 | | 5.4 | 9.8 | 4000[c] | 50 |
| FRG | 2200 | 39 | 49 | 8 | | | 4 | 2.5 | 7.8 | 563[d] | 2 |
| Netherlands | 230 | 60 | 27 | 2 | 11 | | | 4.1 | 5.9 | 362 | 11 |
| Denmark | 130 | 45 | 45 | 10 | | | | 1.2 | 2.7 | 1255[e] | 29 |
| Belgium | 70 | 15 | 83 | 2 | | | | 3.2 | 2.3 | 62 | 2 |
| Norway | 55 | 180 | 82 | | | | | 0.1 | 0.1 | 3115 | 10 |
| Other OSCOM Countries | | | | | | | | | | | |
| France | 840 | 30 | 50 | 20 | | | | 1.0 | 1.2 | 2115[f] | 4 |
| Ireland | 20 | 4 | 51 | | | 45 | | 1.2 | 0.4 | 1212 | 46 |
| Luxembourg | 11 | 90 | 10 | | | | | | | 0 | 0 |
| Spain | 1 | 60 | ←20→ | | | | | 0.1 | 0.1 | 2259[g] | 5 |
| Sweden | 25 | 60 | ←30 | | | | | | 0.5 | 2566[h] | 6 |

Notes: Sources – Sludge production and disposal routes from Vincent and Critchley (1982); Land area from Anon (1986); Length of coastline, US Dept of State (1969).

[a] Population divided by total land area (exclusive of small offshore islands and large lakes); this ignores differences in habitability or usage of land.

[b] The 'Land usage index' is derived from (total sludge production × % diverted to agriculture and tipping) divided by total land areas (units of t $ha^{-1} a^{-1}$). This provides a very crude measure of the significance of land disposal in terms of usage of land area, ignoring variations in suitability of land for agriculture or tipping. Norway is badly treated by this index, because of its very high proportion of mountainous land but, in association with the population density figures, the index provides a guide to the constraints in other countries.

[c] Excluding islands.

[d] Of which ca. 350 km from the North Sea.

[e] Of which ca. 400 km from the North Sea.

[f] Of which ca. 1600 km from the English Channel/Atlantic.

[g] Of which ca. 900 km from the Atlantic.

[h] All in the Kattegat/Baltic.

to over 300,000 t today. The long-term aim of the Netherlands is to provide further treatment of the Hague sludge and to divert it to beneficial usage on land.

In the UK, the major proportion of sludge is disposed of to land (Healey 1984), but in contrast, sea disposal is considered an environmentally, socially and economically sensible disposal option for sludges from large conurbations having good access to the sea and constraints upon land disposal. On the North Sea coastline of Britain, sludges are dumped at sea from sewage treatment works operated by all the English coastal Water Authorities and by the Lothian Regional Council in Scotland. Eight disposal sites (Fig. 1) are used, between the Thames and the Forth.

Trends in disposal are shown in Fig. 2 (note that these figures are in wet figures are dominated by the input to the Thames (three disposal sites, with the bulk of the input being to the Barrow Deep Site), though it should be noted that this tonnage figure is inflated by the unusually high water content of the sludge ( 98%). The Thames load peaked at over 5 mt wet weight in 1979–81 and has thereafter declined slightly; it is likely to remain more or less stable in terms of dry matter inputs (ca. 100 kt) for the long term, although the water content will decrease. The next highest inputs are in the Tyne/Tees area in the north east, which currently account for around 0.5 mt wet weight per year. However, this figure is rising as sewage treatment schemes in the Tyne and Tees areas come on stream, so diverting loads from the rivers to the open sea. In the Forth Estuary, two other sites

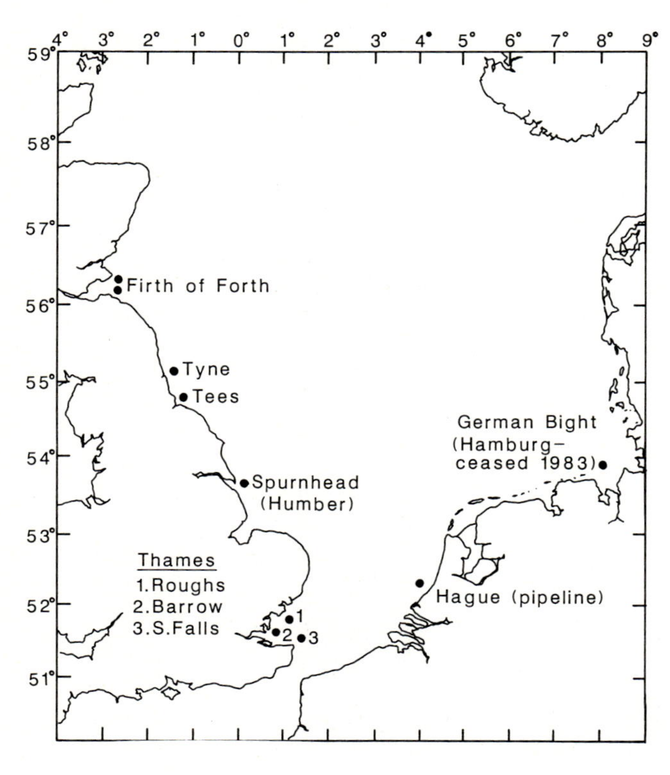


**Fig. 1.** Sewage sludge disposal sites in the North Sea

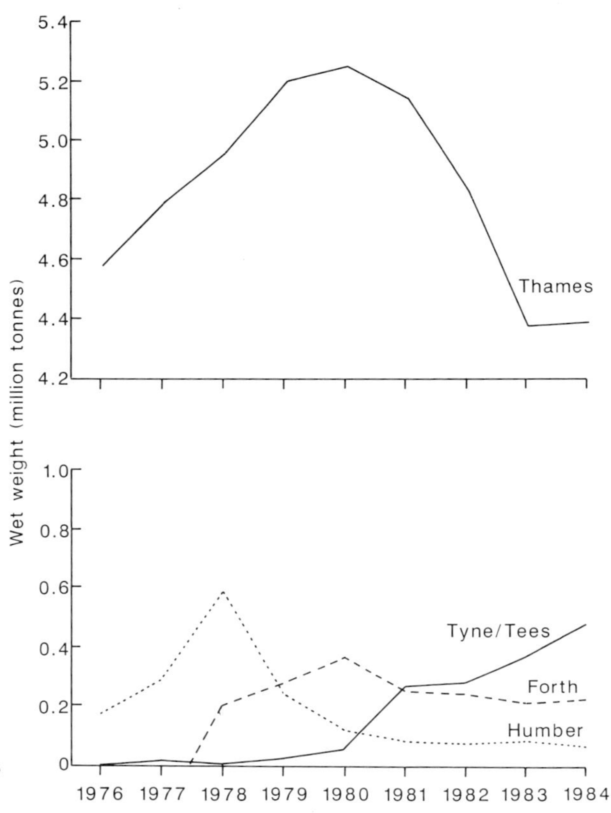


**Fig. 2.** Sewage sludge disposal to areas of the British east coast

are used for sludge from the Edinburgh region (ca. 0.25 mt $a^{-1}$ wet weight) and a smaller input occurs off the mouth of the Humber estuary (0.1 mt $a^{-1}$). The overall quantity *licensed* (as opposed to dumped) for disposal in the British waters in North Sea has remained roughly constant (at around 6 mt wet weight, equivalent to approximately 150–200 kt dry matter) since 1979 and is likely to remain so.

## 4 Nature of the Material

The bulk of the material dumped in British waters of the North Sea, namely the 4–5 mt (wet) dumped annually at the Barrow Deep site in the Thames, is digested sludge. The remainder is primary sludge (i.e. derived from settlement of raw sewage). The gross differences between primary and digested sludge are that digestion results in an overall reduction in Biological Oxygen Demand (BOD) and total dry matter of about 30%; Chemical Oxygen Demand (COD) and grease content are also considerably reduced. The particle size is also reduced, which may affect settling characteristics (Garber et al. 1975). Digestion also leads to a reduction in pathogen levels and a loss of nitrogen from the sludge (Wood 1986). The reduction in carbon content leads to proportional changes in phosphorus and metal concentrations. The bulk of the metal contaminants in sludge remain bound to very small particles (< 10 μm) within the sludge (Chapman 1986) and most organic contaminants also concentrate in the particulate fraction.

Data on nutrient inputs from sewage sludge dumped in UK waters of the North sea in 1981 are given in Table 3; data on inputs from the Netherlands and FRG in that year are not available. These inputs represent about 0.4% of total nitrogen inputs and 2.5% of total phosphorus inputs to the North Sea (Norton 1982). Metal inputs from UK sludges (Table 4) peaked in 1979/80 (the apparent rise before that date partially represents improvement in data gathering) and have since levelled off or declined. The bulk of the inputs are to the Thames, where considerable improvements have been achieved since the passing of the Dumping at Sea Act 1974; in particular, mercury inputs have been reduced substantially. Inputs to the Spurn Head dumping ground off the Humber have also sharply decreased in line with reduction in total inputs, as well as improved sludge quality; this reduction has been offset by the recent rise in inputs to the dumping site off the Tyne. The sludges from the Tyne/Tees area are generally of good quality with respect to contaminants, but the total quantity dumped is rising sharply.

**Table 3.** Nutrient inputs to the North Sea from sewage sludge disposal in the UK (1981 data). (Grogan 1984)

| UK | Nitrogen | | Phosphorus | |
|---|---|---|---|---|
| | Concentration (%) | Max. input ($t\ a^{-1}$) | concentration (%) | max. input ($t\ a^{-1}$) |
| Forth | 3.7 | 353 | 0.5 | 189 |
| Tyne | — | 798 | — | 225 |
| Tees | — | 27 | — | 7 |
| Humber | 2.1–5.1 | 338 | 0.7–2.2 | 93 |
| Thames | 7.5–7.8 | 9210 | 1.5–1.8 | 2125 |
| (Barrow Deep) | | | | |
| Hajwich | {3.1–6.8} | 727 | {0.5–1.4} | 150 |
| S. Falls | | 205 | | |
| | | | | 2 |

**Table 4.** Inputs of metals in sewage sludge to English coastal waters (t)

| Metal | 1976 | 1977 | 1978 | 1979 | 1980 | 1981 | 1982 | 1983 | 1984 |
|---|---|---|---|---|---|---|---|---|---|
| Mercury | 1.4 | 1.6 | 1.6 | 1.7 | 2.7 | 1.5 | 0.9 | 0.4 | 0.4 |
| Cadmium | 7.7 | 8.6 | 5.9 | 6.0 | 6.5 | 4.3 | 3.5 | 2.9 | 2.8 |
| Lead | 98 | 109 | 164 | 162 | 114 | 105 | 99 | 92 | 93 |
| Copper | 105 | 115 | 120 | 118 | 116 | 117 | 111 | 103 | 109 |
| Zinc | 457 | 485 | 608 | 604 | 392 | 367 | 254 | 278 | 188 |
| Nickel | 27 | 25 | 25 | 20 | 18 | 14 | 13 | 13 | 13 |
| Chronium | 56 | 69 | 74 | 64 | 61 | 53 | 50 | 47 | 47 |

As a result of pressure by the licensing authorities on the sewage treatment agencies, much of the achievable reduction in contaminant levels has now occurred; although discussions continue, it is apparent that any further significant reduction is unlikely. This is most noticeable in the cases of copper and lead, where multiple diffuse sources contribute; further reductions in lead input via sludge are likely only as controls on lead in petrol and paint take effect. As a proportion of total inputs to the North Sea (Table 5), sewage sludge provides a relatively small contributor of metals compared with river, atmospheric and dredge spoil inputs in 1981 (Hill et al. 1984; Norton 1982). It is worth noting that there have been improvements in UK sludge quality since 1981 (Table 4), especially with respect to mercury, so that sewage sludge now represents even lower inputs proportionately. Comparable data on inputs of man-made organics in trace amounts in sewage sludge are not available, but there is no reason to believe that the trend would be significantly different.

**Table 5.** Quantities of heavy metals entering the North Sea, 1981 data (t $a^{-1}$), adapted from Hill et al. (1984)

| Metal | Hg | Cd | Pb | Cu | Zn | Ni | Cr |
|---|---|---|---|---|---|---|---|
| Sewage sludge (all sources) | 2 | 5 | 120 | 150 | 480 | 25 | 102 |
| Total input | 20–33 | 150–165 | 1780–3330 | 2430–2470 | 14000–14100 | 2130–280 | 1560–2000 |
| Sewage sludge as % of total | 5–9 | 3 | 4–6.5 | 5.5–6.5 | 3.5 | 1 | 5–6.5 |

# 5 Toxicity

Sludges usually contain a wide range of materials other than the trace contaminants listed above, some of which, e.g. ammonia, might be expected to exert a toxic effect, especially considering the large quantities in which sludges are dumped. However, static tank toxicity tests by Franklin (1983) did not indicate any appreciable toxicity to a variety of adult fish and shellfish, though larvae of shrimps (*Crangon*) were more sensitive. A similar conclusion was reached by Fava et al. (1985), who estimated that in the normal conditions which occurred after dumping in the New York Bight, dilution of the sludge after dumping would rapidly bring it well below the concentration levels at which no effect to the most sensitive crustacea would occur. Franklin's study (1983) included a 60-day bio-accumulation study to assess metal uptake; some uptake of lead, zinc and copper occurred when shrimps were exposed to very highly contaminated sludges, but mercury and cadmium levels were lower than expected; it is possible that these metals in sewage sludge are not readily available to marine organisms. However, field studies indicate that accumulation can occur in organisms in the vicinity of dump sites, underlining the importance of obtaining a good degree of dispersion to overcome both acute and chronic impact.

## 6 Pathogens and Parasites

Sewage is a potent source of pathogens and parasites (Davis 1980; Table 6); this often precludes the more extensive beneficial use of sludge in agriculture. Historically, raw sewage discharged to sea has given rise to problems of contamination at coastal shellfish beds, leading to closures of fisheries, and corresponding problems at bathing beaches. Offshore disposal of sludge from vessels or long-sea-outfalls at sites carefully selected to avoid shoreward transport of sludge materials, and out of contact with significant fisheries, helps to reduce this problem. Many pathogens and parasites cannot survive in seawater, though some (e.g. *Acanthamoeba*, many viruses) are sufficiently resistant to be potentially useful indicators of the extent of contamination of the sea bed by sludge (Vivian 1986).

**Table 6.** Pathogens present in sewage sludge. (After Wood 1986)

1. Bacteria, particularly *Salmonella* species.
2. Viruses, particularly the Enteroviruses.
3. Eggs of metazoan parasites, including tapeworms (*Taenia*) and roundworms (*Ascaris, Trichuris*).
4. Cysts of protozoan parasites (*Giardia, Acanthamoeba*).

## 7 Choice of Disposal Site – Management of Near-field Impact

The main aim in choosing the sea disposal route is to use the natural capacity of the sea to degrade and assimilate the organic and nutrient components of the waste, and to disperse the persistent materials to such an extent that they have no deleterious effect. The main concern in selection of a suitable location for sewage sludge disposal at sea is to obtain sufficient initial dilution at the point of disposal to overcome the possibility of

- deoxygenation in the water column due to high BOD,
- deoxygenation in the sediments due to accumulation of excessive organic matter,
- acute toxicity due to major components (e.g. ammonia).

Subsequent dispersion must overcome chronic effects due to local accumulation of parasites, pathogens or toxic substances.

As with any waste discharge, there is usually an area close to the point of discharge in which the waste will exert measurable local effects. A sensible management policy suggests that the impacted zone should be small with respect to the distribution of significant local resources; it should not affect, for example, shellfish beds or bathing beaches. Within the affected area the impact should not be extreme nor have significant consequences outside the area. Normally, this leads to a limitation of 'acceptable' change in the zone of immediate impact to barely detectable biological change. In the case of piped discharges or dumping of sludge in accumulative sites, more extreme changes in the biology, even to complete abiosis, may be acceptable in a confined area, if this limits the impact outside this zone. An important final criterion is that, having established the scale and extent of acceptable local impact, there should be no deterioration with time, which would suggest inadequate assessment of assimilative capacity.

Site selection is obviously constrained by geographical location, and the rate and scale of disposal must be s7ited to the local dispersal characteristics and the degree of local impact found to be acceptable. Norton and Champ (in press) have reviewed a range of disposal sites and have shown that the degree of local effect can be predicted, at least qualitatively, on the basis of assessment of key site characteristics, namely peak tidal flows at the sea bed, the nature of the sediments and the nature and severity of wave disturbance at the sea bed. In the North Sea, most of the English and Scottish disposal sites are strongly dispersive. In the Thames, where the Barrow Deep disposal site receives some 5 mt (wet weight) per year, Norton et al. (1981), indicate that at this level some areas of accumulation do occur, with localised biological consequences, though tracer studies indicate that on the whole this large load was effectively dispersed (Talbot et al. 1982). The Tyne and Tees sites are somewhat less dispersive and thus more limited in their capacity if significant local effect is to be avoided. The now disused sludge disposal site in the German Bight is also an area of weak tidal current and cohesive sediment (Norton and Champ, in press) and the overlying waters are prone to stratification and subsequent bottom water de-oxygenation in summer. The studies of Caspers (1980) and Rachor (1977) suggest an unstable local biology, in which the instabilities may have been exacerbated by sludge d7mping, by periodic summer low oxygen conditions, and by winter storms.

## 8 Conclusions

Marine disposal of sewage sludge by vessel or pipeline, properly managed, results in only minor, acceptable local impact and, as currently practised, contributes only a small fraction of the total inputs of organic matter, nutrients and persistent contaminants to the North Sea. Sludge disposal is by no means the only route to the sea of sewage derived materials, which enter the sea via rivers and direct treated or untreated discharges. However, these sources are more difficult to quantify, being often individually smaller in quantity, very much larger in number and more diffuse than the large, discrete sludge disposal operations discussed here. Apart from their relatively minor role in contributing to inputs of persistent substances, there is currently no satisfactory evidence that sewage sludge disposal has other than local effects.

## References

Anon (1986) Whitaker's Almanac. Whitaker, Lond

Caspers H (1980) Long term changes in benthic fauna resulting from sewage sludge dumping into the North Sea. Wat Technol 12461–479

Chapman DV (1986) The distribution of metals in sewage sludge and their fate after dumping at sea. Sci Total Environ 48:1–11

Commission of the European Communities (1986) Directive on the protection of the Environment, and in particular the soil, when sewage sludge is used in agriculture, 86/278/EEC. Off J L181:6–12

Davis RD (1980) Control of contamination problems in the treatment and disposal of sewage sludge. Water Res Cent Medmenham Tech Rep TR 156:1–79

Fava JA, McCulloch WL, Gift JJ, Reisinger HJ, Storms SE, Maciorowski AF, Edinger JE, Buchak E (1985) A multi-disciplinary approach to assessment of ocean sewage sludge disposal. Environ Toxicol Chem 4:831–840

Franklin FL (1983) Laboratory tests as a basis for the control of sewage sludge dumping at sea. Mar Pollut Bull 14:217–223

Garber WF, Ohara GT, Colbaugh JE, Raksit SK (1975) Thermophilic digestion at the Hyperion Treatment Plant. J Wat Pollut Contr Fed 47:950–961

Great Britain – Parliament (1974) Dumping at Sea Act 1974. Her Majesty's Stationery Office, Lond, 13 pp

Grogan W (1984) Input of contaminants to the North Sea from the United Kingdom (Report for the Department of the Environment). Heriot-Watt Univ Inst Offshore Eng, Edinburgh, 203 pp (Mimeo)

Healey MG (1984) Guidelines for the utilisation of sewage sludge on land in the United Kingdom. Water Sci Technol 16:461–471

Hill JM, Mance G, o'Donnel AR (1984) The quantities of some heavy metals entering the North Sea. Water Res Cent Medmenham Tech Rep TR 205:1–21

Norton RL (1982) The assessment of pollution loads to the North Sea. Water Res Cent Medmenham Tech Rep TR 182:28

Norton MG, Champ Ma The influence of site-specific characteristics on the effects of sewage sludge dumping. In: Oceanic Processes in Marine Pollution, Volk Ch 15. D. Hood and A Schoner (eds), Robert E Krieger Publ Co, N.Y. (in press)

Norton MG, Eagle RA, Nunny RS, Rolfe MS, Hardiman PA, Hampson BL (1981) The field assessment of effects of dumping wastes at sea: 8. Sewage sludge dumping in the outer Thames Estuary. Fish Res Tech Rep MAFF Direct Fish Res, Lowestoft 62:1–62

Rachor E (1977) Faunenverarmung in einem Schlickgebiet in der Nähe Helgolands. Helgol Wiss Meeresunters 30:633–651

Talbot JW, Harvey BR, Eagle RA, Rolfe MS (1982) The field assessment of effects of dumping wastes at sea: 9. Dispersal and effects on benthos of sewage sludge dumped in the Thames Estuary. Fish Res Tech Rep MAFF Direct Fish Res, Lowestoft 63:1–42

US Dept of State (1969) Sovereignty of the Sea. US Dep State Bureau Intelligence & Res Geogr Bull 3:1–5

Vincent AJ, Critchely RF (1982) A review of sewage sludge treatment and disposal in Europe. Water Res Cent Medmenham Rep 442-M:31

Vivian CMG (1986) Tracers of sewage sludge in the marine environment: A Review. Sci Total Environ 53:5–40

Wood PC (1986) Sewage sludge disposal options. In: Kullenberg G (ed) The Role of the Oceans as a Waste Disposal Option. Reidel, Dordrecht, pp 111–124

# Waste Incineration at Sea

H. COMPAAN[1]

## 1 Introduction

The incineration of chemical waste at sea started in 1969 on the North Sea. Only during the last few years has the technique become hotly debated as a result of the opposition to it by some environmentalist groups.

Incineration at sea is designed specifically for the destruction of liquid organochlorine wastes with a maximum chlorine content of 70% w/w.

I will concentrate here on the present state of the art and the environmental aspects. The main goal of the research that has been carried out during the last 15 years was to study the proper functioning and the environmental impact of the incineration ships. Most of this research took place on the North Sea and in United States waters.

Practically none of the research reports on the subject have been published in scientific journals, but they are all available to the public. A list of publications up to and including 1981 is given by Compaan (1982). Probably the most important discussion of all aspects of waste incineration at sea is the recent OTA report (1986). The studies in US waters have been reviewed by Ackerman and Venezia (1985).

The doubts about environmental safety are caused in particular by two reports by Kleppinger and Bond (1983, 1985). Rebuttals of these reports have been published by Ackerman (1984, 1986).

## 2 The Organochlorine Waste Problem

After the Second World War, the chemical industry developed very rapidly and with it the production of organochlorine compounds. Vinylchloride (VCM) is one of the most important. Provisional estimates of the production of organochlorine products and wastes in Western Europe are given in Table 1.

It is a well-known fact that a considerable amount of the so-called EDC-tar, a black, liquid waste tar, containing 60 to 70% chlorine, has been dumped in the sea. Sperling (1985) states that this amounts to hundreds of thousands of tons of EDC-tar. It was normal practice well into the 1970's on the North Sea as well as in the Gulf of Mexico. After the well-known *Stella Maris* drama in 1971 and the rapid

[1]MT-TNO, Division of Technology for Society, P.O. Box 217, Schoenmakerstraat 97,2628 VK Delft, The Netherlands

**Table 1.** Organochlorine products and waste in western europe in millions of tons

| Product | Period | Production[a] | Waste arising therefrom | Burnt at sea | "Processed" in another way |
|---|---|---|---|---|---|
| VCM[b] | upto–1970 | 16–20 | 0.7–1 | 0.01 | 00.7–1 |
| VCM | 1971–1985 | ca. 65 | 1.5–2 | 0.6–0.8 | ca. 1 |
| Rest | 1945–1985 | ca. 85 | > 2 | 0.5–0.7 | > 1.4 |

[a] Provisional estimates.
[b] Vinyl chloride monomer, material for the production of PVC.

birth of the Oslo Convention and the London Dumping Convention, organohalogen compounds appeared on the "black lists" and the incineration of organochlorine wastes on the North Sea rose within a few years to a fairly steady level of about 90,000 tons per year.

In recent years the waste problem has been made considerably more serious by the fact that increasing quantities of relatively pure organochlorine products have to be destroyed. This is because they are no longer allowed to be used, at least in the western world. Examples are trichlorophenol based herbicides, PCB's and DDT.

## 3 The Ships in Use and the Wastes Incinerated

At the moment of writing there are three incineration ships certified and operational, of which two are in regular use. Some relevant data about these ships are given in Table 2.

Roughly half of the wastes incinerated on the North Sea is EDC-tar, a complicated mixture of saturated and unsaturated C-1 through C-4 chlorinated

**Table 2.** Some data on the incineration ships which are operational at the moment of writing

| | Vulcanus I | Vulcanus II | Vesta |
|---|---|---|---|
| Length | 97 m | 93.5 m | 72 m |
| Width | 16 m | 16.0 m | 11 m |
| Draught (max.) | 6.06 m | 6.12 m | 4.3 m |
| Unladen capacity (t) | 3990 | 4430 | 999 |
| Tank capacity ($m^3$) | 3200 | 3200 | 1300 |
| Incinerators | 2 | 3 | 1 |
| Burners | 3 for each incinerator (rotating cups; Saacke KG) | | |
| Air flow rate | 90,000 $m^3$ $h^{-1}$ for each incinerator | | |
| Capacity (max.) | 25 t $h^{-1}$ | 35 t $h^{-1}$ | 12.5 t $h^{-1}$ |
| Owners | WMI/OCS | WMI/OCS | Lehnkering |
| Construction | Converted | Newly built | Newly built |
| In use since | 1972 | 1982 | 1979 |

aliphatic hydrocarbons. The main components are 1,2-dichloro-ethane and 1,1,2-trichloro-ethane. The other half consists of a very heterogeneous collection of different organochlorine wastes from 500 to 700 different sources in 11 different countries.

About 85% of the waste is collected in Antwerp. Transport takes place by road, rail, rivers, and canals and over the sea. Other ports from which at-sea incineration takes place are Le Havre (France), Newcastle (UK) and Rafnes (Norway) (see Table 3).

The amounts of waste incinerated on different seas up to and including 1986 are shown in Table 4.

**Table 3.** Organochlorine waste transport to the incineration are in the North Sea 1986

| | |
|---|---|
| Total amount | 108,000 |
| via Antwerp (t) | 92,500 |
| of which: | |
| by river | 54,500 |
| by rail | 15,500 |
| by road | 22,500 |
| by sea | 6,000 |
| | |
| From | |
| FRG | 63,000 (58%) |
| France | 5,000 (4.5%) |
| U.K. | 4,500 ( 4%) |

**Table 4.** Amounts of waste incinerated at sea in the world

| | North Sea | Gulf of Mexico | Pacific Ocean | Australia |
|---|---|---|---|---|
| 1969 | 4,000 | | | |
| 1970 | 8,000 | | | |
| 1971 | 28,000 | | | |
| 1972 | 66,000 | | | |
| 1973 | 87,000 | | | |
| 1974 | 85,000 | 12,200 | | |
| 1975 | 85,000 | 4,100 | | |
| 1976 | 85,000 | | | |
| 1977 | 55,440 | 17,800 | 11,200[a] | |
| 1978 | 67,503 | | | |
| 1979 | 107,000 | | | |
| 1980 ca. | 106,000[c] | | | |
| 1981 | 100,673 | 3,350[b] | | |
| 1982 | 94,075 | 3,000[b] | | 4,600 |
| 1983 ca. | 90,000 | | | 220 |
| 1984 ca. | 95,000 | | | |
| 1985 ca. | 100,000 | | | |
| 1986 | 108,000 | | | |

[a] Herbicide "Orange" (esters of 2, 4, 5-trichlorophenoxyacetic acid).
[b] Concentrated PCB wastes.
[c] Nót 166,000 tons, as can be found in some reports (for example OTA 1986).

# 4 Legislation and Control

## 4.1 Regulations

A code of practice containing all the technical aspects of incineration at sea was adopted by the Oslo Commission. Details are given in its publication *The first decade* (1984). Incineration at sea is seen as an interim solution for the organochlorine waste problem and the Commission decided to have a meeting before 1 January 1990 to discuss and establish a date to put an end to incineration at sea. It is sometimes stated erroneously that incineration at sea has to stop at that date, which certainly is not the case. Most probably a time schedule will be discussed in which incineration at sea will have to be phased out.

Some details of the *Technical Guidelines* for the operation of the ships will be discussed later on in this chapter.

## 4.2 Permits

Two permits are required for the incineration of a load of chemical waste at sea. The first concerns the *ship*. This permit is obtained when the ship has proven to be able to incinerate a well-known type of waste, EDC-tar for example, within the limits of efficiency specified by the Oslo Commission. To obtain this permit, which has to be renewed every 2 years, it is necessary to measure not only the combustion efficiency, but also the destruction efficiency. The last one is a rather difficult and specialized type of measurement, which has given rise to many of the investigations mentioned in Section 6 and summarized in Table 6.

The second permit concerns the *type of waste*. Some types of organochlorine waste can give rise to special problems. "Herbicide Orange" is an example, because it contains traces of chlorinated dioxins, which also could be produced during incomplete combustion. PCB's are another example, because they contain traces of chlorinated dibenzofuranes, which could be formed as well during incomplete combustion. (Even the incomplete combustion of dichloro-ethane can be the cause of the formation of chlorinated dioxins (Marklund et al. 1987).) For the incineration of these special types of waste, so-called "research" and/or "special" permits were obtained and it was proven in three, and then two voyages that trichlorophenol-based herbicides and PCB's can be combusted and destroyed with very high efficiency (Ackerman et al. 1978a,b; Compaan 1982; Ackerman et al. 1983). The permits are granted by the authorities of the country in whose territorial waters the burning areas are situated.

## 4.3 The Burning Area

The burning area in the North Sea was moved in 1979 to an area east of the Doggerbank (Fig. 1). This was done after more knowledge was obtained on the distribution of the stack gases. This is discussed in a paper by the Rijkswaterstaat, the Dutch permitting authority (1987).

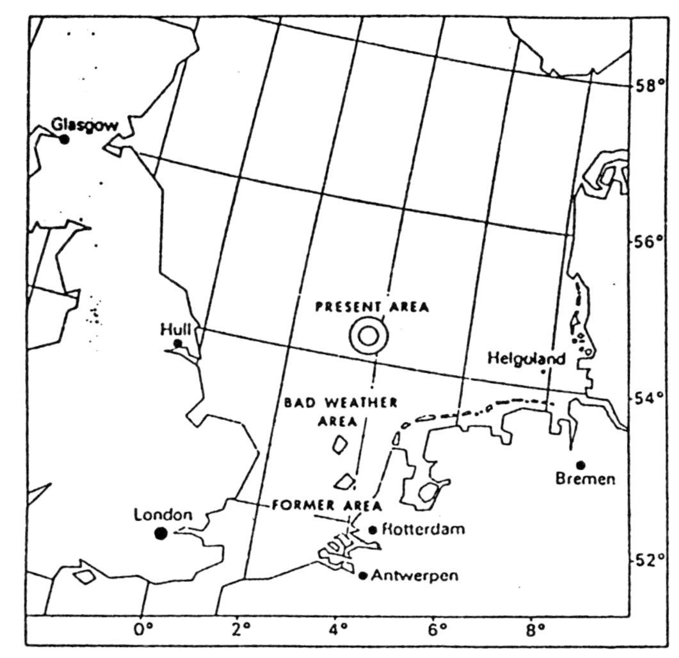


**Fig. 1.** The situation of the burning areas in the North Sea. (After Weitkamp)

## 4.4 Control by the Authorities

All relevant data of an incineration voyage are stored in a datalogger which has been developed and installed on the ships by the Rijkswaterstaat. In this way the control by the authorities, even when they are not on board, has become practically watertight.

There are a number of additional methods by which the authorities can control the proper operation of the ships. The caloric value of the waste to be incinerated is known and the time necessary to incinerate a particular load can be calculated in advance. Nothing can be gained by dumping the waste and burning oil instead.

An employee of the permitting government can travel with the incineration ship or accompany it on another ship. This is done from time to time, especially during research and certification burns ("shiprider").

The ships can be photographed from the air, which is carried out on the North Sea very frequently. A continuous control on shore by satellite relay of the ship's operating data is investigated with regard to its feasibility.

## 5 State of the Art and Routine Operating Procedures

### 5.1 Waste Accepting Procedure

The composition of the waste to be incinerated must be known in order to assess if it is suitable for incineration at sea and that no substances are present that are excluded from the permit. This knowledge is also required to protect personnel and equipment.

Companies offering waste for incineration have to supply full details about its composition and guarantee them. All waste is collected by contractors and analyzed at the OCS laboratory in Antwerp. Special analyses, for example for chlorinated dioxins, are carried out by Belgian, Dutch, and American laboratories specializing in this work. From time to time, the authorities take samples and have them analyzed.

The most important problem in this field is the development of rapid and reliable methods for the determination of PCB's and dioxins in the waste. This is extremely difficult, by the huge and varying amount of numerous interfering organochlorine compounds present in the wastes.

Substances which are excluded from the permits at the moment of writing are: organofluorine compounds in large quantities, PCB's, PCT's, DDT and derivatives and HCH-isomers. The USA authorities have added the herbicide Sylvex to this list. All waste is filtrated before it is incinerated. The residues are disposed of in an official, licensed landincinerator.

### 5.2 Technical Details of the Ships

A list of the presently operational ships was given in Table 2. The following details apply to the "Vulcanus II" in particular and many of them apply to all ships. (Fig. 2) The ships have to meet the requirements specified by the international authorities for chemical tankers. This means, for example, that the part of the ship which contains the tanks has a double hull.

Elaborate precautions have been taken to prevent, contain and clean up spillage during loading. There are no loading or unloading pumps on board. The inert gas which is driven out of the tanks during loading is sent back to the shore tanks by a gas return line.

The waste level in the tanks is measured with ultrasonic instruments. An automatic safety device prevents overloading; it gives alarm signals when the tanks are 85% and 98% filled.

The waste level during incineration must be known accurately for two reasons. There is a special scheme according to which the tanks have to be emptied, in order not to affect the ship's stability due to imbalance. In the second place it is the only way to know the time-averaged waste feed during the measurement of destruction efficiencies. Reliable waste flow meters do not yet exist.

In order to prove that the ship operates only within the prescribed burning area, its position must be known and recorded continuously. For this purpose a

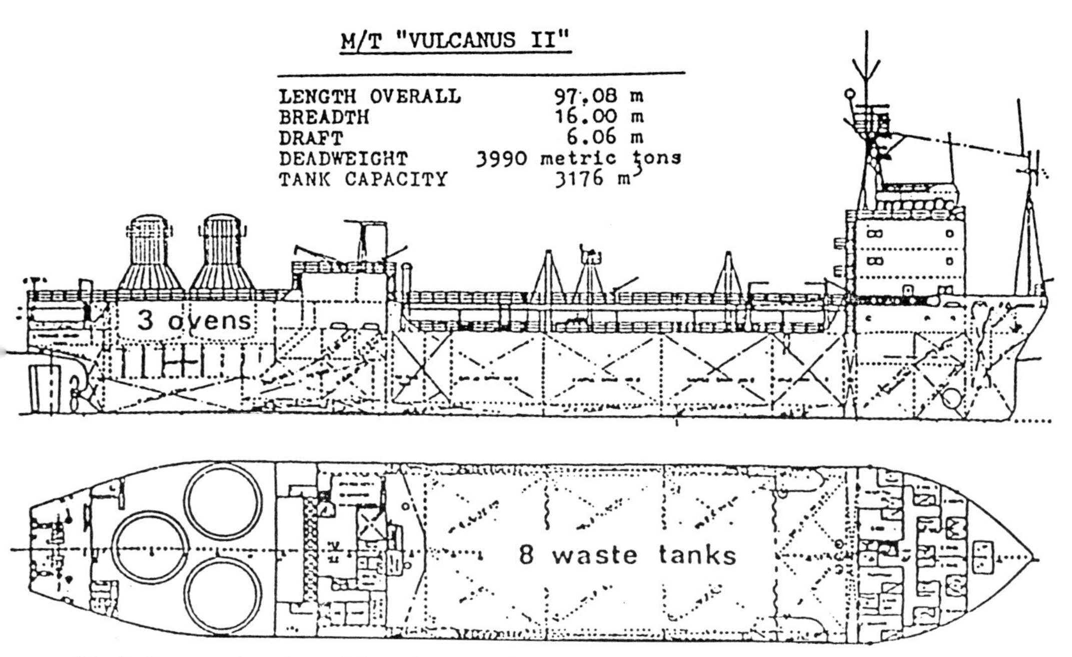


**Fig. 2.** Cross-section views of the *Vulcanus II* (OCS)

satellite navigator and a Decca or Loran navigator are installed on board. The waste feed pumps and magnetic valves are stopped (or closed) automatically by safety devices, mainly actuated by the incinerator thermocouples. The ovens are approximately of the same design on all ships. The temperature of each oven is measured by a number of thermocouples, located at different heights at a depth of a few centimeters in the brick lining. The flame temperature is about 200° to 300° higher than the wall temperature measured in this way. (Ackerman 1984, 1986) The thermocouples are connected to an automatic waste-feed shut-off system. The air feed to the ovens can be measured only with limited accuracy, but it can be calculated from the other combustion parameters (Boubel 1984).

## 5.3 A Routine Incineration Voyage

About 12 h before the ship reaches the prescribed burning area, heating up of the ovens to 800°C (wall temperature) is started by burning fuel oil. When the burning area is reached, the oven temperature is increased up to 1250°C, after which the burners are one by one switched over to waste. The highly chlorinated waste can only maintain its own combustion when the temperature is sufficiently high. The waste pumps cannot be started before a minimum temperature of 1250°C is measured by at least two of the three main thermocouples in each oven. When for some reason the temperature of these thermocouples drops below 1180°C, the waste feed is stopped automatically within 0.5 s. Re-heating with fuel oil then has to take place.

During incineration there is continuous wireless contact with Scheveningen Radio (Dutch Shore Station). Regular warnings are given to other ships in or around the burning area once the ship leaves port.

When all waste has been incinerated, the system of pipes, pumps, and burners is cleaned with fuel oil, which is burned as well. After that, the incineration is stopped. When the ship returns in the port, all monitoring data are collected on board by the relevant authorities.

## 5.4 Combustion Parameters and Stack Gas Composition

A cross section of one oven of the Vulcanus II and the position of the burners near the bottom of the oven are shown in Fig. 3. The burners are positioned in this way to obtain as much turbulence in the ovens as possible. The details of the combustion process are discussed thoroughly by Ackerman (1984, 1986). The flame temperature is 1400 to 1500°C; the residence time is about 1 s. There is an unimportant temperature gradient near the walls of the ovens. The gross composition of the stack gases does not vary much and is given in Table 5.

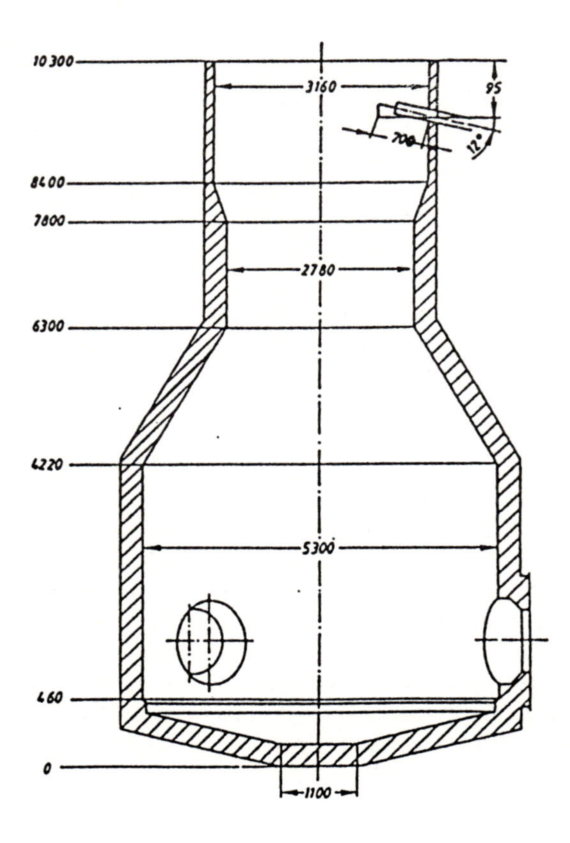


**Fig. 3.** Inside and outside configuration of the three identical incinerators and position of the stack sampling proble and the burners. (After Gielen and Compaan 1983)

**Table 5.** The approximate composition of the stack gases during the incineration of organochlorine wastes with a high chlorine content (50–65% w/w)

| | |
|---|---|
| Oxygen | 10% v/v |
| Carbondioxide | 10% v/v |
| Carbon monoxide | 5–30 ppm |
| Hydrochloric acid | 5% w/v |
| Water | 6% w/v |
| Nitrogen | 70% v/v |
| Organics (total) | ppm level |
| Metals | ppm level |
| Nitrogen oxides | $\geqq 100$ mg $m^{-3}$ |
| Sulfur oxides | Traces? |
| Chlorine | up to 2000 ppm? |
| Phosgene | $<2$ ppm? |

## 5.5 Solid Incineration Residues

Very few ashes remain on the bottom of the ovens. During incineration a black, coke-like material sometimes accumulates on the rotating burner cups. This has to be removed from time to time. In order to do this, the ovens are temporarily switched over on fuel oil and the burners are swung out of the ovens one by one and cleaned. The coke-like material is collected in drums and incinerated on land in a licensed facility.

The brick lining of the ovens must be renewed about every 2 years. Before it is disposed of in a landfill it is examined for the presence of traces of dioxins. Up till now they have not been found (detection limit 10 ppb TCDD + 10 ppb TCDF). Particle sampling in the stack gases has never been carried out on purpose for two reasons. First, at the very high stack gas sampling temperatures (above 1000°C) organic stack gas contaminants can only be present in the vapor phase. Second, all metals that are fed into the ovens within the waste are considered to leave the stack again. Their quantities are limited by the permits (see Sect. 9).

## 5.6 Measuring and Monitoring Nonchemical Data

The following parameters are measured and monitored continuously in such a way that all data are displayed in the burning room and part of them on the bridge as well; they are all stored in a sealed data-logger:

- waste and oil level in the tanks,
- oil and waste feed,
- actions of automatic valves and all pumps,
- air feed (by blower action and valve position),
- temperatures in all ovens,
- actions of automatic safety devices,
- course, speed and position of the ship.

A burning room log-book is kept in which all actions and events during a voyage are noted.

### 5.7 Measuring and Monitoring Combustion Efficiency

A minimum combustion efficiency (CE) of 99.9% is prescribed by the Oslo Commission and the London Dumping Convention. The CE is defined on the basis of the concentrations in the stack gas of carbon dioxide $[CO_2]$ and carbon monoxide $[CO]$:

$$CE = \frac{[CO_2]-[CO]}{[CO_2]} \times 100\%.$$

From Table 5 and the above formula it can be seen that a CE of 99.9% corresponds with a CO concentration of about 100 ppm and at a CE of 99.99% the CO concentration will be about 10 ppm.

During incineration the CO concentration varies in the lower ppm range (5–30 ppm).

Every oven is equipped with a nondispersive infrared-type CO and $CO_2$ monitor. With the aid of the onboard computer the CE is calculated and monitored directly. The CO content is the most sensitive direct control parameter for the combustion process.

The CE is rarely lower than the prescribed 99.9%, and then only for a few minutes and almost exclusively during the combustion of fuel oil.

Every oven has also a magnetic type oxygen monitor installed. Normally the incineration is carried out with 100% air excess, which leads to an oxygen concentration of about 10% by volume.

The nine gas monitors on board are calibrated at least once a day. There is a spare heated suction line for the gas monitors on each oven.

The monitors for CO and oxygen can activate an alarm.

## 6 The Destruction Efficiency

### 6.1 Problems with the DE – Concept

Up till now, no clear-cut relationship has been found between the CE and the DE (Erickson et al. 1985). This means that the DE must be measured separately, which is still a difficult task.

Contrary to the CE, there are formal problems with the definition of the DE and its measurement. In the international regulations the same simple formula as with the CE is used:

$$DE = \frac{[\text{input}] - [\text{output}]}{[\text{input}]} \times 100\%.$$

In this formula [input] is the waste vapor concentration in the stack gases, when no destruction at all has taken place. [Output] then is the waste vapor concentration which is actually measured.

The problems with this concept will be discussed in detail elsewhere (Compaan 1987) but can be summarized as follows.

- The combustion gases of incineration ships are released into the atmosphere without purification. This means that only environmentally acceptable amounts of hazardous substances must be allowed to be emitted. It was the objective of prescribing a minimum DE to guarantee this.
- The DE concept as laid down, however, does not cover the problem of possible emissions of products of incomplete combustion (PIC) or newly formed hazardous substances (e.g., hexachlorobenzene, chlorinated dioxins and dibenzofuranes, etc.). Only negligibly small amounts of such substances must be allowed to be emitted.
- When the determination of the DE is based on the total organochlorine content (TOCl) of wastes and stack gases, the above-mentioned problem still remains. A TOCl measurement has hardly any environmental significance here. [It goes without saying that a total hydrocarbon ($C_xH_y$) measurement has no relevance at all!]
- In many cases the waste incinerated is a mixture of hundreds of chemicals and it is not feasable to determine the DE for every component. It is, of course, impossible to determine the DE for *unknown* components (traces). Some intelligent choice must be made. Probably all parties will have to agree on some form of EPA's POHC-concept (Principal Organic Hazardous Components). A key problem here is the definition of "principal" and "hazardous".
- The request has been made many times to make a "full chemical analysis" of the stack gases. It must be realized, however, that there does not exist such a thing as a "full chemical analysis" and that there are no analytical methods for the determination of "traces of unknown supertoxic compounds" in any material. Toxicity tests on the stack gases might possibly solve this problem, but they will give highly unrealistic, if not useless results if they are carried out after bubbling the stack gases through seawater (Compaan 1987).
- When very high destruction efficiencies are achieved, as has been proved already for many organochlorine compounds (Table 6), the above discussion degenerates practically into discussing a nonevent, certainly when the emissions of the ships are compared with all other pollutant inputs into the North Sea (see also Sect. 9).

## 6.2 Measurement of the Destruction Efficiency

For the determination of the DE it is necessary to have:

- an exact qualitative and quantitative GC-MS waste analysis,
- the measurement of the waste flow rate,
- the measurement of the incineration air flow rate,
- the measurement of the incineration temperature,
- an exact GC-MS determination of organochlorine traces in a series of stack gas samples.

Many determinations have been carried out and most results have been reviewed by Compaan (1982) and Ackerman and Venezia (1985). A short overview of the results is given in Table 6.

**Table 6.** Destruction efficiencies as reported by different workers in the field (Ackerman et al. 1985; Compaan 1982; Gielen et al. 1983; see also Table 7). *Remark*: All DE's with a "larger than" sign are based on the analytical detection limits; the substances were nót detected. This detection limit is higher for trace components in the waste. (TCDD, TCDF, hexachlorobenzene). In reality the DE's are probably much higher, but this has to be proved by more sensitive measurements

| Waste type | DE |
|---|---|
| EDC-tar | 99.995 |
| Aliph. organochlorine | 99.98–99.998 |
| Herbicide Orange | $>$ 99.999 |
| TCDD | $>$ 99.98– $>$ 99.99 |
| PCB's | $>$ 99.99989 |
| Chlorobenzenes | $>$ 99.99993 |
| TCDF | $>$ 99.96 |
| Hexachlorobenzene | $>$ 99.9999 |
| Hexachlorobutadiëne | $>$ 99.985 |
| Carbontetrachloride | $>$ 99.9966 |

The last testburn was the certification burn for the *Vulcanus II* on the North Sea in January 1983. During this burn the US EPA Modified Method 5-stack gas sampler and a newly developed TNO sampler were compared (Gielen and Compaan 1983). Some of the results are given in Table 7.

A study of oven similarity has been made (Bartelds 1983). The results of this study justify carrying out DE-measurements on only one oven.

As long as the DE's measured are orders of magnitude higher than prescribed, the accuracy of the measurements is of little importance.

**Table 7.** Measured DE of four EDC-tar components during the certification voyage of the *Vulcanus II*, January 1983. (Gielen et al. 1983)

| | USEPA | TNO |
|---|---|---|
| 1,1,2-trichlorothane | $>$ 99.9995% | $>$ 99.9996% |
| tetrachlorethylene | $>$ 99.9934% | $>$ 99.9999% |
| 1,1,1,2-tetrachlorethane | $>$ 99.9990% | $>$ 99.9998% |
| 1,1,2,2-tetrachlorethane | $>$ 99.9983% | $>$ 99.9982% |

### 6.3 Recombination Reactions in the Plume

There remains the somewhat hypothetical question of recombination reactions in the rapidly cooling plume, outside the ovens. These reactions could perhaps give rise to traces of hazardous substances, which are not detectable at the top of the ovens. It is very difficult to take samples in the plume at a short distance from the ovens. Another approach may be the simulation of the rapid quenching of the stack gases in a specially constructed sampling apparatus.

## 7 Other Minor Components of the Stack Gases

A number of components not mentioned before can be listed which may occur in the stack gases. They are phosgene ($COCl_2$), chlorine ($Cl_2$), nitrogen oxides ($NO_x$), sulfur dioxide ($SO_2$), volatile hydrocarbons ($C_xH_y$), polycyclic aromatic hydrocarbons (PAH), metals, and solid particles. They are not measured for different reasons:

- they only have a short lifetime in the atmosphere and do not occur or in trace quantities only ($COCl_2$, $Cl_2$);
- the emission can be neglected when compared with other sources ($NO_x$, $SO_2$, $C_xH_y$, PAH);
- the emission can be calculated from the concentration in the waste (metals);
- they do not contain organic substances (solid particles).

Chlorine gas is especially to be expected when the molar hydrogen-chlorine ratio in the waste is smaller than 1. This can happen when highly chlorinated *aromatic* substances are incinerated.

## 8 Plume Measurements

The only quantitatively important substance in the waste gas plume is hydrochloric acid. The dilution and "breakdown" of this acid in the plume has been examined by Weitkamp and co-workers over a number of years by means of an HCl-specific Differential Absorption and Scattering infrared LIDAR (Heinrich et al. 1986). A schematic drawing of the instrument is given in Fig. 4.

Figure 5 shows how HCl concentration cross-sections through the plume are made.

During 3 years of research (1981–83) it turned out that the plume can take very different forms. It can rise almost vertically or lie down on the sea. Splitting of the plume often occurs. All this is determined by the meteorological conditions

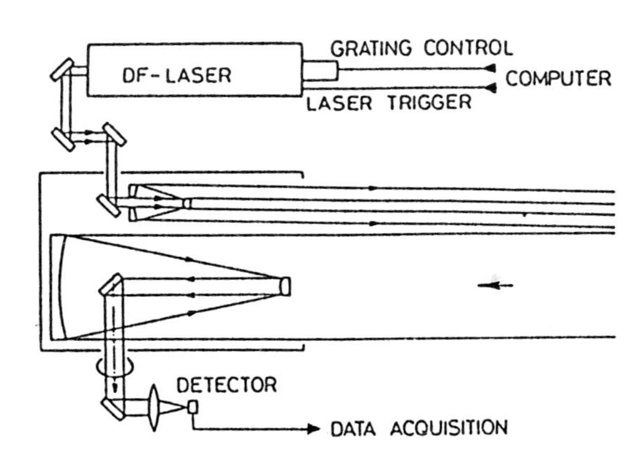


**Fig. 4.** Schematic representation of DAS lidar optics. The angle between the axes of the transmitted beam and the receiver field of view is greatly exaggerated. (After Weitkamp)

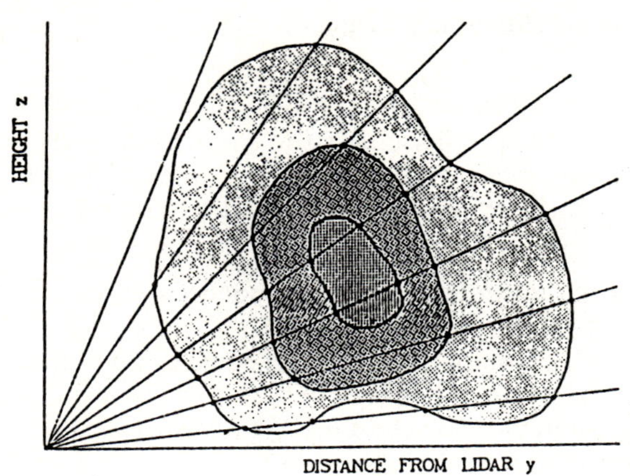


**Fig. 5.** Construction of a plume cross-section from a fan of individual lidar profiles. (After Weitkamp)

prevailing. The visible white plume, largely composed of water, is at the most 1.5 km wide, 1 km high, and can have a length of up to 60 km. Where the plume touches the sea, the slightly alkaline seawater absorbs the HCl without a measurable pH change, due to its pH buffering capacity (Grasshoff 1973).

Part of Weitkamp's results can be summarized as follows:

- In the order of 97.5% of the HCl in the plume is gasiform.
- The maximum HCl-concentrations are 0.5–5 ppm at distances of 1–10 km, with incineration of 10 tons per hour of waste with a chlorine content of 40% w/w.
- The decrease of the HCl-concentration in the plume due to dilution and due to other processes (together called "degradation processes" by Weitkamp et al.) can be measured separately.
- The lifetime (1/e) of the HCl in the plume is in the order of 35 min, rarely longer than about 50 min and depends on the type of plume (Fig. 6).
- Dry deposition is the most important "degradation" process.
- Under certain assumptions it can be calculated that of the HCl that crosses the West European North Sea coast only about 1 thousand millionth part comes from incineration at sea.
- Plume dispersion models that can be used on land cannot be used at sea.

## 9 The Total Emissions from the Ships on the North Sea

The emissions from the combustion of 100,000 tons organochlorine waste per year on the North Sea by the *Vulcanus II* and the *Vesta* together are given below, assuming a DE of 99.995%.

- metals for which the upper limit of the concentration in the waste is 5 ppm (Cd, Hg, As): 0.5 t $a^{-1}$ maximum,
- other metals: 50 t $a^{-1}$ each,
- volatile organochlorine compounds: 5 t $a^{-1}$

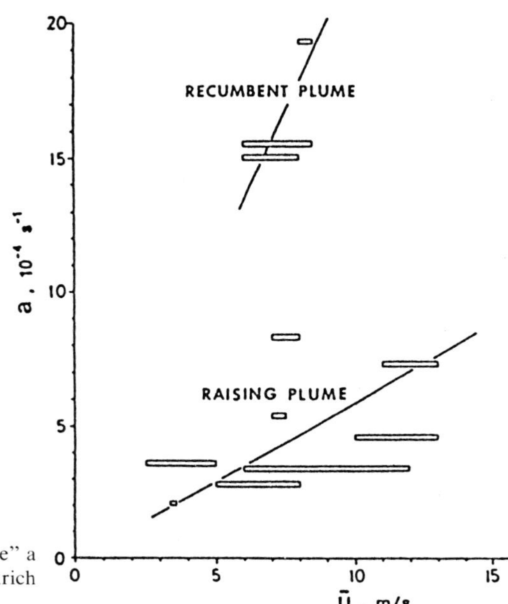


**Fig. 6.** Hydrochloric acid "degradation rate" a as a function of the windspeed ū. (After Heinrich et al. 1986)

– nonvolatile organochlorines: probably less than 0.5 ton per year, but difficult to estimate,
– hydrochloric acid: 40,000–50,000 t $a^{-1}$.

The ships do not clean their tanks and have no slops. Loading spills are very rare and negligible. The reader can easily compare these data with the other pollutant inputs into the North Sea, as given in other Chapters of this book.

## 10 Transport Risks

In 1983 there were about 50,000 tanker movements on the North Sea, 10,000 of which were by chemical tankers. In the same year there were about 70 movements of loaded incineration ships on the North Sea. (Ravenhorst and Van der Tak 1984) There have been more than 1000 incineration voyages on the North Sea since 1969, without one accident. However, the chance of an accident is not zero. When such an accident takes place with a load of EDC-tar in the ship, the North Sea can be expected to survive it, considering its dumping history (see Sect. 2).

This may not be the case when a load of concentrated PCB waste sinks to the seabed. According to Bletchly (1985) at least 250,000 tons of PCB wastes will have to be disposed of in the EC in the near future. It is not known how much of this waste is suitable for destruction at sea. It is not allowed to transport PCB products in bulk. However, the IMO left the possibility open to make an exception for the transport of liquid PCB wastes in incineration ships, which have cargo-tanks with a capacity of about 400 metric tons. A spill with 400 or 800 tons of concentrated PCB waste sinking to the seabed will almost certainly be a very serious environmental accident. If PCB's are going to be incinerated at sea, which is technically feasable, this transport risk must be made much smaller, for example by containerized transport, as proposed in Stolt Nielsen's Sea Burn project (OTA 1986). More than 50% of all wastes to be incinerated is transported on water (see Table 3). The serious risk of bulk transport of liquid waste with high concentrations of substances like PCB, DDT, HCH, etc. seems to have been overlooked until now.

Before the chemical waste reaches an incineration or recycling installation, it has usually been on a long journey with numerous stops on the way. Collection, transfer, storage, further transport by road, by rail, or by ship over inland waters or at sea. The latter also occurs even if the processing does not take place at sea. During these stages there is the risk of spillage, accidents, and fire. A fire in which organochlorine wastes are involved can have very serious consequences because of the risk of dioxin-contamination. A complete risk analysis of these problems has not been made for the European situation.

## 11 Suggestions for Research

- The DE concept must be replaced by a more adequate system.
- To remove all doubts about the environmental safety of the waste incineration at sea, it seems wise to repeat earlier DE measurements with the newest methods available and to search for hazardous PIC's with more sensitive methods.
- The problem of recombination reactions in the plume needs attention.
- Using the results of the new DE measurements and the appropriate dispersion models, the input per unit area of the sea surface of hazardous emissions should be estimated. Only when this estimated input is within the range in which biological effects can be detected significantly with the aid of the present-day methods should an ecotoxicological testing program be developed.
- A thorough risk assessment should be made of the chain of loading, storage, transport and processing of different organochlorine wastes. A result of this assessment can be that for some types of waste on-site processing is the best solution and needs more attention.

## 12 Conclusions

It has been shown by different workers in the field that many organochlorine compounds can be destroyed very efficiently by incineration at sea. To remove all doubts about the environmental safety of the technique some additional research is needed, although up until now no negative environmental effects have been detected. The destruction efficiency concept in the international regulations needs a revision. For some types of waste, PCB's for example, the transport risk must be studied seriously.

## List of Abbreviations

| | |
|---|---|
| AMI | Association of Marine Incinerators |
| CE | Combustion efficiency |
| DAS | Differential Absorption and Scattering (lidar) |
| DDT | Dichloro Diphenyl Trichloroethane |
| DE | Destruction efficiency |
| DF | Deuteriumfluoride (laser) |
| EDC | Ethylene dichloride (= 1,2-dichloroethane) |
| EPA | Environmental Protection Agency (USA) |
| GC-MS | Gaschromatography coupled with mass spectrometry |
| HCH | Hexachlorocyclohexane |
| LDC | London Dumping Convention |
| OCS | Ocean Combustion Service |
| OTA | Office of Technology Assessment (US Congress) |
| PAH | Polycyclic Aromatic Hydrocarbons |
| PCB | Polychlorinated biphenyls |
| PCT | Polychlorinated terphenyls |
| PIC | Products of incomplete combustion |
| POHC | Principal Organic Hazardous Compounds (Components) |
| TCDD | Tetrachlorodibenzodioxins |
| TCDF | Tetrachlorodibenzofuranes |
| TNO | (from the Dutch:) Applied Scientific Research |
| TOCl | Total organic chlorine |
| WMI | Waste Management International (USA) |

## References


Ackerman DG (1986) Ocean incineration. A state-of-the-art disposal technology: an analysis of the 1985 Kleppinger-Bond report. Sitex Consultants East, USA, March

Ackerman DG, Venezia RA (1985) Research on at-sea incineration in the United States. Wastes in the ocean. 5:53–72

Ackerman DG, Fisher HJ, Johnson RJ et al. (1978) At-sea incineration of Herbicide Orange on board the M/T *Vulcanus*. US-EPA, Report No. EPA-600/2-78-086, April

Ackerman DG, McGaughey JF, Wagoner DE (1983a) At-sea incineration of PCB-containing wastes on board the M/T *Vulcanus*. US-EPA Report EPA 600/7-83-024, April

Ackerman DG, Scinto LL, Shih CC, Matthews BJ (1983b) The capability of ocean incineration – a critical review and rebuttal of the Kleppinger report. TRW/CWM, USA, May

Bartelds H (1983) Submission to permitting authorities of data in lieu of trial burn results for future incinerator ships. TNO, Reports 83-09409, 09548, July, Apeldoorn, Netherlands

Bletchly JD (1985) Report to the Commission of the EC on a study of measures to avoid dispersion into the environment of PCBs and PCTs from existing installations. 15 October

Boubel RW (1984) Environ Sci Technol 18:229A

Compaan H (1982) Incineration of chemical wastes at sea. A short review, 2nd edn. TNO, Report CL 82/83, Delft, Netherlands

Compaan H (1987) Paper to be submitted to the Oslo Commission and the LDC. TNO, February, Delft, Netherlands

Erickson MD, Gorman PG, Heggem DT (1985) Relationship of destruction parameters to the destruction/removal efficiency of PCBs. J Air Pollut Control Assoc 35:663–665

Gielen JWJ, Compaan H (1983) Monitoring of combustion efficiency and destruction efficiency during the certification voyage of the incineration vessel *Vulcanus II*, January 1983. TNO, Reports R 83/53 R83/136, March resp July, Delft, Netherlands

Grasshoff K (1973) Gutachten über die möglichen Auswirkungen der Hochtemperaturverbrennung von chlorierten Kohlenwasser-stoffen auf hoher See durch Spezialschiffe

Heinrich H-J, Eck I, Weitkamp C (1986) Ausbreitung von Chlorwasserstoff in Abgasfahnen von Verbrennungsschiffen: Fernmessung von Konzentrationsverteilungen und Bestimmung von Verdünnungs- und Abbauparametern. GKSS Report 86/E/44 (in English; in press)

Kleppinger EW, Bond DH (1983) Ocean incineration of hazardous waste: a critique. EWK Consultants, USA, April, 112 pp

Kleppinger EW, Bond DH (1985) Ocean incineration of hazardous waste: a revisit to the controversy. EWK Consultants, USA March, 38 pp

Marklund S, Rappe C, Tysklind M, Egebäck K (1987) Identification of polychlorinated dibenzofurans and dioxins in exhausts from cars run on leaded gasoline. Chemosphere 16:29–36

Oslo and Paris Commissions (1984) The first decade. New Court London, 387 pp

OTA Office of Technology Assessment (1986) Ocean incineration: its role in managing hazardous waste. US Congress, US Print Off, Washington DC, 231 pp

Ravenhorst EM, Tak C van der (1984) Scheepsbewegingen Noordzee (in dutch). Marin-Marit Res Inst, Rotterdam, Proj. 630269

Sperling K (1985) Möglichkeiten und Grenzen des Monitorings im Meeres-, Küsten- und Astuarbereich. Eine kritische Betrachtung. Mitt Dtsch Ges Meeresforsch 1985 (3–4):3–15

# Input from the Atmosphere

R.M. VAN AALST[1]

## 1 Introduction

Atmospheric inputs into oceans, seas and lakes have been studied extensively in the last decades (NSF 1976; NAS 1978; GESAMP 1980). Special studies have been devoted to the American Great Lakes (Eisenreich 1981), the Baltic Sea (Rohde et al. 1980) and the North Sea (Goldberg 1973; ICES 1978; Cambray et al. 1979; RSU 1980; Norton 1982; van Aalst et al. 1983a; PARCOM 1986). Although estimates of atmospheric inputs into the North Sea are still highly uncertain, it is apparent from these studies that the atmospheric inputs are in the same order of magnitude as inputs by rivers, dumpings and discharges.

In this chapter, a short overview is given of air-sea exchange processes and their measurement, and an estimate is presented of input fluxes of some pollutants, derived from measured data as well as from model calculations.

## 2 Air-Sea Exchange Processes: Measuring and Modelling

Transport fluxes across the air-sea interface occur in two directions. Both atmospheric deposition and emission from the sea surface of gaseous and particulate pollutants may take place (Liss and Slinn 1983).

Wet deposition takes place by uptake of gases and particles by cloud droplets, followed by precipitation (rainout) or by scavenging of gases and particles by falling raindrops or snowflakes (washout). The rate of deposition is dependent on many factors, principally the rain intensity, the size distribution and chemical composition of the rain or cloud droplets, the diffusion constant and Henry's constant of the gases involved. For particles the scavenging is highly dependent on particle size. Wet deposition fluxes are commonly estimated from simple, highly parametrized models. Parameters widely in use are the dimensionless scavenging ratio W and the scavenging coefficient $\Lambda$. A first estimate of the wet deposition flux $F_w$ may be obtained from the concentration in the air $C_a$ of the pollutant by

$$F_w = W \cdot C_a \cdot P$$

here, P is the rain intensity.

[1]National Institute of Public Health and Environmental Protection, P.O. Box 1, 3720 BA Bilthoven, The Netherlands

For many anthropogenic components in aerosol, W takes values in the range $10^5$–$10^6$. The flux of washout of gases and particles may be estimated from

$$F_w = \Lambda \cdot h \cdot P,$$

where $\Lambda$ is the scavenging coefficient ($s^{-1}$) and h the depth of the polluted layer in which precipitation takes place.

Measurement of wet deposition is relatively easy on land. The deposition is determined from the chemical composition and amount of precipitation sampled in rain collectors. Contamination of the sample is to be avoided by closing the collector during dry periods. Samples collected at sea on platforms or research vessels may be strongly contaminated by sea spray and by nearby metal constructions or paint. The amount of precipitation is dependent of the position on the platform or vessel.

Dry deposition is the direct uptake of pollutants by the ground, in this case the sea surface. Depending on the concentrations in air and seawater, the flux of gaseous pollutants may occur both from air to sea (dry deposition) and from sea to air (evaporation). The flux may be estimated from (Liss and Slater 1974):

$$F = K\,(C_a - C_w \cdot H/RT),$$

where $C_a$ and $C_w$ are the pollutant concentrations in air and water, H is the Henry coefficient for the gas (in atm $m^3$/mol), R is the gas constant, T the absolute temperature and K a transfer coefficient. For many gases, $C_w$ H/RT is small with respect to $C_a$, and the flux is given by:

$$F = K\,C_a.$$

The constant K is in this case referred to as the deposition velocity $v_d$. The value of $v_d$ is primarily dependent on wind speed and atmospheric stability, as well as on the diffusion coefficient of the gas. It is usually of the order of 0.01–1.0 cm $s^{-1}$ (Sehmel 1980).

Some pollutant gases, particularly organic lipophilic compounds and some trace metals do not form homogeneous solutions in seawater. These pollutants are absorbed on floating particles or incorporated in organic complexes. Their concentrations in the uppermost layer, the so-called microlayer, may be substantially higher than in the bulk of the seawater (Liss 1975; Hunter 1980). Application of the above mentioned simple model is then complicated. Direct measurement of dry deposition or evaporation fluxes is possible by micrometeorological techniques (Hicks et al. 1980), but very few measurements have been made over sea.

Pollutants in aerosol form may reach the surface by a host of processes, such as sedimentation, diffusion or inertial impaction (van Aalst 1986). The resulting deposition velocity is highly dependent on particle size.

The sea itself is an important source of aerosol. Aerosol is produced by rising and consecutively bursting bubbles and by sea spray at high wind speeds. A likely value of the average emission flux is of the order of 0.1–1 $\mu g\ m^{-2}\ s^{-1}$ (van de Vate and ten Brink 1986). This aerosol may be considerably enriched in trace pollutants, probably as a consequence of the enrichment of the sea surface microlayer.

Direct measurements of particle pollutant fluxes over sea have hardly been published (van Aalst 1986), and comparison of dry deposition fluxes and emission fluxes is not yet possible.

# 3 Input Estimates from Measurements

## 3.1 Procedure

Current estimates are obtained mainly from measurements of wet deposition and air concentrations at coastal stations. The data given in this section refer to inputs to the North Sea with a surface area of 525 000 $km^2$. An average annual amount of precipitation of 685 mm is assumed (van Aalst et al. 1983a). Inputs by wet deposition may be estimated by

$$\text{input (t year}^{-1}) = C_p \cdot 360,$$

where $C_p$ is the concentration in precipitation in $\mu g\ l^{-1}$, and dry deposition by inputs by

$$\text{input (t year}^{-1}) = C_a \cdot v_d \cdot 165,$$

where $C_a$ is the concentration in air (in ng $m^{-3}$), and $v_d$ is the deposition velocity (in cm $s^{-1}$).

These procedures are likely to give overestimates because

1. Concentrations in air and in precipitation are probably higher at coastal stations than at sea due to nearby emissions (see Sect. 4).
2. Fluxes from sea to air by evaporation or by particle production are neglected. This seems to be justified by the fact that for many pollutants wind roses at coastal stations show low concentrations at wind from sea (see e.g. Kretschmar and Cosemans 1979).

## 3.2 Trace Metals

Table 1 shows estimates of inputs by dry and wet deposition of some trace metals (van Aalst et al. 1983a). The estimates were obtained from relatively abundant data obtained from measurements at the British, Belgian and Dutch coastal stations. Values for Hg refer to the aerosol phase only: the concentrations of this element in air and precipitation were often below detection limits.

**Table 1.** Atmospheric input of trace metals into the North Sea (t $year^{-1}$), derived from measurements at coastal stations

| | van Aalst et al. (1983a) | PARCOM (1986) |
|---|---|---|
| As | 220– 720 | 40– 120 |
| Cd | 110– 430 | 45– 240 |
| Cr | 70– 1400 | 300– 900 |
| Cu | 1400–10000 | 400– 1600 |
| Hg | ≤ 36 | 10– 30 |
| Ni | 360– 3600 | 300– 950 |
| Pb | 3600–13000 | 2600– 7400 |
| Zn | 7200–58000 | 4900–11000 |

Dry deposition estimates show wide ranges due to the uncertainty in the dry deposition velocity $v_d$, which was assumed to be 0.1–1 cm $s^{-1}$. This uncertainty is mainly due to lack of knowledge of the size distribution of the aerosol, and very poor knowledge on the dry deposition of aerosols on water surfaces (Slinn and Slinn 1981). A likely value of $v_d$ is 0.1–0.3 cm $s^{-1}$ for most metals.

The range in the values for wet deposition reflects the range in measured concentrations. Generally, the input by dry deposition is less than the wet deposition. Wet deposition, however, is overestimated due to contamination of the samples in open collectors during dry periods. This contamination cannot be considered as representative for dry deposition at sea, as the collector geometry and chemical nature is different from the sea surface, and dry deposition is very sensitive to these factors. If it is assumed that the total deposition may be approximated by the wet deposition as measured with open collectors, an error of less than a factor of two is introduced.

More recent measurements, which are believed to be less contamined and less influenced by local sources indicate lower inputs, as is shown in Table 1 (PARCOM 1986). However, model calculations indicate that actual inputs of trace metals into the North Sea are expected to be lower than these estimates from measurements at coastal stations (see Sect. 4).

### 3.3 Organic Compounds

Estimates of inputs of organic compounds are still more uncertain than those for trace metals. This is due to the scarcity of data on concentrations in air and in precipitation and to the uncertainty in the deposition parameters. The concentration of six polycyclic aromatic hydrocarbons (PAH), viz. fluoranthene, benzo(k)fluoranthene, benzo(b)fluoranthene, benzo(ghi)perylene, benzo(a)pyrene, benzo(e)pyrene and indeno (1,2,3-cd) pyrene were found to sum up to about 2–8 ng $m^{-3}$ from measurements in the Netherlands and in Norway in the period 1976–1981 (van Aalst et al. 1983a). These PAH occur mainly in aerosol particles. Assuming a deposition velocity of 0.1–1 cm $s^{-1}$, a dry deposition input of some 30–1300 t $year^{-1}$ is found. The wet deposition flux is expected to be between 50 and 200 t $year^{-1}$ on the basis of scarce data of concentrations in precipitation (van Aalst et al. 1983). Total atmospheric inputs of these six PAH are probably between 50 and 1300 t $year^{-1}$.

For pesticides and polychlorobiphenyls (PCB), systematic measurements of air concentrations were performed at Delft, the Netherlands (Diederen et al. 1981). During the period 1979–1981, 55 24-h samples were taken and analyzed for a number of pesticides and PCB. Although influence of local sources cannot be excluded in these measurements, the data were not inconsistent with those for the waters around England (Dawson and Riley 1977) and the North Atlantic Ocean (Atlas and Giam 1981). Concentrations in precipitation were measured by Wells and Johnstone (1978) along the English and Scotch east coast. The dry deposition velocity is very uncertain, as some of these substances occur both in gaseous and particulate form. The often strong lipophilic character of these compounds forms a complication in estimating the exchange between air and natural seawater containing biomass. Van Aalst et al. (1983a) estimated the dry deposition velocity

**Table 2.** Estimates of atmospheric inputs of polycyclic aromatic hydrocarbons (PAH), some pesticides and polychlorobiphenyls (PCB) into the North Sea (t $year^{-1}$) (van Aalst et al. 1983a)

| | Dry deposition | Wet deposition |
|---|---|---|
| PAH | 30–1300 | 50–200 |
| Hexachlorocyclohexane | 10– 100 | 2 |
| Hexachlorobenzene | 2– 17 | –[a] |
| DDT, DDE | 2.2– 22 | 1.4 |
| PCB | 10– 160 | 2 |

[a] Unknown.

for pesticides to be 0.1–1 cm $s^{-1}$, and for PCB 0.2–0.5 cm $s^{-1}$, with great uncertainty. Table 2 summarizes the input estimates made by van Aalst et al. from these data. As can be seen, dry deposition may be relatively important for these substances.

## 4 Modelling of Atmospheric Inputs

Several attempts have recently been made to calculate atmospheric inputs into the North Sea by means of atmospheric transport models (van Aalst et al. 1983b; van Jaarsveld et al. 1986). In these models, annually or seasonally averaged concentrations and depositions are calculated from emissions and meteorological data. Wind speed, atmospheric stability, mixing layer depth and precipitation at sea generally differ from those of the surrounding land (Höhn 1973; RSU 1980; Joffre 1985), but to a first approximation these differences may be disregarded. The errors introduced in this way are of the order of some tens of percents, generally acceptable in view of the large uncertainty in emission data. This uncertainty is illustrated by Table 3 where recent estimates for trace metal emissions in countries surrounding the North Sea are given.

**Table 3.** Emission of trace metals (t $year^{-1}$) for some European countries according to van Jaarsveld et al. (1986) (first line) and to Pacyna (1985) (second line)

| | Belgium | Denmark | France | FRG | Lux | Neth. | UK |
|---|---|---|---|---|---|---|---|
| As | 94 | 3.1 | 67 | 230 | 0.11 | 4.9 | 62 |
| | 360 | 7 | 230 | 350 | 0.4 | 38 | 164 |
| Sb | 62 | 12 | 38 | 61 | 0.60 | 13 | 29 |
| | – | – | – | – | – | – | – |
| Cd | 22 | 2.4 | 40 | 78 | 0.26 | 5.4 | 22 |
| | 51 | 9 | 85 | 150 | 1 | 11 | 51 |
| Cr | 79 | 7.1 | 130 | 260 | 24 | 25 | 127 |
| | – | – | – | – | – | – | – |
| Cu | 230 | 17 | 170 | 460 | 10 | 60 | 210 |
| | 610 | 38 | 450 | 1600 | 24 | 105 | 580 |
| Ni | 110 | 57 | 350 | 290 | 15 | 53 | 300 |
| | – | – | – | – | – | – | – |
| Pb | 2200 | 440 | 6200 | 5400 | 176 | 1500 | 5900 |
| | 4000 | 750 | 10500 | 9300 | 300 | 2400 | 10000 |
| Zn | 1000 | 420 | 3600 | 6900 | 93 | 470 | 2100 |
| | 4700 | 710 | 6100 | 12000 | 160 | 1400 | 3500 |

Model calculations can provide independent estimates for atmospheric inputs which can be compared to input estimates from measurements. Moreover, the calculations may indicate the spatial variation of the deposition, and allow the evaluation of the representativeness of measurements at coastal stations. Also, from the calculations estimates of the contributions of source areas and source categories to the deposition may be obtained, and the effectiveness of emission reduction measures to decrease atmospheric inputs may be assessed.

Figure 1 shows deposition maps calculated by van Jaarsveld et al. (1986). There is a rather strong north-south gradient in the deposition flux, and the results indicate that fluxes measured at coastal stations in the Netherlands, Belgium and Great Britain are likely to be a factor of 2–4 higher than the average flux into the North Sea. Taking this into account, the agreement between measured and modelled inputs is reasonable for most trace metals, as shown in Table 4.

It appears that the calculated total depositions are not very sensitive to the choice of model parameters. Variation of the deposition velocity from 0.01 to 1 cm $s^{-1}$ and of the scavenging coefficient from $2 \times 10^{-6}$ to $6 \times 10^{-6}$ resulted in a change in total deposition of a factor of 2 only. This indicates that the main uncertainty in these model estimates is in the emissions. Metal concentrations measured at stations on land are reproduced rather well by these models (van Egmond 1986).

Table 5 shows estimated contributions of some countries to the atmospheric inputs of trace metals into the North Sea. Although major contributions come from countries directly surrounding the North Sea, contributions from somewhat more distant sources cannot be disregarded.

**Table 4.** Atmospheric inputs of trace metals into the North Sea (t $year^{-1}$) as derived from measurements and model calculations

| | Measurements[a] | Model calculations[b] |
|---|---|---|
| As | 13– 40 | 42 |
| Cd | 15– 80 | 14 |
| Cr | 100– 300 | 74 |
| Cu | 130– 530 | 130 |
| Ni | 100– 320 | 150 |
| Pb | 870–2500 | 2600 |
| Zn | 1600–3600 | 1200 |

[a] From Table 1; but reduced by a factor of 3 (see explanation in text).
[b] From van Jaarsveld et al. (1986); rounded values.

**Table 5.** Calculated relative contribution of some countries to the total deposition in the North Sea (in percent)

| | Belgium | Denmark | France | FRG | Neth. | UK | Other |
|---|---|---|---|---|---|---|---|
| As | 20 | 0.6 | 7 | 16 | 1.5 | 38 | 18 |
| Sb | 27 | 4 | 9 | 9 | 7 | 39 | 6 |
| Cd | 12 | 1 | 13 | 15 | 5 | 39 | 15 |
| Cr | 8 | 1 | 8 | 10 | 7 | 57 | 9 |
| Cu | 15 | 1 | 6 | 14 | 7 | 41 | 16 |
| Ni | 6 | 3 | 11 | 7 | 5 | 53 | 15 |
| Pb | 6 | 1 | 11 | 5 | 8 | 60 | 9 |
| Zn | 6 | 2 | 14 | 15 | 5 | 45 | 13 |

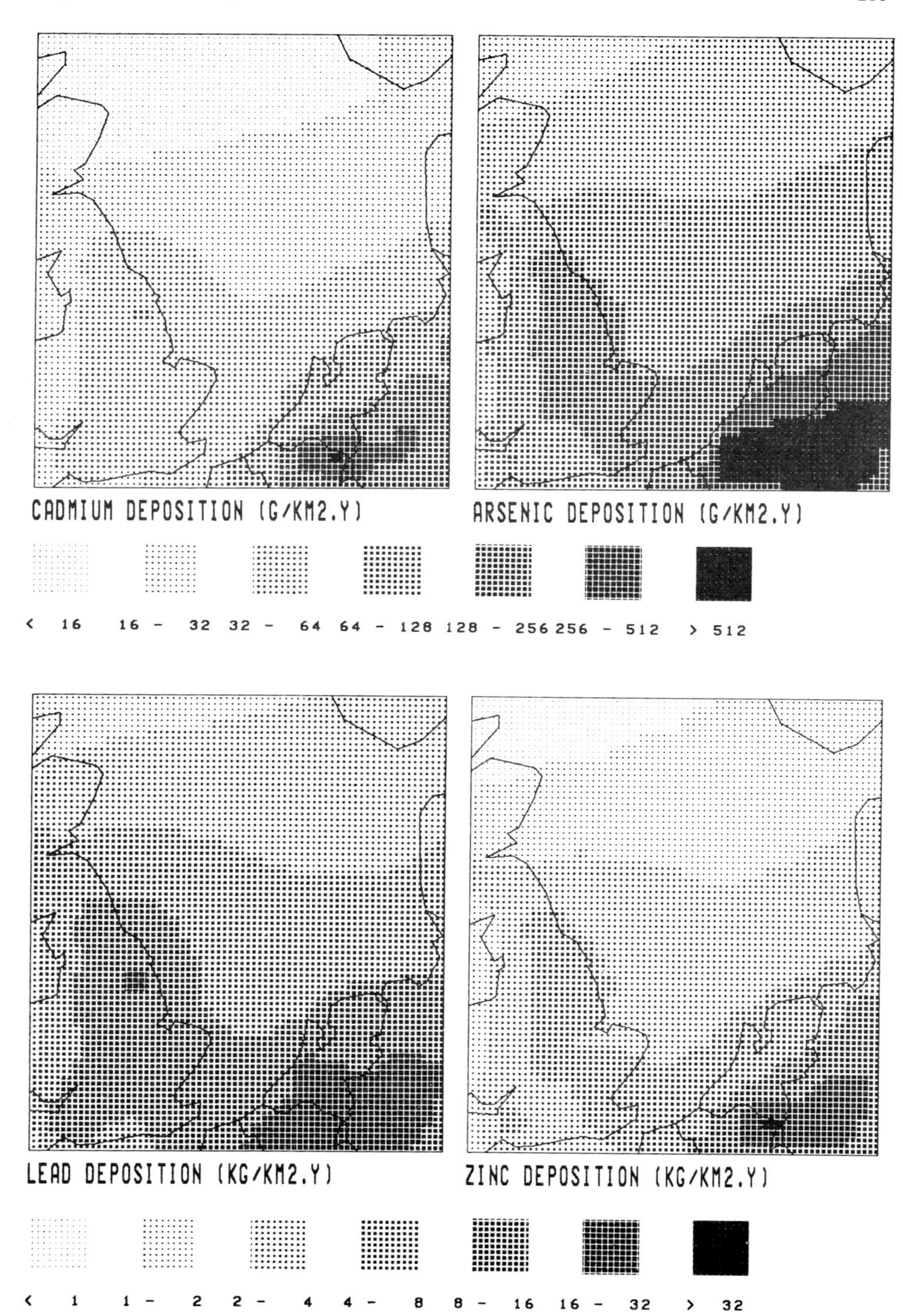


**Fig. 1.** Total (dry plus wet) deposition of Cd and As (in g $km^{-2}$ $year^{-1}$) and Pb and Zn (in kg $km^{-2}$ $year^{-1}$) into the North Sea (van Jaarsveld et al. 1986)

## 5 Major Uncertainties

Our knowledge of atmospheric inputs into the North Sea is presently mainly limited due to lack of:

1. Adequate and representative measuring data of concentrations in air and precipitation;
2. Insight into and quantitative parameters for dry deposition of gases and particles over the sea;
3. Accurate data on emissions in Europe.

Much uncertainty in input estimates could be removed by systematic measurements of concentrations in air over the North Sea. This would allow testing of model calculations and improving estimates of dry and wet deposition. Measurement of the particle size distribution for various pollutants is important. Measurements of concentrations in precipitation at coastal stations and at sea by wet – only samplers are needed. The influence of contamination by sea spray and nearby constructions has to be eliminated. Better estimates of the amount of precipitation at sea are needed.

It is quite difficult to make representative measurements of dry deposition fluxes of gases and particles at sea. However, field studies can provide information on dry deposition parameters. For particles, the dependence on particle size is important. For organic gases, pesticides and PCB, the uncertainty in the flux estimates may be reduced by laboratory studies of the partition coefficient between air and sea, particularly if representative seawater is used.

For many pollutants, better emission estimates are needed, especially for countries surrounding the North Sea. Model calculations on the basis of such emissions, and tested on measured concentrations will provide the best and most representative estimates of atmospheric inputs. Model results will also form a necessary basis for designing strategies to reduce the atmospheric input of pollutants by effective emission reductions.

## References

Aalst RM van, Ardenne RAM van, Kreuk JF de, Lems T (1983a) Pollution of the North Sea from the atmosphere. Report CL 83/152 TNO, Delft, the Netherlands

Aalst RM van, Duyzer JH, Veldt C (1983b) Atmospheric deposition of lead and cadmium in the southern part of the North Sea. Emissions and preliminary model calculations. Report R 83/222 TNO, Delft, the Netherlands

Aalst RM van (1986) Dry deposition of aerosol particles. In: Aerosols (Lee SD, Schneider T, Grant LD, Verkerk PJ eds.), p 933. Lewis Publishers, USA

Atlas E, Giam CS (1981) Global transport of organic pollutants: ambient concentrations in the remote marine atmosphere. Science 211, 163

Cambray RS, Jefferies DF, Topping G (1979) The atmospheric input of trace elements to the North Sea. Marine Science Comm 5, 175

Dawson R, Riley JPC (1977) Chlorine-containing pesticides and polychlorinated biphenyls in British coastal waters. Estuarine and Coastal Marine Science, 4, 55

Diederen HSMA, Hartog JC den, Hollander JCTh, Kaayk J, Schulting FL (1981) Niveaus van luchtverontreiniging gemeten over de periode januari 1979 – maart 1981 (FLAT project) (in Dutch). Report CMP 83/02 TNO, Delft, the Netherlands
Eisenreich JS (1981) Atmospheric pollutants in natural waters. Ann Arbor Science, Mich, USA
Egmond ND van (1986) Air pollution as a result of the emission from coal-fired power stations. Report proj. No. 20.70–004.11, PEO Management Office for Energy Research, Utrecht, the Netherlands
GESAMP (Joing Group of Experts on the Scientific Aspects of Marine Pollution) (1980) Interchange of pollutants between the atmosphere and ocean. Report GESAMP (13) WMO Geneva
Goldberg ED (1973) North Sea science. MIT Press, Cambridge, Mass
Hicks BB, Wesely ML, Durham JL (1980) Critique of methods to measure dry deposition: workshop summary. Report EPA 600/9–80–050 USA
Höhn R (1973) On the climatology of the North Sea. In: North Sea Science (Goldberg ED, ed) p 183 MIT Press, Cambridge, USA
Hunter KA (1980) Processes affecting particulate trace metals in the sea surface microlayer. Marine Chemistry 9, 49
ICES (International Council for the Exploration of the Sea) (1978) Input of pollutants to the Oslo commission area. Report ICES 77, Charlottenlund, Denmark
Jaarsveld JA van, Aalst RM van, Onderdelinden D (1986) Deposition of metals from the atmosphere into the North Sea; model calculations. Report RIVM 842015002, Bilthoven, the Netherlands
Joffre SM (1985) The structure of the marine atmospheric boundary layer: a review from the point of view of diffusivity, transport and deposition processes. Technical report 29, Finnish Meteorological Institute, Helsinki
Kretschmar JG, Cosemans G (1979) A five-year survey of some heavy metal levels in air at the Belgian North Sea coast. Atmospheric Environment 13, 267
Liss PS (1975) Chemistry of the sea surface microlayer. In: Chemical oceanography, vol 2. (Riley JP and Skirrow G eds) Acad Press, London
Liss PS, Slater PG (1974) Fluxes of gases across the air-sea interface. Nature 247, 181
Liss PS, Slinn WGN (1983) Air sea exchange of gases and particles. NATO ASI series. D Reidel
NAS (National Academy of Sciences) (1978) The tropospheric transport of pollutants and other substances to the oceans. Washington, USA
Norton RL (1982) Assessment of pollution loads to the North Sea. Water Research Centre Technical Report 82, Medmemham, UK
NSF (National Science Foundation) (1976) Marine pollutant transfer (Windom NHL, Duce RA eds). Lexington Books
PARCOM (Paris Commission) (1986) Current estimates of atmospheric inputs to the North Sea. Fourth meeting of the Working Group on the atmospheric input of pollutants to convention waters, Oslo, 28–30, October 1986
Rohde H, Söderlund R, Eksted J (1980) Deposition of airborne pollutants in the Baltic. Ambio 9, 168
RSU (Rat für Sachverständigung für Umweltfragen) (1980) Umweltprobleme der Nordsee. Verlag W Kohlhammer, Stuttgart, FRG
Sehmel GA (1980) Particle and gas dry deposition, a review. Atmospheric Environment 14, 983
Slinn SA, Slinn WGN (1981) Modelling of atmospheric particulate deposition to natural waters. In: Atmospheric pollutants in natural waters (Eisenreich SJ ed.). Ann Arbor science
Vate JF van de, Brink HM ten (1986) In: Aerosols (Lee SD, Schneider T, Grant LD, Verkerk PJ, eds). Lewis Publishers, USA, p 3
Wells DE, Johnstone SJ (1978) The occurrence of organochlorine residues in rain water. Water, Air and Soil Pollut 9, 271

# Occurrence and Fate of Organic Micropollutants in the North Sea*

W. ERNST[1], J.P. BOON[2], and K. WEBER[1]

## 1 Introduction

According to Maugh (1978), more than 60,000 organic chemicals are presently in use. Potential sources from which they are likely to reach the marine environment are rivers, the atmosphere, direct dumping into the sea, and shipping activities.

Once entered the sea, a compound takes part in a variety of interactions, involving many transitions between different "compartments" such as water, suspended matter, sediments and organisms.

Depending on their chemical structure, many of these compounds are considerably toxic to marine biota. They can also bioaccumulate in marine species, thus being hazardous to man consuming contaminated seafood. A number of strongly bioaccumulating chemicals, such as PCB's and DDT, have been detected in fish and other marine biota. Only in recent years, however, has it become feasible to identify the sometimes up to $10^7$ times lower concentrations of an array of organic contaminants in seawater, owing to large improvements of the available analytical techniques.

Transformation and degradation processes alter a chemical structure and remove the parent compound from the environment. Regrettably there are only few investigations dealing with the biotic and abiotic transformations and degradations of organics, although these pathways deserve special attention in view of an ecotoxicological evaluation of the compound in question. Degradation processes may occur by physicochemical reactions and on biochemical pathways in microorganisms as well as in higher animals, such as macroinvertebrates, fish, and marine mammals.

This chapter will deal with the partitioning processes between the dissolved phase, suspended particulate matter (SPM) and sediments, as well as with abiotic and biotic degradation processes altering molecular structures. The behavior of a few classes of compounds, about which more than the bare minimum of knowledge exists, will be highlighted in separate paragraphs.

[1]Alfred-Wegener-Institute for Polar and Marine Research, Chemistry Section, Columbusstraße, D-2850 Bremerhaven, FRG
[2]Netherland Institute for Sea Research, P.O. Box 59, 1790 AB Den Burg, Texel, The Netherlands
*Contribution No. 24 of the Alfred-Wegener-Institut for Polar and Marine Research

## 2 Occurrence of Organic Micropollutants in the North Sea

### 2.1 Water

The analyses of water from estuaries, coastal areas and the open North Sea point to rivers as the major sources for many organic pollutants. Not only persistent compounds (i.e., compounds which are hardly degraded by any mechanism), such as chlorinated hydrocarbons, were detected in the North Sea, but also more readily degradable compounds such as phthalic acid esters (e.g., DBP and DEHP) and organic phosphates in use as plasticizers.

Substances identified in estuaries and coastal waters of Germany are listed in Table 1, indicating a large variety of chemical structures (Ernst and Weber 1978; Weber and Ernst 1978, 1983; Ernst et al. 1986). Inputs for selected compounds from rivers and estuaries to the open sea can be taken from Fig. 1.

**Table 1.** Organic chemicals identified in estuaries of the rivers Elbe, Weser, Ems, and in German coastal waters (1977–1984) (After Ernst et al. 1986)

| Pesticides | Technical chemicals intermediates and byproducts |
|---|---|
| Methylparathion | Organophosphates |
| Dichlobenil | Phthalates |
| Hexachlorobenzene | Hexachlorocyclohexanes ($\alpha$, $\beta$, $\delta$, $\varepsilon$) |
| Lindane ($\gamma$-Hexachlorocyclohexane) | Chlorobenzenes |
| Pentachlorophenol | Polychlorinated biphenyls (PCB) |
| | chloronitrobenzene |
| Organotin compounds | Chlorophenols |
| DDT-group | Chlorinated low molecular weight hydrocarbons and ethers |
| Triazines | various chlorinated and non-chlorinated aromatics |
| **Oil components and other hydrocarbons** | |
| Paraffines | |
| Naphthenes | |
| Mononuclear aromatic hydrocarbons | |
| PAH | |

Quantitative data for some components from coastal waters of Germany are shown in Table 2; it should be noted that values for DEHP and bis-(2-chloroisopropyl)ether can sometimes be 1 or 2 orders of magnitude higher.

In Fig. 2 the rapid decline of HCB concentrations from the Elbe estuary to the German Bight is shown. This compound is known to associate strongly with particles (Killer 1986) thus being partly eliminated from the water column. Polychlorinated biphenyls (PCB) with estuarine concentration of 3–10 ng $l^{-1}$ compared to those of 0.4–0.9 ng $l^{-1}$ in the open North Sea do not exhibit such a steep gradient.

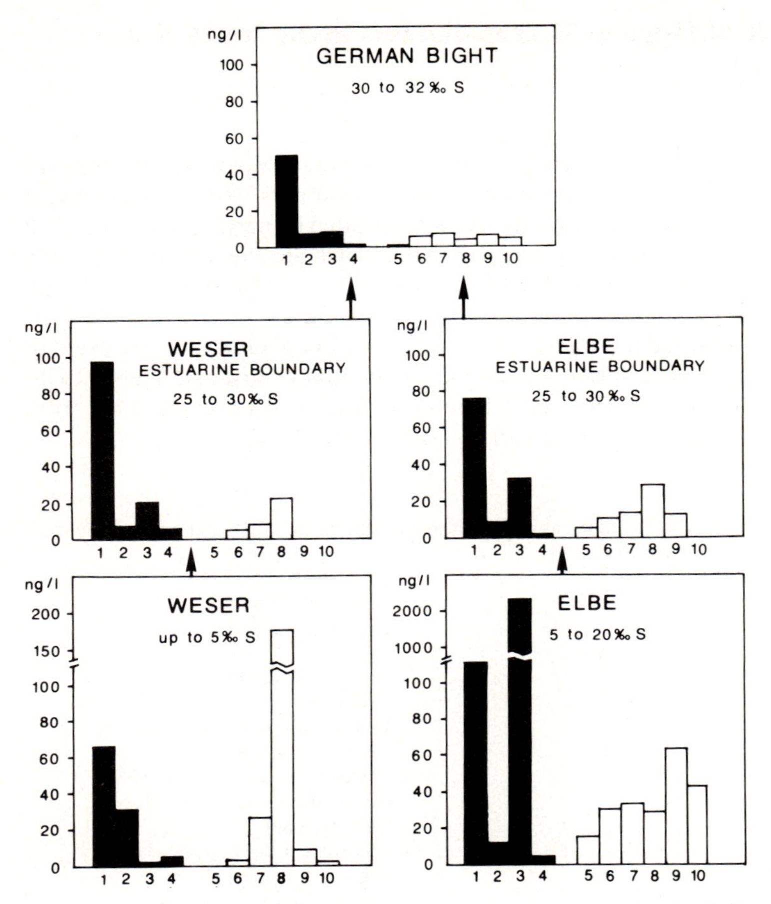


**Fig. 1.** Input of organic pollutants from German estuaries into the North Sea (Weber and Ernst, unpubl.) *1* Phthalates; *2* organophosphates; *3* bis-(2-chloroisopropyl)ether; *4* polychlorinated biphenyls (PCB); *5* hexachlorobenzene; *6* α-hexachlorocyclohexane; *7* γ-hexachlorocyclohexane; *8* pentachlorophenol; *9* prometryne; *10* dimethylparathion

Especially in dynamic estuarine and coastal waters, concentrations of organic micropollutants may vary with tidal movements and emission regimes as well as with microbial degradation potentials present in water. For example, sorption to suspended matter leading to lowered concentrations in solution may especially occur in the turbidity zone of estuaries. Therefore great care must be taken in evaluating a pollution situation when only few measurements of water are available. Results should always be discussed together with the sampling strategy and the used analytical methods.

**Table 2.** Concentration of organic compounds in the German Bight (After Ernst et al. 1986)

| Compounds | Concentration (ng $l^{-1}$) | | |
|---|---|---|---|
| | A | B | C |
| Dimethylparathion | < 0.5 | < 0.5 | 8.5 |
| Dichlobenil | 0.8 | 0.6 | 1.8 |
| p,p'-DDE | 0.09 | 0.07 | ≤ 0.03 |
| γ-Hexachlorocyclohexane (γ-HCH) | 2.9 | 3.6 | 4.5 |
| α-Hexachlorocyclohexane (α-HCH) | 2.9 | 4.5 | 7.1 |
| Tri-butylphosphates | 6 | 8 | 14 |
| Di-n-butylphthala (DBP) | 7.9 | 6.8 | 8.8 |
| Bis-(2-ethylhexyl)phthalate (DEHP) | 31.1 | 35.3 | 24.5 |
| Chloroanthraquinone | 3.8 | 2.9 | 12.0 |
| Dichloroanthraquinones | 5.4 | 5.5 | 8.8 |
| Bis-(2-chloroisopropyl)ether | 2.2 | 1.9 | 36.6 |
| Fluorene | 5.2 | 3.3 | 3.6 |
| Dibenzofurane | 2.5 | 2.9 | 0.9 |
| Fluoranthene | 2.3 | 1.7 | 1.4 |
| Pyrene | 1.3 | 1.1 | 0.8 |
| Chrysene | 2.2 | 1.4 | 1.0 |

A: East Frisian coast, Oct. 1982;
B: North Frisian coast, Oct. 1982;
C: NE-Helgoland, June 1982.

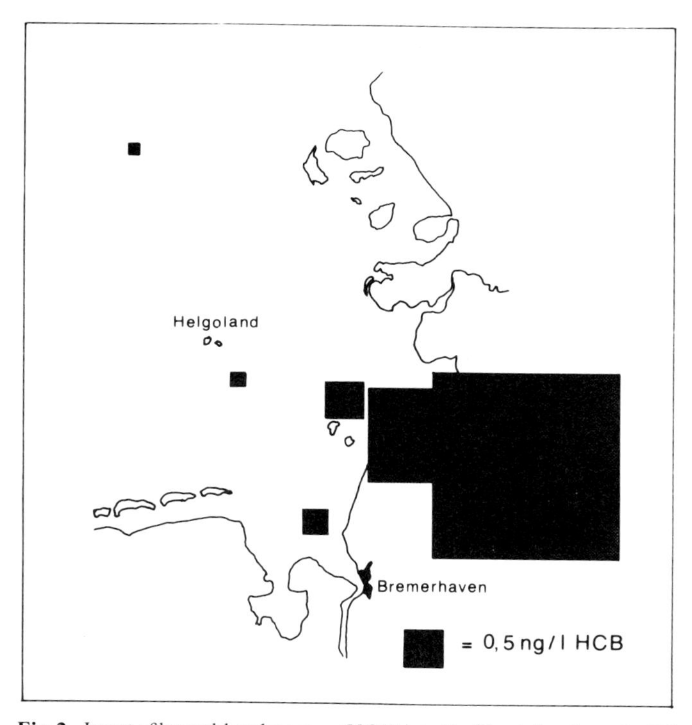


**Fig. 2.** Input of hexachlorobenzene (HCB) into the North Sea from the Elbe estuary. (After Ernst 1986)

## 2.2 Suspended Particulate Matter and Sediments

Organic components in seawater are either dissolved or associated with suspended particulate matter. Distribution depends on the physical properties of the components and the compartments.

Contaminants which are very less soluble in water show a high affinity for being adsorbed onto particulate matter; among the particles, small ones with a high specific surface, and those with a high content of organic matter show the highest sorption capacity (Eder and Weber 1980).

Experimental data obtained in the laboratory is usually in good agreement with in situ data (Killer 1986) and if water solubility of a contaminant is known, its adsorption to particulate matter can be calculated (Kenaga and Goring 1979). The same is generally true for sediments. Thus sandy erosion-type bottoms contain low concentration of hydrocarbons (Law 1981; Law and Fileman 1985) or organochlorines (Duinker et al. 1983; Boon et al. 1985), while the highest concentrations of trace contaminants in the North Sea should be expected in areas characterized by high fractions of silt (Eisma 1981). In silty sediments of the German Bight organic chemicals were determined along a west-east profile, indicating an increase of the concentration approaching the river Elbe outflow (Eder 1984), Table 3.

**Table 3.** Organic compounds in sediments of the German Bight at 53°03′ N and 07°55′ E – 08°13′ E

| Compounds | ng/g wet sediment[a] |
|---|---|
| α-Hexachlorocyclohexane (α-HCH) | 0.03– 0.34 |
| γ-Hexachlorocyclohexane (γ-HCH) | 0.02– 0.12 |
| Bis-(2-ethylhexyl)phthalate (DEHP) | 45.43–222.30 |
| Di-n-butylphthalate (DBP) | 5.25– 16.34 |
| Hexachlorobenzene (HCB) | 0.08– 4.12 |
| Pentachlorobenzene | 7.51– 35.06 |

[a] Water content of sediments: 45–54%.

## 2.3 Specific Compounds: Oil, Polychlorinated Biphenyls (PCB's) Hexachlorocyclohexanes (HCH's)

### 2.3.1 Oil Components

Estimates of the global oil input into the marine environment indicate that about one third of the total oceanic and shelf-sea pollution is due to accidental spills and transport of petroleum derivatives. So, despite the suddenly extremely high oil concentrations occurring after a major spill, continuous input at much lower concentrations via rivers and the atmosphere appear to involve the highest annual quantities (U.S. NAS 1975). Nevertheless, the major spills may well cause the severest ecotoxicological effects because of the high concentrations involved.

The weathering of oil greatly influences its hazard. While birds are in the first place threatened by floating oil, hydrocarbons become bioavailable to gill-breathing aquatic animals only when components dissolve into the water phase, disperse as an oil-in-water emulsion, or adsorb to suspended particles.

The largest oil-spill that has occurred in the North Sea, the Bravo blow-out, caused the escape of $15–23 \times 10^3$ tons of oil in a 2:1 oil:gas mixture through an open production pipe about 20 m above the sea surface. The most volatile components of the oil disappeared rapidly. Because of its low viscosity, the oil spread over the sea surface, increasing volatilization into the atmosphere and dissolution into the seawater. Maximum concentrations of total oil-derived hydrocarbons were about 300 ng $l^{-1}$. Whether such concentrations occur as true solutions or partly as agglomerates, remained uncertain (Audunson 1978; Grahl- Nielsen et al. 1977a,b; Grahl- Nielsen 1978).

Compared to the aliphatic hydrocarbons, the group of poly-aromatic hydrocarbons (PAH's) hardly occur naturally in the marine environment, possess a higher water solubility, are more resistant to microbial breakdown and show a higher toxicity. Less volatile representatives will become enriched in the water phase compared to the original oil mixture. In case of the Bravo, oil residues could be identified as originating from the blow-out with isomers of dimethyl phenanthrenes and di- and trimethyl dibenzothiophenes, since the pattern of their relative concentrations was only slightly affected by weathering processes.

When about the first week after an oil spill has elapsed, the role of evaporation and dissolution decreases and other processes become more important: mousse and tarball formation as well as photochemical and microbial degradation processes (Wolfe 1985).

When oil components become available as a carbon source for microorganisms, populations build up when inorganic nutrients and essential trace elements are also available.

The formation of mousse and tarballs decreases both biodegradation and weathering because of a decreasing surface: volume ratio of the oil residue.

A special occasion is that of a relatively fresh oil layer being left on a tidal flat when the water recedes during the ebb. Such a situation may occur in the shallow estuarine areas bordering the North Sea, such as the Waddensea. In experiments with mesocosms, the oil was rapidly burrowed into the deeper layers of the sediments by the activity of benthic animals, especially the lugworm *Arenicola marina* (Kuiper et al. 1984). The sediments involved may either be aerobic or anaerobic; the latter greatly reduces biodegradation in sediments (Berne and Bodennec 1984).

In comparison with the situation after major disasters, little is known quantitatively about the processes and concentrations involved with the continuous input of oil-derived compounds via rivers and the atmosphere to North Sea waters. Some data on PAH's and a variety of other hydrocarbons in Dutch coastal waters have been reported by van de Meent et al. (1985). However, no distinction between dissolved and particulate fractions were made.

### 2.3.2 Polychlorinated Biphenyls (PCB's)

PCB's belong to the group of cyclic chlorinated hydrocarbons and were or are still in use in a variety of industrial applications. These include their application as cooling fluids in transformers and condensors, in hydraulic oils, additives to paints and technical pesticide formulations, No Carbon Required paper, and plasticizers (US NAS 1979). PCB's represent a mixture of chlorinated biphenyls. All possible chlorine substitutions of the biphenyl molecule together allow for the theoretical existence of 209 components. A typical chromatogram of PCB's in filtered seawater is given in Fig. 3. Concentrations of individual PCB congeners are in the sub-ng $l^{-1}$ range for North Sea waters.

Between environmental compartments and also between different animal species, the PCB patterns, i.e., the relative contribution of each congener to total PCB differ. However, within each environmental compartment, the patterns appear to be largely independent of concentrations. So, within each environmental compartment, PCB concentrations may be compared either on the basis of summation of concentrations of individual congeners prior to comparison, or per single congener. Between different environmental compartments only the latter method is appropriate.

When water is compared with particulates and biota, tri- and tetra-chlorobiphenyls (CB's) dominate in solution, while penta-, hexa- and hepta-CB's dominate in SPM and biota. Mono- and dichlorobiphenyls are usually hardly present, probably as a result of degradation processes and higher volatilities for these lower chlorinated congeners.

Because of the virtual absence of degradation processes in the water column – metabolism was shown in higher organisms by Goerke and Ernst (1986) – PCB distribution between SPM and water can be expressed by a distribution coefficient K(d), presuming equilibrium conditions.

Table 4 indicates that values of K(d) in the open North Sea and in several rivers entering the North Sea, tend to increase with increasing chlorine content and with increasing log $P_{octanol}{}^{\pi}{}_{water}$. Since the latter is a parameter expressing the hydrophobicity of a compound, it shows the tendency of each congener to escape from the highly polar water phase and to "solubilize" into an apolar part of a particle, for which PCB's have a high affinity. Table 4 also shows a higher K(d) for North Sea waters compared to the rivers. This is due to a different nature of the particles dominating at low SPM concentrations in the open North Sea (Duinker 1986). Figure 4 shows that for the North Sea and the rivers investigated, both the dissolved and the particulate phases contribute significantly to transport processes.

### 2.3.3 Hexachlorocyclohexanes

Alpha- and gamma-HCH are found all over the North Sea. The earliest available distribution data for both isomers originate from 1977 and 1979 (Weber and Ernst, in preparation, 1986) Although the data were only derived from investigations of surface water they are highly reliable from an analytical point of view and can be compared with future investigations for trend analyses. Later Gaul and Ziebarth

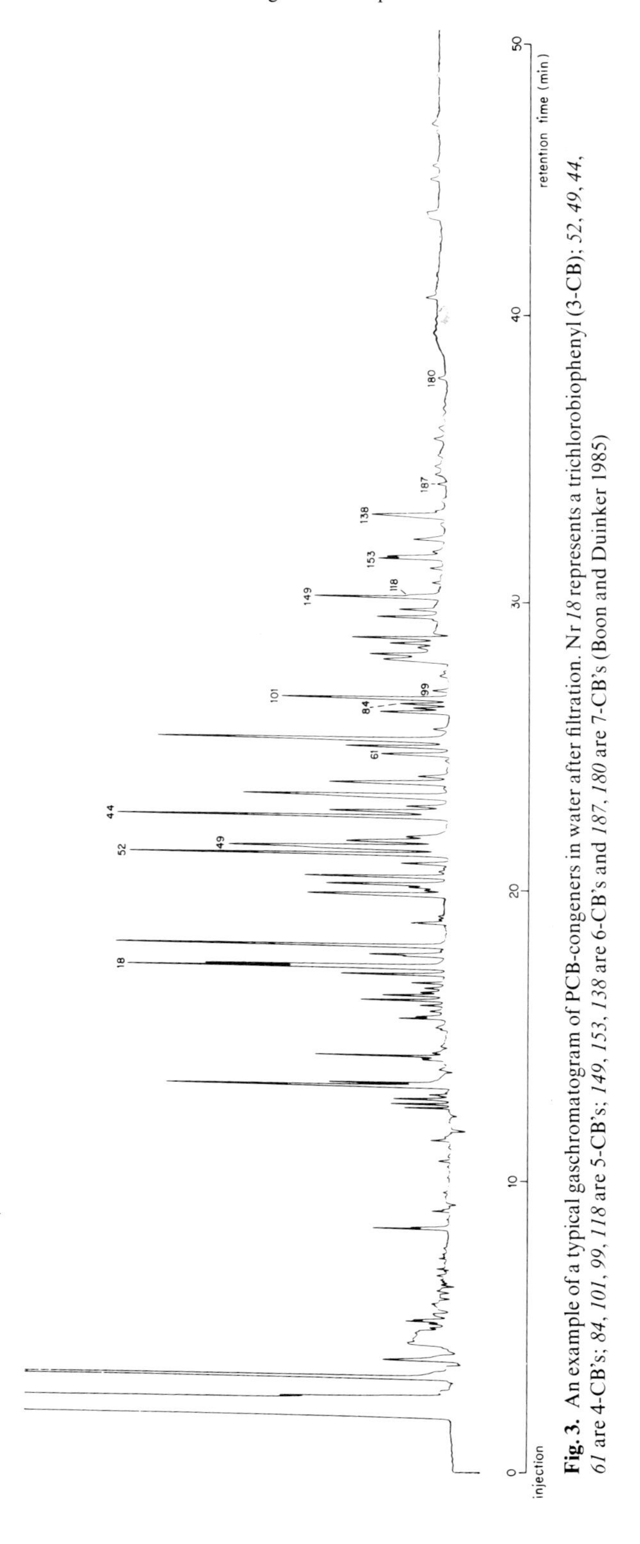


**Fig. 3.** An example of a typical gaschromatogram of PCB-congeners in water after filtration. Nr *18* represents a trichlorobiophenyl (3-CB); *52*, *49*, *44*, *61* are 4-CB's; *84*, *101*, *99*, *118* are 5-CB's; *149*, *153*, *138* are 6-CB's and *187*, *180* are 7-CB's (Boon and Duinker 1985)

**Table 4.** Distribution coefficients for water-suspended matter partition on individual PCB congeners in freshwater of the Scheldt, Rhine, Weser, and Elbe rivers and in the offshore of the North Sea. The latter samples were obtained at positions 52°01′N, 2°47′E, 53°0′N, 1°31′, 5E, 55°20′N, 5°30′E (October 1982) at salinities > 35 × $10^{-3}$. (After Duinker and Boon 1986)

| PCB-congener a) | b) | c) | d) | Scheldt | Rhine | Weser | Elbe | Ems | North Sea |
|---|---|---|---|---|---|---|---|---|---|
| 18 | 2,2′,5 | 3 | 5.55 | $8\times10^3$–$5\times10^4$ | $2\times10^4$ | $2\times10^4$ | $2\times10^4$ | $2\times10^4$ | $4\times10^4$–$6\times10^5$ |
| 15 | 4,4′ | 2 | 4.82 | $6\times10^3$–$6\times10^4$ | $4\times10^4$ | $2\times10^4$ | $2\times10^4$ | $3\times10^4$ | $5\times10^5$ |
| 26 | 2,3′,5 | 3 | 5.76 | $2\times10^4$–$2\times10^5$ | $7\times10^4$ | $2\times10^5$ | $1\times10^5$ | $1\times10^5$ | $8\times10^4$–$2\times10^5$ |
| 31 | 2,4′,5 | 3 | 5.69 | $1\times10^4$–$2\times10^5$ | $7\times10^4$ | –e) | –e) | –e) | $1\times10^5$–$3\times10^5$ |
| 28 | 2,4,4′ | 3 | 5.69 | $2\times10^4$–$1\times10^5$ | $7\times10^4$ | – | – | – | $7\times10^4$–$2\times10^5$ |
| 52 | 2,2′,5,5′ | 4 | 6.09 | $3\times10^4$–$7\times10^4$ | $8\times10^4$ | $4\times10^4$ | $5\times10^4$ | $5\times10^4$ | $3\times10^5$–$2\times10^6$ |
| 84 | 2,2′,3,3′,6 | 5 | 6.04 | $5\times10^4$ | $8\times10^4$ | $8\times10^4$ | $8\times10^4$ | $5\times10^4$ | $7\times10^4$–$4\times10^5$ |
| 101 | 2,2′,4,5,5′ | 5 | 7.07 | $8\times10^4$–$1\times10^5$ | $1\times10^5$ | $1\times10^5$ | $1\times10^5$ | $6\times10^4$ | $8\times10^4$–$1\times10^6$ |
| 99 | 2,2′,4,4′,5 | 5 | 7.21 | $1\times10^4$–$4\times10^5$ | $1\times10^5$ | $1\times10^5$ | $1\times10^5$ | $6\times10^4$ | $6\times10^4$–$2\times10^6$ |
| 153 | 2,2′,4,4′,5,5′ | 6 | 7.75 | $1\times10^5$–$3\times10^5$ | $1\times10^5$ | $2\times10^5$ | $2\times10^5$ | $1\times10^5$ | $5\times10^5$–$5\times10^6$ |
| 138 | 2,2′,3,4,4′,5′ | 6 | 7.44 | $7\times10^4$–$1\times10^5$ | $6\times10^4$ | $2\times10^5$ | $2\times10^5$ | | $2\times10^5$–$5\times10^6$ |
| SPM concentration (mg $dm^{-3}$) | | | | 45 | 19 | 30 | 35 | 40 | 0.5–1.0 |

a) systematic numbering, b) chlorine substitution pattern, c) number of chlorine atoms ($N_{Cl}$) d) Values of $^{10}\log P_{ow}$ as taken from Rapaport and Eisenreich (1984) e): individual peaks of 31 and 28 not separated.

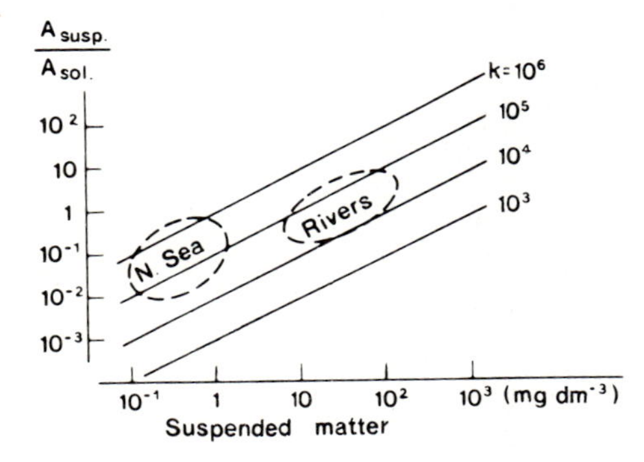


**Fig. 4.** Relation between suspended matter concentration (mg $dm^{-3}$) and the ratio between the amounts of PCB congeners in suspension ($A_{susp}$) and in solution ($A_{sol}$) for various values of the distribution coefficient (K) of any chemical constituent with $10^3 < K < 10^6$. The values for individual PCB congeners in the North Sea and in the rivers given in Table 4 are contained in the encircled regions. (Duinker and Boon, 1986)

(1983) reported similar HCH distributions in the North Sea, including depth profiles.

The important stereoisomers of technically synthesized hexachlorocyclohexanes (HCH) were detected in coastal waters of the North Sea. For instance, water of the Elbe estuary contained a mixture of 27% $\alpha$-, 14% $\beta$-, 31% $\gamma$-, 23% $\delta$- and 5% $\varepsilon$-HCH in October 1982 and river input could be derived from the distribution of the compound levels along the estuary (Weber, unpublished).

In Fig. 5, isolines of HCH isomers ratios are shown for surface water in the German Bight. They demonstrate different input sources in this area. Whereas the river Weser is mainly loaded with $\gamma$-HCH – ten times more of this isomer is usually found compared to $\alpha$-HCH – the river Elbe carried equal portions of both into the

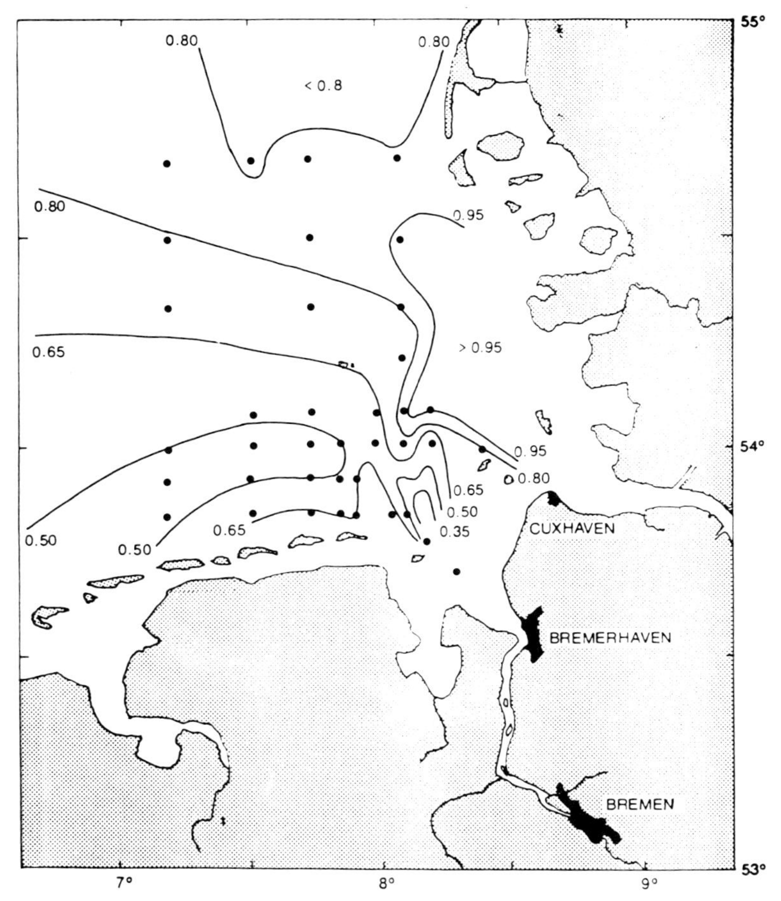


**Fig. 5.** Lines of equal ratios of α- and γ-HCH levels in water of the German Bight in June 1977. (Weber and Ernst, unpublished)

sea at the time of investigation; in East-Frisian coastal waters, directed into the German Bight by residual currents, levels of γ-HCH are about twice as high as levels of α-HCH. The ratio can be used to follow the movements of water masses within the German Bight. However, levels of α-HCH decreased in the last 2 or 3 years compared with average data from 1977 to 1982.

In Fig. 6 α- and γ-HCH in surface water of different parts of the North Sea is compared. The main load for both compounds is found in the German Bight. In remote areas α-HCH levels are higher than those for γ-HCH. The reason for this distribution is not yet understood unambiguously. Environmental photoisomerization or faster degradation of the thermodynamically less stable γ-isomer during long-range transport in the water may be one explanation; favored atmospheric input of the higher volatile α-isomer in remote areas another. In any case, the observed gradients of both isomers within the estuary appear to be considerably less steep than those of HCB.

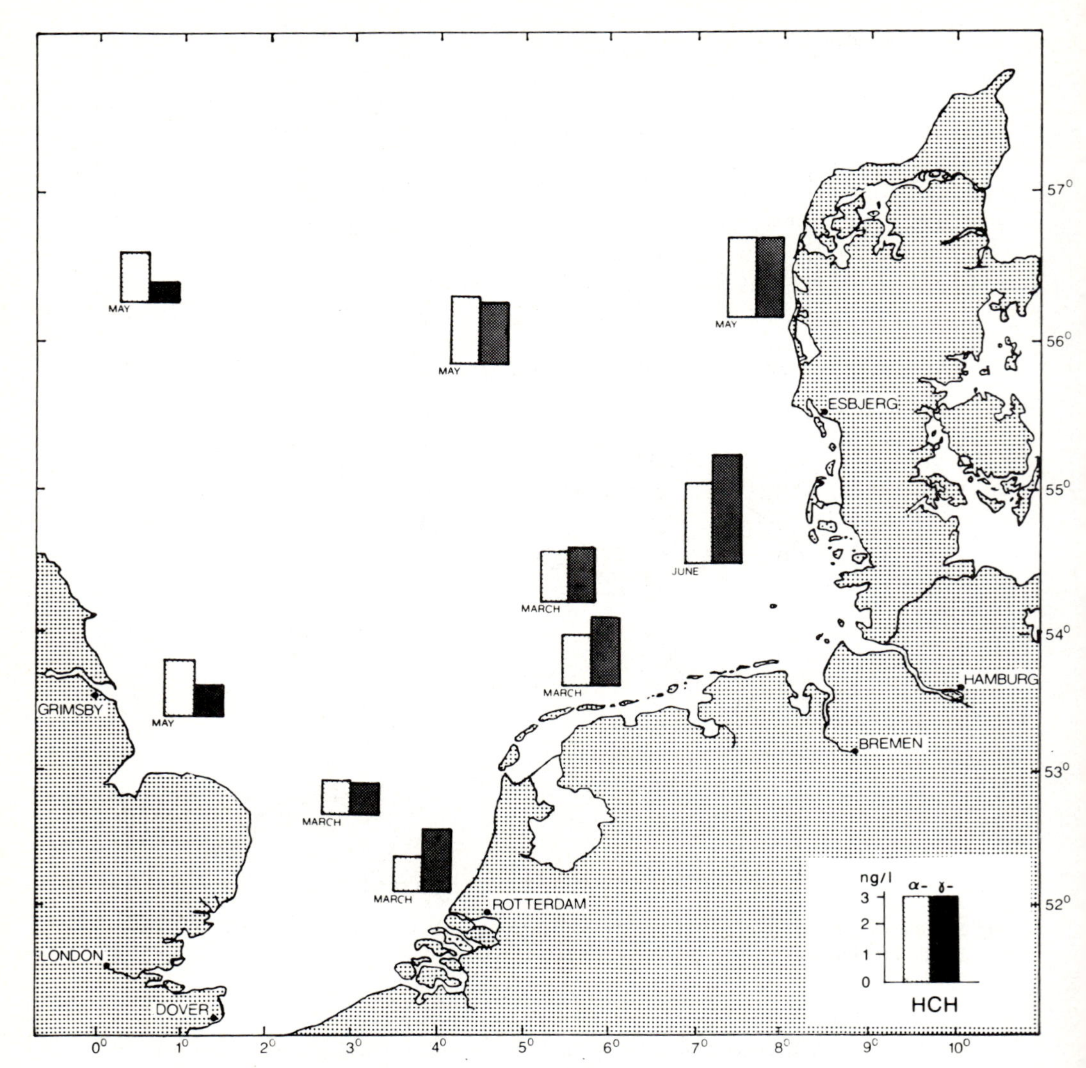


**Fig. 6.** Levels of α- and γ-HCH in water of the North Sea in 1979. (Weber and Ernst, unpublished)

## 3 Fate of Organic Micro Pollutants

Degradation of organic compounds may occur in the marine environment via abiotic processes and along metabolic pathways in the biota. The resulting degradation products and metabolites may be either less or more toxic than the parent compound or practically nontoxic when mineralized, i.e., completely degraded to $CO_2$, $H_2O$ and, in the case of chlorinated compounds, also to HCl.

## 3.1 Biodegradation

Organic compounds, when passing an estuary, may be degraded at different rates according to their chemical structure. In the Weser estuary, for example, biodegradation rates decrease with increasing salinity, probably due to dilution and lower microbial activity (Killer 1986). It is interesting to note that also higher marine animals are capable of transforming anthropogenic organic chemicals. In Fig. 7, examples are shown for hydroxylation of a polychlorinated biphenyl by a marine polychaete and the degradation of a phthalic acid ester in the common

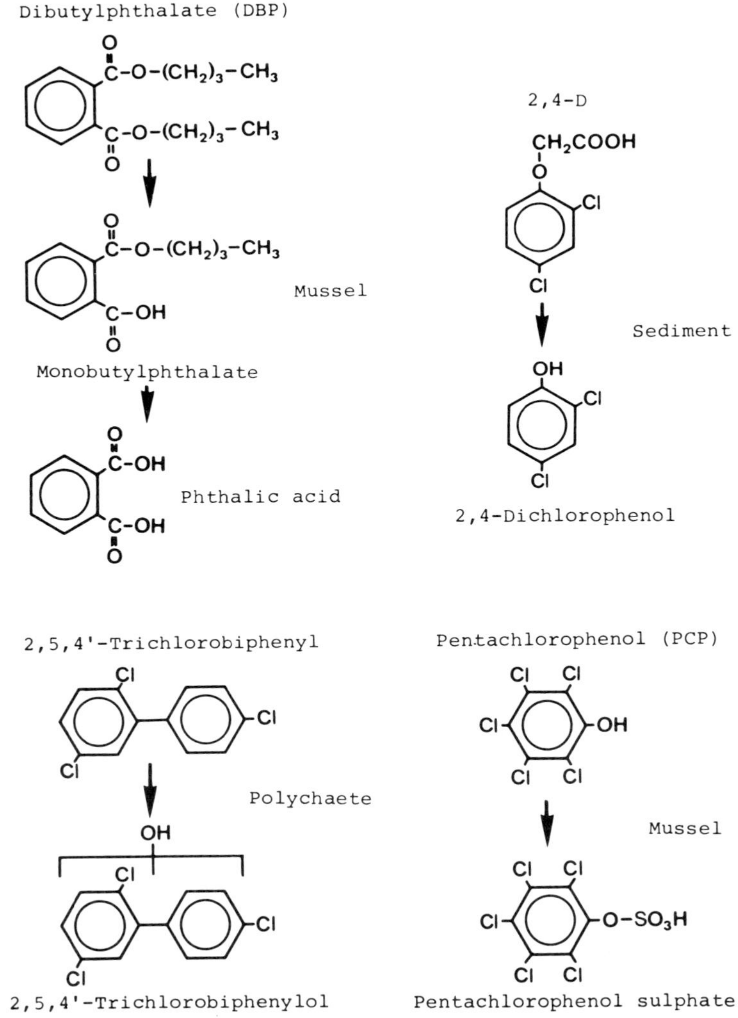


**Fig. 7.** Metabolic degradation of selected organic chemicals in different organisms and in sediment. (After Ernst 1984)

mussel. Mussels can also metabolize such persistent molecules as hexachlorobenzene to pentachlorothioanisole, which was found in natural mussel populations (Quirijins et al. 1979). Phenolic compounds can also be transformed in mussels via a conjugation process, in which sulfuric acid is enzymatically combined with phenols as shown in Fig. 3 for pentachlorophenol. Transformation of 2.4-dichlorophenoxy acetic acid to 2.4-dichlorphenol has been observed in sediments (Eder 1980) obviously due to microbial action. Compared to the unhalogenated aromatic hydrocarbons occurring, for example, in oil, chlorination decreases the susceptibility to biogeochemical reactions in the environment; as a result, PCB components belonging to these compounds, are highly persistent to degradation processes. Their general environmental behavior in the marine environment was recently reviewed (Duinker and Boon 1986).

Biodegradation of oil hydrocarbons occurs in the water column preferably by microbial metabolism. Degradation rates usually decrease in the order n-alkanes > iso-alkanes > naphthenes > mono-aromatics > PAH's, of which degradation decreases with increasing number of rings (Wolfe 1985).

### 3.2 Abiotic Degradation

Among abiotic degradation pathways photochemical processes play a major role and are preferably investigated for hydrocarbons. The importance of photochemical reactions for oil components was recently reviewed by Payne and Phillips (1985). Products formed are alcohols, hydroxylated aromatics, peroxides, sulfoxides, aryl- and alkyl-ethers, carbonyl-compounds and carboxylic acids; they all have in common a much higher water solubility than their parent compounds. Photooxidation reactions often involve chain reactions with highly reactive intermediates such as radicals and peroxides. The rates of photochemical reactions are decreased in turbid waters because of light attenuation.

## 4 Analytical Methods

The crucial point in environmental analysis of trace components, and especially from seawater, is to obtain the samples without contaminations from shipboard or sampling equipment. However, also the further processing by extraction and clean-up steps is vulnerable to laboratory contamination. The sample size is dependent on the location, i.e., the concentration of the compounds to be detected, and on the demand for resolution of the complexity of the compounds. If mass spectrometry is used, 100–500 l of seawater are required for certain identifications of unknown compounds in the open sea. In many cases high resolution gas chromatography with highly sensitive and selective detectors will provide sufficient indication for the identity of the compound also with smaller sample volumes.

The fraction of a water sample that is usually called dissolved, is in fact often an inhomogeneous mixture, since several types of naturally occurring larger molecules are present in this fraction also as so-called "dissolved organic matter"

(DOM). The presence of such naturally occurring organic molecules will influence the partitioning of organic trace contaminants between the "dissolved" phase and SPM.

In general it can be stated that except for oil after spills and certain organochlorines, little knowledge exists on the behavior of organic trace contaminants in marine waters. This is mainly due to the analytical difficulties involved: not many compounds can be analyzed together using one analytical method and with each method the low concentrations involved in seawater present difficulties in obtaining sufficiently accurate data and a good blank procedure.

## 5 Conclusions and Recommendations for Future Research

For the comparatively high concentrations of organic micropollutants in coastal areas, riverine input is mainly responsible. Atmospheric fallout and direct input by shipping activities and dumping of sewage and other materials also contribute to the total load of the North Sea. The concentrations of numerous compounds in water are in the ng- and sub-ng $l^{-1}$ range, which is of ecotoxicological significance for persistant and bioconcentrating compounds such as PCB, HCB, DDD, DDE and HCH.

A long-lasting source of pollutants especially detrimental for the benthic fauna are sediments contaminated, for example, by PCB, HCB, and other persistent chemicals.

There is no analytical method existing to the present to measure the organic chemicals possibly present in the North Sea in their entirety. For an ecotoxicological assessment it has to be considered that, depending on the analytical methods available, preferably the lipophilic compounds will be detected. Analytical methods for the determination of highly polar components have to be developed.

Further improvements in analytical techniques are required for clean sampling and clean-up procedures removing interfering components and for the quantification of atmospheric inputs.

## References

Audunson T (1978) The fate and weathering of surface oil from the Bravo blow out. In: Proceedings of the conference on assessment of ecological impacts of oil spills. American Institute of Biological Sciences, Washington D.C., pp 445–475

Berne S, Bodennec G (1984) Evolution of hydrocarbons after the Tanio oil spill – a comparison with the Amoco Cadiz accident. Ambio 13:109–114

Boon JP, Duinker JC (1985) The kinetics of polychlorinated biphenyl (PCB) components in juvenile sole (*Solea solea*) in relation to their concentrations in water and to lipid metabolism under conditions of starvation. Aquat Toxicol 7:119–134

Boon JP, Duinker JC (1986) Monitoring of cyclic organochlorines in the marine environment. Environ Monit Assess, 7:189–208

Boon JP, Zantvoort MB van, Govaert MJMA, Duinker JC (1985) Organochlorines in benthic polychaetes (*Nephtys* sp.) and sediments from the southern North Sea. Identification of individual PCB components. Neth J Sea Res 19:93–109

Duinker JC (1986) The role of small, low density particles on the partition of selected PCB congeners between water and suspended matter (North Sea area). Neth J Sea Res 20:229–238

Duinker JC, Boon JP (1986) PCB congeners in the marine environment – a review. In: Bjørseth A, Angeletti D (eds) Organic micropollutants in the aquatic environment. Proc 4th Eur Symp, Vienna, Austria, 22–24 October 1985. Reidel, Dordrecht, The Netherlands, pp 187–205

Duinker JC, Hillebrand MTJ, Boon JP (1983) Organochlorines in benthic invertebrates and sediments from the Dutch Wadden Sea; identification of individual PCB components. Neth J Sea Res 17:19–38

Eder G (1980) The formation of chlorophenols from the corresponding chlorooxyacetic acids in estuarine sediment under unaerobic conditions. Veröff Inst Meeresforsch Bremerh 18:217–221

Eder G (1984) Organische Umweltchemikalien in marinen Sedimenten. Veröff Inst Meeresforsch Bremerh, 20:41–48

Eder G, Weber K (1980) Chlorinated phenols in sediments and suspended matter of the Weser Estuary. Chemosphere 9:111–118

Eisma D (1981) The mass-balance of suspended matter and associated pollutants in the North Sea. Rapp P V Réun Cons Int Explor Mer 181:7–14

Ernst W (1984) Pesticides and technical organic compounds in the sea. In: Kinne O (ed) Marine Ecology-Ocean management vol 5. Wiley, New York p 1627–1709

Ernst W (1986) Hexachlorobenzene in the marine environment: distribution, fate and ecotoxicological aspects. In: Hexachlorobenzene: Proceedings of an International Symposium, CR Morris and JRP Cabral (eds) JARC Scientific Publication Lyon. p 211–222

Ernst W, Weber K (1978) The fate of pentachlorophenol in the Wester Estuary and the German Bight. Veröff Inst Meeresforsch Bremerh, 17:45–53

Ernst W, Eder G, Goerke H, Weber K, Weigelt S, Weigelt V (1986) Organische Umweltchemikalien in deutschen Ästuarien und Küstengewässern. Vorkommen, Biotransfer, Abbau. Forschungsbericht M86-001 des Bundesministers für Forschung und Technologie Fachinformationszentrum Karlsruhe FRG

Gaul H, Ziebarth U (1983) Method for the analysis of lipophilic compounds in water and results about the distribution of different organochlorine compounds in the North Sea. Dtsch Hydrogr Z 36:191–212

Goerke H, Ernst W (1986) Elimination of pentachlorobiphenyls by *Nereis virens* (Polychaeta) in the laboratory and the marine environment. Chem Ecol, 2:263–285

Grahl-Nielsen O (1978) The Ekofisk Bravo blowout: Petroleum hydrocarbons in the sea. In: Proceedings of the conference on assessment of ecological impacts of oil spills. American Institute of Biological Sciences, Washington DC pp 476–487

Grahl-Nielsen O, Westrheim K, Wilhelmsen S (1977a) Determination of petroleum hydrocarbons in the water. In: The Ekofisk Bravo blow out; compiled Norwegian Contributions. Int Counc Explor Sea CM/E55

Grahl-Nielsen O, Westrheim K, Wilhelmsen S (1977b) Fate of floating oil. In: The Ekofisk Bravo blow out; compiled Norwegian Contributions. Int Counc Explor Sea CM/E55

Kenaga EE, Goring CAI (1979) Relationship between water solubility, soil sorption, octanol-water partitioning and bioconcentration of chemicals in biota. 3rd Symp Aquatic Toxicology, New Orleans 1978

Killer K (1986) Untersuchungen zum Abbau und zur Sorption anthropogener organischer Stoffe im Weserästuar. (Dissertation Univ Bremen)

Kuiper J, Wilde PAWJ de, Wolff WJ (1984) Oil pollution experiment (OPEX). I. Fate of an oil mousse and effects on macrofauna in a model ecosystem representing a Wadden Sea tidal mudflat. In: Persoone G, Jaspers E, Claus C (eds) Ecotoxicological testing for the marine environment. State Univ Ghent Inst Mar Scient Res 2:331–359

Law RJ (1981) Hydrocarbons concentrations in water and sediments from UK marine waters determined by fluorescence spectroscopy. Mar Pollut Bull 12:153–157

Law RJ, Fileman TW (1985) The distribution of hydrocarbons in surfacial sediments from the central North Sea. Mar Pollut Bull 16:335–337

Maugh TH (1978) Chemicals: how many are there? Science 199:162

Meent D van de, Hollander HA den, Pool WG, Vredenbregt MJ, Oers HAM van, Greef E de, Luijten JA (1985) Organic micro-pollutants in Dutch coastal waters. Paper presented at the IAWPRC-NERC conference on estuarine and coastal pollution. Plymouth, UK, 16–19 July 1985

Payne JR, Phillips CR (1985) Photochemistry of petroleum in water. Photooxidation causes changes in oil and enhances dissolution of its oxidized products. Environ Sci Technol 19:569–579

Quirijns JK, Paauw CG van de, Noever de Brauw MC ten, Vos RH de (1979) Survey of the contamination of dutch coastal waters by chlorinated hydrocarbons, including the occurrence of methylthio-pentachlorobenzene and di(methylthio)tetrachlorobenzene. Sci Total Environ 13:225–233

Rapaport RA, Eisenreich SJ (1984) Chromatographic determination of octanol-water partitioning coefficients ($K_{ow}$S) for 58 polychlorinated biphenyl congeners. Environ Sci Technol 18:163–170; additions and corrections. Environ Sci Technol 19:376 (1985)

US NAS (1975) Petroleum in the marine environment. National Academy of Sciences, Washington DC 107 pp

US NAS (1979) Polychlorinated Biphenyls. Report prepared by the committee on the assessment of polychlorinated biphenyls in the environment. National Academy of Sciences, Washington DC, 182 pp

Weber K, Ernst W (1978) Levels and pattern of chlorophenols in water of the Weser estuary and the German Bight. Chemosphere 7:873–879

Weber K, Ernst W (1983) Vorkommen und Fluktuation von organischen Umweltchemikalien in deutschen Astuarien. Vom Wasser 61:111–123

Wolfe DA (1985) Fossil Fuels: Transportation and marine pollution. In: Duedal IW, Kester DR, Parke PK, Ketchum BH (eds) Energy Wastes in the Ocean. Wastes in the Ocean, vol 4. Wiley New York, pp 46–93

# Distribution and Fate of Heavy Metals in the North Sea

M. KERSTEN[1], M. DICKE[2a], M. KRIEWS[3], K. NAUMANN[3], D. SCHMIDT[2], M. SCHULZ[3], M. SCHWIKOWSKI[3], and M. STEIGER[3]

## 1 Introduction

In recent years coastal and shelf systems such as the North Sea have attracted considerable attention where heavy metal studies are concerned. There has been growing concern over the last decade regarding the effects of industrial discharges on heavy metal levels in nearshore waters and ecosystems, which has been documented in the reports of the first and second International North Sea Conference. As a result of the high input of pollutants from both the British Isles and the European continent, and a limited dilution on account of the shallow waters, the southern North Sea has often been regarded as the most heavily polluted marine area in the world since the early seventies (Weichart 1973). Metal pollution assessments from sediment analysis (e.g., Groot et al. 1971; Banat et al. 1972; Gadow and Schäfer 1973) formed a background of these early discussions.

The International Council for the Exploration of the Sea (ICES), founded in 1902 as one of the oldest intergovernmental scientific organizations of the world and based in Copenhagen, subsequently took the initiative to coordinate investigations for, inter alia, heavy metals in different marine matrices of the North Sea, Baltic Sea, and the North Atlantic Ocean. In 1975, ICES established the "Subgroup on Contaminant Levels in Sea Water" under the chairmanship of one of the authors (D.S.) with the task of investigating heavy metal abundancies in sea water. This Group pioneered the design and execution of a series of five rounds of intercalibration exercises (from 1976 until 1982). Work was then continued in the newly formed Working Group on Marine Chemistry of ICES. Subsequently, ICES was requested by the Oslo and Paris Commissions to collect, statistically handle and evaluate the data submitted by the contracting governments of the Conventions.

One of the most striking facts revealed from more recent studies is the increase in the dissolved nutrient and metal concentrations between NE Atlantic and North Sea waters (Topping et al. 1980). The results presented by Kremling (1983), obtained on a transsect between the open NE Atlantic and the German Bight, were

[1]Technische Universität Hamburg-Harburg, Arbeitsbereich Umweltschutztechnik, Eissendorferstr. 40, D-2100 Hamburg 90, FRG
[2]Deutsches Hydrographisches Institut, Labor Sülldorf, Wüstland 2, D-2000 Hamburg 55, FRG
[2a]present address: Baubehörde-Stadtentwässerung, Stadthausbrücke 12, D-2000 Hamburg 36, FRG
[3]Universität Hamburg, Institut für Anorganische und Angewandte Chemie, Martin-Luther-King-Platz 6, D-2000 Hamburg 13, FRG

one of the first to confirm that the North Sea is exposed to increases in inorganic nutrients (phosphate, silicate) and dissolved trace elements such as Cd and Cu. Despite lack of knowledge on certain aspects, it is possible to obtain an overall impression based on "snapshot" studies on metal concentration and partitioning conducted over the last say 10 years. The majority of these data have been collected as a means of assessing local pollution "hotspots" of particular national concern. Recently a survey has been conducted covering the entire North Sea involving a large multi-disciplinary team. The atmospheric part of this study has been performed by Prof. W. Dannecker's group (University of Hamburg), the water part by the heavy metal group of the Deutsche Hydrographische Institut (Hamburg), and the sediment part by Prof. U. Förstner's group (Technological University of Hamburg). This paper is a result of a rich literature, new data, and manuscripts in review. It will present further evidence that anthropogenic heavy metal impact is not always limited to local pollution hotspots, but may affect the entire North Sea ecosystem.

## 2 Pathways of Heavy Metals in Marine Ecosystem Compartments

To evaluate concentrations found in the different ecosystem compartments of the North Sea, knowledge of sources, the biogeochemical behavior and resulting distribution patterns of the heavy metals is necessary. Thus, before we are presenting concentration data, heavy metal pathways should be outlined in a very much abbreviated form. More indepth discussions of metal behavior in the marine environment are presented by, e.g., Förstner and Wittmann (1979).

In general there are two ways to characterize anthropogenic sources of heavy metals for the North Sea. Point sources are direct discharges from identifiable inputs such as dumping areas or estuaries. Nonpoint sources are diffuse inputs such as from the atmosphere. So far it is only possible to provide range estimates of the heavy metals from each source. Table 1 presents an overview on the latest official data available on inputs of heavy metals to the North Sea as taken from the synthesis of the International North Sea Conference 1987, from which one can imagine to what extent this ecosystem is incurred by metal pollution. The data ranges for the atmospheric input, however, have already been corrected utilizing newest data reported by Michaelis and Stößel (1986) and Krell and Roeckner (1988). The reason for this will be discussed in detail in the atmospheric part of this review. From this overview it is readily obvious that even the best data yet available are still to a large extent uncertain, especially for the atmospheric input.

Apart from atmospheric inputs, contaminated sediments may also become an important nonpoint source of heavy metals, especially in the case of metals characterized by a nutrient-type remobilization (cf. Kersten this Vol.). The sediment, however, can only be viewed conditionally as a "secondary source" of heavy metals, as the metals emerging out of the sediment originate in the first instance from the water column via sedimentation.

Metals are distributed into various compartments or "pools" of the aquatic ecosystem. Figure 1 depicts a conceptual framework which interrelates the major

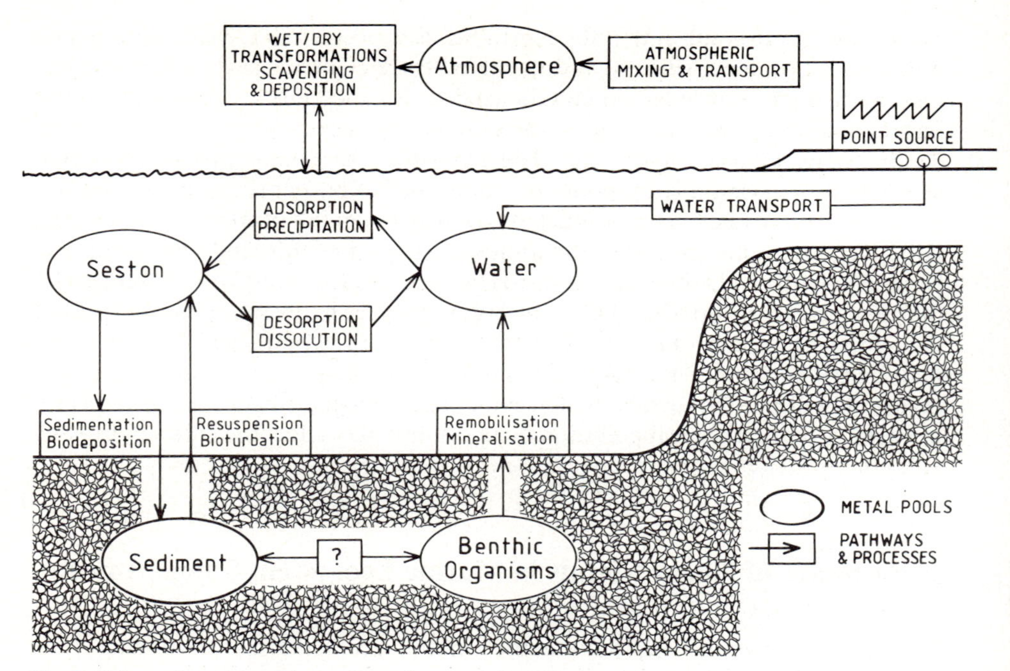


**Fig. 1.** Schematic review of the major pathways and pools of heavy metals in a marine ecosystem

pathways for heavy metals in the marine environment from emission sources to removal processes. This simplified description is based on five abiotic and biotic compartments. The atmosphere, the water column, and the sea bed can be distinguished as important abiotic pools. Main biotic compartments are pelagic and benthic organisms. Since the accumulation of pollutants by various classes of organisms is already treated in other chapters of this volume, we will emphasize here the abiotic compartments.

The pathway which a specific heavy metal follows during its residence time in the different compartments will depend largely upon its physical state, chemical binding forms and existing environmental conditions. In the atmosphere, pollutant dispersion comprises the processes of initial mixing as well as transport and diffusion. Acting in concert, these processes are capable of dispersing heavy metals even to the most remote regions of the North Sea.

On pages 36–58 of this Volume, a description has already been given by one of the authors (M.K.) of a number of processes which may affect the fate of heavy metals in the aquatic environment. In water and sediment, metals tend to undergo chemical and/or biological transformation processes. The ability of aquatic organisms to accumulate metals in their tissues is a well-known phenomenon. Biological activity can therefore represent an important mechanism of transfer of heavy metals from the aqueous to the particulate phase in the water column, and may interact with physical processes (boundary layer shear) and geochemical processes (interstitial water/particle partitioning) at the sea bed to create possibilities of contaminant transport and redistribution.

Understanding the processes regulating heavy metal residence time in the different compartments requires an intensive effort of multi-disciplinary teams and long-term studies. The widespread use of methods which determine only total metal concentrations and the analytical difficulties/costs inherent in obtaining data on specific chemical forms of trace elements, have curtailed the extent of information currently available concerning geobiochemical transformations of metals in the marine environment (Kersten and Förstner 1987). Interactions between the various compartments presented in Fig. 1, as well as balances etc., have in fact yet to be made. Even in well-studied regions like the coastal areas of the North Sea there is still a severe lack of knowledge to predict the impact of heavy metals. Little is known about net fluxes of metals at the water/air and sediment/water boundaries, especially about the amount of wet/dry deposition over the open sea, and about the mobility of heavy metals accumulated in sediments. Adequate models need to be developed from classical approaches and also from physicochemical approaches based on irreversible thermodynamics of nonlinearly coupled geochemical cycles (cf. e.g., Lasaga 1980).

## 3 Atmosphere

There has been increasing concern that considerable quantities of pollutants such as heavy metals and metalloids are reaching the marine environment via the atmosphere. Due to the geographic position of the North Sea, surrounded by highly industrialized countries, this seems to be inevitable. In fact, all published input data confirmed the relative importance of the atmosphere when compared with riverine input or direct dumping. However, there are still great uncertainties in the determination of atmospheric input to the North Sea as a whole (cf. Table 1).

**Table 1.** Summary of heavy metals inputs to the North Sea in tonnes per annum

| Source | Cd Min-Max | Hg Min-Max | Cu Min-Max | Pb Min-Max | Zn Min-Max | Cr Min-Max | Ni Min-Max |
|---|---|---|---|---|---|---|---|
| River inputs | 46–52 | 20–21 | 1290–1330 | 920–980 | 7360–7370 | 590–630 | 240–270 |
| Direct discharges | 40 | ≤ 5 | ≤ 400 | ≤ 165 | ≤ 1220 | ≤ 500 | ≤ 165 |
| Atmospheric | 14–380 | 10–30 | 380–1600 | 1530–6400 | 3900–12000 | 100–530 | ≤ 950 |
| Dumpings | | | | | | | |
| Dredgings | ≤ 20 | ≤ 15 | ≤ 1000 | ≤ 2000 | ≤ 8000 | ≤ 2500 | ≤ 700 |
| Sewage Sludge | ≤ 3 | ≤ 0.6 | ≤ 100 | ≤ 100 | ≤ 220 | ≤ 40 | ≤ 15 |
| Industrial Waste | ≤ 0.3 | ≤ 0.2 | ≤ 160 | ≤ 200 | ≤ 450 | ≤ 350 | ≤ 70 |
| Total (t/a) | 123–350 | 50–70 | 3000–4500 | 4900–11000 | 22000–28000 | 4200–5000 | 1500–2200 |

### 3.1 Field Measurements of Atmospheric Heavy Metal Flux

Three removal mechanisms have to be considered in order to assess the atmospheric input of pollutants to the North Sea. They are dry and wet deposition, and chemical transformations. With regard to the known air chemistry of heavy metals, chemical transformations do not act as a sink but change their state of phase or size spectrum within the aerosol, and by these processes alter the removal characteristics of wet and dry deposition. The current knowledge of dry and wet deposition processes for trace components require certain prerequisites for correct field measurements, which cannot be met, especially on a routine basis, by state-of-the-art techniques. Along with the recent growth in interest in the study of airborne particles, there has evolved a complex terminology to describe an airborne particle's source(s), formation, transformation, transport and removal. The following is a brief overview of the salient features of airborne particles as they relate to North Sea pollution by dry and wet deposition.

The most important parameter used for the characterization of airborne particles is considered to be particle size. Atmospheric particles relatively far removed from specific sources normally have a bimodal size distribution according to mass. These modes are fine (0.1–2.0 $\mu$m) and coarse (2–100 $\mu$m). Consequently, they have lifetimes in the order of days up to several weeks, depending on their size distribution and the parameters determining their removal processes. Coarse particles formed mainly via mechanical processes such as rock weathering are in most cases removed from the atmosphere without being transported long distances. The chemical composition of coarse particles typically reflects the composition of the earth's crust, containing such predominate elements as calcium, aluminum, silicon, and iron. Depending upon the sampling location and meteorological conditions, sea salt may also be present in the coarse mode (Blanchard 1983). In fact, the seas are one of the major natural sources of atmospheric particulate matter (releasing about 5 $mg/m^2/d$ sea salt particles). Sea spray is produced by bubbles bursting at the sea surface due to wave, wind and rainfall action. Consequently, atmospheric transport from sea to land and vice versa results in the marine aerosol being a variable mixture of modified marine and continental source material. Incorporation of sea spray, originating from a microlayer highly enriched in trace metals, into rainwater was invoked by Peirson et al. (1974) to explain the enhanced content of trace metal in rainwater above the North Sea. On the other hand, Dedeurwaerder et al. (1983) showed by resuspension flux measurements that the recycled component can have but a relatively small impact upon the magnitude of the concentrations in rainwater above sea.

By far the most interesting component of the marine aerosol in the pollutional context is contributed from anthropogenic sources releasing the overwhelming amount of toxic trace elements (Pb, Cd, Hg, As, Se, etc.) into the atmosphere. Particles injected into the atmosphere by man-made activities are derived from industrial processes, fuel combustion, solid waste incineration and a host of other processes. The resulting particle sizes of these compounds are mainly in the micron to submicron size range. Such particles strongly adsorb other organic and inorganic air components. One factor that does have a significant effect on the size and chemical composition of primary particles emitted from anthropogenic sources is the type and degree of emission control. Although conventional emission controls are responsible for reducing the overall particle mass loading, the portion of mass loading attributed to fine particle mass contribution predominantly to heavy metal emission has increased (cf. Förstner 1986). Such fine particles are transported through the atmosphere via eddy diffusion and advection and, in the absence of precipitation, may be transported long distances without being deposited (Patterson and Settle 1987). Wet deposition processes, such as rainout and washout, are significant methods of removal of fine particles especially over the sea (Slinn 1983).

Finally the products of the chemical transformations of primary gaseous emissions have to be considered. Such "secondary aerosols" composed of sulfates, ammonium and bromide salts, etc. contribute considerably to the total mass concentrations of the marine aerosol with mass median diameters (mmd's) in the submicron size range (Whitby 1978). Unfortunately, very little information is available on the heavy metal pollution potential of secondary particles. As a result of the contribution of particles produced by the different sources, the mass size distribution of North Sea aerosols is usually bimodal with a coarse particle mode dominated by sea spray particles, and a fine particle mode which mainly consists of long-range transported primary and secondary particles of continental origin. During the last few years, several authors have reported their measurements of ambient concentrations from measuring sites presented in Fig. 2. Compilation of data obtained in this field has shown a great variability in the trace element concentrations as shown in Table 2. Anyhow, since continued laboratory and field research has been devoted to the interrelation of actual ambient concentrations and magnitude of deposition of pollutants, there is a chance that estimates of fluxes can be based on measured as well as on modeled heavy metal concentration patterns.

When dry deposition of a particle spectrum is discussed, its vertical transport down to the sea surface and finally its incorporation at the air/sea interface has to be described correctly. Besides gravitational settling, which is of growing importance for the coarse mode of the aerosol of predominantly crustal origin, turbulent mixing within the planetary boundary layer (of about 1000 m height) and diffusion through the viscous sublayer ($< 0.1$ mm) will provide the vertical flux of particles and gases (Slinn 1983). Scavenging by sea-spray droplets could further enhance the deposition velocity (Williams 1982).

The use of surrogate surfaces to measure dry deposition flux can be misleading. Widely used dry surfaces show reduced sticking probabilities for depositing particles, while the collection efficiency of open vessels is diminished under moderate to high wind speeds. Any artificial collector fails to take into account scavenging phenomena of sea spray droplets, as well as growth of particles in the high humidity environment close above the sea surface. Overestimation may result from the disturbed turbulence spectrum due to the shape of collectors and actual measuring height above the sea surface. These difficulties may explain why only few attempts measuring dry deposition have been made for the North Sea environment (Dedeurwaerder et al. 1983; Stößel 1987).

Since raindroplets are bigger in size, their collection causes less problems. The mechanisms of cloud formation and dynamics of rainfall events, however, are so

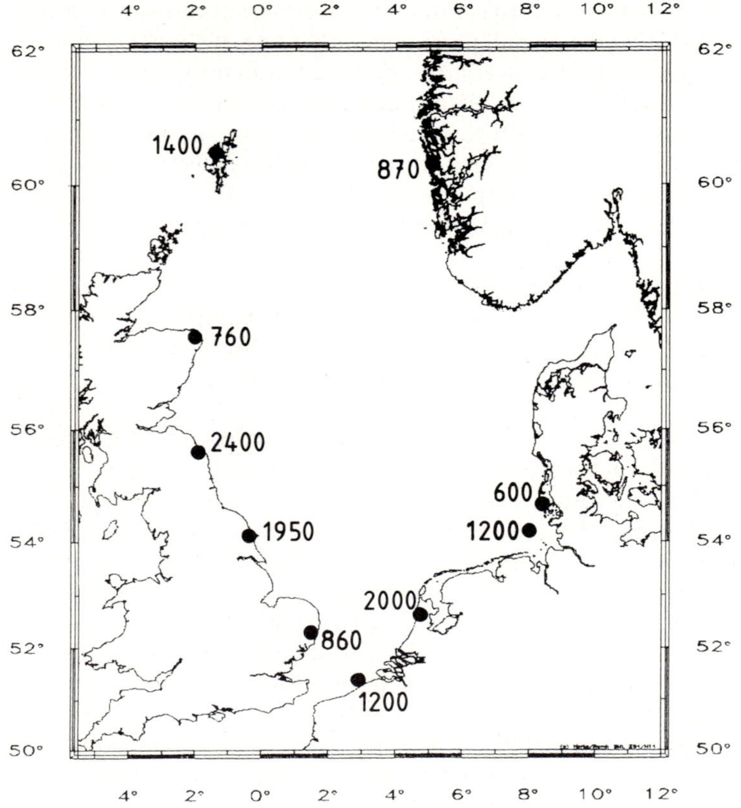


**Fig. 2.** Mean atmospheric lead deposition estimates ($\mu g/m^2$ month) reported so far from all coastal measuring sites around the North Sea. (Data taken from Cambray et al. 1979; Dedeurwaerder et al. 1983; Kretzschmar 1979; Stößel 1987)

complex that it is not well understood how to extrapolate the findings gained from coastal measuring sites to the open North Sea conditions. Mixed layer dynamics and dilution due to convergence by land to sea travel of air parcels as well as rainstorm events complicate the interpretation considerably. A correct yearly wet fall-out flux requires the sampling of all rain events during a full year. To our knowledge this has never been done at sea. Mean fall-out flux estimates calculated as the main rainwater concentration multiplied by the total amount of precipitation, which is believed to amount to about half the value above land, are incorrect when both variables are subject to fluctuations (Dedeurwaerder et al. 1983).

The remaining question is whether wet and dry deposition could be of equal importance for the input of pollutants. Findings from the coastal measuring approaches suggest an important contribution of dry deposition to total deposition (Stößel 1987). Changed size spectra to smaller particles, steady loss of particles by scavenging and rainfall during transport to open waters where no significant pollution sources are to be expected, suggest that wet deposition would become

**Table 2.** Ambient elemental mean concentrations in air as measured by several authors at coastal sites around the southern North Sea (ng/m$^3$)

| | Ostende[a] 1972/77 | Westhinder[b] 1980/85 | Tange[c] 1983 | Pellworm[d] 1984/85 | Helgoland[e] 1985/86 |
|---|---|---|---|---|---|
| Al | – | 394 | 133 | – | 210 (64–600) |
| As | – | – | – | 2.7 | 2.1 (0.3–15.7) |
| B | – | – | – | – | 1.3 (n.n.–3.1) |
| Ba | 19 | – | 5.3 | 4.5 | 3.4 (n.n.–16.5) |
| Be | – | – | – | – | 0.027 (0.003–0.084) |
| Ca | – | – | 147 | 226 | 418 (130–1245) |
| Cd | 5 | 3.9 | 0.6 | 0.67 | 1.4 (0.8–8.1) |
| Co | – | – | – | – | 0.24 (0.017–0.93) |
| Cr | 12 | – | 1.6 | 1.3 | 1.7 (0.34–9.1) |
| Cu | 17 | 16.8 | 3.0 | 3.3 | 3.9 (0.49–20.5) |
| Fe | 1114 | 555 | 275 | 189 | 201 (22–1040) |
| K | – | – | 171 | 234 | 306 (102–754) |
| Mg | – | – | 0.3 | – | 559 (146–1256) |
| Mn | 66 | 28.2 | 9.9 | 8.5 | 7.2 (1.4–42.3) |
| Na | – | 2499 | – | – | 4795 (1222–10707) |
| Ni | 11 | – | 3.0 | 4.4 | 2.6 (n.n.–8.8) |
| Pb | 241 | 147 | 40.9 | 38.8 | 28.9 (3.2–149) |
| S | – | – | 1753 | 3149 | – |
| Sb | – | – | 0.9 | 1.1 | 1.1 (0.19–3.4) |
| Si | – | – | 572 | – | – |
| Se | – | – | 0.7 | 0.9 | 0.9 (0.17–4.1) |
| Sn | – | – | 1.3 | 2.1 | – |
| Sr | – | – | 2.0 | 2.4 | 4.2 (1.1–9.1) |
| Ti | – | – | 16.2 | 14 | 15.2 (1.5–59.8) |
| V | 43 | – | 7.1 | 6.8 | 5.8 (1.2–17.1) |
| Zn | 250 | 150 | 29.7 | 40 | 32.8 (4.7–185) |

[a] Kretzschmar (1979) [b] Dedeurwaerder and Artaxo (1987) [c] Kemp (1984) [d] Stößel (1987) [e] This Chap.

overwhelming. Dedeurwaerder et al. (1983) have shown that the mean annual deposition (wet + dry) of Cu, Zn, Pb, and Fe is larger above coastal sea than above land (Table 3), essentially as the result of the large excess of wet over dry fall-out above sea. Excess contents of these elements in rain water above sea can be due to larger background contents and/or an increased scavenging efficiency of rain above sea. Recently almost simultaneous 12-hourly measurements of total aerosol concentrations and rain sampling on an event basis have been conducted in the German Bight at coastal sites, on the island of Helgoland, and on a German offshore research platform. First results show a decrease in concentration of heavy metals as expected when comparing the land and sea stations, whereby crustal elements appear to show steeper gradients due to their occurrence in particles of larger mmd size. These findings on concentration gradients from land to sea, and predictions on the predominance of wet over dry deposition as well, are consistent with model simulations.

**Table 3.** Results for the deposition rate determination of aerosols sampled from stations around the North Sea. Mean values ($\mu g/m^2 month$)

| | Helgoland[a] | West Hinder[b] | English EMEP stations[c] | Pellworm[d] |
|---|---|---|---|---|
| Al | 141.500[e] | – | 24.000 | – |
| As | 65 | – | 75 | 50 |
| Cd | 22 | 144 | 61 | 37 |
| Co | – | – | 14 | – |
| Cr | 120 | – | 116 | 48 |
| Cu | 550 | 2.200 | 870 | 160 |
| Fe | 74.000[e] | 15.000 | 16.000 | 17.000 |
| Hg | – | – | 1.1 | – |
| K | – | – | – | 24.000 |
| Mn | 1.400 | 520 | 640 | 570 |
| Ni | 219 | – | 290 | 270 |
| Pb | 660 | 1.190 | 920 | 640 |
| Sb | 23 | – | 21 | 22 |
| Se | < 10 | – | 20 | 21 |
| Ti | 7.300[e] | – | – | 1.500 |
| V | 430 | – | 133 | 170 |
| Zn | 9.200 | – | 2.600 | 1.300 |

[a] This Chap.
[b] Dedeurwaerder and Artaxo (1987).
[c] Cambray et al. (1979)
[d] Stößel (1987).
[e] influenced by local dust resuspension.

## 3.2 Modeling of the Atmospheric Impact

Table 4 gives an overview on potential anthropogenic sources for aerosols. Most of the heavy metals of ecotoxicological relevance are emitted into the atmosphere because of their volatility during high-temperature combustion processes (cf. Förstner 1986). Furthermore, Table 4 gives a classification of those sources with respect to their relative contribution to total anthropogenic emissions of the heavy metals. Pacyna (1985) compiled the first European heavy metal emission register based on elemental emission factors and available statistical information on the consumption of ores, fuels, and industrial production data for 1979/80 in Europe

**Table 4.** Important emission sources of trace elements, with its portions in percentage of total anthropogenic emissions on an European average as assessed by Pacyna (1984)

| Source | Element |
|---|---|
| Non-ferrous metal production | As (90%), Cd (80%), Zn (70%) |
| Combustion of fossil fuels | V (≤ 100%), Ni (75%), Se (90%), Be (≤ 100%) |
| Automotive exhaust | Pb (60%) |
| Iron and steel industry | Cr (80%), Mn (60%) |

and the Soviet Union. Although such emission inventories are somewhat questionable due to the considerable uncertainties in the data, they can be regarded as a reasonable first approximation of the actual spatial distribution of heavy metal emissions all over Europe.

Beside the improvements in emission data sets, major progress has been made in the characterization of the source types of airborne particulates using so called "receptor models" (Watson 1984; Kemp 1984). Statistical treatment of data obtained in atmospheric deposition measurements is usually possible, due to the large number of analyses that are generally performed. In practice, this concept means that if a trace element is significantly correlated to one or several major elements, it is possible that the aerosol component containing the major elements can be the carrier. Usually multivariate statistical analysis is at least able to differentiate between the main sources mentioned above: marine, crustal, and anthropogenic. Further classification, especially of the anthropogenic component, is often problematic, when such analysis is applied to data obtained for aerosol particles transported over long distances. Such an aerosol population cannot be expected to be statistically homogeneous, but rather as being composed of geographically distributed subpopulations. The problem at hand is, therefore, to classify the aerosol composition data into statistically homogeneous groups, before attempting a detailed analysis of component associations. Probably the most promising approach in this field is the use of multivariate statistics in combination with modeling of backward wind trajectories. Such an approach has been demonstrated by Kemp (1984) with his data obtained from two measuring sites within the North Sea region (Tange, Jutland; Faroer Islands). The sequence of operation proceeds from a preliminary classification of the observed elemental sample compositions according to the various "transport sectors". Transport sector division, together with trajectory calculations, gives information of the source areas of the air mass passing the station. Fig. 3 points out that obvious differences exist in the Pb concentrations from one sector sampled at different times. Changes in the

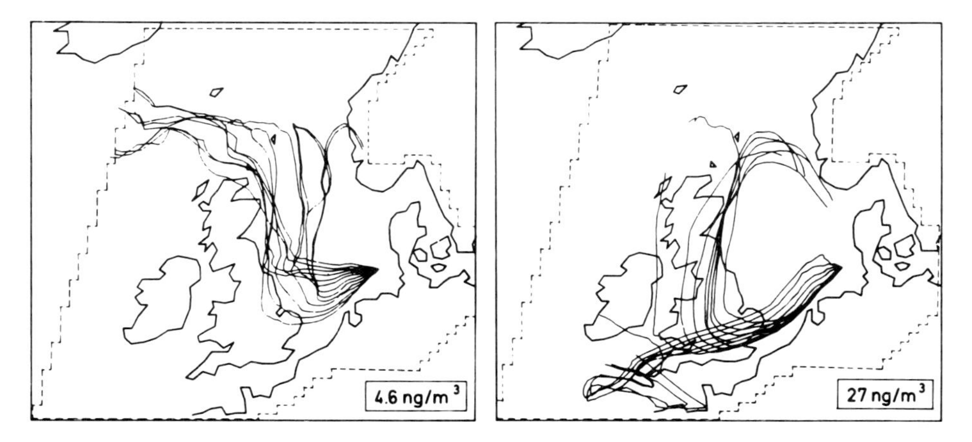


**Fig. 3.** Use of backward wind trajectories for surface and geostrophic winds, calculated for every hour to explanate the difference of lead concentrations measured 12-hourly at the research platform "Nordsee" (NE Helgoland)

receptor model calculated for different sectors and stations could give information concerning the chemical transformations during aerosol transport over land and sea, respectively.

Few attempts have yet been made to assess the atmospheric input of heavy metals to the North Sea by model simulation (Aalst et al. 1983; Jaarsveld et al. 1986). Recently Krell and Roecker (1988) performed stochastic trajectory modeling estimates for wet deposition of Pb and Cd. These trace elements were considered to be best suited as pilot substances because of the amount of reliable information on emission and removal process data available for them. In a three-dimensional approach based on the Monte-Carlo method, the emission data of Pacyna (1985) were adopted and modeled by use of extended meteorological datasets of the European Centre for Medium Range Weather Forecasts in Reading, together with estimates of precipitation, wash-out factors, and the mixed-layer heights given by the Norwegian Weather Service. The model results obtained for two representative months (July 1984 and January 1985) show a considerable total deposition gradient across the coastlines and from south to north irrespective of the element and season concerned (Fig. 4). Main emission sources situated in England, the Netherlands, Belgium, and the Federal Republic of Germany induce high atmospheric depositions in the western and southern part of the North Sea. Total Pb deposition fluxes estimated on a monthly basis decrease from > 500 $\mu g/m^2$ at the southern North Sea coasts to about 100 $\mu g/m^2$ in the central North Sea. An important implication of this study is that previous estimations of deposition data extrapolated from measurements at coastal sites tend to overestimate the atmospheric input of anthropogenically derived heavy metals into the entire North Sea. The lower limits of atmospheric inputs for Cd and Pb shown in Table 1 stem from this work.

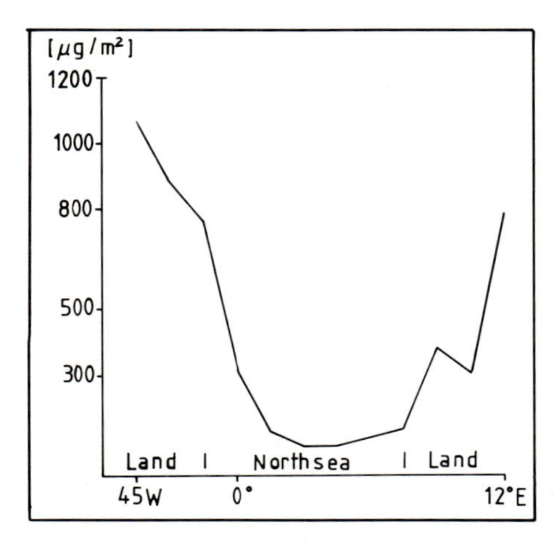


**Fig. 4.** Calculated total deposition of lead along the latitude of 55.5° N. (Krell and Roeckner 1987)

# 4 Water

In the last say, 15 years a small number of laboratories around the world have developed the expertise to produce reliable data that can form the basis for better understanding the behavior of metals in the marine environment. This is not a trivial task, although around 1970 modern analytical techniques for trace determination of heavy metals in seawater became widely available. Work started using graphite furnace atomic absorption spectrometry (GFAAS) with various chemical preconcentration and separation procedures. In this way, mainly Cd, Cu, Fe, Mn, Ni, and Co were determined. Shortly after, anodic stripping voltammetry (ASV) with the hanging mercury drop electrode (HMDE) was introduced with the advantage of direct measurement of Zn, Pb, Cd, and Cu in seawater. This method was subsequently much improved by the introduction of the rotating mercury film electrode that made Pb and Cd determination at the very low level found in open seawaters more accurate. Analyses for Hg in the ppt range (ng/kg) were made possible by the cold vapor atomic absorption spectrometry (CVAAS) with preceding amalgamation on finely dispersed gold.

After the successful completion of the series of intercalibration exercises that were designed by ICES to address the problems of analytical intercomparability, a number of outstanding laboratories around the North Sea could be nominated that were capable of accurately determining ultratrace concentrations of heavy metals in seawater. A detailed description of the success of the various intercalibrations has been given by Schmidt (1983). Even though the concentrations of heavy metals reported by several authors for the North Sea waters exhibit an apparent decrease over the last about 20 years, this is not a real effect, but is now generally accepted as being the result of an improvement in contamination control during sampling, storage, and analysis (e.g., Mart and Nürnberg 1986). Nevertheless there are still unresolved problems with elements like Fe due to its ubiquitousness resulting in contamination risks introduced by, e.g., the research vessel and measuring instruments.

Present knowledge on heavy metal behavior is mainly based on vertical profiles taken in the North Atlantic and Pacific oceans (e.g., Bruland 1983). These data have been explained by a nutrient-like removal from surface waters and a subsequent release in deeper waters. While a deviation from this nutrient-type distribution was reported for Cu (Bruland 1980; Bruland and Franks 1983), a covariation between Cd and Ni to both phosphate and nitrate, and between Zn and silicate has been found (Sclater et al. 1976). Boyle et al. (1976) and Bruland et al. (1978) even suggested a fixed relation between Cd and $PO_4^-$ to be valid throughout all open ocean environments. These relationships, however, could not be confirmed for the North Sea waters (Kremling 1985). The situation in this shelf area with superposed influences like freshwater input, hydrographic peculiarities, strong biological events, and enhanced particulate loads, is still not well understood, although data are available from well-studied areas like the German Bight, the Skagerrak, the Southern Bight, and coastal waters off England. A review of these data is compiled in Table 5, and some results of the first comprehensive study covering the entire North Sea are presented in section 4.2.

### 4.1 Regional Distribution in Coastal Areas

*German Bight.* Monitoring for heavy metals in seawater of the German Bight is one of the official functions of the Deutsches Hydrographisches Institut (DHI, the German Hydrographic Institute in Hamburg). In 1971, one of the authors (D.S.) started to develop a monitoring scheme. First results have been published by Schmidt (1976). During the first decade, monitoring was performed on slightly changing station networks at irregular periods, approximately once per year (Schmidt 1980). In 1980, two new monitoring programs were established: the international program of the Joint Monitoring Group (JMG) of the Oslo and Paris Conventions (JMP), and a German national monitoring program (Gemeinsames Bund/Länder-Meßprogramm für die Küstengewässer der Nordsee: BLMP). Both requested regular monitoring at annual intervals on two overlapping station networks in the German Bight. In addition to Cd, Cu, Fe, Mn, and Ni, determination of Hg was especially introduced for these programs (Schmidt and Freimann 1984).

All monitoring data from 1980 onwards have been published, together with additional results, in the series of annual reports of DHI (Anonymous 1982–1987). Some of the data are also contained in the biannual reports of BLMP and (in condensed form) in JMG publications. Most of the data published so far are listed in Table 5. It is immediately obvious that concentrations often span a relatively wide range. It should be kept in mind, however, that comparison between data sets from different authors obtained at separate time periods with differing analytical techniques is rather problematic; additionally, calculated mean, median, minimum, and maximum are very sensitive to the selection of stations and sampling depths, but also to the individual sampling strategy when surveying a system like the German Bight with steep gradients of metal concentrations. These again are mainly influenced by the complex hydrographic situation, varying point sources, and existence of the estuaries of major rivers draining densely populated countries.

*Southern Bight.* A comparably good data basis exists for heavy metal concentration distributions off the Dutch and Belgian coasts. High concentrations are found in the cross-boundary of rivers Rhine and Scheldt, decreasing toward the center of the Bight, and increasing again toward the English coast (Duinker and Nolting 1982; Nolting 1986; Baeyens et al. 1987 a). In order to gain a better comparison of the data so far reported, the minimum values at maximum salinities reported by all authors are summarized in Table 5. While resulting minimum values for Cd, Cu, and Pb show relatively good correspondence, ranges of concentrations are found to differ from each other. This disagreement may again be attributed to the different choice of sampling stations between the independent investigations. Slightly higher values have been reported by Jones and Jeffries (1983) obtained from a section monitored between the Thames estuary and Hoek of Holland. Results of the station in the middle of this section are enclosed in Table 5. Data reported by Kremling (1985) are added. They exhibit rather varying concentrations and seem to be influenced by freshwater supply to a higher extent than the minimum values given by the authors referred to above. The Hg

**Table 5.** Heavy metal data reported for different parts of the North Sea water

| Element | Cd (ng/l) | Cu (ng/l) | Ni (ng/l) | Zn (ng/l) | Pb (ng/l) | Co (ng/l) | Hg (ng/l) | Mn (μg/l) | Fe (μg/l) |
|---|---|---|---|---|---|---|---|---|---|
| *German Bight* | | | | | | | | | |
| Schmidt (1976) | 80[a,A] | 2300[a,A] | 1200[a,A] | | | | | 4.9[a,A] | 44[a,A] |
| (and this Chap.) | (20–270) | (700–33600) | (400–3400) | | | | | (0.9–53) | (6.8–449) |
| | 70[a,E] | | | 1900[a,E] | | | | | |
| | (20–360) | | | (500–21100) | | | | | |
| Schmidt (1980) | 70[b] | 600[b] | 900[b] | | | | | 3.7[b] | 10[b] |
| (and this Chap.) | (10–180) | (200–2000) | (400–10800) | | | | | (0.9–24) | (1.1–702) |
| Schmidt (1983) | 55[c] | | | | | | | | |
| Mart and Nürnberg (1986) | 28 | 235 | 471 | | 20 | 11 | | | |
| Schönfeld et al. (1988) | 41[d] | 750[d] | 570[d] | | | | 10[d] | 10[d] | 45[d] |
| Sipos et al. (1980) | | | | | | | 6–15 | | |
| May and Stoeppler (1983) | | | | | | | 9.3[e,N] | | |
| | | | | | | | (4.6–15.3) | | |
| | | | | | | | 46.4[e,P] | | |
| | | | | | | | (11.5–50) | | |
| Dicke et al. (1987) | 18[f] | | | | 47[f] | | | | |
| Schmidt and Dicke (1987) | | | | | | | 3[f,g] | | |
| *Southern Bight* | | | | | | | | | |
| Duinker and Nolting (1982) | 20–30[h] | 200–300[h] | | 300–400[h] | | | | | |
| Mart et al. (1982) | 13[i] | 340[i] | | | 41[i] | | | | |
| Jones and Jeffries (1983) | 40[j] | 350[j] | 400[j] | 2000[j] | | | | | |
| Kremling (1985) | 58[k] | 487[k] | 839[k] | | | | | 0.76[k] | |
| | (33–83) | (331–599) | (746–994) | | | | | (0.61–1.01) | |
| Nolting (1986) | 26 | 300 | 241 | | | | | | |
| Baeyens et al. (1987a) | 14[h,l] | 280[h,l] | | 250[h,l] | 45[h,l] | | 6[h,l] | | |
| Baker (1977) | | | | | | | 2–8[m] | | |
| Dicke et al. (1987) | 7[f] | | | | 42[f] | | | | |
| Schmidt and Dicke (1987) | | | | | | | < 0.5–7[n] | | |
| *SE English Coast* | | | | | | | | | |
| Balls (1985a) | 10–60 | 110–580 | | | 10–135 | | | | |
| Baker (1977) | | | | | | | 10–17 | | |
| Dicke et al. (1987) | | | | | 382[f] | | | | |
| *Scottish Coast* | | | | | | | | | |
| Balls (1985b) | 11 | 170 | | | | | | | |
| Kremling and Hydes (1987) | 13 | 191 | 211 | | | 4 | | 0.34 | |
| Baker (1977) | | | | | | | 2–3 | | |
| Kremling (1985) | 17 | 187 | 266 | | | | | 0.52 | |

**Table 5.** *(continued)*

| Element | Cd (ng/l) | Cu (ng/l) | Ni (ng/) | Zn (ng/l) | Pb (ng/l) | Co (ng/l) | Hg (ng/l) | Mn (μg/l) | Fe (μg/l) |
|---|---|---|---|---|---|---|---|---|---|
| *Skagerrak* | | | | | | | | | |
| Magnusson and Westerlund | 22[o] | 330[o,S] | 393[o] | 690[o] | 56[o,S] | | | | 11[o,S] |
| (1983) | | 180[o,B] | | | 190[o,B] | | | | 62[o,B] |
| Danielsson et al. (1985) | 30[p,S] | 410[p,S] | 500[p,S] | 810,390[p,S] | | | | | 0.6,1.4[p,S] |
| | 19[p,B] | 120[p,B] | 200[p,B] | 270[p,B] | | | | | 6[p,B] |
| Brügmann (1986b) | | | | | 33–43[q,S] 25–26[q,B] | | | | |
| Dicke et al. (1987) | 18[f] | | | | 59[f] | | | | |
| Schmidt and Dicke (1987) | | | | | | | 2.5[f] | | |
| *Central and Northern Areas* | | | | | | | | | |
| Jones and Jeffries (1983) | 30[r] | 210[s] | | 1300[r] | | | | 0.31[t] | |
| Baker (1977) | | | | | | | 3 | | |
| Kremling (1985) | 20 | 226 | 320 | | | | | 0.35 | |
| Brügmann (1986) | | | | | 59[u] | | | | |
| Kremling and Hydes (1987) | 16 | 254 | 205 | | | 6 | | 0.37 | |
| Dicke et al. (1987) | 9[f] | | | | 30[f] | | | | |
| Schmidt and Dicke (1987) | | | | | | | 0.5–2[v] | | |

[a]Median values of a survey made in 1973 (18 stations), No. of samples 82–86 per metal, unfiltered, GFAAS[A] and ASV/HMDE[E].
[b]Median values of 1974 (22 stations), No. of samples 66–71 per metal, unfiltered, GFAAS.
[c]Monitoring of 1981, mean of 31 filtered samples, GFAAS.
[d]Median of different stations and depths, unfiltered.
[e]Mean of 10 unfiltered samples collected close to the beach, "nonpolluted area" [N] and "polluted area" [P].
[f]Median of 10 m depth samples.
[g]Including some high spots of 12, 17.5 and 24.5 ng/l.
[h]Minimum values at maximum salinity.
[i]Minimum values.
[j]Mean of different observations at one offshore station.
[k]Stations 88–90 at 6 m water depth.
[l]Samples from 5 m water depth.
[m]Excluding high local spots of 12 and 18 ng/l.
[n]Excluding one high spot of 19 ng/l.
[o]Mean value of surface (S) and bottom (B) water samples, or of different depths.
[p]Mean of two data of surface (S) or bottom (B) water samples.
[q]Mean of two data, range of two methods.
[r]Mean of a survey performed in summer 1976.
[s]Mean of two surveys performed in summer 1974 and 1976.
[t]Mean of a survey performed in summer 1974.
[u]Surface layer (0–10 m) from two stations.
[v]Samples from 10 m water depths, some high spots excluded.

concentrations of different surveys (Baker 1977; Baeyens et al. 1987; and this study) show similar levels, but some inexplicable high spots were found in this region as well as in other parts of the North Sea.

*SE English Coast.* Dissolved Cu, Pb, and Cd values obtained along the British coast have been reported by Balls (1985a). On approaching the coastal area he found an increase in Cu concentrations ranging from 110 to 3300 ng/l. Elevated levels occurred especially in the areas off the Tyne, Tees and Humber estuaries. While a similar pattern was found for total Pb, dissolved Pb (i.e., filtrated $< 0.45$ $\mu$m) revealed a different distribution. Apart from plumes of contaminated coastal waters in the Tyne-Tees area, the concentration values given for dissolved Pb were surprisingly low. The values found for the Humber estuary were comparable to those reported for open ocean waters of the North Atlantic (Schaule and Patterson 1983). Since the lead concentration in suspended matter collected off the Tyne-Tees and Humber estuaries did not vary significantly, the higher total Pb concentrations in the Humber region were explained by the higher particulate load. Cd concentrations showed an increase from north to south along the coastal line, with 95% of the values falling in the range between 10 and 60 ng/l, and reaching values up to 600 ng/l in the Humber estuary.

*Scottish Coast.* In surface waters off W Scotland, Kremling (1985) observed a sharp increase in heavy metal concentrations as high as three- to fivefold compared to the low oceanic values. This finding was confirmed by Balls (1985b), who suggested advection of contaminated Irish Sea water to be the main source for this enrichment. In a more recent study, Kremling and Hydes (1987) found a less significant increase in the same region. This variability in heavy metal concentrations was attributed by the latter authors to occasionally occurring upwelling events redistributing diagenetically remobilized metals from partly reduced sediments to surface waters.

*Skagerrak.* Heavy metal levels found in the Skagerrak are mainly determined by the water exchange between North Sea and Baltic. The outflowing Baltic water in the Skagerrak is restricted to the surficial parts of the water column due to its lower salinity. Magnusson and Westerlund (1983) found accordingly higher Cu concentrations in surface water than in bottom water (Table 5), indicating a contribution of this metal from Baltic waters to the North Sea. No such differences were found, however, in this study for Cd, Ni, and Zn. Enhanced values of Fe and Pb in the deeper waters have been attributed to the influence of suspended matter and bottom sediments originating from the North Sea. In a more recent work, this group found for all metals investigated (Cu, Cd, Ni, Zn, Fe) different concentrations between bottom and surface waters, showing a general decrease with depth except for Fe (Danielsson et al. 1985). An input of Cu, Cd, Ni, Zn, and Hg to North Sea waters by the Baltic inflow was also evidenced and balanced by Brügmann (1986a).

## 4.2 Distributions over the entire North Sea

While the coastal waters of the North Sea have been subjected to intense scientific and monitoring studies, there is a severe lack of information on heavy metal distributions in the central and northern North Sea, especially with respect to spatial and seasonal variations. Most heavy metal concentrations reported so far have been determined at some isolated stations or single transsects from south to north. Two such transsects from the German Bight to the Scottish region have been conducted by Kremling (1983) and Kremling and Hydes (1987). Results reported by these authors are summarized in Table 5. One of the earliest studies covering the whole North Sea was conducted by the British Lowestoft Laboratory in 1974 and 1976. Results for Cd, Cu, Ni, Zn and Mn have been published by Jones and Jeffries (1983), and those for Hg by Baker (1977). A survey covering the entire North Sea area has been conducted in summer 1986 and winter 1987 (Dicke et al. 1987; Schmidt and Dicke 1987). The distributions of Cd, Pb and Hg obtained from samples taken in 10 m water depth during the first summer cruise in 1986 are given in Fig. 5a-c, and will be discussed for each element separately in the following subsections.

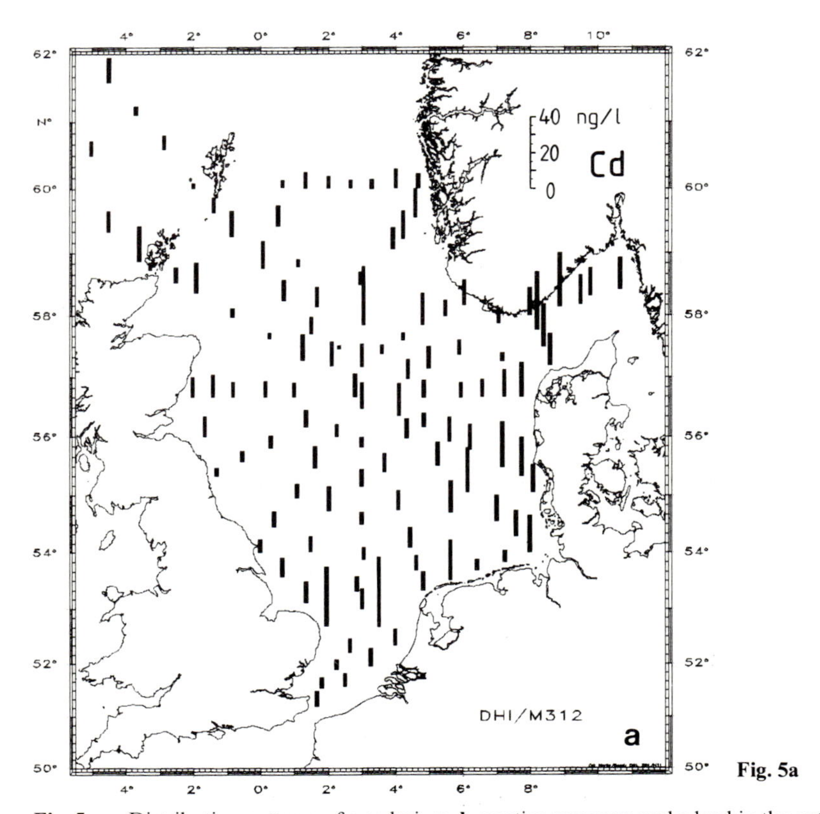


Fig. 5a

**Fig. 5a-c.** Distribution patterns of **a** cadmium, **b** reactive mercury, and **c** lead in the entire North Sea waters. Unfiltered water samples taken in May/June 1986 at 10 m depths. (Schmidt and Dicke 1987; Dicke et al. 1987)

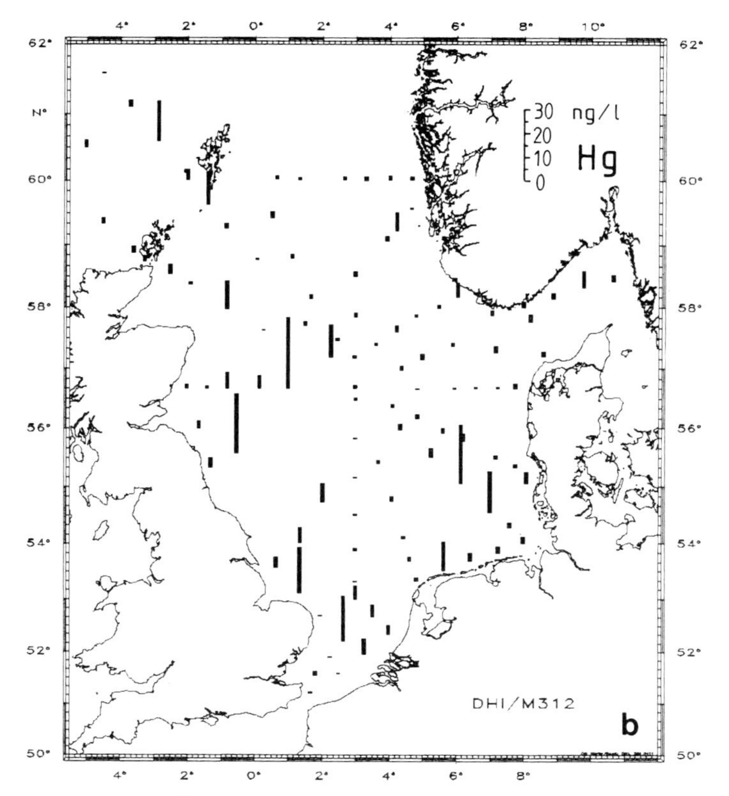


Fig. 5b

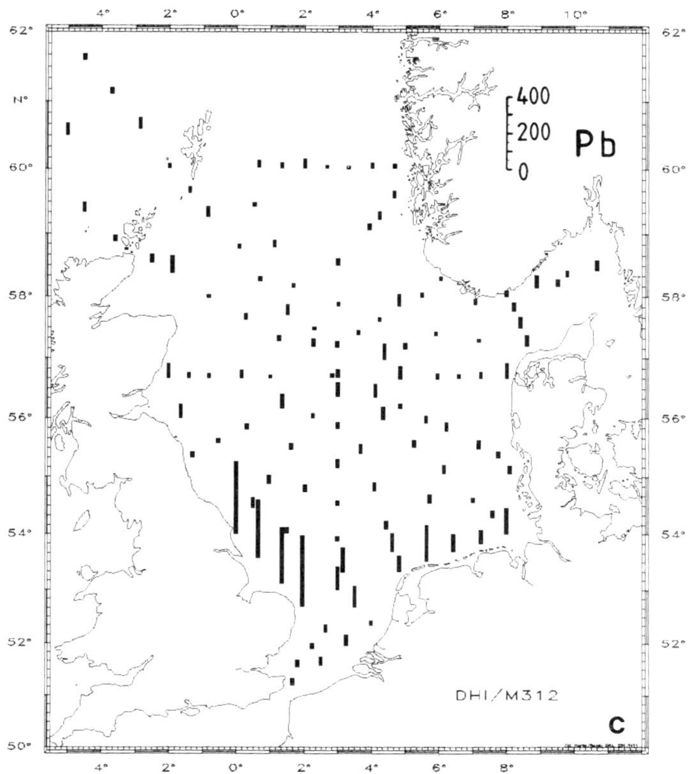


Fig. 5c.

*Cadmium.* Cd concentrations were found to be relatively low in the central and northern North Sea, slightly increasing toward the Southern and German Bights and the Skagerrak (Fig. 5a). In contrast to the findings of Jones and Jeffries (1983) and Balls (1985a) no elevated Cd values could be found near the SE English coast. The highest concentrations (up to 36 ng/l) were found at two stations in the southern North Sea. This maximum value represents an enrichment of only 1:4 compared to a median value of 9 ng/l found in the open North Sea, where Cd concentrations were distributed rather uniformly. Since Cd occurs predominantly in the dissolved phase it is available for long range transport, thus serving as a tracer for metal pollution (Mart and Nürnberg 1986). Compared to data from the open North Atlantic (Kremling 1985),d levels found in the central North Sea are enriched by a factor of 2. It is worth mentioning, however, that the calculation of enrichment factors as an indication of pollution is somewhat questionable in North Sea water, since the natural level in a shallow sea with more intense contact to the sediment may be higher as compared with the deep ocean levels. The median Cd concentration of 9 ng/l found in our survey is, however, the lowest reported up to now for the North Sea. Kremling and Hydes (1987) found an average level of 16 ng/l in surface waters of the central North Sea, and lowest levels of 11 ng/l at two stations off the coast of NE Scotland. This disagreement may be attributed to a mere methodical effect, since our measurements were performed by DPASV (Mart et al. 1980), which usually gives lower concentrations than GFAAS used by other authors. On the other hand, regarding the decrease of mean values given by the more recent studies listed in Table 5, this may still reflect an improvement in contamination control. Irrespective of the disagreements in absolute values, however, the relative difference in the Cd contents of the samples is significant and warrants further investigation.

*Mercury.* As it is uncertain how filtration affects the mercury content of marine water samples, an operational distinction has been established between so-called "reactive" and "total" Hg (Baker 1977). Reactive Hg is defined as the amount reducible by $SnCl_2$ at pH 1 without preceding filtration. Total Hg comprises an oxidation step to degrade the particulate organic matter, in order to analyze both the organic and inorganic fractions. Figure 5b shows the distribution of reactive Hg concentrations in North Sea waters at 10 m depths. The bulk of the values were found to be very low ($< 2$ ng/l), some of them even below the detection limit of 0.5 ng/l. Elevated levels could be detected, however, near the SE English coast and in some southern and northwestern areas. Apart from the elevated values in the northwestern North sea, these findings agree very well with the distribution reported by Baker (1977), including the occurrence of some isolated high spots found in both studies. Baker (1977) attributed these spots to enhanced suspended load or dumping sites. Elevated Hg levels found at a shallow station north of the West Frisian Isles (Fig. 5b) can also be attributed to enhanced loads of resuspended material rather than any pollution effects, whereas high spots observed in the open German Bight appear to reflect an enrichment along a temperature front (Dippner personal communication). The accumulation of Hg found off the coast of E Scotland may be related to local pollution by the outflow of the Firth of Forth receiving an industrial discharge containing elevated Hg loads (Davies et al. 1986). Still little is known about the processes controlling mercury in seawater (Lu et al. 1986). While Mukherji and Kester (1979) found a relation to silicate, other authors (e.g., Olafsson 1983) deny a biological control of Hg distribution. It is generally accepted, however, that sediments may represent an important source at least for the organic fraction of mercury in marine waters (cf. Kersten 1988).

*Lead.* Similar to the Cd situation, total Pb concentrations (i.e., unfiltered and leachable at pH 2) were found to increase from north to south (Fig. 5c). Unlike Cd, however, highest values were found near the SE English coast, which were enriched by a factor of 10 compared to those found in the northern North Sea. This is in sharp contrast to the pattern given by Balls (1985a), who observed plumes of contaminated coastal waters only in the Tyne-Tees area. This disagreement may be explained by the different sample pretreatment. Filtering the water samples as used by the group of Balls may result in a considerable Pb scavenging effect onto the filter cake, with a subsequent depletion of dissolved Pb and a considerable enhancement in particulate Pb concentrations. Unfiltered samples, as used by Dicke et al. (1987), on the other hand, are subjected to a leaching of the suspended matter during storage at pH 2. Since leaching of solids with dilute acids releases the so-called "nonresidual" or anthropogenic metal load (Chester and Voutsinou 1981), the higher values of the unfiltered samples represent a pollution impact of Pb on SE English waters rather than natural sources of elevated Pb contents as assumed by Balls (1985a).

Slightly enhanced values were also found in the Skagerrak area, the Fair Isle Passage, and the Pentland Firth. The most striking feature is a region of enhanced Pb concentrations located in the central North Sea. Median values found in that area (74 ng/l) were more than twice those in the surrounding waters. The higher variability in the Pb concentration distribution as compared to that of Cd may be explained by the different transport and removal mechanisms. While Cd, after leaving the estuarine environment, is transported predominantly in the dissolved phase, Pb remains to a considerable degree associated with the particulate phase. Particle scavenging is thus very effective in diminishing the effect of fluvial input of Pb such as introduced by the Humber estuary and the Firth of Forth (Balls and Topping 1987), resulting in a steep seaward gradient of the dissolved Pb concentrations (Brügmann et al. 1985).

*Relation to Salinity.* A scatterplot of Cd, Hg, and Pb versus salinity is given in Fig. 6a-c. For none of these elements could a significant correlation be found, neither for the data of the entire open North Sea shown in the Figures, nor for the data from separate regions like the Skagerrak or the German Bight. Similarly, no relationship of Cd with salinity and rather wide scattering of some of the other metals around the best-fit dilution line was observed by Kremling et al. (1987) on a cruise through the open North Sea. In contrast to these findings, Kremling and Hydes (1987) found a low but significant correlation of Cd (as well as of Cu, Mn, and Ni) with salinity in surface waters around the British Isles. Deviations from a theoretically linear mixing curve were explained by a rapid remineralization of Cd

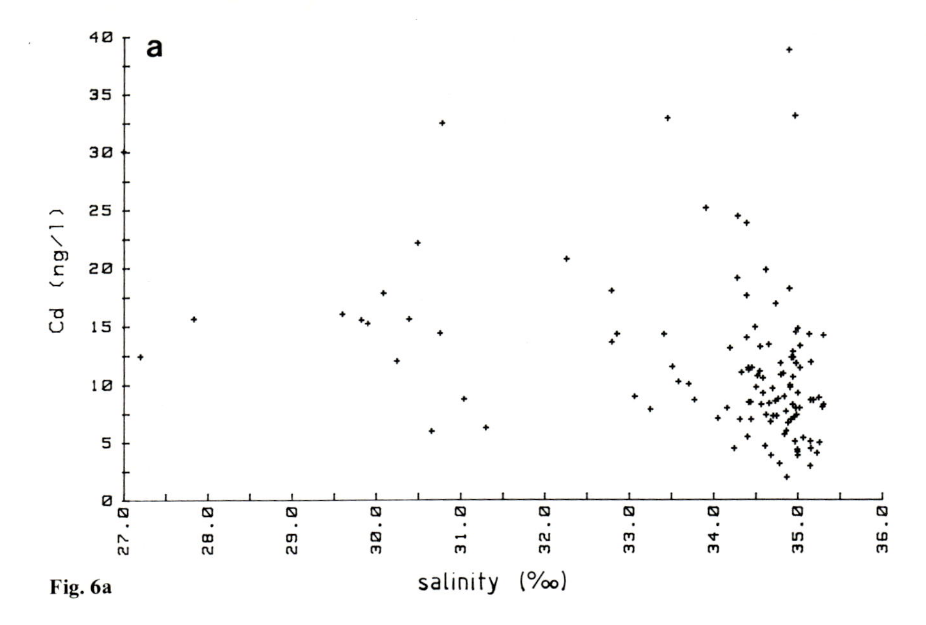


**Fig. 6a**

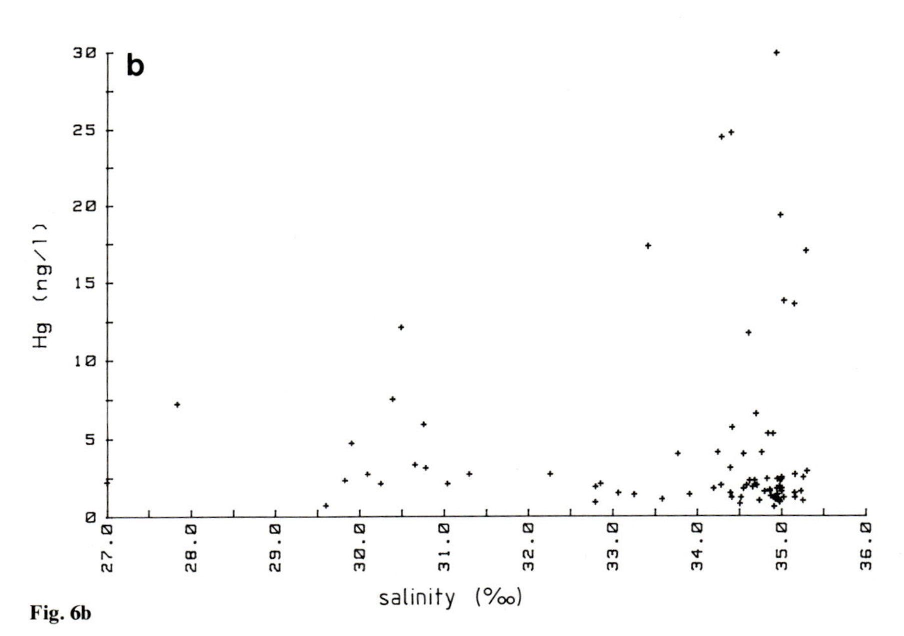


**Fig. 6b**

**Fig. 6a-c.** Scatter plots of **a** cadmium, **b** mercury, and **c** lead versus salinity for water samples taken from the entire open North Sea

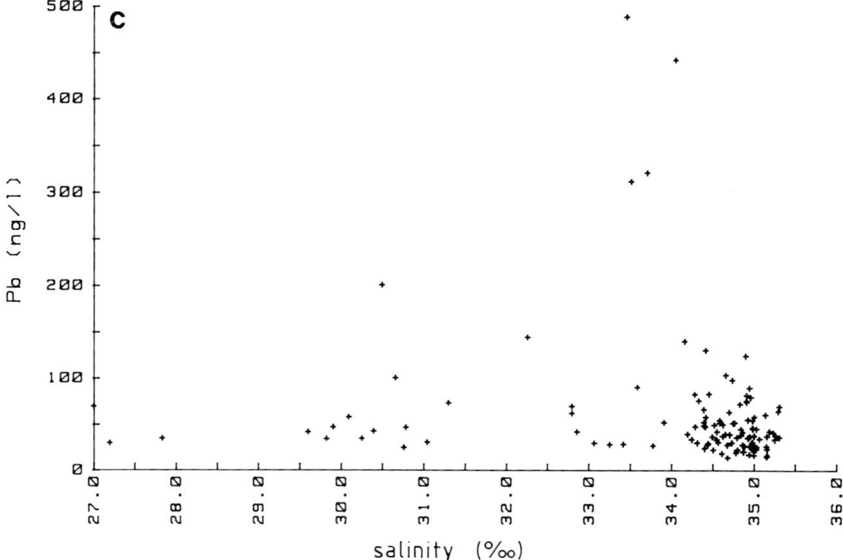


**Fig. 6c**

associated with plankton (Kremling et al. 1987) or changes in adsorption/desorption equilibria (Salomons and Förstner 1984). A poor correlation of heavy metals with salinity on the whole was attributed to different river water "endmember" concentrations which blur simple dilution relationships. Moreover, since Hg and Pb are associated predominantly with the particulate phase, no relationship with salinity may be found due to different transport behavior of water and particulates. Apart from the fact that a statistical treatment of the data in form of a correlation requires a Gaussian distribution of values, which is usually not fulfilled with natural data sets, extrapolating seawater metal concentrations to a salinity of zero to get information on possible sources of contamination is questionable as to the augmenting error of this procedure.

## 4.3 Hydrographical Modeling of the Heavy Metal Distribution

If data on air, water, and sediment movement are available, modeling of the transport and distribution is able to give an integrated view (in time) on the distribution of metals. It can show potentially endangered areas and can help to plan monitoring and sampling programs, especially if modeling and experimental studies are carried out together. A minimum prerequisite for modeling heavy metal distributions in the North Sea waters is the availability of reliable data on the hydrodynamic conditions in order to model the water movement.

Such a model was used to set up an inventory of the annual pollutant loads entering the North sea from coastal inputs based on the situation in 1980 (Klomp et al. 1986). The coastal inputs were mainly derived from rivers with the Rhine as the most important contributant. River inputs are difficult to quantify because of the complex processes occurring in estuaries. However, in the main estuarine part of the river Rhine, continuous dredging takes place and most of the material is discharged in the North Sea off the Dutch coast. Therefore, this estuary does not act as other estuaries as an efficient sink for particulate contaminants. Before 1980 relatively large amounts of contaminated dredged sediments were deposited on land, and thus removed from the aquatic environment (Salomons and Eysink 1981). Contaminated sediments from the river Elbe are still disposed at on land, and therefore withdrawn from entering the North Sea. It should also be added in this context that, because of civil engineering activities, a major part of the pollutants of the river Rhine accumulate in former estuaries which have been converted to freshwater lakes. It has been estimated that more than 50% of the metal pollutant load of the river Rhine is restored in these freshwater basins and therefore do not reach the North Sea (Salomons and Eysink 1981).

To determine the distribution of the trace metals by the input from the various sources, it is necessary to model the hydrodynamics of the North Sea. Vertically integrated two-dimensional hydrodynamic models are in use to simulate wind and tide-induced currents. Klomp et al. (1986) extended an existing numerical grid schematization for the North Sea (Gerritsen 1983) with a grid size of $10 \times 10$ km to include estuarine and coastal waters of the Waddensea area. Average summer and winter wind conditions have been used for the simulation of water movements. Based on the calculated water movement, the mass transport of dissolved and suspended solids by a multi-dimensional water quality model has been used to model the freshwater fraction coming from inflowing waters in the North Sea. In the next step, total input of each heavy metal was treated in the same way as the input of freshwater into the North Sea. The inputs of metals by the rivers were considered thereby as point sources, the atmosphere as a diffuse source (thereby neglecting any atmospheric input gradients from land to sea such as found by Krell and Roeckner 1988). In addition, no distinction was made between particulate and dissolved metals. In Fig. 7 the results of the calculations for Cd, Zn, Cu, and Pb are shown.

This modeling effort also made it possible to calculate the contribution of the Dutch and English coasts to the pollution of the southern North Sea. The results for cadmium are presented in Fig. 8. It can be seen that the relative contribution of sources along the Dutch and Englishs coast to the Cd concentrations in the coastal waters ranges from 20% to over 90%. The remaining contribution can be ascribed to the cross-boundary flow from the Channel and the North Atlantic. For the German Bight, the pattern is slightly different: 10–30% from the Channel, 30–50% from the Dutch coast and 10–70% from the German coast. These results could more or less be confirmed by a more recent study conducted by Baeyens et al. (1987b) for the Belgian coast, in which offshore fluxes of heavy metals based on actual measurements and diffusive as well as mixing processes are calculated. The resulting ratios of the Scheldt output to the offshore flux vary from 38 to 85%, depending on the kind of metal. The contribution of the Scheldt estuary to the flows parallel to the coast, however, ranges only from 1.6 to 3.3%. Quite an opposite situation was found to occur in the German Bight. Model simulations conducted by Müller-Navarra and Mittelstaedt (1987) revealed a negligible metal pollution contributed by the flows into the German Bight as compared to the output flux originating from the estuaries.

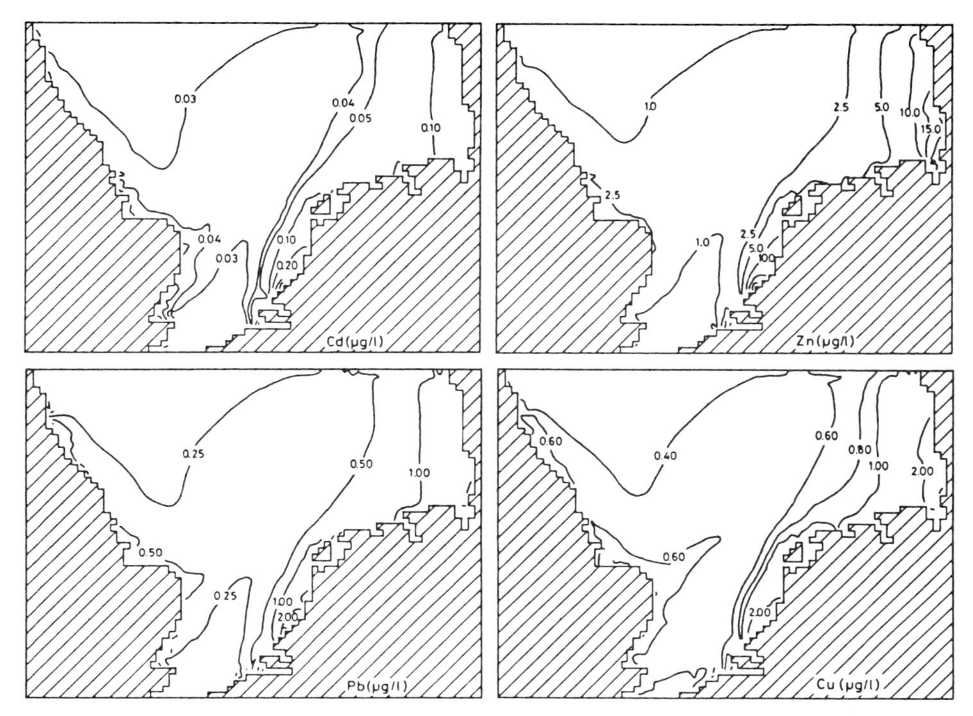


**Fig. 7.** Calculated distribution patterns of cadmium, zinc, lead and copper in the southern North Sea. (Klomp et al. 1986)

Neither the metal concentrations calculated for the Southern Bight area by the model of Klomp et al. (1986), nor the corresponding distributions compared well with field data presented in the former section, especially for Pb. Irrespective of the accuracy of modeled concentrations, however, the advantages of this type of modeling is that it is possible to calculate for different pollution abatement strategies, the results on water quality (sensivity tests). Furthermore, the contribution of the various sources to the pollutant load can easily be calculated. On the other hand, the kind of modeling presented by Klomp et al. (1986) represents only a first-order approximation, because it has to ignore nearly all processes (cf. Fig. 1) leading to a finite residence time and redistribution of pollutants in the various compartments of the North Sea ecosystem.

# 5 Suspended Matter and Deposited Sediments

## 5.1 Enrichment of Heavy Metals in the Particulate Phase

In general, the concentration of heavy metals in the particulate phase, expressed in terms of weight of metal per unit weight of dry solid material (μg/g), is much higher than that of the same metal in the dissolved phase (μg/l). In fact, suspended materials tend to accumulate these trace elements preferentially. It is therefore

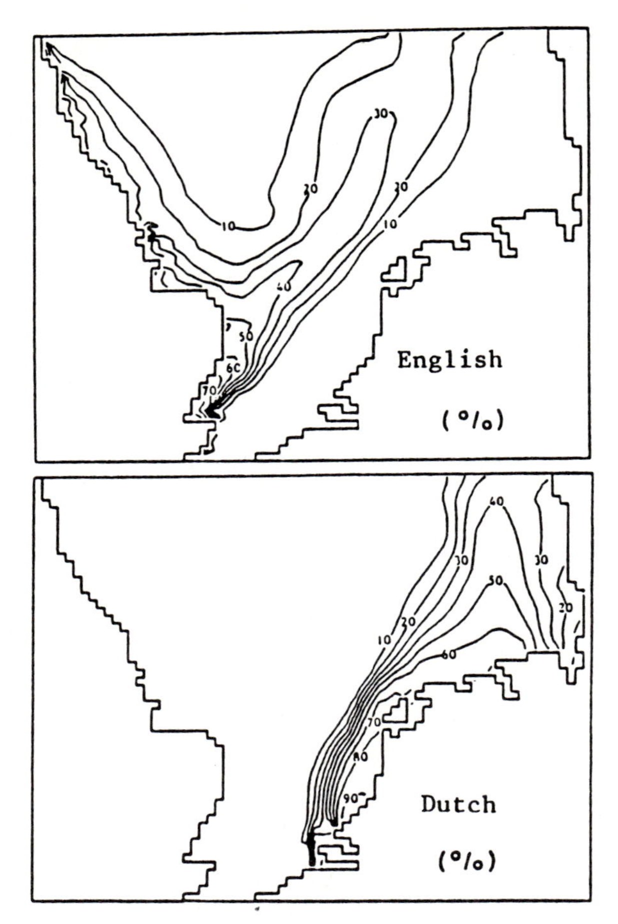


Fig. 8. Relative contribution of the Dutch and English coasts to the cadmium concentration in the southern North Sea. (Klomp et al. 1986)

possible to exploit the relative enrichment of the particulate phase in order to evaluate the level of contamination in an aquatic system. This consideration does not limit itself to the analysis of suspended material, but it also includes the sedimentary deposits which make it possible to establish a geographic map of principal sources and sinks of pollution. Moreover, investigations of sediments as a more or less effective final sink have the advantage of integrating inputs with time. This is particularly important in the case of heavy metals, for which the discharge and removal processes are of a very discontinuous character.

In fact, particles in seawater play a major role in regulating the chemical forms, distributions, and deposition of trace metals released into marine waters as a result of natural processes and human activities (Wollast 1982; Hart 1982; Salomons and Förstner 1984). This is particularly true in coastal waters, where particulate matter in runoff from rivers and coastal discharges interact with seawater, but also in areas where fronts of nutrient-rich water lead to enhanced primary productivity. All the processes of transfer of materials from land to waters are constrained by the removal of reactive elements onto particles. Many of the underlying processes have been elucidated by use of tracers, such as natural and artificial radionuclides in field and model ecosystem experiments (Turekian 1977; Santschi et al. 1980, 1987). While in a conceptual sense these facts are generally recognized, the present state of knowledge is such that detailed information on the processes and mechanisms that result in enrichment of sediments and particulate matter is still lacking. We will not go into details of these processes, which are beyond the scope of this chapter. In addition, our state of knowledge has not yet been developed to the point where accurate quantitative predictions can be made (cf. Förstner et al. 1985).

In brief, particles in coastal waters are the result of ongoing physical, chemical, biological, and geological processes which may vary spatially and seasonally. These processes include the supply of inorganic rock weathering products and organic substances from river runoff, resuspension of previously deposited sediments, aeolian fallout, production and degradation of biological species, but also geochemical adsorption-desorption, precipitation-dissolution, and complexation processes related to surface reactions at the solid-solution interface (Stumm and Morgan 1981).

Particulate organic matter, clays (i.e., aluminosilicate minerals) and the oxides and hydroxides of Fe and Mn represent the most important substrates for trace metal scavenging in the aquatic environment (Gibbs 1973). On the basis of the suspended matter concentration and composition, the North Sea can be divided into a number of separate regions (Eisma and Kalf 1987). The relative contributions of the major components of total particulate material (TPM) vary considerably, showing a most drastic change from the shallow (10–40 m) southern part to the deeper (70–100 m) central and northern area. In the coastal areas, wind and tidal current induce stresses on the sea floor and resuspend bottom sediments into the water column. This detrital material determines significantly the composition of near-coastal TPM. Resuspension and estuarine runoff is the predominant source of clays and oxides in suspended matter. In the deeper central North Sea, erosion of bed sediments affects the composition of the suspended matter much less. Here biomass is the major component of TPM. Since heavy metals are affected differently by uptake or scavenging by these sediment components, and taking into account the different sources of water with its suspended matter load in the North Sea, regional differences in metal accumulation in suspended matter may occur. These regional differences are comprehensively discussed by Nolting and Eisma (1988).

In areas of nondeposition, i.e., the deeper central parts of the North Sea, organic material derived from plankton and associated pollutants may be the most important heavy metal removal mechanism. Reworking of what is left after decomposition will result in burial of at least part of the pollutants supplied to the bottom in this way. This type of "biodeposition" process (cf. Fig. 1), whereby organic matter is gradually accumulated in coarse-grained bottom sediments, is going on in the Fladen Grounds and the Great Fisher Bank in the northern North Sea (Eisma 1981). Here only a very small amount of suspended matter has been deposited for about 9000 years (Jansen et al. 1979), but $^{14}C$ data indicate the deposition of more recent organic material and reworking of the top 10–15 cm (Erlenkeuser 1978). Reworking in this area can be caused by bottom fauna ("bioturbation", cf. Fig. 1), or bottom trawling.

When assessing heavy metal concentrations in sediments, one major problem is differentiation between natural (e.g., weathering of local rorks) and anthropogenic source of metal accumulation. Metal concentrations measured on total sediment basis are usually without value for pollution reconnaissance studies, because there are also natural fractionation processes that determine metal accumulation in sediments. The coarser sediment fraction containing mainly quartz, carbonate shell debris and heavy minerals usually contain low concentrations of heavy metals, while the finer fractions contain the predominant proportion of the (especially anthropogeneous) heavy metals, because the large surface area of the fine particles adsorbs the excess fraction of heavy metals. In areas where sand or gravel forms the sea bed, this fine-grained material is trapped in the pore space between the grains. Although it may contribute only a very small fraction of the total sediment, it may contain nearly the whole proportion of the pollutants.

Consequently, any analysis of whole sediments must be viewed in the context of the portion of the sediments which contain the contaminant. These effects, as well as the methods of normalization, were comprehensively investigated in an earlier study on sediments of the south-eastern North Sea and adjoining estuaries (Förstner et al. 1982). One of the more widely used approaches is to normalize sediment data by analyzing only the fine fractions. This has the effect of reducing particle size effects, but as yet there is no international consensus on which size fraction to use. As long as methods of grain size fractionation are not standardized, however, data cannot be readily compared with each other.

While variations in grain size, which are significant over the entire North Sea bed, are readily compensated for by analyzing the finer size fraction, there is also a considerable variation in sediment mineralogy and geochemistry. Förstner et al. (1982) showed that there is a great internal variability of metal concentrations even after separation of the fine-grained sediment fraction. Natural variability in heavy metal contents must be compensated additionally by a geochemical normalization in order to separate the natural and anthropogenic contribution, since it is not possible to separate them chemically. This geochemical normalization involves the concept of a "carrier substance". A carrier substance can be defined as a major constituent of the sediment fines into which are incorporated, or onto which are attached, one or several trace elements. If a heavy metal is significantly correlated to one or several major elements, it is possible that the mineral phase containing the major elements can be the carrier. Trefry and Presley (1976) assumed for background sediment conditions that the relationship between Fe and a trace metal will be linear. That means, if the concentration of Fe changes because of changing mineralogy or natural processes such as diagenetic alterations, the concentration of the associated heavy metal will change with a constant relation to Fe. The population that occurred within the $\pm 95\%$ prediction interval about the least-squares linear regression was therefore defined as natural, or unpolluted. Metal concentrations occurring above the 95% interval were postulated as indicating possible anthropogenic sources.

Enrichment (or pollution) factors (EF) are used to characterize the heavy metal status of sediments relative to natural back-ground levels (Förstner and Wittmann 1979). Estimation of the contamination level of a sediment by a certain element requires therefore knowledge of the preindustrial concentration. Quite often background values used to compute these factors are simply taken from a compilation of the mean composition of crustal material. The advantage of this standard is that it is essentially free from any anthropogenic admixtures; disadvantages are, however, that regional geochemical characteristics are not taken into account, which warrants again relating the metal to iron or aluminum contents. Background values can also be estimated from remote regions not influenced by significant anthropogeneous impact, such as from samples collected at the shelf edge off the Hebrides or from dated deep sediment cores.

## 5.2 Suspended Sediments

In winter 1987 a sampling cruise over the entire North Sea was performed to collect suspended matter samples for major and trace element analysis. Sampling and analysis has been performed according to the methods presented earlier (Kersten and Förstner 1985). It is worth mentioning that the suspended matter samples have not been collected by filtering due to the severe distortion of adsorption/desorption equilibria and arising enhancement of especially particulate Pb concentrations, but by continuous flow centrifugation. The advantages of this sampling method has been discussed in detuil by Horowitz (1986).

Before going on to discuss here briefly the distributions of particulate Cd and Pb, it seems necessary to obtain some background information on their carrier, the suspended matter. The relation between element contents and total suspended matter (TPM) concentrations displayed in the series of Fig. 9a-f can provide information on particle characteristics when element contents are represented on a relative weight basis, rather than on a volume basis. Most element concentrations are related to the TPM concentrations: Al (and Fe, Ti, K, Ca and V) tend to increase with suspended matter concentration, whereas POC (Cu and Cd as well) increase strongly at low TPM. This reflects the predominance of resuspended mineral particles at high TPM, as occurring in the coastal areas, and the predominance of organic matter at low suspended matter concentrations.

Based on these relationships, Duinker (1983) has shown that three different types of dependence can be distinguished. He has explained these patterns by a model that considers suspended matter to be composed of small/low density particles that are permanently in suspension and larger/denser particles derived from resuspended bottom sediment. The relative amount of the permanently suspended fraction is larger at low than at high suspended matter concentrations. The larger/denser particles dominate at high concentrations (typically in the order of several mg/l). The fractions have different settling velocities and also different compositions. The permanently suspended fraction has higher contents of organic C, Cu and Cd (group I elements, Fig. 9a). The larger/denser fraction has higher contents of Al, Ti, Fe and V (group II elements). These elements correlate well ($R \geq +0.67$) with each other as exemplified in Fig. 10a for Fe versus Al, and Fig. 10b for V versus Fe. The Al/Fe ratios measured fell very close to 0.5, which is the value assumed to represent terrigenic aluminosilicate minerals (Price and Skei 1975). No preference was found for Mn, Zn, Pb, and Cr (group III elements). At higher seston concentrations, the major element concentrations tend to approach the values

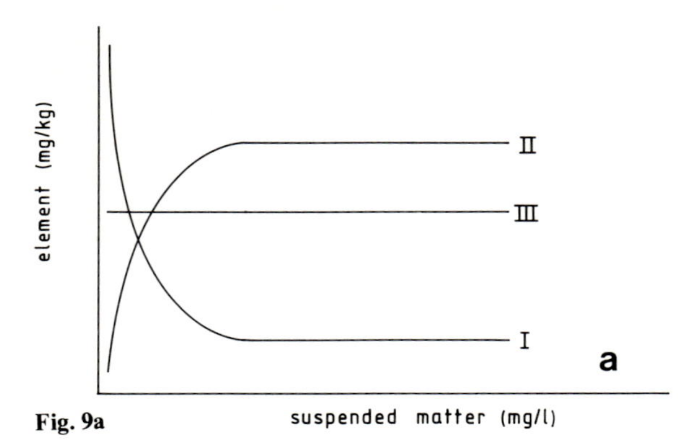


Fig. 9a

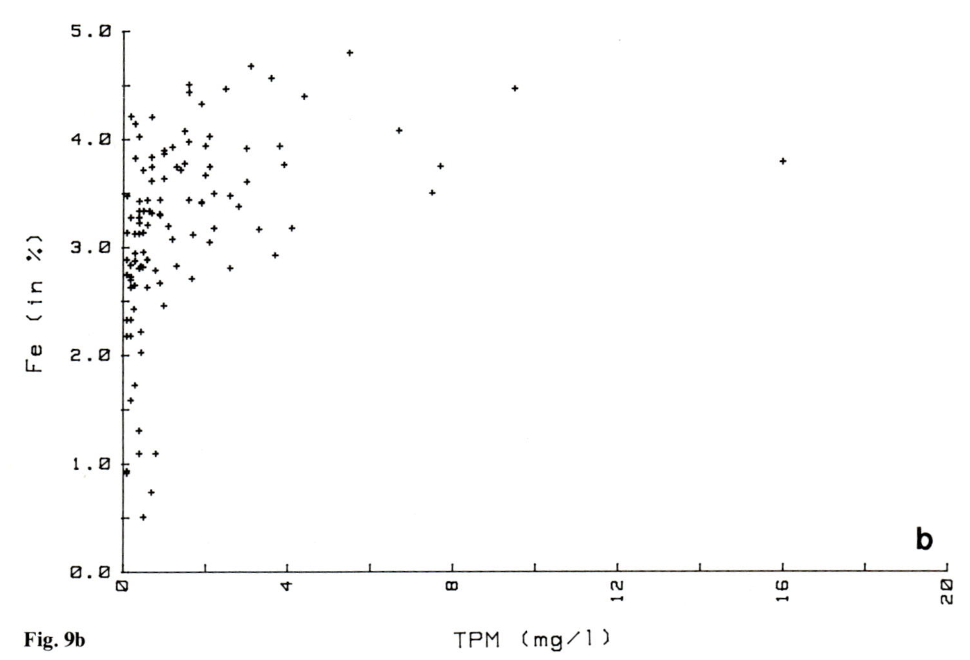


Fig. 9b

**Fig. 9a-f.** Schematic relation between the element contents in suspended matter and particulate load for group I, II and III elements (see text; Duinker 1983). Actual data for iron, particulate organic carbon, cadmium, chromium, and lead in suspended matter from the entire North Sea area. (Figs. 9e and f see page 330)

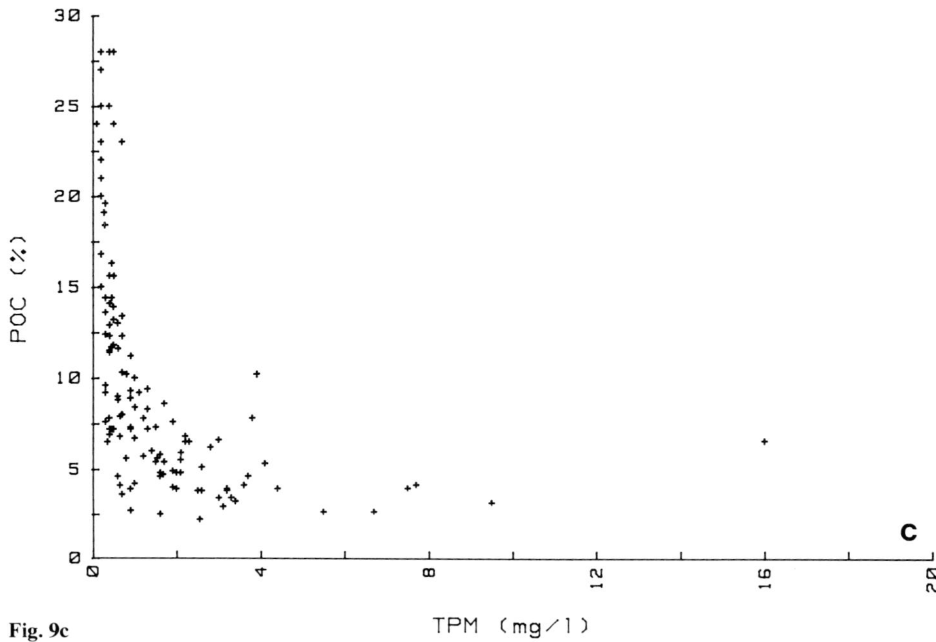
30
25
20
15
10
5
0
POC (%)
c
0
4
8
12
16
20
TPM (mg/l)

Fig. 9c

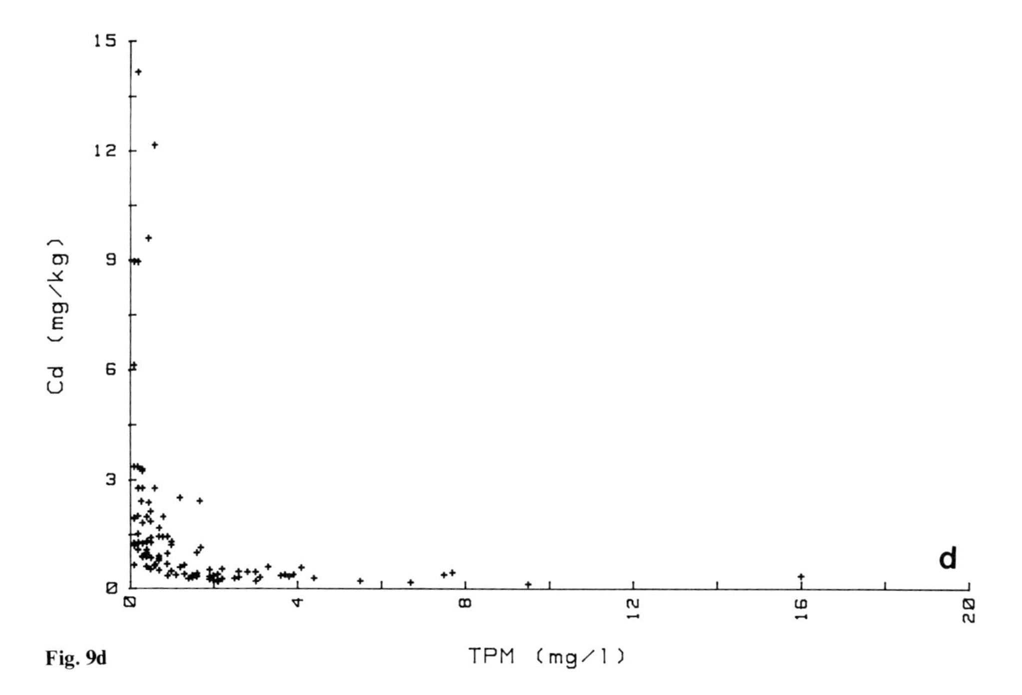
15
12
9
6
3
0
Cd (mg/kg)
d
0
4
8
12
16
20
TPM (mg/l)

Fig. 9d

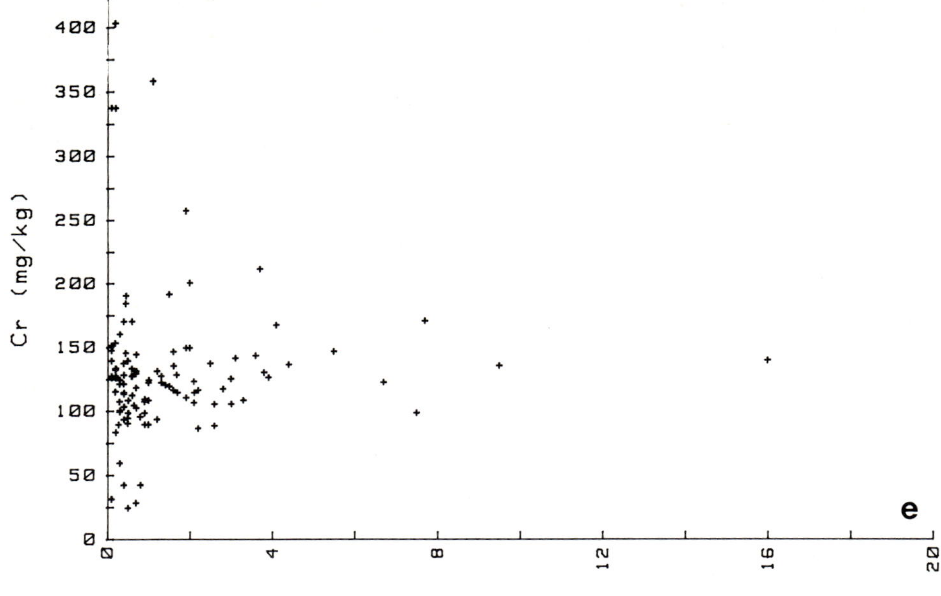
Cr (mg/kg)
400
350
300
250
200
150
100
50
0
0
4
8
12
16
20
TPM (mg/l)
e

Fig. 9e

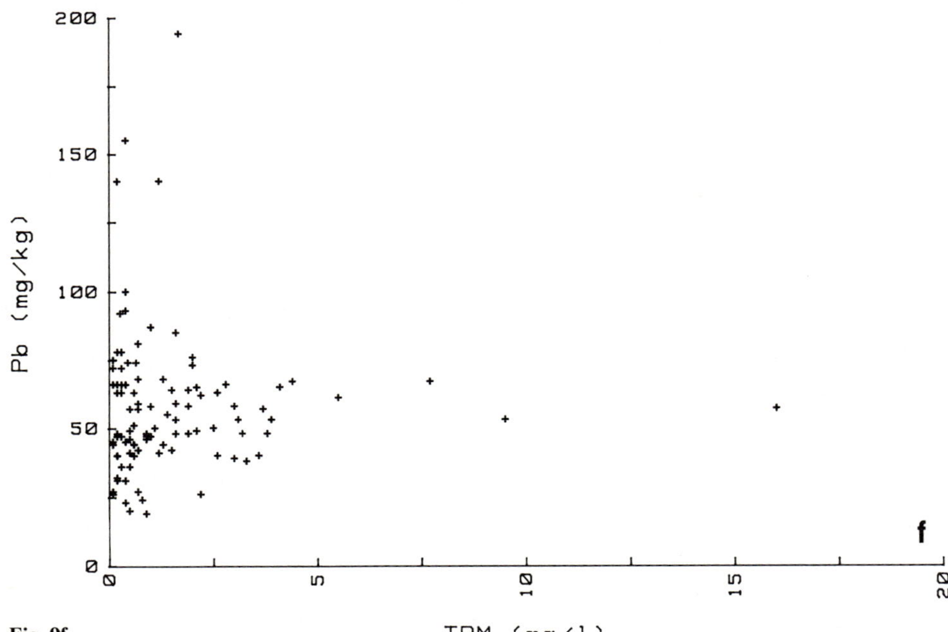
Pb (mg/kg)
200
150
100
50
0
0
5
10
15
20
TPM (mg/l)
f

Fig. 9f

reported for mean sediment composition given by Bowen (1979). The appropriate values for the heavy metals are somewhat elevated in nearly all areas, especially for Cr and Zn. This suggests that suspended matter at high seston concentrations contains a high fraction of bottom derived terrigenous material. The deviations for the heavy metals can partly be explained by a relatively higher content of organic matter in the suspended matter as compared with the mean sediment composition.

The highest element contents for group I elements are found to occur with very low seston concentrations mainly composed of small/low density particles. The relative importance of small/low density particles on the accumulation of copper is well demonstrated in Fig. 10c, in which the concentration of this element in the particulate phase is shown in relation to the specific surface area of the suspended matter as determined directly by the single-point BET gas adsorption technique. Since the main reason for the higher metal content in the small/low density particles is their larger surface area and, therefore, their higher adsorption capacity, it is not surprising to find a significant correlation between surface area and Cu content of the suspended matter. It is not without interest to note that the highest values of surface area are observed in the central part of the North Sea where the relative contribution of detrital organic matter and plankton cell walls with their reactive surface binding sites is the most important, and that the concentration of particulate Cu diminishes near the coasts where the contribution of inorganic material of continental origin reduces considerably the surface area of the seston.

The strong positive correlation between organic carbon and group I elements, and the negative correlation with group II elements, might be used to suggest the association of group I elements (i.e., Cu and Cd) with the organic matter fraction of the suspended sediment. Dehairs et al. (1986) have shown accordingly that both living plankton and detrital organic matter are potentially a sorption site for these metals. Group III elements (i.e. Zn, Cr, and Pb), on the other hand, may be associated with Mn and Fe oxide coatings, occurring in association with organic matter and clay minerals. This is supported by the positive correlation between Mn and these elements as exemplified in Figure 10d for lead.

The above brief considerations give a sufficient background for the interpretation of the distributions of particulate Cd, V, and Pb concentrations in the North Sea waters shown in Fig. 11a-c. In spite of the low density of sampling sites, certain areas of high concentrations indicate that the concentration distributions are not even, and that mixing of suspended matter in the North Sea is far from being complete. Particulate Cd and V concentrations indicate a division in the composition of suspended matter between a northern part including the northern and central North Sea and most of the Skagerrak and Norwegian Trench, and a southern part including the coastal areas. Cd contents vary between 0.1 and 14.2 $\mu g/g$ (mean 1.57 $\mu g/g$), which coincide well with the range found by Nolting and Eisma (1988). In the southern North Sea, Cd contents are generally less than 2 $\mu g/g$, whereas highest concentrations were found in the northern North Sea. Quite an opposite concentration distribution is found with vanadium (Fig. 11c).

The reason must be seen mainly in the different suspended matter concentrations found in these areas. The central and northern North Sea and the Skagerrak are strongly influenced by the inflow from the North Atlantic with little resuspension from the bottom because of the large waterdepth (Eisma 1981).

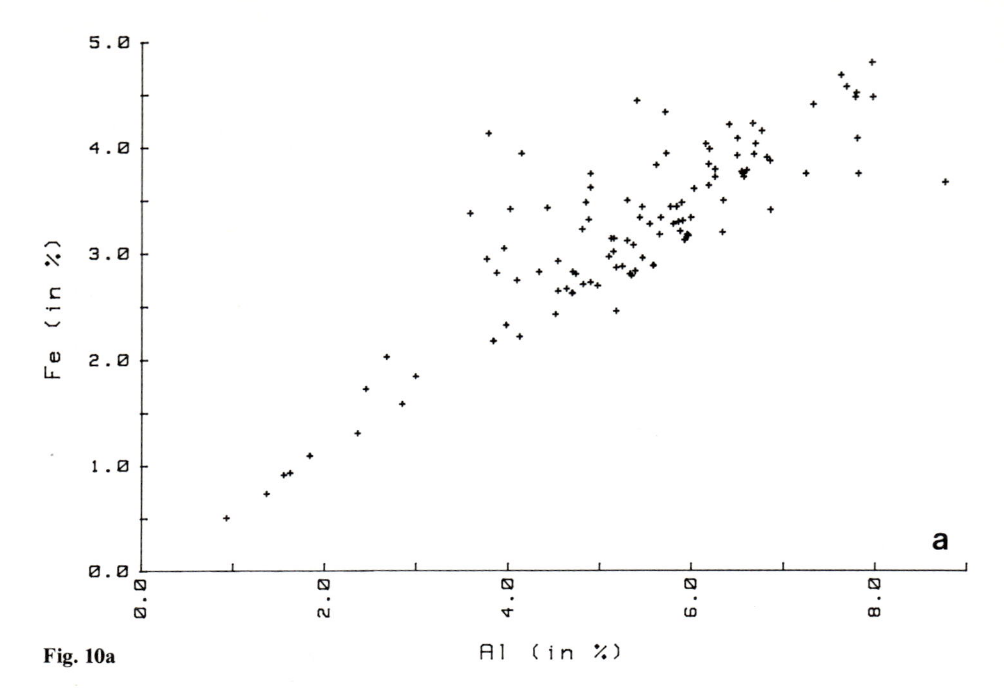


Fig. 10a

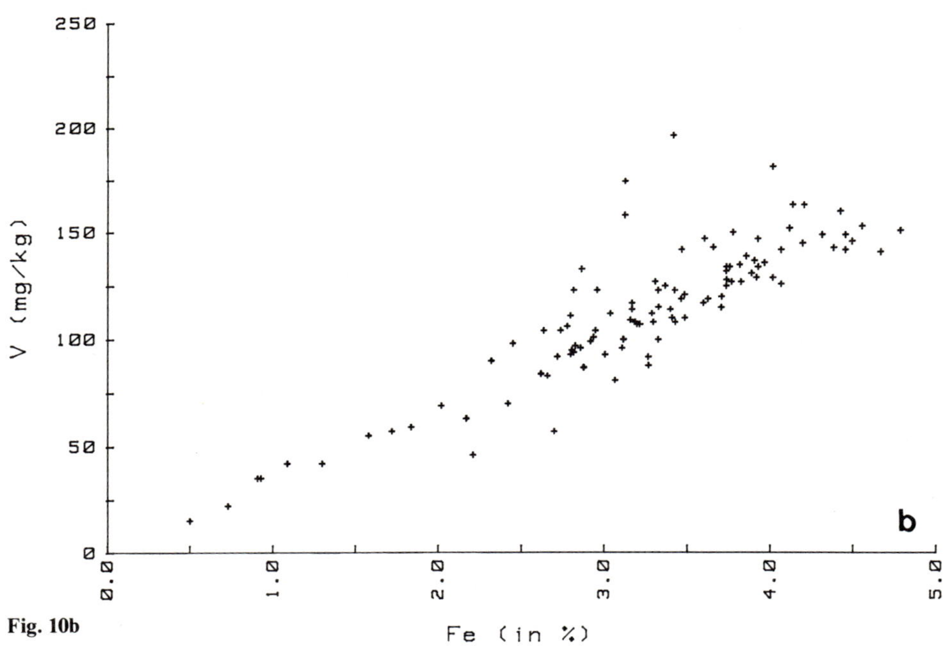


Fig. 10b

**Fig. 10a-d.** Scatter plots of **a** aluminum versus iron, **b** vanadium versus iron, **c** copper versus specific surface area, and **d** lead versus manganese ratios for suspended matter taken from North Sea waters in winter 1987

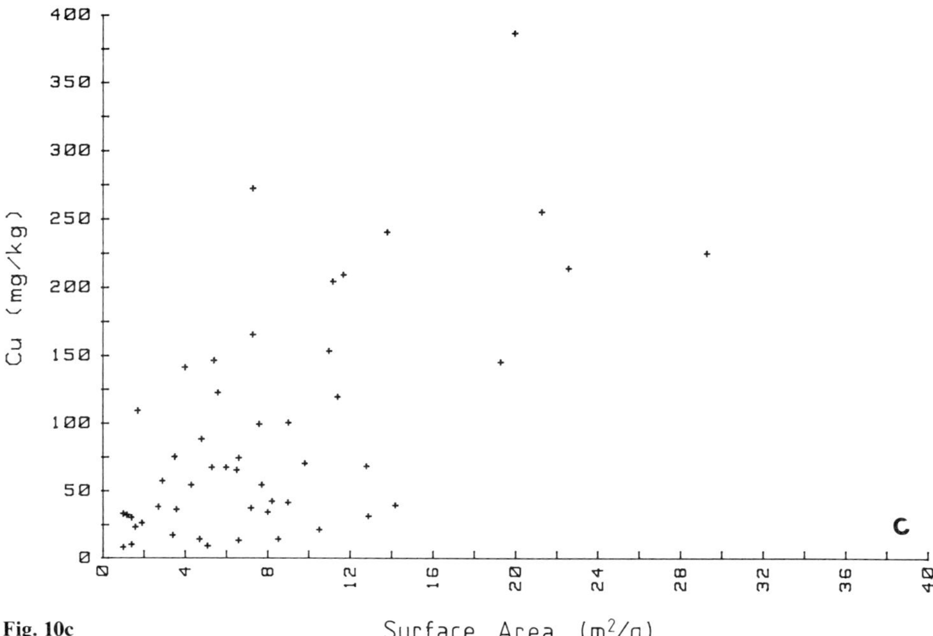

Fig. 10c

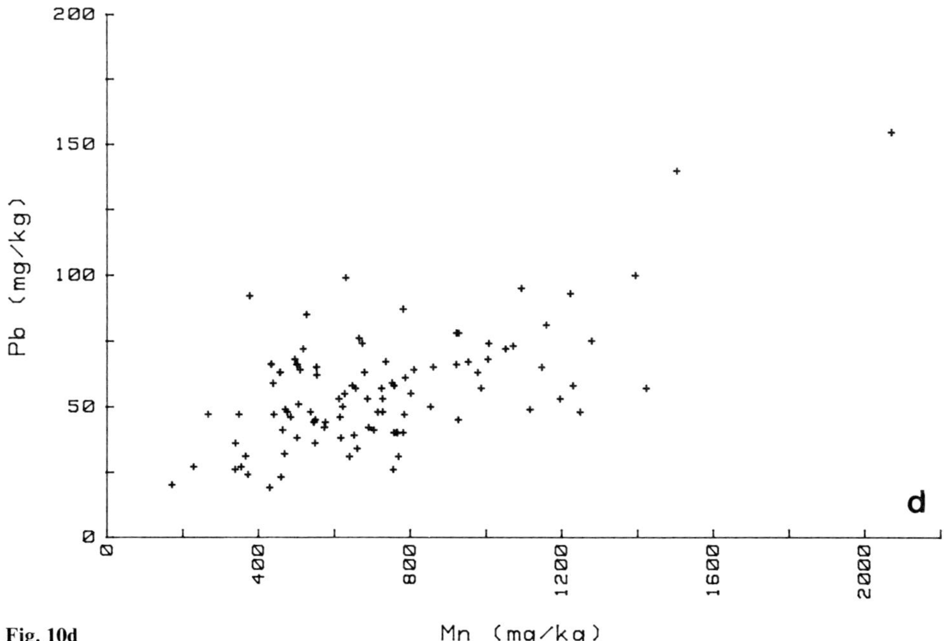


Fig. 10d

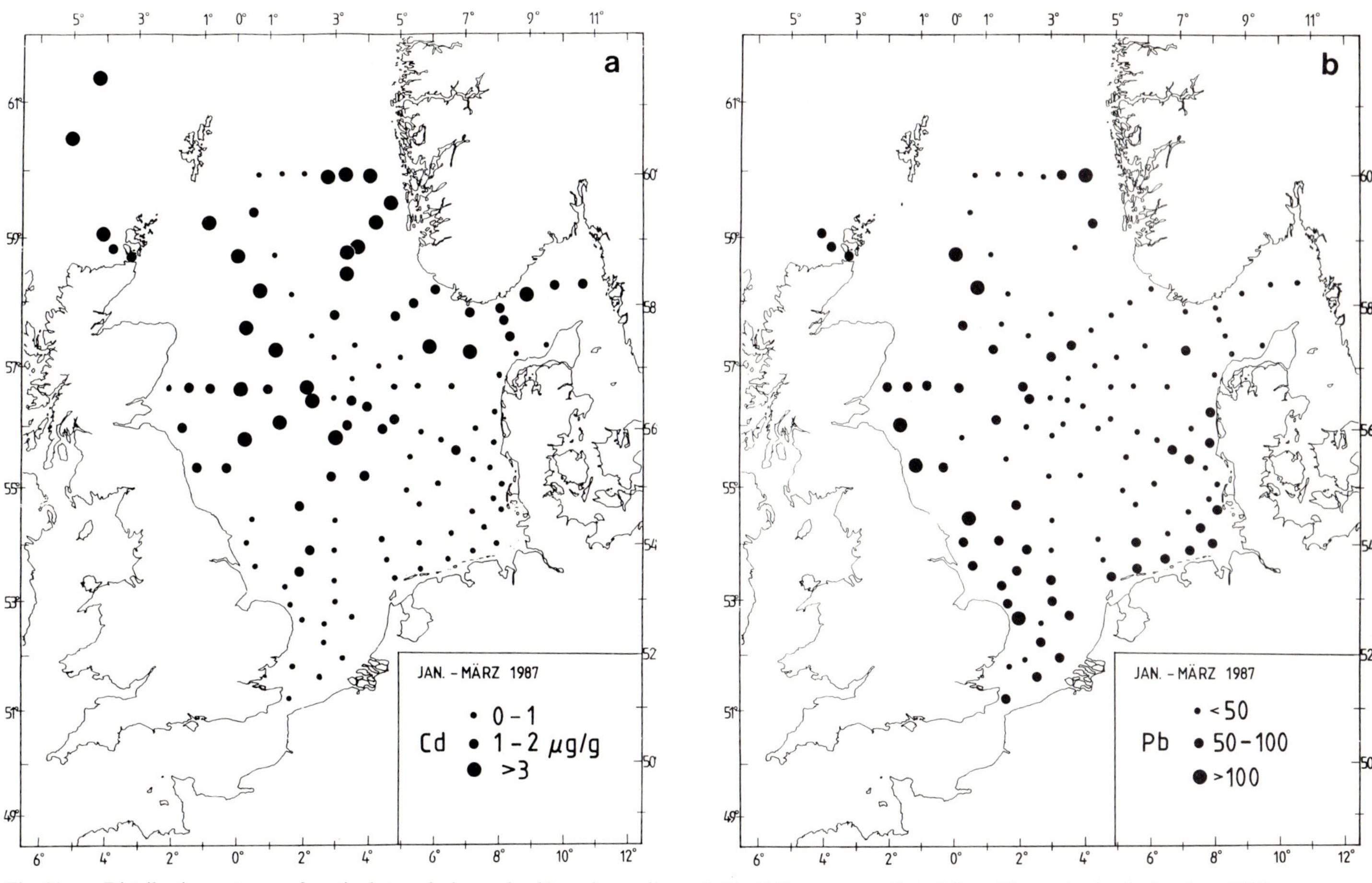


**Fig. 11a-c.** Distribution patterns of particulate cadmium **a**, lead **b**, and vanadium **c** in North Sea waters collected from 10 m water depths in winter 1987

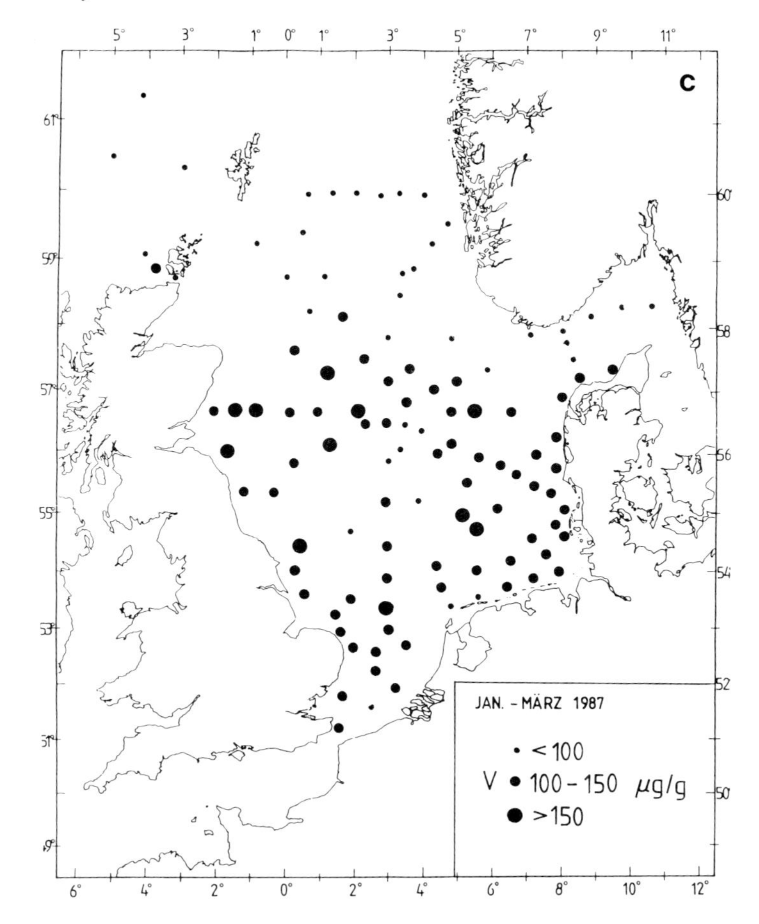


Fig. 11c

Further explanations for these different patterns can at present only be speculative as the nature of the particles in suspension and of the source material is insufficiently known. Since the inflowing Atlantic and outflowing Baltic waters are rich in nutrients (cf. Brockmann et al. this Vol.), elimination of the nutrients from the water by phytoplankton productivity subsequent to "upwelling" of deep or bottom waters into the surface layers may lead to an effective removal of dissolved Cd into the particulate form. Evidence for such "nutrient pumps" produced by increased vertical mixing in the vicinity of shelf edge areas has been discussed by several authors (e.g., Sandstrom and Elliott 1984). Hence, enhanced particulate Cd concentrations in the northern part of the North Sea may be the result of low particulate loads and patchy biological removal rather than simply being an input process (from the sediment or atmosphere). The mixing zone of both Atlantic and North Sea waters may therefore represent a last "biogeochemical barrier" to prevent dissolved Cd from escaping into the ocean. This process is, in general, more effective in summer than in winter (Kremling and Hydes 1987).

Enhanced particulate Pb concentrations were found along the coasts, especially off the British east coast, but also in the northwestern part of the North Sea (Fig. 11b). In the central North Sea, Pb contents are generally low, showing values around 50 μg/g. Since Pb, in contrast to Cd, concentrations do not depend significantly on the particulate load, this pattern represents evidence of considerable Pb contamination of these areas. The concentration range is 15–200 μg/g (mean 59 μg/g). These data are somewhat lower than earlier data reported by Brügmann (1986: range 5–430, mean 88 μg/g), who also found along their south-north profiles enhanced particulate Pb concentrations in the northwestern part of the North Sea. The large inputs delivered by the Belgian-Dutch-German rivers do not seem to reach far beyond the coastal zone. Particle scavenging by sedimentation in these estuaries is in fact very effective in diminishing the impact of the fluvial input (Salomons and Förstner 1984).

In our above consideration, atmospheric input was suspected to be the major source of Pb in the North Sea. From atmospheric deposition modeling, a significant decrease of the total Pb deposition is found from land to sea (Krell and Roeckner 1988, cf. Fig. 4), which is in good agreement with the particulate Pb distribution found in the southern North Sea. Brügmann et al. (1985) discussed water residence time as a factor which may additionally explain the enhanced Pb concentrations found in the northwestern North Sea. Although atmospheric deposition rates are lower, this would give a higher accumulation rate of lead. A simple calculation using the deposition rate of 100 $\mu g/m^2$/month reported by Krell and Roeckner (1988) would result within a year for a 10-m-deep water column in an accumulation of 10 ng/l Pb. This is less than one order of magnitude lower than the values found in this region, which is a reasonably good correspondence bearing in mind the considerable uncertainties in the data basis on which the deposition rate simulations are based upon (Krell personal communication).

Furthermore, in the shallower southern part of the open North Sea, the atmospheric deposition is taken up by a higher particulate load, which increases the removal of anthropogenic lead by relatively short particle residence times. Particulate lead in the northern parts of the North Sea seems to be bound to material of autochton hydroxidic or biogenic origin, whereas for the coastal area part the scavenging by terrigenic material with aluminosilicates, poor in lead, seems to be more important. This conclusion could be drawn from Table 6, which compares the mean enrichment factors of particulate Pb for North Sea aerosols, suspended matter, sediments and the earth's crust. Taking the composition of Bowen's "mean sediment" as reference value, these factors clearly show the decrease of Pb enrichment following the most likely pathway from waste combustion as a potential source to the sediments as the intermediate or final sink. Suspended matter already shows considerably reduced relative lead contents compared with aerosols, which confirm the rapid leaching of the loosely held Pb associations from marine aerosols by seawater as evidenced by Chester et al. (1986), and "dilution" of the remaining particulate Pb into the marine suspended matter.

**Table 6.** Mean enrichment factors[a] for Pb, Cd, and Zn in municipal incinerator fly-ash and different compartments of the North Sea calculated on the basis of the mean sediment composition reported by Bowen (1979)

| Material | Cd | Pb | Zn |
|---|---|---|---|
| Municipal incinerator fly-ash | | | |
| mean of three plants | | | |
| collected above the roof[b] | 45380 | 21920 | 6500 |
| Aerosols (> 0.1 μm) | | | |
| Helgoland[f] | 3025 | 450 | 127 |
| Suspended matter (> 0.4 μm) | | | |
| NE Atlantic[c] | 764 | 58 | 18 |
| N North Sea (north of 56°N)[d] | 19.4 | 4.7 | 2.2 |
| S North Sea (south of 56°N)[d] | 4.3 | 4.0 | 1.8 |
| Sediment | | | |
| entire North Sea[e] | 1.8 | 2.8 | 1.6 |
| Earth's crust | | | |
| (Bowen 1979) | 0.6 | 0.6 | 0.7 |

[a] $EF = \frac{(\text{Element/Al})_{\text{Material}}}{(\text{Element/Al})_{\text{Mean Sediment}}}$

[b] Data taken from Greenberg et al. (1978).

[c] Data taken from Brügmann (1986b)

[d] Samples collected in January/February 1987.

[e] Data taken from Irion and Müller (1987).

[f] Samples collected weekly in 1986/87

## 5.3 Deposited Sediments

As discussed above, sediments can reflect the current quality of the system, as well as the historical development of pollution. The study of dated sediment cores has proved to be especially useful, as it provides a historical record of the natural background levels and the man-induced accumulation of metals. Clear records of increasing heavy metal contamination in the sediments of the inner German Bight have been documented earlier (Förstner and Reineck 1974). The sedimentary profiles given in Fig. 12 are basically the same for all investigated heavy metals. The minimum concentrations of the metals in the deeper sections of the sediment core may be considered as regional background values for this area. From the deepest sections the metal concentrations display a uniformly and more or less steep increase towards the top horizon, where maximum values are reached at a sediment depth of 15–20 cm, indicating a severe increase of contamination of the sediments deposited after the mid-1950's according to radiometric age data (Dominik et al. 1978). The concentrations of Pb, Hg, Cd, and Zn were four to ten times as high as the natural background concentrations.

Similar local sediment pollution "hotspots" have been found in other coastal areas, namely in the Dutch Waddensea (Kramer and Vlies 1983), the Dutch coastal zone (Salomons and Eysink 1981), the German Bight (Irion and Schwedhelm 1983), and off the NE English coast (Nicholson and Moore 1981). The highest metal values on a total sediment basis were found in samples containing a large portion of fine particulate matter, which in turn is generally associated with the inputs from

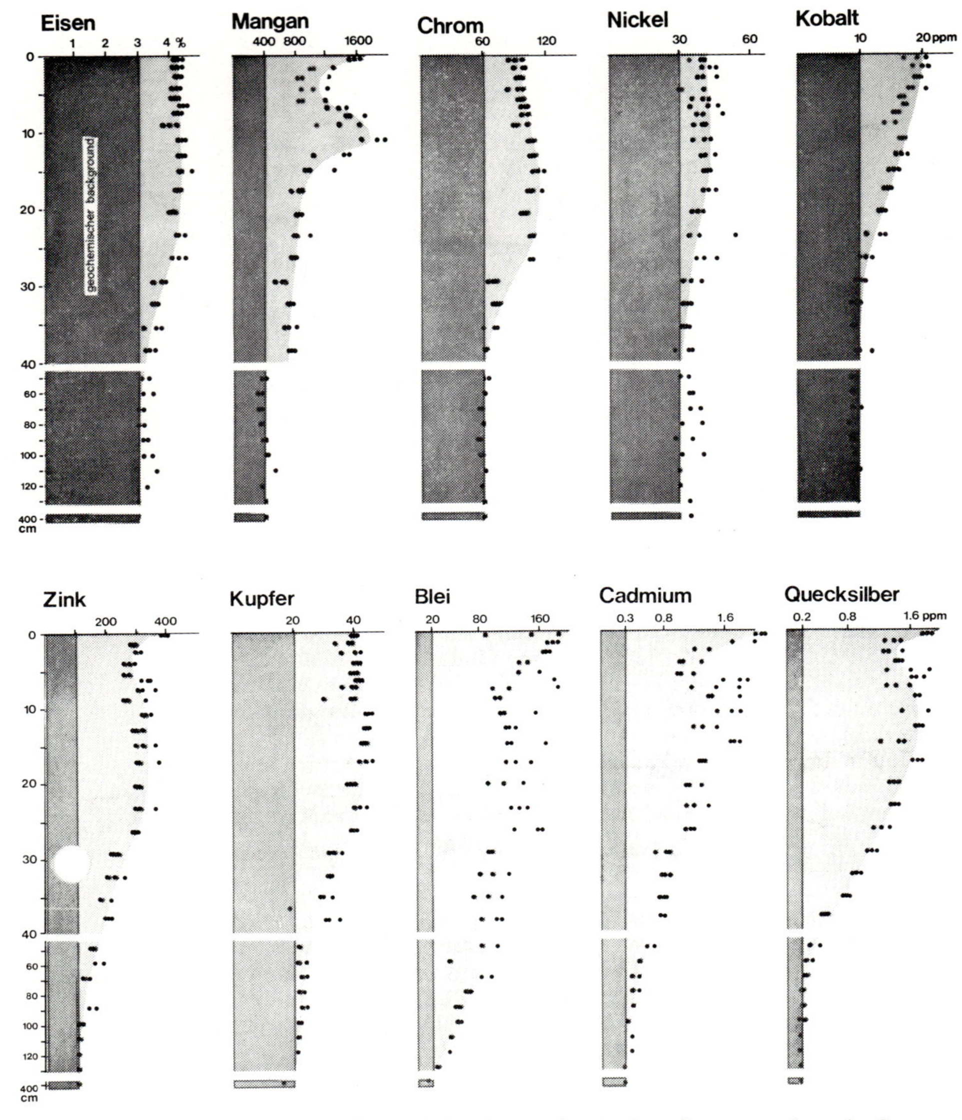


**Fig. 12.** Heavy metal contents in the pelitic fraction (< 2 μm) of a sediment core from the German Bight. (Förstner and Reineck 1974)

the major estuaries. This indicates that the areas in the North Sea where suspended material is being deposited are also the principal areas where heavy metals in particulate form are accumulated in the sea bed. Based on the situation of 1980, an inventory was made of the yearly pollutant load entering the North Sea concerning discharges from the coastal zones between 56° N and the Strait of Dover (Klomp et al. 1986; cf. Sect. 4.3). Figure 13 shows that the Dutch coastal zone contributes

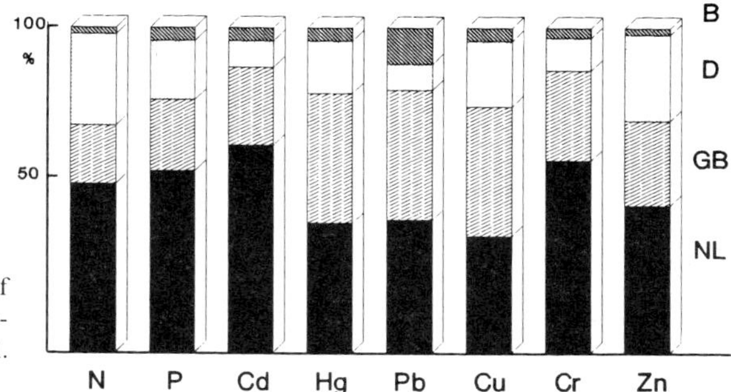


**Fig. 13.** Modeled coastal inputs of trace metals from countries surrounding the North Sea. (Klomp et al. 1986)

considerably (40–50% of all sources from coastal zones) to the pollution of the North Sea. This is closely connected with the cross-boundary of rivers Rhine, Meuse, Ems, and Scheld.

Little or no information could yet be obtained, however, on the possible spreading of the coastal contamination over wider areas of the North Sea. In 1986 bottom samples were taken from the entire North Sea (Kersten and Klatt 1988). This study is based on the analysis of altogether 122 surface sediment samples between the Straits of Dover and 60° N by use of energy-dispersive X-ray fluorescence, i.e., direct analysis of total metal contents. This method has been described and evaluated in a previous paper (Grossmann et al. 1985). Abundance of the $< 20\,\mu$m fraction amounted to between 0.09% and 80% of the total sediment. A coarse sand bed covers most of the survey area. Mud areas are restricted to nearshore locations, notably off major river estuaries, the Skagerrak Basin, and the Norwegian Trench.

Geographical distribution of lead concentrations in total sediment and in the fraction $< 20\,\mu$m are given in Fig. 14a, b. In spite of the low density of sampling sites certain areas of high concentrations stand out. Starting from the shelf edge area off the Hebrides, the general pattern shows an increase in metal concentrations of the sediment fines toward the central North Sea. North of 56° N and between 2° and 6° E, i.e. in the Fisher Bank area, surprisingly high values for Pb were found. Lower concentrations occurred near the coasts except for an area off the NE English coast. In total sediments, on the other hand, high Pb contents corresponds with high proportions of fine-grained sediment as occurring in coastal areas and the Skagerrak basin (Fig. 14b).

The heavy metal concentrations found in the sediment fines were evaluated by the statistical technique of Trefry and Presley (1976) mentioned before. Most anomalously high Pb and Cr values found above the 95% interval are from the central North Sea area. Some of the high Pb points are also located in an elongated area along the NE coast of England off the Tyne and the Tees, but north of the Humber estuary. Webb (1978) has pointed out, however, that mineralized catchment areas drained by these rivers give rise to sediments with an anomalously high metal burden, especially of lead. Most of the anomalously high Zn values were found in a strip of nearshore sediments between the mouth of the Scheld and the SW Danish coast.

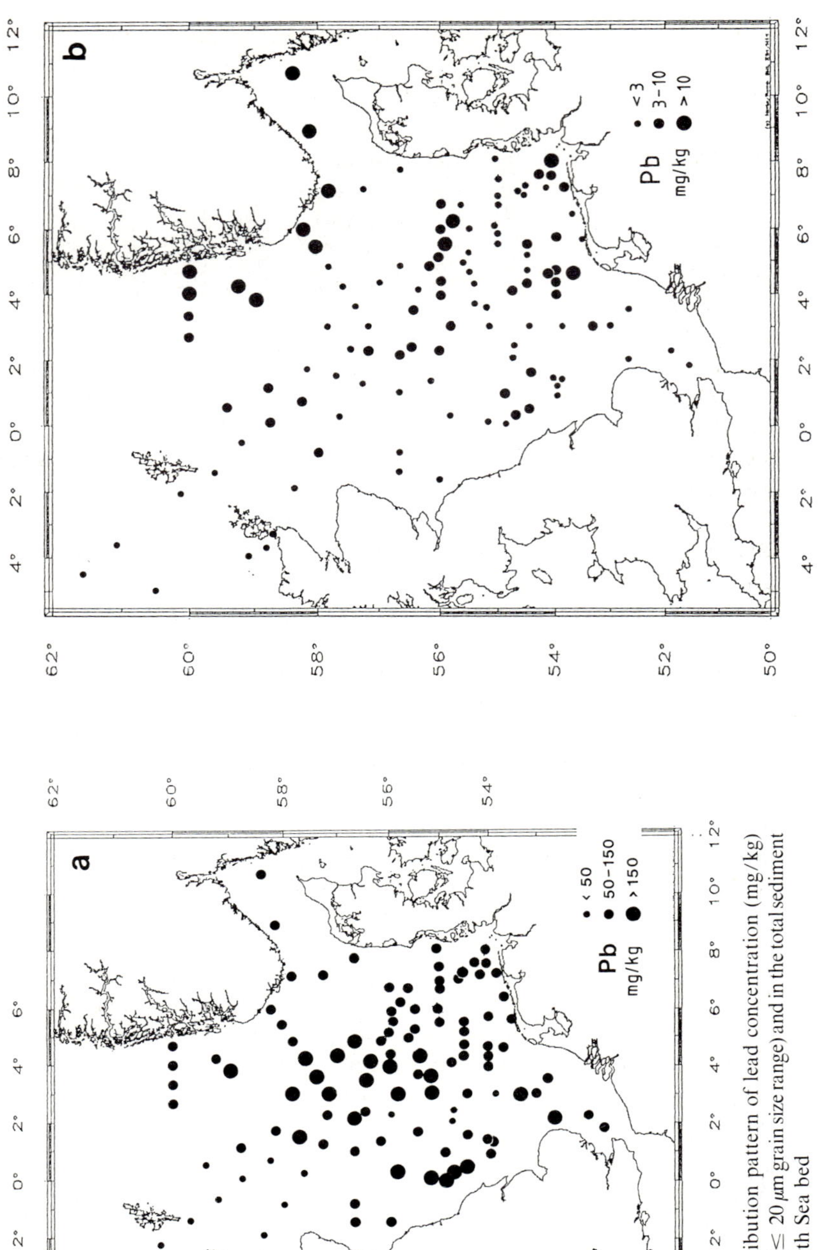


**Fig. 14a,b.** Distribution pattern of lead concentration (mg/kg) in sediment fines ($\leq$ 20 $\mu$m grain size range) and in the total sediment of the entire North Sea bed

Heavy metal enrichment factors were calculated based on the composition of the sediment fines and the low metal concentrations found at the shelf edge off the Hebrides (Kersten and Klatt 1988). They are highest for Pb (up to 5.8 in an area NE off England), followed by Zn (3.7 in the German Bight), Cr (2.7 at the Fisher Bank) and Cu (1.5 in the German Bight). Average enrichment factors calculated for sediment fines of the entire North Sea on basis of pre-industrial sediment composition as reported by Irion and Müller (1987) are somewhat lower: Pb 2.78, Zn 1.63, Cu 1.55, Cd 1.81, Cr 1.17, Ni 0.96, and Co 1.21.

Other factors than known sources will have to be taken into account in order to describe or even to understand the measured distribution. Since the Fisher Bank area is not affected by waste discharges or dumping activities, one may consider atmospheric transport and deposition as a source, at least in the case of lead as discussed above. This area is, however, additionally affected by erosion rather than sedimentation as indicated by the low portions of fines in the sediments (less than 1% by weight). Hence the mode of incorporation of fine particles with their heavy metal load into these surficial sediments is also open for discussion.

As pointed out in the previous section, primary productivity enhanced by upwelling effects in that area may be responsible for heavy metal accumulation on particulates. The fines from the water column are possibly admixed to the underlying coarse sediment bed by mechanisms of biodeposition and bioentrainment rather than by the physical regime alone (cf. Kersten this Vol.). Significant correlations of metal contents and organic carbon, which shows highest concentrations in the sediment fines of the central area, corroborate biological control of heavy metal input into the surficial sediment layers. The high Mn concentrations in sediment fines of that region and the significant correlations between Pb and Mn concentrations on suspended particles indicates also scavenging of dissolved Pb by enlarged formation of hydrous manganese oxides due to an enlarged cycling of Mn between sediments and the water column (Kersten and Klatt 1988). Moreover, in this area the supply of fine-grained lithogenic sediments is low leading to low "dilution effects" with less contaminated sediment fines. Metal concentrations found in surficial fine-grained sediments of the Skagerrak and Norwegian Trench, which are the most important sediment sinks within the North Sea, were found to be less than those of the central North Sea, probably as a result of intermixing with less polluted fines eroded from the slopes and shores surrounding the basins (Weering et al. 1987).

## 6 Conclusion

The conclusions resulting from the present review can be summarized as follows:

1. The most significant result is that clear regional differences exist in the concentrations of heavy metals such as Cd and Pb in the atmosphere, waters, suspended and deposited sediments of the entire North Sea. Atmospheric metal concentrations tend to decrease from land to sea. Metal concentrations in water and metal contents in total sediment tend to decrease from coastal areas to open sea

as well. Metal contents of suspended matter, however, are generally higher for the northern part than for the southern and coastal part of the North Sea, mostly due to relatively lower suspended matter concentrations, and higher contribution of organic matter.

2. A generally low correlation between different major and trace elements in the particulate phase, and a failing correlation of dissolved heavy metals with salinity, confirms the assumption that heavy metals in the open North Sea stem from different (natural and anthropogenic) sources, and are transported by different carrier phases. The influence of the composition of such carrier phases on the heavy metal behavior have been interpreted by a model describing the relationship between particulate metal contents and suspended matter concentrations.

3. Though lead contents on the basis of the whole sediment are relatively low in the central North Sea, the environmental significance (in terms of bioavailability and toxicity) of the observed anomalous elevations in metal content in the sediment fines ($< 20\,\mu m$) is high. These fines found in the bed-forming coarse sands originate possibly from biodeposition and bioentrainment of organic-rich particles. Filter- and deposit-feeding macrobenthos exposed to the contaminated fines may ultimately contribute to dietary uptake of heavy metals in demersal fish resulting in food web transfer and enhanced ecosystem stress far remote from any anthropogenic metal sources.

4. An enrichment is found for lead, and to a lower degree also for cadmium and mercury in form of isolated "hot spots", in several regions of the North Sea waters. Since the processes of aquatic heavy metal transport are yet not well understood, we cannot simply explain these enrichments only in terms of anthropogenic enrichment factors. Progress in this field, and of biogeochemical cycles of trace contaminants in general, demands further development of analytical techniques such as speciation of trace elements in water and particulate phases.

5. To elucidate the question as to what extent biological events determine the level of heavy metals found in North Sea waters, the next step warranted is a comparison of both a winter and a summer data set. As a first approach in this direction the evaluation of seasonal variations found in two cruises of summer 1986 and winter 1987 is in preparation.

6. If the atmospheric deposition gradient is as steep as suggested by several authors, all the sensitive coastal environments like, e.g., the Waddensea area would have to face most of the atmospheric pollution. Furthermore, concerning the strong indications that enhanced concentrations of especially lead in water, suspended and deposited sediments in wide areas of the central North Sea may be due to aeolian transport and deposition, then action should be taken immediately on further air pollution control.

*Acknowledgements.* The authors gratefully acknowledge the skilful and kind assistance given to them during the project by the following: the captains and crews of the RV Gauß and Planet for support during the cruises; K. Backhaus, B. Hussel, V. Klatt, S. Koelling, A. Michel, B. Pohle, K. Przygodda, M. Rieck, H. Sonnenberg, and U. Wendlandt for difficult and tedious sampling and analytical work; P. König for performance of salinity profiles; M. Schroeder for providing us with yet unpublished data of suspended matter and POC concentrations; H. Luthardt for calculation of backward wind trajectories of Fig. 3; and last not least P. Sinclair for improving the English version. Special thanks are due to Profs. Dr. W. Dannecker and Dr. U. Förstner for support, encouragement and critical comments. Prof. Dr. J. Sündermann is thanked for the opportunity to join the multidisciplinary German project Zirkulation und Schadstoffumsatz in der Nordsee (ZISCH) granted by the Bundesministerium für Forschung und Technologie.

# References

Aalst RM van, Ardenne RAM van, Kruk JF de, Lems T (1983) Pollution of the North Sea from the atmosphere. TNO Rep No C182/152, Delft

Anonymous (1982–1987) Überwachung des Meeres. Bericht für das Jahr 1980/1981/1982/1983/1984/1985. Teil II: Daten. 6 DHI Reports, Hamburg

Baeyens W, Gillain G, Decadt G, Elskens I (1987a) Trace metals in the eastern part of the North Sea. Part I: Analyses and short-term distributions. Oceanol Acta 10:169–179

Baeyens W, Gillain G, Ronday F, Dehairs F (1987b) Trace metals in the eastern part of the North Sea. Part II: Flows of Cd, Cu, Hg, Pb and Zn through the coastal area. Oceanol Acta 10:301–309

Baker CW (1977) Mercury in surface waters of seas around the United Kingdom. Nature (Lond) 270:230–232

Balls PW (1985a) Copper, lead and cadmium in coastal waters of the western North Sea. Mar Chem 15:363–378

Balls PW (1985b) Trace metal fronts in Scottish coastal waters. Estuar Coastal Shelf Sci 20:717–728

Balls PW, Topping G (1987) The influence of inputs to the Firth of Forth on the concentrations of trace metals in coastal waters. Environ Pollut 45:159–172

Banat K, Förstner U, Müller G (1972) Schwermetalle in Sedimenten von Donau, Rhein, Ems, Weser und Elbe im Bereich der Bundesrepublik Deutschland. Naturwissenschaften 12:525–528

Blanchard DC (1983) The production, distribution and bacterial enrichment of sea salt aerosol. In: Liss PS, Slinn WGN (eds) Air-to-Sea Exchange of Gases and Particles. Reidel, Doordrecht, pp 407–454

Bowen HJM (1979) Environmental chemistry of the elements. Academic Press, London

Boyle EA, Sclater F, Edmond JM (1976) On the marine geochemistry of cadmium. Nature (Lond) 263:42–44

Brügmann L (1986a) The influence of coastal zone processes on mass balances for trace metals in the Baltic Sea. Rapp P V Réun Cons Int Explor Mer 186:329–342

Brügmann L (1986b) Particulate trace metals in waters of the Baltic Sea and parts of the adjacent NE Atlantic. Beitr Meereskunde 55:3–18

Brügmann L, Danietsson, LG, Magnusson B, Westerlund S (1985) Lead in the North Sea and the North East Atlantic ocean. Mar Chem 16:47–60

Bruland KW (1980) Oceanographic distributions of cadmium, zinc, nickel and copper in the north Pacific. Earth Planet Sci Lett 47:x76–198

Bruland KW (1983) Trace elements in sea-water. In: Riley JP, Chester R (eds) Chemical Oceanography, Vol 8. Academic Press, London, pp 157–220

Bruland KW, Franks RP (1983) Mn, Ni, Cu, Zn, and Cd in the western North Atlantic. In: Wong CS, Boyle E, Bruland KW, Burton JD, Goldberg ED (eds) Trace metals in sea water. Plenum Press, New York, pp 395–414

Bruland KW, Knauer GA, Martin JH (1978) Cadmium in northeast Pacific waters. Limnol Oceanogr 23:618–625

Cambray RS, Jeffries DF, Topping G (1979) The atmospheric input of trace elements to the North sea. Mar Sci Communic 5:175–194

Chester R, Voutsinou FG (1981) The initial assessment of trace metal pollution in coastal sediments. Mar Pollut Bull 12:84–91

Chester R, Murphy KJT, Towner J, Thomas A (1986) The partitioning of elements in crust-dominated marine aerosols. Chem Geol 54:1–15

Danielsson LG, Westerlund S (1985) Short-term variations in trace metal concentrations in the Baltic. Mar Chem 15:273–277

Danielsson LG, Magnusson B, Westerlund S (1985) Cadmium, copper, iron, nickel and zinc in the north-east Atlantic ocean. Mar Chem 17:23–41

Davies JM, Griffiths AH, Leatherland TM, Metcalfe AP (1986) Particulate mercury fluxes in the Forth estuary, Scottland. Papp P V Réun Cons Int Explor Mer 186:301–305

Dedeurwaerder HL, Artaxo P (1987) Composition of ambient aerosols above the North Sea. In: Lindberg SE, Hutchinson TC (eds) Proc Int Conf Heavy metals in the environment Vol 2, New Orleans 1987. CEP Consultants, Edinburgh, pp 131–133

Dedeurwaerder HL, Dehairs FA, Decadt GG, Baeyens WF (1983) Estimates of dry and wet deposition and resuspension fluxes of several trace metals in the Southern Bight of the North Sea. In: Pruppacher HR (ed) Precipitation scavenging, dry deposition, and resuspension. Elsevier, Amsterdam, pp 1219–1231

Dehairs FA, Gillain G, Debondt M, Vandenhout E (1986) The distribution of trace and major elements in Channel and North Sea suspended matter. In: Grieken R van, Wollast R (eds) Proc Progress in Belgian Oceanogr Res, Brussels, March 1985. University Press, Antwerp, pp 136–146

Dicke M, Schmidt D, Michel A (1987) Trace metal distribution in the North Sea. In: Lindberg SE, Hutchinson TC (eds) Proc Int Conf Heavy metals in the environment Vol 2, New Orleans 1987. CEP Consultants, Edinburgh, pp 312–314

Dominik J, Förstner U, Mangini A, Reineck HE (1978) $^{210}$Pb and $^{137}$Cs chronology of heavy metal pollution in a sediment core from the German Bight (North Sea). Senckenb Marit 10:213–227

Duinker JC (1983) Effects of particle size and density on the transport of metals to the oceans. In: Wong CS, Boyle E, Bruland KW, Burton JD, Goldberg ED (eds) Trace metals in sea water. Plenum Press, New York, pp 209–228

Duinker JC, Nolting RF (1982) Dissolved copper, zinc and cadmium in the Southern Bight of the North Sea. Mar Pollut Bull 13:93–96

Eisma D (1981) The mass-balance of suspended matter and associated pollutants in the North Sea. Rapp P V Réun Cons Int Explor Mer 181:7–14

Eisma D, Kalf J (1987) Dispersal, concentration and deposition of suspended matter in the North Sea. J Geol Soc Lond 144:161–178

Erlenkeuser H (1978) The use of radiocarbon in estuarine research. In: Goldberg ED (ed) Biogeochemistry in estuarine sediments. UNESCO, Paris, pp 140–153

Förstner U (1986) Chemical forms and environmental effects of critical elements in solid-waste materials – combustion residues. In: Bernhard M, Brinckman FE, Sadler PJ (eds) The importance of chemical speciation in environmental processes. Springer, Berlin, Heidelberg, New York, Tokyo, pp 465–491

Förstner U, Reineck HE (1974) Die Anreicherung von Spurenelementen in den rezenten Sedimenten eines Profilkerns aus der Deutschen Bucht. Senckenb Marit 6:175–184

Förstner U, Wittmann GTW (1979) Metal pollution in the aquatic environment. Springer, Berlin, Heidelberg, New York, Tokyo

Förstner U, Calmano W, Schoer J (1982) Heavy metals in bottom sediments and suspended material from the Elbe, Weser and Ems estuaries and from the German Bight (south eastern North Sea). Thalassia Jugosl 18:97–122

Förstner U, Calmano W, Schoer J (1985) Verteilung von Spurenmetallen zwischen Lösung und Feststoffen – aktuelle Fragen der Gewässergüte-Praxis an die Sedimentforschung. Vom Wasser 64:1–16

Gadow S, Schäfer A (1973) Die Sedimente der Deutschen Bucht: Korngrößen, Tonmineralien und Schwermetalle. Senckenb Marit 5:165–178

Gerritsen H (1983) Residual currents: a comparison of two methods for the computation of residual currents in the southern half of the North Sea. Delft Hydraulics Lab Rep R1469-III, Delft

Gibbs R (1973) Mechanisms of trace metal transport in rivers. Science 180:71–73

Greenberg RR, Gordon GE, Zoller WH, Jacko RB, Neuendorf DW, Yost KJ (1978) Composition of particles emitted from the Nicosia municipal incinerator. Environ Sci Technol 12:1329–1332

Groot AJ de, Goeij JJM, Zegers C (1971) Contents and behaviour of mercury as compared with other heavy metals in sediments from the rivers Rhine and Ems. Geol Mijnbouw 50:393–398

Grossmann D, Kersten M, Niecke M, Puskeppel A, Voigt R (1985) Determination of trace elements in membrane filter samples of suspended matter by the Hamburg proton microprobe. In: Degens ET, Kempe S, Herrera R (eds) Transport of carbon and minerals in major world rivers, Part 3. Mitt Geol-Paläontol Inst Univ Hamburg, SCOPE/UNEP Sonderbd 58:619–630

Hart BT (1982) Uptake of trace metals by sediments and suspended particulates: a review. Hydrobiologia 91:299–313

Horowitz AJ (1986) Comparison of methods for the concentration of suspended sediment in river water for subsequent chemical analysis. Environ Sci Technol 20:155–160

Irion G, Müller G (1987) Heavy metals in surficial sediments of the North Sea. In: Lindberg SE, Hutchinson TC (eds) Proc Int Conf Heavy metals in the environment Vol 2, New Orleans 1987. CEP Consultants, Edinburgh, pp 38–41

Irion G, Schwedhelm E (1983) Heavy metals in surface sediments of the German Bight and adjoining areas. In: Müller G (ed) Proc Int Conf Heavy metals in the environment Vol 2, Heidelberg 1983. CEP Consultants, Edinburgh, pp 892–895

Jaarsveld JA van, Aalst RM van, Onderdelinden D (1986) Deposition of metals from the atmosphere into the North Sea: model calculations. Nat Inst Public Health Environ Hygiene Rep 842015002, Bilthoven

Jansen JHF, Weering TCE van, Eisma D (1979) Late quaternary sedimentation in the North Sea. In: Oele E, Schüttenhelm RTE, Wiggers AJ (eds) The Quaternary history of the North Sea. Acta Univ Ups, Uppsala, pp 175–187

Jones PGW, Jeffries DF (1983) The distribution of selected trace metals in United Kingdom shelf waters and the North Atlantic. Can J Fish Aquat Sci 40 (Suppl. 2):111–123

Kemp K (1984) Multivariate analysis of elements and $SO_2$ measured at the Danish EMEP stations. Nat Agency Environ Protect Rep MST LUFT-A88, Roskilde

Kersten M (1988) Geochemistry of priority pollutants in anoxic sludges: cadmium, arsenic, methyl mercury, and chlorinated organics. In: Salomons W, Förstner U (eds) Chemistry and biology of solid waste. Springer, Berlin, Heidelberg, New York, Tokyo, pp 170–213

Kersten M, Förstner U (1985) Trace metal partitioning in suspended matter with special reference to pollution in the southeastern North Sea. In: Degens ET, Kempe S, Herrera R (eds) Transport of carbon and minerals in major world rivers, Part 3. Mitt Geol-Paläontol Inst Univ Hamburg, SCOPE/UNEP Sonderbd 58:631–645

Kersten M. Förstner U (1987) Effect of sample pretreatment on the reliability of solid speciation data – implications for the study of early diagenetic processes. Mar Chem 22:299–312

Kersten M, Klatt V (1988) Trace metal inventory and geochemistry of the North Sea shelf sediments. In: Kempe S, Liebezeit G, Dethlefsen V, Harms U (eds) Biogeochemistry and distribution of suspended matter in the North Sea and application to fish biology. Mitt Geol-Paläontol Inst Univ Hamburg 65: (in press)

Klomp R, Pagee JA van, Glas PCG (1986) An integrated approach to analyse the North Sea ecosystem behaviour in relation to waste disposal. In: Kullenberg G (ed) The role of the oceans as a waste disposal option. Reidel, Doordrecht, pp 205–231

Kramer CJM, Vlies LM van der (1983) Heavy metals in sediments of the Dutch Wadden Sea. In: Müller G (ed) Proc Int Conf Heavy metals in the environment Vol 2, Heidelberg 1983. CEP Consultants, Edinburgh, pp 892–895

Krell U, Roeckner E (1987) Simulations of the atmospheric transport and deposition of heavy metals. In: Lindberg SE, Hutchinson TC (eds) Proc Int Conf Heavy metals in the environment Vol 2, New Orleans 1987. CEP Consultants, Edinburgh, pp 26–28

Krell U, Roeckner E (1988) Model simulation of the atmospheric input of lead and cadmium into the North Sea. Atmos Environ 22:375–381

Kremling K (1983) Trace metal fronts in European shelf waters. Nature (Lond) 303:225–227

Kremling K (1985) The distribution of cadmium, copper, nickel, manganese, and aluminium in surface waters of the open Atlantic and European shelf area. Deep-Sea Res 32:531–555

Kremling K, Hydes D (1988) Summer distribution of dissolved Al, Cd, Co, Cu, Mn and Ni in surface waters around the British Isles. Cont Shelf Res 8:89–105

Kremling K, Wenk A, Pohl C (1987) Summer distribution of dissolved Cd, Co, Cu, Mn and Ni in central North Sea. DHZ 40:103–114

Kretzschmar JG (1979) A five-year survey of some heavy metal levels in air at the Belgian North Sea coast. Atmos Environ 13:267

Lasaga AC (1980) The kinetic treatment of geochemical cycles. Geochim Cosmochim Acta 44:815–828

Lu X, Johnson WK, Wong CS (1986) Seasonal replenishment of mercury in a coastal fjord by its intermittent anoxicity. Mar Pollut Bull 17:263–267

Magnusson B, Westerlund S (1983) Trace metal levels in sea water from the Skagerrak and the Kattegat. In: Wong CS, Boyle E, Bruland KW, Burton JD, Goldberg ED (eds) Trace metals in sea water. Plenum Press, New York, pp 467–474
Mart L, Nürnberg HW (1986) Cd, Pb, Cu, Ni and Co distribution in the German Bight. Mar Chem 18:197–213
Mart L. Nürnberg HW, Valenta P (1980) Prevention of contamination and other accuracy risks in voltammetric trace metal analysis of natural waters. III: Voltammetric ultratrace analysis with a multicell system designed for clean bench working. Fresenius Z Anal Chem 300:350–362
Mart L, Rützel H, Klahre P, Sipos L, Platzek U, Valenta P, Nürnberg HW (1982) Comparative studies on the distribution of heavy metals in the oceans and coastal waters. Sci Total Environ 26:1–17
May K, Stoeppler M (1983) Studies on the biogeochemical cycle of mercury I. Mercury in sea and inland water and food products. In: Müller G (ed) Proc Int Conf Heavy metals in the environment Vol 1, Heidelberg 1983. CEP Consultants, Edinburgh, pp 241–244
Michaelis W, Stößel RP (1986) Untersuchungen zur Schwermetalldeposition auf der Insel Pellworm – Ein Beitrag zur Ermittlung des atmosphärischen Schadstoffeintrags in die Nordsee. GKSS-Jahresber 1986, Geesthacht, pp 8–23
Mukherji P, Kester DR (1979) Mercury distribution in the gulf stream. Science 204:64–66
Müller-Navarra S, Mittelstaedt E (1987) Schadstoffausbreitung und Schadstoffbelastung in der Nordsee – Eine Modellstudie. DHZ Ergänzungsheft 18, Deutsch Hydrogr Inst Hamburg
Nicholson RA, Moore PJ (1981) The distribution of heavy metals in the superficial sediments of the North Sea. Rapp P V Réun Cons Int Explor Mer 181:35–48
Nolting RF (1986) Copper, zinc, cadmium, nickel, iron and manganese in the Southern Bight of the North Sea. Mar Pollut Bull 17:113–117
Nolting RF, Eisma D (1988) Elementary composition of suspended matter in the North Sea. Neth J Sea Res 22: (in press)
Olafsson J (1983) Mercury concentrations in the North Atlantic in relation to cadmium, aluminium and oceanographic parameters. In: Wong CS, Boyle E, Bruland KW, Burton JD, Goldberg ED (eds) Trace metals in sea water. Plenum Press, New York, pp 475–485
Pacyna JM (1984) Estimation of the atmospheric emissions of trace elements from anthropogenic sources in Europe. Atmos Environ 18: 41–50
Pacyna JM (1985) Spatial distributions of the As, Cd, Cu, Pb, and Zn emissions in Europe within a 1.5° grid net. Norw Inst Air Res Rep No 60–85, Zillestrom
Patterson CC, Settle DM (1987) Review of data on eolian fluxes of industrial and natural lead to the lands and seas in remote regions on a global scale. Mar Chem 22:137–162
Peirson DH, Cawse PA, Cambray RS (1974) Trace elements in the atmospheric environment. Nature (Lond) 251:675
Price NB, Skei JM (1975) Areal and seasonal variations in the chemistry of suspended particulate matter in a deep water fjord. Estuar Coastal Mar Sci 3:349–369
Salomons W, Eysink W (1981) Pathways of mud and particulate trace metals from rivers to the Southern North Sea. In: Nio SD, Schuettenhelm RTE, Weering TCE van (eds) Holocene marine sedimentation in the North Sea Basin. Spec Publ Int Assoc Sedimentol 5:429–450
Salomons W, Förstner U (1984) Metals in the hydrocycle. Springer, Berlin, Heidelberg, New York, Tokyo
Sandstrom H, Elliott JA (1984) Internal tide and solitons on the Scotian shelf: a nutrient pump at work. J Geophys Res 89:6415–6426
Santschi PH, Yuan Hui Li, Carson SR (1980) The fate of trace metals in Narragansett Bay, Rhode Island: Radiotracer experiments in microcosms. Estuar Coastal Mar Sci 10:635–654
Santschi PH, Amdurer M, Adler D, O'Hara P, Yuan Hui Li, Doering P (1987) Relative mobility of radioactive trace elements across the sediment-water interface in the MERL model ecosystems of Narragansett Bay. J Mar Res 45:1007–1048
Schaule BK, Patterson CC (1983) Perturbations of the natural depth profile in the Sargasso Sea by industrial lead. Proc NATO Adv Res Inst Trace metals in seawater, Erice 1981. Plenum Press, New York, pp 487–504
Schmidt D (1976) Distribution of seven trace metals in sea water of the inner German Bight. ICES C.M. 1976/C:10
Schmidt D (1980) Comparison of trace heavy-metal levels from monitoring in the German Bight and the southwestern Baltic Sea. Helgol Meeresunters 33:576–586

Schmidt D (1983) Neuere Erkenntnisse und Methoden zur Probenahme von Meerwasser für die Ultraspurenanalyse von Schwermetallen. Fresenius Z Anal Chem 316:566–571
Schmidt D, Dicke M (1987) Trace determination of mercury in the water column: two surveys covering the entire North Sea. In: Lindberg SE, Hutchinson TC (eds) Proc Int Conf Heavy metals in the environment Vol 2, New Orleans 1987. CEP Consultants, Edinburgh, pp 315–317
Schmidt D, Freimann P (1984) AAS-Ultraspurenbestimmung von Quecksilber im Meerwasser der Nordsee, der Ostsee und des Nordmeers. Fresenius Z Anal Chem 317:385–387
Schmidt D, Freimann P, Zehle H (1986) Changes in trace metal levels in the coastal zone of the German Bight. Rapp P V Réun Cons Int Explor Mer 186:321–328
Schönfeld W, Schmidt D, Radach G (1988) Spatial and temporal variability of heavy metal concentrations in the German Bight. DHZ 41: (in press)
Sclater FR, Boyle EA, Edmond JM (1976) On the marine geochemistry of nickel. Earth Planet Sci Lett 31:119–128
Sipos L, Nürnberg HW, Valenta P, Braniça M (1980) The reliable determination of mercury traces in sea water by subtractive differential pulse voltammetry at the twin gold electrode. Anal Chim Acta 115:25–42
Slinn WGN (1983) Air-to-sea transfer of particles. In: Liss PS, Slinn WGN (eds) Air-to-sea exchange of gases and particles. Reidel, Doordrecht, pp 299–405
Stößel RP (1987) Untersuchungen zur Naß- und Trockendeposition von Schwermetallen auf der Insel Pellworm. GKSS Rep 87/E/34, Geesthacht
Stumm W, Morgan JJ (1981) Aquatic chemistry. Wiley, New York
Topping G, Bewers JM, Jones PGW (1980) A review of past and present measurements of selected trace metal in sea water in the Oslo Commission and ICNAF (NAFO) areas. Coop Res Rep Cons Int Explor Mer 97, Copenhagen
Trefry JH, Presley JB (1976) Heavy metals in sediments from San Antonio Bay and the northwest Gulf of Mexico. Environ Geol 1:282–294
Turekian KK (1977) The fate of metals in the oceans. Geochim Cosmochim Acta 41:1139–1144
Watson JG (1984) Overview of receptor model principles. J Air Pollut Control Assoc 34:619–623
Webb JS (1978) The Wolfson Geochemical Atlas of England and Wales. Clarendon, Oxford
Weering TCE van, Berger GW, Kalf J (1987) Recent sediment accumulation in the Skagerrak, northeastern North Sea. Neth J Sea Res 21:177–189
Weichart G (1973) Pollution of the North Sea. Ambio 2:99–106
Whitby KT (1978) The physical characteristics of sulfur aerosols. Atmos Env 12:135–159
Williams RM (1982) Model for the dry deposition of particles to natural water surfaces. Atmos Env 16:1933–1938
Wollast R (1982) Methodology of research in micropollutants – heavy metals. Wat Sci Tech 14:107–125

# North Sea Nutrients and Eutrophication

U. BROCKMANN[1], G. BILLEN[2], and W.W.C. GIESKES[3]

## 1 Introduction

The term "nutrient" refers to the biologically available inorganic forms of nitrogen, phosphorus, and silica. Unlike carbon, these elements, which are among the major constituents of living organisms, are often present in short supply with respect to the needs of petoplankton in marine ecosystems. Because of this, primary production and phytoplankton standing crop are highly dependent on nutrient concentrations and the processes regenerating nutrients from the organic compounds where they have been incorporated. Input of nutrients from external sources also considerably affects primary producers and the whole ecosystem, sometimes in an undesirable way causing "eutrophication" effects.

The seasonal variation of nutrient concentrations observed in the open North Sea, with maximum values during winter and exhaustion in the upper layer during the growth season, indicates that phytoplankton cells come into contact during the period of one season with nearly all molecular ensembles in the mixed layer. By this way they will have interacted also with dissolved pesticides and heavy metals which can be particularized by uptake, sorption to biomass or coagulation with released organic substances. Pollutants which are fixed to particles will be transported differently in comparison to the dissolved fraction, part of them will sink to the lower layer, or reach the sediment.

Horizontal nutrient gradients, observed in coastal areas from spring to autumn, are indicative of the spreading of river plumes and coastal inputs. During this time of the year, when nutrients are more or less exhausted in the mixed layer offshore, the nutrient load of river plumes remains available to algae. At this time of the year, but also during autumn when thermocline breakup enables nutrient input from deep layers, exceptional phytoplankton blooms can occur. Eutrophication effects caused by such events as well as by permanent production enhancement will be discussed.

[1]Institut für Biochemie und Lebensmittelchemie der Universität Hamburg, Martin-Luther-King-Platz 6, D-2000 Hamburg 13, FRG
[2]Université Libre de Bruxelles, Campus de la Plaine CP221, B-1050 Bruxelles, Belgium
[3]State University of Groningen, Dept. Marine Biology, Biological Center, Box 14, 9750 AA Haren (GN.), The Netherlands

## 2 Processes Affecting Nutrient Concentrations Within the Pelagic North Sea Ecosystem

A simplified diagram of the fluxes of nutrients (mainly N and P) through marine ecosystems is represented in Fig. 1. The quantity of fluxes within the system is, of course, not the same in different regions, and is also variable from season to season. The interactions described in the diagram affect nutrient distribution in the sea. The main pathways of the elements nitrogen and phosphorus within the planktic ecosystem are from nutrients to phytoplankton, and then either to sediments directly to micro- and macro-zooplankton (including bacteria) or to dissolved organic substances.

Uptake by phytoplankton is the main process causing nutrient concentrations to decrease. Accordingly, nutrients levels are generally highest in late winter, when primary production is minimal mainly due to low light intensity. The onset of spring phytoplankton development corresponds to the moment when its specific growth rate in the photic zone (which is dependent on available light, duration of photo period and temperature) matches sinking and dilution in the lower aphotic zone due to turbulence of the water column. Gieskes and Kraay (1975) have found that the growing season starts when the mean light intensity in the mixed water column is more than 0.03 gcal $cm^{-2}$ $min^{-1}$. In the deeper regions this level can only be reached after stratification has become established. Generally speaking, the phytoplankton crop starts to increase in February-March in shallow regions, in April in the open North Sea, and in May in the most turbid zones. Thereafter, i.e., when sufficient light is available, the rapid growth of phytoplankton (the "spring bloom") usually results in exhaustion of at least one of the required nutrients. From

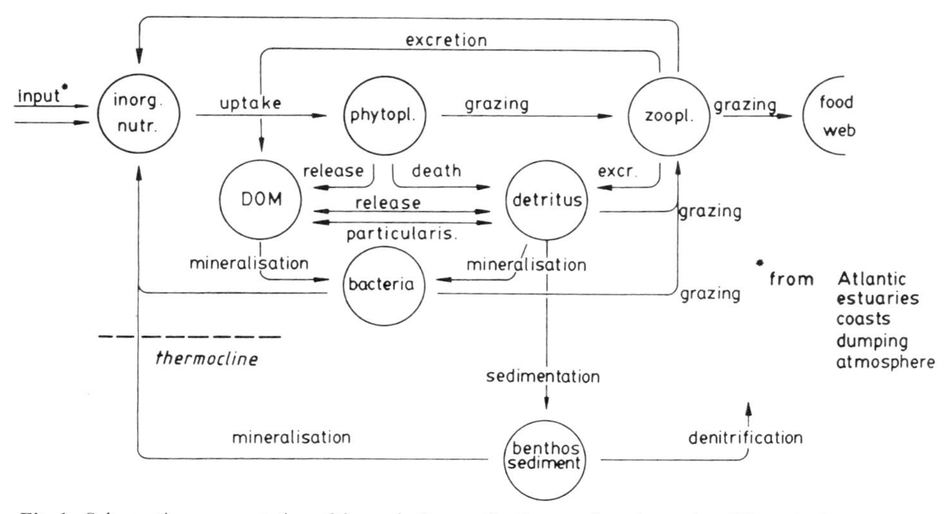


**Fig. 1.** Schematic representation of the main fluxes affecting nutrient dynamics of the pelagic ecosystem

that moment on, and during the whole summer, growth of phytoplankton is controlled by the rate of nutrient regeneration and input to the photic zone.

Concerning the abundance and succession of phytoplankton species it is of interest to know which among the different nutrients is actually limiting phytoplankton growth. N and P are used as a mean in a molar ratio 16:1 (Redfield et al. 1963). Silicate is only required by diatoms which take it up in a molar ratio 1.4:1 with respect to nitrogen (Stefánson and Richards 1963). Comparison of the N:P:Si ratios in the standing stock of nutrients with these ideal Redfield ratios has often been used for assessing the limiting nutrient. As has been discussed by Ryther and Dunstan (1971), however, the ratios between the fluxes of nutrient supply are much more indicative.

Part of the primary production is released as dissolved organic matter either by direct excretion or secretion of extracellular organic molecules (Whittle 1977; Fogg 1977; Gagosian and Lee 1981; Lancelot 1983, 1984) or by cellular lysis. This material is usually rapidly taken up and mineralized by bacterioplankton, which follows the blooms of phytoplankton with only a few days delay (Lancelot and Billen 1984; Billen and Fontigny, 1987). Owing to this rapid response of bacteria, dissolved organic nitrogen and phosphorus are most of the time maintained near steady-state concentration (Billen et al. 1980) and peaks of these substances, when they were observed, were generally short-lived (Riley and Segar 1970; Wafar et al. 1984). The important pool of dissolved organic bound nutrient elements (Jackson and Williams 1985) has been analyzed in no more than a few investigations.

Zooplankton release dissolved organic substances too, namely during grazing. Zooplankton grazing starts immediately after the beginning of growth of phytoplankton, but zooplankton blooming occurs days or weeks later than the phytoplankton maximum peak (Fransz and Gieskes 1984). As shown by several authors (Fransz and Gieskes 1984; Radach et al. 1984; Joiris et al. 1982) for coastal as well as for open sea areas, zooplankton grazing matches phytoplankton production only during summer months. In the spring and autumn the primary production is not kept under control by herbivores. About one third of the grazed biomass will be partly mineralized by the zooplankton and excreted as urea, ammonia or phosphate (Steele 1974). By this way nutrients will be made available to phytoplankton growth within minutes, so that short-cycled nitrogen and phosphorus fluxes will be established during summer. A well-balanced phytoplankton growth results, with a steady and usually rather low biomass level.

Phytoplankton blooms may also occur in the autumn, as a response to the disruption of stratification so that nutrient-rich bottom waters reach the photic zone. They are also observed at any time during the flowering period at frontal systems, or in coastal regions as a response to terrestrial nutrient discharge. These blooms, sometimes involving very high levels of biomass, constitute the main visible aspect of eutrophication.

Finally, much of the particulate material produced by photosynthesis may reach the sediment. A characteristic feature of the North Sea ecosystem in many areas is the significant sedimentation of phytoplankton cells during the spring blooms. This was observed for the northern deep areas (Fladen Ground), where 20–25% of primary production reached the bottom (Davies and Payne 1984). Also in the shallower southern zones sedimentation occurs (Billen 1978; Joiris et al.

1982; Peinert et al. 1982; Rutgers van der Loeff 1980b). These processes are very variable. For example, during 1983 most of the primarily produced material in the Fladen Ground area ended in the DOC pool and only 1% sank out (Cadée 1986).

Zones of minimum bottom erosion energy are sites of mud accumulation (Adam et al. 1981) and preferential sites for organic matter deposition even from adjacent zones. However, observations by Jenness and Duineveld (1985) showed that considerable amounts of phytoplankton material are also incorporated into sandy sediments, in tidally active places where no mud deposition occurs, by a mechanism involving transient deposition during slack current periods, followed by burying down to at least 5 cm through ripple formation during ebb current periods. This explains that up to 35–50% of the nutrient taken up by primary production is recycled in the sediments in the Belgian and Dutch coastal zone (Billen 1978; Rutgers van der Loeff 1980b).

Recent studies on the kinetics of organic matter mineralization in the sediments indicate a rapid response (within a day) to deposition of fresh planktonic material during or at the end of blooms, superimposed to a much steadier level of activity, remaining significant even during the winter. In unstratified areas, benthos therefore contributes nutrients to the pelagic systems directly by rapid remineralization mechanisms as well as by slow release following long-term storage (Nixon 1981; Billen and Lancelot 1987).

Regeneration of dissolved silica is a much slower process in the water column, mainly dependent on the rate of dissolution of diatoms frustules. Van Bennekom et al. (1974) presented evidence that benthos is more important than the planktonic phase for the regeneration of silica.

In stratified areas, the thermocline acts as a barrier between the benthos and the photic zone for the exchange of dissolved substances. However, rapid dislocations of water masses by gales, combined with a break-up of the thermocline during autumn, sometimes cause a sudden mixture of bottom and surface water, sometimes even enhancing interstitial water exchanges (Smetacek et al. 1976). Such events can lead to sudden increases of nutrient concentrations in surface water as observed, for instance, by Brockmann and Eberlein (1986) in the German Bight in August, 1982.

## 3 External Sources of Nutrients

Input of nutrients into the North Sea is from the Atlantic, the estuaries, from coastal discharges, and from the atmosphere. Assuming a net residual transport through the Channel of $5.3 \times 10^3$ km$^3$ a$^{-1}$ (Prandle 1984) (Table 1) and mean annual concentrations of total P of 0.5 $\mu$g at dm$^{-3}$, total N of 9.5 $\mu$g at dm$^{-3}$, and Si of 2.5 $\mu$g at dm$^{-3}$ (calculated from Pingree et al. 1977), inputs to the Southern bight are 82,000 t P, 705,000 t N and 371,000 t Si.

The in and output at the northern boundary has been calculated to be about $50 \times 10^3$ km$^3$ a$^{-1}$ each (Anonymous 1986; ICES 1983; see Table 1). These values are different from those calculated by Lee (1980). For the inflow and outflow of nutrients different layers must be distinguished, at least during summer when

**Table 1.** North Sea in- and outflows. (Anonymous 1986 p 42, ICES 1983)

| | Inflows | | Outflows |
|---|---|---|---|
| | | $10^3$ km$^3$ a$^{-1}$ | |
| Orkney – Shetland | 9.5 | | – |
| Shetland – Norway | 40.0 | | > 57 |
| Channel | 5.3[a] | | – |
| Baltic | 0.5 | | 0.5 |

[a] 1973–1980 means from Prandle 1984.

surface layers are nutrient-poor, while bottom layers are much richer in nutrient contents. Hydrographic data are not available for a transport calculation of nutrient elements in different layers, which are rather variable depending on season, location, and nutrient concentration (Dooley 1983). Most of the water masses that enter the North Sea will reach the Atlantic again within a short time, but a certain proportion will be integrated in current systems moving anticlockwise round half of the North Sea. The fate of water masses is dependent on meteorological events (Backhaus 1984). For budget studies it would therefore be advisable to calculate fluxes within defined time steps with actual real-time data sets.

The North Sea receives the freshwater of rivers draining one of the most densely populated and industrialized areas in the world. A compilation of the most recent estimates of freshwater, N, P, and Si inputs of the major rivers discharging into the North Sea is presented in Table 2. Due to differences in calculation between the different sources cited, these figures are not always consistent. Also, due to variations in freshwater discharge of the rivers, the input figures can vary by up to a factor of 2 from year to year.

As shown in Table 2, the terrestrial input of nutrients into the North Sea is higher than the input from the Atlantic through the Straits of Dover. Terrestrial input must therefore be expected to be of great significance to both function and structure of the ecosystem, particularly in the Southern Bight and the German Bight, where most of the total discharge is concentrated. The N:P:Si ratio in the terrestrial input is also indicated in Table 2. As a whole, terrestrial inputs are severely depleted in Si with respect to the requirements of diatoms.

A regular increase of nutrient discharge by estuaries has been observed for the last 20 years. This increase has been particularly rapid for phosphorus and quite significant also for nitrogen, while silica, which is not subject to human discharge, has remained essentially constant or decreased due to increased uptake in the river systems. These trends are especially well documented for the river Rhine (van Bennekom et al. 1975; van Bennekom and Salomons 1981; Postma 1978; Gerlach 1985). The outflow of phosphorus by the Rhine increased from 3000 t a$^{-1}$ in 1932 to 30,000 t a$^{-1}$ in 1970 and 48,000 t a$^{-1}$ in the 1980's. For nitrogen the corresponding figures are 77,000, 340,000 and 398,000 t a$^{-1}$.

The input of nutrients from the atmosphere is not well known. A value of 3.2 $\mu$g at N m$^{-3}$ s$^{-1}$ is calculated by Rutgers van der Loeff (1980b); this is about $0.8 \times 10^6$ tN a$^{-1}$ for the whole North Sea. Goldberg (1973) cites a figure of $1 \times 10^6$ tN a$^{-1}$. Atmospheric input of P and Si is generally considered negligible.

**Table 2.** Terrestrial nutrient input into the North Sea

| | | Water discharge $m^3 s^{-1}$ | $SiO_2$ $10^3$ | N $t a^{-1}$ | P | Reference | Atomic ratio Si : N : P |
|---|---|---|---|---|---|---|---|
| B | Scheldt estuary | 100 | 42 | 27 | 2.2 | 1;5;6 | 10 : 27 : 1 |
| | Schipdonk canal | | | 8 | | | |
| | Yser estuary | 5 | | 4 | | | |
| | Coastal watershed | | | 9 | | | |
| | | | | 48 | | | |
| NL | New Waterway | 1525 | | 260 | 34 | 2 | |
| | Haringvliet | 900 | | 120 | 12 | 2 | |
| | North Sea canal | 81 | | 18 | 2 | 2 | |
| | Direct discharges | | | 10 | 2 | 2 | |
| | | | 410 | 408 | 50 | | 4.4 : 18 : 1 |
| | Lake Ijssel | 606 | | 66 | 4 | 2 | 36 : 1 |
| | Ems-Dollard | 120 | | 42 | 3 | 2 | 31 : 1 |
| D | Weser | 500 | | 42 | 8.5 | 3 | 11 : 1 |
| | Elbe | 1150 | | 250 | 14 | 3 | 40 : 1 |
| DK | | | | 22 | 2.4 | 4 | 20 : 1 |
| UK | North Sea inputs | | | | | | |
| | Rivers | 1223 | | 111 | 3.5 | 3 | 70 : 1 |
| | Sewage, industrial discharges | 266 | | 73 | 21 | | 8 : 1 |
| | Dumping | 260,000 | | 11[a] | 2.5[a] | | 10 : 1 |
| | | | | 195 | 27 | | |
| | Channel inputs | | | | | | |
| | Rivers | 163 | | 26 | 0.8 | 3 | 72 : 1 |
| | Sewage, industrial discharges | 16 | | 15 | [b] | | |
| | Dumping | 14,132 | | [b] | [b] | | |

[a] Sewage sludge figures only.
[b] No data.

1 Billen et al. (1985).
2 Data communicated to the consultation meeting on nutrients in the Eastern and Southern North Sea. Skagerrak and Kattegat under art. 9 of the Paris convention (Copenhagen 28–29 Nov. 1985) by the Dutch delegation.
3 Anonymous 1986 in Carlson 1986.
4 id. by the Danish delegation.
5 Wollast R (1982).
6 van Bennekom et al. (1975).

Nitrogen fixation by algae is of little significance for the nutrient budget of North Sea marine ecosystems. Recent measurements showed significant $N_2$ fixation rates in organic-rich marine sediments (Capone 1988), but unfortunately no data are available for assessing the rate of this process in the North Sea.

## 4 Sinks of Nutrients

Besides the outflow and the recirculation at the northern boundary, long-term losses of nutrients from the water column occur mainly through burial of sedimented material in mud accumulation areas and through exchange with the atmosphere after transformation into gaseous forms. Remineralization processes have been well studied (e.g., Krom and Berner 1981), and also release of nutrients to the water column has been described (Smetacek et al. 1976; Rutgers van der Leoff et al. 1981; Nixon 1981), but it is still difficult – if not impossible – to establish budgets of nutrient element cycling including fluxes to and from the sediment (Kelderman 1980; Mommaerts et al. 1984).

Mass balances of supply and deposition of suspended material in the North Sea have been performed recently (Eisma 1981a,b) giving broad estimates of fluxes. The supply from primary production processes was assumed to be $10^6$ t (dry weight) $a^{-1}$. Total supply and outflow with deposition was in the range of $35–50 \times 10^6$ t per year. Deposition was found to be limited to a number of small areas: southwest of Doggerbank, in estuaries, tidal flats, and small areas in the German Bight, but mainly in the Skagerrak/Kattegat area. The fate of nutrient elements that are deposited is not known for the different sites. The Skagerrak acts as a sink for recent sediments which are transported in suspension or by traction currents along the bottom (van Weering 1981). The amount of carbon in the clay and silt fraction of the Skagerrak is much higher than can be expected from sedimentation of biomass produced in this area. The high concentrations indicate therefore a net transport from the southern North Sea, where the nutrient elements were fixed in biomass.

Gaseous losses of nutrients have only to be taken in consideration for nitrogen. Photolytic degradation of nitrate and nitrite into nitrous or nitric oxide has been described by Scranton (1983). The process of denitrification is without doubt of more significance for the nitrogen budget of marine ecosystems (Hattori 1983). Rates of denitrification in the range of 5–70 mg N $m^{-2}$ $day^{-1}$ were determined in Belgian, Dutch, and Danish coastal sediments representing 8–23% of the amount of nitrogen mineralization in the benthos (see review by Billen and Lancelot 1986).

Rutgers van der Loeff (1980b) found a denitrification rate of about 10 mg N $m^{-2}$ $day^{-1}$ in offshore sandy sediments. Extrapolating this value to the whole bottom of the North Sea would give $2 \times 10^6$ tN $a^{-1}$. This process, which increases with increase of discharge (Seitzinger and Nixon 1985), apparently causes a very significant loss of nitrogen, which could partly offset the impact of terrestrial nitrogen input to the North Sea. A similar situation is found in the Baltic sea, where it has been estimated that denitrification in the deeper water and sediments of the Baltic proper eliminates about half the total input of nitrogen to this area (Rönner 1985). Note that denitrification in estuaries and river systems is also of significance in reducing the nitrogen load carried to the sea (Nixon 1981). An extreme case is that of the Scheldt hydrographical network, where 50% of the total input is eliminated by denitrification in anoxic river sediments and in the water column before reaching the sea (Billen et al. 1985, 1986).

## 5 Distribution of Nutrients: Hydrographic Characterization of the North Sea

The large differences in hydrographical regime of the regional ecosystems in the North Sea strongly affect the nutrient distribution. In the central North Sea a thermocline develops during summer, interrupted only by occasional upwelling which brings nutrient-rich water to the surface. In flat and coastal areas the water column remains well mixed throughout the summer, mostly by tidal effects (Pingree et al. 1968). Frontal systems develop at the boundaries between well-mixed and stratified areas. Near the estuaries the influx of river plumes influences coastal areas. The coastal zone along the Wadden Sea is influenced by irregular interaction with the mud flats that are covered with water in a tidal rhythm. Coastal currents induce additional variability in nutrient distribution. Following the mean residual circulation pattern and considering the different flushing times (ICES 1983), nutrients and ecological aspects will be discussed in the various regions of the North Sea.

### 5.1 The Central North Sea

The winter situation in the central North Sea is characterized by a vertically well-mixed water column with high nutrient concentrations. The distributions observed in February, 1984, by Brockmann and Wegner (1985) (Fig. 2) are in general agreement with the maps published earlier by Johnston (1973) and Postma (1978), although these are based on means of several years of winter investigations. One striking difference with the values documented earlier is the Doggerbank area in which minima for all nutrients were found in 1984. Since a chlorophyll maximum was detected in the same area (Brockmann and Wegner 1985), we suggest that primary production continues throughout the winter in this shallow region with clear water, because here the vertical turbulence is limited by water depths of less than 40 m, allowing the phytoplankton a longer stay in the euphotic zone and exposure to sufficient light. This nutrient minimum in the Doggerbank area was documented only for nitrate by Johnston (1973) more westerly for phosphate, but not for silicate; Postma (1978) did not find this minimum because his presentation was based on the data set of Johnston and Jones (1965). We may safely state that this nutrient minimum is characteristic for the Doggerbank in the winter season; it was again found in February 1985, again coupled with a chlo-

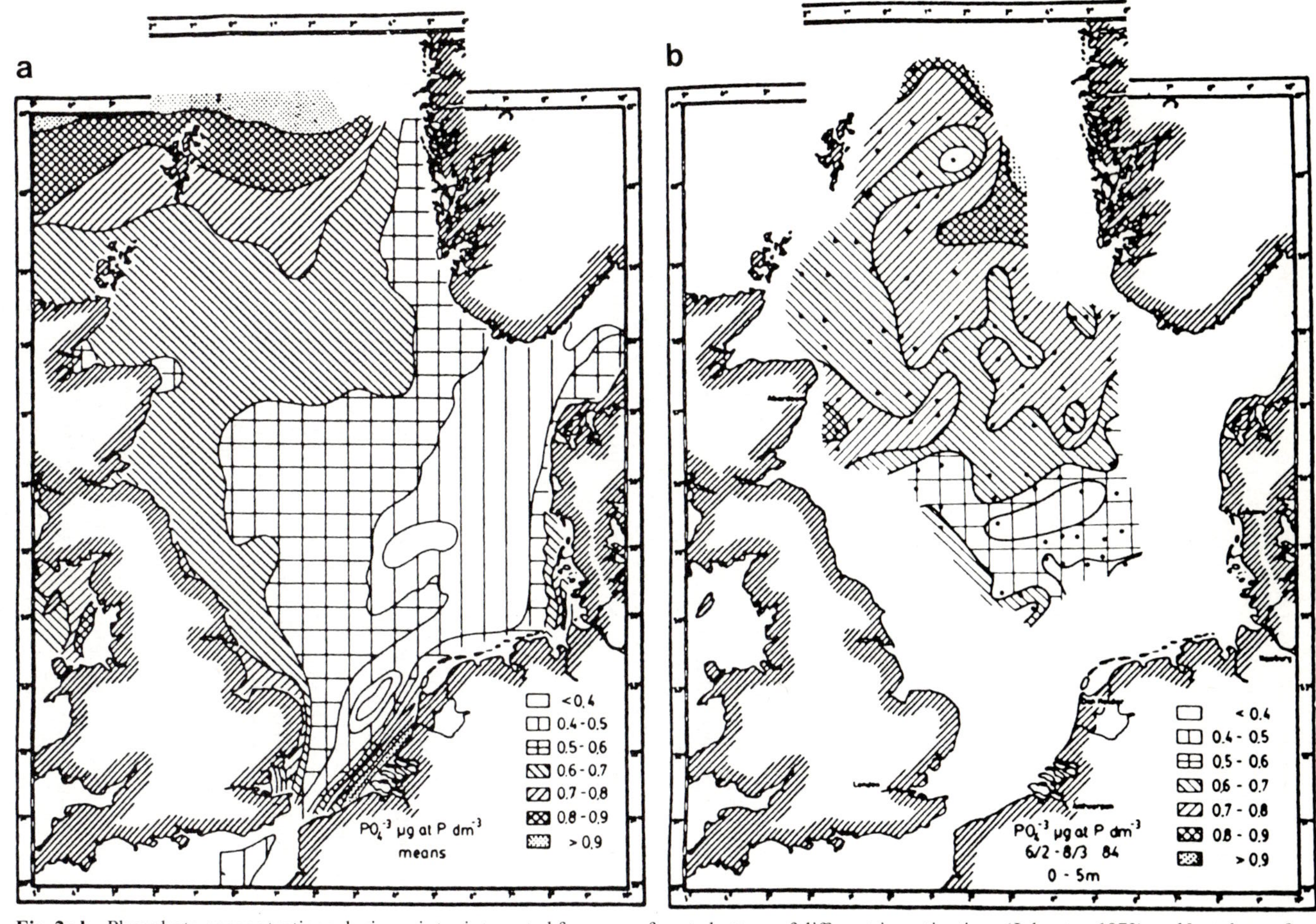


**Fig. 2a,b.** Phosphate concentrations during winter integrated from **a** surface to bottom of different investigations (Johnston 1973) and **b** at the surface during one cruise in February, 1984. Sampling stations are indicated by *points*. (Brockmann and Wegner 1985)

rophyll maximum. As the nutrient analysis in 1984 was combined with a hydrographic investigation, hydrodynamic reasons for the inhomogenous nutrient distribution can be given as well.

Backhaus (1984) has shown that the North Sea circulation is very variable and dependent on the actual wind stress during a few days. Therefore snapshots of nutrient distribution during one cruise in combination with hydrographic measurements are a valuable tool to recognize the variability of horizontal distribution of nutrients and accordingly areas of different potential productivity.

The phosphate and nitrate concentrations were highest in the Atlantic influx from the Norwegian Deep, spreading east into the northern Central North Sea (Brockmann and Wegner 1985). The Fair Isles current brought less nutrient-rich water into the North Sea, as also observed by Johnston (1973). The reason for this lower nutrient content is the origin of these water masses in coastal areas west of Britain. This origin is very well documented by $^{137}Cs$ measurements (Kautsky 1985) tracking the way of radioactive waste from the Irish Sea (Sellafield or Windscale) around Scotland along the east coast reaching the Central North Sea.

Around mid-April, heating of surface water creates a temperature gradient which results in the formation of a thermocline. The biological events initiated at that period of time are illustrated by Fig. 3, combining results obtained in the north of the Central North Sea during FLEX 1976 that were within the range of natural variability, observed in this area before (Cushing 1983). Simultaneous with the formation of the thermocline, an increase of chlorophyll concentration was observed. A few days later, phosphate and nitrate decreased (Eberlein et al. 1980) and were even depleted after about 10 days, while stagnation of chlorophyll contents indicated the end of phytoplankton growth. A second phytoplankton bloom was observed by late May. During the first bloom, diatoms such as *Chaetoceros* dominated, during the second bloom flagellates formed the main part of the phytoplankton standing crop (Wandschneider 1980, 1983). Maxima of about 12 $\mu g$ Chl a $dm^{-3}$ (i.e., about 4000 mg C $m^{-2}$) were reached (Gieskes and Kraay 1980). Zooplankton came to maximum concentrations during the end of May (Krause and Trahms 1983). The bacteria increased already during the phytoplankton maximum, in response to the release of dissolved organic substances from the phytoplankton (Hentzschel 1980; Wandschneider 1983; Brockmann et al. 1983).

Already during early May a slight increase of phosphate of about 0.05 $\mu g$ at P $dm^{-3}$ was observed in the bottom layer in comparison to the values found in March (Eberlein et al. 1980). Since there was also a significant increase of ammonia during this time at all depths from 0.2 to 1.2 $\mu g$ at N $dm^{-3}$ (see also Fig. 4), these nutrient increases can be attributed to remineralization processes starting in the lower layer immediately after the first phytoplankton maximum.

The fate of nitrogen was followed by investigation of vertical profiles at selected dates (Fig. 4) (Brockmann et al. 1983). Before the beginning of the 1976 phytoplankton spring bloom the dominating nitrogen compartment was nitrate, next to dissolved amino acid compounds. During the first phytoplankton spring bloom of 1976 nitrogen was mainly fixed in particulate material. About 30% of primary production was sedimenting between 24 April and 19 May, 1976 (Davies and Payne 1984). However, in 1983 only 1% of the primary production of the spring bloom reached the bottom layer (Cadée 1986). It is possible that the different

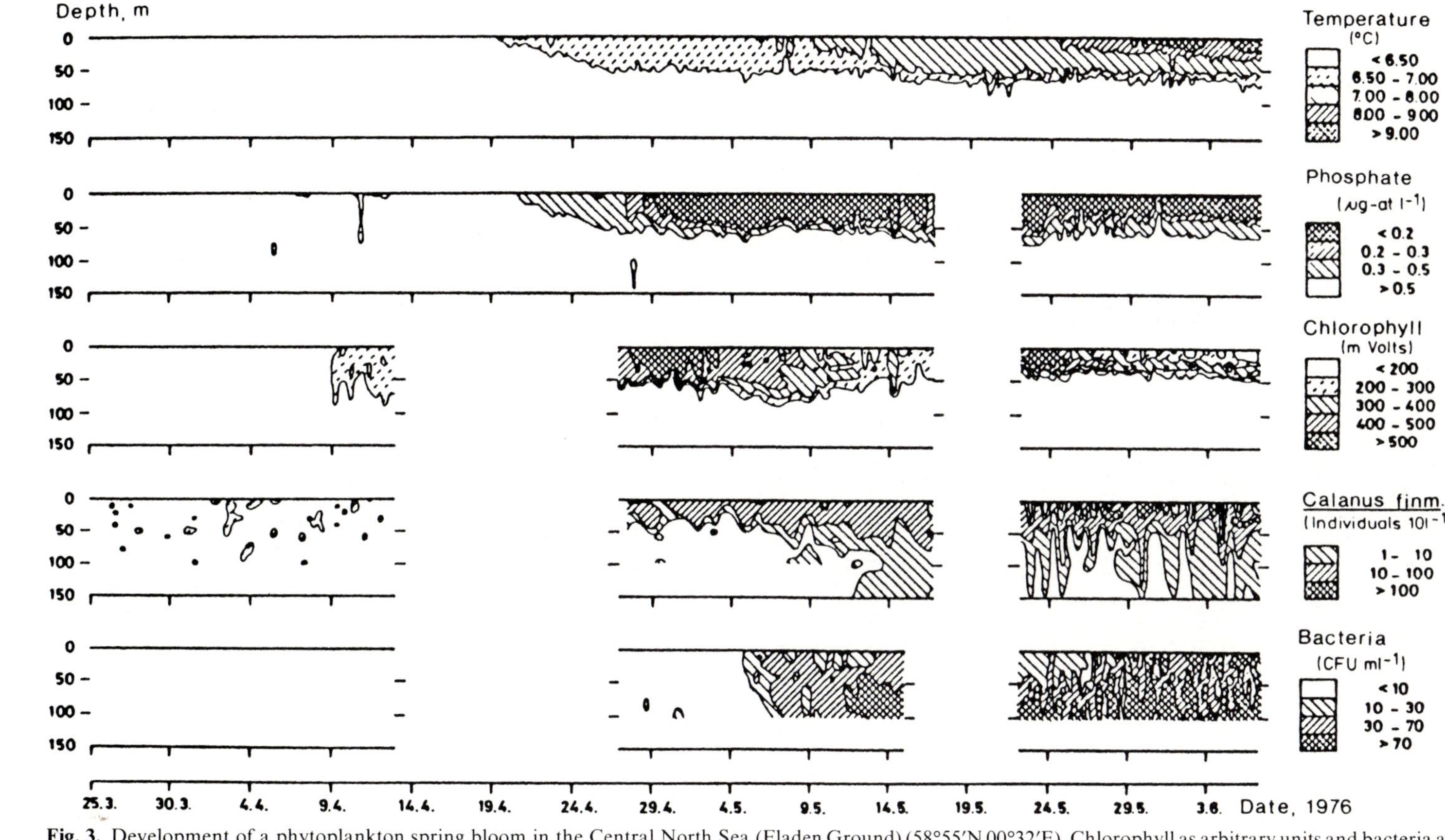


**Fig. 3.** Development of a phytoplankton spring bloom in the Central North Sea (Fladen Ground) (58°55′N 00°32′E). Chlorophyll as arbitrary units and bacteria as colony forming units (CFU). For graphical reasons the *hatching* of phosphate concentrations is *inverse*: light at high and dark at low concentrations. (Ittekkot et al. 1981. Data from Brockmann et al. 1985)

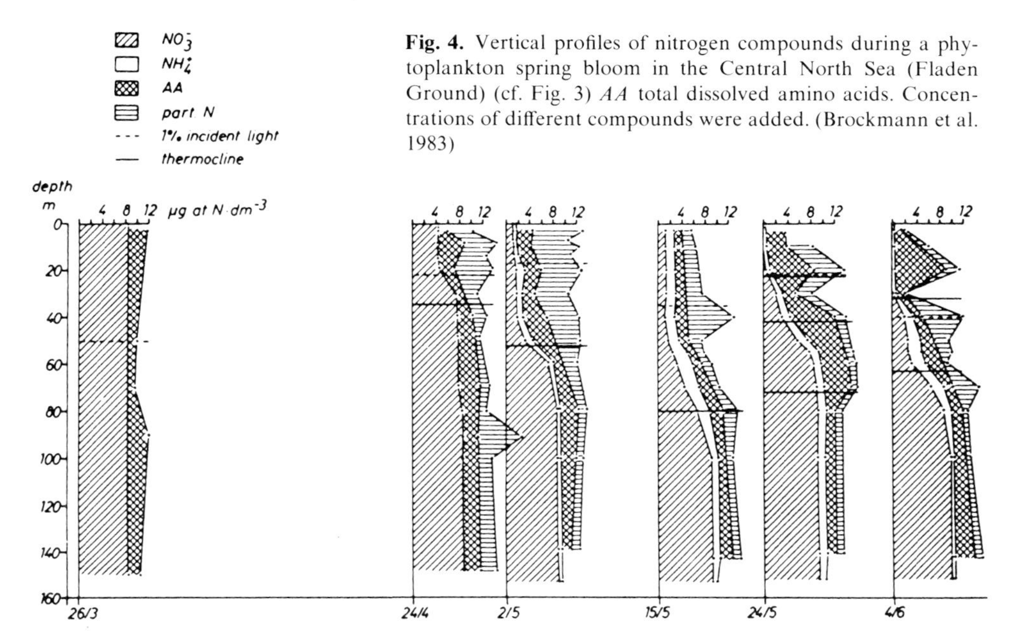


**Fig. 4.** Vertical profiles of nitrogen compounds during a phytoplankton spring bloom in the Central North Sea (Fladen Ground) (cf. Fig. 3) *AA* total dissolved amino acids. Concentrations of different compounds were added. (Brockmann et al. 1983)

species composition of the 1983 bloom (*Thalassiosira conferta* and *Corymbellus aureus*) (see Gieskes and Kraay 1986) caused this difference in sedimentation. Within a few days after the phytoplankton bloom of 1976, detectable ammonia concentrations appeared, released by remineralization processes. Nitrate concentrations in the mixed layer were reduced and exhausted already on 24 May. During the following "oligotrophic" period, dissolved organic substances were probably the most important sources of nutrient elements. In the course of primary production processes a considerable amount of biosynthesized products are released (Brockmann et al. 1983). In the mixed layer of the last two vertical profiles (Fig. 4) most nitrogen was bound in amino acid compounds. After degradation of particulate nitrogen or dissolved proteinaceous substances, which were enriched at density gradients, rapid utilization of released nutrients in the mixed layer is likely, so an increase of nutrients cannot be detected.

Organic-bound nitrogen can also be taken up as nitrogen source directly by phytoplankton, e.g., urea (Hellebust and Lewin 1977), but the uptake of other organic nitrogeneous compounds seems to be ecologically insignificant under natural conditions (Paul 1983). The transitory dominance of the dissolved organic nitrogen fraction indicates the importance of these compounds within the nitrogen cycle.

It is a pity that data sets including both nutrients and dissolved and particulate organic nitrogen and phosphorus are very rare. Therefore, during summer when organic-bound nutrient elements are essential in ecological balances, the nutrient situation is difficult to assess in the absence of information on the organic stock or rate measurements.

The period of stratification in the central and northern North Sea lasts from spring to autumn when the thermocline is broken by surface cooling and gales.

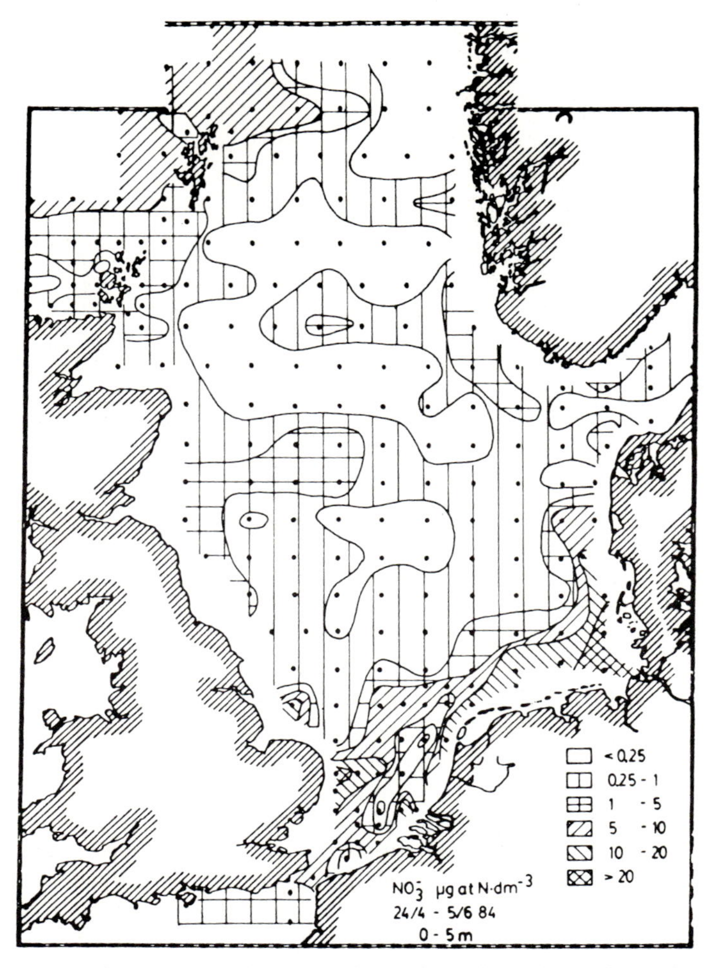


**Fig. 5.** Nitrate concentrations during late spring at the surface of the North Sea during one cruise from 24/4–5/6/84. Sampling stations are indicated by *points*. (Brockmann and Kattner 1985)

During this period the mixed layer in large areas of the North Sea is nutrient-exhausted as shown by late spring nitrate distribution, which dropped below 1 μg at dm$^{-3}$ in most of the North Sea (Fig. 5). Besides the influx of nutrient-rich Atlantic water the coastal and estuarine runoff will become more evident after May, detected at the estuaries of Humber and Thames, Rhine and Elbe/Weser, e.g., in May 1984. Primary production now depends on regenerated sources of nutrient and intermittent supply of "new" nutrients into the surface layer from the lower layer which occurs by turbulent diffusion (King and Devol 1979; Klein and Coste 1984), or by breaking of internal waves, upwelling of nutrient-rich water at fronts (Pingree and Griffiths 1978) or shelf breaks (Pingree et al. 1968). Some spots of increased nutrient concentrations within the exhausted mixed layer in the surroundings indicate such events (Fig. 5).

The nitrate/phosphate ratio, which was close to the Redfield ratio of 16 in the whole Central North Sea in the winter (except in the Doggerbank area, with values of about 8) drops during the flowering period to values smaller than 3 in surface water of large areas of the Central North Sea (Fig. 6). Only in the adjacent Atlantic and small upwelling areas along the Norwegian coast were atomic ratios above 10 found. These ratios were also detected in small areas in the Channel and the Humber estuary as well as in the frontal system between water masses coming through the Channel and re-circulating Central North Sea water. Due to the high nitrate load from the estuaries, nitrate to phosphate ratios increased along the continental coast to values of more than 40.

The situation had not changed very much during July, when only some small areas with N:P ratios higher than 10 were detected, perhaps formed by vertical exchange at spots of interrupted thermocline. Also near the bottom (within 10 m distance) the atomic ratio of nitrate to phosphate was reduced during spring and even more so during summer in large areas of the southern North Sea. This process started from the Doggerbank area, spreading during July.

Other evidence of the dominating biological influence on nutrient distribution is the ammonia concentration, analyzed in samples near the bottom (Fig. 7). In the Central North Sea only concentrations below 1 $\mu g$ at $dm^{-3}$ were detected during February 1984 with maxima near the Shetland and Orkney Islands (Brockmann and Wegner 1985). During May 1984 the bottom concentrations of ammonia had increased to more than 3 $\mu g$ at $dm^{-3}$ (Brockmann and Kattner 1985). Besides the maximum in the northern North Sea, which had increased and spread easterly along 60° latitude N, a second maximum was found in the Central North Sea covering the bottom from Britain to Norway with the contour of a wedge. The latter maximum was found in the circulating Central North Sea water passing the British east coast and turning northeast to south Norway (Wegner 1985). Both maxima were formed by remineralization of biogenic material which had sedimented to the bottom layer. In the Doggerbank area a minimum of ammonium concentrations was observed which indicates short-cycled nutrient fluxes between remineralizing organisms and phytoplankton.

## 5.2 The Scottish Coast

The Scottish coast is characterized by relatively deep water ($>$ 50 m) with a southerly drift, transporting water masses from the Atlantic and the British west coast (Johnston 1973; Kautsky 1985). The annual inflow is calculated as $9.5 \times 10^3$ $km^3$ (ICES 1983; see Table 1). There is only little fresh water inflow: about 65 $m^3$ $s^{-1}$ from the Forth (Anonymous 1986). In summer the coastal water is stratified and frontal systems occur at the boundaries with the mixed near-shore water (Pingree et al. 1968). Due to tidal mixing near the Orkney Islands, relatively high nutrient concentrations were observed here throughout the year (Burns 1967; Burns and Johnston 1968; Johnston and Jones 1965), but further to the south nutrients are depleted during summer, with gradients from east to west showing lowest concentrations along the shore. Apparently, nutrient discharge from the coast is significant in summer. The discharge of the Firth of Forth could clearly be detected in 1984 (Brockmann and Kattner 1985).

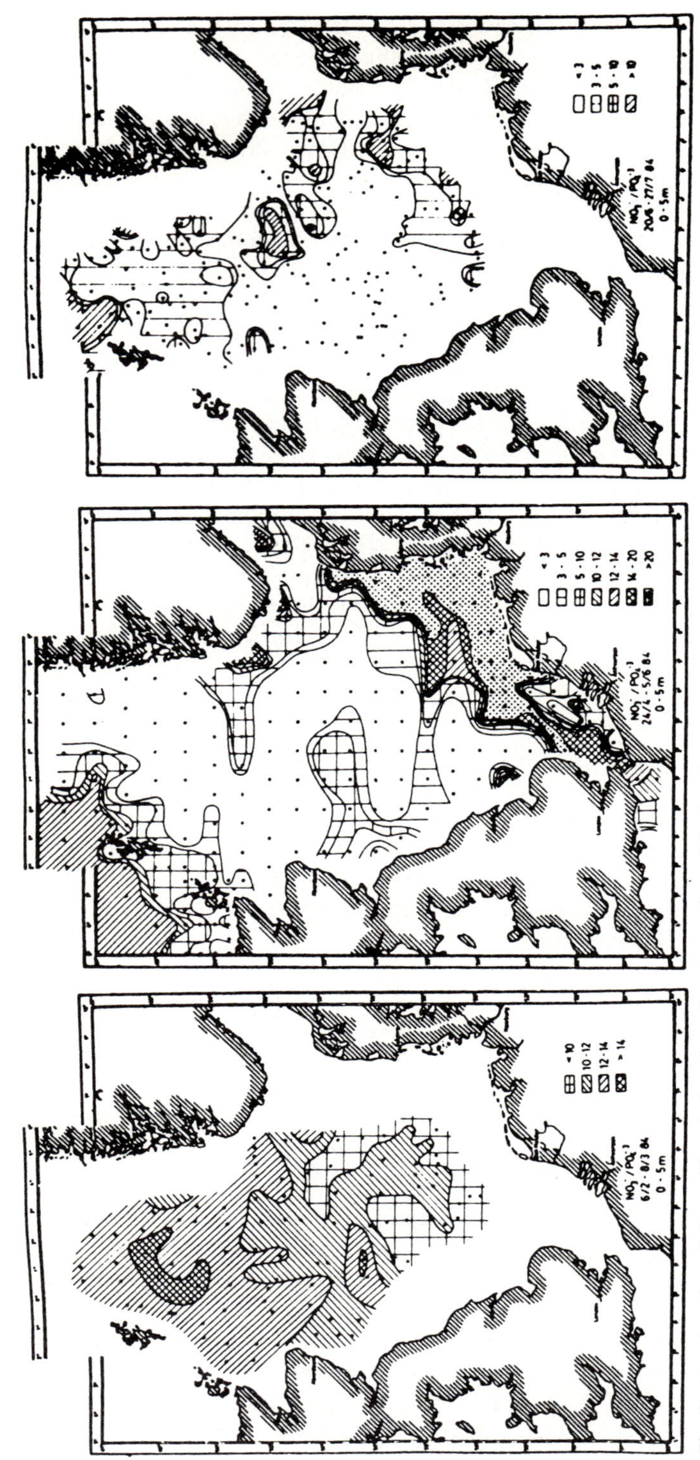

**Fig. 6.** Atomic ratios of nitrate-N and phosphate-P at the surface of the North Sea during three cruises in 1984. Sampling stations are indicated by *points*. (Brockmann and Kattner 1985)

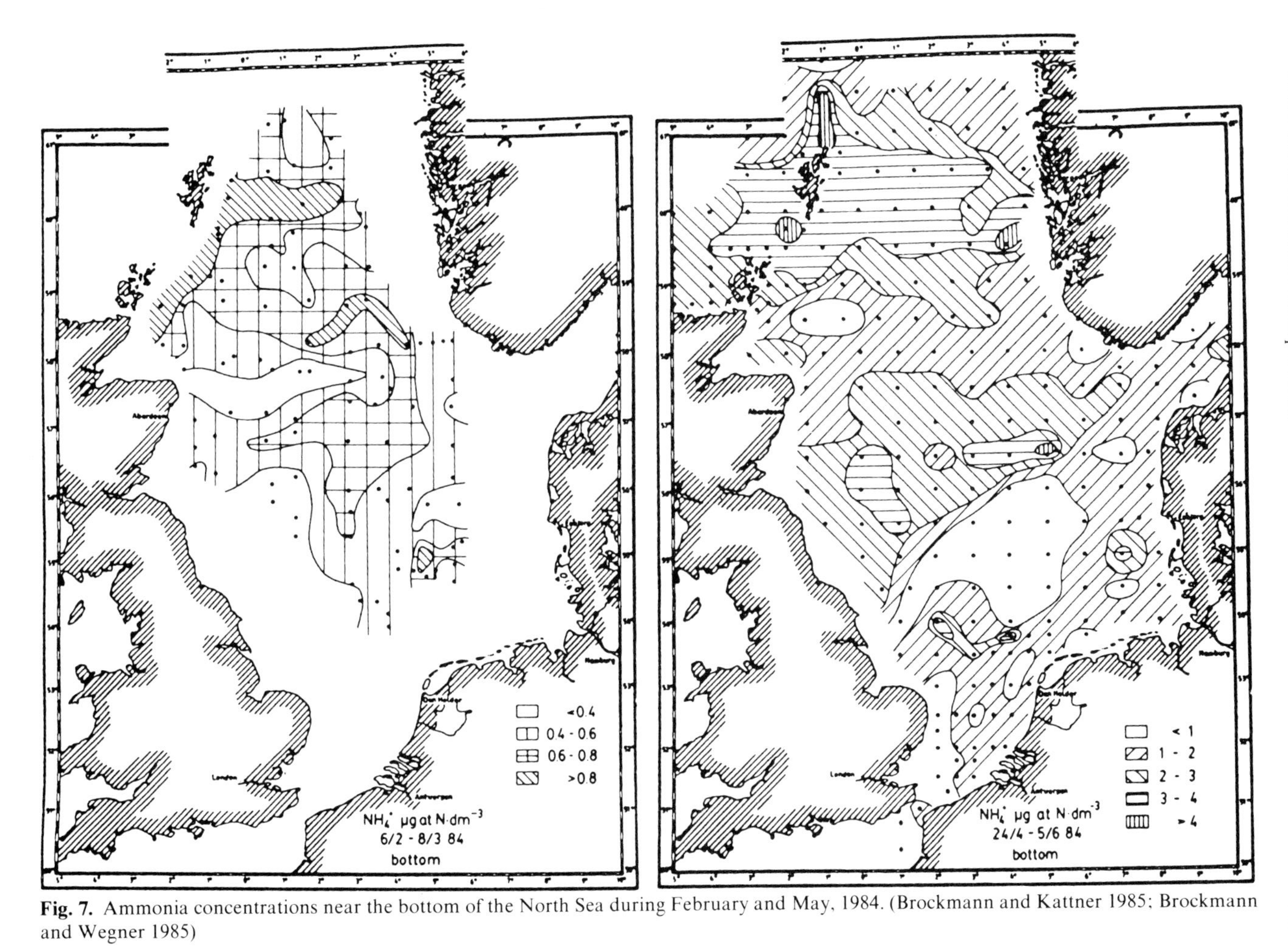


**Fig. 7.** Ammonia concentrations near the bottom of the North Sea during February and May, 1984. (Brockmann and Kattner 1985; Brockmann and Wegner 1985)

### 5.3 The English Coast

The English coastal water is more shallow to the south ($<$ 50 m) with increasing fresh water influx dominated by Humber (280 $m^3 s^{-1}$) and Thames (154 $m^3 s^{-1}$) (Anonymous 1986) with discharges rich in nutrients. Means of tidal limit for the Thames are 900 $\mu g$ at N $dm^{-3}$ and 100 $\mu g$ at P $dm^{-3}$ (Jones 1982). Gradients of nutrients from coast to sea are often observed (Johnston and Jones 1965; Brockmann and Kattner 1985).

Frontal systems between mixed coastal water and stratified regions also occur at the English east coast, stretching across the southern North Sea (Pingree et al. 1968). Pingree et al. (1968) described frontal systems at Flamborough Head with nutrient gradients following the temperature and salinity gradients. The meandering fronts were also reported by Harding et al. (1978).

### 5.4 The Channel Inflow

Due to the eastward residual current the nutrient situation in the western Channel is important for the Southern Bight of the North Sea. During summer the concentrations west of the Channel (50°2′N, 4°22′W) are reduced: Phosphate from 0.45 to 0.2 $\mu g$ at P $dm^{-3}$, silicate from 3 to 1.5 $\mu g$ at Si $dm^{-3}$ and nitrate from 5 to 0.5 $\mu g$ at N $dm^{-3}$ as has been estimated from long-time means (Butler et al. 1979), but during summer the dissolved organic phosphorus increased from 0.15 to 0.25 $\mu g$ at P $dm^{-3}$ the dissolved organic nitrogen from 4.5 to 9 $\mu g$ at N $dm^{-3}$; so total dissolved nitrogen remained at a constant level. Since the nutrient concentrations in this area are distributed patchily (Armstrong et al. 1974) and differ considerably from year to year (Armstrong and Butler 1968), the nutrient input from the Channel is very variable, taking into account the scale of monthly mean residual flows between 114 and $205 \times 10^3$ $m^3 s^{-1}$ (Prandle 1984).

Another variation is introduced by changing coastal discharges which are transported along the shorelines, whereas the Atlantic inflow can be followed in the center of the Channel to the southern Bight (Holligan et al. 1978).

### 5.5 The Southern Continental Coastal Waters

The continental coastal area from the Straits of Dover to the German Bight and the western Danish coast is characterized by a residual circulation of the water masses directed to the northeast, and high tidal mixing preventing stratification of the shallow water column. The Channel outlet is characterized during spring to autumn by inflow of oligotrophic Atlantic water (Armstrong and Butler 1968; van Bennekom et al. 1975; Pingree et al. 1977) which is only slightly enriched by the discharge of the Seine and Somme estuaries. This nutrient-poor water meets nutrient-loaded, less haline water in the opposite regions of the Thames and Rhine estuaries, where strong horizontal nutrient gradients are found (van Bennekom et al. 1975; Holligan et al. 1978) (Figs. 5, 6). The nutrient distribution is very patchy in this area (Johnston and Jones 1965; Tijssen 1968, 1969, 1970; Gieskes 1974). Some values of dissolved organic phosphorus were reported for this area (Tijssen 1968) which were in the same range as phosphate-phosphorus.

In front of the Belgian coast, the water masses from the Channel are mixed with the highly polluted Scheldt waters. Although the discharge of the Scheldt is rather low, an extended area is influenced by this estuary, caused by a residual gyre (Nihoul and Ronday 1975), which elongates the residence time of the water masses. As a result, mean winter concentrations up to 30–40 $\mu$mol $dm^{-3}$ nitrate and 3 $\mu$mol $dm^{-3}$ phosphate are reached, and a considerable phytoplankton spring bloom develops in many years, reaching higher levels and lasting for a longer period than in the Channel (Fig. 8).

The Dutch coastal zone is dominated by the inflow from the Rhine. Winter levels of nitrate and phosphorus are typically 60–70 and 3–4 $\mu$mol $dm^{-3}$, respectively. Phytoplankton biomass reaches high values, often up to 30 $\mu$g Chl a $dm^{-3}$, and sometimes even 50, in the spring. The biomass remains around 10 $\mu$g Chl a $dm^{-3}$ during the summer (Fig. 8; cf. Gieskes and Kraay 1975, 1977a). Both the Belgian and the Dutch coastal waters, however, are characterized by high turbidity (Gieskes and Kraay 1975, 1977a) which strongly limits primary production (Gieskes and Kraay 1975) and hence causes spreading of the nutrient discharges from the estuaries to larger areas due to reduced uptake. Postma (1978, 1981) and Gieskes and Kraay (1975, 1977a) reported the start of the early spring bloom in a persistent turbidity minimum which is located between the salinity maximum of Atlantic water and the Dutch coast (Fig. 9). The nutrient minima move to the west later, corresponding with the minimum of suspended matter. The high turbidity in the Belgian and southern Dutch coastal zone may partly explain that phytoplankton development is often highest in the Northern Dutch zone (also influenced by the Wadden Sea) although the nutrient levels are lower there (Fig. 8).

## 5.6 Wadden Sea

From the Dutch coast to Esbjerg, exchange of North Sea water with the Wadden Sea influences the coastal ecosystem. Due to the complicated morphology, tidal currents, changing wind stress and river runoff cause complex hydrodynamic processes affecting not only the exchange of water masses but also changes of sediment structure by variation of resuspension and sedimentation. Concerning the nutrient regime in the North Sea the most important function of the Wadden Sea is the transient and permanent storage of particulate nutrient elements.

As a whole, the Wadden Sea acts as a trap for particulate organic material produced in the coastal zone or brought there from the estuaries. This material represents as a mean 18% of the total input of particulate organic matter in the Wadden Sea (Postma 1981). Intense mineralization occurs in the tidal flat area, particularly in the summer, causing enrichment of phosphate, ammonia and silicate of the adjacent coastal water (Helder 1974; De Jonge and Postma 1974; Postma 1981). Because of the presence of organic-rich sediments, intense benthic denitrification takes place causing a net loss of nitrate, particularly during winter

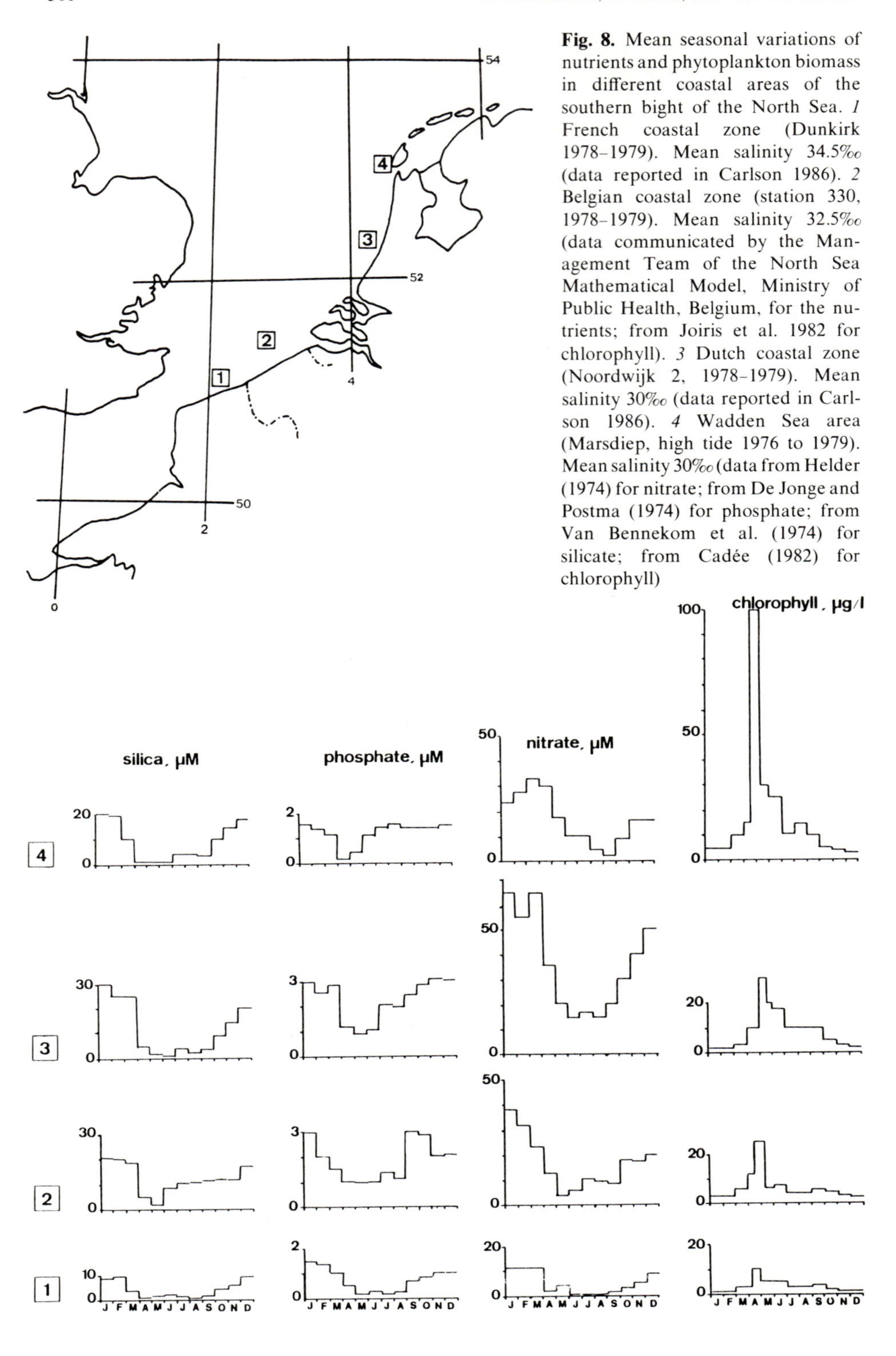


**Fig. 8.** Mean seasonal variations of nutrients and phytoplankton biomass in different coastal areas of the southern bight of the North Sea. *1* French coastal zone (Dunkirk 1978–1979). Mean salinity 34.5‰ (data reported in Carlson 1986). *2* Belgian coastal zone (station 330, 1978–1979). Mean salinity 32.5‰ (data communicated by the Management Team of the North Sea Mathematical Model, Ministry of Public Health, Belgium, for the nutrients; from Joiris et al. 1982 for chlorophyll). *3* Dutch coastal zone (Noordwijk 2, 1978–1979). Mean salinity 30‰ (data reported in Carlson 1986). *4* Wadden Sea area (Marsdiep, high tide 1976 to 1979). Mean salinity 30‰ (data from Helder (1974) for nitrate; from De Jonge and Postma (1974) for phosphate; from Van Bennekom et al. (1974) for silicate; from Cadée (1982) for chlorophyll)

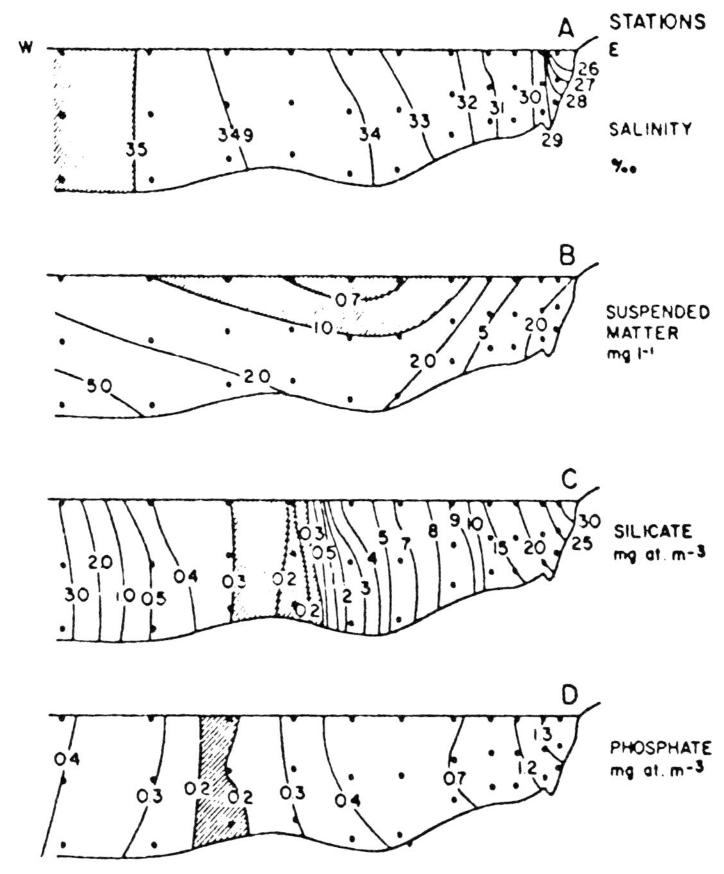


**Fig. 9.** Turbidity minimum off the Dutch coast in which phytoplankton spring bloom starts as indicated by nutrient minima. (Postma 1981)

and spring when nitrate concentration is the highest in the water column (Rutgers van der Loeff 1980b). Similar situations are found in Danish tidal basins (Henriksen et al. 1984).

Not only particulate, but also much dissolved organic matter produced in the adjacent marine areas is preferentially decomposed in the Wadden Sea areas, as shown by Hickel (1984). Most of this dissolved organic matter is set free in the water after the end of phytoplankton blooms, the mortality being induced either by nitrogen depletion (Lancelot 1983, 1984), phosphate limitation (van Bennekom et al. 1975; Veldhuis et al. 1986a,b) or silicate limitation (Gieskes and van Bennekom 1973).

## 5.7 German Bight and Danish West Coast

This area is characterized by a northward drift with stratification during summer time in areas deeper than 30 m. In the German Bight the discharge from the Rhine and other Dutch waterways, transported by the residual current, is still felt, but direct discharges by rivers dominated by the Elbe and Weser influence the nutrient concentrations most. The Elbe, Weser and Ems have a total inflow of about 1800 $m^3\ s^{-1}$ (Table 2). Frontal areas which were observed during summer between different water masses complicate the hydrographic system also in the German Bight (Martens 1978; Becker and Prahm-Rodewald 1980; Becker et al. 1983; Krause et al. 1986). Gradients of nutrients, with increases towards the estuaries, are characteristic for the German Bight (Kalle 1956; Krey 1956; Brockmann 1985; Weichart 1985; Eberlein et al. 1985; Brockmann and Eberlein 1986). Shape and extension of the nitrate-rich Elbe river plume is strongly influenced by wind stress and runoff (Brockmann and Eberlein 1986).

The high charge and coastal gradients of dissolved organic nitrogen are demonstrated in an example of the German Bight during June 1981 (Fig. 10; Eberlein et al. 1985). The estuarine input was evident (high nitrate concentrations in the Elbe river plume). Ammonia gradients showed an increase in the plume as well as towards the coast, as did dissolved carbohydrates, suggesting an influx of these compounds from the Wadden Sea. Besides the estuary, dissolved organic nitrogen compounds with 20–40 $\mu$g at N $dm^{-3}$ were dominating the inorganic nutrient concentrations. These results show again that it is necessary to include measurements of dissolved organic compounds in nutrient element analysis.

Rapid exchanges of water masses by gales, combined with a breakup of the thermocline during autumn, sometimes cause a sudden exchange of bottom water in these shallow coastal areas. Therewith, interstitial water is exchanged very rapidly. This has been demonstrated in enclosure experiments (Smetacek et al. 1976). Nutrients which have been produced in the sediments by degradation of organic materials (Krom and Berner 1981) or in the Wadden Sea near the shore (Rutgers van der Loeff 1980a,b; Rutgers van der Loeff et al. 1981) will be released suddenly, increasing the nutrient content in the coastal zone. Such an event, which can be very local, was actually observed in the German Bight in August 1982 (Brockmann and Eberlein 1986) (Fig. 11), causing an increase of phosphate concentration amounting to about 0.3 $\mu$g at P $dm^{-3}$.

At the Biological Station of Helgoland nutrients have been analyzed several times a week since 1956 (Fig. 8) (Lucht and Gillbricht 1978; Gillbricht 1981, 1983; Radach and Berg 1984). Lucht and Gillbricht (1978) calculated significant correlations between phosphate concentrations and freshwater runoff from the river Elbe during winter. Gillbricht (1981) documented a steady increase of phosphate-phosphorus since 1962 at Helgoland Roads of about 0.3 $\mu$g at $dm^{-3}$ and a corresponding decrease of the nitrate-to-phosphate ratio. Berg and Radach (1985) calculated a phosphate increase from 0.5 $\mu$g at $dm^{-3}$ (1962) to 1 $\mu$g at $dm^{-3}$ (1981). A parallel increase of phytoplankton biomass during the first decade has been considered to be the result of this increase of phosphate concentrations (Gillbricht 1981). The increase was slower during the last decade due to the decreasing N:P ratio, nitrogen becoming the limiting nutrient. However, also influences of climatic

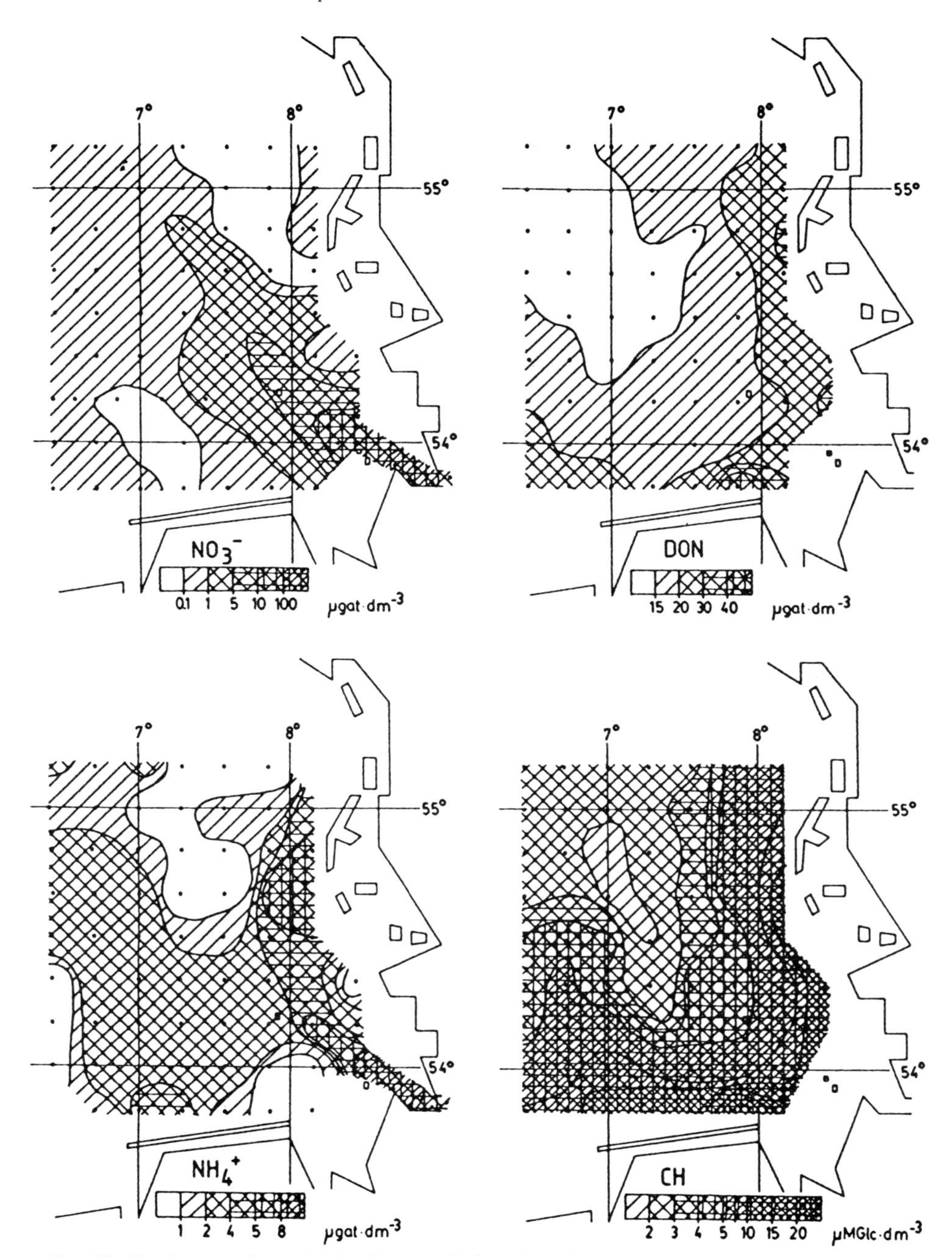


**Fig. 10.** Nutrient gradients in the German Bight (nitrate and ammonia) and dissolved organic substances (dissolved organic nitrogen and carbohydrates). (Eberlein et al. 1985)

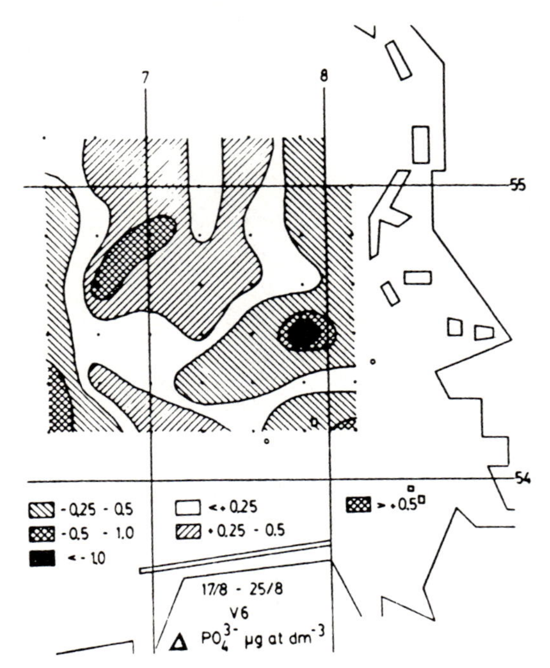


**Fig. 11.** Increase of phosphate concentrations in the German Bight following a thermocline breakup. Stations were investigated within a 4-day period. (Brockmann and Eberlein 1986)

changes and salinity changes cannot be neglected (Gieskes and Kraay 1977b; Reid 1978; Radach 1984). A decrease of salinity caused by higher amounts of freshwater runoff stabilizes the water column in coastal areas, which also can enhance primary production (Gillbricht 1983).

## 5.8 The Norwegian Deep

The Norwegian Deep is characterized by two layers. The surface layer moving within the Norwegian coastal current northward, transports Skagerrak and Norwegian coastal water which is less haline and, during summer, nutrient-exhausted (Dahl and Danielson 1981; Føyn and Rey 1981; Brockmann et al. 1981). Into this nutrient-poor surface layer nutrient-rich Atlantic water from the deep layer is frequently brought up into the euphotic zone. The tides are extremely low at the shore of south Norway, yet mixing processes often occur, namely by exchange of Skagerrak and Atlantic water in the fjord systems caused by the meandering Norwegian Coastal Current (Svendsen 1981; Brockmann et al. 1981). Thus, the nutrient distribution in the surface layer is influenced strongly by this meandering current, which sometimes forms eddies with quick alterations due to hydrological processes (Audunson et al. 1981). The deep layer, characterized by Atlantic water, flows southwards in the west and northwards in the east. The water exchange between Shetland and Norway of about $50 \times 10^3$ km$^{-3}$ a$^{-1}$ (Table 1) affects the vertical structure of nutrients, as documented by measurements in sections through the Norwegian Deep (Franck et al. 1966; Olsen 1967; Føyn and Rey 1981) (Fig. 12).

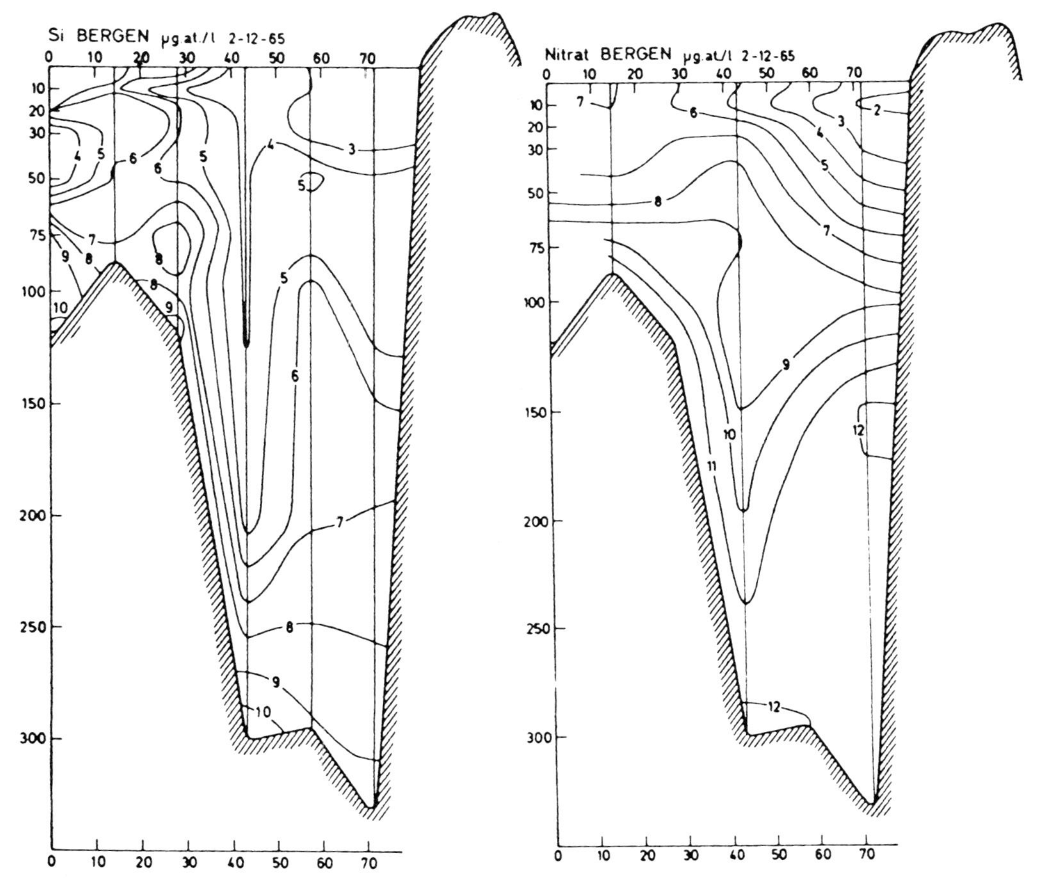


**Fig. 12.** Vertical distribution of silicate and nitrate, Bergen section on the 2 December, 1965. (Olsen 1967)

The Norwegian Coastal Current is for example at the Bergen section (December 1965) characterized by low silicate concentrations and salinity ($S < 33.5$). On the east as well as on the west side of the Norwegian Deep, silicate-rich, more haline water bodies can be recognized indicating the lateral inhomogenities caused by hydrodynamic events.

## 6 Eutrophication

The term eutrophication refers to the changes induced at the ecosystem level by an increased load of nutrients from external sources. These changes may include general enhancement of biomass and/or production, oxygen depletion, reduction of species diversity, shortening of the trophic chains. This theoretical definition, however, does not provide an operational criterion for ascribing observed effects to specific eutrophication events. Krey (1956) defined trophic stages on the base of protein concentrations. "Eutrophic" are in his definition concentrations of albumin, as an indicator for biologically active substances, in the range of 400–800 $\mu g\ dm^{-3}$ which corresponds to a biomass of 1–2 $mg\ dm^{-3}$. Following this criterion, Krey found already in August 1956 eutrophic water masses in the Elbe estuary with total protein concentrations of more than 700 $\mu g\ dm^{-3}$. This approach, however, does not distinguish between external enrichment and internal processes leading to biomass or production enhancement.

Conversely, it is very difficult to predict eutrophication effects of given external enrichments on the complex coastal ecosystems including variable plankton populations and patchy benthic communities, all subject to the influence of hydrodynamic and other physical conditions.

Johnston (1973) summarized a range of effects of nutrient enrichment in coastal ecosystems, starting with general enhancement, up to complete "kill-off", but the operational value of his discussion is limited for real systems.

For predictive purposes, it can only be stated that systems which are naturally subject to conditions promoting intense algal development or oxygen depletion are particularly sensitive to additional external inputs of nutrients, which still reinforce the natural trends.

### 6.1 Oxygen Deficiency

Recently (1981, 1982, 1983) oxygen deficiency was observed in the German Bight and off the Danish west coast in bottom layers isolated by a thermocline from the oxygen-saturated mixed layer (Rachor and Albrecht 1983; von Westernhagen and Dethlefsen 1983; Brockmann and Eberlein 1986) (Fig. 13). Areas up to 15,000 $km^2$ including the submarine valley "Helgoland Channel" with depths below 30 m were covered with low-oxygen ($< 3\ mg\ dm^{-3}$) but nutrient-rich water indicating well-advanced remineralization processes (Fig. 14).

During 1982, oxygen depletion was particularly severe regarding the degree (oxygen concentrations below 3 $mg\ dm^{-3}$), the duration (several weeks) and the extent (from the German Bight to Danish coastal waters) (von Westernhagen and Dethlefsen 1983).

Ammonia concentrations, presumably produced in the bottom layer, reached a manifold (5–9 $\mu g\ dm^{-3}$) of winter values (0.2–2 $\mu g\ dm^{-3}$; see Fig. 14). Silicate, which is remineralized most slowly of all nutrients, increased to higher values in the bottom layer during August 1982 (9–11 $\mu g\ dm^{-3}$) than in winter (7–10 $\mu g\ dm^{-3}$). The phosphate concentrations did not increase to a level above 0.5–1.5 $\mu g\ dm^{-3}$, which should, according to Johnston (1973), cause a "general enhancement of biological production". However, due to the coupling to benthic remineralization, part of the nutrients was stored in the sediment, and only another – probably smaller – part was released with exchange of interstitial water.

The finding of oxygen depletion in 1981 was just a matter of coincidence. It is unknown how often such events have occurred before; oxygen depletion has been observed already in 1902 (Gehrke 1916), long before the nutrient discharge of the

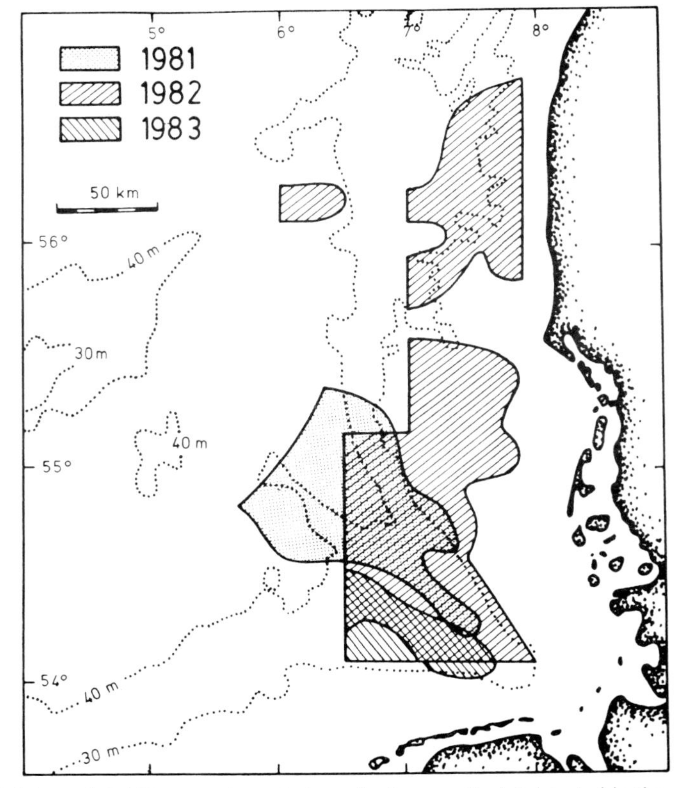


**Fig. 13.** Areas of oxygen deficiency (< 40% saturation; < 4 mg $dm^{-3}$, respectively) detected in the German Bight 1981–1983. (van Westernhagen and Dethlefsen 1983; Rachor and Albrecht 1983; Brockmann and Eberlein 1986)

estuaries increased. This indicates that also in some North Sea areas oxygen depletion may be a natural phenomenon. The hydrodynamic situation, with a thermocline and enriched nutrients in the bottom layer, was also found during June in the years before 1981; oxygen measurements were not performed then.

At least two conditions must be fulfilled for oxygen depletion to take place:

1. A stable thermocline with nearly stationary bottom water,
2. accumulation or sedimentation of organic material, either high concentrations within a short time or low concentrations during a longer period.

This situation can occur in regions where organic matter accumulates, such as the Wadden Sea; but coastal zones with underwater valleys are also sensitive. The probability of oxygen depletion will increase when nutrients are discharged from rivers and estuaries into a nutrient-poor euphotic zone. Dumping of large amounts of manure in the open North Sea, as has been proposed as a solution to the manure waste problem of some EEC countries, should certainly be discouraged, especially during the growing season. In the German Bight, oxygen depletion is most likely

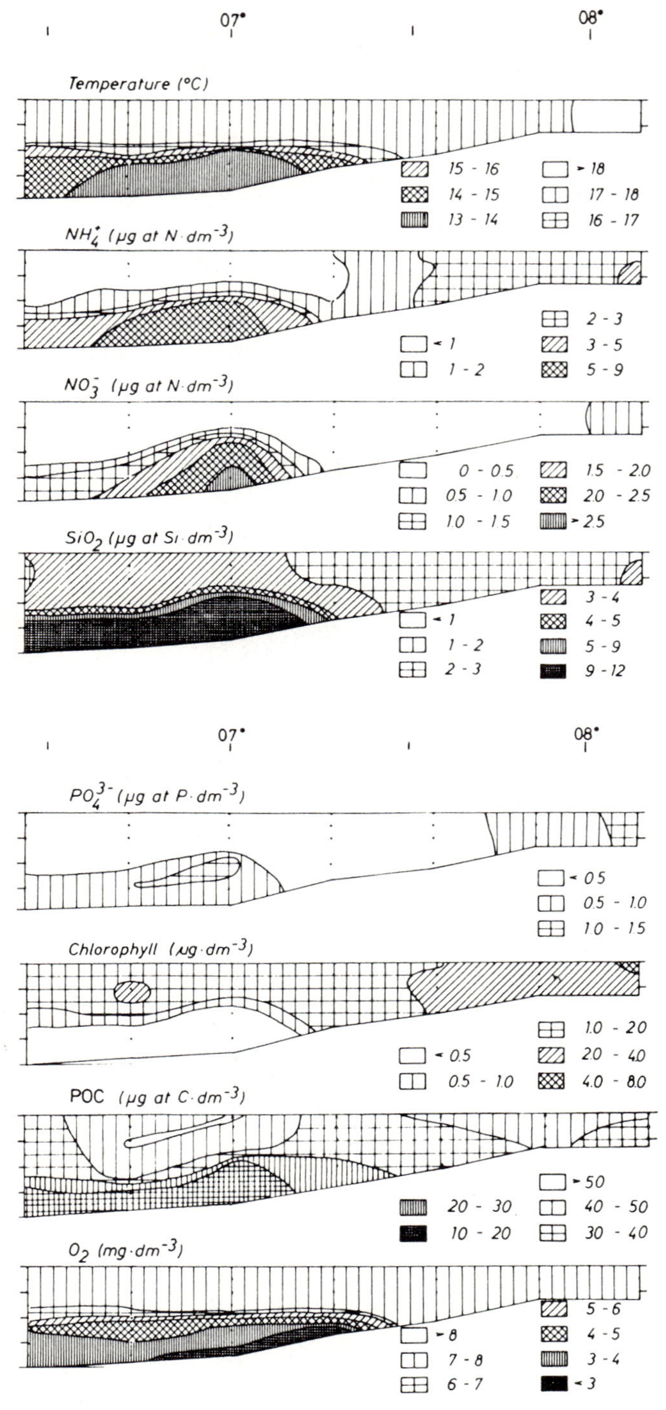


**Fig. 14.** Vertical profiles in the German Bight with oxygen depleted and nutrient enriched bottom layer (east-west profile 53°30′N) on 19/8/82. *Points* indicate sampling positions (Brockmann and Eberlein 1986)

due to the long flushing times (Maier-Reimer 1979) and high nutrient discharge in spreading river plumes covering a region of about 5000 $km^2$ above the "Helgoland Channel" (Brockmann and Eberlein 1986). Also in other North Sea areas, such as the Oyster Grounds (central-southern North Sea) oxygen concentrations may become low beneath the thermocline. De Wilde et al. (1984) have warned that areas where oxygen concentrations drop to low values due to natural conditions (i.e., a large part of the North Sea) are susceptible to eutrophication: A further increase in organic matter loading could become fatal, and extensive mass mortality of benthic and pelagic organisms could be a result.

The consumption of about 4 mg $dm^{-3}$ oxygen is in a 20-m bottom layer of a 5000 $km^2$ area equivalent to 154,000 t biomass or 30,000 t nitrogen which is about 12% of the mean annual Elbe-Weser discharge. Due to the fluctuation of river runoff and variable retarded fixation of nutrient elements in the coastal ecosystem, quantitative interactions cannot be established, but the relation of nutrient discharge and oxygen depletion is significant. Additionally, water masses from the nutrient-rich mixed shore (e.g., Wadden Sea, estuaries) move offshore, crossing the large sediment trap "Helgoland Channel", enriching the bottom layer by sedimenting biomass and detritus.

In oxygen-deficient areas fish catches are usually low. Dead fish were caught and observed by television in the German Bight during oxygen depletion; also dead benthic organisms were seen (von Westernhagen and Dethlefsen 1983). During early benthos investigations in the inner German Bight sudden breakdowns in population densities were observed, especially during late summer (Rachor 1977). A long-term trend of impoverishment of the fauna has been attributed to the development of anaerobic conditions in the muddy sediments. The fauna in this area was shown to be very susceptible to environmental stress like oxygen deficiency (Rachor 1980), supporting the suggestion of a correlation between oxygen depletion and numbers of dead benthic organisms (von Westernhagen and Dethlefsen 1983). Von Westernhagen and Dethlefsen (1983) draw attention to the fact that lethal or sublethal thresholds of toxic substances are lowered significantly when oxygen concentrations are low.

Eutrophication effects have also been observed at the shores of the southern North Sea: Thames estuary, Dutch and German coast (Anonymous 1986) and in the Wadden Sea, but also in estuaries of Tyne and Tees reduced oxygen concentrations (65–75% of saturation) were found. In Norwegian fjords (Skei 1981) eutrophication effects have been detected as well.

### 6.2 Other Eutrophication Effects

Oxygen deficiency can easily be detected, but other eutrophic effects, e.g., changes in species composition or an increase in productivity, are usually difficult to recognize due to the complexity of coastal ecosystems (Fransz and Verhagen 1985) and to natural fluctuations (Gieskes and Kraay 1977b).

Enrichment of nutrients, including organic bound nitrogen and phosphorus, will enhance primary production directly only when these elements are limiting factors. As we have seen, this is the case in large North Sea areas from late spring to autumn. In the estuaries with high nutrient load an increase of production will mostly not happen due to light limitation in the turbid environment (Gieskes and Kraay 1975). Also at weak boundaries, or in frontal systems between nutrient-depleted mixed layers in offshore regions and river plumes, coastal currents and well-mixed inshore waters, increased production can lead to high standing crops, including exceptional blooms, or extension of production in time or space (e.g., Cadée and Hegeman 1986). When nutrient elements are fixed in particulate matter, deposition far from the original place may be the ultimate fate. Mineralized nutrients, released from the sediment, may be transported within the bottom layer far away, until they will be mixed up into the euphotic zone and cause increased production, so the effect of discharge may sometimes be felt in distant regions. Therefore not only high biomass concentrations near the eutrophication source, but also extension of production processes in space and time can indicate eutrophication. It should be kept in mind that nutrient elements, once added to coastal ecosystems, can be utilized several times before they are deposited finally. Of course, nutrients may also eventually leave the North Sea by advective transport, either of the inorganic nutrients, or after fixation in particulate or dissolved organic matter.

An increase of phytoplankton from 25 to about 50 $\mu g$ C $dm^{-3}$ (yearly means) in the German Bight from 1972–1974 has been reported, based on measurements performed three times per week at Helgoland Roads (Hagmeier 1978); the increase was attributed to phosphate increase ($0.5 \rightarrow 0.85$ $\mu g$ at $dm^{-3}$). This biomass increase continued to 1980 (Gillbricht 1981) and 1981 (Gillbricht 1983), reaching 200 $\mu g$ C $dm^{-3}$ (see also Sect. 5.7).

In the southern North Sea the time series of observations extended over a larger number of years. A phytoplankton increase was observed between 1948 and 1975, seen as silk color increase of continuous plankton recorder nets (Gieskes and Kraay 1977b). A further analysis of CPR data revealed an increase of silk color in the other North Sea areas as well (Reid 1975, 1978) but this increase was less pronounced than in Dutch coastal waters (Gieskes and Kraay 1977b). In the northwestern North Sea the distinct spring and autumn bloom maxima overlap since 1962 due to extension of the blooming season (Reid 1975). This may be an eutrophication effect, since the spring bloom in this region, and also elsewhere in the North Sea, is normally limited more by nutrient depletion than by grazing (Fransz and Gieskes 1984; Radach et al. 1984).

The zooplankton caught by the CPR at a standard depth of 10 m decreased from 1948 to 1980 in the whole North Sea, but least in the most eutrophic regions (Gieskes and Kraay 1977b; Colebrook 1978a,b, 1982). The decrease has been related to climatic changes, such as a gradual decrease of sea surface temperatures (Colebrook 1978a, 1982). The annual fluctuation of zooplankton was correlated with sea surface temperature as well as frequency of westerly winds. A feed-back by nutrient regeneration to phytoplankton growth has been mentioned, but significant correlations cannot be given since information on nutrient dynamics and pathways as extended in space and time as the CPR data set is not available.

Due to natural fluctuations, shifts in species composition become evident only in a few cases and it will very seldom be possible to attribute such changes to eutrophication by nitrogen or phosphorus because the set of physical, chemical, and biological factors that influences the annual species succession is hardly ever investigated completely. One general consequence of nitrogen and phosphorus eutrophication is probably the shift from diatoms to flagellates due to the limitation of diatoms by silicate, a nutrient that usually remains constant (van Bennekom et al. 1975; Gieskes and van Bennekom 1973; Fransz and Verhagen 1985).

At Helgoland Roads not only a phosphate increase was observed (Hagmeier 1978; Gillbricht 1981, 1983) but also a nitrate increase since 1962 (Radach and Berg 1984; Jahresberichte BAH 1967–1985), parallel to a significant silicate decrease in the less haline (0.5‰ difference) estuary-influenced water masses. Also Gieskes and van Bennekom (1973) and Officer and Ryther (1980) have stressed the importance of silicon to marine eutrophication effects. Due to potentially high growth rates, diatoms can gain dominance over dinoflagellates in a well-balanced nutrient ecosystem with sufficient concentrations. It has been suggested that diatoms dominate spring blooms in the North Sea for this reason (Fig. 15): nitrogen and phosphorus are remineralized faster than silicon, and silicon becomes the limiting factor for diatoms during summer and early autumn. In the central and northern North Sea the phytoplankton during summer – early autumn is dominated by dinoflagellates represented by *Ceratium* sp. while diatoms are only a minor component at that time of the year (e.g., Gieskes and Kraay 1984). This succession can be observed nearly every year (cf. Fig. 15). However, besides the large dinoflagellates, also small $\mu$-flagellates appear, sometimes dominating the biomass and also the primary production in the North Sea (e.g., Gieskes and Kraay 1984).

In the North Sea a significant decrease of diatoms was observed during the last decades (Gieskes and Kraay 1977b; Reid 1977, 1978) parallel to an increase of phytoplankton color which has been attributed by Gieskes and Kraay (1977b) to an increase of small phytoplankton species ("nano phytoplankton") that cannot be detected by the CPR method. Since the decrease of diatoms was observed also in the unpolluted north-east Atlantic (Reid 1977), it has been suggested that at least part of the changes was not caused by eutrophication, but by climatic fluctuations (Gieskes and Kraay 1977b; Radach 1984).

However, because in the papers just quoted only standing plankton stocks have been dealt with, not processes, eutrophication effects causing the decrease of diatoms in the North Sea cannot simply be excluded. The increase of phosphorus and nitrogen in the North Sea with silicate remaining at constant or even decreased levels due to increased silicate utilization in the eutrophic rivers themselves, will shift species composition in large areas due to "memory" effects caused by long-term enrichment of phosphorus and nitrogen fluxes in the nutrient cycles. Indeed, the depletion of silicate at the end of the spring bloom (Gieskes and van Bennekom 1973) causes a breakdown of diatom populations. Later on, at low nutrient concentrations, flagellates may have ecological advantages over diatoms: mobility for migration to the euphotic zone or to nutrient-richer boundary layers, different growth strategy, adaptability to different nutrient levels (Holligan 1985) and avoidance of grazing zooplankton by colony formation (*Phaeocystis* and *Corymbellus*) or toxicity.

Some final remarks: it is extremely difficult to show significant and direct relations between eutrophication (enrichment of nutrients) and increased biomass or even production in open marine areas. The reason is that there is so much natural

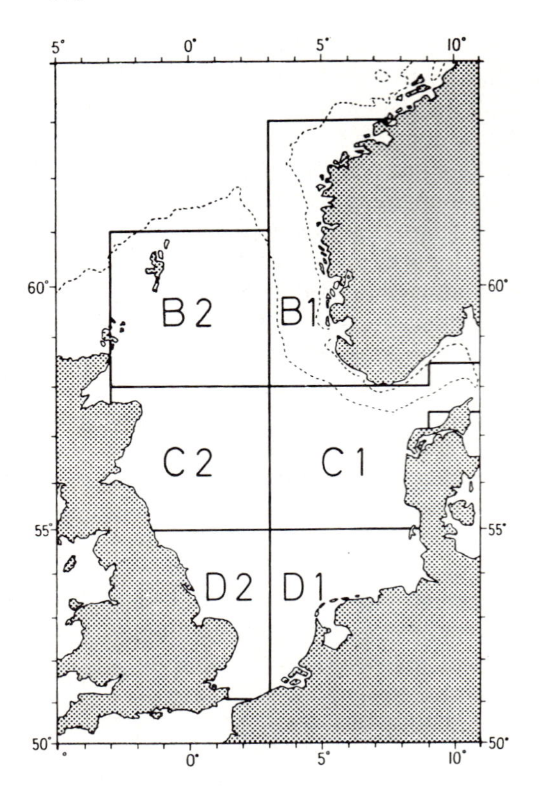


**Fig. 15.** Phytoplankton measured as silk color, diatoms, and *Ceratium* sp., sampled by continuous plankton recorder in the southern North Sea. (Reid 1978)

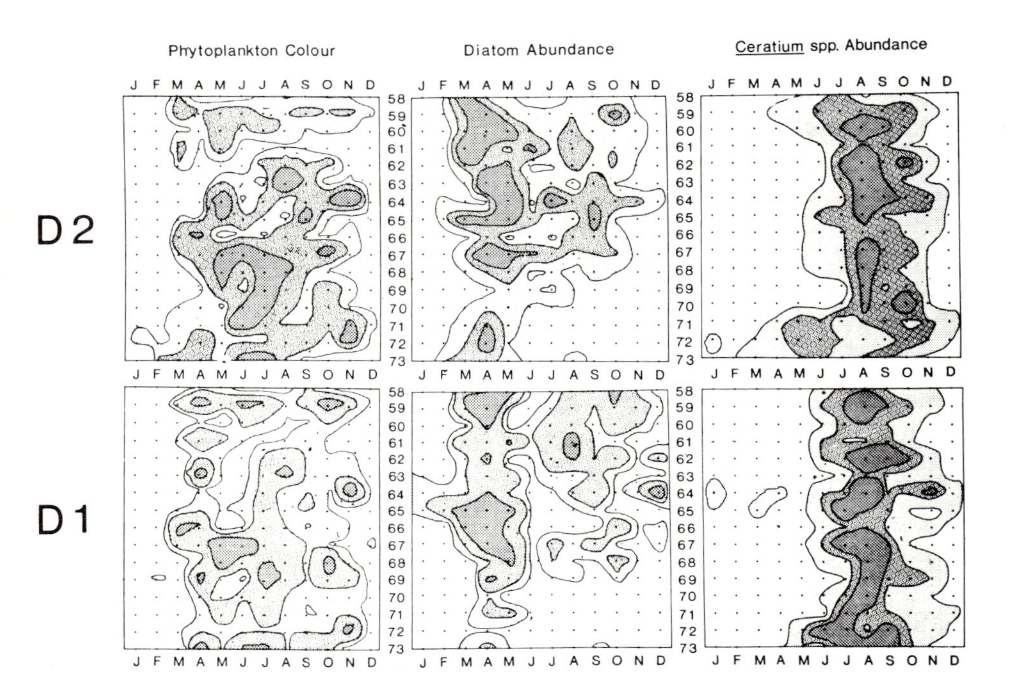

fluctuation in current patterns and in climatic events; moreover, natural long-term changes have been reported, while the transfer of nutrient elements within food webs is highly complex (Reid 1978; Gieskes and Kraay 1977b; Radach 1984; Goldman and Caron 1985). The function in material conversion and fluxes of many important members of shelf sea ecosystems is not well known. This is especially true for algal nicoplankton and heterotrophic nanoplankton (bacteria, protozoa, fungi, heterotrophic $\mu$-flagellates, ciliates) sometimes forming aggregates (Goldman 1984). All these groups (the "small food web") have been overlooked largely until recently; their importance in the system has now finally been recognized. Also dissolved organic compounds which are included in the fluxes of nitrogen and phosphorus are hardly ever involved in eutrophication studies.

### 6.3 Exceptional Phytoplankton Blooms

Due to the importance of toxic blooms to fisheries and recreation, some interdisciplinary meetings have been organized during the last few years (Taylor and Seliger 1979; ICES 1984; Anderson et al. 1985) where also correlations between eutrophication and red tide occurrence were discussed. Only in some cases was evidence of eutrophication effects on toxic blooms reported for semi-land-locked bights (Anonymous 1986, p. 332, 333).

As early as in 1849, long before the present-day eutrophication level was reached, blooms of phytoplankton were reported (see van Bennekom et al. 1975). The fact that reports on blooms are published more frequently nowadays may merely mean that the number of observations has increased. Diatom blooms tentatively ascribed to eutrophication have been reported in the North Sea (Gieskes 1973), but nearly all excessive marine phytoplankton growth due to eutrophication effects are flagellate blooms (Officer and Ryther 1980). In the North Sea, exceptional blooms have been observed of *Phaeocystis* sp. (Gieskes and Kraay 1975, 1977a; Tijssen and Eygenraam 1982; Lancelot 1982), *Mesodinium rubrum* (Fonds and Eisma 1967; Gieskes and Kraay 1983), *Ceratium* sp., *Prorocentrum* sp., and toxic flagellates like *Gonyaulax* sp. (White 1984; Ayres 1975) and *Gyrodinium aureolum* (Tangen 1977; Dahl and Tangen 1983); a really very exceptional bloom, namely of a species never before found in the North Sea, was reported by Gieskes and Kraay (1986): mass occurrence of *Corymbellus aureus* in the open North Sea, in the Fladen Ground, in May 1983.

The relation between algal abundance and eutrophication is not clear either for dinoflagellates. For example, the occurrence of "Red Tides" was already reported in the Bible (Gillbricht 1983) and Indians who were living along the (now) Canadian west coast had the good custom of not eating shellfish during months when red tides were observed (Kremer 1981; Seliger and Holligan 1985).

*Phaeocystis* is one of the most common and regular bloomers in the coastal North Sea, especially in April and May (Gieskes and Kraay 1975, 1977a; Lancelot 1982). Long-term series observations on *Phaeocystis* in Dutch coastal waters have been available since 1948. The species seemed to decrease in abundance from 1948 to 1975 (Gieskes and Kraay 1977b). However, in recent years, the duration of the growth season and the maximum abundance reached have increased considerably (Cadée and Hegeman 1986). In spite of eutrophication, which has been going on since the 1930's, phosphate may still become limiting to further expansion of the *Phaeocystis* crop (van Bennekom et al. 1975; Veldhuis et al. 1986a,b,), at least in Dutch coastal waters; in Belgian coastal waters nitrogen has been claimed as a limiting nutrient for *Phaeocystis* (Lancelot 1984; Lancelot and Billen 1984). In Fig. 16 a typical distribution pattern of *Phaeocystis* is plotted in the Southern Bight. The species is often most abundant along fronts, although this distribution is not always as obvious as in the Figure presented here.

*Gyrodinium aureolum* blooms were observed for the first time in the North Sea in 1966 (Tangen 1977). Already during this event fish kills at the coast of south Norway were found, caused by a still unidentified toxin. Since 1966 *Gyrodinium* blooms have been documented on the Danish west coast, the German Bight, the Irish Sea, the southwestern Channel, and most frequently in Norwegian waters (Tangen 1977; Dahl et al. 1982; Dahl and Tangen 1983). Hydrodynamic events like frontal systems (Holligan 1979) or concentration of offshore populations at the coastal area (Lindahl 1985) are assumed to be the key factors enhancing *Gyrodinium* blooms. Ecological advantages of this species have been reported such as nitrate uptake coupled with photosynthesis occurring also at low light intensities (Paasche et al. 1984), slow growth rates (0.3 divisions per day), migration (Dahl and Brockmann 1985) (no diurnal migration), and keeping constant stocks in spite of nutrient exhaustion for some days (Brockmann et al. 1985a).

*Gonyaulax excavata* blooms causing paralytic shellfish poisoning were observed in 1984 in the Faroe Islands (Mortensen 1985). Also an epidemic paralytic shellfish poisoning which occurred in western Europe in 1976 is attributed to *Gonyaulax* sp. (Lüthy 1979).

In some cases the toxic effects observed in shellfish cannot be related to single dinoflagellate species. Enrichment of toxins in shellfish have been observed along the Norwegian and Swedish coasts which are probably caused by *Dinophysis* species which appeared at the same time (Underdal et al. 1985; Dahl and Yndestat 1985). Shellfish poisoning was still observed at low cell concentrations (100 cells $dm^{-3}$) of *Dinophysis* sp.

## 7 Future Research

The research effort to study nutrients and eutrophication effects will increase in the next future. Plans exist to establish a working group on nutrients within the Technical Working Group of the Paris Convention and a symposium on North Sea studies was recently held in May 1988 in the Netherlands. ICES is stepping up its activities in stimulating interdisciplinary field projects (e.g., SCAPINS). National monitoring programs will no doubt be continued or intensified. In 1986 at least ten cruises covering large areas of the North Sea will be performed by Belgian, British, Dutch, and German scientists for analysis of nutrients, phytoplankton stocks, grazers, and benthic organisms. An international approach to the *Phaeocystis* problem is now being followed. Scientists from all North Sea countries have held

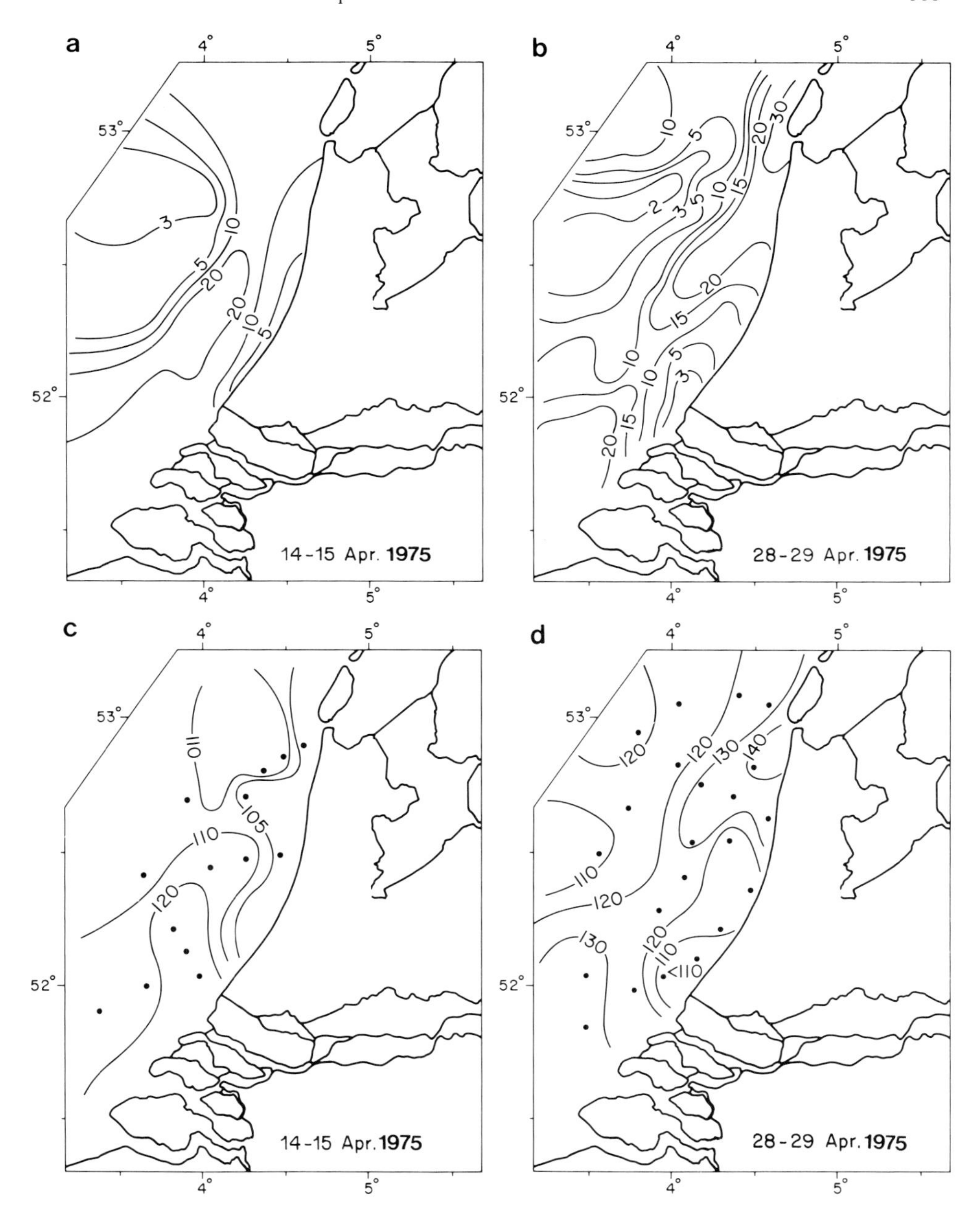


**Fig. 16a-d.** Distribution of *Phaeocystis pouchetii* (**a,b**) and of oxygen saturation values (%) (**c,d**) in Dutch coastal waters in the spring of 1975. *Phaeocystis* concentration in μg chl. a $dm^{-3}$; values over 10: more than 90% of phytoplankton crop is *Phaeocystis* (Gieskes and Kraay 1977a)

**Fig. 17.** Foam on the beach of North Holland, June 1986, caused by organic matter remaining in the water after mass occurrence of *Phaeocystis pouchetii*. Foam masses can reach a height of more than 2 m when a deteriorating bloom is accumulated towards the coast during persistent northwesterly winds (Bätje and Michaelis 1986)

workshops in the Netherlands in 1985 and 1986, and Lancelot et al. (1987) have prepared a review published in "Ambio". That this species can interfere with recreation facilities is well illustrated in Fig. 17.

In order to understand eutrophication processes, it is necessary to consider all relevant parameters, which should include dissolved organic compounds, microphytoplankton and microheterotrophs (the "small food chain") and other aspects of structure and function of ecosystems that received too little attention in the past. It is further of importance to combine concentration analysis with flux measurements. Due to the complexity of interaction processes between physical, chemical, and biological parameters, interdisciplinary research must be intensified. Modelling and simulation of model ecosystems (Ursin and Andersen 1978; Fransz and Verhagen 1985) should (and will) receive much attention in the future, not only to increase our understanding of the systems and to investigate which essential parts or functions should be studied more, but also to be able to predict the consequence of natural or man-made changes and interferences with the North Sea ecosystem.

*Acknowledgements.* This work, submitted 1986, was supported partly by the German Federal Research Board (DFG) SFB 94 and the Ministry of Research and Technology (MFU 0541). G. Billen is research associate of the Belgian National Funds for Scientific Research.

## References

Adam Y, Bayens W, Mommaerts JP, Pichot G (1981) Mathematical modelling of recent sedimentology in the shallow waters along the Belgian coast. In: Nihoul JCJ (ed) Ecohydrodynamics. Elsevier, Amsterdam
Anderson DM, White AW, Baden DG (eds) (1985) Toxic dinoflagellates. Elsevier, New York
Anonymus (1986) In: Carlson H (ed) Quality status of the North Sea. Dtsch Hydrogr Z Reihe B Nr 16
Armstrong FAJ, Butler EI (1968) Chemical changes in sea water off Plymouth during the years 1962 to 1965. J Mar Biol Assoc UK 48:153–160
Armstrong FAJ, Butler EI, Boalch GT (1974) Hydrographic and nutrient chemistry surveys in the western English Channel during 1965 and 1966. J Mar Biol Assoc UK 54:895–914
Audunson T, Dalen V, Krogstad H, Lie HN, Steinbakke P (1981) Some observations of ocean fronts, waves and currents in the surface along the Norwegian coast from satellite images and drifting buoys. In: Saetre R, Mork M (eds) The Norwegian coastal current, vol 1. University Bergen, pp 20–56
Ayres PA (1975) Mussel poisoning in Britain with special reference to paralytic shellfish poisoning. Environ Health 83:261–265
Backhaus JO (1984) Estimates of the variability of low frequency currents and flushing times of the North Sea. ICES C24, pp 23
Bätje M, Michaelis H (1986) Phaeocystis pouchetii blooms in the East Frisian coastal waters (German Bight, North Sea). Mar Biol 93:21–27
Becker GA, Prahm-Rodewald G (1980) Fronten im Meer. Salzgehaltsfronten in der Deutschen Bucht. Seewart 41:12–21
Becker GA, Fiuza AFG, James ID (1983) Water mass analysis in the German Bight during Marsen, phase I. J Geophys Res 88:9865–9870
Bennekom AJ van, Salomons W (1981) Pathways of nutrients and organic matter from land to oceans through rivers. In: River input to ocean systems. Proc Workshop United Nations Environment Program, Rome (1979)
Bennekom AJ van, Krijgsman-Van Hartingsveld E, Veer GC van der, Voorst HFJ van (1974) The seasonal cycle of reactive silicate and suspended diatoms in the Dutch wadden sea. Neth J Sea Res 8:174–207
Bennekom AJ van, Gieskes WWC, Tijssen SB (1975) Eutrophication of Dutch coastal waters. Proc R Soc Lond B Biol Sci 189:359–374
Berg J, Radach G (1985) Trends in nutrient and phytoplankton concentrations at Helgoland Reede (German Bight) since 1962. ICES CM 1985/L2/SessR, 16 pp
Billen G (1978) A budget of nitrogen recycling in North Sea sediments off the Belgian coast. Est Coast Mar Sci 7:127–146
Billen G, Fontigny A (1987) Dynamics of Phaeocystis-dominated spring bloom in the Belgian coastal waters. II Bacterioplankton dynamics. Mar Ecol Progr Ser 37:249–257
Billen G, Lancelot C (1987) Modelling benthic microbial processes and their role in nitrogen cycling of temperate coastal ecosystems. In: Blackburn HT, Sørensen J (eds) Nitrogen in coastal marine environments. Scope Wiley, New York, pp 341–378
Billen G, Joiris C, Wijnants J, Gillain G (1980) Concentration and microbial utilization of small organic molecules in the Scheldt estuary, the Belgian coastal zone of the North Sea and the English Channel. Est Coast Mar Sci, 2:279–294
Billen G, Somville M, Becker E de, Servais P (1985) A nitrogen budget of the Scheldt hydrographical basin. Neth J Sea Res 19:223–230
Billen G, Lancelot C, Becker E de, Servais P (1986) The terrestrial-marine interface: modelling nitrogen transformations during its transfer through the Scheldt river system and its estuarine zone. In: Nihoul JCJ (ed) Marine interfaces ecohydrodynamics. Elsevier, Amsterdam, New York, Oceanography ser 42 pp 429–452
Brockmann U (1985) Summer concentrations of nutrients in the German Bight 1979–1982. In: Gerlach SA (ed) Oxygen depletion 1980–1983 in coastal waters of the Federal Republic of Germany. Ber Inst Meereskd Kiel 130:22–23, 25

Brockmann UH, Eberlein K (1986) River input of nutrients into the German Bight. In: Skreslet S (ed) The role of freshwater outflow in coastal marine ecosystems. NATO ASI Ser G7. Springer, Berlin Heidelberg New York, pp 231–240

Brockmann UH, Kattner G (1985) Distribution of nutrients in the North Sea during February, May and July, 1984. ICES C 46/SessR, pp 19

Brockmann UH, Wegner G (1985) Hydrography, nutrient and chlorophyll distribution in the North Sea in February 1984. Arch Fischereiwiss 36:27–45

Brockmann UH, Koltermann KP, Dahl E et al. (1981) Water exchange in Rosfjorden during spring '79, a detailed account of physical, chemical and biological variations. In: Saetre R, Mork M (eds) The Norwegian coastal current, vol 1. University Bergen, pp 93–130

Brockmann UH, Ittekkot V, Kattner G, Eberlein K, Hammer KD (1983) Release of dissolved organic substances in the course of phytoplankton blooms. In: Sündermann J, Lenz W (eds) North Sea dynamics. Springer, Berlin Heidelberg New York, pp 530–548

Brockmann UH, Dahl E, Eberlein K (1985a) Nutrient dynamics during a *Gyrodinium aureolum* bloom. In: Anderson DM, White AW, Baden DG (eds) Toxic dinoflagellates. Elsevier, New York, pp 239–244

Brockmann UH, Eberlein K, Huber K, Neubert HJ, Radach G, Schulze K (eds) (1985b) JONSDAP'76: FLEX/Inout Atlas. ICES oceanographic data lists and inventories 63

Burns RB (1967) Chemical observations in the North Sea in 1965. Ann Biol (Copenh) 22:29–31

Burns RB, Johnston R (1968) Chemical observations from 1964 in the northern North Sea. Ann Biol (Copenh) 21:29–36

Butler EJ, Knox S, Liddicoat MJ (1979) The relationship between inorganic and organic nutrients in sea water. J Mar Biol Assoc UK 59:239–250

Cadée GC (1982) Tidal and seasonal variation in particulate and dissolved organic carbon in the western Dutch Wadden Sea and Marsdiep tidal inlet. Neth J Sea Res 15:228–249

Cadée GC (1986) Organic carbon in the water column and its sedimentation, Fladen Ground (North Sea), May 1983. Neth J Sea Res 20:347–358

Cadée GC, Hegeman J (1986) Seasonal and annual variation of *Phaeocystis pouchetii* (Haptophyceae) in the westernmost inlet of the Wadden Sea during the 1973 to 1985 period. Neth J Sea Res 20:27–36

Capone DG (1988) Benthic nitrogen fixation: Microbiology, physiology and ecology. In: Blackburn HT, Sørensen J (eds) Nitrogen cycling in marine coastal environments. Proc SCOPE Symp 33. Wiley

Carlson H (ed) (1986) Quality status of the North Sea. Dtsch Hydrogr Z Reihe B Nr. 16

Colebrook JM (1978a) Continuous plankton records: Zooplankton and environments, North-East Atlantic and North Sea, 1948-1975. Oceanol Acta 1:9–23

Colebrook JM (1978b) Changes in the zooplankton of the North Sea, 1948 to 1973. Rapp P V Reun Cons Int Explor Mer 1972:390–396

Colebrook JM (1982) Continuous plankton records: Phytoplankton, zooplankton and environment, north-east Atlantic and North Sea, 1958–1980. Oceanol Acta 5:473–480

Cushing DH (1983) Sources of variability in the North Sea ecosystems. In: Sündermann J, Lenz W (eds) North Sea dynamics. Springer, Berlin Heidelberg New York, pp 498–516

Dahl E, Brockmann UH (1985) The growth of *Gyrodinium aureolum* Hulburt in in situ experimental bags. In: Anderson DM, White AW, Baden DG (eds) Toxic dinoflagellates. Elsevier, New York, pp 233–238

Dahl E, Danielssen DS (1981) Hydrography, nutrients and phytoplankton in the Skagerrak along the section Torungen-Hirtshals, January-June 1980. In: Saetre R, Mork M (eds) The Norwegian coastal current, vol 1. University Bergen, pp 294–310

Dahl E, Tangen K (1983) Forekomsten av *Gyrodinium aureolum* høsten 1982. Nor Fiskeoppdrett 1:17–19

Dahl E, Yndestad M (1985) Diarrhetic shellfish poisoning (DSP) in Norway in the autumn 1984 related to the occurrence of *Dinophysis* spp. In: Anderson DM, White AW, Baden DG (eds) Toxic dinoflagellates. Elsevier, New York, pp 495–500

Dahl E, Danielssen DS, Bøhle B (1982) Mass occurrence of *Gyrodinium aureolum* Hulburt and fish mortality along the southern coast of Norway in September-October 1981. Flødevigen Rapportserie 4, pp 15

Davies JM, Payne R (1984) Supply of organic matter to the sediment in the northern North Sea during a spring phytoplankton bloom. Mar Biol 78:315–324

Dooley HD (1983) Seasonal variability in the position and strength of the Fair Isle Current. In: Sündermann J, Lenz W (eds) North Sea dynamics. Springer, Berlin Heidelberg New York, pp 108–119

Eberlein K, Kattner G, Brockmann U, Hammer KD (1980) Nitrogen and phosphorus in different water layers at the central station during FLEX '76. Meteor Forschungsergeb A 22:87–98

Eberlein K, Leal MT, Hammer KD, Hickel W (1985) Dissolved organic substances during a *Phaeocystis* bloom in the German Bight (North Sea). Mar Biol 89:311–316

Eisma D (1981a) Supply and deposition of suspended matter in the North Sea. Spec Publ Int Assoc Sediment 5:415–428

Eisma D (1981b) The mass-balance of suspended matter and associated pollutants in the North Sea. Rapp P V Reun Cons Int Explor Mer 181:7–14

Fogg GE (1977) Excretion of organic matter by phytoplankton. Limnol Oceanogr 22:576–577

Fonds M, Eisma D (1967) Upwelling water as a possible cause of red plankton bloom along the Dutch coast. Neth J Sea Res 3:458–464

Føyn L, Rey F (1981) Nutrient distribution along the Norwegian coastal current. In: Saetre R, Mork M (eds) The Norwegian coastal current, vol 2. University Bergen, pp 629–639

Franck H, Nehring D, Rohde K-H (1966) Hydrographische und chemische Untersuchungen in der Norwegischen Rinne im April 1965. Z Fisch Hilfswiss 14:111–122

Fransz HG, Gieskes WWC (1984) The unbalance of phytoplankton and copepods in the North Sea. Rapp P V Reun Cons Int Explor Mer 183:218–225

Fransz HG, Verhagen JHG (1985) Modelling research on the production cycle of phytoplankton in the southern bight of the North Sea in relation to riverborne nutrient loads. Neth J Sea Res 19:241–256

Gagosian RB, Lee C (1981) Processes controlling the distribution of biogenic organic compounds in seawater. In: Duursma EK, Dawson R (eds) Marine organic chemistry. Elsevier, Amsterdam, pp 91–123

Gehrke J (1916) Uber die Sauerstoffverhältnisse der Nordsee. Ann Hydrogr Marit Meteorol 44:177–193

Gerlach SA (1984) Oxygen depletion 1980–1983 in coastal waters of the Federal Republic of Germany. Ber Inst Meereskd Kiel 130:87

Gerlach SA (1985) Wurde der 1981 in der Deutschen Bucht beobachtete Sauerstoffmangel durch anthropogene Nährstoff-Franchten begünstigt? In: AMK (Hrsg) Wasser Berlin '85, Kongreß-vorträge. Wissenschaftsverlag V Spiess, Berlin, pp 430–451

Gieskes WWC (1973) De massale planktonsterfte op de Kust by Callantsoorg, Juli 1972. Wadden-bulletin 8:36–37

Gieskes WWC (1974) Phytoplankton and primary production studies in the southern bight of the North Sea, eastern part, in 1972. Ann Biol (Copenh) 29:54–61

Gieskes WWC, Bennekom AJ van (1973) Unreliability of the $^{14}$C-method for estimating primary production in eutrophic Dutch coastal waters. Limnol Oceanogr 18:494–495

Gieskes WWC, Kraay GW (1975) The phytoplankton spring bloom in Dutch coastal waters of the North Sea. Neth J Sea Res 9:166–196

Gieskes WWC, Kraay GW (1977a) Primary production and consumption of organic matter in the southern North Sea during the spring bloom of 1975. Neth J Sea Res 11:146–167

Gieskes WWC, Kraay GW (1977b) Continuous plankton records: Changes in the plankton of the North Sea and its eutrophic southern bight from 1947 to 1975. Neth J Sea Res 11:334–364

Gieskes WWC, Kraay GW (1980) Primary productivity and phytoplankton pigment measurements in the northern North Sea during FLEX '76. Meteor Forschungsergeb A 22:105–112

Gieskes WWC, Kraay GW (1983) Dominance of Cryptophyceae during the phytoplankton spring bloom in the central North Sea detected by HPLC analysis of pigments. Mar Biol 75:179–185

Gieskes WWC, Kraay GW (1984) Phytoplankton, its pigments, and primary production at a central North Sea station in May, July and September 1981. Neth J Sea Res 18:51–70

Gieskes WWC, Kraay GW (1986) Analysis of phytoplankton pigments by HPLC before, during and after mass occurrence of the microflagellate *Corymbellus aureus* during the spring bloom in the open northern North Sea in 1983. Mar Biol 92:45–52

Gillbricht M (1981) Hydrographie, Nährstoffe und Phytoplankton bei Helgoland. Jahresbericht 1980 Biologische Anstalt Helgoland, Hamburg, pp 23–27

Gillbricht M (1983) Eine "red tide" in der südlichen Nordsee und ihre Beziehungen zur Umwelt. Helgol Wiss Meeresunters 36:393–426

Goldberg ED (1973) Introduction. In: Goldberg ED (ed) North sea science. MIT Press, Cambridge, pp 1–14

Goldman JC (1984) Oceanic nutrient cycles. In: Fasham MJR (ed) Flows of energy and materials in marine ecosystems. Plenum, New York, pp 137–170

Goldman JC, Caron DA (1985) Experimental studies on an omnivorous microflagellate: implications for grazing and nutrient regeneration in the marine microbioal food chain. Deep Sea Res 32:899–915

Hagmeier E (1978) Variations in phytoplankton near Helgoland. Rapp P V Reun Cons Int Explor Mer 172:361–363

Harding D, Ramster J, Nichols JH, Folkard AR (1978) Studies on planktonic fish eggs and larvae in relation to environmental conditions in the west central North Sea. Ann Biol (Copenh) 33:62–69

Hattori A (1983) Denitrification and dissimilatory nitrate reduction. In: Carpenter EJ, Capone DG (eds) Nitrogen in the marine environment. Academic Press, London, pp 191–232

Helder W (1974) The cycle of dissolved inorganic nitrogen compounds in the Dutch Wadden Sea. Neth J Sea Res 8:154–173

Hellebust JA, Lewin J (1977) Heterotrophic nutrition. In: Werner D (ed) The biology of diatoms, vol 13. Blackwell Scient, Oxford, pp 169–197

Henriksen K, Jensen A, Rasmussen MC (1984) Aspects of nitrogen and phosphorus mineralization and recycling in the northern part of the Danish wadden sea. Neth Inst Sea Res Publ Ser 10:51–69

Hentzschel G (1980) Wechselwirkungen bakteriolytischer und saprophytischer Bakterien aus der Nordsee. Mitt Inst Allg Bot Hamb 17:113–124

Hickel W (1984) Particulate nitrogen in the German Bight and its potential oxygen demand. In: Gerlach SA (ed) Oxygen depletion 1980–1983 in coastal waters of the Federal Republic of Germany. Ber Inst Meereskd Kiel, pp 48–51

Holligan PM (1979) Dinoflagellate associated with tidal fronts around the British Isles. In: Taylor DL, Seliger HH (eds) Toxic dinoflagellate blooms. Elsevier, New York, pp 249–256

Holligan PM (1985) Marine dinoflagellate blooms – growth strategies and environmental exploitation. In: Anderson DM, White AW, Baden DG (eds) Toxic dinoflagellates. Elsevier, New York, pp 133–139

Holligan PM, Pingree RD, Pugh PR, Mardell GT (1978) The hydrography and plankton of the eastern English Channel in March 1976. Ann Biol (Copenh) 33:69–71

ICES (1983) International Council for the Exploration of the Sea. Cooperative research report 123. Flushing times of the North Sea

ICES (1984) International Council for the Exploration of the Sea. Report of the ICES Special Meeting on the causes, dynamics and effects of exceptional marine blooms and related events. CM 1984/E42

Ittekkot V, Brockmann U, Michaelis W, Degens ET (1981) Dissolved free and combined carbohydrates during a phytoplankton bloom in the northern North Sea. Mar Ecol Prog Ser 4:299–305

Jackson GA, Williams PM (1985) Importance of dissolved organic nitrogen and phosphorus to biological nutrient cycling. Deep Sea Res 32:223–235

Jahresberichte BAH (Biologische Anstalt Helgoland) 1967–1985 Helgoland

Jenness MI, Duineveld GCA (1985) Effects of tidal currents on chlorophyll *a* content of sandy sediments in the southern North Sea. Mar Ecol Prog Ser 21:283–287

Johnston R (1973) Nutrients and metals in the North Sea. In: Goldberg ED (ed) North Sea science. MIT Press, Cambridge, pp 293–307

Johnston R, Jones PGW (1965) Inorganic nutrients in the North Sea. Serial Atlas Mar Env 11 Am Geogr Soc, New York

Joiris C, Billen G, Lancelot C et al. (1982) A budget of carbon cycling in the Belgian coastal zone: relative roles of zooplankton, bacterioplankton and benthos in the utilization of primary production. Neth J Sea Res 16:260–275

Jones PGW (1982) A review of nutrient salt and trace metal data in UK tidal waters. Aquat Environ Monit Rep MAFF Direct Fish Res Lowestoft (7), 22 pp

Jonge de VN, Postma H (1974) Phosphorus compounds in the Dutch Wadden Sea. Neth J Sea Res 8:139–153

Kalle K (1956) Chemisch-hydrographische Untersuchungen in der inneren Deutschen Bucht. Dtsch Hydrogr Z 9:55–65

Kautsky H (1985) Distribution and content of different artificial radio nuclides in the water of the North Sea during the years 1977 to 1981 (complemented with some results from 1982 to 1984) Dtsch Hydrogr Z 38:193–224

Kelderman P (1980) Phosphate budget and sediment-water exchange in lake Grevelingen (SW Netherlands). Neth J Sea Res 14:229–236
King FD, Devol AH (1979) Estimates of vertical eddy diffusion through the thermocline from phytoplankton nitrate uptake rates in the mixed layer of the eastern tropical Pacific. Limnol Oceanogr 24:645–651
Klein P, Coste B (1984) Effects of wind-stress variability on nutrient transport into the mixed layer. Deep Sea Res 31:21–37
Krause G, Budeus G, Gerdes D, Schaumann K, Hesse K (1986) Frontal systems in the German Bight and their physical and biological effects. Marine Interfaces Ecohydrodynamics, JCJ Nikoul (ed), Elsevier, Amsterdam, pp 119–140
Krause M, Trahms J (1983) Zooplankton dynamics during FLEX '76. In: Sündermann J, Lenz W (eds) North Sea dynamics. Springer, Berlin Heidelberg New York, pp 632–661
Kremer BP (1981) Toxische Planktonalgen. Naturwissenschaften 68:101–109
Krey J (1956) Die Trophie küstennaher Meeresgebiete. Kiel Meeresforsch 12:46–64
Krom MD, Berner RA (1981) The diagenesis of phosphorus in a nearshore marine sediment. Geochim Cosmochim Acta 45:207–216
Lancelot C (1982) Etude écophysiologique du phytoplankton de la zone côtière belge. Ph D Thesis, Univ Brussels
Lancelot C (1983) Factors affecting phytoplankton extracellular release in the southern bight of the North Sea. Mar Ecol Prog Ser 12:115–121
Lancelot C (1984) Metabolic change in *Phaeocystis pouchetii* during the spring bloom in Belgian coastal waters. Est Coast Shelf Sci 18:593–600
Lancelot C, Billen G (1984) Activity of heterotrophic bacteria and its coupling to primary production during the spring phytoplankton bloom in the southern bight of the North Sea. Limnol Oceanogr 29:721–730
Lancelot C, Billen G, Sournia A, Weisse T, Colijn F, Veldhuis MJW, Davies A, Wassman P (1987) Phaeocystis blooms and nutrient enrichment in the continental coastal zones of the North Sea. Ambio 16:38–46
Lee AJ (1980) North Sea: Physical oceanography. In: Banner FT et al. (eds) The northwest European Shelf Seas: The sea-bed and the sea in motion. II. Physical and chemical oceanography, and physical resources. Elsevier, Amsterdam, pp 467–493
Lindahl O (1985) Blooms of *Gyrodinium aureolum* along the Skagerrak coast – a result of the concentration of offshore populations? In: Anderson DM, White AW, Baden DG (eds) Toxic dinoflagellates. Elsevier, New York, pp 231–232
Lucht F, Gillbricht M (1978) Long-term observations on nutrient contents near Helgoland in relation to nutrient input of the river Elbe. Rapp P V Reun Cons Int Explor Mer 172:358–360
Lüthy J (1979) Epidemic paralytic shellfish poisoning in western Europe, 1976. In: Taylor DL, Seliger HH (eds) Toxic dinoflagellate blooms. Elsevier, New York, pp 15–22
Maier-Reimer E (1979) Some effects of the Atlantic circulation and of river discharges on the residual circulation of the North Sea. Dtsch Hydrogr Z 32:126–130
Martens P (1978) Contributions to the hydrographical structure of the eastern German Bight. Helgol Wiss Meeresunters 31:414–424
Mommaerts JP, Pichot G, Ozer J, Adam Y (1984) Nitrogen cycling and budget in Belgian coastal waters: North Sea areas with and without river inputs. Rapp P V Reun Cons Int Explor Mer 183:57–69
Mortensen AM (1985) Massive fish mortalities in the Faroe Islands caused by a *Gonyaulax excavata* red tide. In: Anderson DM, White AW, Baden DG (eds) Toxic dinoflagellates. Elsevier, New York, pp 165–170
Nihoul JCJ, Ronday FC (1975) The influence of the tidal stress on the residual circulation. Tellus 27:484–489
Nixon SW (1981) Remineralization and nutrient cycling in coastal marine ecosystems. In: Neilson BJ, Cronin LE (eds) Estuaries and nutrients. Humana, pp 111–138
Officer CG, Ryther JH (1980) The possible importance of silicon in marine eutrophication. Mar Ecol Prog Ser 3:83–91
Olsen OV (1967) Hydrographic investigations in the Skagerrak and northern North Sea. Ann Biol (Copenh) 22:40–45
Paasche E, Bryceson I, Tangen K (1984) Interspecific variation in dark nitrogen uptake by dinoflagellates. J Phycol 20:394–401

Paul JH (1983) Uptake of organic nitrogen. In: Carpenter EJ, Capone DG (eds) Nitrogen in the marine environment. Academic Press, London, pp 275–308

Peinert R, Saure A, Stegman P, Steinen C, Haardt H, Smatacek V (1982) Dynamics of primary production and sedimentation in a coastal ecosystem. Neth J Sea Res 16:276–289

Pingree RD, Griffiths DK (1978) Tidal fronts on the shelf seas around the British Isles. J Geophys Res 83:4615–4622

Pingree RD, Holligan PM, Mardell GT (1968) The effects of vertical stability on phytoplankton distributions in the summer on the northwest European Shelf. Deep Sea Res 25:1011–1028

Pingree RD, Maddock L, Butler EI (1977) The influence of biological activity and physical stability in determining the chemical distributions of inorganic phosphate, silicate and nitrate. J Mar Biol Assoc UK 57:1065–1073

Postma H (1978) The nutrient contents of North Sea water: Changes in recent years, particularly in the southern bight. Rapp P V Reun Cons Int Explor Mer 172:350–357

Postma H (1981) Exchange of materials between the North Sea and the wadden sea. Mar Geol 40:199–213

Prandle D (1984) Monthly-mean residual flows through the Dover Strait, 1949–1980. J Mar Biol Assoc UK 64:722–724

Rachor E (1977) Faunenverarmung in einem Schlickgebiet in der Nähe Helgolands. Helgol Wiss Meeresunters 30:633–651

Rachor E (1980) The inner German Bight – an ecologically sensitive area as indicated by the bottom fauna. Helgol Wiss Meeresunters 33:522–530

Rachor E, Albrecht H (1983) Sauerstoff-Mangel im Bodenwasser der Deutschen Bucht. Veröff Inst Meeresforsch Bremerhaven 19:209–227

Radach G (1984) Variations in the plankton in relation to climate. Rapp R V Reun Cons Int Explor Mer 185:234–254

Radach G, Berg J (1984) Trends in concentrations of plant nutrients at Helgoland Reede 1962–1981. In: Gerlach SA (ed) Oxygen depletion 1980–1983 in coastal waters of the Federal Republic of Germany. Ber Inst Meereskol Kiel 130:28–31

Radach G, Berg J, Heinemann B, Krause M (1984) On the relation of primary production to grazing during the Fladen Ground Experiment 1976 (FLEX '76). In: Fasham MJR (ed) Flows of energy and materials in marine ecosystems. Plenum, New York, pp 596–625

Redfield AC, Ketchum BH, Richards FA (1963) The influence of organisms on the composition of sea water. In: Hill MN (ed) The sea, vol 2. Wiley, New York, pp 26–77

Reid PC (1975) Large scale changes in North Sea phytoplankton. Nature 257:217–219

Reid PC (1977) Continuous plankton records: Changes in the composition and abundance of the phytoplankton of the north-eastern Atlantic Ocean and North Sea 1958–1971. Mar Biol 40:337–339

Reid PC (1978) Continuous plankton records: Large-scale changes in the abundance of phytoplankton in the North Sea from 1958 to 1973. Rapp P V Reun Cons Int Explor Mer 172:384–389

Riley JP, Segar DA (1970) The seasonal variation of the free and combined dissolved amino-acids in the Irish Sea. J Mar Biol Assoc UK 50:713–720

Rönner U (1985) Nitrogen transformations in the Baltic proper: Denitrification counteracts eutrophication. Ambio 14:134–138

Rutgers Loeff MM van der (1980a) Time variation in interstitial nutrient concentrations at an exposed subtidal station in the Dutch wadden sea. Neth J Sea Res 14:123–143

Rutgers Loeff MM van der (1980b) Nutrients in the interstitial waters of the southern bight of the North Sea. Neth J Sea Res 14:144–171

Rutgers Loeff MM van der, Es FB van, Helder W, Vries TP de (1981) Sediment water exchange of nutrients and oxygen on tidal flats in the Ems-Dollard estuary. Neth J Sea Res 15:113–129

Ryther JH, Dunstan WN (1971) Nitrogen, phosphorus and eutrophication in the coastal marine environment. Science 171:1008–1013

Scranton MI (1983) Gaseous nitrogen compounds in the marine environment. In: Carpenter EJ, Capone DG (eds) Nitrogen in the marine environment. Academic Press, London, pp 37–64

Seitzinger SP, Nixon SW (1985) Eutrophication and the rate of denitrification and $N_2O$ production in coastal marine sediments. Limnol Oceanogr 30:1332–1339

Seliger HH, Holligan PM (1985) Sampling criteria in natural phytoplankton populations. In: Anderson DM, White AW, Baden DG (eds) Toxic dinoflagellates. Elsevier, New York, pp 540–544

Skei J (1981) The entrapment of pollutants in Norwegian fjord sediments – a beneficial situation for the North Sea. Spec Publ Int Assoc Sediment 5:461–468

Smetacek V, Bodungen B von, Bröckel K von, Zeitzschel B (1976) The plankton tower. II. Release of nutrients from sediments due to changes in the density of bottom water. Mar Biol 34:373–378

Steele JH (1974) The structure of marine ecosystems. Harvard University Press, Cambridge, Massachusetts

Stefanson U, Richards FA (1963) Processes contributing to the nutrient distributions off the Columbia river and Strait of Juan de Fuca. Limnol Oceanogr 8:394–410

Svendsen H (1981) Wind-induced variations of circulation and water level in coupled fjord-coast systems. In: Saetre R, Mork M (eds) The Norwegian coastal current, vol 1. University Bergen, pp 229–262

Tangen K (1977) Blooms of *Gyrodinium aureolum* (Dinophyceae) in north European waters, accompanied by mortality in marine organisms. Sarsia 63:123–133

Taylor DL, Seliger HH (eds) (1979) Toxic dinoflagellate blooms. Elsevier, New York

Tijssen SB (1968) Hydrographic and chemical observations in the southern bight, August and November 1967. In: Ann Biol (Copenh) 24:52–56

Tijssen SB (1969) Hydrographic and chemical observations in the southern bight, February, May, August and November. Ann Biol (Copenh) 25:51–59

Tijssen SB (1970) Hydrographic and chemical observations in the southern bight, July and October, 1969. Ann Biol (Copenh) 26:73–81

Tijssen SB, Eygenraam A (1982) Primary and community production in the southern bight of the North Sea, deduced from oxygen concentration variations in the spring of 1980. Neth J Sea Res 16:247–259

Underdal B, Yndestad M, Aune T (1985) DSP intoxication in Norway and Sweden, autumn 1984 – spring 1985. In: Anderson DM, White AW, Baden DG (eds) Toxic dinoflagellates. Elsevier, New York, pp 489–494

Ursin E, Andersen KP (1978) A model of the biological effects of eutrophication in the North Sea. Rapp P V Reun Cons Int Explor Mer 1972:366–377

Veldhuis MJW, Admiraal W, Colijn F (1986a) Chemical and physiological changes of phytoplankton during the spring bloom dominated by *Phaeocystis pouchetii* (Haptophyceae): observations in Dutch coastal waters of the North Sea. Neth J Sea Res 20:49–60

Veldhuis MJW, Colijn F, Venekamp LAH (1986b) The spring bloom of *Phaeocystis pouchetii* (Haptophyceae) in Dutch coastal waters. Neth J Sea Res 20:37–48

Wafar M, Le Corre P, Birrien JL (1984) Seasonal changes of dissolved organic matter (C, N, P) in permanently well mixed temperate waters. Limnol Oceanogr 29:1127–1132

Wandschneider K (1980) Die Artensukzession des Phytoplanktons während der Frühjahrsblüte im Fladengrundgebiet (nördliche Nordsee). Mitt Inst Allg Bot Hamb 17:39–48

Wandschneider K (1983) Some biotic factors influencing the succession of diatom species during FLEX '76. In: Sündermann J, Lenz W (eds) North Sea dynamics.Springer, Berlin Heidelberg New York, pp 573–583

Weering TCE van (1981) Recent sediments and sediment transport in the northern North Sea: surface sediments of the Skagerrak. Spec Publ Int Assoc Sediment 5:335–359

Wegner G (1985) Changes in water mass distributions in the North Sea during spring 1984. ICES C 20/SessR, pp 17

Weichart G (1985) High pH values in the German Bight as an indication of intensive primary production. Dtsch Hydrogr Z 38:93–117

Westernhagen H von, Dethlefsen V (1983) North Sea oxygen deficiency 1982 and its effects on the bottom fauna. Ambio 12:264–267

White AW (1984) Paralytic shellfish toxins and finfish. In: Regalis EP (ed) ACS Symposium series 262. American Chemical Society, pp 171–180

Whittle KJ (1977) Marine organisms and their contribution to organic matter in the ocean. Mar Chem 5:381–411

Wilde PAWJ de, Berghuis EM, Kok A (1984) Structure and energy demand of the bentic community of the Oyster Ground, central North Sea. Neth J Sea Res 18:143–159

Wollast R (1982) Behaviour of organic carbon, nitrogen and phosphorus in the Scheldt estuary and the adjacent coastal zone. In: Nihoul JCJ, Wollast R (eds) Hydrodynamic and dispersion models: boundary fluxes and boundary conditions. ICES CM 1982/E40, pp 199–257

# Radioactive Substances

H. KAUTSKY

## 1 Introduction

In the limited space available, one can only consider and summarize into a comparatively rough picture the, in part, very complex processes in the North Sea and the behavior of the radio nuclides in seawater. In the references also, for reasons of space economy, in part, only summarized overview references are cited, which contain further detailed literature information.

## 2 Origin and Content of Radio Nuclides in Seawater

In nature, considerable quantities of radio nuclides are contained in seawater. To them belong especially potassium 40, rubidium 87, uranium, and radium (Table 1), as well as the nuclides of the decay series of uranium and thorium.

As the result of the utilization of the nuclear fission process by man, further so-called artificial or man-made radio nuclides are to be found in measurable quantities in seawater. These nuclides, in principle, are also natural nuclides but, because of their short half-life times compared to geological temporal periods, have decayed in the course of the Earth's development and have disappeared from nature. Only the renewed starting up of the corresponding nuclear fission processes by man has permitted their re-birth.

The first large introduction of these artificial radio nuclides into the sea took place until the middle of the 1960's – as the result of the atmospheric atomic bomb tests – via the global fallout. Particularly the large tests in the years 1961/62 led to easily measurable activity concentrations of some radio nuclides in seawater.

In the second half of the 1960's, in which – after practical cessation of the atmospheric atomic bomb tests in 1962 – the introduction of the radio nuclides via the fallout was still only slight, the activity concentrations of those radio nuclides decreased clearly also in the North Sea. Today, for $^{137}Cs$ for example, from this source they lie at a maximum of about 3 Bq $m^{-3}$.

In 1970, in the southern North Sea and in 1971 in the northwestern North Sea, a renewed increase of the $^{137}Cs$ concentration could again be observed. Thorough

[1]Vogt Wellsstraße 24A, D-2000 Hamburg 54, FRG

**Table 1.** Content and mean activity concentration of different radio nuclides in the water of the North Sea between 51°N to 61°N and 4°W to 9°E (Kautsky 1985, 1986)

| Isotope | Year | Total content in TBq[a] referred to [b]32,720 $km^3$ | [c]42,444 $km^3$ | Mean activity concentration Bq $m^{-3}$ |
|---|---|---|---|---|
| K 40 | – | ~387,500 | ~503,000 | ~11,800 |
| Rb 87 | – | ~4 250 | ~5 520 | ~130 |
| U 238 | – | ~629 | ~785 | ~18.5 |
| Ra | – | ~100 | ~127 | ~3.0 |
| Cs 137 | 1978 | 5 836 | – | 174 |
| | 1979 | 5 209 | – | 155 |
| | 1980 | 3 950 | – | 117 |
| | 1981 | 2 775 | – | 82 |
| | 1982 | 3 034 | 3 424 | 90/81 |
| | 1984 | 2 219 | 2 621 | 68/62 |
| Sr 90 | 1979 | 925 | – | 28 |
| Pu 239 + 240 | 1980 | 1.5 | – | 0.044 |
| H 3 | 1981 | 25 766 | – | 760 |

[a] TBq = Terabecquerel = $10^{12}$ Bq (= 27 Ci).
[b] In the region of the Norwegian Deep is only the upper 100 m water layer included (measurement values of the radio nuclides are normally available for this layer only).
[c] Total water column.

investigations carried out by the Deutsches Hydrographisches Institut in Hamburg and the MAFF Fisheries Laboratory in Lowestoft, England, proved clearly that these nuclides originated from the nuclear fuel reprocessing plants Sellafield Works near the Irish Sea and La Hague near the English Channel.

The discharge into the sea of radio nuclides with the waste waters of normal nuclear power plants is so slight that the activity concentrations in the water resulting therefrom can practically not be distinguished from those of the nuclides still present from the time of the atomic bomb tests.

The North Sea is the only sea area on the earth in the vicinity of which three nuclear fuel reprocessing plants are situated, the radioactive waste waters of which reach the sea directly and flow into it. These are Sellafield Works near the Irish Sea, Dounreay in North Scotland near the Pentland Firth, and La Hague near the English Channel.

The largest activity quantities by far are discharged into the sea from the Sellafield Works. The discharge by La Hague, of the nearest corresponding nuclides, contributes comparatively only a few percent. The plant at Dounreay discharges only insignificant quantities of radio nuclides into the sea, which do not make themselves conspicuous in the North Sea.

The main quantity of the radioactivity discharged – which in all cases is well below the quantities permitted for the individual works – is restricted to only a few radio nuclides within a broad spectrum. The isotopes caesium 137, caesium 134, strontium 90, ruthenium 106, antimony 125, and tritium belong particularly to them. The quantities of transurancis discharged, and especially the plutonium 239 + 240, lie at more than 2 orders of magnitude below that of the Cs 137. In general, in the water of the North Sea activity concentrations of kBq $m^{-3}$ are met with for $^{3}H$; Bq $m^{-3}$ for Cs, Sr, Ru, Sb; and mBq $m^{-3}$ for the transurancis $^{239+240}Pu$, $^{238}Pu$, and $^{241}Am$.

In Table 2, the discharge rates for different radio nuclides, insofar as they were available, are given for the Sellafield Works and La Hague over several years. Thereby, the essentially lesser discharge from La Hague and the different percentual mixture of the quantities of nuclides discharged are clearly recognizable.

Only Ru 106 was discharged from both plants in comparable quantities. In 1983, by gamma spectroscopy, from 50 l seawater not only Ru 106 but also Sb 125 could be detected in North Sea water from Dover to as far as the German Bight. Both radio nuclides originated from the waste waters from La Hague. The Ru 106 discharged from Sellafield Works could no longer be detected in the North Sea.

Besides these sources for the artificial radio nuclides present today in the North Sea, there is still a smaller constant supply of Cs 137 and Sr 90. This takes place via the water masses flowing into the North Sea from the Atlantic Ocean and the Baltic Sea, as well as the freshwater from the land via the rivers and the runoff (Table 3).

In total, the discharge rates of the Sellafield Works have receded heavily since 1978 (Table 2). Parallel to this, from 1978 to 1984, the quantity of Cs 137 present in the North Sea has decreased by around 62% (Table 1).

Unfortunately, from the entire region of the North Sea, there are only sufficient measurement data available from individual years for Sr 90 (1979), Pu 239 + 240 (1980), and tritium (1981) (Wedekind 1982) in order to be able to undertake an estimate of the total content of those nuclides in the water of the North Sea. These values, together with those of the Cs 137, for which corresponding data are available over several years, are summarized in Table 1.

In contrast to those of the Sellafield Works, the discharge quantities of radio nuclides from La Hague, although on an essentially lower level, have remained largely the same between 1972 and 1982. In the case of Sr 90 only, one can recognize clear variations in the discharge rates (to about a factor of 3) (Table 2).

The artificial radio nuclides present in the seawater, according to their different sources, are in no way distributed evenly over the region of the North Sea. The highest activity concentrations for Cs 137, Sr 90, and Pu 239 + 240 are generally to be found in the northwestern North Sea near the Pentland Firth and the adjoining regions of the British East coast towards the south. This can be traced back to the discharge of the radio nuclides originating from the Sellafield Works with the water flowing into the North Sea in the region of the Orkney Isles.

In the southern and southeastern North Sea, which is influenced by the water flowing in near Dover, as a result of the essentially lower nuclide discharge from La Hague, correspondingly lower activity concentrations are present.

Thus, for example, in February/March 1978, the year of the heaviest introduction of radio nuclides into the North Sea, in the regions lying southwards of the Pentland Firth between 58°N and 56°N Cs 137 activity concentrations up to 555 Bq $m^{-3}$ were measured. In the same period of time, on the other hand, in water near Dover only 16 Bq $m^{-3}$ were present. In May 1985, the corresponding values lay at around 110 Bq $m^{-3}$ and 16 Bq $m^{-3}$.

For Sr 90, in 1978 southwards of the Pentland Firth between 58°N and 56°N activity concentrations of between 60 and 80 Bq $m^{-3}$ were measured and near Dover

**Table 2.** Discharge rates of different radio nuclides (TBq $a^{-1}$) in different years from the Sellafield Works and La Hague (Cambray 1982; Calmet and Gueguéniat 1985; Hunt 1986 and previous reports)

| Year | Sellafield Works | | | | | La Hague | | | | | |
|---|---|---|---|---|---|---|---|---|---|---|---|
| | Cs137 | Cs134 | Sr90 | Ru106 | Pu239 + 240 | Cs137 | Cs134 | Sr90 | Ru106 | Pu239 + 240 | Sb125 |
| 1970 | 1154 | 251 | 232 | – | 34 | 89 | 14 | 4 | 200 | 0.023 | 0.9 |
| 1971 | 1325 | 236 | 456 | – | 42 | 242 | 48 | 17 | 285 | 0.145 | 2.3 |
| 1972 | 1289 | 215 | 561 | – | 57 | 33 | 6.1 | 33 | 280 | 0.066 | 18 |
| 1973 | 768 | 166 | 275 | – | 66 | 69 | 8.4 | 19 | 263 | 0.081 | 66 |
| 1974 | 4061 | 997 | 394 | – | 46 | 56 | 9.0 | 104 | 537 | 0.551 | 69 |
| 1975 | 5231 | 1081 | 467 | 761 | 44 | 34 | 4.3 | 75 | 829 | 0.261 | 72 |
| 1976 | 4289 | 738 | 383 | 766 | – | 35 | 6.5 | 40 | 555 | 0.156 | 36 |
| 1977 | 4478 | 594 | 427 | 816 | 36 | 51 | 9.5 | 73 | 539 | 0.239 | 55 |
| 1978 | 4088 | 404 | 598 | 810 | 46 | 39 | 7.8 | 140 | 801 | 0.215 | 62 |
| 1979 | 2562 | 235 | 252 | 393 | 37 | 23 | 3.6 | 117 | 747 | 0.236 | 53 |
| 1980 | 2966 | 239 | 352 | 344 | 20 | 27 | 3.9 | 59 | 774 | 0.156 | 51 |
| 1981 | 2357 | 168 | 277 | 530 | 16 | 39 | 6.0 | 54 | 639 | 0.165 | 48 |
| 1982 | 2000 | 138 | 319 | 419 | – | 51 | 8.4 | 73 | 738 | 1.900 | 74 |
| 1983 | 1200 | 89 | 204 | 553 | – | | | | | | |
| 1984 | 434 | 35 | 72 | 348 | – | | | | | | |
| 1985 | – | – | 52 | 81 | – | | | | | | |

**Table 3.** Estimation of the yearly influx of Cs 137 and Sr 90 into the North Sea from other sources than nuclear facilities (Kautsky and Murray 1981; Kautsky 1983)

| Assumptions for calculation | | | |
|---|---|---|---|
| | Average activity concentration in Bq $m^{-3}$ | | |
| Medium | Cs 137 | Sr 90 | |
| Atlantic water | 4.4 | 3.7 | |
| Baltic Sea water | 22 | 22 | |
| Runoff | 1.9 | 19 | |
| Whole North Sea | | Content in TBq | |
| inflowing water | $Km^3$ | Cs 137 | Sr 90 |
| Runoff incl. rivers | 290 | 0.55 | 5.3 |
| Atlantic from NW | 9000 | 40 | 33 |
| Atlantic from N | 8700 | 39 | 32 |
| Near Dover | 4500 | 20 | 17 |
| Baltic Sea | 500 | 11 | 11 |
| Fallout | — | 5.5 | 5.5 |
| Total | 22,990 | 116 | 104 |

11 Bq $m^{-3}$. In 1985, the corresponding values lay at around 20 Bq $m^{-3}$ in the north-west, and 49 Bq $m^{-3}$ near Dover (in June 1984, 22 to 29 Bq $m^{-3}$ and 8.6 Bq $m^{-3}$).

Normally, the activity concentrations of Sr 90 in the water of the North Sea lie clearly below those of Cs 137. Only in the region of Dover and the southeastern North Sea, as the result of the annually strongly varying discharge quantities of Sr 90 from La Hague, can intermittent clearly higher Sr 90 activity concentration values occur in the water.

For Pu 239 + 240, in August 1980 between Pentland Firth and 57°30′N, activity concentration values between 135 and 160 mBq $m^{-3}$ were measured; near Dover 34 mBq $m^{-3}$. In May 1985, the corresponding values lay at 117 mBq $m^{-3}$ and 36 mBq $m^{-3}$.

On the other hand, tritium showed another distribution pattern, because its main input takes place over the precipitation, the runoff from land, and the rivers. The highest activity concentration values can therefore be measured normally near the land, and especially in the German Bight. Thus, in August 1980, southwards of the Pentland Firth near 58° to 57°N, tritium activity concentrations from 0.8 to 0.9 kBq $m^{-3}$ were measured, near Dover around 1.2 kBq $m^{-3}$, and in the German Bight 2.45 kBq $m^{-3}$. In 1981, the corresponding values have decreased to 0.6 to 0.7 kBq $m^{-3}$, 1.0 kBq $m^{-3}$, and 1.9 kBq $m^{-3}$.

In general, one can observe a gradual reduction of the activity concentrations of artificial radio nuclides in the North Sea.

The behavior of each individual radio nuclide in seawater is different according to the element (Pentreath 1985).

Cs 137 and Cs 134 are predominantly present in solution. Our own estimations, as well as calculations carried out by other scientists (Jefferies et al. 1982;

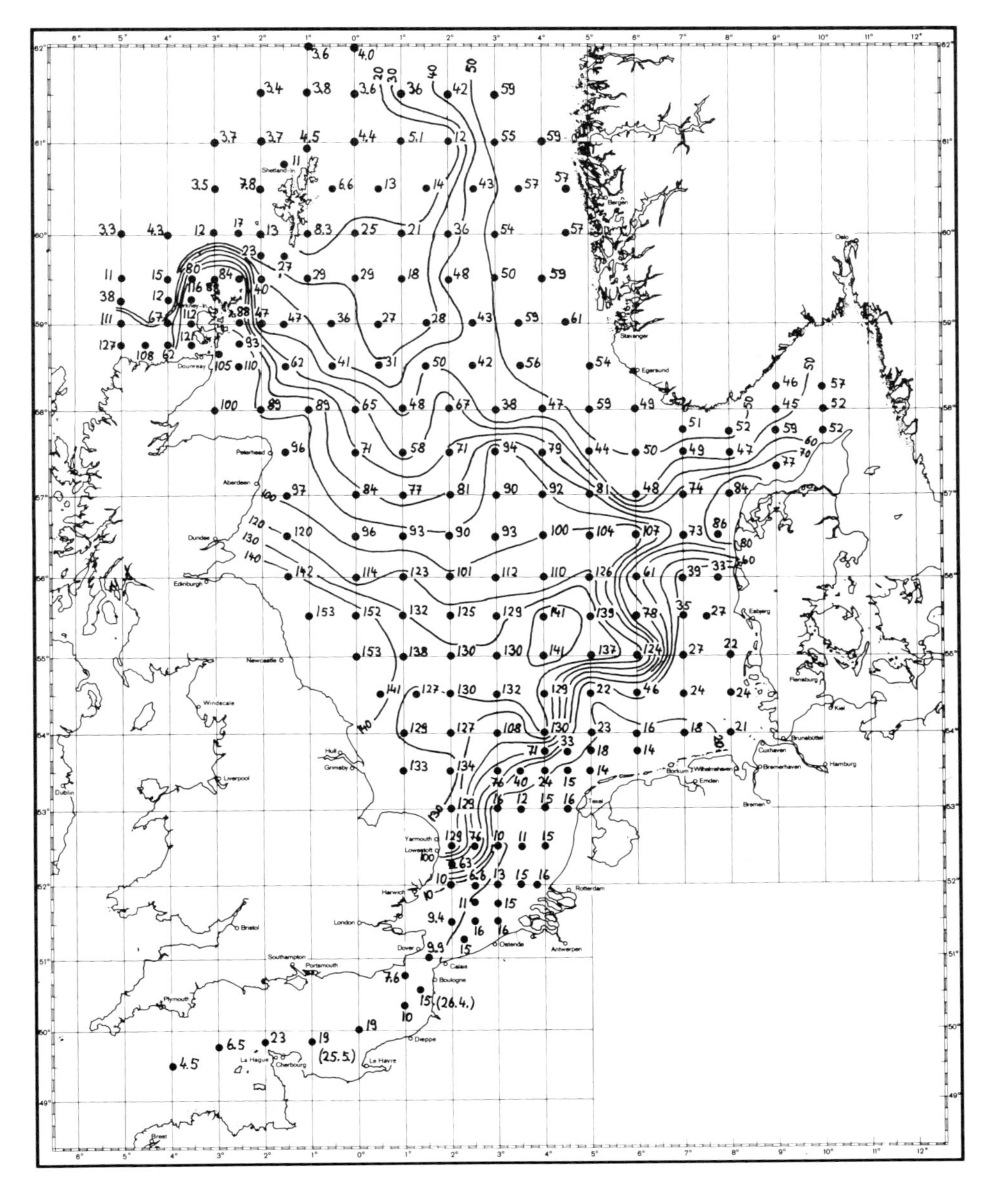


**Fig. 1.** Cs 137 + 134 in surface water (Bq $m^{-3}$) April/June 1984

Livingston et al. 1981) indicate a loss of Cs 137 from the seawater by absorption into particulate matter or bottom sediments to be in the range of no more than about 5 to 10%.

The measurement data of Cs 137 + 134 from 1984 given in Fig. 1 should provide an idea of the distribution pattern of artificial radio nuclides in the surface water of the North Sea.

Sr 90, under the conditions present in seawater, is easily soluble. An absorption worth mentioning on suspended particles or sediments is not known. Seawater itself contains around 8 mg $kg^{-1}$ of inactive strontium.

The Pu 239 + 240 (likewise Pu 238) is predominantly already enriched in the bottom sediments in the vicinity of the waste water discharge of the reprocessing plants. The part still remaining in solution shows practically the same distribution behavior in water as Cs 137. However, its behavior in the marine environment must be regarded as being extraordinarily complex.

The compounds of the elements Ru, Sb, Ce, Zr, and Nb are the same as those of the heavy metals mostly present for only a limited period in soluble form in the alkaline milieu of the seawater. They are transformed into hydroxides or oxides, and are then present in the form of fine, insoluble particles, which can sediment.

Ruthenium has a certain exceptional position, which in part is discharged from the reprocessing plants in the form of rather stable chemical complex compounds, and thereby can remain for a long time in solution.

Tritium is present in the form of tritiated water molecules, and is without limit mixable with the rest of the water.

The accident at Chernobyl had led, at the end of May beginning of June 1986, to Cs 137 values of over 100 to up to 205 Bq $m^{-3}$ in the German Bight and along the coast of Jutland (DHI 1987). As a result of the constant water exchange processes, the radio nuclides which entered the water of the North Sea from the fallout have in the meantime largely been transported away again. Only in the sediments especially near the coasts, can one expect a certain enrichment of different nuclides resulting from this accident.

## 3 Transport Paths and Times

The distribution picture of the radiocaesium in the North Sea, measured over many years (1971 to 1984) and in principle only slightly variable, permits a clear estimation of the existing transport paths (Fig. 2).

The water flowing into the North Sea in the region of the Orkney Islands moves along the British east coast towards the south. In the course of this route, one can recognize in three regions, at ca. 57° N to 58° N, 55° N to 56° N, and 53° N to 54° N, transport currents directed towards the east or northeast.

The water near Dover, flowing into the North Sea from the south, moves along the mainland coast, in a relatively narrow strip, towards the north.

These two water masses flow between about 52° 20′ N 2° E and 56° N 6° 30′ E, alongside but separate from one another. Westwards of Jütland, between 54° 30′ N and 56° N, the boundary between the two waterbodies can be observed practically constantly at around 6° 30′ E.

A mixing of these two waterbodies and that of the water masses on the northernmost transport route towards the east, first takes place at about 57° N and particularly in the region of the Skagerrak.

The further transport then takes place mainly along the Norwegian coast towards the north; small quantities reach the Baltic Sea (Aarkrog et al. 1982; Kautsky et al. 1980; Kautsky 1981).

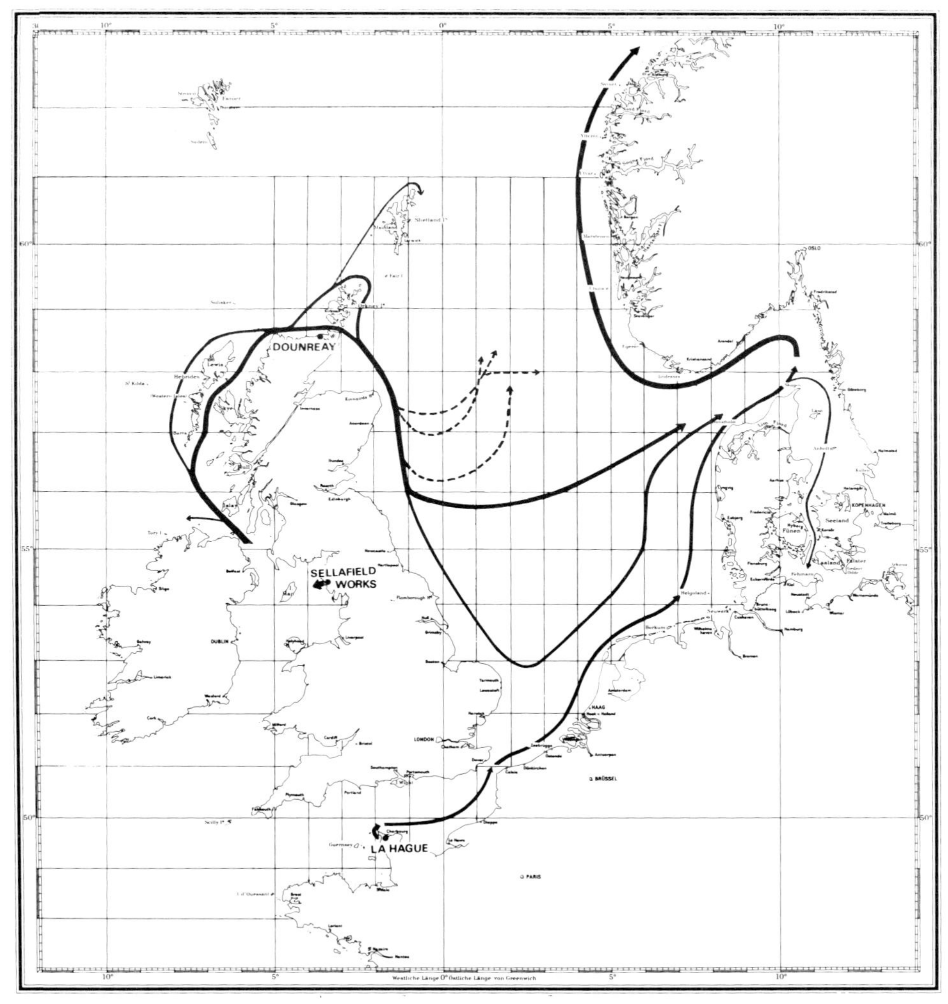


**Fig. 2.** Transport routes of Cs 137 deduced from the measurements of the activity concentration distribution in the years 1971 to 1984. *Dotted lines* indicate temporary different transport routes

The main quantity of the radio nuclides originating from the Sellafield Works is transported through the central and southwesterly North Sea in the direction of the Skagerrak.

The course of the two curves of the Cs:Sr ratio in the discharge of the Sellafield Works and in the region of the Pentland Firth, with a tendency of a 3- years' shift, can be viewed as being parallel. That indicates a transport time of about 3 years from the Sellafield Works as far as the Pentland Firth (Fig. 3). If one compares further the corresponding ratio figures westwards of Jütland at 55° 30′ N and 6° E, this then results in a practically parallel curve, but offset by a further year. That means that the time of transit of the water mass between the Pentland Firth and this region must lie at about 1 year.

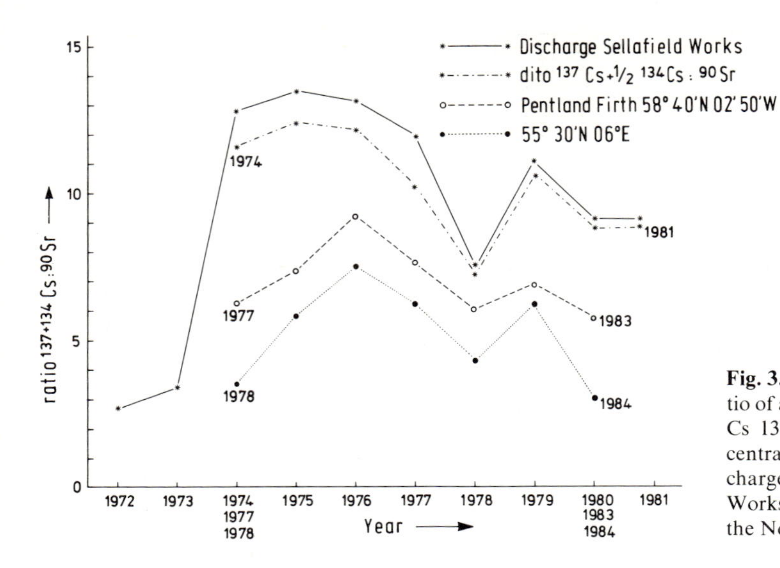


**Fig. 3.** Comparison of the ratio of activity concentration of Cs 137 + 134 to activity concentration of Sr 90 in the discharge from the Sellafield Works and at two positions in the North Sea

Measurements of the ratio Cs 134:Cs 137 (from 2000 l water) in the years 1983 and 1984 near 58° 30′ N 2° 30′ W, 53° 30′ N 1° 00′ E, and 57° 00′ N 7° 00′ E in comparison with the Cs ratio in the discharge from Sellafield Works resulted in transport times of 3 years as far as the Pentland Firth and 4 to 4.5 years as far as into the southwesterly North Sea resp. the entrance to the Skagerrak. Between La Hague and the German Bight, on the basis of measurements taken in the years 1971/72 and 1974/75, transport times of around 15 months were ascertained. In accordance with recent investigations, using the irregular Sr 90 discharges from La Hague, amongst other things shorter transport times for this route of less than 1 year are indicated (Gabriel, pers. commun.).

## 4 Further Research Needed

Even if a hazard for the biosphere caused by the quantities of artificial radio nuclides present in the seawater at the moment, is certainly not to be expected, the further development of their spreading should still be carefully observed. Their path through the biocycle, and especially their final deposition in the sea cannot yet be viewed as being satisfactorily clarified.

As it is to be assumed that the radio nuclides are ultimately deposited in the bottom sediments, particular attention should be directed to the investigation of the sediments.

Substances which are firmly integrated in the bottom, possibly still in its deeper layers, are practically removed from the biosphere.

On the other hand, there always exists solution equilibrium between substances particulate-bound and in solution, so that deposited nuclides can always return again in solution in measurable quantities.

# References

Aarkrog A, Bøtter-Jensen L, Dahlgaard H, Hansen H, Lippert J, Nielsen SP, Nilsson K (1982) Environmental radioactivity in Denmark 1981. Risø-Rep 469

Calmet D, Guegueniat P (1985) Les rejets d'effluents liquides radioactifs du centre de traitement des combustibles irradiés de La Hague (France) et l'évolution radiologique du domaine marin. Vienna Intl Atomic Energy Agency, TECDOC-329

Cambray RS (1982) Annual discharge of certain long-lived radio nuclides to the sea and the atmosphere from Sellafield Works, Cumbria 1957 to 1981. Harwell: U K Atomic Energy Authority, AERE-M 3269

DHI (1987) Meereskundl.Beobacht.und Ergebnisse Nr.62: Die Auswirkungen des Kernkraftwerksunfalls von Tschernobyl auf Nord- und Ostsee, August 1986 überarbeitet und ergänzt Januar 1987. Dtsch Hydrogr Inst Hamburg, Nr 2149/34

Hunt GT (1986) Radioactivity in surface and coastal waters of the British Isles. Aquat Environ Monitoring Rep, MAFF Direct Fish Res, Lowestoft (see also the same reports of previous years)

Jefferies, DF, Steele AK, Preston A (1982) Further studies on the distribution of Cs137 in British coastal waters, Irish Sea. Deep Sea Res 29:713–738

Kautsky H (1981) Radiological investigations in the western Baltic Sea including Kattegat during the years 1975 to 1980. Dtsch Hydrogr Z 34:125–149

Kautsky H (1983) The introduction of radio nuclides over the atmosphere into the North Sea. Dtsch Hydrogr Z 36:145–155

Kautsky H (1985) Distribution and content of different artificial radio nuclides in the water of the North Sea during the years 1977 to 1981. Dtsch Hydrogr Z 38:193–224

Kautsky H (1986) Distribution and content of Cs137 + 134 in the water of the North Sea during the years 1982 to 1984 Dtsch Hydrogr 239:139–159

Kautsky H, Murray CN (1981) Artificial radioactivity in the North Sea. Atomic Energy Rev Suppl 2:63–105

Kautsky H, Jefferies DF, Steele AK (1980) Results of the Radiological North Sea Programme RANOSP 1974 to 1976. Dtsch Hydrogr Z 33:152–157

Livingston HD, Bowen VT, Kupferman SL (1981) Radio nuclides from Windscale discharges, 1. Nonequilibrium tracer experiments in high latitude oceanography. J Mar Res 40:253–272

Pentreath RJ (1985) General review of literature relevant to coastal waste discharges. In: Behaviour of radio nuclides released into coastal waters. IAEA TECDOC 329:17–99

Wedekind Ch (1982) Tritium distribution and spreading in the North Sea and the Baltic Sea in 1980/81 as well as in the surface water of the North Atlantic in 1979. Dtsch Hydrogr Z 35:117–183

# Mathematical Modelling as a Tool for Assessment of North Sea Pollution

J.A. van Pagee, P.C.G. Glas, A.A. Markus, and L. Postma[1]

## 1 Introduction

During the last decades the North Sea has been exposed to a seriously increased input of various pollutants. Various solid and liquid waste materials are deposed into the North Sea, directly from outfalls, ships and offshore activities, and indirectly through river discharges, atmospheric deposition and cross-boundary inflows from the English Channel, the North Atlantic Ocean, and the Baltic Sea.

In order to assess the rate and risk of North Sea pollution, an integrated approach of physics, chemistry and biology is needed. In such an approach attention should be paid to the input of pollutants, the transport of pollutants through the North Sea, the subsequent pollution of water and sediments and the (potential) impact of the different pollutants on both individual organisms and population dynamics.

Within the framework of drawing up a Water Quality Management plan for the Dutch part of the North Sea, studies have been carried out to quantify the impact of anthropogenic (human-derived) pollution inputs on coastal water quality, sediments and biota (Rijkswaterstaat 1985; Beukema et al. 1986). As the Dutch part of the North Sea cannot be studied separately, the study area comprised the southern North Sea between the Straits of Dover and 56°N.

To study the effects of anthropogenic pollution inputs, attention was given to the impact of increased inputs of nutrients and hazardous substances. Nutrients like phosphorus (P) and nitrogen (N) are studied because the input of these substances has shown a drastic increase during recent decades. The increased availability of these nutrients for phytoplankton growth might result in changes of algal biomass and species composition, and consequently influence the North Sea ecosystem as a whole. Hazardous substances are studied because of their threat to marine organisms. Due to a lack of data for organic micro-compounds, the studies were mainly concerned with heavy metals, namely cadmium (Cd), mercury (Hg), copper (Cu), zinc (Zn), chromium (Cr) and lead (Pb).

Figure 1 roughly presents the framework of analysis as used in these studies. The figure illustrates the relation between hydrodynamics, sediment transport, water quality, sediment pollution and the impact on marine resources by eutrophication and/or toxicological effects. As the transport of pollutants and their

[1]Delft Hydraulics, Water Resources and Environment Division, P.O. Box 177, 2600 MH Delft, The Netherlands

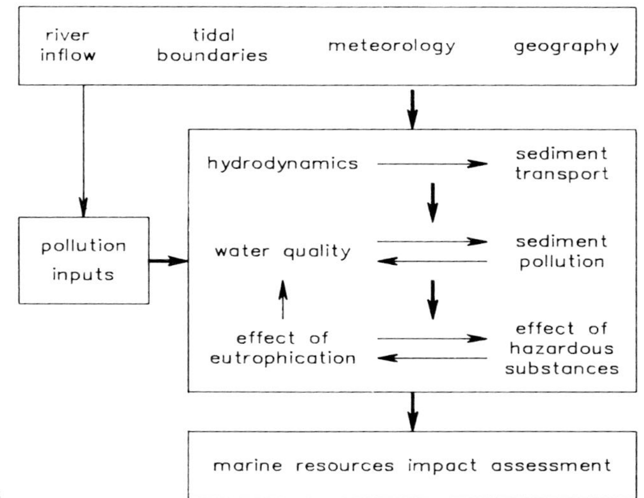


**Fig. 1.** Framework of analysis

impact on biota depend on a complex interaction of various processes, mathematical modelling techniques are used to determine the impact of pollution inputs.

Due to the wide variety in spatial and temporal scales, the analysis cannot be carried out by one single method. In estuarine and coastal regions the transport phenomena are highly influenced by the tidal motion, while in the North Sea the transport over time periods up to more than one year are relevant as well. Time scales for accumulation of pollutants in sediments can even reach more than 100 years. Within the studies to prepare the Water Quality Management Plan the main objective was to obtain an overall view of the impact of various pollution inputs in the Dutch part of the North Sea for yearly average or, if relevant, seasonal conditions. The results have been used for a tentative assessment of North Sea pollution and to defined priorities in coastal pollution research and control in the Netherlands.

Recently additional studies have been carried out using similar modelling techniques. In these studies special attention was given to (1) the long-term characteristics of transport of dissolved constituents in the southern North Sea, (2) the transport of suspended solids and adsorbed contaminants, (3) the impact of increased nutrient supply on primary production and subsequent risk of oxygen depletion in stratified regions. Although some of the results of these studies should be qualified as tentative results, they have already increased the insight into the complex functioning of the North Sea ecosystem in relation to waste disposal.

In this paper an overview is given with respect to the role of mathematical modelling in assessment of North Sea pollution. In accordance with the framework of analysis, the steps described include the modelling of hydrodynamics and transport phenomena, the transport of dissolved and suspended pollutants, the risk of increased heavy metal and nutrient inputs.

## 2 Hydrodynamics and Transport Phenomena

In order to analyze the influence of pollution inputs on the water quality of the North Sea, mathematical modelling techniques are used to quantify the transport of pollutants. In the shallow southern part of the North Sea, the influence of density gradients is relatively small in comparison with the influence of tide, wind and topography. Therefore a set of vertically integrated two-dimensional hydrodynamic models is used to simulate wind and tide-induced currents. The areas covered by the models are depicted in Fig. 2. The largest model covers the whole continental shelf sea around the British Isles and is mainly used for storm surge predictions

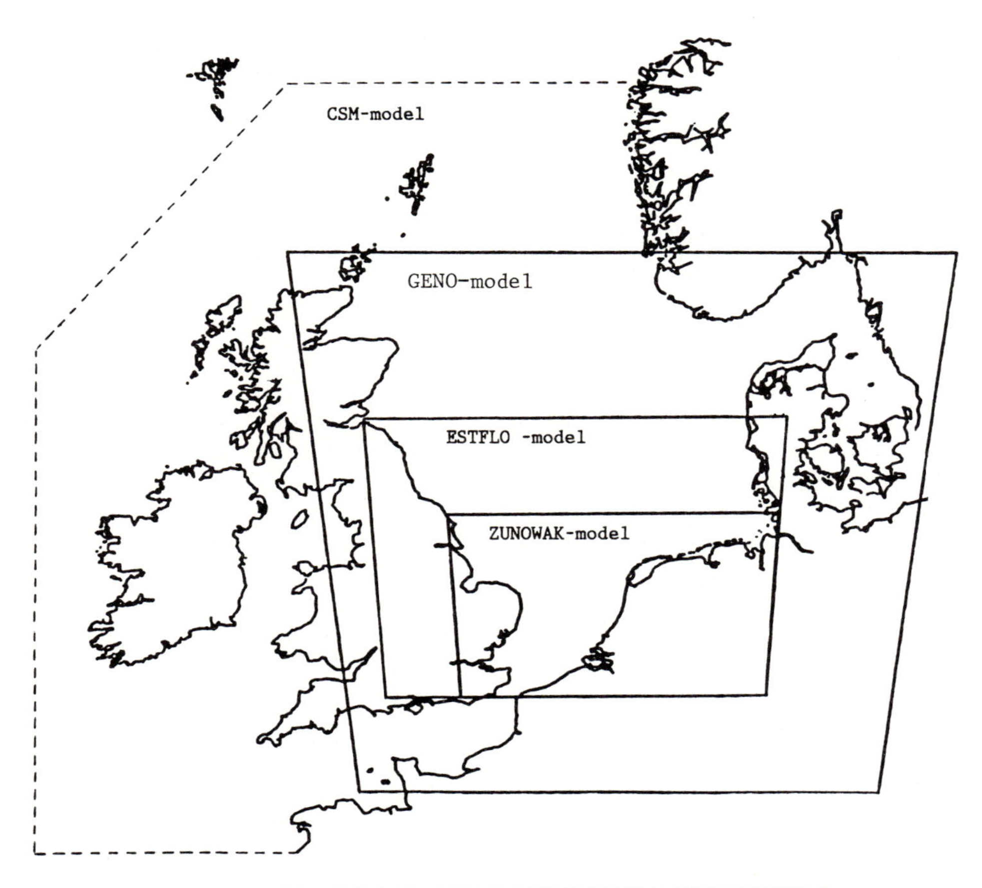


| Model | Gridsize | Comp. elements | Reference |
|---|---|---|---|
| CSM | 1/8° x 1/12° | ca. 20.000 | Verboom (1986) |
| GENO | 8 x 8 km | 10.561 | Voogt (1984) |
| ZUNOWAK | 3.2 x 3.2 km | 12.527 | Dijkzeul (1985) |
| ESTFLO | 10 x 10 km | 2.199 | Gerritsen (1986) |

**Fig. 2.** Boundaries of the models used for this study

(Verboom et al. 1986). From this Continental Shelf Model (CSM) the hydrodynamic boundary conditions are derived for the GENO model, which in its turn provides the boundary conditions for the ESTFLO and ZUNOWAK models. To use the results as a base for transport and water quality modelling the boundary conditions represent a mean cyclical tidal period of 24 h 50 min. As shown from the model specifications, the ZUNOWAK model is the most detailed. Its application, however, is limited to the area between the Straits of Dover and 54° 15′ N.

In order to quantify the influence of various pollution inputs on the long-term water quality of the North Sea, average wind conditions and river inflows are used for the simulation of water movements. With the boundary conditions given in terms of Fourier components, a fully cyclic flow situation is reached as soon as the initial state disturbance has disappeared. Integration of the computed tidal flows over the diurnal tidal period then yields the (time-independent) residual flow pattern.

Using average south-westerly winds, the streamlines of residual transport as calculated by the models show a counter-clockwise circulation from Scotland along the British coast towards the Danish coast where its joins the net inflow from the English Channel (Fig. 3a). Although this circulation pattern is less pronounced with average north-westerly winds, the overall pattern still holds (Fig. 3b). Similar circulation patterns were found by other models (Prandl 1978; Backhaus 1985). It should be noticed, however, that this pattern can be disturbed as a result of changing windfields (Backhaus 1987).

As shown in Fig. 4, the calculated residual transport velocities for a 4.5 m/s south-westerly windfield range between 0.01 and 0.13 m/s.

To assess the transport of substances in the North Sea, a transport and water quality model is used, based on the vertically averaged advection-diffusion equation (Pagee et al. 1986; Ruijter et al. 1987; Postma 1988). The same grid is applied as for the corresponding hydrodynamic models. The equation is solved with an implicit central difference scheme, which has no numerical diffusion up to the second order.

The steady residual transport velocity field, calculated by the hydrodynamical models, is used as advective term in the transport model. A mixing term accounts for turbulent diffusion and the combined effects of integrating over varying wind fields, tides and depth. The dispersion coefficient was calibrated roughly to reproduce global salinity patterns as given by ICES (1962). A uniform value of 150 $m^2/s$ for the GENO and ESTFLO model and 120 $m^2/s$ for the more detailed ZUNOWAK model provided acceptable results (Fig. 5).

The general behaviour of mass transport in the southern North Sea was studied based on various simulations for both conservative and non-conservative inputs of dissolved tracers. An overview of the transport characteristics of cross-boundary inflows from the English Channel and various rivers is presented in the Transport Atlas of the Southern North Sea (Ruijter et al. 1987). As the results can be used to analyze the impact of discharged (dissolved) constituents, the atlas contains a display programme for discharge-related concentration distributions which can be used on a personal computer.

Figure 6 shows the results of individual point loads (1 kg/s) deposed near the mouth of the Humber, Thames, Rhine and Elbe. These results show that, as a

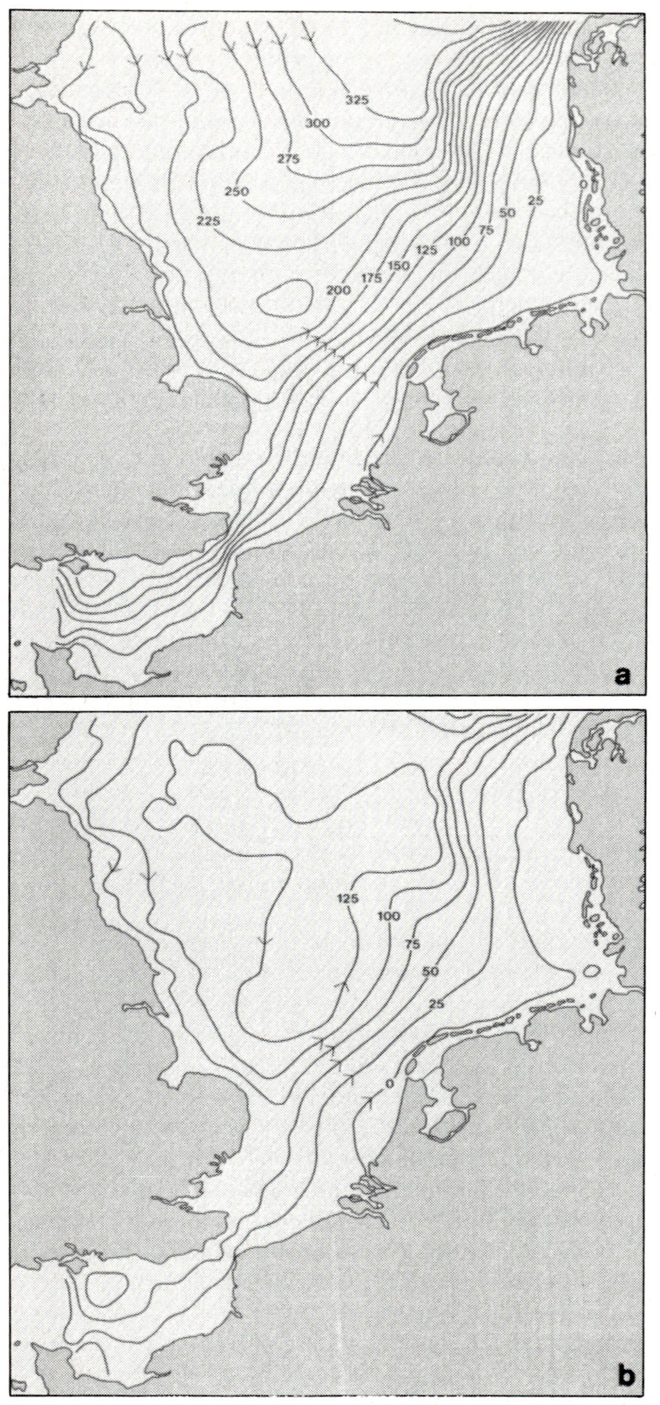


**Fig. 3a,b.** Stream lines of residual transport calculated for 4.5 m/s SW-wind **(a)** and 3.5 m/s NW-wind **(b)** (contour interval 25000 $m^3/s$)

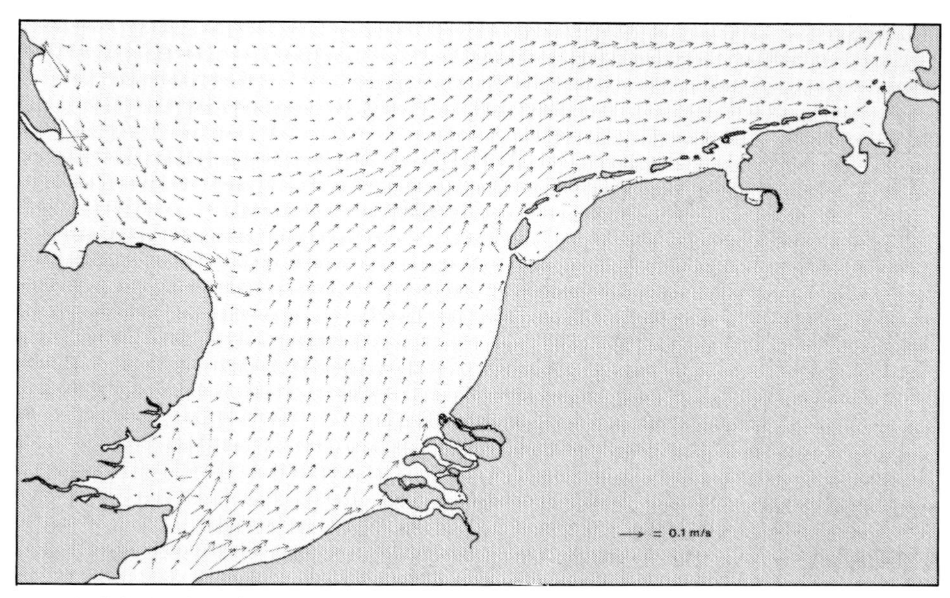


**Fig. 4.** Calculated residual transport velocities for a 4.5 m/s southwesterly wind field

consequence of transport patterns, the British inputs follow the general residual circulation into the open sea, crossing the North Sea towards the German Bight. Constituents deposed from the Continent remain trapped along the coast and also pass through the German Bight into the Danish coastal waters.

## 3 Rate of (Anthropogenic) Pollution of Water and Sediments

To analyze the rate of North Sea pollution, inventories were made of the annual pollutant loads entering the North Sea between the Straits of Dover and 56°N, based on the situation in 1980 (Pagee and Postma 1987). The data on pollutant loads from coastal zones are primarily based on ICES (1978), updated with information from the Governmental contributions to the 1984 North Sea Ministerial Conference in Bremen (Carlson 1986). Estimates of atmospheric deposition are based on observations at land sites close to the North Sea (van Aalst 1982). As these data show a wide range of deposition rates, their average has been derived and applied for the whole area under consideration (ca. 220,000 km$^2$). Cross-boundary inputs from the English Channel and the North Atlantic are calculated from measured concentrations and inflows of water masses as quantified by hydrodynamic modelling.

A summarized overview of land-based inputs from each bordering country is given in Table 1 together with the estimated input from the atmosphere and cross-boundary inflows.

From a review of "natural" river concentrations (Van Eck et al. 1983) an estimate was made for non-polluted river inputs. Based on this information, the

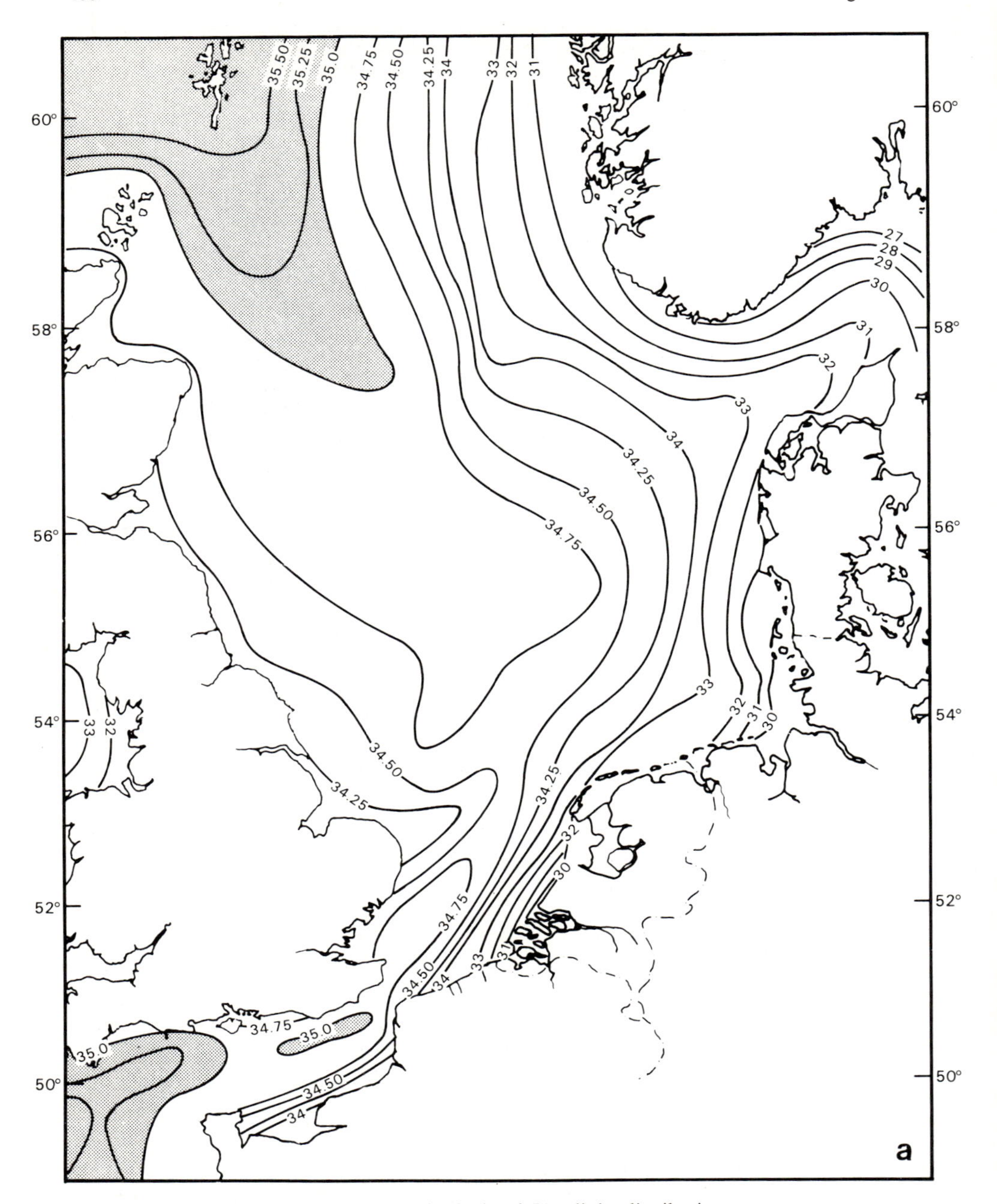


**Fig. 5a,b.** Measured (**a**), (ICES, 1962) and calculated (**b**) salinity distribution

**Fig. 6a-d.** Mass transport calculations for conservative tracer-inputs of 1 kg/s from Thames (**a**), Humber (**b**), Rhine (**c**), Elbe (**d**)

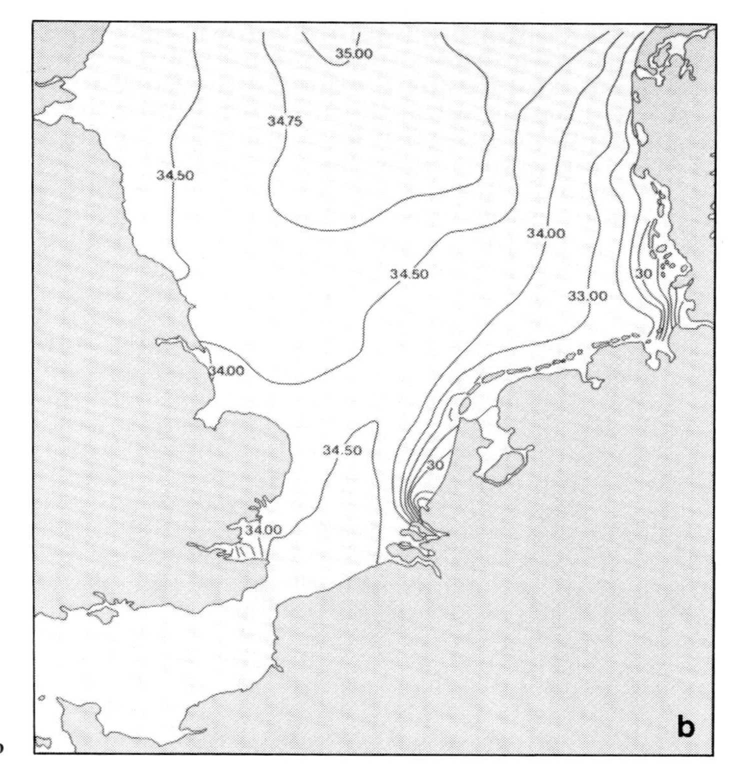
35.00
34.75
34.50
34.00
34.50
33.00
30
34.00
34.50
30
34.00
b

Fig. 5b

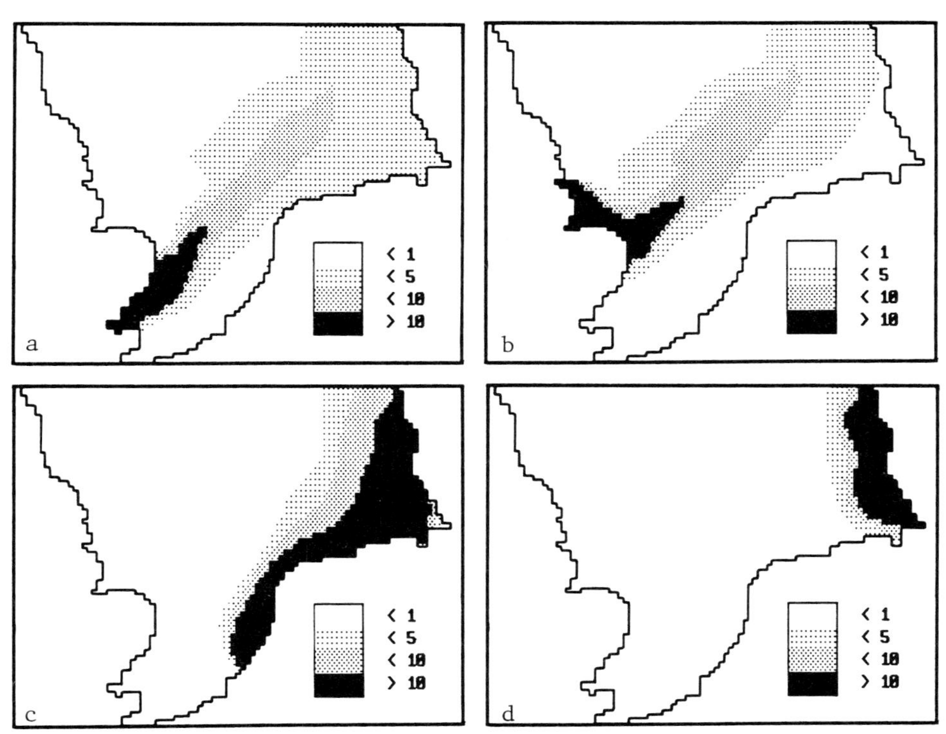
< 1
< 5
< 10
> 10
a
b
c
d

Fig. 6

**Table 1.** Overview of sources in the southern part of the North Sea (area 219,900 $km^2$) and an estimate of the anthropogenic fraction, 1980

| Sources of pollution | Flow $km^3/a$ | N 1000 t/a | P 1000 t/a | Cd t/a | Hg t/a | Pb t/a | Cu t/a | Cr t/a | Zn t/a |
|---|---|---|---|---|---|---|---|---|---|
| Coastal inputs: | | | | | | | | | |
| United Kingdom | 19 | 196 | 28 | 48.4 | 20.2 | 1106 | 1152 | 695 | 4152 |
| Belgium | 0.6 | 7 | 1.2 | 0.1 | 1.1 | 246 | 5 | 4 | 131 |
| Netherlands | 110 | 583 | 72.5 | 134.8 | 20.9 | 1165 | 1125 | 1523 | 7629 |
| W. Germany | 52 | 309 | 25.1 | 20.7 | 9.1 | 244 | 640 | 266 | 4428 |
| Denmark | – | 10 | 0.3 | 3.6 | – | 32 | 21 | 21 | 140 |
| Total (1980) | 182 | 1105 | 127 | 208 | 51.2 | 2793 | 2943 | 2509 | 16480 |
| Total (reference) | (182) | (273) | (18.2) | (12.7) | (7.3) | (619) | (510) | (546) | (2548) |
| Atmosphere | | 220 | 10 | 66 | 8.8 | 2420 | 1320 | 66 | 8806 |
| Channel | 4786 | 957 | 120 | 129 | 15.8 | 813 | 1914 | 3254 | 3493 |
| N. Atlantic O. | 5899 | 767 | 112 | 147 | 15.9 | 295 | 1652 | 2242 | 2537 |
| Total (1980) | 10867 | 3049 | 369 | 550 | 91.7 | 6321 | 7829 | 8071 | 31316 |
| Total (reference) | | (2217) | (260) | (355) | (47.8) | (4147) | (5396) | (6108) | (17384) |
| Anthropogenic fract. | | 27% | 30% | 35% | 48% | 34% | 31% | 24% | 44% |

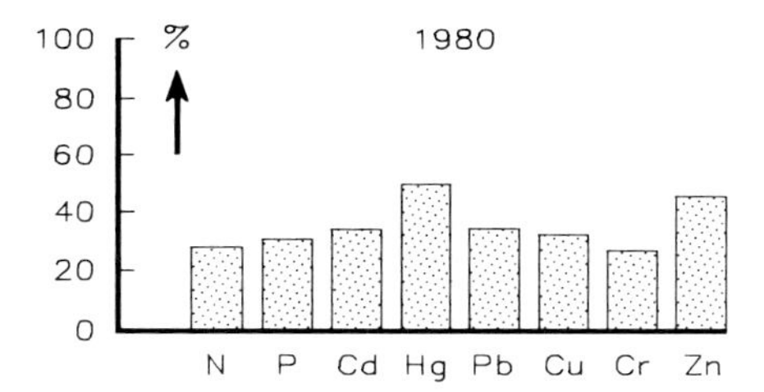


**Fig. 7.** Anthropogenic fraction (%) of total input from all sources

total land-based reference input was derived and compared with the total 1980 input into the considered part of the North Sea. The difference between the 1980 and reference input is defined as the anthropogenic fraction. Figure 7 shows that the anthropogenic influence thus defined reaches between 25–50% for the considered nutrients and heavy metals.

As this percentage represents the average influence for this part of the North Sea, the influence will regionally be higher.

It should be noticed that the anthropogenic contribution by atmospheric deposition and cross-boundary inflows is not considered. The anthropogenic influence in the total input therefore represents a lower limit.

Based on the data of pollution inputs, concentrations of pollutants are calculated with the same model as is used to simulate salinity distributions and transport of water masses. Although the model is able to consider all kinds of physical, chemical and biological processes, the first application of the model is restricted to a conservative behaviour of pollutants. As only total (dissolved + particulate) concentrations of nutrients (N, P) and heavy metals (Cd, Hg, Pb, Cu, Cr, Zn) are considered, this limitation means ignoring the exchange of pollutants with the bottom sediment. Another limitation of the model's application is the use of annually averaged pollution inputs, ignoring temporal variations. As a consequence, these results should be interpreted as a first attempt to quantify the influence of various pollution inputs and to assess the rate of pollution caused by human activities.

Figure 8 shows that the highest concentrations are calculated for the Dutch coastal zone and the German Bight. As the calculated concentrations are based on steady annual pollution inputs for 1980 and a steady residual water circulation for the winter period, a comparison with coherently measured water quality data is not possible. However, an overall comparison of calculated and measured concentrations in the Dutch coastal zone (for 1975–1980) shows remarkable agreement (Table 2).

Another way to analyze the origin of North Sea pollution is to consider the building up of concentrations from west to east over the North Sea along the 53°N, sea Fig. 9. This figure illustrates the increase of concentrations from the open sea towards the coastal region. This increase results partly from the input of higher concentrations from rivers caused by the "natural" weathering process in the drainage area. Due to anthropogenic inputs, however, the concentrations in the coastal zones are increased significantly, as shown by the difference in the calculated reference ($C_{ref}$) and 1980 ($C_{80}$) concentrations. Along the 53°N, the

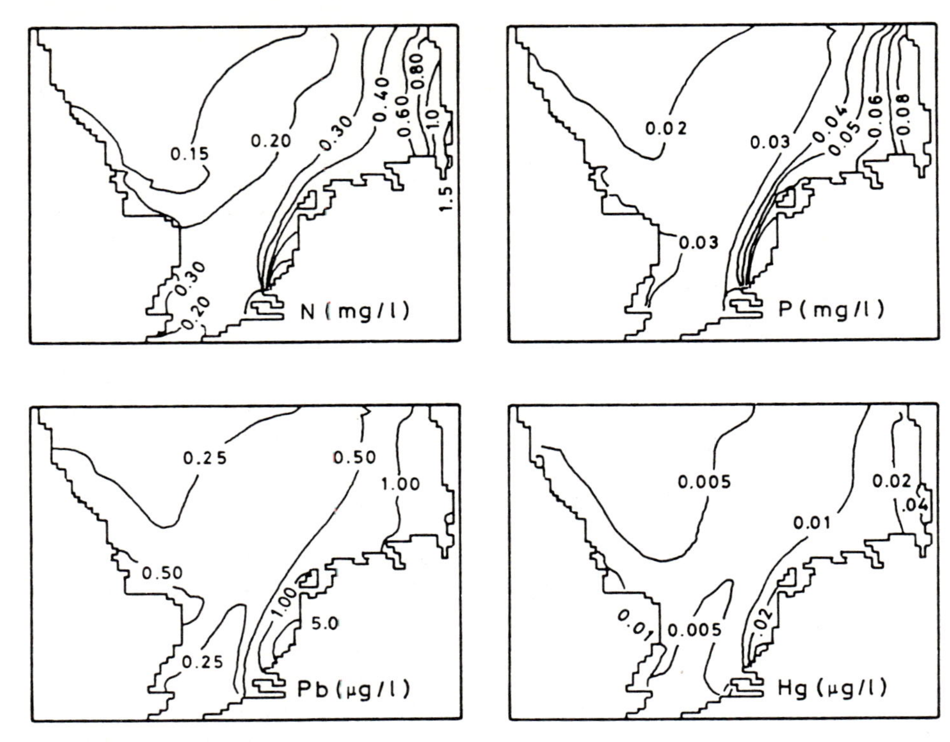


**Fig. 8.** Calculated concentration of N, P, Pb and Hg (pollution input 1980)

**Table 2.** Comparison between observed (Kooy 1983) and calculated pollutant concentrations in Dutch coastal waters

| Pollutant | Observed | Calculated |
|---|---|---|
| N mg/l | 0.2 -1.0 | 0.25 - 1.0 |
| P | 0.03-0.2 | 0.03 - 0.2 |
| Cd µg/l | 0.09-0.15 | 0.04 - 0.20 |
| Hg | 0.01-0.06 | 0.005- 0.03 |
| Pb | 1.0 -2.2 | 0.5 - 2.0 |
| Zn | 2.1 -9.0 | 2 -10 |
| Cr | 0.5 -2.3 | 1 - 3 |

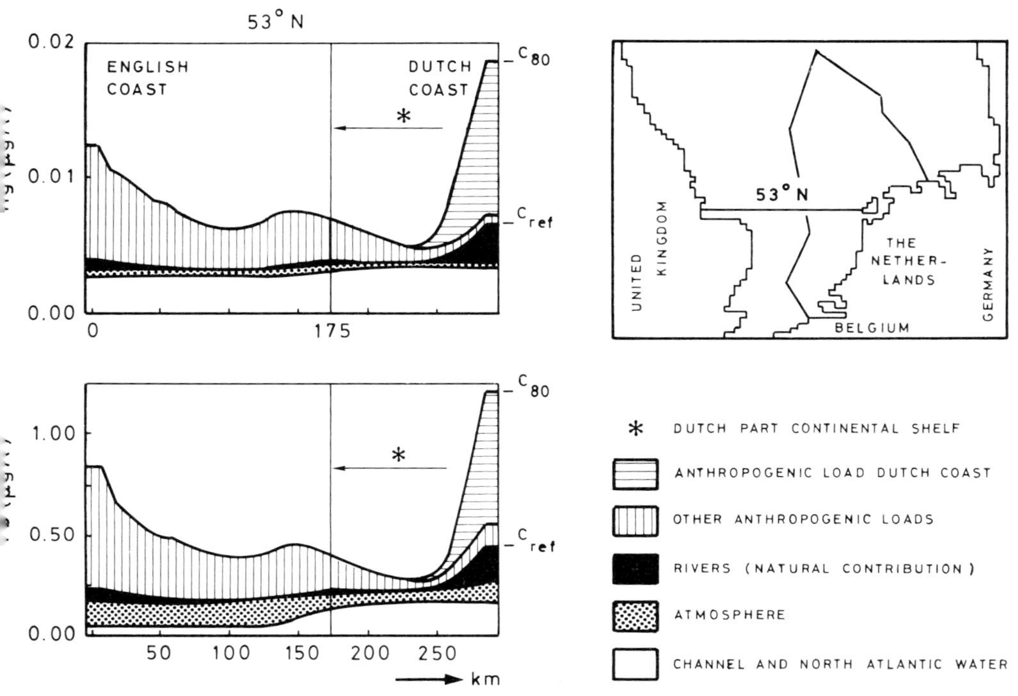


**Fig. 9.** Overview of the Hg and Pb concentration build up in a cross-section between the Dutch and English coast (53°N)

crossing of pollutants originating from the NE and SE coast of the UK in NE direction results in a concentration maximum near the UK-Dutch boundary of the continental shelf.

In order to quantify the contamination due to human activities, estimates have been made of "natural" (uncontaminated) river inputs. Together with "natural" cross-boundary inflows from the Channel and North Atlantic, these "natural" inputs have been used to calculate reference concentrations for the southern North Sea using the water quality model. From the calculated concentrations for the pollution inputs for 1980 ($C_{80}$) and these reference concentrations ($C_{ref}$), so-called anthropogenic fractions (A) can be calculated using the relation:

$$A_{(x,y)} = \frac{C\,80_{(x,y)} - C\,ref_{(x,y)}}{C\,80_{(x,y)}}.$$

As the anthropogenic contribution to atmospheric deposition and cross-boundary inflows is not considered in this analysis, it is likely that the above method provides a relatively low estimate of the anthropogenic influence. Figure 10 shows that in large parts of the southern North Sea the anthropogenic fraction of Cd, Hg, Pb and Zn exceeds 50%. Although the anthropogenic influence in nutrient concentrations is less than for these heavy metals, about 50% of the concentrations of N and P in the Dutch coastal zone and the German Bight originate from human activities.

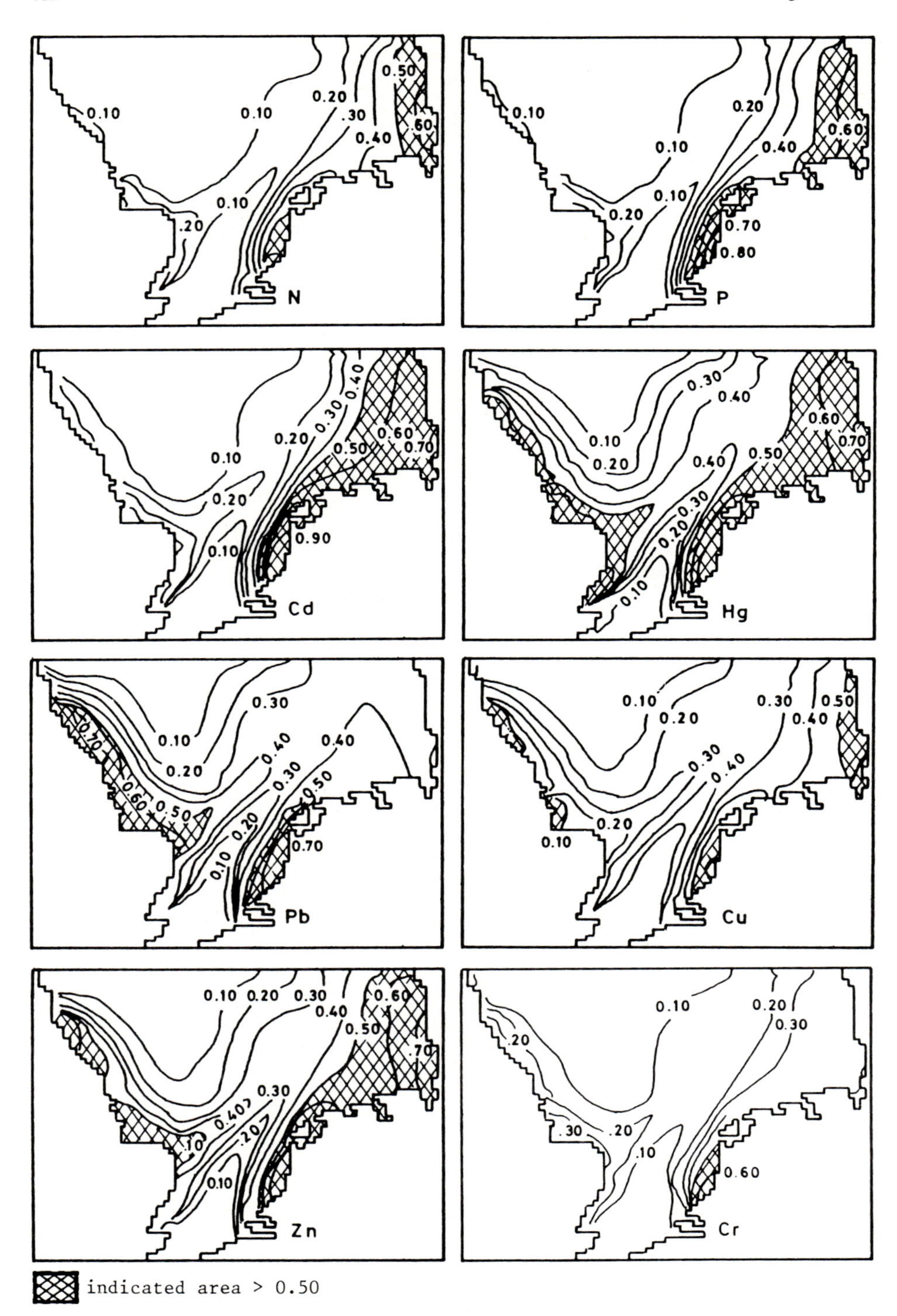


**Fig. 10.** Calculated anthropogenic fractions for nutrients and heavy metals (1980)

## 4 Risk Assessment of Heavy Metals

The risk of heavy metals for marine organisms depends highly on its speciation and related bio-availability. Unfortunately, the chemical speciation cannot be measured in full detail, whereas the measurement of specific fractions needs special attention as a result of the low concentrations in marine waters. Monitoring of metal concentrations in the North Sea is mainly restricted to:

- dissolved concentrations in water;
- particulate concentrations in water;
- particulate concentrations in suspended sediments;
- particulate concentrations in bottom sediments;
- accumulated concentrations in organisms.

The uptake of metals by organisms can be ascribed to both direct uptake from the water and the foodweb. For organisms at the beginning of the foodweb, the uptake is dominated by the direct uptake from the water, whereas for birds and mammals the uptake mainly originates from their food. For pelagic organisms the dissolved fraction mostly determines the uptake from the water. For demersial and benthic organisms, the concentration of interstitial water in the bottom and therefore the quality of bottom sediment seems to be relevant as well. As the pollution of heavy metals increased especially in the period from 1960 till 1980 (see Fig. 11), while the timescale of the biologically active bottom layer is 50 years or more, the accumulation in the bottom is only beginning (Fig. 12).

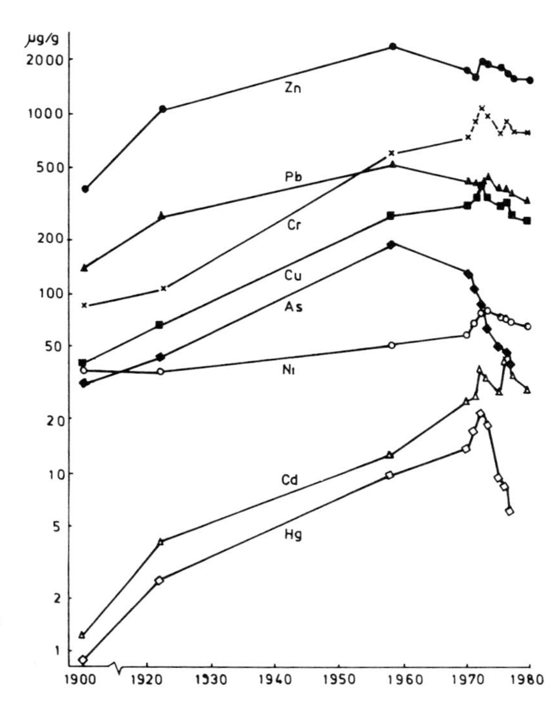


**Fig. 11.** Trends in contamination of Rhine sediments 1900–1980 (Salomons 1984)

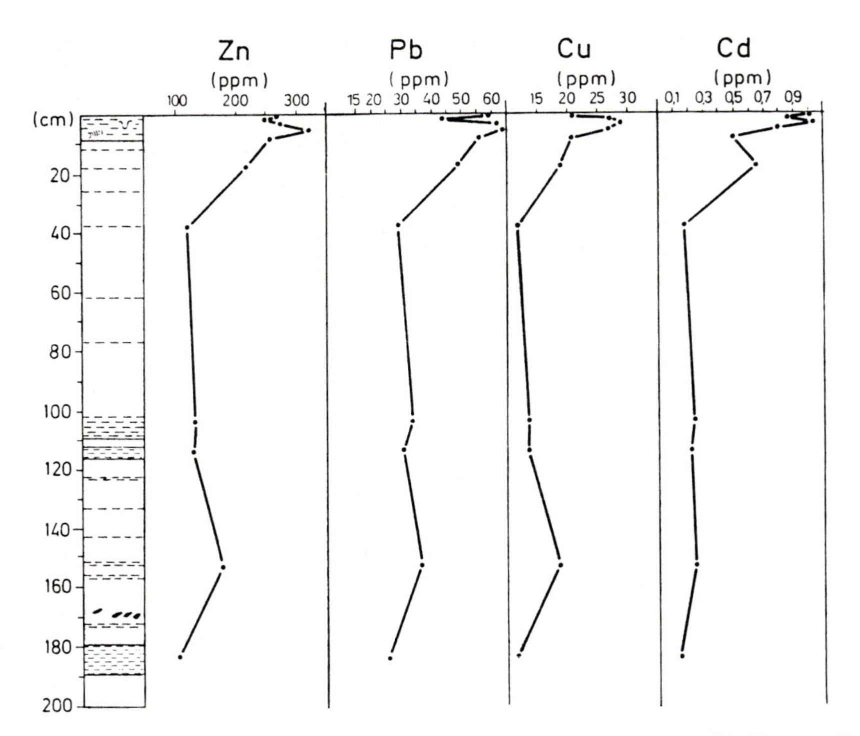


**Fig. 12.** Measured contaminants in bottom sediments in the German Waddensea, Carolinensiel (Schwedhelm and Irion 1985)

In order to analyze the impact of increased heavy metal inputs more clearly, tentative modelling calculations are being carried out to simulate both dissolved and particulate fractions. Ratios between dissolved and particulate fractions are estimated from measured ratios in the river Rhine estuary. To simulate the dissolved fraction the same transport phenomena are used as for the simulation of salinity. The simulation of particulate fractions additionally takes into account the characteristics of erosion and sedimentation in different regions of the North Sea. Net erosion and sedimentation regions are derived from field data (Eisma 1983; Hopstaken 1986; Alphen 1987). Together with a coastwards transport in Dutch coastal waters, the net settling velocities of suspended sediments are derived from a comparison between measured and calculated suspended sediments.

Figure 13 shows the simulated results of dissolved cadmium concentrations and pollution of suspended sediments. The cadmium pollution of suspended sediments represents the ratio between the simulated particulate cadmium concentrations and the simulated suspended sediment concentrations. In net sedimentation regions these concentrations can be considered as a basis for pollution of local bottom sediments.

To analyze the risk of these concentrations for marine organisms, information is needed on accepted standards for ambient water quality in marine waters or critical values related to observed negative effects on marine organisms. Water

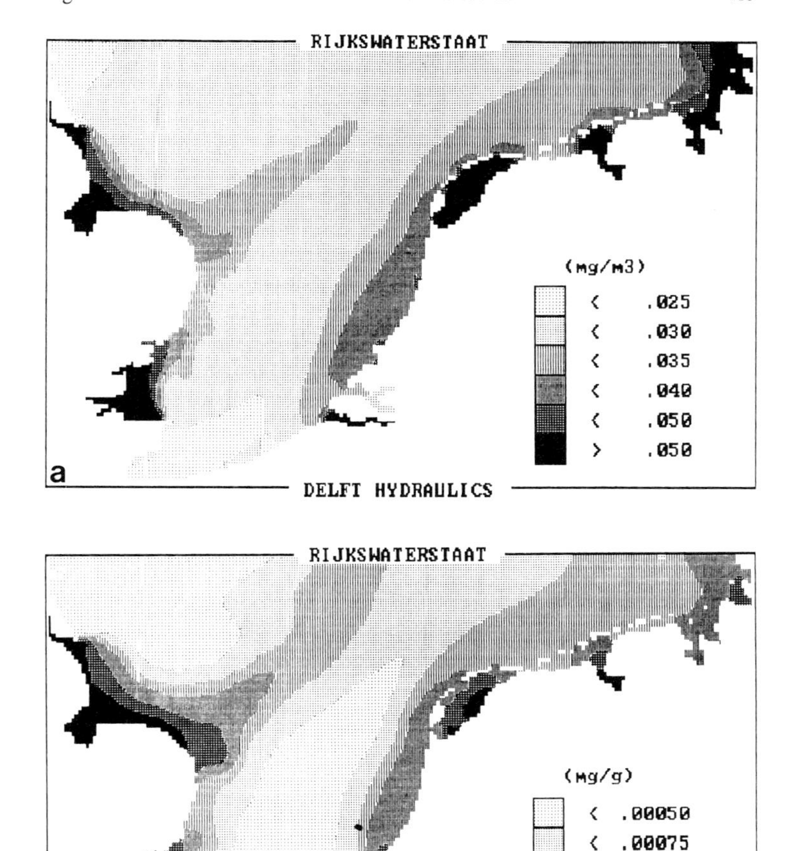


**Fig. 13a,b.** Calculated dissolved cadmium concentrations **(a)** and cadmium concentrations **(b)** in suspended sediments (1984/1985)

quality standards to protect marine life have been derived for a number of metals using the results of a toxicological exposure test under laboratory conditions and field observations. An overview of standards and assessment levels is given in Table 3. The use of standards and assessment levels as presented in this table is not generally adopted and needs further consideration. In this respect special attention should be given to the safety factors included in these levels, which account for uncertainties in using the results of laboratory tests as a basis for assessment of concentrations in the field.

**Table 3.** Water quality data for protection of marine life in μg/l (STWG 1987)

| | EC Directives (a) (c) | (d) | UK National Standard (a) | USEPA (m) | NL Assessment levels (b) |
|---|---|---|---|---|---|
| Cadmium | 2.5 | 5.0 | 2.5 | – | 0.05 |
| Copper | – | | 5 | 4 | – |
| Lead | – | | 25 | 25 (r) | 3.4 |
| Zinc | – | | 40 | 58 | 250 |
| Nickel | – | | 30 | 7.1 | – |
| Mercury | 0.3 | 0.5 | 0.3 | – | 0.01 |
| Chromium | – | | 15 | 18 (r) | 300 |

(a) All values average. (m) Maximum 24h average values. (r) Recoverable.
(b) Assessment levels – These are the highest concentrations which are believed to have no effect on marine organisms and are taken as 1% of the lethal concentrations for the most sentitive organisms.
(c) Territorial and internal coastal waters.
(d) Estuary waters affected by discharges.

## 5 Risk Analysis of Increased Nutrient Inputs

During the last decades the nutrient content in rivers like the Rhine and Meuse has seriously increased as a result of urban, industrial and agricultural developments. Although this nutrient enrichment has resulted in serious eutrophication problems in the shallow lakes in the Netherlands, so far the impact on the coastal waters has been less pronounced. In order to study the (potential) impact of increased nutrient inputs into the Dutch coastal waters on phytoplankton biomass and species composition, a nutrient and phytoplankton submodel has been developed and applied to this region (Fransz and Verhagen 1985).

The model (SEAWAQ) describes the carbon and nutrient (N, P, Si) cycles schematized in Fig. 14. As is shown in this schematization, the following state variables have been simulated:

- phytoplankton: specified as "diatoms" and "other phytoplankton" with a distinction into young and old populations (differing in mortality rates) and a resting stage (spores);
- suspended detritus;
- bottom detritus;
- dissolved nutrients.

External steering variables and forcing functions include light, temperature, background extinction and grazing by zooplankton and benthic consumers.

The seasonal changes in the various nutrient pools and fluxes are simulated by a set of coupled differential equations for each state variable. Due to the lack of validation data for the larger part of the southern Bight of the North Sea, the use of SEAWAQ was restricted to the Dutch coastal zone. In its first application, a simplified spatial discretization of seven compartments parallel to the Dutch coastline was considered. This area roughly coincides with the area covered by a monitoring programme of the Dutch Ministry of Public Works (Rijkswaterstaat). Monitoring data from the period 1975–1982 have been used to calibrate the model. Flow conditions used for defining the advective inflow and outflow of the com-

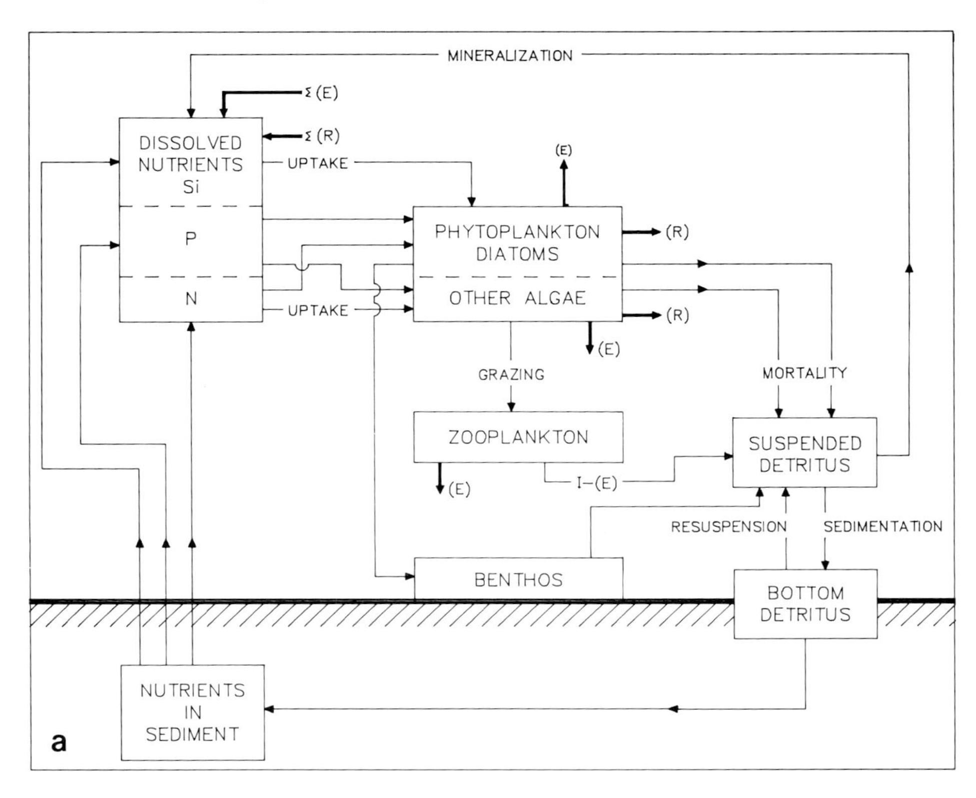


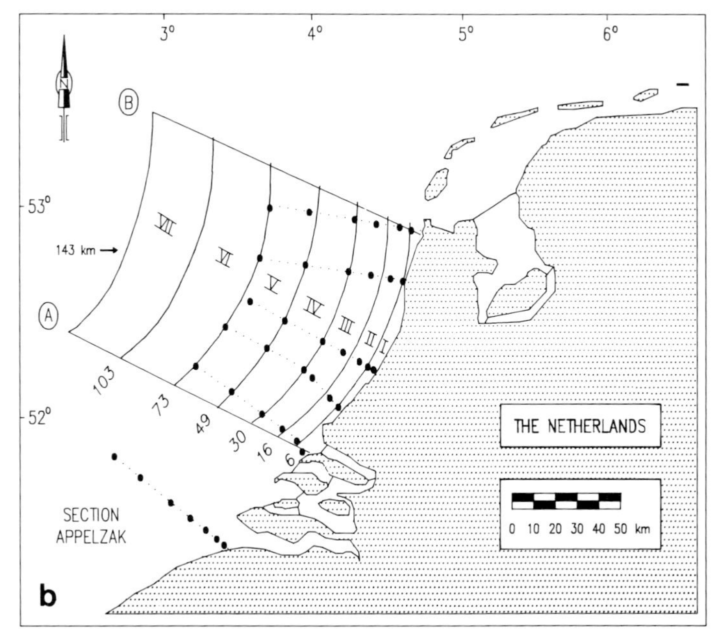


**Fig. 14a,b.** Model structure of nutrient and phytoplankton model SEAWAQ **(a)** and its application for the Dutch coastal zone **(b)**

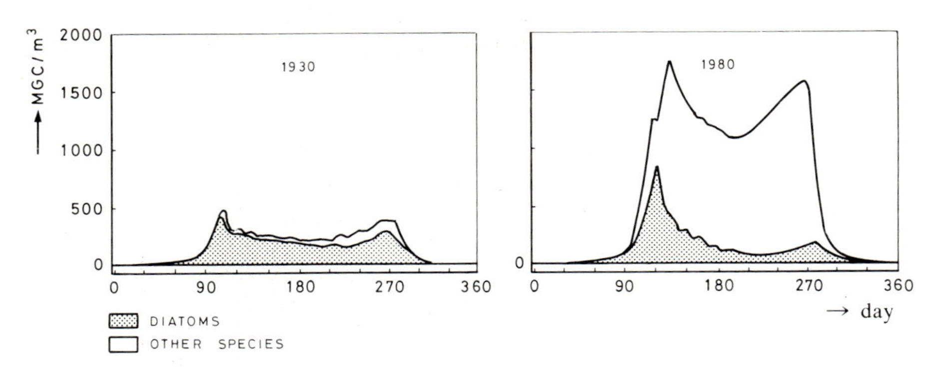


**Fig. 15.** Simulated phytoplankton biomass in the Dutch coastal zone for 1930 and 1980 nutrient inputs

partments referred to season-dependent average water transport parallel to the coast. Perpendicular to the coast only dispersive transport was considered. Flows and dispersion specifications were calibrated in an earlier stage by comparison of calculated and measured salinity figures for the same area. Riverine inputs from Rhine and Meuse were attributed to compartment I. This was done for a 1930 situation and for averaged figures of the period 1975–1985, which is here referred to as the 1980 situation.

From the results (Fig. 15) it was concluded that the eutrophication of the Dutch coastal waters has had little effect on the total biomass of diatoms, probably because Si is the limiting factor and Si-inputs have decreased since 1930. This Si-decrease may have been caused by eutrophication of freshwater systems in the catchment area of the river Rhine. Increased availability of P promotes higher freshwater diatom biomasses and thus a higher level of Si-fixation before the river Rhine enters the North Sea.

A major effect was calculated for the other phytoplankton group. Where diatoms were calculated to have been dominant in 1930, the ratio of diatoms to other phytoplankton species has since reversed from more than 1–2 to 0.2–0.3 in 1980.

Although there are no reliable data on phytoplankton change since 1930 the model results roughly correspond to recent findings based on measurements for the period 1970–1985 (Beukema 1986).

Whether these changes may in the long run be beneficial to a higher productivity and related increase of potentials for fisheries, or in contrast may eventually lead to the deterioration of the ecosystem, needs to be studied more extensively. In this respect special attention should be given to the risk of increased oxygen depletion in the stratified region of the North Sea. As shown by Gerlach (1984) the oxygen content in the German Bight can fall to very low values during the (thermally stratified) summer period. The analyze the risk of oxygen depletions as a results of increased nutrient inputs, a worst case analysis was carried out. Based upon the simulated nutrient concentrations (see Sect. 3) the maximum produced biomass was calculated and subsequent settling and oxygen demand in the bottom layers (Glas et al. 1988). As shown by the results (Fig. 16), the most critical areas for

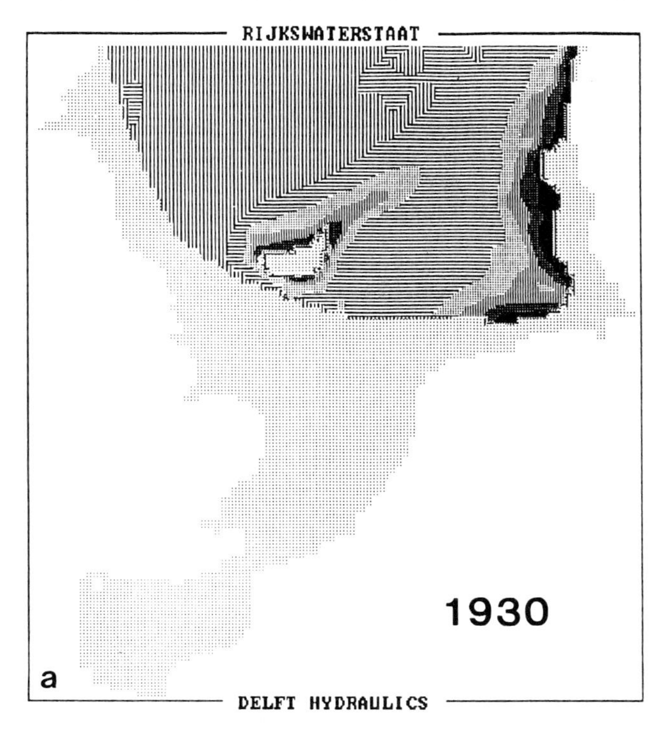


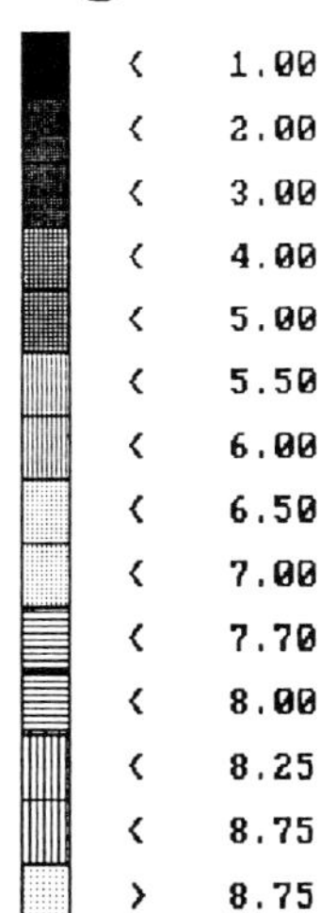


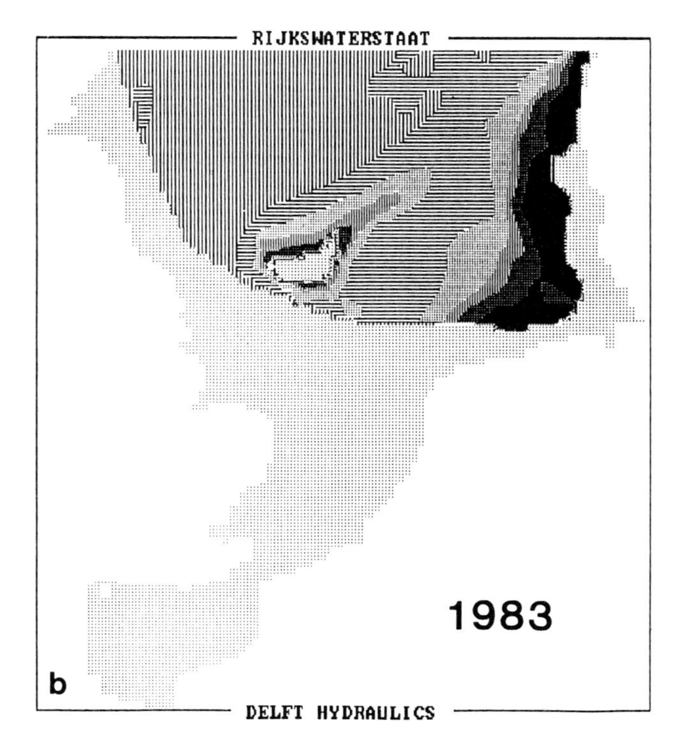


**Fig. 16a,b.** Calculated oxygen contents in the bottom layer after 120 days of stratification, for 1930 (**a**) and 1983 (**b**) nutrient inputs (Glas et al. 1988)

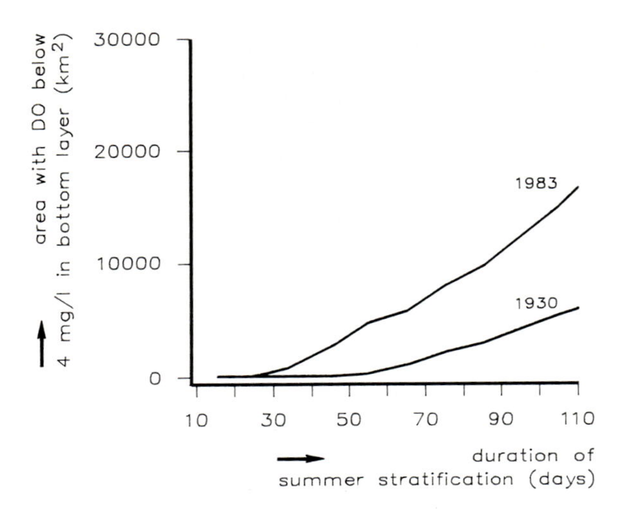


**Fig. 17.** Calculated area where dissolved oxygen concentration in the bottom layer drops below 4 mg/l since the start of a continuous stratification

low oxygen contents are found in the German Bight as a result of high nutrient concentrations and relatively small water volume between the bottom and thermocline. The comparison of the simulated oxygen contents in the bottom layer for the 1930 and 1985 situation shows a significant difference in the area of oxygen deficits after 120 days of continuous stratification without any vertical mixing.

The calculated increase of the area with oxygen contents below 4 mg/l, is presented in Fig. 17 as a function of the duration of stratified conditions. As can be concluded from this figure, the increase of nutrients since 1930 might have caused a serious increase in the frequency of low oxygen contents in the stratified region of the North Sea. In 1930 calculated oxygen concentrations below 4 mg/l only occur after 55 days of continuous stratification. In 1985 this period is shortened to 25 days.

Measured temperatures in the German Bight show that 60 days seems to be a "normal" duration of stratified conditions in the summer period from mid-July to mid-September (Tomczak and Goedecke 1962).

## 6 Discussion and Conclusions

Although the presented studies are only a first step in the development of analytical tools for research and management, the results of the studies have proved to be very useful for the governmental decision-making process in the Netherlands (Beukema et al. 1986). The integral approach using available data interlinked with mathematical modelling techniques enabled a spatial evaluation of the present status of North Sea pollution. Especially in large water bodies like the North Sea, it is very hard to obtain reliable and consistent field data of pollutants such as heavy metals and organic micro-compounds. Not only the low concentrations, but also the variability in time, often prevent an overall view on (long-term) spatial distributions of pollutants. The variability in transport phenomena in the North Sea, the boundary inflows from the English Channel, North Atlantic and rivers also prevents an early determination of long-term trends in pollution based on field data only. It was shown by these studies that the interlinked use of measured river loads, mathematical modelling and field data enables a more detailed evaluation of the present and future North Sea pollution.

## References

Aalst RM van, Kreuk JF de (1982) Pollution of the North Sea from the atmosphere; executive summary. Report CL 82/151, TNO, The Hague

Alphen (1987) Het slibtransport in de Belgisch-Nederlandse kustzone in de periode 1970–1986. RWS, Directie Noordzee nota NZ-N-87 19

Backhaus JO, Maier-Reimer E (1983) On seasonal circulation patterns in the North Sea. pp 63–84. In: North Sea dynamics. J Sündermann et al (eds.) Springer Berlin Heidelberg New York Tokyo 693 pp

Backhaus JO, Hainbucher D, Quadfasel D, Bartsch J (1985) North Sea circulation anomalies in response to varying atmospheric forcing. International Council for the Exploration of the Sea, CM 1985/C:29

Beukema AA, Hekstra GP, Venema C (1986) The Netherlands environmental policy for the North Sea and Wadden Sea environmental monitoring and assessment 7, 117–155

Beukema J (1986) Eutrophication: Reasons for satisfaction or concern. Proceedings 2nd North Sea Conference, Rotterdam, Vol 2:27–38

Carlson H (1986) Quality status of the North Sea. International Conference on the Protection of the North Sea, Bremen, 1984. Dt hydrogr Z, Erg -HB, nr 16

Dijkzeul J (1985) Rijkswaterstaat notitie GWIO-85, 113 (in Dutch)

Eisma D (1981) Supply and deposition of suspended matter in the North Sea. Spec Publ Int Ass Sediment 5, pp 415–428

Eck B van, Turkstra E, Sant H van't (1983) Voorstel referentie waarden Nederlandse zoute wateren. Ministerie van VROM, Leidschendam (in Dutch)

Fransz HG, Verhagen JHG (1985) Modelling research on the production of phytoplankton in the southern bight of the North Sea in relation to riverborne nutrient loads. Neth J Sea Res 19(3/4): pp 241–250

Gerlach SA (1984) Oxygen depletion 1980–1983 in coastal waters of the Federal Republic of Germany. Ber Inst f Meeresk Kiel, 130 pp 1–87

Gerritsen H (1986) Residual currents: A comparison of two modelling approaches. Lecture notes on coastal and estuarine studies. Vol 16. Physics of shallow estuaries and bays, pp 81–101

Glas PCG, Markus AA, Briten (1988) Modelling primary production and oxygen deficiencies in the southern bight of the North Sea – a management context. Neth J Sea Research (in press)

Hopstaken CF (1986) Transport of suspended matter and associated sedimental pollution in the southern North Sea. 16th EBSA symposium. Dynamics of turbid coastal environments, Plymouth

ICES (1962) Mean monthly temperature and salinity of the surface layer of the North Sea and adjacent waters from 1905–1954. Copenhagen

ICES (1978) Input of pollutants to the Oslo Commission area. International council for the exploration of the sea. Cooperative Research Report No. 77

Kooy LA van de, Stokman GNM (1983) De waterwaliteit van de Noordzee 1975–1982. Rijksinstituut voor Zuivering van Afvalwater, nota nr 83.084 (in Dutch)

Pagee JA van, Gerritsen H, Ruijter WPM de (1986) Transport and water quality modelling in the southern North Sea in relation to coastal pollution research and control. Wat Sci Tech Vol 18, Plymouth, pp 245–256

Pagee JA van, Postma L (1987) North Sea Pollution, the use of modelling techniques for impact assessment of waste inputs. Proc 2nd North Sea Conference, Rotterdam

Postma L (1988) Transport and water quality model DELWAQ, users manual. Delft Hydraulics report T327

Prandle D (1978) Residual flows and elevations in the southern North Sea. Proc Roy Soc London, A 359, 189–228

Rijkswaterstaat and Delft Hydraulics (1985) Waterkwaliteitsplan Noordzee. Staatsuitgeverij The Hague. 85 p + 5 additional reports (partly in Dutch)

Ruijter WPM, Postma L, Kok JM de (1987) Transport atlas of the Southern North Sea, Rijkswaterstaat. Delft Hydraulics, Delft, Netherlands

Salomons W, Förstner U (1984) Metals in the hydrocycle. Springer Berlin Heidelberg New York Tokyo

Schwedhelm E, Irion G (1985) Schwermetalle und Naturelementen in den Sedimenten der deutschen Nordseewatten. CFS, Frankfurt, ISBN 3-924500-09-6

STWG (1987) Second international conference on the protection of the North Sea. Quality status of the North Sea, London

Tomczak G, Goedecke E (1962) Monatsharter der Temperatur der Nordsee, dargestellt für verschiedene Tiefenhorizonte. Erganzungsh dt Hydrog Z, B Nr 7, 16 pp

Verboom GK, Dijk RP van, Dijkzeul JCM (1986) A fine grid tidal flow and storm surge model of the North Sea. IAHR meeting, Taiwan

Voogt L (1984) Een getijmodel van de Noordzee gebaseerd op JONSDAP 1976 meeting. Rijkswaterstaat, WWKZ-84 G 006, 25 pp (in Dutch)

# Part III
# Impacts on Selected Areas and by Human Activities

# German Bight

V. DETHLEFSEN[1]

## 1 Introduction

Much of the concern which is presently expressed in the context with North Sea pollution has its origin in studies carried out in the German Bight. In this area, located off the coastlines of Schleswig-Holstein and Niedersachsen of the Federal Republic of Germany and limited by a line of 55°N and 07°E, first systematic studies of fish diseases were carried out, first indications for oxygen deficiency were found, long-term studies indicated the increase of nutrient levels in the water column, benthic changes occurred in estuaries and changes in species composition of Wadden Sea organisms including fishes were demonstrated. In the following chapters some remarks will be made on the hydrography, topography, and chemistry of the German Bight, followed by information on the above-mentioned biological changes.

### 1.1 General Remarks on Hydrography and Chemical Pollution

The German Bight is receiving a high pollution load from various inputs. The first and most significant introduction of wastes is via major rivers like the Elbe, Weser, and Ems. The influence of the first two river systems joins in the Helgoland Bight, which constitutes one of the biggest sedimentation areas within the North Sea as a whole. For example, the Elbe and Weser carry 16% of the total riverine North Sea input of cadmium.

Coastal parallel currents are transporting discharges from the river Rhine into the German Bight. The Rhine adds a further 11% of the total riverine North Sea input and carries 67% of mercury, 32% of the total chromium, etc.

In the center of the German Bight, 12 nm northwest of the island of Helgoland, wastes from titanium dioxide production have been dumped since 1969. Quantities were 750,000 t per year of 12% $H_2SO_4$, and they were changed to 450,000 t of 23% $H_2SO_4$. The dumping area is located at

54° 20′N / 07° 35′ E
54° 25′N / 07° 35′ E
54° 20′N / 07° 42,5′E
54° 25′N / 07° 52,5′E.

[1]Bundesforschungsanstalt für Fischerei, Institut für Küsten und Binnenfischerei, Außenstelle Cuxhaven, Niedersachsenstraße, 2190 Cuxhaven

**Table 1.** Composition and quantities of wastes from titanium dioxide production dumped in the German Bight 1980–1989

| | | 1980 | 1982 | 1984 | 1987 | 1988 | 1989 |
|---|---|---|---|---|---|---|---|
| Total Amount | (t) | 725 000 | 481 700 | 415 000 | 390 000 | 234 000 | 195 000 |
| Copperas | (t) | 55 200 | 54 370 | Termination of dumping | 0 | 0 | 0 |
| Cr | (t) | 116 | 73.7 | 73.2 | 68.8 | 41.3 | 34.4 |
| Cu | (kg) | 870 | 213 | 244 | 229 | 138 | 114.7 |
| V | (t) | | 192.7 | 192.2 | 180.7 | 108.4 | 90.3 |
| Pb | (t) | 1.16 | 0.2 | 0.21 | 0.2 | 0.12 | 0.1 |
| Hg | (kg) | 5.8 | 7.1 | 6.1 | 5.7 | 3.4 | 2.9 |
| Cd | (kg) | 58.0 | 35.4 | 9.15 | 8.6 | 5.16 | 4.3 |
| Fe | (t) | 37 700 | 26 564 | 19 926 | 18 726 | 11 235 | 9 363 |
| Zn | (t) | 23.2 | 7.8 | 6.7 | 6.3 | 3.8 | 3.15 |
| Ni | (t) | 8.7 | 9.56 | 9.46 | 8.89 | 5.33 | 4.44 |

The composition of the wastes and quantities dumped are given in Table 1.

From 1961 to 1980 annually 350,000 t of sewage sludge of the city of Hamburg were dumped into an area located in the outer stretches of the Weser and Elbe estuary. Afterwards the dumping was shifted to a location further offshore in the German Bight and was finally terminated in 1983 (Dethlefsen 1986). Within the given geographic lines the German Bight is shallow, with a medium depth between 20 and 30 m. The sediment structure is dominated by sand and pebbles and it is only in the Helgoland Bight, and from here extending in a northwesterly direction, that significant fractions of silt and clay can be found in the sediments (Figge 1981). Although rapid tidal water movement occurs in onshore waters, a high retention time of water masses in the German Bight is typical.

Already in 1922 Böhnecke found the existence of a large-scale eddy in the German Bight, the location of which is wind-dependent and not stationary (Böhnecke 1922).

According to Maier-Reimer (1979), the body of water from the central German Bight would be totally exchanged, i.e., reach the geographical borders of the North Sea after 36 months. Due to its shallowness, the German Bight as a whole can be considered to represent features normally found in estuaries with repercussions on sedimentation and remineralization of pollutants (Eisma 1981).

## 2 Oxygen Deficiency and Effects

Although already in 1902 Gehrke found low oxygen values in the central North Sea (48% oxygen saturation at 68 m depth at 65°N 03°E, Gehrke 1916), the presence of low dissolved oxygen waters in the German Bight was not described before August 1982, when Rachor and Albrecht (1983) and Dethlefsen and von Westernhagen

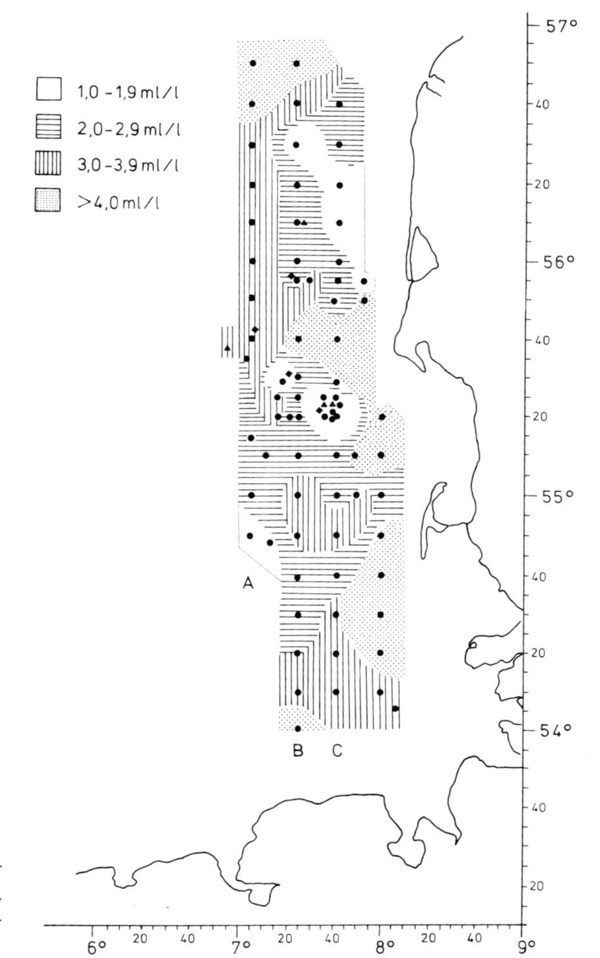


**Fig. 1.** Oxygen in near bottom water in the German Bight August 1982. (After Dethlefsen and von Westernhagen 1983)

(1983) found low dissolved oxygen concentrations in bottom waters of the German Bight (Fig. 1). In August and September of this year in two thirds of more than 15,000 km$^2$ oxygen content was less than 5 ml l$^{-1}$ corresponding to 60% saturation. Lowest values were near 1 ml l$^{-1}$. In oxygen deficiency areas (oxygen saturation around 10%) fish catches were low containing dead *Agonus cataphractus, Pleuronectes platessa* and *Limanda limanda.*

When operating underwater television cameras, *Agonus cataphractus, Callionymus lyra, Ammodytes* sp., and flatfish were detected lying dead on the bottom. Evaluation of underwater photography revealed the occurrence of dead benthic organisms in areas with low dissolved oxygen. The species were *Ophiura albida* and *Venus striatula* (Dethlefsen and von Westernhagen 1983).

In continuing these studies, von Westernhagen et al. (1986) found extremely low oxygen concentrations in summer 1983 and higher oxygen concentrations in summer 1984, when 40 to 70% oxygen saturation was found in some areas in subthermocline waters.

When considering biological effects related to low dissolved oxygen it must be mentioned that Dyer et al. (1983) detected unusually low catches in Danish coastal waters, which were later interpreted to be due to low dissolved oxygen in bottom water.

Kröncke (1985) investigated the benthic community structures in low dissolved oxygen areas in years with and without oxygen deficiencies in bottom waters. She concludes, using cluster analyses, diversity index and eveness index, that effects on benthos occurred in August 1983, the period with lowest oxygen in bottom water. Three species dominated, which were described as being particularly tolerant to unfavorable oxygen conditions, while in August 1984, with much better oxygen conditions, the benthos was dominated by species which are considered to be recolonizers, occurring under improving environmental conditions (von Westernhagen et al. 1986).

Considering possible causes for the occurrence of low dissolved oxygen in bottom waters in the German Bight, the first guess would be eutrophication, all the more so as the prerequisites, i.e., increased nitrate, phosphate, and phytoplankton concentrations are given for the area under consideration. Since 1962 concentrations of winter nitrate have increased nearly fourfold at stations in the vicinity of the island of Helgoland, those of phosphate by a factor of more than 1.5-fold. Total inorganic nitrogen increased 1.7-fold. There was a fourfold phytoplankton-C increase consisting of flagellates alone. They increased to the 16-fold biomass concentration compared to 1962 (Radach and Berg 1984). Weichart (1984) demonstrated that between 1936 and 1978 in certain areas of the German Bight an increase of concentrations of phosphate by a factor 3 has occurred. One of the further prerequisites to producing low dissolved oxygen conditions in near bottom water is the formation of a stratification which needs a prolonged calm weather period. All attempts undertaken so far to establish cause-effect relationships were negative, so that some uncertainty exists whether the phenomena observed in German and Danish coastal waters might be called eutrophication of hypertrophication or not (Gerlach 1985).

It is therefore uncertain whether reduction measures to decrease the input of nutrients would improve the situation in the German Bight.

## 3 Benthic Changes

Salzwedel et al. (1985) compared benthic data from the German Bight from 1975 with earlier studies and found higher biomasses in the newer samples, which is interpreted to be due to better sampling methods. Dörjes (1986) states that in comparison to older studies on macrozoobenthos in the German Bight (Hagmeier 1925; Hagmeier and Kändler 1927; Caspers 1938), a general decrease for many species can be found. Dörjes subdivides the changes in macrobenthos according to species with different behavior.

**1.** Species no longer represented. To these he counts *Astropecten irregularis, Solaster papposus, Thyone fusus,* and *Cucumaria elongata,* species which Caspers considered to be abundant in areas south of Helgoland. The crustaceans *Upogebia deltaura* and *Callianassa subterranea,* which, prior to the 1960's were very abundant, are no longer present in their former habitats south of Helgoland. Also molluscan species like *Ostrea edulis* and *Helcion pellucidus,* which were present in the 1930's and 1960's in substantial populations, are no longer present on the German coast (Dörjes 1986).

**2.** Species with strong population decreases. Here Dörjes (1986) mentioned the molluscs *Venus ovata, Cultellus pellucilus, Crepidula fornicata,* and *Cardium fasciatum,* which were considered to be the most common molluscs of the area of Helgoland (Heincke 1894). Also polychaetes like *Glycera alba, Lubrinereis tetraura (impatiens),* and *Sabellaria spinulosa,* echinoderms *Echinocyamus pusellis,* the sponge *Halichondria panicea,* and the crustacean *Porcellana longicornis* and *Pisidia longicornis* show significant reductions in their populations.

**3.** Species with strong fluctuations. To this group Dörjes counts species with reduced numbers of individuals for prolonged periods, developing into rich populations, which after a certain period are reduced to their former niveau. Polychaetes *Pectinaria koreni* and *Pholoe minuta,* molluscs *Mysella bidentata, Montacuta ferruginosa, Abra alba,* and *Donax vitatus* are counted, as well as the echiurid *Echiurus echiurus,* the decapod crustacean *Corystes cassivellaunus* and the fish *Callionymus lyra.*

Rachor (1985) investigated long-term benthic changes in two different areas in the German Bight, one being located in the inner German Bight, the former sewage sludge dumping ground which within the period 1961 and 1980 received 350,000 $m^3$ sewage sludge annually (Dethlefsen 1986), and one located in the central German Bight, being the dumping area for wastes from titanium dioxide production.

Results were different with regard to the areas. For *Nucula nitidosa,* for example, he found that a significant decrease for this species could be stated for the former sewage sludge dumping area in the period from 1969 to 1983, while in the area in the central German Bight an increase for this species could be found. The situation was similar for the species *Abra nitida.* Also Caspers (1979) and Mühlenhardt-Siegel (1981; 1985) demonstrated clear effects of the dumping of sewage sludge on macrobenthos resulting in impoverished communities, which at times totally collapsed.

Rachor (1982) did not see a connection between fluctuations in the benthic communities in the dumping area for wastes from titanium dioxide production northwest of the island of Helgoland and the input of the wastes, but he states that the changes in macrobenthos communities might not be the most sensitive criteria towards heavy metal exposure. Long-term comparison of benthic communities in the Schleswig Holstein Wadden Sea area led Riesen and Reise (1982) to the conclusion that the disappearance of oysters and the polychaetes *Sabellaria* have to be interpreted as effects due to intensive fisheries. The growth of banks of *Mytilus edulis* is interpreted to be an indication for eutrophication (Reise 1986).

The invasion of the polychaete *Heteromastus* in sandy areas (Forschungsstelle Norderney 1984) indicates an extension of reducing processes in Wadden Sea areas.

Grotjahn and Michaelis (1985) investigated benthos in effluents of a titanium dioxide production site in the Weser estuary. In comparison to data from 1968 it is stated that the alga *Vaucheria* is reduced in population density and zoobenthos that opportunistic species are taking over from brackish water specialists.

The interpretation of long-term changes in macrobenthos in the German Bight is complex. While eutrophication-related problems should lead to an increase of macrobenthos, toxic effects could counteract this development. A third and significant impact has to be seen in the activities of fisheries.

Rauck (1985) summarized the catching activities of Dutch sole fisheries in these waters and it is evident that impact on benthos through heavy fishing gear must be significant.

Statements by Dörjes (1986) concerning large-scale reduction of macrobenthos species are new and should give rise to concern. They indicate that the sum of impact constituted by waste loads on one side, changes in the productivity on the other, and the fishery on the third exert an influence which results in a stress which can no longer be compensated for by the populations.

## 4 Reproduction of Fish

First indications for a possible impairment of reproduction of fish through accumulative substances was given by von Westernhagen et al. (1981) for flounders of the western Baltic, and by Hansen et al. (1985) for spring spawning herring of the same region. It could be shown that gonad residues of organochlorine substances were correlated with hatching success of these species. These investigations were extended to fishes from Dutch and German coastal waters, and first results of hatching experiments with whiting of the southern North Sea are reported by Dethlefsen et al. (1986). Fishes were caught off the Dutch coast in March 1984. Eggs of running ripe whitings were artificially inseminated and reared until hatching. Organochlorine residues in the respective gonads were analyzed. Early embryos were taken for investigation of chromosomal aberrations. Hatching success was generally very low (Fig. 2). High mortality rates were found, which were higher than those for fishes from the Baltic from preceding experiments. Also the contamination of different organs of whiting with organochlorine substances, in comparison to those of flounder and herring and cod from the Baltic, was much higher. Increased hatching success was only detectable when gonads showed lower comparison to those of flounder herring and cod from the Baltic, was much 3,800 ng $g^{-1}$ PCB fat weight. The threshold which has been found to depress hatching success in flounders was 4,800 ng $g^{-1}$. In two thirds of the female whiting used for these experiments the PCB contamination of gonads was more than 6,000

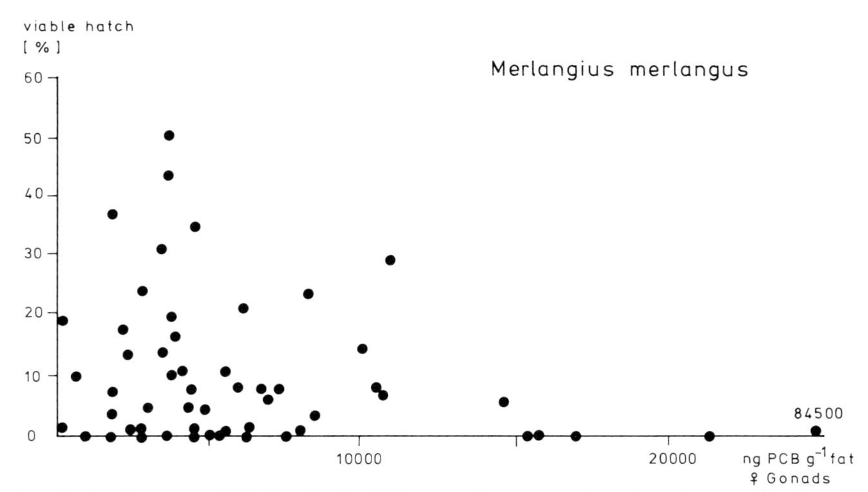


**Fig. 2.** Correlation between viable hatch of whiting (*Merlangius merlangus*) and contamination of gonads with PCBs (fat weight). (After Dethlefsen et al. 1986)

ng $g^{-1}$ fat weight. This contamination level led to severe mortalities of larvae of the Atlantic salmon (*Salmo salar*). Frequencies of chromosomal aberrations were positively correlated with the contamination of gonads and livers of female whiting with polychlorinated biphenyls and with the contamination of gonads with DDT, DDE, and $\alpha$-HCH.

In March 1984 and February 1985, embryos of dab (*Limanda limanda*), flounder (*Platichthys flesus*), plaice (*Pleuronectes platessa*), cod (*Gadus morhua*), long rough dab (*Hippoglossoides platessoides*), and whiting (*Merlangius merlangus*) were caught and investigated on board a research vessel for the presence of malformations. Results are given by Dethlefsen et al. (1985, 1986). For 1984 it was found that 38% of the early embryos of flounders of the Schleswig-Holstein coast were malformed. Highest malformation rates were found in an area west of the island of Sylt. Off the Dutch coast, malformation rate of early embryos of dab was 36%, in the German Bight 29%. Maximum values of malformations of whiting during a later developmental stage were 47%. When malformation frequencies of all developmental stages of different fish embryos were summarized for the year 1984, it was obvious that the highest malformation rates were found in the center of the German Bight and on stations off the East Frisian Islands on the main shipping routes leading into Dutch coastal waters.

The picture obtained for February and March 1985 was generally similar. Malformation rates of fish embryos off the Rhine were 50%. Further high malformation rates of early embryos of dab were found off the Schleswig-Holstein coast, with a striking regional coincidence with the dumping area for wastes from titanium dioxide production (Fig. 3). Also during this investigation, embryos of whiting displayed highest malformation rates. An evaluation of these findings is not attempted, since the material is considered to be too preliminary.

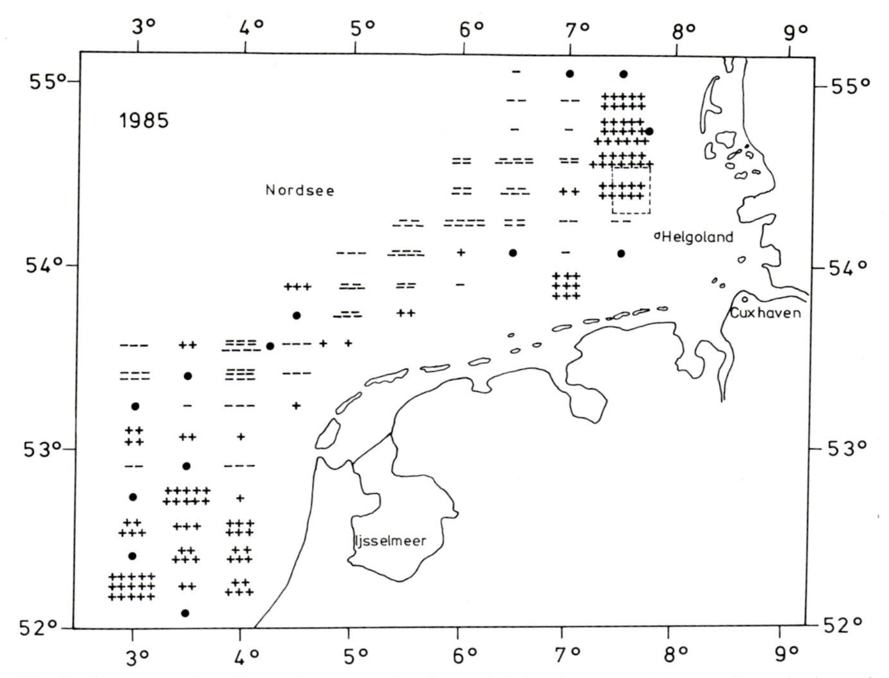


**Fig. 3.** Summary of malformation rates of embryos (all developmental stages, all species investigated) February/March 1985. Stations with higher (+) and lower (–) malformation rates related to weighted mean □ dumping area for wastes from titanium dioxide production. (After Dethlefsen et al. 1986)

## 5 Studies on Fish Diseases

First systematic studies on occurrence and abundance of fish diseases in the German Bight were done in 1977. Möller carried out studies in February 1977, 1978 and August 1978 and 1980 (Möller 1979; 1981). While during the first three cruises he investigated diseases of flatfish in the southern North Sea, in his cruise in August 1980 he concentrated on fish from German and Danish coastal waters.

Dethlefsen started his cruises in 1977. He published first results in 1980. Main results of these early studies were that two species were afflicted with externally visible diseases in significant quantities.

**1.** Cod (*Gadus morhua*). The major diseases were ulcerations, skeletal deformities, and fin rot;

**2.** Dab (*Limanda limanda*). The major diseases were lymphocystis, hyperplasia and epidermal papilloma, and ulcerations in various stages of healing.

In the meantime more than 200,000 dab and more than 80,000 cod, only to consider the more important species, have been investigated. The results published so far are to be found in Dethlefsen et al. (1984) on sources of variance in data from fish disease surveys considering the variability of the occurrence of three external diseases of dab. Results on histological investigations of different diseases of dab and cod are published by Watermann (1982, 1984), Watermann and Dethlefsen (1982). The epidemiology of pseudobranchial tumors of cod of the North Sea is described by Watermann et al. (1982). Watermann and Dethlefsen (1985) give information on epidermal hyperplasia and degenerative skin diseases of gadoid fishes of the North Sea. Dethlefsen (1980) provides more general information on the occurrence of different diseases of cod, dab, plaice and flounder, where 31 different fish species were investigated on the occurrence of externally visible diseases. Further details on different diseases of dab and cod are to be found in Dethlefsen (1984a,b, 1985).

Based on a material of 180,000 dab, Wolthaus (1984) investigated seasonal fluctuations of disease frequencies of dab with lymphocystis, epidermal papilloma, and ulcerations on three different stations in the German Bight.

Knust and Dethlefsen (1986) described the influence of x-cell gills of dab on condition and growth of the fishes, and Dethlefsen et al. (1986) summarized all epidemiological and chemical investigations in relation to diseases of dab in the southern North Sea. Dethlefsen (1986) reports on a possible connection between the dumping of wastes from titanium dioxide production in the German Bight and the occurrence of high disease prevalence of dab within that area, and finally Dethlefsen and Huschenbeth (1986) provide information on organochlorine contamination of dab in the southern North Sea. Further studies on disease prevalence in the German Bight have also been carried out by Danish scientists, who started their systematic investigation in 1983, concentrating on diseases of dab and plaice (Mellergaard and Nielsen, 1984a,b, 1985a,b). Another study which has been extended to the southern border of the German Bight is carried out by Dutch colleagues, and first results, as published by Vethaak (1985), seem to indicate higher disease prevalence in polluted areas as compared to less polluted reference sites.

During all these studies, dab proved to be the species most frequently afflicted with externally visible diseases, and it could be shown that for three major diseases, epidermal papilloma, lymphocystis, and ulceration, congruently areas of increased disease prevalences could be detected.

1. An area starting in the center of the German Bight, extending from 54°N to 56°N on a northwesterly transect;
2. an area starting off the Humber estuary at 54°N extending on a northeasterly transect, ending at 55°30′N;
3. an area off the British coast at 56°N off the Firth of Forth.

During all cruises centering on these phenomena especially high disease rates were found in the middle of the German Bight (Figs. 4 and 5). This area coincides with the dumping area for wastes from titanium dioxide production, in operation since 1969. For location and quantities of wastes dumped see above. Dethlefsen and Watermann (1980) therefore suspected that a connection might exist between high disease rates within that area and toxic components of the wastes. This assumption received some criticism and the data were reviewed in 1986 after a recalculation of the epidemiological data using a statistical procedure which takes different lengths and sexes of fish at the various locations into consideration.

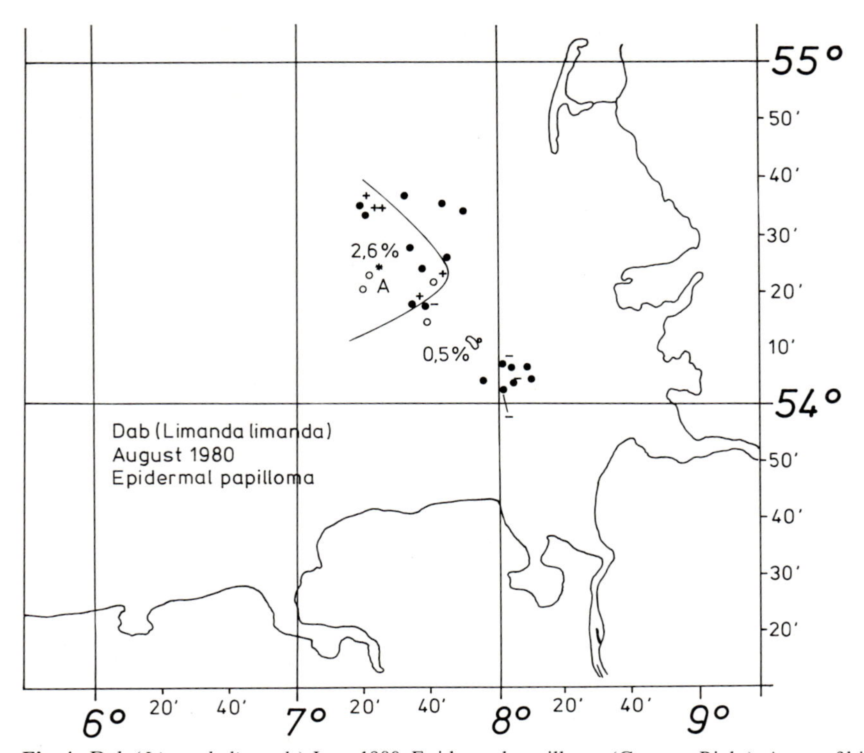


**Fig. 4.** Dab (*Limanda limanda*) June 1980. Epidermal papilloma (German Bight). Areas of higher (*A* and *B*) and lower (*C*) prevalence of epidermal papilloma. + • stations outside the 1 sigma limit; + + • stations outside the upper 2 sigma limit; • stations where ≤ 14 dab were caught (not considered in analysis; • stations with low number of dab (disregarded in drawing boundary lines but taken into account for calculation of significance). *Open circles* marked by an *asterisk*: stations omitted from consideration because length was not measured in all fishes caught. (After Dethlefsen 1984b)

Furthermore data on chemical analysis were added, including the measurement of iron in the water column, of heavy metals in the sediments and of chromium in diseased and healthy fishes. The most striking finding in this context was, that there was a positive correlation, though not significant, between the size of papilloma of infected fish and their liver residues of chromium (Fig. 6). This finding indicates a possible cause-effect relationship between the two phenomena. It was therefore concluded that it cannot be excluded that a cause-effect relationship between the dumping and the high disease rates in the area exists and that, as a precautionary measure, in the absence of the final proof for a connection, the dumping should be stopped. Officials responsible for this decision in the Federal Republic of Germany therefore decided to stop dumping in the German Bight by 1989. From then on wastes will be recirculated.

One of the major shortcomings in the attempts to prove that the occurrence of fish diseases in the southern North Sea is related to pollution has to be seen in the fact that disease maxima occur in offshore regions, especially on the Dogger Bank, an area which, due to its distance from the coast, was considered to be nonpolluted (Figs. 7 and 8). Newer results of investigations on residues of organochlorine and

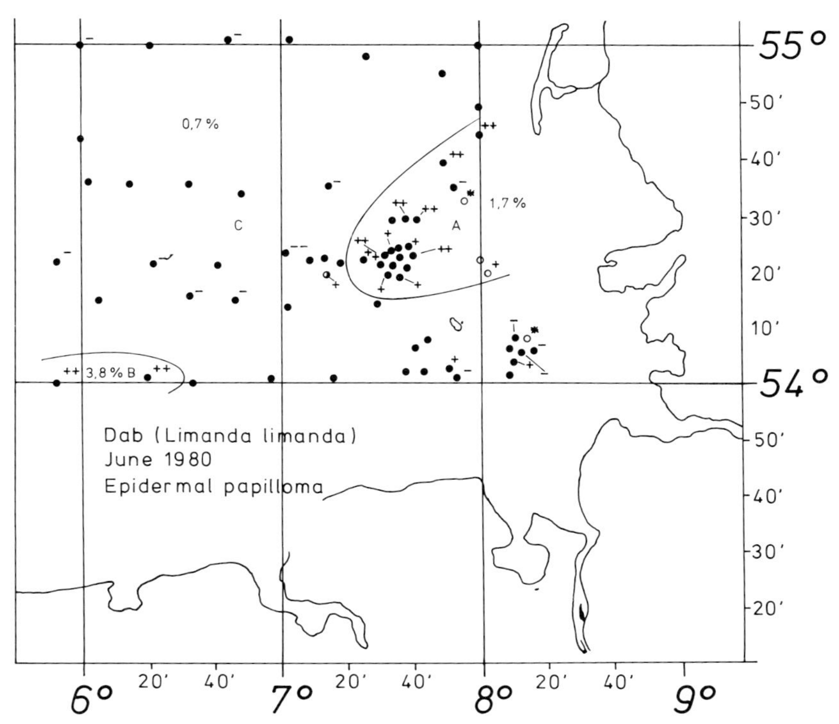


**Fig. 5.** August 1980. Areas of higher and lower prevalence of epidermal papilloma. For explanation of symbols see Fig. 4. (After Dethlefsen 1984a)

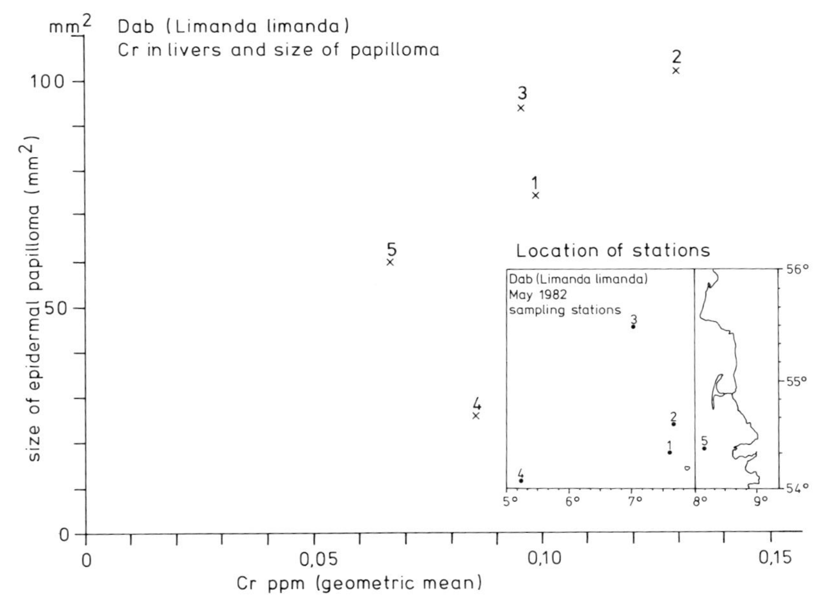


**Fig. 6.** Dab (*Limanda limanda*) May 1982. Relation between size of epidermal papilloma and chromium in livers of diseased fishes (both sets of data length corrected). (After Dethlefsen 1986)

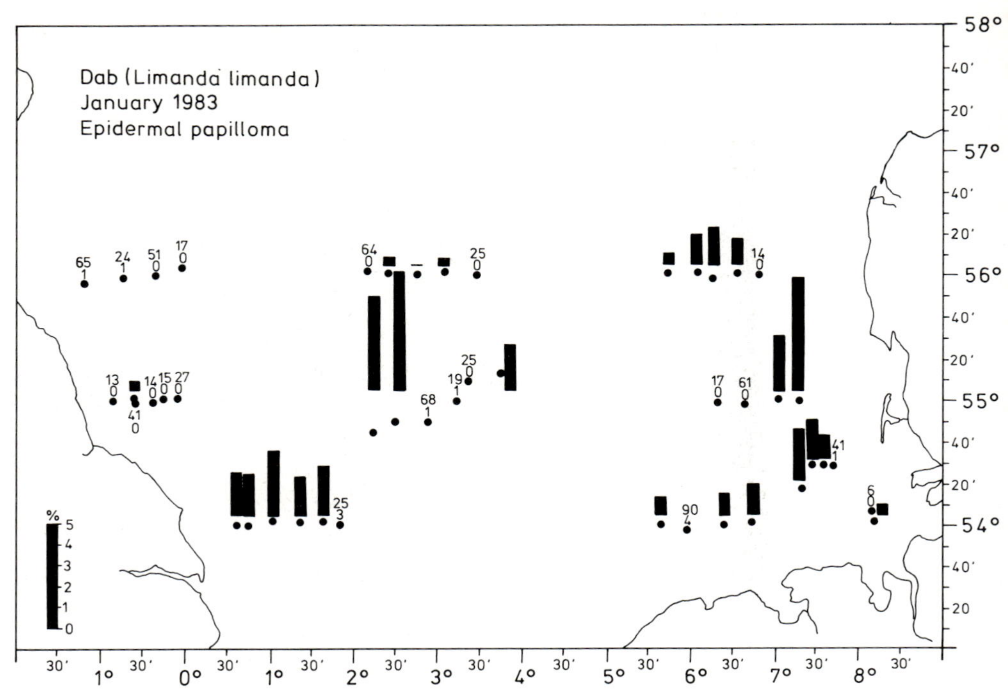


**Fig. 7.** Dab (*Limanda limanda*). January 1983. Epidermal papilloma, percentage of infection. For stations with less than 100 individuals investigated, upper figure gives numbers inspected, lower figure number of diseased fish at these station

heavy metals in livers of dab from the German Bight and the southern North Sea reveal that for certain heavy metals and for a number of organochlorines, residues were highest at stations where highest disease rates were found, especially in the central part of the southern North Sea (Harms and Claußen, Büther, pers. commun.).

## 6 Changes in Fish Populations

In a long-term study from 1954 to 1981, bycatch data of German shrimpers were analyzed for the occurrence of fish and a crustacean species. Seven inhabitants of the Wadden Sea area have decreased in abundance, some drastically: shorecrab, butterfish, gobies, eel, little sole, and gurnard, while three of the facultative inhabitants, i.e., dab, sprat, and cod, have increased in numbers to equal the former total biomass. For two species, changes in hydrographic-climatological factors could be used to explain the changes in the species composition. For other species, no such explanation was possible, leading to the conclusion that pollution problems could be amongst the responsible factors (Tiews 1983).

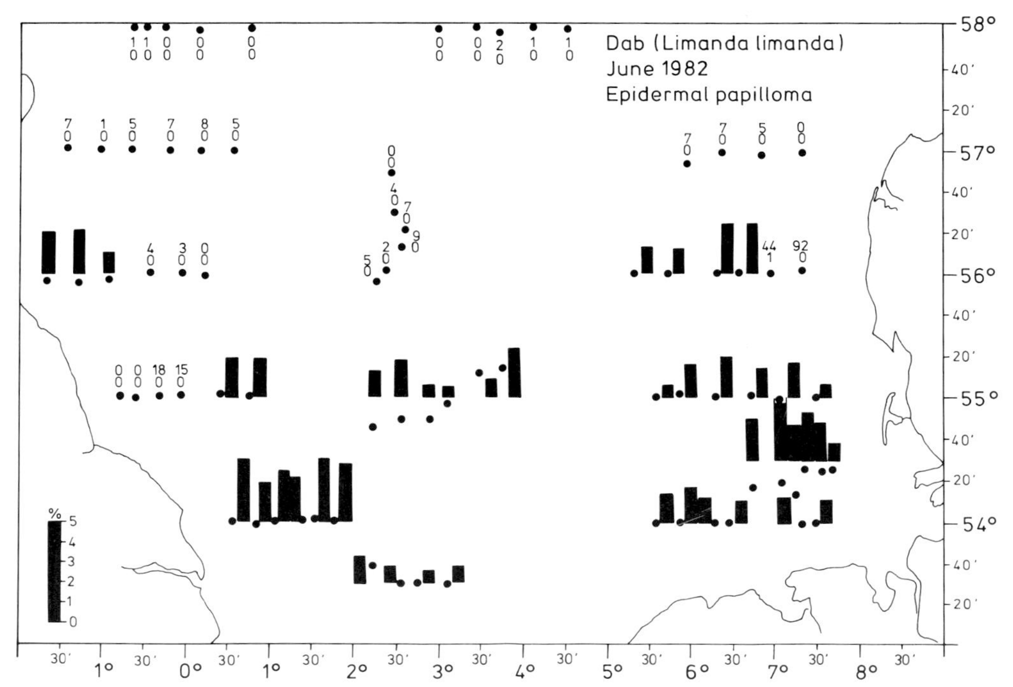


**Fig. 8.** Dab (*Limanda limanda*). June 1982. Epidermal papilloma, percentage of infection. For stations with less than 100 individuals investigated, *upper figure* gives numbers inspected, *lower figure* gives number of diseased fish at these stations

## 7 Conclusions

Ecological changes in the German Bight are manifold and they comprise various compartments of the ecosystem. It is striking that longer-term systematic studies on a substantial basis always revealed deviations from normal. Most striking changes were seen in phytoplankton communities, benthic populations, fish populations, and the most obvious signs of biological dysfunctions were fish diseases and impairment of reproductive capacity of fishes. The changes in macrobenthos populations are somewhat obscured by partly counteracting changes, i.e., eutrophication-related increases on the one hand and decreases, possibly due to pollution, on the other. When looking at causative factors, no final answer is possible. It has to be accepted that, in the absence of long-term time series to characterize the development of these changes and changes in the degree of pollution, the interpretation has to remain incomplete. However, even in the future with the expectation that better time series will be available, the evidence produced will be only circumstantial, giving a certain degree of probability that a cause-effect relationship exists between certain pollutants and certain dysfunctions. Due to the complexity of ecosystem changes and interrelationships, one cannot expect to produce clear answers to these questions. It appears, therefore, that the precautionary concept is justified, and seems to be the only tool to protect marine ecosystems in the light of otherwise increasing amounts of wastes introduced in marine areas.

Long-term investigations on various compartments of marine ecosystems are nevertheless fully justified because they will produce signs for an amelioration of the situation after the cessation of dumping or the decrease of the input of pollutants into the sea.

## References

Böhnecke G (1922) Salzgehalt und Strömungen der Nordsee. Veröff Inst Meereskd Berlin, NF (A) 10:1-34

Caspers H (1938) Die Bodenfauna der Helgoländer Tiefen Rinne. Helgoländer Wiss Meeresunters 2:1-112

Caspers H (1979) Die Entwicklung der Bodenfauna im Klärschlamm-Verklappungsgebiet vor der Elbe-Mündung. Arb Dtsch Fischerei Verb 27:109-134

Dethlefsen V (1980) Observations on fish diseases in the German Bight and their possible relation to pollution. Rapp P Réun Const Int Explor Mer 179:110-117

Dethlefsen V (1984a) Untersuchungen zur Erkennung subletaler Schadstoffeffekte an Fischen in der Deutschen Bucht: Krankheiten, Biochemie, Physiologie, Rückstände. Forschungsbericht BMFT 1-427

Dethlefsen V (1984b) Diseases in North Sea fishes. Helgoländer Meeresunters 37:353-374

Dethlefsen V (1985) Krankheiten von Nordseefischen als Ausdruck der Gewässerbelastung. Abh Naturwiss Verein Bremen 40/3:233-252

Dethlefsen V, Tiews K (1986) Wirkung der Verschmutzung der Nordsee auf Fische und die Fischerei. Veröff Inst Küsten Binnenfisch 93:1-51

Dethlefsen V (1986) Überblick über Auswirkungen der Verklappung von Abfällen aus der Titandioxidproduktion in der Deutschen Bucht. Veröff Inst Küsten Binnenfisch 95:1-42

Dethlefsen V, Huschenbeth E (1986) Regional differences in organochlorine residues in livers of dab (Limanda limanda) of the southern North Sea. Arch Fisch Wiss 37, 1/2:25-42

Dethlefsen V, Watermann B (1980) Epidermal papilloma of North Sea dab (Limanda limanda): histology, epidemiology and relation to dumping of wastes from $TiO_2$ industry. ICES Special Meeting on Diseases of Commercially Important Marine Fish and Shellfish, 8

Dethlefsen V, Westernhagen H von (1983) Oxygen deficiency and effects on bottom fauna in the eastern German Bight 1982. Meeresforschung 30:42-53

Dethlefsen V, Watermann B, Hoppenheit M (1984) Sources of variance in data from fish disease surveys. Arch Fisch Wiss 34, 2/3:155-173

Dethlefsen V, Cameron P, Westernhagen H von (1985) Untersuchungen über die Häufigkeit von Mißbildungen in Fischembryonen der südlichen Nordsee. Inf Fischwiss 32 (1):22-27

Dethlefsen V, Cameron P, Westernhagen H von, Janssen D (1986) Morphologische und chromosomale Untersuchungen an Fischembryonen der südlichen Nordsee in Zusammenhang mit der Organochlorkontamination der Elterntiere. Veröff Inst Küsten Binnenfisch 96:1-56

Dörjes J (1986) Langfristige Entwicklungstendenzen des Makrozoobenthos der Deutschen Bucht. Data to be submitted by FRG to the Sci Tech Work Group, Int North Sea Conf, 1987

Dyer MF, Pope JG, Fry PD, Law RJ, Portmann JE (1983) Changes in fish and benthos catches off the Danish coast in September 1981. J Mar Biol Assoc UK 63:767-775

Eisma D (1981) Suspended matter as a carrier for pollutants in estuaries and the sea. Elsevier Oceanogr Ser 27B:281-295

Figge K (1981) Sedimentverteilung in der Deutschen Bucht. Karte 2900 m Beih. Dtsch Hydrogr Inst

Forschungsstelle Norderney (ed) (1984) Bestandsschwankungen der Bodenfauna im ostfriesischen Watt bei Norderney, 1976-1980. In: Gewässergütemessungen im Küstenbereich der Bundesrepublik Deutschland, Gemeinsames Bund/Länder-Meßprogramm für die Nordsee, Hannover, pp 54-59

Gehrke J (1916) Über die Sauerstoffverhältnisse in der Nordsee. Ann Hydrogr Mar Meteorol 44:177–193
Gerlach SA (ed) (1985) Oxygen Depletion 1980–1983 in Coastal Waters of the Federal Republic of Germany. Berichte aus dem Inst f Meereskunde der Christian-Albrechts-Univ Kiel, Nr. 130
Grotjahn M, Michaelis H (1985) Das Benthos der Wesermündung im Einleitungsbereich säure- und eisenhaltiger Abwässer – Vergleich 1968 und 1980. Jahresber 1984 Forschungsstelle Küste 36:113–140
Hagmeier A (1925) Vorläufiger Bericht über die vorbereitenden Untersuchungen der Bodenfauna der Deutschen Bucht mit dem Petersen-Bodengreifer. Ber Dtsch Wiss Komm Meeresforsch 1:247–272
Hagmeier A, Kändler R (1927) Neue Untersuchungen im nordfriesischen Wattenmeer und auf den fiskalischen Austernbänken. Wiss Meeresunters 16:1–90
Hansen P-D, Westernhagen H von, Rosenthal H (1985) Chlorinated hydrocarbons and hatching success in Baltic herring spring spawners. Mar Environ Res 15:59–76
Heinke F (1894) Die Mollusken Helgolands. Wiss Meeresunters NF 1:1–92
Knust R, Dethlefsen V (1986) X-cells in gills of North Sea dab (Limanda limanda L.), epizootiology and impact on condition. Arch Fisch Wiss 37, 1/2:11–24
Kröncke I (1985) Makrofaunahäufigkeiten in Abhängigkeit von der Sauerstoffkonzentration im Bodenwasser der östlichen Nordsee. Diplomarb, Univ Hamburg, pp 1–127
Maier-Reimer E (1979) Some effects of the Atlantic circulation and of river discharges on the residual circulation of the North Sea. Dtsch Hydrogr Z 32 (3):126–130
Mellergaard S, Nielsen E (1984a) Preliminary investigations on the eastern North Sea and the Skagerrak plaice (*Pleuronectes platessa*) population and their disease. ICES CM/E:29
Mellergaard S, Nielsen E (1984b) Myxobolus infections in plaice (*Pleuronectes platessa*). ICES CM/E:15
Mellergaard S, Nielsen E (1985a) Fish diseases in the eastern North Sea dab (*Limanda limanda*) population with special reference to the epidemiology of epidermal hyperplasia/papillomas. ICES CM/E:14
Mellergaard S, Nielsen E (1985b) Preliminary investigations on the eastern North Sea and the Skagerrak dab (*Limanda limanda*) populations and their diseases ICES CM/E:28
Möller H (1979) Geographical distribution of fish diseases in the Atlantic. Meeresforschung 27:217–235
Möller H (1981) Fish diseases in German and Danish coastal waters in summer 1980. Meeresforschung 27:215–235
Mühlenhardt-Siegel U (1981) Die Biomasse mariner Makrobenthos-Gesellschaften im Einflußbereich der Klärschlammverklappung vor der Elbemündung. Helgoländer Meeresunters. 34:427–437
Mühlenhardt-Siegel U (1985) Die Weichbodengemeinschaft vor der Elbemündung unter dem Einfluß der Klärschlammverklappung. Diss, Univ Hamburg, pp 1–177
Rachor E (1982) Indikatorarten für Umweltbelastungen im Meer. Decheniana Beih 26:128–137
Rachor E (1985) Eutrophierung in der Nordsee – Bedrohung durch Sauerstoffmangel. Abh Naturwiss Verein Bremen 40:283–292
Rachor E, Albrecht H (1983) Sauerstoffmangel im Bodenwasser der Deutschen Bucht. Veröff Inst Meeresforsch Bremerhaven 19:209–227
Radach G, Berg J (1985) Trends in nutrient and phytoplankton concentrations at Helgoland Reede (German Bight) since 1962. ICES CM/L:2
Rauck G (1985) Wie schädlich ist die Seezungenbaumkurre für Bodentiere? Inf Fischwirtsch 32, 4:165–167
Reise K (1986) Gütezustand der Nordsee: Teilbereich Benthos. Data to be submitted by FRG to the Sci Tech Work Group, Int North Sea Conf, 1987
Riesen W, Reise K (1982) Macrobenthos of the subtidal Wadden Sea: revisited after 55 years. Helgoländer Meeresunters. 35:409–423
Salzwedel H, Rachor E, Gerdes D (1985) Benthic macrofauna communities in the German Bight. Veröff Inst Meeresforsch Bremerhaven 20:199–267
Tiews K (1983) Über die Veränderungen im Auftreten von Fischen und Krebsen im Beifang der deutschen Garnelenfischerei während der Jahre 1954–1981. Ein Beitrag zur Ökologie des deutschen Wattenmeeres und zum biologischen Monitoring von Ökosystemen im Meer. Arch Fisch Wiss 34:1–156
Vethaak AD (1985) Inventarisatie naar het Voorkomen van Visziekten in Relatie tot Vervuiling in Nederlandse Kustwateren. Rijksinst Visserijonderzoek, RIZA 59:1–61

Watermann B (1982) An unidentified cell type associated with an inflammatory condition of the subcutaneous connective tissue in dab, *Limanda limanda* L. Short communication. J Fish Dis 5:257–261

Watermann B (1984) Untersuchungen zur Histologie und Pathogenese von Hautwucherungen der Kliesche (*Limanda limanda* L.) aus der Nordsee. Diss, Univ Hamburg, pp 1–79

Watermann B, Dethlefsen V (1982) Histology of pseudobranchial tumours in Atlantic cod (*Gadus morhua*) from the North Sea and the Baltic Sea. Helgoländer Meeresunters 35:231–242

Watermann B, Dethlefsen V (1985) Epidermal hyperplasia and dermal degenerative changes as cell damage effects in gadoid skin. Arch Fisch Wiss 35, 3:205–221

Watermann B, Dethlefsen V, Hoppenheit M (1982) Epidemiology of pseudobranchial tumours in Atlantic cod (*Gadus morhua*) from the North Sea and the Baltic Sea. Helgoländer Meeresunters 35:231–242

Weichart G (1984) Unusually high pH values in the German Bight as an indication of strong primary production. Special Meeting on the causes, dynamics and effects of exceptional marine blooms and related events. ICES CM/B:6

Westernhagen H von, Rosenthal H, Dethlefsen V, Ernst W, Harms U, Hansen PD (1981) Bioaccumulating substances and reproductive success in Baltic flounder Platichthys flesus. Aqu Toxicol 1:85–99

Westernhagen H von, Hickel W, Bauerfeind E, Niermann U, Kröncke I (1986) Sources and effects of oxygen deficiencies in the southeastern North Sea. (in press)

Wolthaus BG (1984) Seasonal changes in frequency of diseases in dab Limanda limanda from the southern North Sea. Helgoländer Meeresunters 37:375–387

# Impact of Pollution on the Wadden Sea

W.J. WOLFF[1]

## 1 Introduction

In 1978 a multi-authored volume on the pollution of the Wadden Sea was published (Essink and Wolff 1978). Since then our knowledge has increased considerably, but basically the situation is still the same. A large number of contaminants have been demonstrated to occur in the Wadden Sea, but only for a small number has an impact on organisms or ecosystems been demonstrated. This chapter will focus on the latter cases for which sufficient scientific proof is available. A few cases in which grave suspicions exist will be mentioned as well.

This chapter limits itself strictly to the Wadden Sea (Fig. 1), which is defined as the area between the chain of the Frisian Islands and the mainland. Areas above average high-tide level, e.g., salt marshes, have not been considered.

## 2 Main Characteristics of the Wadden Sea

The Wadden Sea (Vadehavet, Wattenmeer, Waddenzee) is a shallow coastal sea extending along the coasts of Denmark, the Federal Republic of Germany and the Netherlands (Fig. 1). Its area is about 8000 km$^2$, of which 10% belong to Denmark, 55% to Germany and 35% to the Netherlands.

The average tidal amplitude varies between 1.36 m near Den Helder, The Netherlands, and 3.43 m at Husum in Germany. In general there is a gradient of increasing tidal amplitude both from the westernmost and the northernmost part of the area towards the Elbe estuary. Because of these tidal ranges the Wadden Sea geomorphology may be characterized as mesotidal. This appears from the more or less continuous series of barrier islands and the large areas of tidal flats. On average about 50% of the area emerges at low tide.

Sediments in the area are largely sandy, but near high-water mark and at tidal divides more muddy sediments occur as well.

Vegetation in the Wadden Sea consists largely of unicellular algae, planktonic as well as benthic. On tidal flat areas eelgrass meadows and sparse vegetations of macroalgae occur, but the only other vegetation occurs in the salt marshes.

Especially on the tidal flats, high numbers and biomasses of benthic invertebrates occur. These form the major food source for many fish species and millions of

[1]Research Institute for Nature Management, P.O. Box 46, 3956 ZR Leersum, The Netherlands

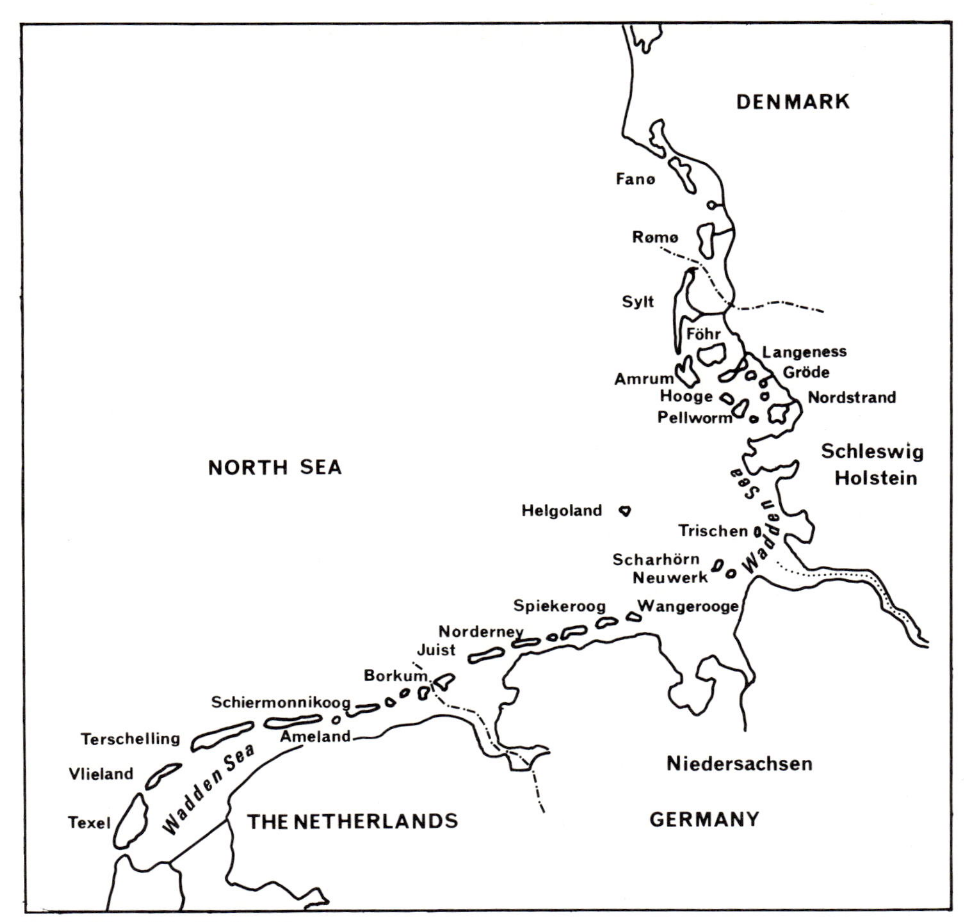


**Fig. 1.** Map of the Wadden Sea

shorebirds. Marine mammals occurring in the Wadden Sea are harbor seals (*Phoca vitulina*) and harbor porpoises (*Phocaena phocaena*).

An extensive account of the ecology of the Wadden Sea was published recently (Wolff 1983).

## 3 Impact of Pollutants

### 3.1 Micropollutants

#### 3.1.1 Presence of Micropollutants

Duinker and Koeman (1978) summarized the presence of micropollutants in the Wadden Sea. Since then numerous papers have given additional data (Dittmer 1982; Duinker and Hillebrand 1979; Duinker et al. 1983, 1984; Drescher et al. 1977; Essink

1977; Essink 1980, 1985; Goede 1985; Goede and De Voogt 1985; Luten et al. 1986; Reijnders 1980; De Wit et al. 1982; several chapters in this Volume).

As a general conclusion it may be stated that the Wadden Sea is contaminated by a large number of micropollutants and that for many compounds the levels are relatively high.

### 3.1.2 Impact of Micropollutants

*Heavy Metals.* Effects of mercury on the Wadden Sea biota have been suggested by Koeman et al. (1971) for seals and by Essink (1980) for subtidal benthic fauna, but for neither case does firm evidence exist.

Goede (1985) makes similar suggestions about the effects of arsenic on knots (*Calidris canutus*), but also in this case firm evidence is lacking.

*Organochlorines.* Koeman (1971; see also Koeman and Van Genderen 1972; Koeman et al. 1967; Swennen 1972) demonstrated that discharges of the pesticides telodrin and dieldrin from a pesticide factory at the Rhine estuary near Rotterdam led to large-scale bird mortality in the Dutch Wadden Sea. It resulted in a decline of the breeding populations of spoonbill (*Platalea leucorodia*), eider duck (*Somateria mollissima*), herring gull (*Larus argentatus*), common tern (*Sterna hirundo*), arctic tern (*Sterna macrura*), sandwich tern (*S. sandvicensis*), and little tern (*S. albifrons*). Telodrin, manufactured in Rotterdam only, and applied nowhere in Europe, was demonstrated to occur even in the eggs of terns breeding at the German island of Minsener Oldoog. The decrease in population size was most serious for sandwich and little tern, which were reduced to 5% or less of their original population size. Common tern, spoonbill, and eider declined to 15–25% of their original numbers and herring gulls were reduced to about 65% (Rooth 1980; Swennen 1982). Becker and Erdelen (1987) showed that a similar although smaller decrease occurred at the same time in the German Wadden Sea. Since then the factory has been partly closed and partly reorganized, resulting in an increase of the population size of all species concerned. Eider duck and herring gull are back at their original levels, but spoonbill and terns after 20 years (!) have not completely recovered (Figs. 2, 3). Conrad (1976) found relatively high amounts of organochlorines in the eggs of little terns (*Sterna albifrons*) from the Elbe estuary. He suspects a relation between these high levels and the decrease of the German population of this species.

Koeman et al. (1973) attributed the death of Cormorants (*Phalacrocorax carbo*) to the effects of polychlorinated biphenyls (PCB's), although he did not exclude the possibility that chlorinated dibenzofurans were involved.

Drescher (1978), Drescher et al. (1977) and Reijnders (1980, 1982) have correlated the death of seals or the decline of the seal population in the Dutch Wadden Sea with the presence of high levels of PCB's in seal tissues. A number of mechanisms has been proposed to explain these correlations. Drescher (1978) postulated that the immuno-suppressive effect of PCB's inhibited the healing of the umbilicus of newly born seals and ultimately led to the death of the animals through inflammation of the umbilical region (Fig. 4). Reijnders (1980, 1982) developed the hypothesis that PCB's interfere in some way with the reproductive

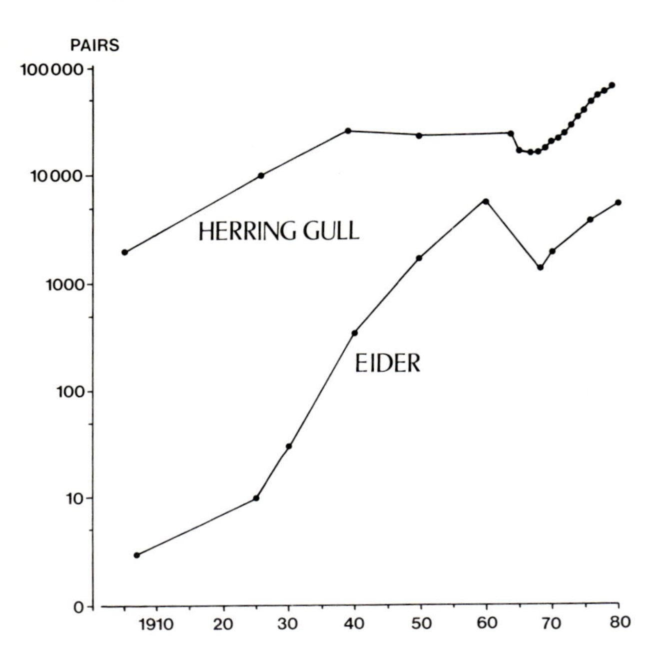


**Fig. 2.** Population size of eider duck and herring gull in the Dutch Wadden Sea. After a decrease of the breeding population due to insecticides in the 1960's the populations have recovered completely. (After Swennen 1982)

process in female seals and thus cause a lowered reproductive potential (Fig. 5). Experimental work with mink (*Mustela vison*) and seals led to the conclusion that PCB's do interfere with reproduction in harbor seals and are thus the cause of the decline of the Dutch seal population (De Boer 1984; Reijnders 1986).

Shortly before the Wadden Sea harbor seal population declined strongly, also the harbor porpoise decreased dramatically. In the 1960's, the species became virtually extinct in the Dutch and German parts of the Wadden Sea. PCB's and other organochlorines have been invoked as an explanation (Koeman et al. 1972; Verwey 1975; Van Bree 1977; Duinker and Hillebrand 1979; Verwey and Wolff 1981), although other explanations have been forwarded as well. Nevertheless, very high levels of PCB's and other contaminants have been demonstrated to occur in harbor porpoises from Dutch coastal waters (Koeman et al. 1972; Duinker and Hillebrand 1979).

A common feature of all cases described above is the heavy influence of the River Rhine on the biota of the Wadden Sea. Indeed, De Groot (1963), Roskam (1966), Koeman (1971), Zimmerman and Rommets (1974), Van Bennekom et al. (1975) and Zimmerman (1976) reveal that a significant part of the Rhine water, and probably an even larger part of the suspended matter carried by this river reaches the westernmost part of the Dutch Wadden Sea about 1 month after discharge at

**Fig. 3.** Population size of spoonbill and sandwich tern in the Dutch Wadden Sea. After a decrease of the breeding population due to insecticides in the 1960's the populations have not yet recovered completely. (After Rooth 1980)

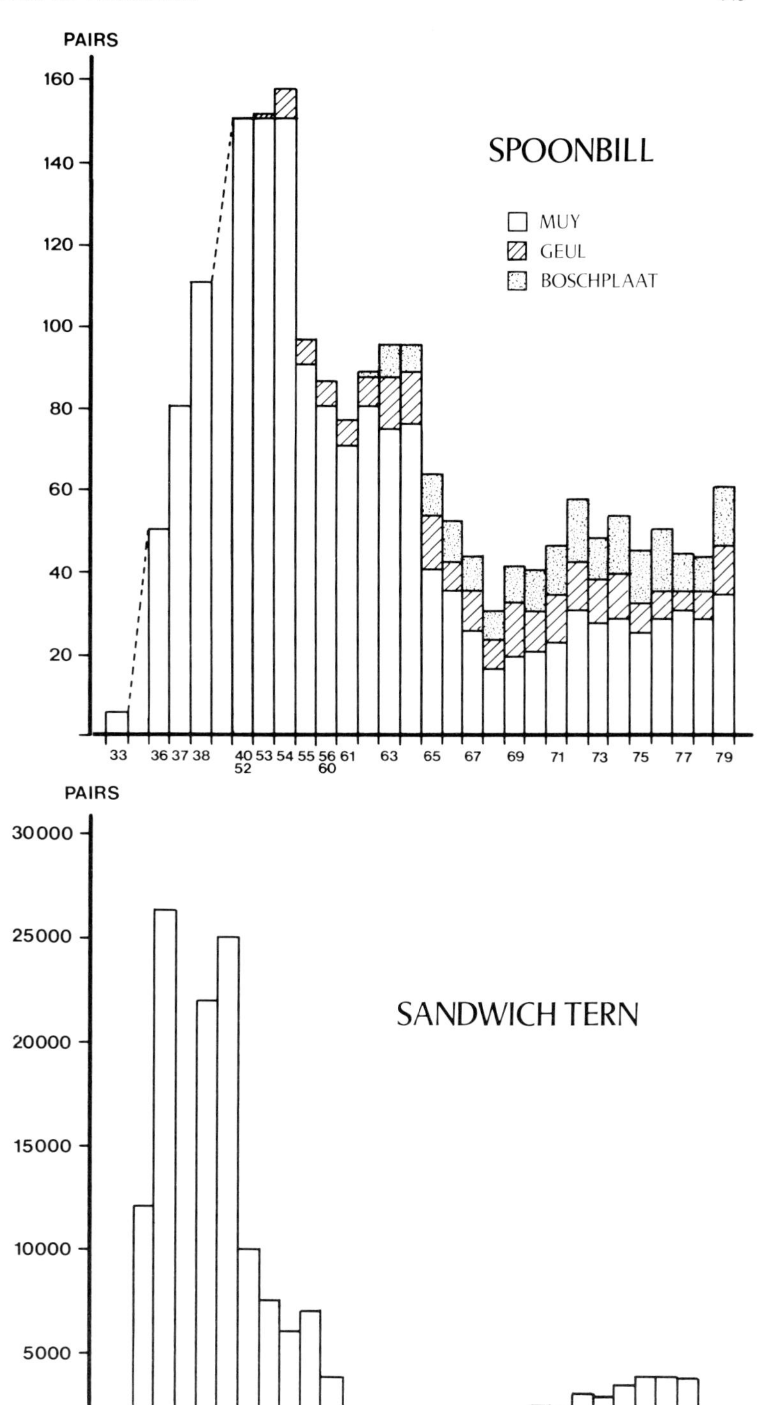
PAIRS
160
140
120
100
80
60
40
20
SPOONBILL
MUY
GEUL
BOSCHPLAAT
33
36 37 38
40 52
53 54 55 56 60
61
63
65
67
69
71
73
75
77
79
PAIRS
30000
25000
20000
15000
10000
5000
SANDWICH TERN
?
1953
55
57
59
61
63
65
67
69
71
73
75
77
79

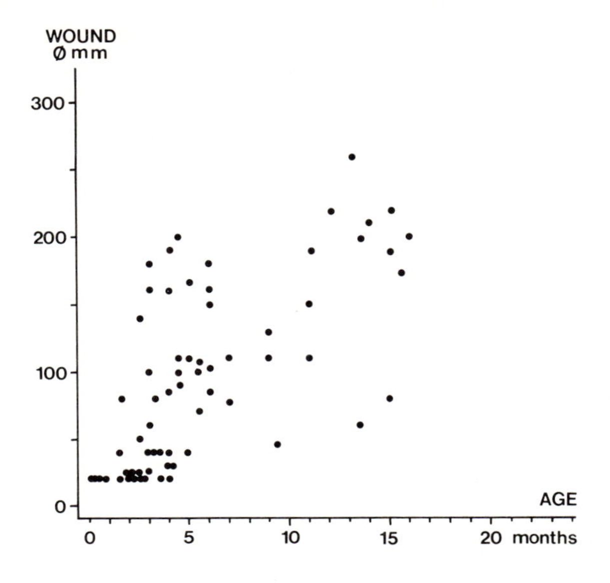


**Fig. 4.** Relationship between age of young harbor seals and size of wounds on the belly. (After Drescher 1978)

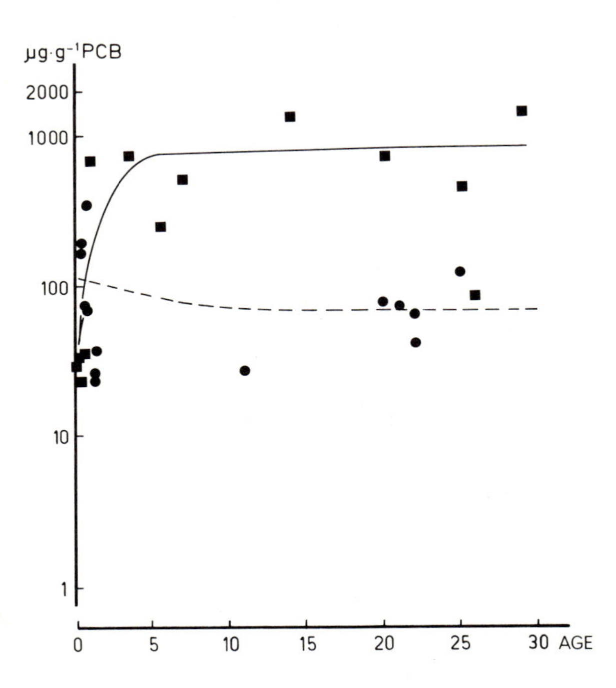


**Fig. 5.** Relationship between age of harbor seals and PCB content of the blubber. *Squares* denote seals from the Dutch Wadden Sea, *dots* from the Wadden Sea in Schleswig-Holstein and Denmark. (After Reijnders 1980)

the Rhine estuary near the Hook of Holland. From West to East the amount of Rhine water in the Wadden Sea decreases (Zimmerman 1976), as do the ecological effects of pollutants (Koeman 1971; Reijnders 1980).

## 3.2 Eutrophication

### 3.2.1 Increase of Nutrient Levels

De Jonge and Postma (1974) estimated that phosphate levels in the Dutch western Wadden Sea had increased threefold in the period 1950–1970. Helder (1974) concluded a twofold increase of dissolved nitrogen compounds (especially ammonia) in the Dutch Wadden Sea in the period 1960–1970. De Wit et al. (1982) observed for the same area a further increase of phosphate levels in winter in the period 1971–1981, but found no increase for summer phosphate levels and levels of dissolved nitrogen compounds over the same period. Increased nutrient levels in the Wadden Sea have been attributed to increased mineralization of organic matter imported into the Wadden Sea from the North Sea (De Jonge and Postma 1974), increased nutrient levels in coastal North Sea water exchanged with Wadden Sea water (De Wit et al. 1982) and increased nutrient levels in freshwater discharged from the mainland. Ultimately, however, increased nutrient levels in the River Rhine are held responsible for these changes.

Also for other parts of the Wadden Sea increased nutrient levels have been reported, e.g. Ems estuary (Colijn 1983) and Gradyb near Esbjerg (Henriksen et al. 1984).

### 3.2.2 Impact of Increased Nutrient Levels

Increased levels of nutrients might be reflected in the first place in increased levels of primary production. In the westernmost part of the Dutch Wadden Sea phytoplankton production did not increase in the period 1950–1976, but did so after 1978 (Postma and Rommets 1970; Cadée and Hegeman 1979; Cadée 1984). The latter author also showed an increasing production by the microphytobenthos at one particular station in the western Wadden Sea over the period 1960–1981 (Fig. 6). Reise (1984) suggests that the recent spread of mats of macroalgae on the tidal flats of the Wadden Sea is a response to increased nutrient levels. Such mats were observed in the Bay Königshafen at Sylt since 1979.

Increased primary production might be reflected in increased levels of consumer production. Beukema and Cadée (1986) provide evidence that macrobenthic animals doubled both in biomass and annual production (Fig. 7) on the tidal flats of the western Wadden Sea in the period 1970–1984. Also growth rate and reproductive success increased in this period. These authors suggest that these phenomena can be explained as a response to eutrophication. Although this hypothesis is supported strongly by the available data, it cannot be proven so far.

So far these increased densities of plants and animals have not been accompanied by mass mortalities due to oxygen deficits, although rather low oxygen values have been observed a few times (Tijssen and Van Bennekom 1976).

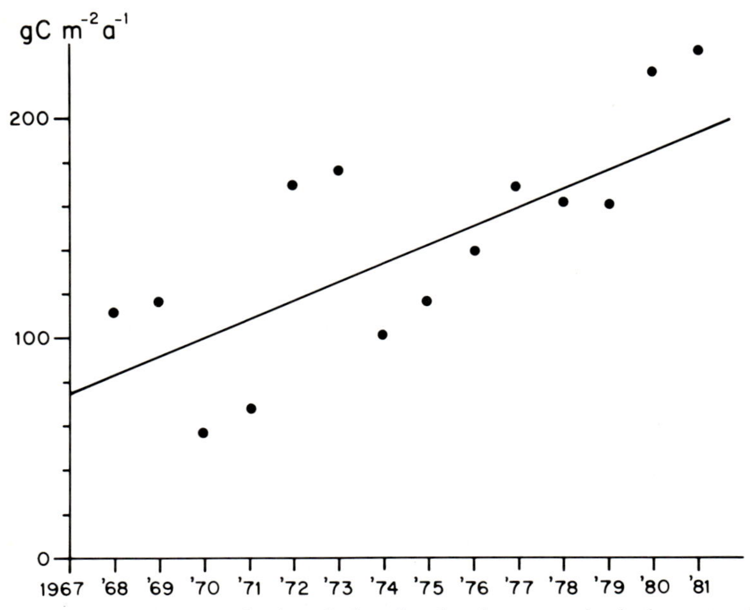


**Fig. 6.** Annual production of microphytobenthos at a station in the western Dutch Wadden Sea in the period 1960–1981. (After Cadée 1984)

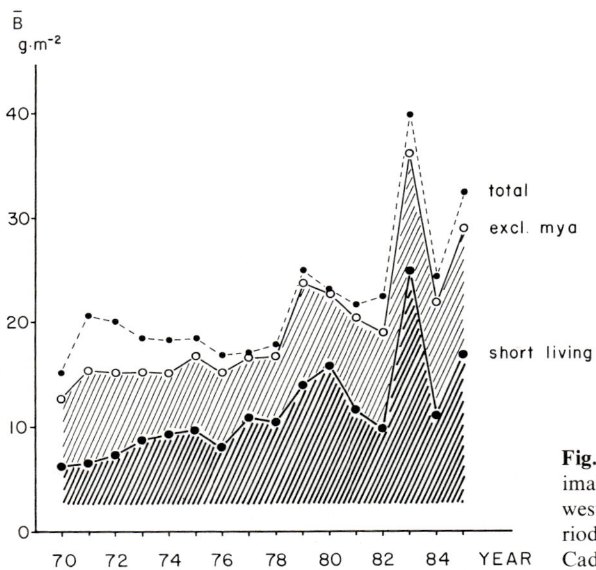


**Fig. 7.** Biomass of macrobenthic animals at the Balgzand tidal flats in the western Dutch Wadden Sea in the period 1970–1984. (After Beukema and Cadée 1986)

## 3.3 Organic Waste

### 3.3.1 Quantities of Organic Waste Discharged

Essink (1978a) lists the discharges of organic waste along the entire Wadden Sea. Quantities discharged are generally low except for the vicinity of Esbjerg (see also Henriksen et al. 1984), the Elbe estuary, the Weser estuary, the Dollard embayment and the Ems estuary, and the vicinity of Noordpolderzijl, the Netherlands. Essink (1984) discusses the history of direct introduction of organic waste into the Dutch Wadden Sea. Next to a decreasing number of minor discharges, major discharges occurred only at Noordpolderzijl, in the Ems estuary and in the Dollard embayment. The total quantity discharged at the latter three points has been reduced from over 300 tons $BOD_5^{20}$ per day in the early 1970's to some 140 tons recently, and further reductions are planned. This decreasing tendency applies to all discharges of organic waste into the Wadden Sea.

### 3.3.2 Impact of Discharges of Organic Waste

Essink (1978b, 1984) and Essink and Beukema (1986) describe the effects of the discharge of large quantities of organic waste. This primarily affects the oxygen balance, and eventually the levels of inorganic nutrients (cf. Sect. 3.2). Oxygen deficits lead to mortality of mainly benthic invertebrates when other conditions are adverse. Thus around the outfall of a waste pipeline near Noordpolderzijl all macrobenthic animals disappeared from about 15–20 ha, when 40–55 ton BOD per day were discharged. Discharge of about 250 ton BOD of organic waste per day into the Dollard embayment led to an impoverished fauna in an area of about 1500 ha. Since the latter discharge was among the largest in the world, it can be concluded that the negative direct effects of discharge of organic waste into the Wadden Sea are relatively restricted. Positive direct effects have not been demonstrated.

Next to the direct effects on the area where the waste is discharged, indirect effects may occur through mineralization of the organic material and subsequent eutrophication of the coastal waters (see Sect. 3.2). Thus, Reise (1984) ascribes the development of mats of green algae on the tidal flats of Königshafen, Sylt, to discharge of organic waste.

## 3.4 Oil Pollution

### 3.4.1 Occurrence of Oil Pollution

Potential sources of oil pollution in the Wadden Sea are several. Exploitation of a small oil field is developing near the island of Trischen north of the Elbe estuary. A pipeline, also to be used for transport of oil, may be constructed between the North Sea and the coast of the province of Groningen, The Netherlands. Small tankers transport various oil products across the western Wadden Sea in the Netherlands. Large-scale transport of oil occurs off the Wadden Sea islands across the North Sea and part of this transport is directed towards the estuaries of Ems, Jade, Weser and Elbe, and to Esbjerg. Particularly the transport to Wilhelmshaven, Germany, is important. All over the Wadden Sea small vessels, e.g., fishermen and pleasure craft, use the channels. The actual occurrence of oil pollution is frequent. Many times per year small oil slicks (< 10 m$^3$) may be observed in the Wadden Sea, the tidal inlets and the major estuaries. Larger oil occurrences (10–1000 m$^3$) are seen much less frequently; *Torrey Canyon*-size spills have not occurred so far. However, a standardized system for reporting oil occurrences in the Wadden Sea has not been developed yet.

### 3.4.2 Impact of Oil Pollution

The effect of "real" oil pollution on aquatic organisms has not been studied in the Wadden Sea itself. However, Bergman (1982, 1983) has made an elaborate prediction of the effects of an oil spill in the Wadden Sea, based on a thorough study of the international literature, including experimental work done in the Wadden Sea (Van Bernem 1976). The effects of an oil spill will vary with the species of organism considered, life stage, time of the year, temperature etc. Some species, e.g., the amphipod *Corophium* sp., are extremely vulnerable, others, e.g., the mussel *Mytilus edulis*, are relatively resistant. Kuiper et al. (1984) demonstrated experimentally that effects may still occur more than half a year after a small oil spill. In general, however, it may be predicted that the effects will have disappeared after a period of time. On the sandy beaches and flats around the tidal inlets effects may last a few months only, at the extensive tidal flats further inward the impact may last 1–4 years and only at the muddy flats near the tidal divides and the mainland coast and in the salt marshes may the damage last for many years (Gundlach and Hayes 1978; Bergman 1983). This prediction is made for large-scale (> 1000 m$^3$) oil spills; it is, however, very difficult to distinguish the effects of regularly occurring small-scale spills from the natural variability of the environment.

Much better information is available on the effects of oil on birds. Goethe (1978), Joensen (1978) and Swennen (1978) present data still characteristic for the situation in the Wadden Sea. It is estimated that several thousands of birds are killed annually by the "normal" small-scale oil pollution.

The spill of 8000 tons of oil after a tanker collision off the Elbe estuary in January 1955 led to large-scale mortality of coastal birds. Over 2200 dead birds, mainly common scoters, were found along the coast of Schleswig-Holstein, Germany, and over 3000 oiled birds were killed by hunters in Denmark (Goethe 1961; Joensen 1972). The same authors estimate that the real mortality was much larger: 200,000 (Goethe 1961) – 500,000 birds (Goethe 1978) in Germany and over 20,000 birds in Denmark (Joensen 1972, 1978).

Another incident was described by Swennen and Spaans (1970) for the westernmost part of the Wadden Sea in February 1969. A few hundreds of tons of oil drifted in from an unknown source at the North Sea and resulted in 14,500 birds found dead, mainly eider ducks and common scoters. The authors estimated that the total number of victims will have amounted to about 40,000.

Goethe (1978) and Joensen (1978) list several other cases, but from the two examples described it will be clear that oil can have a disastrous impact on birds in the Wadden Sea. Effects on seals seem to be relatively minor (Bergman 1982, 1983).

### 3.5 Flotsam and Jetsam

The beaches and salt marshes of the Wadden Sea area are characterized by large quantities of flotsam and jetsam, which for a considerable part is of human origin (Nassauer 1981).

Although the impact on the landscape is considerable, ecological effects are relatively small. Hartwig et al. (1985) give examples of effects on fishes, birds, and seals.

### 3.6 Thermal Pollution

Essink (1978c) listed the sources of thermal pollution in the Wadden Sea. Since that publication the picture has hardly changed. Only at Esbjerg, Denmark, and Texel, The Netherlands, power stations (500 and 15 megawatt, respectively) discharge directly into the Wadden Sea. All other discharges of heated cooling water are concentrated at the estuaries of Ems, Jade, Weser, and Elbe. Nevertheless, the impact on temperature regimes in these estuaries is slight.

No ecological effects of thermal pollution have been demonstrated in the Wadden Sea area.

### 3.7 Acid Wastewater from a Titanium Dioxide Factory

Wienecke (1982) and Grotjahn and Michaelis (1985) describe the ecological effects of the discharge into the Weser estuary of 20–30,000 $m^3$ per day of wastewater of a titanium dioxide factory. The main constituents are $H_2SO_4$ and $FeSO_4$ and the pH is 2–2.5. No clearcut effects on diversity and biomass of flora and fauna could be observed, but it should be mentioned that this discharge was situated in the brackish zone of the estuary which is biologically poor by natural causes and because of several other human interventions.

### 3.8 Radioactive Pollution

Levels of radioactive compounds in the Wadden Sea are low. In only a few cases are levels slightly elevated, notably for tritium (3H).

No effects of radioactive pollution have been demonstrated in the Wadden Sea.

# 4 Conclusion

At present pollution problems in the Wadden Sea arise from organic micropollutants, eutrophication, and oil. Eutrophication appears to be an increasing problem, whereas the situation with regard to organic micropollutants and oil seems to be more or less stable. The presence and effects of inorganic micropollutants and organic waste are declining thanks to an active anti-pollution policy. Ecological effects of flotssam and jetsam, discharge of heated cooling water and radioactive substances are negligible so far, but may become more significant in future.

These conclusions apply to the Wadden Sea as a whole. Locally and temporarily the situation can be different.

# References

Becker PH, Erdelen M (1987) Die Bestandsentwicklung von Brutvögeln der deutschen Noordseeküste 1950–1979. J Orn 128:1–32

Bennekom AJ van, Gieskes WWC, Tijssen SB (1975) Eutrophication of Dutch coastal waters. Proc R Soc Lond B 189:359–374

Bergman M (1982) Gedrag, bestrijding en biologische effecten van olie in estuariene gebieden. I. Literatuuroverzicht. RIN-rapport 18:420

Bergman M (1983) Gedrag, bestrijding en biologische effecten van olie in estuariene gebieden. II Olie in de Waddenzee. RIN-rapport 22:97

Bernem KH van (1976) Beobachtungen zur Einwirkung von Rohölen auf ausgewählte Wattenorganismen in mesohalinen Bereichen des Elbe-Aestuars. Diplom-Arbeit Univ Hamburg Fachbereich Biol, 143 pp

Beukema JJ, Cadée, GC (1986) Zoobenthos responses to eutrophication of the Dutch Wadden Sea. Ophelia 26:55–64

Boer MH de (1984) Reproduction decline of harbour seals: PCBs in the food and their effect on mink. Ann Rep Res Inst Nat Management 1983:77–86

Bree, PJH van (1977) On former and recent strandings of cetaceans on the coast of the Netherlands. Z Säugetierk 42:101–107

Cadée GC (1984) Has input of organic matter into the western part of the Dutch Wadden Sea increased during the last decades? Neth Inst Sea Res Publ Ser 10:71–82

Cadée GC, Hegeman J (1979) Phytoplankton primary production, chlorophyll and composition in an inlet of the western Wadden Sea (Marsdiep). Neth J Sea Res 13:224–241

Colijn F (1983) Primary production in the Ems-Dollard estuary. Thesis Univ Groningen, 123 pp

Conrad B (1976) Die Belastung der freilebenden Vogelwelt der Bundesrepublik Deutschland mit chlorierten Kohlenwasserstoffen und PCB und deren mögliche Auswirkungen. Thesis, Köln, 90 pp

Dittmer, JD (1982) Untersuchungen über Schwermetallkonzentrationen in eulitoralen Sedimenten und Organismen des Jadegebietes. Jahresber Forsch stelle Insel Küstenschutz 32:69–98

Drescher HE (1978) Hautkrankheiten beim Seehund, *Phoca vitulina* Linné 1758, in der Nordsee. Säugetierk Mitt 26:50–59

Drescher HE, Harms U, Huschenbeth E (1977) Organochlorines and heavy metals in the harbour seal (*Phoca vitulina*) from the German North Sea coast. Mar Biol 41:99–106

Duinker JC, Hillebrand MTJ (1979) Behaviour of PCB, pentachlorobenzene, hexachlorobenzene, $\alpha$-HCH, $\beta$-HCH, $\gamma$-HCH, dieldrin, endrin and p,p′-DDD in the Rhine-Meuse estuary and the adjacent coastal area. Neth J Sea Res 13:256–281

Duinker JC, Koeman JH (1978) Summary report on the distribution and effects of toxic pollutants (metals and chlorinated hydrocarbons) in the Wadden Sea. In: Essink K, Wolff WJ (eds) Pollution of the Wadden Sea area. Balkema Rotterdam, pp 45–54

Duinker JC, Hillebrand MTJ, Boon JP (1983) Organochlorines in benthic invertebrates and sediments from the Dutch Wadden Sea: identification of individual PCB components. Neth J Sea Res 17:19–38

Duinker JC, Boon JP, Hillebrand MTJ (1984) Organochlorines in the Dutch Wadden Sea. Neth Inst Sea Res Publ Ser 10:211–228

Essink K (1978a) Inventory of the quantities of waste discharged into the Wadden Sea from terrestrial sources. In: Essink K, Wolff WJ (eds) Pollution of the Wadden Sea area. Balkema Rotterdam, pp 25–36

Essink K (1978b) The effects of pollution by organic waste on macro-fauna in the eastern Dutch Wadden Sea. Neth Inst Sea Res Publ Ser 1:135

Essink K (1978c) Sources of thermal pollution. In: Essink K, Wolff WJ (eds) Pollution of the Wadden Sea area. Balkema Rotterdam, pp 36–38

Essink K (1980) Mercury pollution in the Ems estuary. Helgol Wiss Meeresunters 33:111–121

Essink K (1984) The discharge of organic waste into the Wadden Sea – local effects. Neth Inst Sea Res Publ Ser 10:165–177

Essink K (1985) Monitoring of mercury pollution in Dutch coastal waters by means of the teleostean fish *Zoarces viviparus*. Neth J Sea Res 19:177–182

Essink K, Beukema JJ (1986) Long-term changes in tidal flat zoobenthos as indicators of stress by organic pollution. Hydrobiologia 142:209–215

Essink K, Wolff WJ (1978) Pollution of the Wadden Sea area. Balkema Rotterdam, 61 pp

Goede AA (1985) Mercury, selenium, arsenic and zinc in waders from the Dutch Wadden Sea. Environ Pollut Ser A 37:287–309

Goede AA, Voogt P de (1985) Lead and cadmium in waders from the Dutch Wadden Sea. Environ Pollut Ser A37:287–309

Goethe F (1961) Deutscher Olpestbericht 1953–1961. Internat Rat Vogelschutz Deutsche Sektion Ber 1:1–12

Goethe F (1978) Oil pollution affecting birds along the German North Sea coast. In: Essink K, Wolff WJ (eds) Pollution of the Wadden Sea area. Balkema Rotterdam, pp 58–59

Groot AJ de (1963) Mangaantoestand van Nederlandse en Duitse Holocene sedimenten in verband met slibtransport en bodemgenese. Versl. Landbouwk Onderz Wageningen 69.7:1–164

Grotjahn M, Michaelis M (1985) Das Benthos der Wesermündung im Einleitungsbereich säure- und eisenhaltiger Abwässer-Vergleich 1968 und 1980. Jahresber 1984 Forschstelle Küste, Norderney 36:113–140

Gundlach ER, Hayes MO (1978) Vulnerability of coastal environments to oil spill impacts. Mar Techn Soc J 12:18–27

Hartwig E, Reineking B, Schrey E, Vauk-Hentzelt E (1985) Auswirkungen der Noordsee-Vermüllung auf Seevögel, Robben und Fische. Seevögel Sonderb 6:57–62

Helder W (1974) The cycle of dissolved inorganic nitrogen compounds in the Dutch Wadden Sea. Neth J Sea Res 8:154–173

Henriksen K, Jensen A, Rasmussen MB (1984) aspects of nitrogen and phosphorus mineralization and recycling in the northern part of the Danish Wadden Sea. Neth Inst Sea Res Publ Ser 10:51–69

Joensen AH (1972) Oil pollution and sea birds in Denmark 1935–1968. Dan Rev Game Biol 6, 8:1–24

Joensen AH (1978) The Danish Wadden Sea. In: Essink K, Wolff WJ (eds) Pollution of the Wadden Sea area. Balkema Rotterdam, pp 59–60

Jonge VN de, Postma H (1974) Phosphorus compounds in the Dutch Wadden Sea. Neth J Sea Res 8:139–155

Koeman JH (1971) Het voorkomen en de toxicologische betekenis van enkele chloorkoolwaterstoff aan de nederlandse kust in de periode van 1965 tot 1970. Thesis Univ Utrecht, 136 pp

Koeman JH, Genderen H van (1972) Tissue levels in animals and effects caused by chlorinated hydrocarbon insecticides, chlorinated biphenyls and mercury in the marine environment along the Netherlands coast. In: Marine Pollution and Sea Life Fishing News (Books), Survey, pp 1–8

Koeman JH, Oskamp AAG, Veen J, Brouwer E, Rooth J, Zwart P, Broek E van de, Genderen H van (1967) Insecticides as a factor in the mortality of the Sandwich tern. A preliminary communication. Med Rijksfac Landbouwwet, Gent 32:841–854

Koeman JH, Canton JH, Woudstra A, Peeters WHM, Goeij JJM de, Zeegers C, Haaften JL van (1971) Kwik in het nederlandse kustmilieu. TNO-nieuws 26:402–409

Koeman JH, Peeters WHM, Smit CJ, Tjioe PS, Goeij JJM de (1972) Persistent chemicals in marine mammals. TNO-nieuws 27:570–578

Koeman JH, Velzen-Blad HCW van, Vries R de, Vos JG (1973) Effects of PCB and DDE in cormorants and evaluation of PCB residues from an experimental study. J Reprod Fert Suppl 19:353–364

Kuiper J, Wilde PAWJ de, Wolff WJ (1984) Oil pollution experiment (OPEX). I. Fate of an oil mousse and effects on macrofauna in a model ecosystem representing a Wadden Sea tidal mud flat. In: Persoone G, Jaspers E, Claus C (eds). State Univ, Gent, Ecotoxicological testing for the marine environment. 2:331–359

Luten JB, Bouquet W, Burggraaf MM, Rauchbaar AB, Rus J (1986) Trace metals in mussels (*Mytilus edulis*) from the Wadden Sea, coastal North Sea and the estuaries of Ems, Western and Eastern Scheldt. Bull Environ Contam Toxicol 36:770–777

Nassauer G (1981) Untersuchungen zur Müllbelastung von Stränden der deutschen Nordseeküste. Seevögel 3:53–57

Postma H, Rommets J (1970) Primary production in the Wadden Sea. Neth J Sea Res 4:470–493

Reijnders PJH (1980) Organochlorine and heavy metal residues in harbour seals from the Wadden Sea and their possible effects on reproduction. Neth J Sea Res 14:30–65

Reijnders PJH (1982) On the ecology of the harbour seal (*Phoca vitulina*) in the Wadden Sea: population dynamics, residue levels and management. Veter Quart 4:36–42

Reijnders PJH (1986) Reproductive failure in common seals feeding on fish from polluted coastal waters. Nature (Lond) 324:456–457

Reise K (1984) Indirect effects of sewage on a sandy tidal flat in the Wadden Sea. Neth Inst Sea Res Publ Ser 10:159–164

Rooth J (1980) Sandwich tern (*Sterna sandvicensis*). In: Smit CJ, Wolff WJ (eds) Birds of the Wadden Sea. Balkema Rotterdam, pp 250–258

Roskam (1966) Kopervergiftiging in zee. Water, bodem, lucht 56:19–23

Swennen C (1972) Chlorinated hydrocarbons attacked the Eider population in the Netherlands. TNO-nieuws 27:556–560

Swennen C (1978) The Dutch Wadden Sea. In: Essink K, Wolff WJ (eds) Pollution of the Wadden Sea area. Balkema Rotterdam, pp 56–58

Swennen C (1982) De vogels langs onze kust. In: Wadden, duinen, delta. Pudoc, Wageningen, pp 78–100

Swennen C, Spaans AL (1970) De sterfte van zeevogels door olie in Februari 1969 in het Waddengebied. Vogeljaar 18:233–245

Tijssen SB, Bennekom AJ van (1976) Lage zuurstofgehaltes in het water op het Balgzand. $H_2O$ 9:28–31

Verwey J (1975) The cetaceans *Phocaena phocaena* and *Tursiops truncatus* in the Marsdiep area (Dutch Wadden Sea) in the years 1931–1973, I + II. Ned Inst Onderz Zee Publ Versl 17a + b:153

Verwey J, Wolff WJ (1981) The common or harbour porpoise (*Phocaena phocaena*). In: Reijnders PJH, Wolff WJ (eds) Marine mammals of the Wadden Sea. Balkema Rotterdam, pp 51–58

Wienecke G (1982) Untersuchungen von Sediment und Bodenfauna in der Wesermündung in Zusammenhang mit säure- und eisenhaltigen Abwässern. Jahresber Forsch stelle Insel Küstenschutz, Norderney 32:119–171

Wit JAW de, Schotel FM, Bekkers LEJ (1982) De waterkwaliteit van de Waddenzee 1971–1981. Rijkswaterstaat Rijksinst Zuivering Afvalwater Nota 82 065–67

Wolff WJ (1983) Ecology of the Wadden Sea. Balkema Rotterdam, 2000 pp

Zimmerman JTF (1976) Mixing and flushing of tidal embayments in the western Dutch Wadden Sea, I + II. Neth J Sea Res 10:149–191, 397–439

Zimmerman JTF, Rommets J (1974) Natural fluorescence as a tracer in the Dutch Wadden Sea and the adjacent North Sea. Neth J Sea Res 8:117–125

# The Impact of Anthropogenic Activities on the Coastal Wetlands of the North Sea

A.H.L. HUISKES[1] and J. ROZEMA[2]

## 1 Introduction

The North Sea is surrounded by one of the most industrialized areas of the world. The industrial activities and harbor facilities are mainly found along the mouths of the major rivers flowing into the North Sea (see chapter 2, this Vol.), where also the main estuarine wetland areas are concentrated (Beeftink and Rozema, this Vol.).

The wetlands surrounding the North Sea are under heavy pressure from man's activities: civil engineering operations (dredging, diking, and other building operations); discharge of large quantities of industrial processing water, sometimes changing temperature and salinity gradients in the estuary; eutrophication by the discharge of human and agricultural waste water, and the contaminations by substances of anthropogenic origin discharged as sewage from industrial, domestic, or agricultural sources. This anthropogenic source of contamination of the estuarine system is the subject dealt with in this paper.

Most of the organic and inorganic contaminating agents in the estuarine system are generally adsorbed to the particulate matter transported with the river water, a minor part is of airborne origin. It is a well-known fact that the estuarine part of the river acts as a sink for particulate matter (Lauff 1967; Wiley 1976) except for a few very big deltaic rivers (e.g., Mississippi, Ganges etc.).

All major rivers flowing into the North Sea have larger or smaller estuaries, which means that most wetland areas around them are under the influence of contaminants of some kind or the other.

The effects these contaminants can have on the wetlands, and more in particular on the organisms living on it and in it are manifold, depending on the kind of contaminant, the concentration, and the physicochemical environment.

In this chapter the effects of anthropogenic substances are divided in the following categories: trace metals, organic microcontaminant, oil, and radioactive materials.

[1]Delta Institute for Hydrobiological Research, Vierstraat 28, 4401 EA Yerseke, The Netherlands*
[2]Department of Ecology and Ecotoxicology, Free University, P.O. Box 7161, 1007 MC Amsterdam, The Netherlands
*Communication Nr 378

## 2 Effects of Trace Metals on the Functioning of the Wetland Ecosystem

The occurrence of trace metals is not an unnatural phenomenon in the wetlands of the estuaries. Many rivers flow through formations containing metal ores resulting in a natural background of heavy metal ions. Förstner and Wittmann (1979) discuss the natural levels of metal ions in fluviatile deposits. In the study of anthropogenic contamination of sediments, shales are commonly used as a standard, as the levels do not differ greatly from those found in sediments usually regarded as uncontaminated (Table 1). Adopting the shale standard as an unpolluted background level, all other enrichments found in sediments should be regarded by definition as pollution. A measure for the state of contaminant is the factor of enrichment (Beeftink and Rozema, this Vol.).

Table 1 shows the situation in estuaries in the southwest of the Netherlands (data from Beeftink et al. 1982), which indicates a notable pollution of the wetlands along the Westerschelde estuary as compared with the Oosterschelde, a former estuarine branch of the river Scheldt but closed off from the riverine system by railway dams constructed around 1870. In the same study it was shown that the levels of heavy metals in the shoots of the various salt-marsh species could differ considerably, depending on the species and metals concerned, and on the state of pollution.

Table 2 and Fig. 1 give a compilation of the results whereby the levels of the contaminating metals are subdivided into ten classes based on the average level in the shoots of the plants collected in three marshes in the eastern part of the Westerschelde estuary with the species ranking highest given an extra class number (11). A statistical analysis indicates that the species that express distinctly a different behavior towards Cd-Cu uptake on the one hand and towards Pb-Hg-Ni on the other belong to different clusters. Another cluster of species shows low levels of heavy metals in the aerial tissues, which indicates supposedly either an active exclusion by the roots (known for certain grasses: *Agrostis tenuis* and *A. stolonifera* (Peterson 1975; Wainwright and Woolhouse 1975) or excretion by leaf glands, or that certain metals (e.g., Hg) are not transported from the roots to the shoots (Gardner et al. 1978; Rozema et al. 1985c).

Rozema and Roosenstein (1985), however, argue that no such exclusion or excretion mechanism exists for heavy metals and ascribe the differences in the plant tissue to different levels of availability of heavy metals in the soil or to differences in the speciation of the metals. If in salt-marsh plants the general translocation pattern in the plant is acropetal from the root to the shoot, there is also an uptake route via the leaves (Rozema et al. 1985a), especially when the plants are inundated. For *Zostera marina* (a submerged plant from the tidal flats) such a basipetal transport was shown (Faraday and Churchill 1979) while Brinkhuis et al. (1980) and Schroeder and Thorlaug (1980) found a two-way transport between leaves and roots. The translocation processes (of Cd) in *Zostera* seem to be dependent on variation of environmental salinity (Brinkhuis et al. 1980).

Not only does the uptake of heavy metals by salt-marsh plants differ from species to species, but it is also strongly influenced by soil conditions like pH, redox

**Table 1.** Heavy-metal concentrations (ppm) in sediments of Dutch estuarine and coastal waters compared with concentrations of uncontaminated sediments. Concentration of Western (WS) and Eastern (ES) Scheldt extrapolated to 50% fraction < 16 $\mu$m, in comparison with recent data from Rhine and Meuse to 100% fraction < 16$\mu$m. (After Beeftink et al. 1982, from various sources)

| | Year | Cr | Ni | Cu | Zn | As | Cd | Hg | Pb |
|---|---|---|---|---|---|---|---|---|---|
| Western Scheldt (50% clay) | | | | | | | | | |
| Mouth | 1974 | 98 | 22 | 26 | 165 | 17 | 0.9 | 0.6 | 59 |
| Saaftinge | 1974 | 170 | 34 | 80 | 393 | 48 | 7.6 | 1.9 | 114 |
| Waarde | 1977 | | 4 | 48 | 442 | 74 | 2.5 | 1.4 | 134 |
| Eastern Scheldt (50% clay) | | | | | | | | | |
| Mouth | 1975 | 88 | 23 | 23 | 157 | | 0.9 | | 55 |
| Rattekaai | 1975 | 81 | 26 | 29 | 143 | | 1.1 | | 49 |
| Stroodorpepolder salt marsh | 1977 | | 35 | 22 | 145 | 27 | 0.5 | 0.4 | 72 |
| Dutch sediments (100% clay) | | | | | | | | | |
| Western Scheldt (Saaftinge) | 1974 | 268 | 62 | 155 | 715 | 87 | 1.5 | 3.5 | 210 |
| Eastern Scheldt (Rattekaai) | 1975 | 109 | 49 | 58 | 257 | | 2.0 | | 97 |
| Rhine | 1970 | 1240 | 103 | 600 | 2900 | 220 | 45 | 25 | 800 |
| Meuse | 1970 | 620 | 83 | 340 | 2500 | — | 45 | — | 600 |
| Haringvliet | 1970 | 370 | 56 | 120 | 1200 | 160 | 7 | 10 | 330 |
| Standard references | | | | | | | | | |
| Shales | | 90 | 68 | 45 | 95 | 13 | 0.3 | 0.4 | 20 |
| Near-shore sediments | | 100 | 55 | 48 | 95 | — | — | — | 20 |
| 16th Century Rhine deposits | | 63 | 33 | 21 | 93 | 11 | 0.3 | 0.1 | 31 |
| Fossil Rhine deposits | | 47 | 46 | 51 | 115 | ? | 0.3 | 0.2 | 30 |
| Estimated contamination rate | | | | | | | | | |
| WS marshes | | 2–3 | — | 2 | 4 | 4–5 | 8–25 | 3–10 | 4–5 |
| ES marshes | | — | — | — | 1–2 | 1–2 | 2–3 | — | 2 |

**Table 2.** Scores in the range of heavy-metal concentrations in the plant species. The range of the concentrations of each metal is subdivided into 10 classes (score 1–10), and attaches a score of 11 to the highest concentration. (After Beeftink et al. 1982)

| Species | Cd | Cu | Pb | Hg | Ni |
|---|---|---|---|---|---|
| Atriplex hastata | 4 | 1 | 1 | 2 | 3 |
| Elytrigia pungens | 1 | 1 | 1 | 1 | 1 |
| Halimione portulacoides | 4 | 1 | 2 | 4 | 3 |
| Spartina anglica | 1 | 2 | 1 | 2 | 3 |
| Artemisia maritima | 11 | 11 | 3 | 4 | 5 |
| Aster tripolium | 10 | 4 | 2 | 4 | 3 |
| Limonium vulgare | 1 | 3 | 1 | 5 | 7 |
| Salicornia europaea | 3 | 3 | 4 | 7 | 8 |
| Suaeda maritima | 2 | 2 | 4 | 9 | 7 |
| Festuca rubra | 1 | 4 | 9 | 11 | 9 |
| Plantago maritima | 1 | 5 | 9 | 9 | 11 |
| Puccinellia maritima | 1 | 2 | 11 | 9 | 10 |
| Triglochin maritima | 1 | 3 | 10 | 8 | 9 |

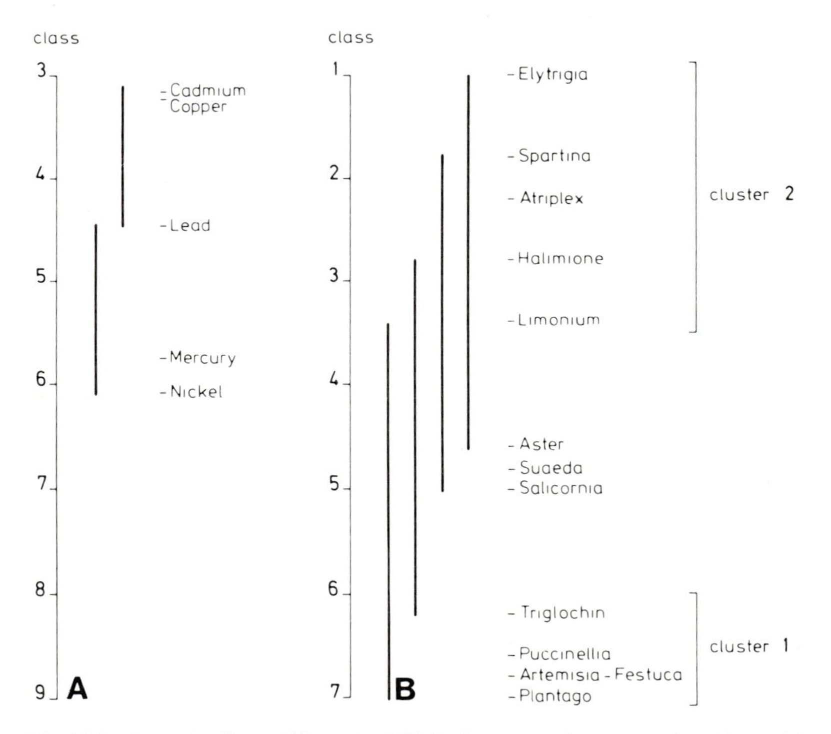


**Fig. 1A,B.** Least significant differences (LSD) in heavy-metal concentrations (dry weight) of the young aerial tissues of plant species expressed as scores in the range of metal concentrations (see Table 5), calculated by two-way analysis of variance. **A** All plant species together. **B** All metal concentrations together. Averages of classes of scores joined by *underlinings* are not significantly different. (After Beeftink et al. 1982)

potential, and sulfide content. The Cd content in the plants, for instance, is greatly increased with increasing redox potential and with decreasing pH (Cooper 1984; Gambrell et al. 1977; Lee et al. 1982; Huiskes and Nieuwenhuize 1985; Rozema et al. 1985b). A correlation between salinity levels of the soil, the uptake of heavy metals and Na excretion has also been shown (Cooper 1984; Rozema et al. 1985b).

No obvious effects on growth, reproduction, and germination in salt-marsh plants have been detected, nor does there seem to be any change in plant species composition in the salt marshes around the North Sea, such as is reported from salt marshes along the Kalu river (India) (Mhatre et al. 1980). On the experimental level, however, effects of heavy metals were shown to have impact on, e.g., germination (Mrozek 1980; Wollan et al. 1978).

Chronic exposure to subtoxic levels of heavy metals is known to produce shifts in the gene pool of populations of species towards more tolerant genotypes, as has happened with species growing for instance, in mining areas, (Ernst 1974; Lefèbvre 1974; Antonovics 1972). Although this is an extreme case whereby certain genotypes of the species simply cannot exist in metal mining areas, it may be occurring in a much more obscured way in the salt marsh with the species growing there.

Heavy metals can enter the body of wetland animals in various ways. Most of them are ingested with food, but they may also enter the body by the skin. The intake of heavy metals with the food is not necessarily bound to the levels of trace metals in the food itself, but also to the sediment particles that are attached to it. Especially in the case of filter-feeding animals, large quantities of sediment pass their ducts.

## 3 Organic Micropollutants

### 3.1 Polychlorobiphenyls

#### 3.1.1 Introduction

Polychlorobiphenyls (PCB's) comprises a group of 209 compounds with one or more (10 maximum) of the hydrogen atoms of the biphenyl molecule substituted by chloride compounds. PCB's are persistent, stable, and lipophilic substances, and have been produced since 1929 (Ballschmiter and Zell 1980). Mixtures of PCB's are applied in hydraulic systems, dieleetric fluids, heat transfer fluids, paints and dyes, and a wide variety of other industrial systems. The PCB mixture Aroclor is produced by Monsanto (U.S.A.) and Clophen by Bayer (F.R.D.). The composition of a PCB mixture may be characterized by the degree of chlorination expressed as the weight percentage of the chlorine atoms of the total molecular weight (Hutzinger et al. 1974). There is no PCB production in the Netherlands.

The content of chlorine of PCB's varies from 18 to 79%. In the series Aroclor 1242, 1254, 1260 the number of chlorine atoms per atom and the number of PCB molecules with chlorine increases. The capacity of microorganisms to degrade PCB is decreased with increasing chlorination.

### 3.1.2 Measurement of PCB's in the Estuarine Environment: Water, Sediment, Atmosphere, and Animals and Plants

*PCB's in Water and Sediment.* PCB's reach the estuarine waters and sediment from the supply of sewage and run off from industries in the river waters, but also by direct disposition of sewage sludge and mud spoil from harbors and waterways. The solubility of PCB's in water is low, but PCB's absorb well to soil and organic particles, therefore PCB's accumulate in sewage sludge and mud spoil. Because of the low volatilization of PCB's, only a small part of the PCB's in estuarine waters and sediment reaches the atmosphere. However, atmospheric PCB's may be rapidly transported for long distances. Also, despite the low atmospheric content of PCB's, the uptake rate of airborne PCB's by plant shoots is higher than via the root system (Nash and Beal 1970). Recently, it has been suggested to take PCB levels in plant foliage as an indication of tropospheric contamination level (Gaggi et al. 1985).

PCB's are highly persistent and adsorb well to river and estuarine sediment (Saylor et al. 1978); they are almost insoluble in water, but highly soluble in organic fluids, lipids in particular.

Research into the levels of organochlorines in water, sediment, invertebrates, and vertebrates in the Dutch Wadden Sea and other Dutch estuarine areas has shown that bioaccumulation of PCB's in marine mammals (harbor seal, *Phoca vitulina*) has led to decreased reproduction and an increased death rate of juvenile and adult individuals of the harbor seal (Reijnders 1980; Duinker et al. 1982). PCB's reach rivers, lakes, and estuarine and coastal waters as industrial and sewage effluents.

In upland sites along a landward-coastal transect of the former river Rhine-Meuse estuary Amer, Nieuwe Merwede, Hollandsch Diep, Haringvliet, there was a decrease of PCB, HCB, and $\gamma$HCH content in the sediment. Also the total of extractable organochlorines (EOCL) decreased in this direction (Rijkswaterstaat 1986). A similar gradient was observed of decreasing contamination of the sediment with heavy metals indicating a joint origin, transport, deposition, and association (adsorption) to organic and clay particles.

Duinker and Hillebrand (1985) studied the content of PCB (Aroclor 1254) equivalents in the shrimp from different parts in the Wadden Sea, and found the highest values in the Balgzand population and in that of the Eems-Dollard estuary. Shrimps collected from the Wadden Islands Terschelling and Schiermonnikoog had the lowest PCB levels in their tissues. PCB's in the soil appear to be highly immobile (Hutzinger et al. 1974) due to adsorption of PCB's to organic matter and clay particles in the soil (Strek and Weber 1982).

In general, and when PCB contamination is rather constant over a long period, such as with the river Rhine, the distribution of PCB components is rather constant between water and suspended matter, and water and fat in organisms for a large range of concentrations (Duursma et al. 1986). Distribution coefficients ($K_d$'s) were found to be $10^4$–$10^5$ for suspended matter and $10^5$–$10^6$ for fat in organisms for samples from the Dutch Delta, the Wadden Sea and the North Sea. For macrophytes like *Enteromorpha* sp., *Fucus vesiculosus*, *Ulva lactuca* and *Zostera marina* $K_d$'s ranged from $(0.6–2.5) \times 10^4$.

### 3.1.3 PCB's in the Atmosphere

PCB's in the atmosphere may originate from industrial sites. Södergren (1972) measured highest PCB levels on those locations that were under the direct influence of factory chimneys and other industrial discharges.

Unlike PCB's in the sediment, atmospheric PCB's may be transported rapidly and for long distances. Although the atmospheric PCB levels are low compared to those of the sediments, airborne PCB's have contaminated distant ecosystems rather than soilborne sources. Since the volatibility of PCB's decreases with increasing degree of chlorination, lower chlorinated PCB's reach higher levels in atmospheric samples. It has been estimated that 85% of the total PCB input of Lake Superior (U.S.A.) is from the atmosphere (Eisenreich et al. 1981). For the North Sea this atmospheric input of PCB amounts to 96% of the total.

### 3.1.4 Uptake of PCB's by Terrestric Coastal Organisms, Plants in Particular

Because of the relatively extensive and detailed knowledge of PCB's in aquatic ecosystems, in comparison to terrestrial ecosystems, it is not surprising that one of the first PCB-uptake studies on higher plants was done on the marshland species (*Veronica beccabunga*). Of the $^{14}C$-2,2$^{1}$-dichlorobifenyl dissolved in aceton and applied as droplets to the leaves 1.89% was found on the leaf, and 93.25% was measured in the atmosphere. This experiment does not, of course, indicate whether or not higher plants take up PCB's from the soil (Moza et al. 1973). Iwata and Gunther (1976) demonstrated uptake of low chlorinated PCB's from the technical mixture Aroclor 1254 in (or on) the roots of the carrot (*Daucus carota*) while higher chlorinated biphenyls remained in the soil and were not taken up by the plant roots. Similar results of relatively enhanced uptake of lower chlorinated PCB's has been reported for radish (*Raphanus sativus*), sugar beet (*Beta vulgaris*), soybean (*Glycine max*), (Susuki et al. 1977), rape (*Brassica rapa*), beans (*Phaseolus vulgaris*) (Sawney and Hankin 1984) and for the coastal halophyte *Spartina anglica* (Mrozek and Leidy 1981). The relative high uptake of the lower chlorinated PCB's has been explained by the higher solubility in water of these components. Uptake of PCB's in carrots was higher than in sugar beet (Moza et al. 1976), which may be ascribed to a relatively high lipid content of the carrot root tissue.

It may be questioned whether the above results really reflect the uptake of PCB's *into* the plant root tissue or adsorption to the root surface. Iwata and Gunther (1976) reported that 97% of the total amount of PCB was in the "peel" of carrots and only 3% in the carrot tissue. The rate of translocation of PCB from the root to the shoot seems to be extremely low. Four years after application of only 0.8% of $^{14}C$-tri-tetra and pentachlorolbiphenyl to the soil with prevention of evaporation resulted in only 5% of the PCB levels in the shoot of soybean plants compared to soybean plants in pots without inhibited evaporation (Fries and Marrow 1981).

Other reports confirm the hypothesis that translocation of PCB's from root to shoot is extremely slow, and that a major part of the PCB's in the shoot has been taken up directly from the atmosphere. Thomas et al. (1984) report PCB levels in the leaves of six plant species in response to atmospheric PCB pollution, that correlated well with the leaf tissue content of lipids and aromatic compounds. Three of the species studied (*Juniperus, Vaccinium,* and *Pinus*) have a wax layer on the cuticle and it may well be that a major part of the PCB content is on the leaf surface, rather than inside the leaf.

In an experimental study, Mrozek et al. (1983) studied the effect of labelled $^{14}$C-polychlorinated biphenyls on *Spartina alterniflora* plants on salt marsh and river bottom sand. There was no significant reduction of the total plant biomass production in the presence of PCB's in the growth medium, although there was an inhibition of the belowground biomass development.

More generally, it seems to be particularly difficult to analyze the effects of contaminants such as heavy metals, PCB's, and PAH's on the growth and physiology of coastal wetland species, since variation of natural environmental factors, such as salinity and anaerobic conditions, in itself has profound effects on the growth of salt-marsh plants, masking the effects of inorganic and organic micropollutants (Rozema et al. 1985a,b,c,d).

To sum up, the pollution with heavy metals and organic micropollutants of the North Sea and coastal wetlands and estuaries has been relatively well documented (Salomons and Förstner 1984; Harmonisatie Noordzeebeleid 1985), reporting high levels of contamination in the (former) river Rhine-Meuse estuary and the Westerschelde and lower levels in the Oosterschelde, Eems-Dollard and the Waddensea (Rozema et al. 1985c; Rozema et al. 1986).

Animals take up heavy metals, PCB's, and PAH's and the concentration factors (ratio of tissue concentration of pollutant and concentration in sea water) vary from $10^3$–$10^4$ for heavy metal uptake by phytoplankton and animals to $10^8$ for PPT and PCB uptake by these organisms (Harmonisatie Noordzeebeleid 1985).

For animals lethal effects have been found next to other components of reduced fitness (such as reduced fertility of the harbor seal, Reijnders 1980). In comparison with zoological studies, there is only scanty information on the relationship between PCB's (and PAH's) and plants in the coastal environment.

PCB's do occur in coastal waters and sediment, and are found in belowground and aerial parts of coastal plants. A significant part of PCB's in plant roots appears to be localized in the outer root cell layers. Also PCB's in the leaves seem to refer mainly to adsorption of PCB's to the wax layer on the leaf cuticle.

The toxic effects of PCB's and PAH's on plants are less obvious than those on animals (Mrozek and Leidy 1981). On the other hand, plants or, more precisely, the plant root environment may help to remobilize (PCB's (and PAH's) from the sediment. There is further research needed to reach more conclusive results on the occurrence, effects, and fate of PCB's and other organic micropollutants in the coastal sediments in relation to plants and animals of coastal ecosystems.

## 3.2 Polycyclic Aromatic Hydrocarbons (PAH's)

### 3.2.1 The Occurrence of PAH Compounds in the Environment

Polycyclic aromatic hydrocarbons (PAH's) consist of three or more fused benzene rings in linear, angular, or cluster arrangement and contain only carbon and hydrogen (Blumer 1976). PAH's are formed when organic substances are exposed to high temperature and are also synthesized by plants and micro organisms (bacteria). Most PAH's are water-insoluble, the boiling point of 14 common PAH's ranged from 150–525°C and the melting point from 101°C–438°C (Edwards 1983). Many PAH's are weakly or strongly carcinogenic or mutagenic. Some of the PAH's, like benzo(a)pyrene, are not toxic by themselves, but biological activitation by enzymes produces epoxides that are carcinogenic and mutagenic. Conversely, plant substances like ellagic acid may inactivate the harmful epoxide form of benzo(a)pyrene. (Sayer et al. 1982).

Unlike PCB's, which are all of anthropogenic origin, PAH's may be formed naturally (fires in dry ecosystems, volcanoes, and also by direct synthesis by plants and micro organisms. Anthropogenic sources of PAH's consist of the combustion of fossil fuels by electric plants, car exhausts, refuse burning, and agricultural burning. It should be emphasized that the amounts of PAH's formed by natural processes are generally very small compared to those from anthropogenic origin. According to Andelman and Suess (1980) the origin of most PAH's is through high temperature pyrolysis of various naturally occurring organic materials like coal. Sources of PAH's to the water environment include coal tar pitch, shale oil, C black, cracked mineral oil effluents from various industrial processes and sewage sludge (Harmonisatie Noordzeebeleid 1985). Of these sources, PAH's of sewage sludge form an important potential input to the terrestrial vegetation. However, a major source of contamination of the soil is by deposition from the air (Suess 1976).

Photo-oxidation products of some PAH's may be more toxic than the original compounds. For algae (*Chlorella* and *Scenedesmus*) and higher plants, it was found that the degree of carcinogenic effects of the PAH-compounds benzo(a)pyrene, 1,2,5,6-dibenzanthracene and 1,2-benzantracene corresponded well with PAH-induced growth stimulation.

### 3.2.2 Uptake of PAH's by Organisms

While more than 160 different polycyclic aromatic hydrocarbon compounds appeared to occur in sewage sludge, waste compost, and in agricultural soil amended with this municipal waste compost, there was no significant uptake of any of the PAH by the roots of plant species tested.

Nevertheless there was an increased level of PAH's in (on?) the aerial plant parts, this level being higher in plants with a large leaf surface (potatoes) than in those with a smaller leaf surface (cereals). This indicated uptake of polycyclic aromatic hydrocarbons from the atmosphere (Ellwardt 1977). Evidence for precipitation or adsorption of PAH's on the leaf surface is provided by the fact that after intensive rinsing a major part of the PAH's may be removed (Shabad 1980). Similarly, it has been shown for benzo(a)pyrene (BaP) that in soils with concentrations of BaP not exceeding 1000 mg BaP per kg soil, there was no uptake by the plant roots. Only in model tests for several years was there significant uptake of PAH's in carrots and radish (with a relatively high oil and lipid content) but far less significantly in potatoes, tomatoes, peas and beans (Dorr 1970; Fritz 1971). Studies with the mud snail suggest this organism to be a good indicator of long-term bioaccumulation of many PAH's, especially those which are more persistent (Kay et al. 1986).

## 4 Effects of Oil and Emulsifiers on the Functioning of the Wetland Ecosystem

Dudley (1971) lists the major causes for oil pollution on wetlands with reference to Milford Haven:

1. Inshore facilities and refineries. Drainage from failures and ruptures and overflow of tanks; failure of automatic gauges although this is collected in inshore central collecting points, but natural seepage will occur.
2. Poor design of ships and terminals, especially poor design of valves allowing oil to escape past closed valves, which would suggest that the design of these valves has not kept pace with the discharging rules and pressures currently used.
3. Mechanical failure of hoses, pipelines; corrosion.
4. Operating procedures, although this is not an important cause of pollution.
5. Human error: the main cause of pollution from shore and ship: ignorance, language difficulties, poor communication, tired personnel, and carelessness resulting in overflowing of tanks, failing to ensure that the sea valves are closed, wrongly connected hoses and pipelines, and at sea, navigation errors.

Oil pollution due to poor design and failing operating procedures is in general of a chronic character, causing effects different from single spillages (Baker 1971a). Single spillages have mostly the character of a catastrophe, an uncontrollable (large) amount of oil affects the estuarine ecosystem. These oil spill disasters are in general well documented, contrary to other pollution problems, mostly because they have dramatic effects on ecosystem and landscape, and cause direct problems to man. Two recent oil spill disasters, although outside the North Sea, viz. the wreckage of the *Torrey Canyon* off the Cornwall coast and the *Amoco Cadiz* off the coast of Brittany have been described in detail (Smith 1968; Nelson-Smith 1968; O'Sullivan and Richardson 1968; Bellamy et al. 1967; Conan et al. 1978; and others) and give a list of data on the effects of oil and oil-cleaning chemicals on plants and animals.

The effect of an oil slick on salt-marsh vegetation is a serious reduction or complete die-off of the aboveground parts of the plants (Baker 1971a-f; Baker et al. 1984; d'Ozouville et al. 1978). The extent of the damage depends on the amount and type of oil, the time of the year, and the plant species (Baker 1971b). Mostly, however, recovery takes place in the years following the oil spill. Baker et al. (1984) show in experiments on this matter that *Spartina anglica* stands are seriously reduced in number of plants or tillers after the application of oil, but a recovery takes place in about 2 years. Thus, Bender et al. (1980) report a complete recovery of marsh grasses in American salt marshes 3 years after the application of oil (so do De Laune et al. 1979). Webb et al. (1981) found a complete restoration of the *Spartina alterniflora* vegetation (Galveston, Texas) 1 year after the spill. Géhu (1979, 1981) and Géhu and Géhu-Franck (1979) describe in detail the recovery of salt-marsh vegetation after the *Amoco Cadiz* oil disaster. Of the 76 salt-marsh plant species, none has been completely destroyed, not even rare endemic species and associations. The recovery time depended on the species, *Triglochin maritimum, Juncus maritimus,* and *Limonium vulgare* seem to be more tolerant than other species.

Although oil seems to have little, or at least a reversible effect on existing salt-marsh vegetation, the effects on flowering, seed production, and germination are strongly negative, which also demonstrates that oil effects depend strongly on the time of year in which the spill occurs (Baker 1971c). The general negative effect on reproductive processes makes the annual species more vulnerable than the perennial species.

The difference in effect by the various types of oil depends mainly on the composition of the oil. The low boiling oil-fractions have the most toxic effects (Baker 1971d). The heavier compounds reduce the transpiration rate by blocking the stomata, thereby reducing the photosynthesis. Disruption of chloroplast membranes, mitochondrial damage, and leakage of cell membranes (Boyles 1967 in Baker 1971d) have also been reported.

The influence of oil in the soil proved harmful, beneficial, or indifferent to salt-marsh plants. Harmful effects are reported by Kloke and Leh (1966), who found that oil interfered with water uptake and N, P, and K uptake. Light oil had a short-term harmful effect as compared with heavier oil, which could last many years. Germination processes were affected more by lighter oils in the soil due to penetration in the seeds (Baker 1971e). Several authors describe beneficial effects of oil in the soil to plant growth (Carr 1919; Galtsoff et al. 1935; Mackin 1950; Stibbings 1968).

It is suggested that the plants were deriving some nutritional benefit from the breakdown products of oil due to bacterial action and to the mineralization of oil-killed animals (Baker 1971e). In general, it can be concluded that the effects of a single oil spill on higher plant vegetation are in the long run not negative. The effects of oil on other organisms are different.

Plante-Cuny et al. (1979) report a rapid recovery of populations of heterotrophic bacteria, diatoms and chlorophytes on heavily soaked soil after the *Amoco Cadiz* disaster. Benthic fauna is in general heavily affected by oil, as is fish. Meiofauna seems to be much less affected, although the reactions of different species vary greatly (Boucher 1985).

The most dramatic influence of oil pollution, especially to the general public, is that to birds. Monat (1978) estimated a death toll of about 20,000 birds after the *Amoco Cadiz* oil disaster, which is rather low as compared with other oil spills.

Apart from direct influences of oil on birds, like soiling the feathers or causing internal problems, Monat (1978) also cited publications reporting a decreased reproductivity or even sterility. All the facts described above are the consequences of a single oil spill. Chronic oil pollution has in general more serious effects especially on plants. Baker (1971a) reports on the results of successive spraying with oil on experimental plots. If one or a few sprayings allow a recovery, more sprayings have the result that there is hardly any recovery of the species, or at least not in 1 or 2 years (Figs. 2 and 3). Not all salt-marsh species are equally susceptable to successive oil sprayings. Baker (1971a) classifies the salt-marsh plants in six categories from very susceptible to very tolerant.

It appears that not only the animals mentioned earlier are very susceptible to successive spraying, but also the perennial species *Halimione portulacoides* (L.) Aellen. Algae are also in the susceptible category in Baker's scale. The perennial salt-marsh grasses form an intermediate group while most of the dicotyledonous perennial species are resistent. The perennial umbelliferous species, *Oenanthe*

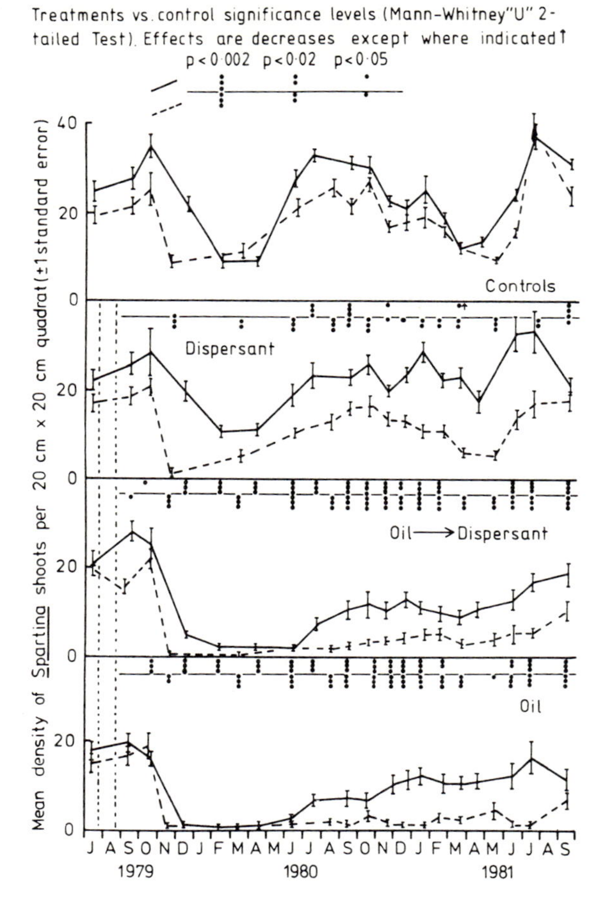


**Fig. 2.** Steart Salt Marsh 2 experiment. Effects of Forties crude oil and BP 1100 WD dispersant on the density of *Spartina anglica* shoots (results are given for two duplicate plots per treatment). *Vertical broken lines* indicate the two treatment dates. (After Baker et al. 1984)

*lachenalii*, survived 12 oil sprayings in Baker's experiment. This difference in susceptibility may result in a change in the vegetation pattern in the marsh after successive oilings. The effect of a refinery effluent can, according to Baker (1971f) also be regarded as chronic oil pollution. Hershner and Lake (1980) made observations on *Spartina alterniflora* after repeated oiling. The results were that a substantial part of the vegetation was completely killed and the remaining live plants showed delayed development in spring, reduced mean weight, and a lower leaf production.

The life of microorganisms in soil is not really affected by chronic oil pollution (Thomson and Webb 1984), but on higher organisms like benthic fauna the effect is serious: even with discharges of "safe" levels of oil in the water coming from refineries, populations of animals can be reduced (Crapp 1971a).

Far more serious than the influence of oil and oil products on the organisms living in and on estuarine wetlands is the use of emulsifiers in the cleaning process of an oil spill. The emulsifiers, used to – as the word says – make an emulsion of the oil in order to wash it away either with the tides or with water being brought

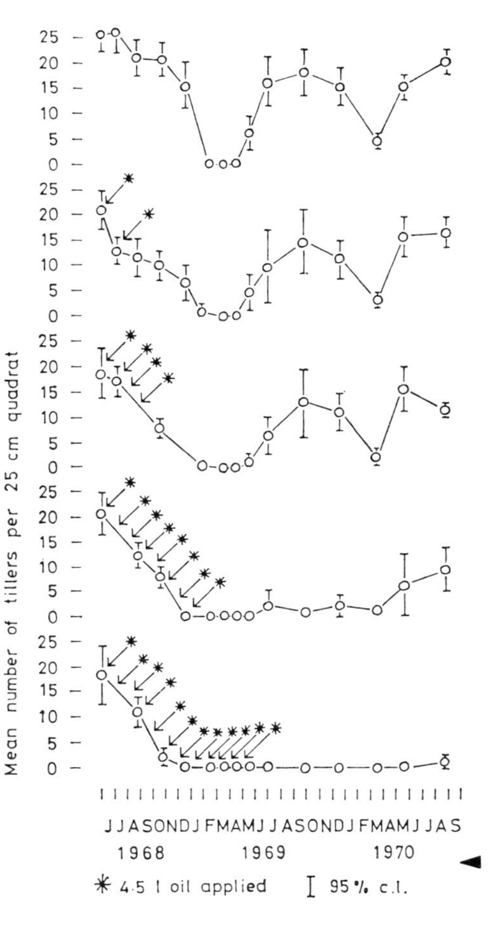


Fig. 3. Effects of successive oil sprayings on *Spartina anglica*. (After Baker 1971a)

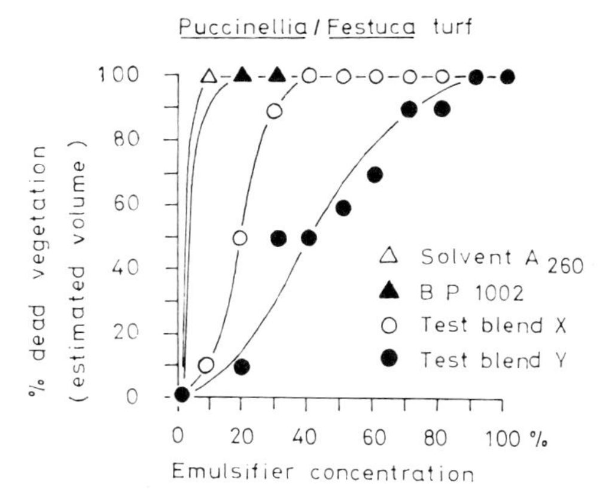


**Fig. 4.** Effects of emulsifiers on Puccinellia turf. (After Baker 1971d)

from elsewhere. These compounds are generally rich in aromatics and therefore likely to be toxic to plants and animals. Emulsifiers can easily penetrate into oil, but also into plants due to their lipophilic plant surfaces (Baker 1971d). Baker (1971d) studied the influence of various emulsifiers on salt-marsh plants, and concluded that – although differing in levels – the compounds are invariably toxic to them (Fig. 4).

The emulsifiers are also toxic to very toxic to animals (Simpson 1968; George 1961) (Fig. 5). Crapp (1971b) confirmed these findings in laboratory experiments on benthic animals (molluscs) in which low concentrations of emulsifiers already caused high mortalities in species like *Gibbula umbilicalis, Littorina obtusata, Thais lapillus* and *Mytilus edulis*. The animals that stayed alive after application of the emulsifier were inactivated for shorter or longer periods, which made them vulnerable to tidal currents that could wash them away from the intertidal areas to deeper water or higher areas of the shore, thus causing a secondary mortality risk.

All authors conclude that mechanical cleaning of intertidal areas after an oil spill is far preferable to the use of emulsifiers in cleaning operations, especially when action is taken immediately after the pollution. Cutting and removing of the

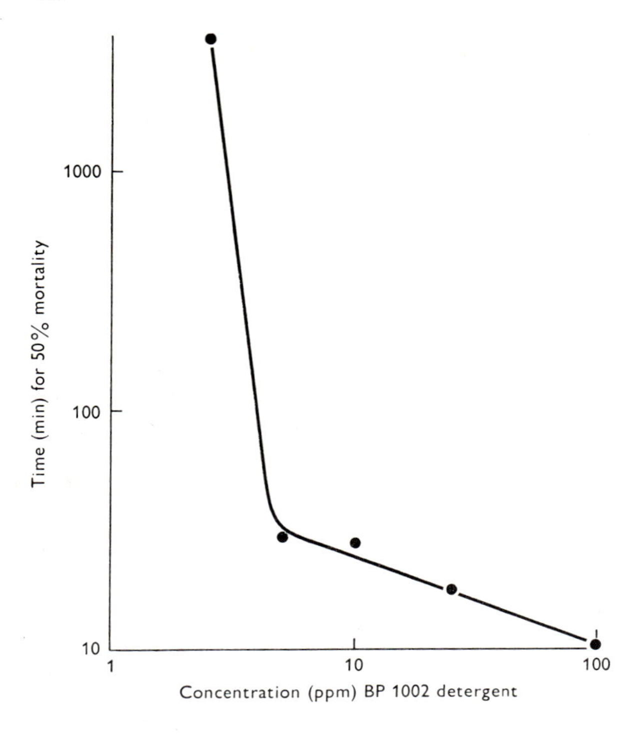


**Fig. 5.** Effect of different concentrations of BP 1002 on swimming activity of cyprids of *Elminius modestus* at 20°C. (After Smith 1968)

soiled salt marsh turf, for instance, has a much less harmful effect. Immediate cleaning of freshly deposited oil on intertidal sand- and mudflats may save much of the fauna, and reduces the mortality caused by the oil when left for longer periods. In general it makes the application of emulsifiers superfluous. In cases of chronic oil pollution, like efflux from refineries, the use of bars to guide the oil away from intertidal areas is far preferable (and cheaper) than a regular application of emulsifying compounds (Baker and Crapp 1971).

## 5 The Occurrence of Radionuclides in Intertidal Wetlands

Compared with the polluting effects of heavy metals, organic micropollutants, and oil on intertidal ecosystems, the pollution and effects of radionuclides are negligible. The reason for this is the generally understood harmful character of these materials that have resulted in stringent regulations for use, transport, storage, and discharge in most countries using them. Radionuclides supplied to estuaries can originate from atmospheric deposition and land run-off. They may come from nuclear weapon use (tests) and the failure of the launching of the SNAP-9A satellite in 1969 that added 17kCI of $^{238}$Pu to the world's atmosphere (Duursma et al. 1985). Discharge of nuclear power plants and reprocessing plants of nuclear power fuel are also sources of radionuclide pollution to wetlands. The discharge of radionuclides of nuclear power plants is generally negligible, more significant are the effects of atmospheric sources and those of the reprocessing plants. The nuclear plants located at Sellafield (G.B.) and La Hague (Normandy, France) give detectable levels of radionuclides in, e.g., the Westerschelde estuary (Duursma et al. 1985). So far the effects of the radioactive behavior of the nuclear compounds are negligible due to their low levels. Their chemical behavior can be compared with that of other trace metals).

## 6 Conclusions

The wetlands around the North Sea are subject to various sources of pollution, the short-term and long-term effects of the various pollutants are largely unknown or were small. Visible changes in the community pattern in wetlands have so far not occurred and where they have, e.g., after oil disasters, the effects were mostly reversible. Because of variation of environmental factors in the coastal wetland such as salinity, frequency of immergence, redox potential, and organic matter content, it is difficult to discern the effects of the various contaminants. Considering the contamination of salt marshes with heavy metals, anaerobic conditions cause many of the heavy metals to remain insoluble as sulfide salts (Rozema et al. 1986). On sandy salt marshes and estuaries the sediment easily becomes drained, and due to a limited number of adsorption sites on sand grains, heavy metals are easily available and may cause severe growth reduction of plants (Rozema et al. 1986).

It is unknown to what extent the availability of heavy metals will increase when coastal sand and mud flats become drained when the water level drops as a result of construction of dams and changes of the tidal regime and amplitude.

Much less well-documented are the influences of chronic pollution on the communities, causing slow but sure variation. The influence of chronic oil pollution has been reported in this paper; not reported are the effects of an increasing eutrophication by discharge of various compounds, which may – although very slowly – alter the species composition of salt marshes and intertidal flats. An influence by chronic pollution of heavy metals may be the variation in genetic composition of species populations as has been reported in other circumstances (Ernst 1974). In the long run this change in the gene pool of certain species may evoke a changing response to physiological and competitive processes resulting in gradual alteration of community composition. It is these obscured effects of pollutants that endanger the wetlands around the North Sea, much more than the single-event disasters that are dealt with immediately and effectively. Because visible effects are difficult to detect, human action tends to be neglected. Legal action is impaired by the lack of adverse effects on a short-term base and the difficulty of making a proper assessment of the present or future damage. In cost-benefit analyses these damages are therefore easily forgotten or neglected. Potential hazards have to be treated very seriously to prevent irreversible damage to the coastal wetland ecosystems.

*Acknowledgments.* Drs. M.L. Otte, Dept. Ecology & Ecotoxicology, Free University is acknowledged for his contribution to the section on organic micropollutants in this chapter.

## References

Andelman JB, Suess MJ (1980) Polynuclear aromatic hydrocarbons in the water environment. Bull WHO 43:479–508

Antonovics J (1972) Population dynamics of the grass *Anthoxanthum odoratum* on a zinc mine. J Ecol 60:351–366

Baker JM (1971a) Successive spillages. In: Corvell EB (ed) The ecological effects of oil pollution on littoral communities. Institute of Petroleum, London, pp 21–32

Baker JM (1971b) The effects of a single oil spillage. In: Cowell EB (ed) The ecological effects of oil pollution on littoral communities. Institute of Petroleum, London, pp 16–20

Baker JM (1971c) Seasonal Effects. In: Cowell EB (ed) The ecological effects of oil pollution on littoral communities. Institute of Petroleum, London, pp 44–51

Baker JM (1971d) Comparative toxicities of oils, oil fractions and emusifiers. In: Cowell EB (ed) The ecological effects of oil pollution on littoral communities. Institute of Petroleum, London, pp 78–87

Baker JM (1971e) Growth stimulation following oil pollution. In: Cowell EB (ed) The ecological effects of oil pollution on littoral communities. Institute of Petroleum, London, pp 72–77

Baker JM (1971f) Refinery effluent. In: Cowell EB (ed) The ecological effects of air pollution in littoral communities. Institute of Petroleum, London, pp 33–43

Baker JM, Crapp GB (1971) Predictions and recommendations. In: Cowell EB (ed) The ecological effects of oil pollution on littoral communities. Institute of Petroleum, London, pp 217–220

Baker JM, Crothers JH, Little DI, Oldham JH, Wilson CM (1984) Comparison of the fate of ecological effects of dispersed and nondispersed oil in a variety of intertidal habitats. In: Allen TE (ed) Oil Spill chemical dispersants: research, experience and recommendations, STP 840. Am Soc Testing Materials, pp 239–279

Ballschmiter K, Zell M (1980) Analysis of polychlorinated biphenyls (PCB) by glass capillary gas-chromatography. Fresen Z Anal Chem 302:20–31

Beeftink WG, Stoeppler M, Nieuwenhuize J, Mohl C (1982) Heavy metal accumulation in salt marshes from the Western and Eastern Scheldt. Sci Total Environ 25:199–223

Bellamy DJ, Clarke PH, John DM, Jones D, Whittick A, Darke T (1967) Effects of pollution from the Torrey Canyon and Littoral and Sublittoral Ecosystems. Nature 216:1170–1173

Bender ME, Shearls EA, Murray L, Hugget RJ (1980) Ecological effects of experimental oil spills in eastern coastal plain estuaries. Environ Int 3:121–133

Blumer N (1976) Polycyclic aromatic compounds in nature. Sci Am 234:35–45

Boucher G (1985) Long-term monitoring of meiofauna densities after the *Amoco Cadiz* oil spill. Mar Pollut Bull 16:328–333

Brinkhuis BH, Penello WF, Churchill AL (1980) Cadmium and manganese flux in eelgrass *Zostera marina*. II. Metal uptake by leaf and root-rhizome tissues. Mar Biol 58:187–196

Carr RH (1919) Vegetative growth in soil containing crude petroleum. Soil Sci. 8: 67–68

Conan G, d'Ozouville L, Marchand M (1978) Amoco Cadiz. Premières observations des consequences a court terme de la pollution par hydrocarbures sur l'environnement marin. Centre National Pour l'Exploitation des Oceans, Paris, 238 pp

Cooper A (1984) A comparative study of the tolerance of salt marsh plants to manganese. Plant Soil 81:47–59

Crapp GB (1971a) Chronic oil pollution. In: Cowell EB (ed) The ecological effects of oil pollution on littoral communities. Institute of Petroleum, London, pp 187–203

Crapp GB (1971b) Laboratory experiments with emulsifiers. In: Cowell EB (ed) The ecological effects of oil pollution on littoral communities. Institute of Petroleum, London, pp 129–149

Dorr R (1970) Die Aufnahme von 3,4 Benzpyren durch Pflanzenwurzeln. Landwirtsch Forsch 23:371–379

Dudley G (1971) Oil pollution in a Major Oil Port: The incidence, Behaviour and Treatment of Oil Spills. In: Corvell EB (ed) The ecological effects of oil pollution on littoral communities. Institute of Petroleum, London, pp 5–12

Duinker JC, Hillebrand MTJ (1985) Milieuverontreiniging. In: Inleiding tot de Oecologie. Bohn, Scheltema & Holkema, Amsterdam, p 501

Duinker JC, Hillebrand MTJ, Nolting RF, Wellershaus S (1982) The river Elbe: processes affecting the behaviour of metals and organochlorines during estuarine mixing. Neth J Sea Res 141–169

Duursma EK, Frissel MJ, Guary C, Martin MM, Nieuwenhuize J, Pennders RMJ, Thomas AJ (1985) Plutonium in sediments and mussels of the Rhine-Meuse-Scheldt estuary. In: The Behaviour of Radionuclides in Estuaries. CEC 1985 XII/380/85, EN, pp 71–106

Duursma EK, Nieuwenhuize J, Liere J van, Hillebrand MTJ (1986) Partitioning organochlorines between water, particulate matter, and some organisms in estuarine and marine systems of the Netherlands. Neth J Sea Res 20:239–251

Edwards CA, Lofty (1977) Biology of Earthworms. Chapman & Hall, London

Edwards NT (1983) Polycyclic Aromatic Hydrocarbons (PAH's) in the terrestrial environment. A review. J Environ Qual 12:427–441

Eisenreich SJ, Looney BB, Thornton JD (1981) Airborne organic contaminants in the Great Lakes ecosystem. Environ Sci Technol 15:30–38

Ellwardt PC (1977) Variation in content of polycyclic aromatic hydrocarbons in soil and plants by using municipal waste composts in agriculture. In: Proc Symp on soil organic Matter Studies, Vol 2. Int Atomic Energy Agency, Vienna

Ernst WHO (1974) Schwermetallvegetation der Erde. Fischer, Stuttgart, 194 pp

Faraday WE, Churchill AC (1979) Uptake of cadmium by the eelgrass *Zostera marina*. Mar Biol 53:293–298

Förstner U, Wittmann GTW (1979) Metal pollution in the Aquatic Environment, vol 8. Springer, Berlin Heidelberg New York, 486 pp

Fries FG, Marrow GS (1981) Chlorobiphenyl movement from soil to soybean plants. J Agric Food Chem 29:757–759

Fritz W (1971) Extent and sources of contamination of our food with carcinogenic hydrocarbons. Ernährungsforschung 16:547–557

Gaggi C, Bacci E, Calamari D, Fanelli R (1985) Chlorinated hydrocarbons in plant foliage: an indication of tropospheric contamination level. Chemosphere 14:1673–1686

Galtsoff PS, Pryterch HF, Smith RO, Koehring V (1935) Effects of crude oil pollution on oysters in Louisiana waters. Bull Bur Fish Wash 18:143–209

Gambrell RP, Khalid RA, Verloo MG, Patrick WH (1977) Transformations of heavy metals and plant nutrients in dredged sediments as affected by oxidation-reduction potential and pH, vol 2. Materials and methods/results and discussion. Office, Chief of Engineers, US Army, Washington DC, 309 pp

Gardner WS, Kendall DR, Odom RR, Windom HL, Stephens JA (1978) The distribution of methyl mercury in a contaminated salt marsh ecosystem. Environ Pollut 15:243–251

Géhu J-M (1979) Suivi phytocoenologique de l'impact de la marée noire sur les prés-sales de la côte nord-américaine. Inst Eur d'Ecologie, Lille, 20 pp

Géhu J-M (1981) Suivi phytocoenologique de l'impact de la marée sur les prés-sales de la côte nord-américaine. Inst Eur d'Ecologie, Metz, 43 pp

Géhu J-M, Géhu-Franck J (1979) Evolution des prés-sales nord-américaine sous l'impact de la marée noire. In: Conséquences d'une pollution accidentelle par les hydrocarbures. Centre National pour l'Exploitation des Oceans, Paris, pp 443–453

George M (1961) Oil pollution of marine organisms. Nature 192:1209

Harmonisatie Noordzeebeleid (1985) (Waterkwaliteitsplan Noordzee, Noordzeecommissie Den Haag), 85 pp

Hershner C, Lake J (1980) Effects of chronic oil pollution on a salt-marsh grass community. Mar Biol, 56:163–173

Huiskes AHL, Nieuwenhuize J (1985) Uptake of heavy metals from contaminated soil by salt-marsh plants. In: Lekkas TD (ed) Heavy metals in the environment, vol 2. CEP Consultants, Edinburgh, pp 307–309

Hutzinger O, Safe S, Zitko V (1974) The chemistry of PCB's. CRC, Cleveland, Ohio

Iwata Y, Gunther FA (1976) Translocation of polychlorinated biphenyl Aroclor 1254 from soil into carrots under field conditions. Arch Environ Contam Toxicol 4:44–59

Kay SH, Marquenie JM, Simmers JW (1986) Time-dependent uptake of PAHS by mud snails *Ilyanassa obsoleta*. In: Environmental contamination. Proc 2nd Int Conf, Amsterdam, pp 70–72

Kloke A, Leh H (1966) Untersuchungen über die Beeinflussung der Pflanzenentwicklung durch Veruntreinigungen des Bodens mit Heiz- und Treibölen. Wasser Boden 9:324–328

Lauff GH (ed) (1967) Estuaries. Am Assoc Adv Sci, publication 83.757 pp

Laune RD de, Patrick WH Jr, Buresh RJ (1979) Effect of crude oil on a Louisiana *Spartina alterniflora* salt marsh. Environ Pollut 20:21–31

Lee CR, Folsom BL Jr, Engler RM (1982) Availability and plant uptake of heavy metals from contaminated dredged material placed in flooded and upland disposal environments. Environ 7:76–71

Lefèbvre C (1974) Population variation and taxonomy in *Armeria maritima* with special reference to heavy-metal-tolerant populations. New Phytol 73:209–219

Lion LW, Altmann RS, Leckie JO (1982) Trace metal adsorption characteristics of estuarine particulate matter: evaluation of contributions of Fe/Mn oxide and organic surface coatings. Environ Sci Technol 16:660–666

Mackin JG (1950) A comparison of effects of application of crude petroleum on saltgrass *Distichlis specata* (L.) Greene. Texas A & M Res Found Project 9 Rep, 8

Mhatre GN, Chaphekar SB, Ramani Rao JV, Patil MR, Haldar BC (1980) Effect of industrial pollution on the Kalu river ecosystem. Environ Pollut Ser A Ecol Biol 22–67–78

Monat JY (1978) Effet du pétrole de l'Amoco Cadiz sur les oiseaux de mer. Bilan provisoire. In: Conan G, d'Ozouville L, Marchand M (eds) Premières observations des consequences à court terme de la pollution par hydrocarbures sur l'environnement marin. Centre National pour l'Exploitation des Oceans, Paris, pp 135–142

Moza P, Weisgerber I, Klein W, Korte F (1973) Verteilung und Metabolismus von 2,2'-dichlorobiphenyl -$^{14}$C in der höheren Sumpfpflanze *Veronica beccabunga*. Chemosphere 2:217–222

Moza P, Weisgerber I, Klein Q (1976) Fate of 2,2'-dichlorobiphenyl -$^{14}$C in carrots, sugarbeets and soil under outdoor conditions. J Agric Food Chem 24:881–885

Mrozek E (1980) Effect of mercury and cadmium on germination of *Spartina alterniflora* Loisel seeds at various salinities. Environ Exp Bot 20:367–377

Mrozek E, Leidy RB (1981) Investigation of selective uptake of polychlorinated biphenyls by *Spartina alterniflora*. Loisel Bull Environ Contam Toxicol 27:481–488

Mrozek E, Queen WH, Hobbs LL (1983) Effects of polychlorinated biphenyls on growth of *Spartina alterniflora*. Loisel Water Air Soil Pollut 17:3–15

Nash RG, Beal ML (1970) Chlorinated hydrocarbon insecticides: root uptake versus vapor contamination of soybean foliage. Science 168:1109–1111

Nelson-Smith A (1968) The effects of oil pollution and emulsifier cleansing on shore life in south-west Britain. J Appl Ecol 5:97–107

O'Sullivan AJ, Richardson AJ (1968) The "Torrey Canyon" disaster and intertidal marine life. Nature 24:448, 541–542

d'Ozouville L, Gundlach ER, Hayes MO (1978) Effect of coastal processes on the distribution and persistence of oil spilled by the *Amoco Cadiz*. Preliminary conclusions. In: Conan G, d'Ozouville L, Marchand M (eds) Amoco Cadiz. Premières observations des consequences a court terme de la pollution par hydrocarbures sur l'environnement marin. Centre National Pour l'Exploitation des Oceans, Paris, pp 69–96

Peterson PJ (1975) Element accumulation by plants and their tolerance of toxic mineral soils. Proc Int Conf Heavy Metals Environ, Toronto, pp 39–54

Plante-Cuny MR, Campion-Alsumard T le, Vacelet E (1979) Influence de la pollution due a l'Amoco Cadez sur les peuplements bactériens et microphytiques des marais maritimes de l'Ile Grande. II. Peuplements microphytiques. In: Amoco Cadiz: Conséquences d'une pollution accidentelle par les hydrocarbures. Centre National pour l'Exploitation des Oceans, Paris, pp 429–442

Reijnders PJH (1980) Organochlorine and heavy metal residues in harbour seals from the Wadden Sea and their possible effects on reproduction. Neth J Sea Res 14:30–65

Rozema J, Roosenstein J, Broekman R (1985a) Effects of Zinc, Copper and Cadmium on the mineral nutrition and ion secretion of salt secreting halophytes. In: Beeftink WG, Rozema J, Huiskes AHL (eds) Ecology of Coastal Vegetation. Junk, Dordrecht, pp 554–556

Rozema J, Roosenstein J (1985b) Effects of zinc, copper and cadmium on the growth and mineral composition of some salt-marsh halophytes. Vegetatio 62:551–553

Rozema J, Otte R, Broekman R, Punte A (1985c) Accumulation of heavy metals in estuarine salt marsh sediment and uptake of heavy metals by salt marsh halophytes. Proc Int Symp Heavy Metals Environ, Athens, Greece, pp 545–547

Rozema J, Bijwaard P, Prast G, Broekman R (1985d) Ecophysiological adaptations of coastal plants from foredunes and salt marshes. In: Beeftink WG, Rozema J, Huiskes AHL (eds) Ecology of Coastal Vegetation. Junk, The Hague, pp 499–521

Rozema J, Otte ML, Broekman RA, Wezenbeek JM (1986) The uptake and translocation of heavy metals by salt marsh plant from contaminated estuarine salt marsh sediment: possibilities of bio-indication. Environmental Contamination, Amsterdam, pp 123–125

Rijkswaterstaat (1986) Ecosysteemonderzoek Noordelijk Deltabekken (An)organische microverontreinigingen. Div Benedenrivieren Uitgev door Vakgr Oecologie en Oecotoxicologie. Vrije Universiteit, 67 pp

Salomons W, Förstner U (1984) Metals in the hydrocycle. Springer, Berlin Heidelberg New York, 349 pp

Sawney BL, Hankin L (1984) Plant contamination by PCB's from amended soils. J Food Prot 47:232–236

Sayer JM, Yagi H, Wood AW, Conney AH, Jerina DM (1982) Extremely facile reaction between the ultimate carcinogen benzo(a)pyrene – 7,8-diol 9,10-epoxide and ellagic acid. J Am Chem Soc 104:5562–5564

Sayler GS, Thomas R, Colwell RR (1978) Polychlorinated biphenyl (PCB) degrading bacteria and PCB in estuarine and marine environments. Estuarine Coastal Mar Sci 6:553–567

Schoefer HF, Schmitz R (1986) Occurrence of chlorinated hydrocarbons in the river Sieg (West Germany). In: Environmental Contamination. Proc 2nd Int Conf, Amsterdam, pp 89–92

Schroeder PB, Thorlaug A (1980) Trace metal cycling in tropical-subtropical estuaries dominated by the seagrass *Thalassia testudinum*. Am J Bot 67:1075–1088

Shabad LM (1968) The distribution and the fate of the carcinogenic hydrocarbon benzo(a)pyrene (3.4-benzopyrene) in the soil. Z Krebsforsch 70:204–210

Shabad LM (1980) Circulation of carcinogenic polycyclic aromatic hydrocarbons uin the human environment and cancer prevention. J Natl Cancer Inst 64:405–410

Simpson AC (1968) Oil, emulsifiers and commercial shelfish. Field Studies 2 (suppl), pp 81–90

Smith JE (1968) "Torrey Canyon", Pollution and Marine Life. Cambridge Univ Press, Cambridge, pp 14 + 196

Södergren A (1972) Chlorinated hydrocarbon residues in airborne fallout. Nature 236:395–397

Stibbings RE (1968) *Torrey Canyon* oil pollution on salt marshes and a shingle beach in Brittany 16 months after. Nature Concervancy, Furzebrook Research Station, 12 pp

Strek HJ, Weber JB (1982) Behaviour of polychlorinated biphenyls (PCB's) in soils and plants. Environ Pollut Ser A Ecol Biol 291–312

Suess MJ (1976) The environmental load and cycli of polycyclic aromatic hydrocarbons. Sci Total Environ 6:239–250

Susuki M, Aizawa N, Okano G, Takahashi T (1977) Translocation of polychlorinated biphenyls from soil into plants: a study by a method of culture of soybean sprouts. Arch Environ Contam Toxicol 5:343–352

Thomas W, Rühling A, Simon H (1984) Accumulation of airborne pollutants (PAH, chlorinated hydrocarbons) in various plant species and humus. Environ Pollut Ser A Ecol Biol 36:295–311

Thomson AD, Webb KL (1984) The effect of chronic oil pollution on salt-marsh nitrogen fixation (acetylene reduction). Estuaries 7:2–11

Wainwright SJ, Woolhouse HW (1975) Physiological mechanisms of heavy metal tolerance in plants. In: Chadwick MJ, Goodman GT (eds) The Ecology of Resource degradation and renewal. 15th Symp British Ecological Society, Blackwell, Oxford, pp 231–259

Wallhöfer P, Königer M, Engelhardt G (1975) Verhalten von xenobiotischen chlorierten Kohlenwasserstoffen (HCB und PCB's) in Kulturpflanzen und Böden. Z Pflanzenkr Pflanzenschulz 82:91–100

Walsh GE, Hollister TA, Forester J (1974) Translocation of four organochlorine compounds by Red Mangrove (*Rhizophora mangle L.*) seedlings. Bull Environ Contam Toxicol 12:129–135

Webb JW, Tanner GT, Koerth BH (1981) Oil spill effects on smooth cordgrass in Galveston Bay, Texas. Contrib Mar Sci 24:107–114

Wiley M (ed) (1976) Estuarine processes. II: Circulation sediments and transfer of material in the estuary. Academic Press, London, 428 pp

Wollan E, Davis RD, Jenner S (1978) Effects of sewage sludge on seed germination. Environ Pollut 17:195–205

# Fjords

J. MOLVAER and J.M. SKEI[1]

## 1 Introduction

The impact of pollutants on fjords in Scandinavia is highly variable due to different pollutant load, differences in fjord morphology, water exchange, and type of pollutant. To exemplify these variations three case histories are presented, representing three morphologically different fjords with different pollutant load, amounts as well as quality. However, they all have in common a long history of industrial pollution.

## 2 Frierfjord, Southeast Norway

### 2.1 Environmental Setting

The Frierfjord represents the inner part of a fjord system on the Norwegian Skagerrak-coast (Fig. 1). The fjord is about 10 km long, with a maximum depth of 98 m and connected to the Langesundsfjord via a narrow and shallow sill (23 m). The surface area is 17.5 $km^2$. The Langesundsfjord is 90–130 m deep, with a 50-m-deep sill facing the coastal water at the south end.

The freshwater runoff is dominated by the Skien river, varying from 50 $m^3 s^{-1}$ to 1000 $m^3 s^{-1}$. The annual average is 270 $m^3 s^{-1}$. This freshwater and the sills create a marked three-layer structure in the fjord water. Down to 10–15 m depth there is a strong estuarine circulation, and the average residence time for the surface layer (0–4 m) is 2–4 days. Below the sill level, the residence time of the basin water is 1–3 years, and 6–10 months in the Langesundsfjord. During stagnation periods, the basin water below 45–50 m depth in the Frierfjord will contain hydrogen sulfide.

### 2.2 Pollutant Load

The area around the Frierfjord is strongly industrialized, and the surface water receives effluents from a pulp factory, a chlor-alkali plant, a magnesium plant, fertilizer plant, ferro-manganese smelter, petrochemical industry, etc. in addition to mainly untreated municipal wastewater from a population of 60,000.

[1]Norwegian Institute for Water Research, P.O. Box 33 Blindern, 0313 Oslo 3, Norway

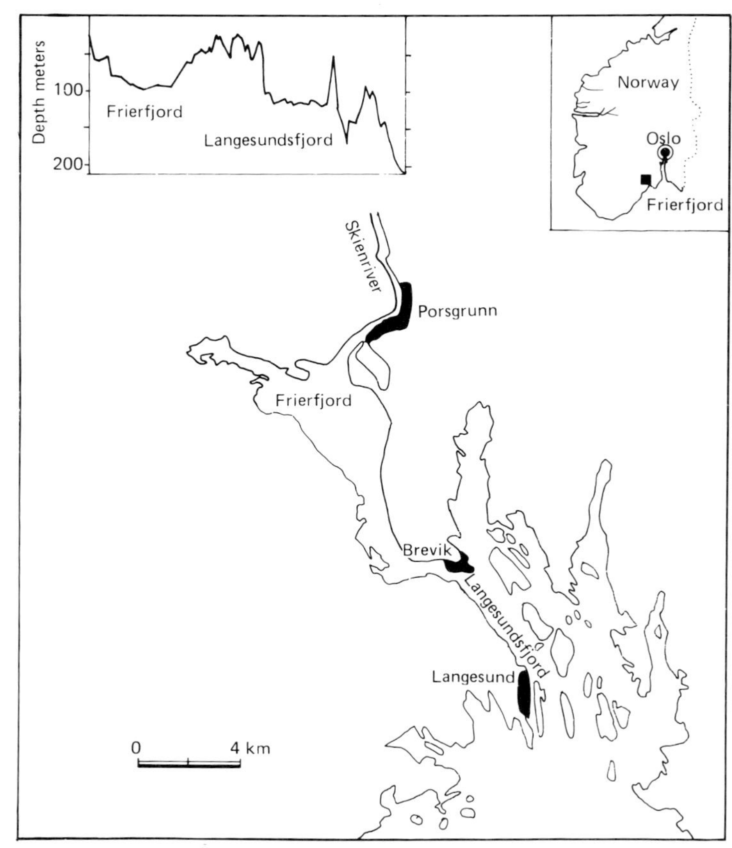


**Fig. 1.** Map of the Frierfjord and the fjord system outside the sill

The pollution load has been significantly reduced during the last 8–10 years. At present the main problems are connected with the load of organic micropollutants and nutrients (Fig. 2).

## 2.3 Environmental Problems

Due to reduced discharges from the magnesium plant, the concentrations of hexachlorobenzene and octachlorostyrene declined until 1976, and have been relatively stable since (Fig. 3). For decachlorbiphenyl the trend is more uncertain. The levels in the Langesundsfjord are typically 10–20% of the levels in the Frierfjord. There are no further data indicating the size of the transport of chlorinated hydrocarbons into the coastal water.

In response to reduced discharges, also the *mercury* concentration in cod (*Gadus morhua*) has fallen almost by a factor of 10 since 1970 (Fig. 4). The apparently strong increase in 1974 and 1975 has attracted much interest, as

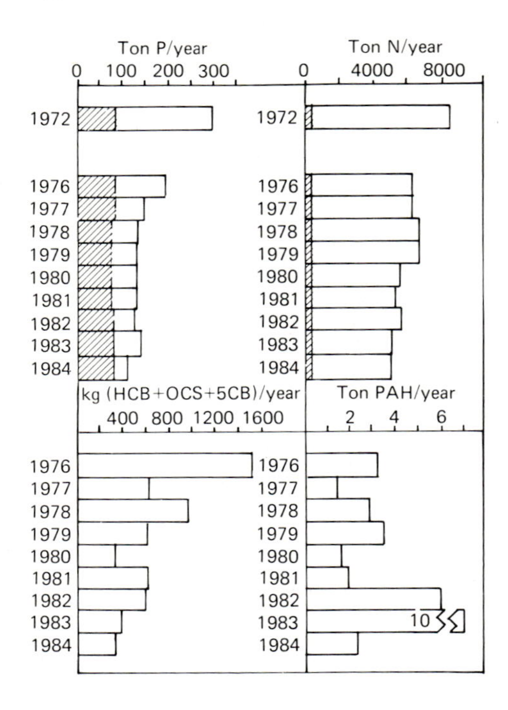


**Fig. 2.** Annual discharge of phosphorous, nitrogen, chlorinated hydrocarbons (HCB, OCS, 5CB) and polycyclic aromatic hydrocarbons to the Frierfjord

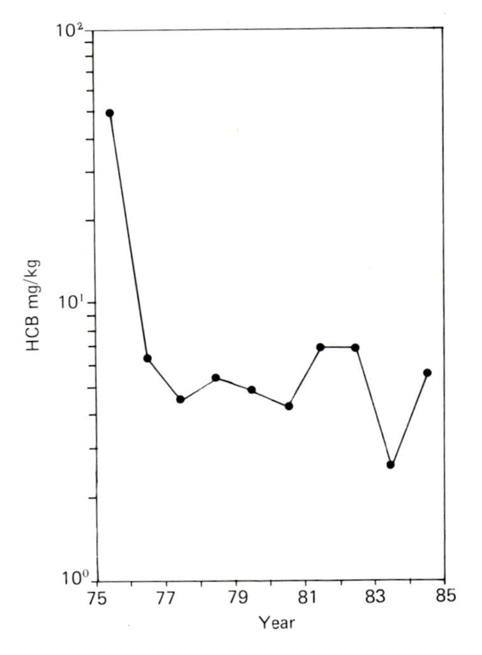


**Fig. 3.** Concentrations of hexachlorobenzene in cod liver from the Frierfjord (mg $kg^{-1}$ wet weight). Annual means calculated for a "standard" cod of 1 kg. (After Rygg et al. 1985)

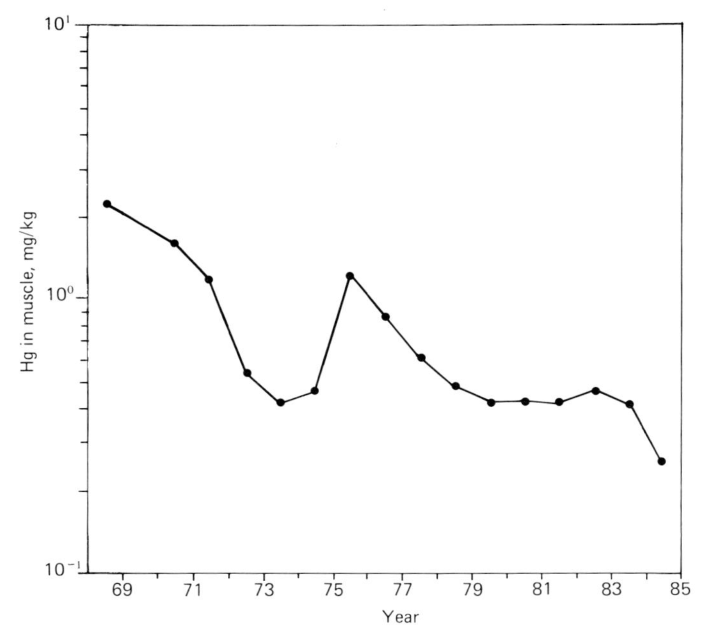


**Fig. 4.** Concentrations of mercury in cod muscle from the Frierfjord (mg $kg^{-1}$ wet weight). Annual means calculated for a "standard" cod of 1 kg. (After Rygg et al. 1985)

approximately 500,000 $m^3$ of sand, silt, and clay from dredging were dumped in the Frierfjord from October 1974 to May 1975. However, statistical analyses support the theory that the variations were caused by subpopulations of fish from the less contaminated Langesundsfjord being included in the samples (Gramme et al. 1984). At present, leakage of mercury from the strongly polluted Gunnekleivfjord (Skei 1978) probably represents the main mercury load on the Frierfjord. The data from 1984 showed improved conditions relative to earlier years.

In 1980–81 the local health authorities warned that due to high concentrations of chlorinated hydrocarbons and mercury, liver of fish caught in the Frierfjord was unfit for human consumption. The muscle could be eaten twice a week. Fish from the Langesundsfjord could be eaten four to five times a week.

Since 1980 high contents of *PAH* have been found in the common blue mussel (Fig. 5). The main source is the ferro-manganese smelter at the mouth of Skien river. In 1984 the state health authorities concluded that mussels in the Frierfjord and Langesundsfjord are unfit for human consumption. Data from 1985 (NIVA, unpublished) further indicate concentrations in the order 5–20 times background levels in samples from a mussel farm on the coast, situated about 25 km from the smelter. These preliminary results may thus indicate a significant transport of PAH from the fjord system into the coastal water, with serious consequences for the local aquaculture industry.

The daily discharge of *nutrients* to surface water of the Frierfjord is 14 t of nitrogen, mainly $NH_4$ -N and $NO_3$ -N from the fertilizer plant, and 250 kg of phosphorous, mainly as $PO_4$ -P.

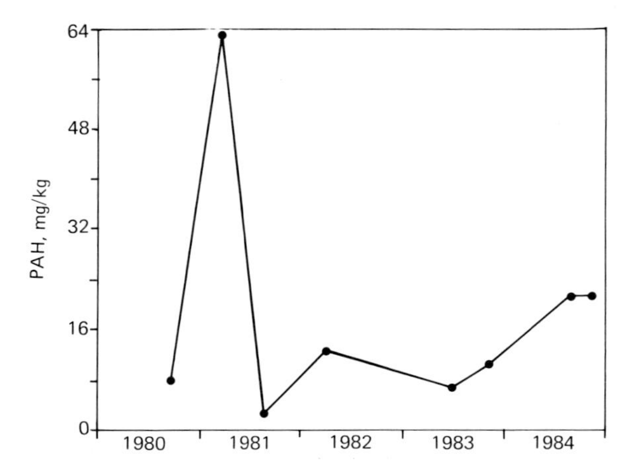


**Fig. 5.** Concentrations of polycyclic aromatic hydrocarbons in *Mytilus edulis* from the Langesundsfjord 1980–84 (mg $kg^{-1}$ dry weight)

The turbid surface water in the Frierfjord (secchi-depth 2–4 m both summer and winter), salinity as low as 3–6‰ and short residence time limits the pelagial eutrophication effects. The fjord system outside the sill, with longer residence time and higher salinity, are more affected.

In this context the transport of nitrogen from the Frierfjord into the coastal water is of special interest, as nitrogen generally is considered a limiting factor for the primary production in marine waters. This aspect has not yet been studied in detail, but a rough calculation may indicate the size of the transport. As the estuarine circulation is a dominant feature, one may use the Knudsens relations (Defant 1961) to calculate the average surface layer flow into the coastal water. Data from the Langesundsfjord indicate that for an average salinity of 12‰for the surface layer, a salinity for 30‰ of the deep return current, and a freshwater discharge of 270 $m^3s^{-1}$, the outgoing surface current amounts to 450 $m^3s^{-1}$. Between 1978 and 1981 the average concentration of total nitrogen (n = 22) in the outflowing surface layer was 960 mg $m^{-3}$, and 230 mg $m^{-3}$ in the return current at 12 m depth. This indicates a net daily transport of 20–30 t of nitrogen into the coastal water. The difference between this number and the 14 t direct discharge into the Frierfjord, constitute the net transport from the Skien river into the fjord (4–8 t $day^{-1}$) and runoff to the Langesundsfjord.

The impact on the coastal water quality and plankton production from this nitrogen load has not been studied. As there is growing concern of serious eutrophication effects in the coastal water, with increased occurrence of toxic dinoflagellates as the most conspicuous effect, the possible negative effects of the nitrogen transport from the Frierfjord into the coastal water should be thoroughly investigated.

### 2.4 Summary

Due to present and earlier discharges of nutrients, organic micropollutants, and metals, the Frierfjord and the fjord system outside the sill are seriously polluted. The potential negative effects from a substantial transport of nitrogen (and possibly also PAH) into the coastal water should be studied.

## 3 Kristiansandsfjord, South Norway

### 3.1 Environmental Setting

The Kristiansandsfjord is situated far south on the Norwegian Skagerrak-coast (Fig. 6). It has an open connection to the Skagerrak, with approximately 200 m as maximum depth at the south end.

The freshwater runoff to the fjord is 150 $m^3 s^{-1}$ as an annual mean, of which the Otra river supplies 95%. In addition 60–70 $m^3 s^{-1}$ of freshwater is supplied through the brackish surface current from the Topdalsfjord. In the Kristiansandsfjord this creates a 1–3-m-deep brackish surface layer, with salinity in the range of 1–32‰, with approximately 12‰ as an annual mean.

The circulation and water exchange is dominated by estuarine circulation, effects from local wind and density currents from wind-generated up- and downwelling in the coastal water.

### 3.2 Pollutant Load

The Kristiansandsfjord receives untreated municipal wastewater from a population of 80,000. From industrial effluents the present and earlier discharges of metals and organic micropollutants from a nickel plant are of most concern.

The daily discharge of metals during the last 5 years is summarized in Table 1. For organic micropollutants only few data are available, but for 1984 they indicate that the nickel factory discharged in the order of 7 kg EPOCl $week^{-1}$ (extractable, persistent organic chlorine) into the fjord. This does not include quantified amounts of hazardous substances such as hexachlorobenzene (HCB), octachlorostyrene, and alkylbenzenes. However, only a minor part of the EPOCl has been identified.

### 3.3 Environmental Problems

The different effects from the metal load have been studied frequently since 1973. In 1975 high content of HCB was found in cod caught in the harbor (Brevik et al. 1978), but the source for the HCB was not discovered at that time.

In 1982–84 the Norwegian State Pollution Authority sponsored a baseline study of the Kristiansandsfjord, and the nickel plant was identified as the main

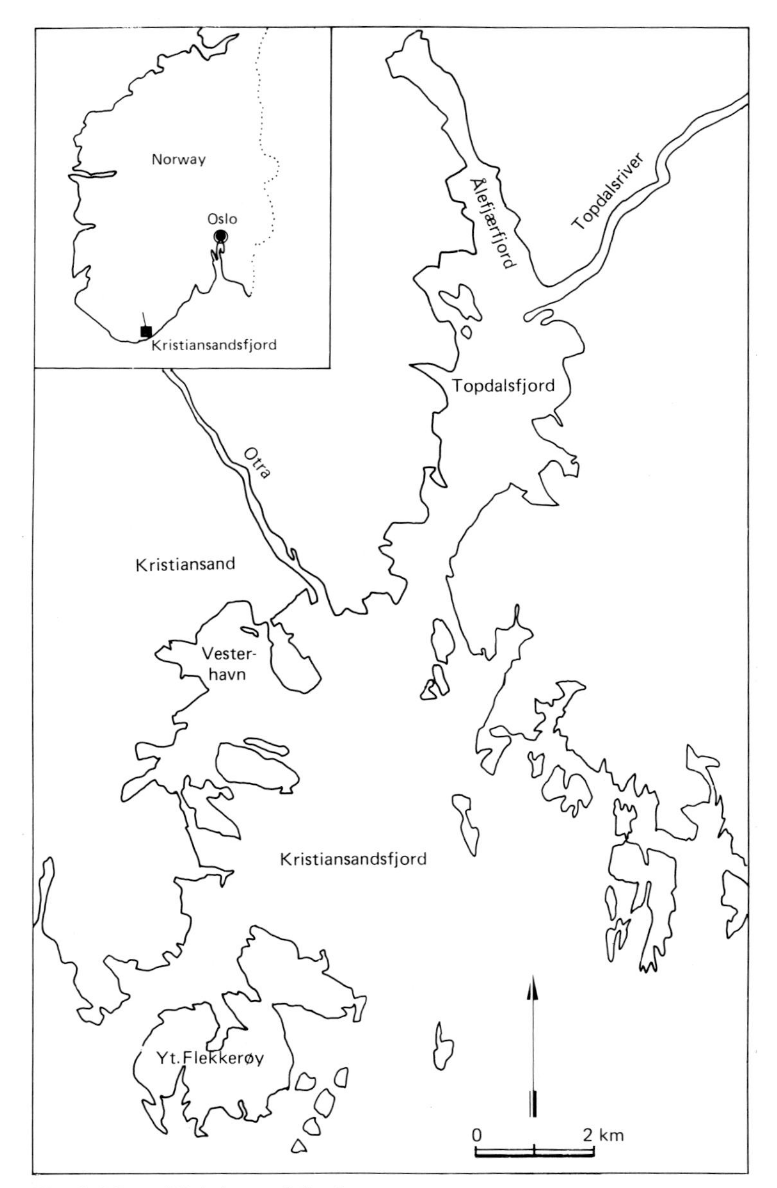


**Fig. 6.** Map of Kristiansandsfjord

source for the organic micropollutants (Naes 1985). A process which probably had contributed significantly to the discharge of HCB had then been stopped a few months earlier.

Studies of the sediments in 1983 (Naes 1985) showed extremely high concentrations of nickel, copper, arsenic, and cobalt in the harbor area (Fig. 7). Even in the outer fjord, 6 km from the nickel plant, the levels of nickel and copper in the 0–1 cm layer was three to five times that of the estimated background levels.

**Table 1.** Discharge of metals to the Vesterhavn, Kristiansand (kg day$^{-1}$)

| | Before 1.7.82 | Until 1.1.85 | After 1.1.85 |
|---|---|---|---|
| Iron | 6000 | 120 | 120 |
| Nickel | 500 | 210 | 50 |
| Lead | 100 | 1 | 1 |
| Zinc | 10 | 6 | 6 |
| Copper | 160 | 53 | 20 |
| Cobalt | 20 | 10 | 4 |
| Arsenic | 450 | 3 | 3 |

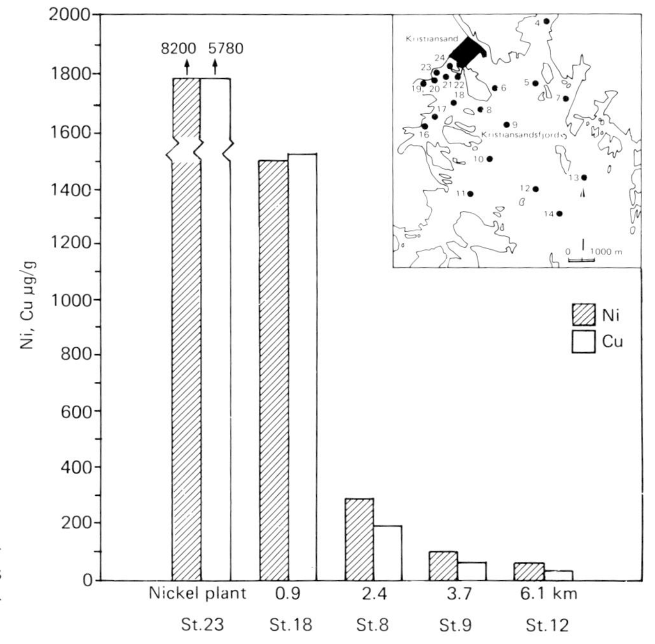


**Fig. 7.** Concentrations of nickel and copper in sediments (0–1 cm) in the Kristiansandsfjord. (After Naes 1985)

A similar picture appeared from analysis of HCB and EPOCI. In the harbor area concentrations up to 10,000 times estimated background levels were observed in the bottom sediments.

Studies of the softbottom fauna (Rygg 1985) showed corresponding serious biological effects. In the harbor area the ecosystem in an area of 3–5 km$^2$ was partly destroyed. The negative effect is probably caused both by the toxic effect of the copper in the sediments, and the earlier hypersedimentation of particulate materials.

Analysis of cod (*Gadus morhua*), flounder (*Platichthys flesus*) and mussels (*M. edulis*), both from Vesterhavn and the outer part of the Kristiansandsfjord showed very high content of HCB and EPOCI (Knutzen et al. 1984). However, the reduced discharge since 1982 appears to have resulted in somewhat lower levels in cod and flounder (Fig. 8), and also in mussels. However, in many of the samples only

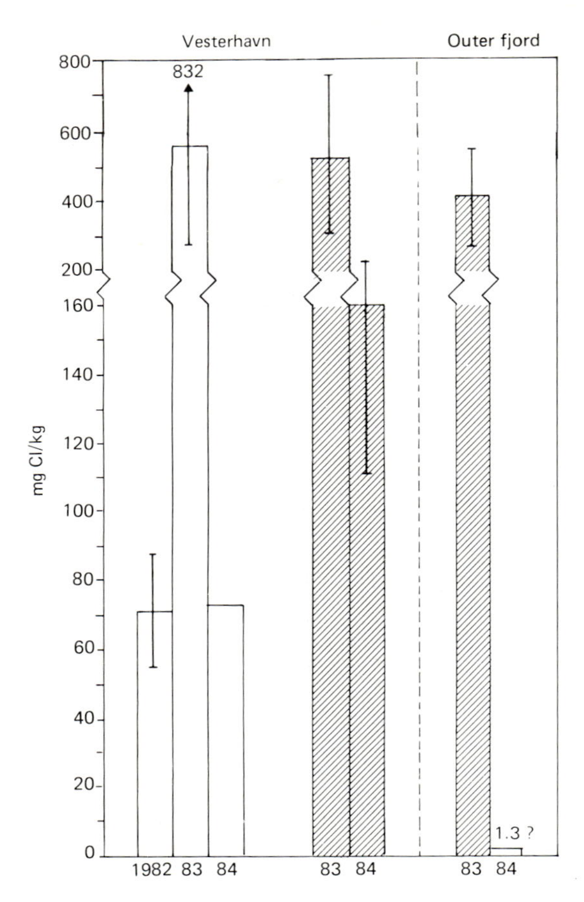


**Fig. 8.** EPOCl in flounder (*Platichtys fl.*) □ and cod (*Gadus morhua*) ▨ from the Kristiansandsfjord 1984–84, mg $kg^{-1}$ fat. Mean and variation ⊢⊣ for two parallel samples. (After Knutzen et al. 1986)

5–20% of the organic chlorine has been explained by identifiable and quantifiable compounds. The high content of dangerous substances indicated by the high EPOCl-levels has therefore resulted in a warning against consumption of fish from the area, including the outer part of the Kristiansandsfjord.

The eutrophication problems in the area are small and connected to local discharges of municipal effluents. However, estimates similar to those for the Frierfjord indicate that 2–8 t of nitrogen daily escape the fjord system into the coastal water, rivers being the most important source.

### 3.4 Summary

Due to present and earlier discharges of metals and organic micropollutants, the inner part of the Kristiansandsfjord is severely polluted. While probably insignificant amounts of metals escape into the coastal water, the situation regarding the organic micropollutants is more uncertain. The nitrogen input to the coastal water is in the order of 2–8 t $day^{-1}$.

# 4 Sørfjord, West Norway

## 4.1 Environmental Setting

Sørfjord is a north-south-trending extension of Hardangerfjord, 1 to 2 km wide and 40 km long (Fig. 9). The maximum depth is 390 m and the sill at the mouth is 225 m deep. Most of the runoff enters the fjord at its head (50 $m^3 s^{-1}$), partly as glacial runoff. The majority of the silt is deposited within Sørfjord while the finest clay fraction is transported further into Hardangerfjord. In the central parts of Sørfjord an average sediment accumulation rate of 2 mm $a^{-1}$ or 1.28 kg $m^{-2} a^{-1}$ has been estimated from excess lead-210 profiles (Skei 1981).

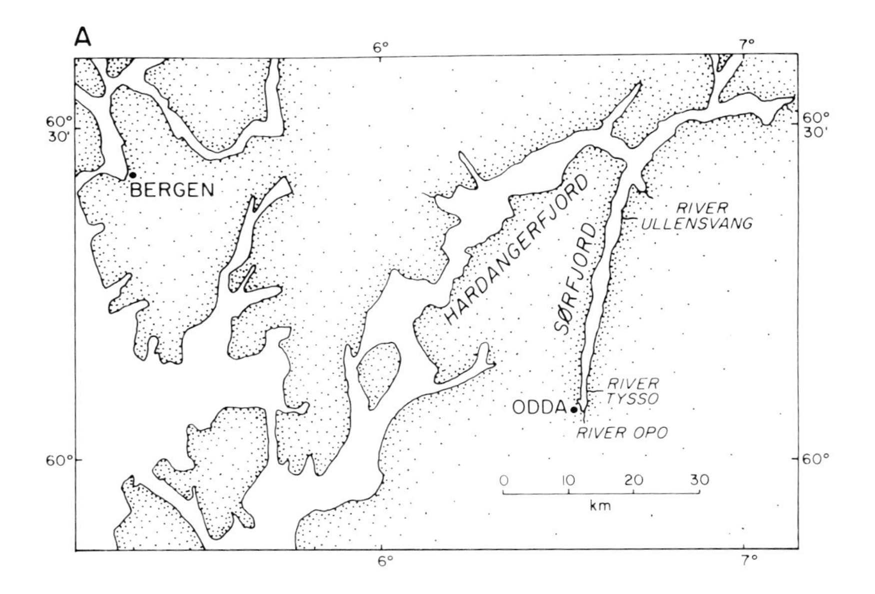


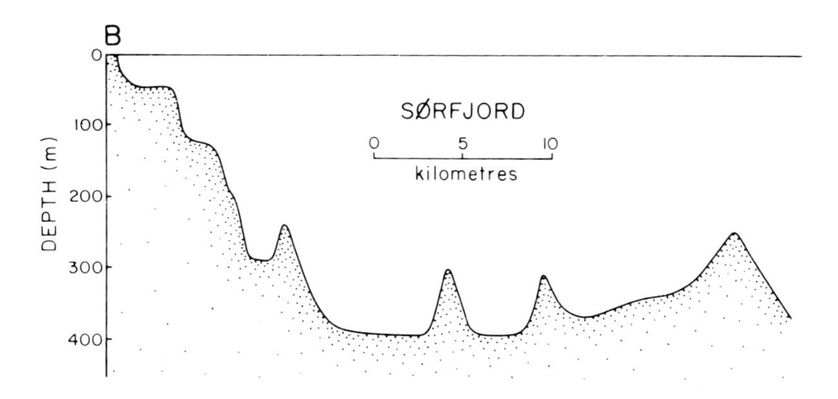


**Fig. 9.** **A** Map of Sørfjord and Hardangerfjord. **B** Depth profile of Sørfjord. (After Syvitski et al. 1987)

Circulation in the fjord is predominantly two-layer estuarine and frequent deep water exchange maintains oxygen concentrations in the bottom water above critical levels ($> 4$ ml $l^{-1}$). The sediments are generally oxic and do not contain hydrogen sulphide, except locally in the harbor basin of Odda.

## 4.2 Pollutant Load

Sørfjord has been considered the most metal-polluted fjord in the world, with several percent of heavy metals in the sediments (Skei et al. 1972), and several thousands of $\mu g l^{-1}$ of metals in the water (Skei 1981). The main source of the metals is a zinc plant situated at the head of the fjord at the industrial town of Odda. Additionally, Sørfjord has received industrial effluents from an aluminium smelter (shut down in 1981) and a carbide-dicyanamide smelter. The history of smelting at the town of Odda extends some 60–70 years back, which implies that Sørfjord has acted as a recipient of industrial waste for a long time. All compartments of the fjord environment are therefore severely polluted.

The daily discharge of metals to Sørfjord in 1972 is shown in Table 2. Only minor changes in discharge rates have occurred in recent years. The most important metal source today is a jarosite residue that is discharged at 20 m depth as a slurry with density 1090 kg $m^{-3}$. As the density of this suspension is higher than the receiving seawater, it travels as a density current seawards. This is very similar to the behavior of mine tailings being discharged into the marine environment (Tesaker 1978; Nyholm et al. 1983; Hay 1982). An additional metal source is seepage from old residue masses from zinc production dumped on shore and exposed to tidal water.

**Table 2.** Metals in industrial waste discharged into Sörfjorden (1972). (After Environmental Committee Report 1973)

| Metal | Compound in waste | Discharged weight (t $day^{-1}$) |
|---|---|---|
| Zn | $ZnO.Fe_2O_3$, $ZnSO_4$ | 6 |
| Cu | $CuSO_4$, $CuO_2$, Cu-jarosite | 0.3 |
| Cd | $CdSO_4$, CdS, $CdO.Fe_2O_3$ | 0.03 |
| Pb | $PbSO_4$ | 4.5 |
| As | $FeAsO_4$ | 0.09 |
| Sb | $FeSbO_4$ | 0.06 |
| Hg | HgS, HgSe, $Hg_2Cl_2$, Hg, $Hg^{2+}$ | 0.003 |
| Fe | $(NH_4)_2$ $Fe_6(SO_4)_4$ $(OH)_{12}$, $ZnO.Fe_2O_3$, $Fe_2O_3$ | 23 |
| Mn | $MnSO_4$, $MnO_2$ | 0.15 |

### 4.3 Environmental Problems

The jarosite consists of particles varying in size between 1 and 20 $\mu$m, and also contains soluble metal salts. Upon entering the fjord, the soluble fraction of the metals is released immediately and with the finest solid fraction, transported long distance. The coarse-grained part of the solid waste is deposited close to the discharge point, causing an exponential increase of heavy metals in the sediments towards the source (Fig. 10). The fine fraction of the jarosite slurry behaves like a plume, gradually mixing into the surrounding water, eventually to loose its identity (Fig. 11).

The severe pollution of the surface water, despite the fact that the regular discharges from the zinc plant today occur at 20 m depth or deeper, has led to intensified investigations of other potential sources. Attention has focused recently on land dump sites near the zinc plant, flooded during high tide. These sites are acidic residue dumps from the earlier days of zinc production, containing 10–15% zinc and lead, as well as high concentrations of associated metals (i.e., cadmium). During ebb tide, seawater with low pH (3–5) and extremely high metal content (830 mg $l^{-1}$ zinc, 22 mg $l^{-1}$ cadmium, 2.4 mg $l^{-1}$ lead and 5.2 mg $l^{-1}$ copper, all maximum values) enters the surface water of the fjord. This undoubtedly contributes significantly to the surface pollution of Sørfjord and presumably Hardangerfjord. High levels of cadmium and zinc in mussels and seaweeds 100 km from the source may be attributed to the leaching from the old residue. Measurements of metals and pH near the shore outside the residue dump over a tidal period, indicates the efficiency of the tidal pump (Fig. 12). In 1986 a barrier was built in front of the dump site to prevent seawater to penetrate the old residue.

One of the main issues of metal pollution in Sørfjord is to what extent the pollutants create environmental problems in Hardangerfjord and possibly in the coastal water outside. Investigations so far have revealed that the major part of Hardangerfjord have elevated metal content in mussels, seaweeds and the bottom sediments. Due to the long distance from the source of pollution to open coastal water (150–200 km), only minor amounts of metals are likely to reach the North Sea. Rough calculation of the retention of zinc, lead, and mercury in the Sørfjord bottom sediments suggested that 25% of zinc, 35% of lead, and 85% of mercury discharged from the smelter could be accounted for in the sediments within 40 km from the source (Skei 1981). This clearly indicates that Hardangerfjord must be expected to be considerably affected. Finally, it should be added that in 1986 the jarosite residue from the zinc smelter will no longer be discharged into the fjord. Instead the effluent will be stored in tunnels on shore. Consequently, a rehabilitation of the environment of Sørfjord and Hardangerfjord is expected in the years to come.

### 4.4 Summary

Due to discharge of substantial amounts of heavy metals to Sørfjord for a period of 50–60 years, water, sediments, and biota are severely polluted. As a consequence, restrictions on human consumption of fish and shellfish have been imposed in the

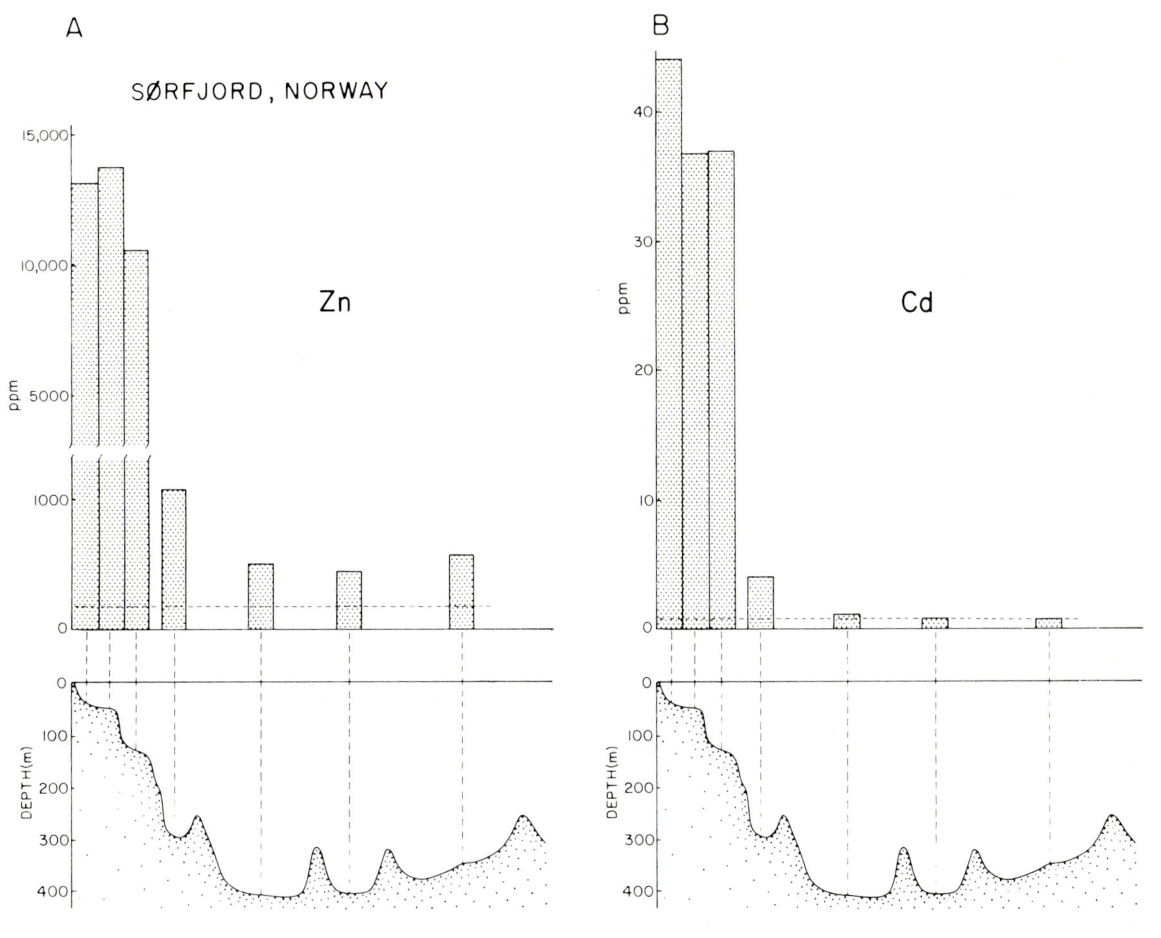


**Fig. 10A,B.** Zinc (**A**) and cadmium (**B**) in the surface sediments (0–2 cm) of the Sørfjord. (After Syvitski et al. 1987)

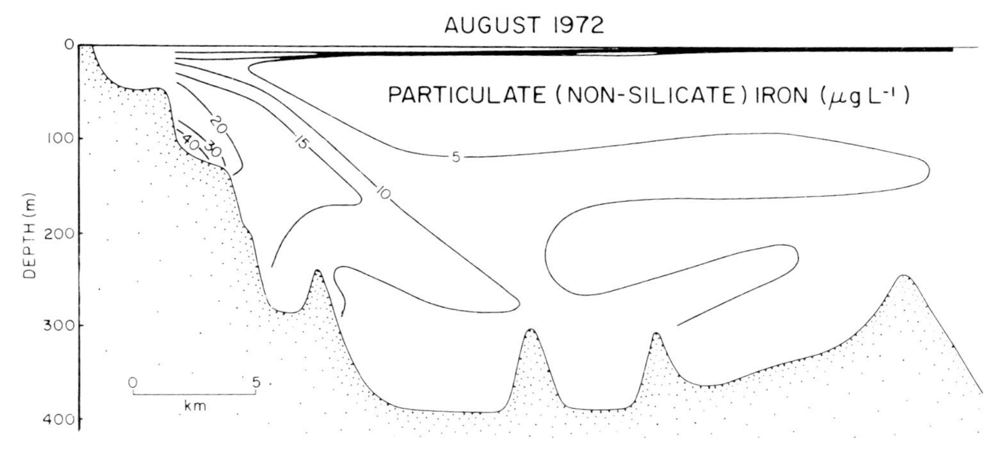


**Fig. 11.** Particulate iron (> 0.4 μm) in the water masses of Sørfjord. (After Syvitski et al. 1987)

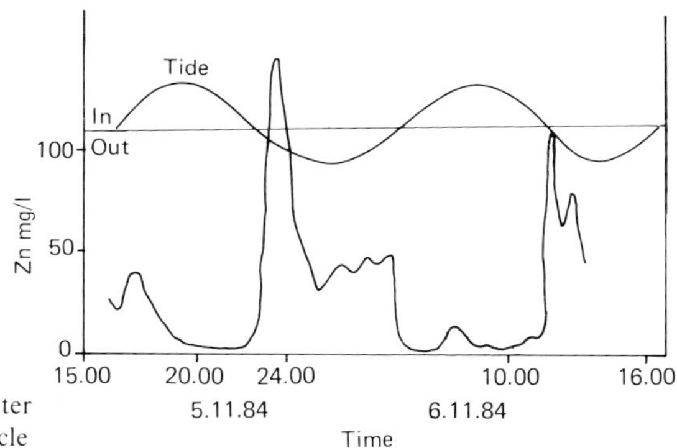


**Fig. 12.** The variation of zinc in the water outside the residue dumps over a tidal cycle

area. The impact may at least be traced 100 km from the source. There is another 50–100 km to the open coastal water and it is assumed that only minor amounts of pollutants are escaping Hardangerfjord and influencing the water quality of the North Sea.

## References

Brevik EM, Bjerk JE, Kveseth NJ (1978) Organochlorines in codfish from harbors along the Norwegian coast. Bull Environ Contam Toxicol 20:715–720

Defant A (1961) Physical Oceanography, vol 1. Pergamon, London

Environmental Committee Report (1973) Resipientundersøkelser i Sørfjorden 1972 (unpublished in Norwegian)

Gramme PE, Norheim G, Bøe G, Underdal B, Bøckman OC (1984) Detection of cod (*Gadus morhua*) subpopulations by chemical and statistical analysis of pollutants. Arch Environ Contam Toxicol 13:433–440

Hay AE (1982) The effects of submarine channels of mine tailings disposal in Rupert Inlet, BC. In: Ellis DU (ed) Marine Tailings Disposal. Ann Arbor Science, Ann Arbor, Michigan, pp 139–181

Knutzen J, Martinsen K, Naes K (1984) Om observasjoner av klororganiske stoffer i organismer og sedimenter fra Kristiansandsfjorden. (Preliminary note on organochlorines and sediments from the Kristiansandsfjord (S Norway) in 1982–83. VANN 3(1984), pp 392–400 (English summary)

Knutzen J, Martinsen K, Enger B (1986) Basisundersøkelse av Kristiansandsfjorden. Delrapport 4. Miljøgifter i fisk og andre organismer. Norwegian Institute for Water Research, Oslo, 100 pp (in Norwegian)

Naes K (1985) Basisundersøkelse av Kristiansandsfjorden. Delrapport 2 Metaller i vannmassene, metaller og organiske miljøgifter i sedimentene 1983. Norwegian Institute for Water Research, Oslo, 62 pp (in Norwegian)

Nyholm N, Nielsen TK, Pedersen K (1983) Modeling heavy metals transport in an arctic fjord system polluted from mine tailings. Modelling the rate and effects of toxic substances in the environment. ISEM Conference, Copenhagen

Rygg B (1985) Basisundersøkelse av Kristiansandsfjorden. Delrapport 1 Bløtbunnsundersøkelser 1983. Norwegian Institute for Water Research, Oslo, 66 pp (in Norwegian)

Rygg B, Bjerkeng B, Molvaer J (1985) Grenlandsfjorden og Skienselva 1984. Norwegian Institute for Water Research, Oslo, 66 pp (in Norwegian)

Skei JM (1978) Serious mercury contamination of sediments in a Norwegian semi-enclosed bay. Mar Pollut Bull 9:191–193

Skei JM (1981) Dispersal and retention of pollutants in Norwegian fjords. Rapp P V Reun Cons Int Explor Mer 181:78–86

Skei JM, Price NB, Calvert SE, Holthedal H (1972) The distribution of heavy metals in sediments of Sørfjord, West Norway. Water Air Soil Pollut 1:452–461

Syvitski JPM, Burell DC, Skei JM (1987) Fjords: processes and products. Springer, Berlin Heidelberg New York

Tesaker E (1978) Sedimentation in recipients. Disposal of particulate mine waste. VHL-report No STF 60 A78105, 72 pp

# Impact of Sewage Sludge

T. AP RHEINALLT[1]

## 1 Introduction

In the United Kingdom, disposal of sewage sludge to the sea is almost entirely by dumping from ships into estuarine and coastal waters. Only very small quantities are discharged by pipeline. In other countries bordering the North Sea, sludge disposal is almost exclusively to land, although some countries, such as the Federal Republic of Germany, have practised disposal to the North Sea in the past (Collinge and Bruce 1981). The present chapter, therefore, is mainly concerned with the impact of disposal from ships in the marine environment, and impact is considered to be measured largely in terms of biological effects.

Actual and potential effects of sludge on the marine environment will first be reviewed. Studies carried out at North Sea disposal sites (Fig. 1) will then be described and discussed. Finally, the research methods presently employed will be reviewed and new approaches suggested.

## 2 Effects of Sewage Sludge on the Environment

A distinction can be made between two types of disposal ground: "accumulating" and "dispersing" (McIntyre 1981). These can be considered as extremes of a continuum, the determining parameter being the amount of dispersion to which the sludge is subjected. On accumulating grounds, of which the Garroch Head site (Strathclyde) is an example, currents are weak and sedimentation is rapid. Material accumulates on the seabed in the vicinity of the disposal site, and ecological effects on the benthos may be considerable, but are likely to be confined to a small area. On dispersing grounds, such as the Thames Estuary, currents are frequently strong and/or the water column is strongly stratified, and material may be widely dispersed before settling out at a distance from the disposal site. Its residence time in the water column can be as long as 8 h (Talbot et al. 1982), which may be sufficient to affect pelagic communities significantly. Effects on the benthos are likely to be less intense but more widespread than at accumulating grounds, and in practice difficult to evaluate, especially if pollution from other sources is present in the same

[1]WRc Environment, Medmenham Laboratory, PO Box 16, Henley Road, Medmenham, Marlow, Bucks, SL7 2HD, United Kingdom

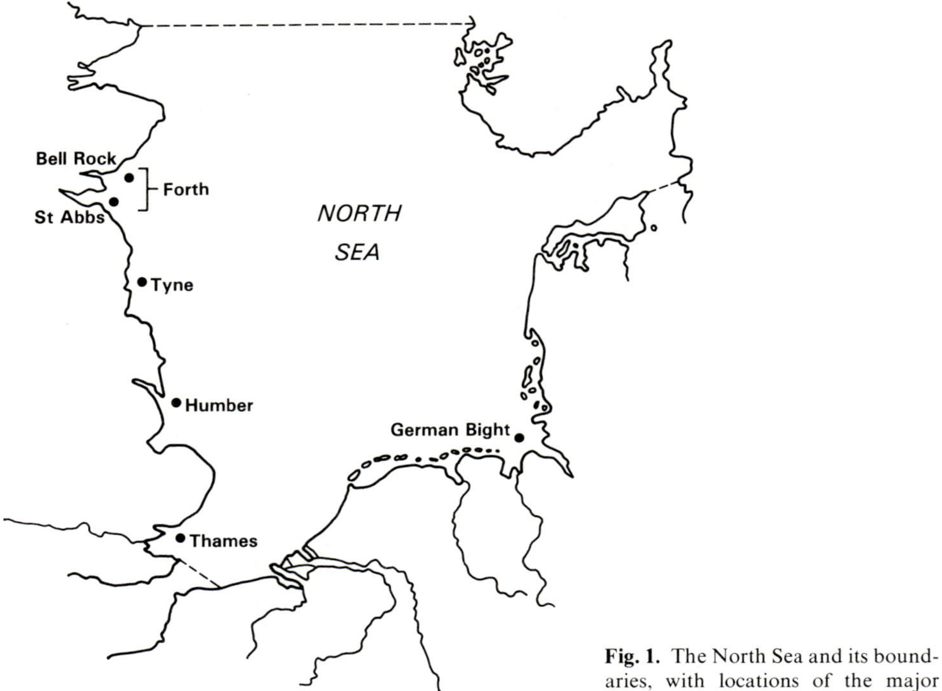


**Fig. 1.** The North Sea and its boundaries, with locations of the major sewage sludge disposal grounds

general area. It has been suggested that contamination of a seabed by sewage sludge can be predicted from the nature of water movements at the seabed and that of sediments at the particular site (Devine et al. 1986). The authors calculated factors which described (a) the local and regional dispersiveness, and (b) the local and regional accumulation rates for each of eight disposal sites: six around the coasts of the UK, one in the German Bight and one in the New York Bight.

The release of sewage sludge from a ship results in the presence of a surface slick, which may be visible for up to several hours. As sludge begins to disperse horizontally and vertically, a plume of turbid water is formed. Finally, the solids settle on the seabed.

For the purposes of discussion, the effects of sewage sludge on the surface layer, in the water column, and on the seabed can be considered separately, although in practice these effects overlap (Rowe et al. 1976). Ecological effects of sludge were reviewed in O'Sullivan (1971), Gould (1976) and McIntyre (1981), on which the following general discussion is based.

## 2.1 Effects at the Surface

The slick that forms at the surface will be accompanied by a marked odour and possibly some surface debris: since disposal grounds are well out to sea, there should be no amenity threat (DoE 1972a). The slick alters the characteristics of the surface film, and may affect the neuston (the community at the air/water interface) although there is no evidence for this.

## 2.2 Effects in the Water Column

### 2.2.1 Nutrients

Large quantities of nutrients, in particular nitrogen and phosphorus, are added to the water column, providing favourable conditions for phytoplankton growth. In the absence of other limitations, an increased supply of nutrients can be expected to lead to increased primary productivity, and thus a more abundant food supply for consumers. Particular species of primary producer may be favoured, resulting in blooms of *Phaeocystis* or *Noctiluca.* These may be directly toxic to marine organisms, or have indirect effects by their decay and subsequent deoxygenation of the water. They can accumulate in shellfish that filter them, causing gastrointestinal disorders or paralytic poisoning if consumed by man. There is, however, no evidence that directly links sludge-derived nutrients to these effects.

No detailed studies of nutrients in the water column have been carried out on the North Sea disposal grounds, but in the Liverpool Bay disposal area, sewage sludge was probably the major source of nutrients in summer. High nutrient levels seemed to have little effect on phytoplankton biomass. Some dense localised patches of ammonia were at concentrations likely to be toxic to some marine organisms, but these patches were apparently short-lived and confined to the immediate vicinity of the disposal point (DoE 1972a,b). Further studies have shown that on occasion higher standing crops of phytoplankton occur in the sub-surface water, but the extent of these has not been fully evaluated (Head 1980).

Any tendency of sludge-derived nutrients to enhance productivity could be countered by factors such as contaminant toxicity or increased turbidity.

### 2.2.2 Organic Matter

Organic matter in sludge can act as a food source for planktonic organisms, which may feed directly on fine organic flocs, or on detritophagous bacteria and fungi (Chapman 1986). The importance of this effect, relative to the increase in phytoplankton promoted by the addition of sludge to the water column, is not known. However, it is known that zooplankton biomass is sometimes high at disposal sites, e.g. at Garroch Head (McIntyre and Johnston 1975). Perhaps the most important potential effect of organic matter in sludge is the oxygen demand caused by its degradation. Whether or not this leads to oxygen depletion in the water column at a particular site will depend on the rapidity of dilution and dispersion, and it would seem that on most disposal grounds, oxygen depletion will not occur. In Liverpool Bay, for example, departures from full saturation at the disposal location are rare (DOE 1972a,b; Head 1980).

#### 2.2.3 Inert Particulates

The effects of inert particulates cannot be separated entirely from those of organic particulates, but in suspension their main effect is an increase in turbidity. This may attenuate light and thus decrease photosynthetic activity, but evidence that these effects occur on disposal grounds is lacking. The same applies to the possibility that increased turbidity could decrease the efficiency of filtering in some planktonic organisms.

#### 2.2.4 Persistent Materials

It is widely recognised that metals and other contaminants may be incorporated into benthic food chains after accumulating on the seabed. Less attention has been paid to the behaviour of these contaminants in the water column, however. Small quantities of metals probably dissolve and are available for accumulation by phytoplankton, but most are bound to small organic flocs and may be directly ingested by zooplankton. In this way metals and pesticide residues can be incorporated into the planktonic food chain and ultimately be accumulated by fish (Chapman 1985, 1986). However, the significance of this process in the context of sludge disposal is unknown. Similarly, while uptake and accumulation of organochlorine residues and PCBs by marine organisms has been studied the importance of uptake from sewage sludge in the water column is unknown.

### 2.3 Effects on the Seabed

#### 2.3.1 Nutrients

As in the water column, it is to be expected that nutrient input from sludge will increase primary productivity in the benthic community. However, attached algae may not be present in the disposal area, in which case the major potential effects of sludge-derived nutrients are increased availability of planktonic food to benthic consumers, and deoxygenation as a result of phytoplankton decay.

#### 2.3.2 Organic Matter

Addition of sludge-derived organic material to sediments on the seabed may have significant effects. Expected changes include an increase in organic carbon content of the sediments, accompanied by benthic oxygen depletion.

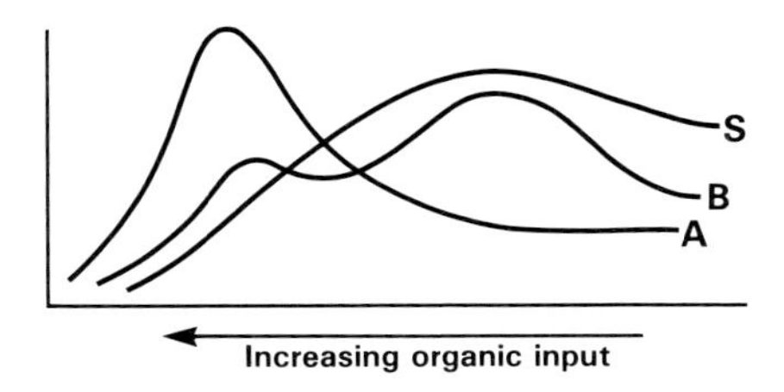


**Fig. 2.** Diagram of changes in species number *(S)*, total biomass *(B)*, and total abundance *(A)*, along a gradient of increasing organic input. (Reproduced by permission from Pearson and Rosenberg 1978. Copyright 1978 Aberdeen University Press Ltd)

The effects of organic enrichment in general on the macrobenthos are reviewed by Pearson and Rosenberg (1978), who also evaluate the usefulness of different measures of community structure in quantifying such effects. Increasing organic input is characterised by well-defined changes in species number, biomass, and total abundance (Fig. 2). Where organic enrichment is greatest, the sediment may be anoxic, leading to a complete absence of life. As it declines, a community composed of a few small opportunistic species, usually polychaetes such as *Capitella*, appears and increases to high abundance. At still lower levels of enrichment, these species disappear, while both species number and biomass rise to a maximum. Here, organic enrichment is sufficient to provide a rich food source, but does not significantly deplete oxygen levels.

Changes over time with increasing or decreasing organic input are similar to changes with distance relative to the source of the input.

In general, organic enrichment leads to an increase in deposit-feeders with an associated decline in suspension-feeders. The responses of mobile epifaunal organisms, on the other hand, are not well known (Pearson and Rosenberg 1978), although organic debris was considered to attract shrimps to a disposal ground (Halcrow et al. 1973).

Clearly, the effects of a given input of organic matter will vary with the hydrographic conditions (Pearson and Rosenberg 1978; Devine et al. 1986). At the Garroch Head (Firth of Clyde) sludge disposal site, which is of the accumulating kind, the various stages of the response to organic enrichment are clearly defined: an initial increase in numbers of all animals present, followed by a reduction in species number accompanied by an increase in the biomass of the few remaining species. The last stage of the progression, the complete absence of life, is not found (Mackay 1986).

### 2.3.3 Inert Particulates

Deposition of sludge-derived particulates may change the sediment structure, leading ultimately to the development of a muddy substratum. This can affect benthic organisms in several ways, the net result being a change in community structure from that typical of the original sediment type. Benthic filter-feeders are adversely affected by very high levels of suspended sediment near the seabed, while deposition of material may smother some organisms and provide an unsuitable substrate for the settlement of others. This effect takes place concurrently with those of organic enrichment and oxygen depletion.

### 2.3.4 Persistent Materials

Most metals and pesticide residues in sludge settle out fairly quickly with the particulate matter and are incorporated into the sediment. The degree of accumulation of these contaminants will obviously depend on the dispersion characteristics of the disposal site (Devine et al. 1986). Some studies have shown increased levels of metals in the sediments of disposal areas, and also in benthic organisms.

However, it is difficult to separate the effects of metal accumulation due to sludge from those due to other sources. Furthermore, while the uptake of contaminants by, and their effects on, organisms, are the subject of many studies, the question of the significance of these effects in the field remains unanswered. For example, laboratory experiments showed that various organisms (fish, shrimp, molluscs) did not accumulate cadmium or mercury from sewage sludge over a 60-day period, but these results were considered to be at variance with field studies showing that fish in the Thames Estuary had rather high body burdens of mercury in the early 1970's, possibly derived from sewage sludge (Franklin 1983).

Of interest in recent years has been the possibility that diseases and abnormalities in marine organisms, particularly fish, could be caused by contaminants. At least two studies in the North Sea (Dethlefsen 1980; Bucke et al. 1983) have specifically investigated disease in fish populations of sludge disposal grounds and will be described below. In general, it can be said that there is no universal agreement on the role of pollutants in inducing fish disease (Dethlefsen 1986).

### 2.3.5 Pathogens

Bacteria, viruses, protozoans, and helminths are present in sewage sludge, and some, such as enteric viruses, can survive in the sea for several days or more if adsorbed onto particles. While direct contamination of humans is unlikely due to the offshore location of disposal grounds, it is possible that pathogens could accumulate in benthic filter-feeders such as shellfish, which are consumed by humans. However, while the effect of sewage from coastal outfalls on shellfish is widely recognised, there seems to be little information on the effect of sludge-derived pathogens, although faecal bacteria are widely used as indicators of sludge (Ayres 1977). Pathogens may be of little importance unless shellfish are exploited near the disposal sites, which are, however, usually chosen to preclude bacteriological contamination of shell fisheries.

# 3 Impact of Sludge Disposal: Studies in the North Sea

## 3.1 Macrobenthic Surveys

### 3.1.1 Thames Estuary

Three separate studies of the effects of sludge disposal on the benthic fauna of the Thames Estuary site, the largest in the United Kingdom, receiving 5 million tonnes per year, have been carried out by the Ministry of Agriculture, Fisheries and Food (MAFF). MAFF conluded from their studies of dispersion and sediment physico-chemical characteristics that, in spite of the generally good dispersive characteristics of the area, some regions of metal and organic matter accumulation nevertheless exist (Norton et al. 1981).

The first survey was carried out in April 1970. It was found that the benthos was fairly normal for an estuary with strong tidal currents, with no areas of anaerobic sediments and impoverished fauna. It was suggested that the high numbers and diversity of polychaetes in organically-enriched sediments were a consequence of sludge disposal (Shelton 1971).

In 1972 a more detailed survey was carried out over a wider area. Nine faunal associations were identified. Poorly-sorted sediments adjacent to the disposal ground were found to support a rich and diverse fauna compared to stations in the outer region of the outer estuary, considered to be unaffected by sludge (Talbot et al. 1982).

Results of the 1977 survey are, unfortunately, not directly comparable with those of the 1972 survey because of the larger sieve size used in the later survey. Also, a much higher proportion of organisms in the samples were identified to species level in 1977, with consequent effects on diversity and associated indices. A dense sampling grid was used, and 17 faunal associations were identified. Initial analysis attempted to classify these associations into those typical of the sediment type and others considered as "unusual". Two stations contained a very sparse fauna dominated by the capitellid worm *Notomastus latericeus* and had high concentrations of organic carbon: these impoverished sites were, however, far away from the main sites of sludge settlement and may have reflected localised conditions. The presence of *N. latericeus* in another association was also thought to indicate pollution and possible reduction in oxygen tension in the sediments. Other associations on mixed sediments with high densities and diversities of fauna were considered to show enrichment in response to sludge, but no 'clean' mixed-sediment stations were available for comparison.

Partial correlation analysis was then used to relate faunal and sediment characteristics within individual associations. Some correlations significant at the 80% to 95% level between sediment organic content and abundance or species richness were found (Table 1). Most but not all of these correlations were negative. Some of the negative correlations were within associations considered in the previous analysis to show faunal enrichment.

Thus, there was some rather weak evidence that associations responded to increased organic input, usually negatively, although the fauna was largely determined by the natural substratum and hydrographic conditions (Norton et al. 1981).

**Table 1.** Thames Estuary: correlation between faunal data (abundance, species richness) and sediment organic carbon levels in different species associations. (Adapted from Tables 8 and 10, Norton et al. 1981. Crown copyright 1981)

| Association | Sediment type | Commonest species (> 25 ind. $m^{-2}$) | Faunal variable | Correlation +/− | P |
|---|---|---|---|---|---|
| 1 | Fine sand | *Bathyporeia elegans* | Species richness | − | 0.10 |
| 4 | Sand + some gravel | *Spiophanes bombyx* | Species richness | − | 0.15 |
| | | *Nephthys hombergi* | Abundance | − | 0.07 |
| 6 | Fine sand + clay | *Nephthys hombergi* | Species richness | + | 0.09 |
| | | | Abundance | + | 0.18 |
| 10 | Mixed + gravel | *Pomatoceros triqueter* | Abundance | − | 0.12 |
| | | Anthozoa (attached) | | | |
| | | *Scoloplos armiger* | | | |

The apparent large reduction in faunal density from 1972 to 1977 was probably due to the increase in mesh size of the sieves used to sort the samples.

The MAFF results do not allow any unambiguous conclusions to be drawn, since both positive and negative responses (in terms of species richness and diversity) were identified. This is as expected (Fig. 2). Nevertheless, it was considered that sludge disposal had placed the natural fauna under stress and promoted the growth of pollution indicator species, current rates of disposal exceeding the dispersive capacities of the area (Norton et al. 1981).

Data collected by Thames Water in 1983 and 1984, and by the Water Research Centre (WRc) in 1985, are at present being analysed. Preliminary analyses indicate that stations in the disposal area support a rich and fairly diverse fauna, consisting mostly of polychaetes, in contrast to sandier stations in the surrounding area, which support a sparser fauna. It is possible that the sites near to the disposal ground are showing faunal enrichment in response to sludge. No obvious adverse effects have been identified.

### 3.1.2 German Bight

The sludge disposal ground at 20 m depth in the German Bight was in use from 1961 to 1980, receiving about 250,000 tonnes per year. Here, larger particles were thought to be deposited over a limited area, while fine material was distributed more widely by tidal currents. The natural sediment of this area was muddy and almost indistinguishable from sewage sludge by particle size and other analyses.

Sampling, which started in 1970, showed that the deeper layers were sometimes fully anaerobic, with the formation of hydrogen sulphide taking place. However, tidal movement ensured the presence of at least a thin oxidised surface over the whole disposal ground. Similar anaerobic sediments, receiving deposits of planktonic detritus, occurred naturally in the area. Sediment analyses for heavy metals showed that there was no difference between the disposal ground and its surroundings, the dominant source of metals being outflow from the Elbe.

The benthos was sampled frequently from 1970 to 1979. Populations of individual species varied greatly from year to year, but the dominant species in the disposal area was the bivalve *Abra alba*, whose abundance increased with increased sludge deposition until it was limited by the space available. The success of *A. alba* in areas affected by sewage sludge was ascribed to its being a deposit-feeder, in an area of heavy sedimentation unfavourable to filter-feeders. Considerable regional and temporal fluctuations occurred in the size of populations, and greatly reduced numbers in 1978 and 1979 were possibly related to this.

The surrounding area was characterised by the bivalve *Nucula turgida*, which occurred only in small numbers in the central disposal area.

Species associations were fairly constant: *A. alba* was associated with the polychaete *Pectinaria koreni*, while *N. turgida* was associated with the cumacean *Diastylis rathkei*. The number of species present in the disposal area was no lower than that found outside, but fluctuations in abundance were considerable.

This example provides a very clear illustration of the impact of sludge disposal on the benthos (Fig. 3), the relation between the conditions created and the ecology of the most successful species being obvious. Whether or not the changes induced by sludge disposal should be classified as ecologically negative is open to debate (Caspers 1980). In a more general context, however, it was stated by Rachor (1980) that the whole German Bight area had suffered depletion in species numbers during the period 1969 to 1979, to which the addition of sewage sludge had perhaps

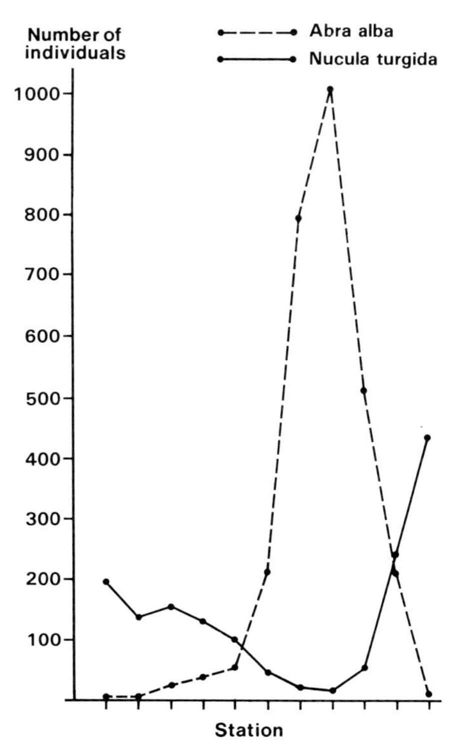


**Fig. 3.** Cross-section through the German Bight sludge disposal ground with the numbers of *Abra alba* and *Nucula turgida* in 1973. (Reproduced by permission from Caspers 1980. Copyright 1980 Pergamon Journals Ltd)

contributed, decreasing oxygen saturation values by an estimated 10% in particular conditions. This area was considered to be naturally susceptible to oxygen depletion because of the nature of the sediment (clay with organic matter) and the hydrographic conditions, the latter permitting the establishment of a thermocline during calm summer periods. The fauna would therefore be sensitive to any additional influence of pollution, which could lead to faunal instability and impoverishment (Rachor 1980).

### 3.1.3 Humber Estuary

Since 1971, sludge has been disposed of outside the mouth of the River Humber, 20 km east of Spurn Head. In 1976, the input of sludge was 171,000 tonnes. Sewage sludge contributes only a small proportion of nutrient and metal inputs to the estuary, but in the disposal area itself, materials entering in sludge are more significant.

The disposal area is located in water less than 25 m deep, with rapid tidal currents, reaching 1 m $s^{-1}$ near the surface. Dispersal of sludge is rapid, with little deposition of fine sediments. Surveys in 1975 and 1977 showed that there was an area of organic enrichment in the sediments, considered to correspond to sludge deposition. Metal distributions were complex, with the influence of sludge mostly at a local level.

Benthic sampling was impeded by the hardness of the substratum, but the dominant faunal association was characterised by the mussel *Modiolus modiolus*. This association, commonly found on coarse sediments in moderately deep water, was ubiquitous over the sampling area, with no evidence for any effect of disposal (Murray et al. 1980).

### 3.1.4 Tyne

This ground, situated 9 km from the mouth of the Tyne, has been in use only since 1980, and in 1984 received 500,000 tonnes of sludge.

Monitoring of the benthos at seven stations between 1980 and 1983 showed no changes that could be attributed to sludge disposal in a community dominated by several species of polychaete. Species abundances varied greatly, the most significant feature being a dramatic increase in some species at all stations during 1981/2, followed by a decrease. Similarly, the results of chemical monitoring at the same time showed no significant effect of sludge disposal up to 1983 (Pomfret and McHugh 1983).

In 1984, MAFF carried out a programme of sampling at a series of 10 stations, predominantly on fine sand and running SSE from the disposal ground, in the direction of the residual near-surface current flow. It was found that the concentration of sediment metals fell with distance from the disposal site: there was probably an influence of inputs other than sludge disposal. Carbon values were higher in the disposal site than elsewhere, but there was a great deal of background variability due to the presence of coal.

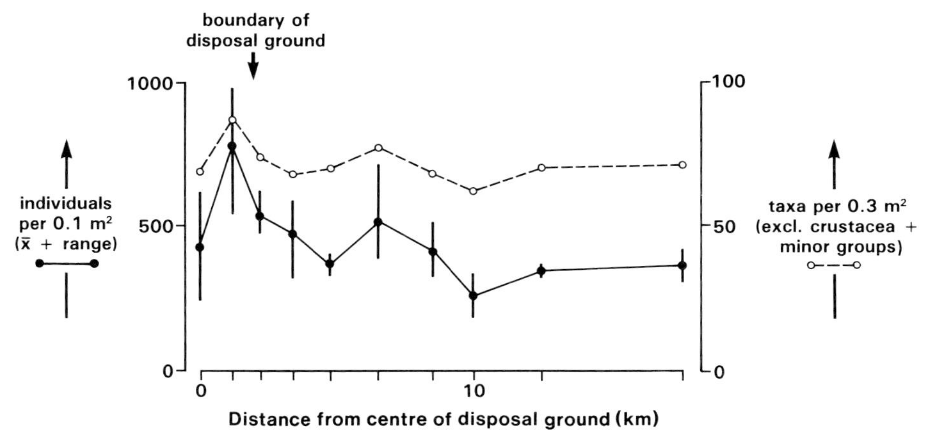


**Fig. 4.** Trends in total abundance and numbers of taxa at the Tyne disposal ground in 1984. (Reproduced by permission from Rees et al. 1985)

At all stations, there was a diverse macrobenthic community dominated by small polychaetes, particularly the capitellid *Heteromastus filiformis* and spionids (*Prionospio* spp.). A peak in macrobenthic abundance occurred at the southern edge of the disposal site, corresponding to a peak in the density of sludge indicators (faecal streptococci and tomato seeds). Outside the disposal site there was a gradual fall in macrobenthic density with distance (Fig. 4). Thus there appeared to be a positive but rather limited effect of sludge, the increase in density being unaccompanied by any significant changes in species composition, possibly because the natural community of this area contains species which might in other circumstances be associated with anthropogenic inputs of organic matter (Rees et al. 1985).

### 3.1.5 Firth of Forth

Two sites, Bell Rock and St. Abbs Head, are designated as sludge disposal sites off the Firth of Forth. They have been in use since 1978, on a 6-monthly rotation, Bell Rock receiving 115,000 tonnes and St. Abbs 170,000 tonnes of sludge in 1981.

Benthic and physico-chemical surveys were carried out by the Department of Agriculture and Fisheries for Scotland (DAFS) in 1978 and 1979. During this period, the mean particle size of sediments fell, while elevated carbon levels were detected in sediments at the disposal sites, suggesting an effect of sludge input. During the 6-month non-disposal phases, a degree of recovery was exhibited. In spite of these changes, however, no statistically significant changes in the fauna attributable to sludge were detected (Moore 1981).

From 1981, benthic surveys were carried out by the Institute of Offshore Engineering (IOE). In 1981, both sites were found to have a rich macrofauna dominated by polychaetes, deposit-feeding species being the most abundant. Values of diversity and evenness indices showed little variation among stations,

and were generally high and comparable with values from uncontaminated offshore areas of the North Sea. At some stations, breaks in the straight line of log-normal species distribution, potentially indicative of pollution (Gray and Mirza 1979), were considered instead to be due to the high numbers of juveniles present in the autumn. The communities were considered to be representative of the sediment type sampled, and dominance by a small number of species from a restricted range of feeding types was not observed (IOE 1983a).

In 1982, results were similar, but at St Abbs, cluster analysis produced evidence that stations near the centre of the disposal site, where concentrations of faecal indicators (tomato seeds) were highest, differed slightly from surrounding stations (IOE 1983b).

In 1983, the existence of these differences was not borne out in any clear manner, but other evidence showed that at St Abbs there was a change in community structure. This was manifested as a decrease in species richness and a fall in the values of diversity and evenness indices, certain polychaete species such as *Spiophanes bombyx* becoming particularly abundant. Breaks in the lognormal distribution may, on this occasion, have represented a pollution effect. These changes could indicate the beginnings of a detrimental trend due to disposal, but the overall species richness remained high, and continued monitoring was recommended. At Bell Rock no changes were observed. Organic carbon and silt levels in 1983 did not differ significantly from those found in 1981 and 1982 at either site (IOE 1985).

Data on metal concentrations in the sediments in 1981, 1982, and 1983 showed no evidence of accumulation due to disposal (Lothian Regional Council 1983a,b, 1984).

### 3.2 Contaminants in Fish and Shellfish

Between 1970 and 1977, accumulation of metals (mercury, cadmium, lead, zinc, and copper), $\alpha$-HCH, $\gamma$-HCH, dieldrin, DDT (including DDD and DDE), and PCBs by fish and shellfish in the Thames Estuary and Humber was studied by MAFF, with specific reference to sludge disposal.

In the Thames, concentrations of lead, zinc, and copper were no higher than those typical of coastal waters in general, but concentrations of cadmium were slightly elevated in some samples of brown shrimp and oysters. Mercury concentrations, generally high in the earlier years of the study, fell by 1977 to normal levels, as did pesticide and PCB residues in fish. These reductions seemed to be associated with a reduction in inputs of contaminants over the same period. A more detailed analysis of mercury levels in flounder did not reveal any tendency for body burdens to be higher in fish captured in the immediate vicinity of the disposal ground than in surrounding areas.

In and near the Humber disposal ground, elevated levels of zinc and copper were found in herring and some shellfish, with some indication that shellfish from the disposal ground itself had higher body burdens than those from the surrounding area. In some fish species, DDT concentrations were elevated, and a single sample of hermit crabs had elevated dieldrin levels. The contribution of sludge to these high concentrations is difficult to assess, since many other sources of contaminants are important in this area (Murray and Norton 1982).

Data from the period 1977 to 1984 showed, in general, a continued reduction in mercury and organics in fish from the Thames estuary, and a reduction in dieldrin levels in fish from the Humber (Franklin 1987).

In 1983, fish and prawns trawled at the Bell Rock (Firth of Forth) disposal site were analysed for metals and organohalogens. Concentrations of zinc and copper in sole were similar to those reported in fish from uncontaminated areas, whereas concentrations in prawns were slightly lower than those reported in hermit crabs from the same area in 1980. Cadmium and PCBs were not detected. DDT, DDE, and dieldrin were present at very low levels, but their analysis was perhaps subject to some interference (Lothian Regional Council 1983b).

### 3.3 Sludge Disposal and Fish Disease

A detailed study of epidermal anomalies in demersal fish was carried out by MAFF in 1980 as part of their programme of investigations into the biological effects of sludge disposal in the Thames Estuary. Of the abnormalities observed, fin-rot was the most frequent. About 8% of all the fish trawled had some external abnormality. In general, there was little evidence of a relation between the incidence of abnormalities and the sampling location, either within the Thames Estuary or when the Thames Estuary as a whole was compared with the control areas (Rye Bay and Southwold). However, fin-rot was slightly more frequent ($P < 0.025$) in cod, flounder, plaice and dab from the Thames Estuary than in those from the two control areas combined, although the authors did not consider that any firm conclusions could be drawn from the data. Similarly, there was some evidence that the incidence of ulcers, haemorrhages and pigment abnormalities was highest in the sampling area furthest into the estuary (Bucke et al. 1983).

The absence of marked spatial trends in this study, as in the study of body burdens in fish from the Thames Estuary (Murray and Norton 1982), may reflect the fact that fish, even those species considered to be rather sedentary, are wide-ranging, so that few or no inferences as to the amount of exposure to pollution can be made from the capture location. It should be noted, however, that when data from these studies were analysed, little or no allowance was made for the possibility that body burdens and disease incidence could be related to fish size or age, as they certainly are in reality.

A further study into the incidence of fish disease in the Thames estuary is currently being undertaken by the Water Research Centre.

A survey of fish off the German coast found that, whilst the incidence of fin-rot and ulcers in cod and dab from the German sewage sludge disposal ground was higher than that in a titanium dioxide disposal area and a control site (the Dogger Bank), this formed part of a general trend of falling disease incidence with distance offshore. There was no evidence that fish captured in the central sludge disposal site exhibited a higher incidence of abnormalities than those from the immediately surrounding area, and it was therefore not possible to state that sludge had an effect on disease (Dethlefsen 1980).

Another survey of fish disease in the North Sea performed by Möller (1981) failed to find evidence that fish abnormalities were more prevalent in the German Bight sludge disposal area than elsewhere.

### 3.4 Effects of Sludge on Plankton

Some preliminary results on the possible effects of sludge on plankton in the Thames Estuary have been published. Sampling for water quality and zooplankton was carried out at the sludge disposal ground (Barrow Deep) and in an adjacent channel (Middle Deep). In addition, laboratory experiments were performed to study the effects of sludge on copepods. In spring, chlorophyll *a* levels were higher in the Middle Deep than in the Barrow Deep, as were copepod densities, with the exception of *Calanus finmarchicus.* Addition of sewage sludge in the laboratory at concentrations of 1 to 100 in seawater led to mortality of the copepod *Acartia,* either by direct toxicity or by inhibition of feeding. At higher dilutions (1 in 1000 or 1 in 2000) mortality was lower and some copepods appeared to feed on sludge (Chapman 1985).

It was suggested that sludge disposal may initially cause some plankton mortality and settlement by trapping within sludge flocs and fibres immediately after the operation takes place. However, once the sludge has dispersed, the finer suspended material may provide a food source for some species of zooplankton. There is a possibility that these may accumulate metals, either directly or by consuming algae which have themselves accumulated dissolved metals. As stated above, this may provide a pathway for the incorporation of sludge-derived contaminants into the marine planktonic food chain (Chapman 1985, 1986).

## 4 Discussion

### 4.1 The Impact of Sludge Disposal in the North Sea

The majority of research into the biological effects of sludge disposal on the marine environment has concentrated on the macrobenthos. Analysis of data from macrobenthic surveys is aimed at the detection of spatial or temporal trends that are attributable to sludge; this is inherently difficult due to the considerable natural variability of the communities studied.

In the North Sea, a variety of effects have been detected (Table 2). This reflects the dependence of the effects of organic enrichment on the intensity of the input (Fig. 2). The fact that impact varies depending on the species association, for example in the Thames Estuary where abundance and species richness appeared to respond positively to organic carbon input in one association and negatively in others (Table 1), further complicates analysis. Only at accumulating sites, such as Garroch Head, can a clearly defined progression of effects be expected.

It is possible that interactions occur between the effects of organic enrichment and those of contaminants, the overall impact perhaps varying with sludge

**Table 2.** Summary of the effects of sewage sludge on macrobenthos in the North Sea

| Site | Date | Effect | Reference |
|---|---|---|---|
| Thames estuary | 1970 | Increase in numbers and diversity of polychaetes | Shelton (1971) |
| | 1972 | Increased diversity and faunal abundance | Talbot et al. (1982) |
| | 1977 | Impoverished fauna dominated by *N. latericeus* (2 stations) Increased density and/or diversity at mixed sediment sites? Organic carbon levels correlated with abundance/species richness (−ve or +ve in different associations) | Norton et al. (1981) |
| German Bight | 1970–1979 | Drastic change in species composition, *A. alba* abundant, no change in species number to 1977 Fall in abundance and species richness 1978–9 (part of normal fluctuations?) | Caspers (1980) |
| Humber estuary | 1975, 1977 | No effect detected | Murray et al. (1980) |
| Tyne | 1980–1983 | No effect detected | Pomfret and McHugh (1983) |
| | 1984 | Increase in abundance and species richness | Rees et al. (1985) |
| Forth | 1978–1979 | No effect detected | Moore (1981) |
| | 1981 | No effect detected | IOE (1983a) |
| | 1982 | Possible small change in species composition at St. Abbs | IOE (1983b) |
| | 1983 | Fall in diversity, some species more abundant, break in log-normal species abundance distribution (St. Abbs) | IOE (1985) |

composition and the physico-chemical characteristics of the disposal ground. For example, high faunal abundance and species richness at the Tyne disposal ground were ascribed to an effect of enrichment by a sludge which, being largely domestic rather than industrial in origin, contained low quantities of contaminants, and also to the recent history of disposal (Rees et al. 1985). Effects of disposal are easier to detect where the sediment is uniform over a wide area, as in the German Bight, rather than where it is variable and harbours a range of species associations, as in the Thames Estuary.

At present, it appears that the only North Sea disposal ground showing characteristics of an accumulating area, with drastic ecological changes directly attributable to the input of sewage sludge, is that in the German Bight, the use of which is now discontinued. For all other sites, further monitoring on a uniform basis will be required before any detrimental effect of disposal on the macrobenthos can be unambiguously identified.

Other than effects on the macrobenthos, some increaseaminant body burdens in fish and shellfish from the Thames and Humber disposal grounds have been observed, while no effects of sludge on the incidence of disease in fish have been positively identified. In addition, limited evidence suggests that sludge disposed of in the Thames may have adverse effects on plankton.

The concentration of research effort in the field on benthos rather than on plankton is favoured by a number of factors. First, disposal rarely takes place twice in the same body of water, so that cumulative effects on plankton are unlikely. Secondly, while natural variations in the benthos are considerable, variations in the plankton with respect to a fixed point such as a disposal ground are even greater (Norton and Rolfe 1978).

With regard to mobile animals such as fish, similar considerations apply: fish may not be exposed to the cumulative effects of any single contaminant source, and the capture location may not be typical of the area where the animal has spent most of its life. This may explain why most studies have failed to find small-scale relationships between sludge disposal and body burdens of contaminants or the incidence of diseases.

## 4.2 Research Strategies and Research Requirements

### 4.2.1 Plankton

As stated above, the study of the impact of sewage sludge on plankton faces certain difficulties, and is at present little advanced in the North Sea (Chapman 1985, 1986). Clearly this is a field which would repay further research, and in particular it is important to know the relative importance of planktonic and benthic food chains in the accumulation of contaminants.

### 4.2.2 Benthic Surveys

Macrobenthic work is an integral part of most monitoring programmes on sludge disposal grounds and it will probably continue. Whereas its usefulness can be questioned on the grounds that detection of effects is difficult and can take several years, and that the expense involved is often great, it is nevertheless true that no other currently employed technique examines the cumulative long-term effects of pollution on communities as a whole.

Many of the problems associated with benthic monitoring arise from a lack of clearly-defined objectives, both at the conceptual and design stages. For example, what constitutes undesirable change in the benthic community? The problem of objectives has been repeatedly pointed out in the past, but in practice is still frequently ignored.

To date, comparisons among surveys carried out at different times and by different organisations have been hampered by the use of different sampling sites, sampling methods, and sieve sizes. Furthermore, there can be great variability among different organisations in the accuracy of species identifications (Ellis and Cross 1981; Ellis 1985).

Smaller benthic organisms (meiobenthos) have not been studied on North Sea sludge disposal grounds, but were surveyed in Liverpool Bay, where some tentative relationships between nematode feeding types and the presence of sludge were identified (DoE 1972b). Trawling of mobile epibenthic organisms, which tend to be under-represented in samples collected with grabs, could also be used to examine effects of disposal.

### 4.2.3 Shorter-Term Effects on Marine Organisms

While macrobenthic surveys examine the long-term impact of sludge disposal at the community level, a number of other techniques are designed to investigate effects in the shorter term on individuals and populations. A general discussion of such techniques is outside the scope of the present chapter (but see, for example, Sheehan et al. (1984) for a recent review of pollutant effects). However, some physiological and cellular indices of fitness in mussels, *Mytilus edulis*, have recently been employed to assess contaminant-induced stress at the Thames Estuary ground (Whitelaw and Andrews 1987). These tests provide a means of assessing stress within communities that is much more rapid and easier to evaluate than benthic surveys, while at the same time being directly related to the field situation, unlike laboratory bioassays. The main limitations are that results may only be applicable to the populations tested (often not indigenous), and that biochemical or physiological stress may not translate directly into reductions in population size.

Studies of diseases and abnormalities, and of contaminant body burdens, can be difficult to relate to specific inputs, as previously discussed. However, current work being carried out on North Sea fish shows considerable improvement in design over early studies. An interesting approach is to study body burdens and disease incidence together in the same populations.

If a relation exists between sludge disposal and disease then this may be easier to detect in less mobile organisms than fish: crustaceans could be suitable for study as these have been shown in the laboratory to develop gill erosion when exposed to sediments from American sewage sludge disposal sites (Young and Pearce 1975).

Another potentially interesting approach to the study of the biological impact of sewage sludge is that of population genetics: it is well known that allele frequencies can respond to environmental change (Berry 1980) and such changes at polymorphic loci have been related to pollution (Nevo et al. 1978; Battaglia et al. 1980). Allele frequencies would be expected to differ in populations at clean sites and at sites affected by sludge if the latter are under some form of genetic "stress".

### 4.2.4 Laboratory and Experimental Work

The main problem with existing laboratory studies into the effects of sewage sludge on the survival, growth and development, and reproduction of marine organisms is the lack of knowledge about the relation between sludge concentrations and contaminant behaviour under experimental conditions and in the field.

Bioassay techniques can help to clarify this point. Test solutions of sewage sludge (Franklin 1983; Chapman 1985) can be compared with water samples collected in the wake of the sludge vessel, as in work in progress at WRc Environment. Bioassays using sediment collected near the disposal site (Whitelaw and Andrews 1987) are also currently employed.

A related approach is that of Eleftheriou et al. (1982) who observed the effect of experimental addition of sludge at different concentrations on benthic communities in the field. Further development of this method would provide a means of relating laboratory bioassays to the effects observed in the field.

## 5 Conclusions

Assessment of available data suggests that at present no consistent overall picture of the impact of sewage sludge in the North Sea, on which scientifically-based management decisions could be made, has emerged. The difficulty of identifying effects is associated with the dispersing nature of most North Sea disposal grounds. It can be said, however, that with the possible exception of the German Bight, no gross negative ecological effects have yet been detected, and there would appear to be no firm justification for advocating the cessation of sea disposal.

It is to be hoped that new methods of study will shortly improve our knowledge of the biological effects of sludge. The approaches which seem most promising are physiological, biochemical and genetic indices of stress, as well as experimental determination of the effects of sludge-contaminated sediments. At the same time, macrobenthic monitoring should continue, but only after careful consideration has been given to the objectives of such monitoring.

Monitoring at sludge disposal grounds has recently been reviewed by the Marine Pollution Monitoring Management Group in the UK. The conclusions and recommendations of this working group are about to be published (April 1987).

In the event of a decrease or increase in the rate of disposal being advocated, it is important to know the relative contribution of sludge to observed effects in an area subject to pollution from several sources. Studies of the biological impact of sludge should therefore proceed at the same time as studies of the behaviour of sludge in the sea (see Parker, this Vol.).

It appears that most of the negative effects of sewage sludge arise from the presence of contaminants and from excessive organic enrichment. A recent, unorthodox, suggestion is that with appropriate pre-treatment to reduce levels of trace metals and persistent organics, sludge could be applied to the sea in such a manner as to enhance productivity without the occurrence of adverse effects once an appropriate low application rate has been identified (Segar et al. 1985). A rather different viewpoint is that, if disposal at sea is practised, then adverse effects are acceptable as long as they are restricted in space, and that accumulating sites are actually preferable to dispersing sites where harmful substances, perhaps not initially recognised as dangerous, can spread over a wide area in an uncontrolled and uncontrollable manner (Mackay 1986).

## 6 Summary

The main input of sewage sludge into the North Sea is from disposal by ships, carried out by the United Kingdom.

Potential effects of sludge in the water column include stimulation of phytoplankton growth by nutrients, oxygen depletion caused by the degradation of organic matter, increased turbidity, and assimilation of contaminants by plankton. Effects on the seabed include an increased food supply for consumers, a fall in oxygen level, deposition of inert particulates as silt, and accumulation of contaminants by benthic organisms.

Macrobenthic surveys at the main North Sea disposal sites (Thames Estuary, German Bight, Humber Estuary, Tyne, Firth of Forth) suggest that the effects of sludge addition are varied: an increase in faunal density may or may not be accompanied by a rise or fall in diversity, while decreases in density and diversity can occur together. Only the German Bight disposal site, no longer in use, appears to act as an accumulating area with drastic changes in faunal composition.

Some studies have reported high contaminant body burdens in fish and shellfish from certain sites. No influence of sludge disposal on fish diseases has been positively identified. Work on the effects of sludge on plankton is only at a preliminary stage.

At present, there would appear to be no sound scientific basis for advocating the cessation of disposal. Opinion is divided as to the relative merits of dispersing and accumulating sites.

While macrobenthic surveys continue, other approaches to the study of the impact of sludge disposal show promise: these include physiological, biochemical, and genetic indices of pollution-induced stress, and the experimental determination of the effects of sludge-contaminated sediments.

## References


Ayres PA (1977) The use of faecal bacteria as a tracer for sewage sludge disposal in the sea. Mar Pollut Bull 8:283–286

Battaglia B, Bisol PM, Rodino E (1980) Experimental studies on some genetic effects of marine pollution. Helgol Meeresunters 33:587–595

Berry RJ (1980) Genes, pollution, and monitoring. Rapp P V Réun Cons Int Explor Mer 179:253–257

Bucke D, Norton MG, Rolfe MS (1983) The field assessment of effects of dumping wastes at sea: 11 Epidermal lesions and abnormalities of fish in the outer Thames Estuary. Fish Res Tech Rep MAFF Direct Fish Res, Lowestoft 72:16

Caspers H (1980) Long-term changes in benthic fauna resulting from sewage sludge dumping into the North Sea. Prog Wat Tech Toronto 12:461–479

Chapman DV (1985) Preliminary observations on the interaction between plankton and sewage sludge dumped at sea. ICES CM 1985/ E:26

Chapman DV (1986) The distribution of metals in sewage sludge and their fate after dumping at sea. Sci Total Environ 48:1–11

Collinge VK, Bruce AM (1981) Sewage sludge disposal: a strategic review and assessment of research needs. WRc Tech Rep TR 166:31

Department of the Environment (1972a) Out of sight, out of mind. Report of a working party on sludge disposal in Liverpool Bay Volume 1: main report. HMSO Lond, 36p

Department of the Environment (1972b) Out of sight, out of mind. Report of a working party on sludge disposal in Liverpool Bay Volume 2: Appendices. HMSO Lond, 486p
Dethlefsen V (1980) Observations on fish diseases in the German Bight and their possible relation to pollution. Rapp P V Réun Cons Int Explor Mer 179:110–117
Dethlefsen V (1986) Marine pollution mismanagement: towards the precautionary concept. Mar Pollut Bull 17:54–57
Devine MF, Norton MG, Champ MA (1986) Estimating particulate dispersiveness and accumulation at nearshore ocean dumpsites. Mar Pollut Bull 17:447–452
Eleftheriou A, Moore DC, Basford DJ, Robertson MR (1982) Underwater experiments on the effects of sewage sludge on a marine ecosystem. Neth J Sea Res 16:465–473
Ellis DV (1985) Taxonomic sufficiency in pollution assessment. Mar Pollut Bull 16:459
Ellis DV, Cross SF (1981) A protocol for inter-laboratory calibrations of biological species identifications (Ring tests). Water Res 15:1107–1108
Franklin A (1987) The concentration of metals, organochlorine pesticide and PCB residues in marine fish and shellfish: results from MAFF fish and shellfish monitoring programmes, 1977–1984. Aquat Environ Monit Rep MAFF Direct Fish Res, Lowestoft 16:38
Franklin FL (1983) Laboratory tests as a basis for the control of sewage sludge dumping at sea. Mar Pollut Bull 14:217–223
Gould DJ (1976) Ecological effects of sewage discharges to the sea: an assessment of research needs. Water Res Cent Tech Rep TR 26:34
Gray JS, Mirza FB (1979) A possible method for the detection of pollution-induced disturbance on marine benthic communities. Mar Pollut Bull 10:142–146
Halcrow W, Mackay DW, Thornton I (1973) The distribution of trace metals and fauna in the Firth of Clyde in relation to the disposal of sewage sludge. J Mar Biol Ass UK 53:721–739
Head PC (1980) The environmental impact of the disposal of sewage sludge in Liverpool Bay. Prog Wat Tech Brighton 13:27–38
Institute of Offshore Engineering (1983a) Biological monitoring of the St Abbs and Bell Rock sewage sludge dumping grounds, Autumn 1981 surveys, 77 p
Institute of Offshore Engineering (1983b) Biological monitoring of the St Abbs and Bell Rock sewage sludge dumping grounds, May 1982 survey, 85p
Institute of Offshore Engineering (1985) Biological monitoring of the St Abbs and Bell Rock sewage sludge dumping grounds, June 1983 survey, 57p
Lothian Regional Council (1983a) Chemical monitoring of the St Abbs and Bell Rock sewage sludge dumping grounds Autumn 1981 surveys, 38p
Lothian Regional Council (1983b) Chemical monitoring of the St Abbs and Bell Rock sewage sludge dumping grounds 1983 surveys, 27p
Lothian Regional Council (1984) Chemical monitoring of the St Abbs and Bell Rock sewage sludge dumping grounds 1982 surveys, 34p
Mackay DW (1986) Sludge dumping in the Firth of Clyde – a containment site. Mar Pollut Bull 17:91–95
McIntyre AD (1981) Effects on the ecosystem of sewage sludge disposal by dumping from ships. Wat Sci Tech Lond 14:137–143
McIntyre AD, Johnston R (1975) Effects of nutrient enrichment from sewage in the sea. In: Gameson ALH (ed) Discharge of sewage from sea outfalls. Pergamon Press, Oxford, pp 131–141
Möller H (1981) Fish diseases in German and Danish coastal waters in summer 1980. Meeresforsch 29:1–16
Moore DC (1981) Studies of environmental effects of sewage sludge dumping off the Firth of Forth. ICES CM 1981/E:42
Murray LA, Norton MG (1982) The field assessments of effects of dumping wastes at sea: 10 Analysis of chemical residues in fish and shellfish from selected coastal regions around England and Wales. Fish Res Tech Rep MAFF Direct Fish Res, Lowestoft 69:42
Murray LA, Norton MG, Nunny RS, Rolfe MS (1980) The field assessment of effects of dumping wastes at sea: 6 The disposal of sewage sludge and industrial waste off the River Humber. Fish Res Tech Rep MAFF Direct Fish Res, Lowestoft 55:35
Nevo E, Shimony T, Libni M (1978) Pollution selection of allozyme polymorphism in barnacles. Experientia 34:1562–1564

Norton MG, Rolfe MS (1978) The field assessment of effects of dumping wastes at sea: 1 An introduction. Fish Res Tech Rep MAFF Direct Fish Res, Lowestoft 45:9

Norton MG, Eagle RA, Nunny RS, Rolfe MS, Hardiman PA, Hampson BL (1981) The field assessment of effects of dumping wastes at sea: 8 Sewage sludge dumping in the outer Thames Estuary. Fish Res Tech Rep MAFF Direct Fish Res, Lowestoft 62:62

O'Sullivan AJ (1971) Ecological effects of sewage discharge in the marine environment. Proc Roy Soc Lond B 177:331–351

Pearson TH, Rosenberg R (1978). Macrobenthic succession in relation to organic enrichment and pollution of the marine environment. Oceanogr Mar Biol Ann Rev 16:229–311

Pomfret JR, McHugh P (1983) Monitoring the effects of sewage sludge disposal in a designated area of the North Sea. The situation to April 1983. Northumbrian Water Rep 7, 10 p

Rachor E (1980) The inner German Bight – an ecologically sensitive area as indicated by the bottom fauna. Helgol Meeresunters 33:522–530

Rees HL, Rowlatt S, West PA, Shakespeare N, Limpenny D, Parker MM (1985) Benthic studies at an offshore sewage sludge disposal site. ICES CM 1985/E:27

Rowe GT, Smith KL, Clifford CH (1976) Benthic-pelagic coupling in the New York Bight. In: Gross MG (ed) Middle Atlantic continental shelf and the New York Bight. Am Soc Limnol Oceanogr Spec Symp 2, pp 370–375

Segar DA, Stamman E, Davis PG (1985) Beneficial use of municipal sludge in the ocean. Mar Pollut Bull 16:186–191

Sheehan PJ, Miller DR, Butler GC, Bourdeau P (1984) Effects of pollutants at the ecosystem level SCOPE 22. Wiley and Sons, Chichester, 443p

Shelton RGJ (1971) Sludge dumping in the Thames estuary. Mar Pollut Bull 2:24–27

Talbot JW, Harvey BR, Eagle RA, Rolfe MS (1982) The field assessment of effects of dumping wastes at sea: 9 Dispersal and effects on benthos of sewage sludge dumped in the Thames Estuary. Fish Res Tech Rep MAFF Direct Fish Res, Lowestoft 63:42

Whitelaw K, Andrews MJ (1987) The effects of sewage sludge disposal on the outer Thames Estuary. Presented at the International Conference on Environmental Protection of the North Sea, 24–27 March, 1987, Lond, UK

Young JS, Pearce JB (1975) Shell disease in crabs and lobsters from the New York Bight. Mar Pollut Bull 6:101–105

# Impact of Contaminants Mobilized from Sediment Upon Disposal

J.M. MARQUENIE[1] and L. TENT[2]

## 1 Introduction

Contaminated sediments are often regarded as a nuisance, although not much is known about the effects of the contaminants on organisms exposed to them. High concentrations of contaminants in sediments are a general cause of concern. The problem becomes acute when large amounts of contaminated dredgings have to be disposed of. Harbor authorities worldwide are faced with the conflict between environmental concern on one hand and economical, feasible solutions on the other, a conflict that raises such questions as why and how sediments should be disposed of, what uses they can be put to, and how they should be treated (Tent 1984). Strangely enough we are far from answering the crucial questions in this context: whether the ecosystems living on top of contaminated sediments are "good" or "bad", and whether disposal of contaminated sediments poses a threat to ecosystems elsewhere. Hardly any research has been done to answer these questions. This creates the incongruous situation where great efforts are being made towards solving the problem of transporting, confining and treating contaminated sediments when a sound ecotoxicological basis is lacking. This chapter summarizes and evaluates research attempting the basic question, whether, and if so, why, contaminated sediment poses a threat to the environment.

## 2 Uptake of Contaminants by Benthic Invertebrates

There have been a great deal of studies on increased levels of contaminants in marine invertebrates living in or feeding on sediments. They usually aim to assess gradients in contamination within the same hydrographic system or to compare different areas (comparative surveys). The two types of study share the same disadvantages when it comes to evaluating the role of sediments as a direct source of contaminants to organisms. These contaminants may be taken up via different routes from the main abiotic compartments, via water, suspended matter, sediment particles, and interstitial water. All these compartments are interdependent as

[1]Ministry of Transport and Public Works, Tidal Waters Division, P.O. Box 20904, NL-2500 EX The Hague, The Netherlands

[2]Behörde für Wirtschaft, Verkehr und Landwirtschaft, Strom- und Hafenbau, Dalmannstrasse 1–3, D-2000 Hamburg 11, FRG

regards contamination, with the result that pollution gradients run parallel in all of them. It should therefore be borne in mind that an organism can pick up a contaminant from any of these different compartments or from any combination of them.

The complex situation we are dealing with is illustrated in Fig. 1, which shows that the concentration of cadmium in a benthic organism such as an alga (*Fucus*) attached to a rocky substrate and in contact only with surface water follows the same gradient as does that in sediment and deposit feeders (*Nereis*, *Macoma*).

For purposes of evaluation, the studies reported in the literature can be categorized as follows:

1. Studies on concentrations in tissues of benthic invertebrates only.
2. Studies on concentrations in tissues and one or more abiotic compartments.
3. Correlation studies relating concentrations in tissues to concentrations in one or more abiotic compartments.
4. Modified correlation studies of specific physicochemical forms of contaminants.
5. Experimental studies under controlled conditions.

Although there are a great many studies in the first two categories (1 and 2), they are only of minor importance as regards attempts to relate contaminants in sediments with those in organisms. Their value lies in providing background concentrations, which may serve as references, as well as the maximum concentrations that some organisms are apparently able to survive in the field. Studies dealing with correlations (categories 3 and 4) give or attempt to give estimates of qualitative and quantitative relationships, although they share the above-mentioned disadvantage of failing to provide direct proof of a causal relationship between a specific abiotic compartment and contaminant concentrations in the organism. The last type of study (5) concerns both field and laboratory studies affording conclusive evidence for the direct transfer of contaminants from the sediment interstitial water complex to the organisms it harbors.

## 3 Correlations and Regressions

Many authors have attempted to relate concentrations of contaminants in sediments to concentrations of contaminants in organisms. These attempts have been successful for data from a limited area. A good example is given by Ray et al. (1979), who calculated regressions between the concentration factor (ratio C-animal/C-sediment) and concentrations in sediments on log-transformed data (Table 1).

Although most correlation coefficients are significant at a level of 5%, the number of samples is rather small. In a later study, Ray and McLeese (1980) had to conclude that: "The analysis of bulk sediment provides only limited indications of its potential to be a source of chlorinated hydrocarbons and heavy metals to benthic fauna". Other workers came to the same conclusion. They believe that within sediments, only specific chemical forms of contaminants are bioavailable. For this reason, the sediments were subjected to more moderate extraction. Luoma

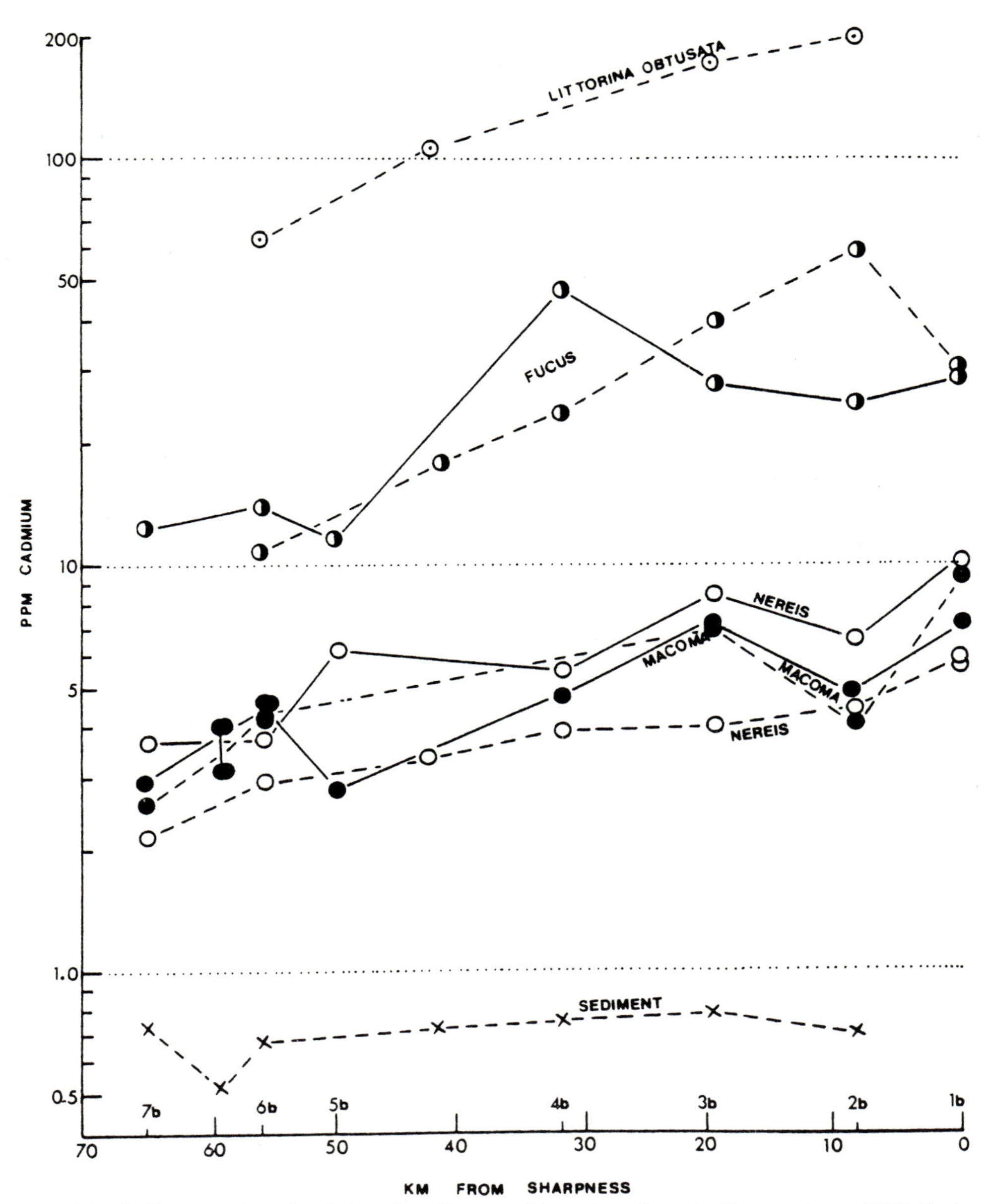


**Fig. 1.** Concentration of cadmium in sediment and organisms from the Severn estuary (UK). *Broken line* is Nov. 1976 and *continuous line* April 1977. Multiple points show that replicates of different-sized organisms were analyzed. (After Bryan et al. 1980)

**Table 1.** Relationships between log concentration factor of metal in animals (log Y) and log metal concentration in sediment (log X) for *Crangon* and *Macoma*. (After Ray et al. 1979)

| | | | |
|---|---|---|---|
| *Copper* | | | |
| | Crangon: | log Y = 0.912–0.999 log X | (r = –0.995[a], n = 5) |
| | Macoma: | log Y = 0.645–1.018 log X | (r = –0.981[a], n = 5) |
| *Zinc* | | | |
| | Crangon: | log Y = 1.301–0.865 log X | (r = –0.754 , n = 5) |
| | Macoma: | log Y = 1.635–0.883 log X | (r = –0.878[a], n = 5) |
| *Lead* | | | |
| | Crangon: | log Y = 0.595–1.107 log X | (r = –0.974[a], n = 5) |
| | Macoma: | log Y = 0.337–0.542 log X | (r = –0.300 , n = 5) |
| *Cadmium* | | | |
| | Crangon: | log Y = –0.282–1.004 log X | (r = –0.993[a], n = 5) |
| | Macoma: | log Y = –0.601–0.985 log X | (r = –0.957[a], n = 5) |

[a] Significant at $p = 0.05$.

and Jenne (1976) were the first to devise an empirical chemical extraction scheme for determining the bioavailability to *Macoma balthica* of Ag, Cd, Co, and Zn sorbed to sediments. They concluded that Ag was bioavailable to the organism in all types of sediments studied. The best estimates of bioavailable fractions of the other metals were, however, related to specific extractions:

- Cd and Co extracted with 70% ethanol or 1 N ammonium acetate.
- Zn extracted with 1 N ammonium acetate or with 1 N NaOH plus EDTA.

Prompted by the somewhat improved correlations, and eager to find relations of predictive value, investigators have focused further research on complexating substances in the sediments. They have shown, for example, that the biological availability of sediment-bound Pb and As to the deposit-feeding bivalve *Scrobicularia plana* is strongly influenced by the level of readily extractable Fe in the sediment.

Luoma and Bryan (1978) used the Pb/Fe ratio in sediments to predict Pb concentrations in *S. plana.* They collected sediment samples from the oxidized surface layer of intertidal sediments, and wet-sieved these through 0.1 mm polythene mesh using diluted seawater. Subsamples of 1 g of air-dried sediment were extracted for 2 h with 10 ml of 1 N HCl. Correlation coefficients for Pb-animal/Pb-sediment were found to be low ($r = 0.69$). Much better coefficients were found when the Pb concentrations in tissues were correlated with the Pb/Fe ratios in sediments. The log-transformed data afforded slightly more accurate predictions of Pb concentrations in *S. plana* at sites with very low Pb/Fe ratios (Fig. 2). Concentrations of As in *S. plana* also seem to be influenced by Fe concentrations in the sediment. This was shown by Langston (1980), who followed the same approach. The author determined As in the alga *Fucus* and the polychaete worm *Nereis*, as well as in *S. plana*. Correlations between concentrations of As in these tissues and As/Fe ratios in sediments were, however, poor (*Fucus* $r^2 = 0.62$; *Nereis*

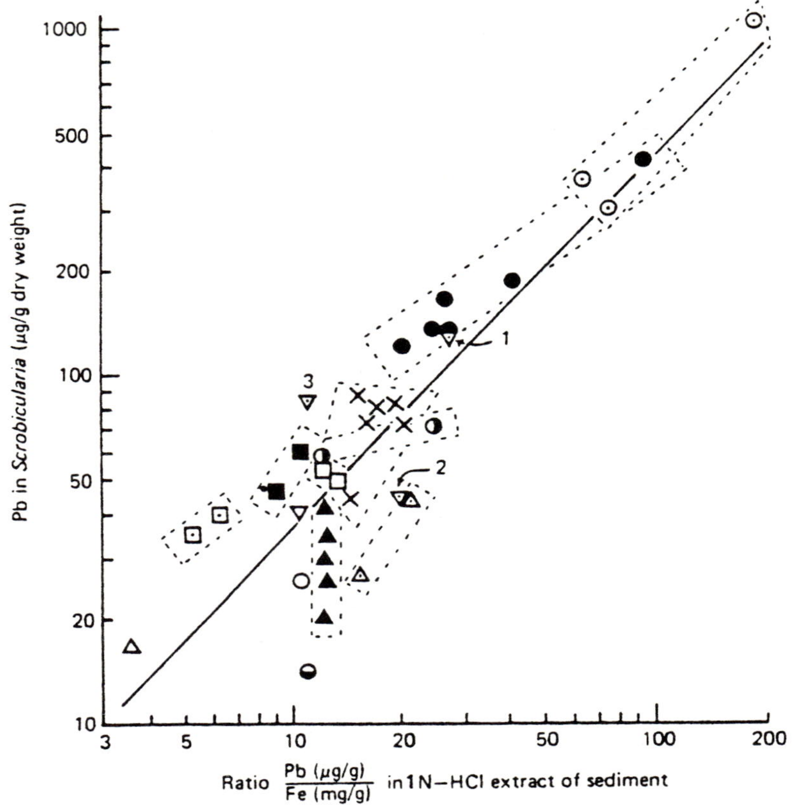


**Fig. 2.** Correlation between concentration of lead in soft tissues of *Scrobicularia plana* and the ratio Pb/Fe extracted with 1N hydrochloric acid. Symbols stand for different sites within the same estuary. The equation for results on logarithmic scales is: log 10 Pb-animal = 1.071 log 10 Pb/Fe + 0.496, n = 37, r = 0.88. (After Luoma and Bryan 1978)

$r^2 = 0.55$) compared to those in *S. plana* ($r^2 = 0.82$). Removing Fe from these equations did not improve the correlation coefficients.

Two remarks are in order at this point. First, attempts to correlate trace metal concentrations in *Fucus* with those in sediments seem to be of dubious value, because of the natural association of this alga with rocky substrates. Secondly, correlations for the polychaete *Nereis* might be low because of its ability to regulate metals, which has been recognized for iron (Jennings and Fowler 1980) and zinc (Bryan 1976; Bryan et al. 1980).

Factors influencing the bioavailability of zinc in both British and U.S. estuaries have been described by Luoma and Bryan (1979). These authors had initially concluded that a correlation existed over a wide range of concentrations between Zn concentrations in *S. plana* and sediment concentrations of Zn extractable with ammonium acetate. This correlation was, however, not precise enough ($r = 0.62$) to be of predictive value for *S. plana*, and did not apply to *Macoma balthica*. Further attempts at correlation indicated that Zn concentrations in *M. balthica* from San Francisco Bay were strongly related to the product:

[hydroxylamine soluble iron] / [ammonium acetate soluble manganese] × 1 / [organic carbon]

The term for organic carbon can be neglected if it is 1% or less. For Zn concentrations in *S. plana* from Britain, it was found that log-transformed concentrations in *S. plana* were related to the product:

[ammonium acetate soluble zinc] × [oxalate soluble manganese] × [humic substances (absorbance in 1 N ammonia)] / [total organic carbon].

Langston (1982) found that the bioavailability of mercury from sediments to these two species of deposit-feeders is strongly related to total organic carbon. He derived the following empirical relations (concentrations in mg/kg):

[Hg-organism] = a + b [Hg-sediment] / [% organic matter]
*S. plana* a = 0.22, b = 1.9
*M. balthica* a = 0.34, b = 2.7

In summary, it was shown that increasing concentrations of Fe led to increasing availability of Zn to *M. balthica*, whereas Fe in sediments showed no detectable effect in *S. plana* for this metal by the multiple regression technique applied. For Pb and As, however, increased Fe decreased bioavailability to *S. plana*. Increasing concentrations of Mn led to a decreasing availability of Zn to *M. balthica* (USA) and increased availability of Zn to *S. plana* (UK). The only material that has the same effect on all organisms in all situations is organic matter, which presence always coincides with a reduction in bioavailability.

Typical of the confusing empirical relationships derived so far is that for Zn and *Scrobicularia plana*:

$$\log \text{Zn-}S.\ plana = 0.03 + 0.30(\log H) - 0.59(\log C) + 0.19(\log \text{Zn-AmAc}) + 1.22(\log \text{Zn-solute}) + 0.17(\log \text{Mn-oxal}) - 0.08(\log CO_3)$$

Where H = concentration of humic substances
C = total organic carbon

It is to be noticed that stronger correlations are found between the concentrations in organisms and in sediments when the major complexing components like iron, manganese, humic substances and total organic carbon of the sediments are taken into account. Remarkably, the effects can be very different for different species of organisms and sediments from different areas. Clearly, in retrospect, factors can be found that may explain differences in accumulated levels of contaminants in certain organisms. However, the relations are too weak and too little understood to be of predictive value.

## 4 Laboratory and Semi-Field Assessments of Bioavailability

Since data from field surveys fail to yield a clear understanding of the processes involved, or to yield criteria based on sediment contamination that can be used in the handling of dredged material, another approach was followed in later years. In this approach, the emphasis was on direct assessment of bioavailability of contaminants in sediments through the controlled exposure of organisms. The techniques employed, e.g., bioassays, rapidly became very popular. As a result, research was no longer aimed at revealing the mechanisms of uptake, but at the development of techniques and their application. Unexpectedly, however, the development and application of bioassays led to a tremendous gain in information.

First, it was shown that the deposit-feeding bivalves *Scrobicularia plana* and *Macoma balthica* were very capable of separately accumulating metals both from sediment and surface water (Marquenie et al. 1983). The authors describe a series of experiments involving field exposure of marked individuals that had been collected from an unpolluted estuary. The animals were taken to a contaminated estuary, and recollected after various periods of time. The authors concluded that, for active biomonitoring purposes, exposure periods of 40 to 60 days suffice. In a next stage, they studied the influence of sediment particle size on the accumulation of metals dosed to the surface water in a flow-through system. The accumulation rates proved to be higher when the animals were kept in gravel rather than fine muds, leading to mortality for *S. plana* in 20 days (Table 2). Furthermore, the animals in the mud rather than the gravel were still accumulating metals after 100 days. These results are in agreement with those of Pesch (1979), who studied the influence of three types of sediment on the toxicity of copper to the polychaete *Neanthes arenaceodentata* (Fig. 3).

Both Pesch's experiments and our own clearly showed that accumulation of contaminants in deposit-feeding organisms may ensue when surface water is contaminated, as is often the case in polluted estuaries. They also showed that uncontaminated sediment may provide organisms with a certain degree of protection when contaminated wastes are disposed of, e.g., at sea. This protection is probably due to adsorption of contaminants to sediment complexation sites. Since

**Table 2.** Concentrations of Cd, Cu and Zn in soft tissues of *M. balthica* and *S. plana* kept in sediments with different grain size classes after 100 days of exposure to overlying water concentrations of 5 (Cd), 20 (Cu) and 100 (Zn) $\mu g\ l^{-1}$. Concentrations in $\mu g\ g^{-1}$ ash-free dry weight

| | *M. balthica* | | | *S. plana* | | |
|---|---|---|---|---|---|---|
| Grain size | Cd | Cu | Zn | Cd | Cu | Zn |
| Gravel | 3.7 | 59 | 1180 | - died within 20 days - | | |
| 0.1% < 16 μm | 4.3 | 52 | 1040 | 13.2 | 209 | 1910 |
| 8 % < 16 μm | 3.1 | 105 | 950 | 6.5 | 75 | 1270 |
| 41 % < 16 μm | 1.7 | 42 | 660 | 4.5 | 52 | 1030 |
| control | | | | | | |
| 41 % < 16 μm | 0.3 | 28 | 990 | 0.8 | 42 | 1130 |

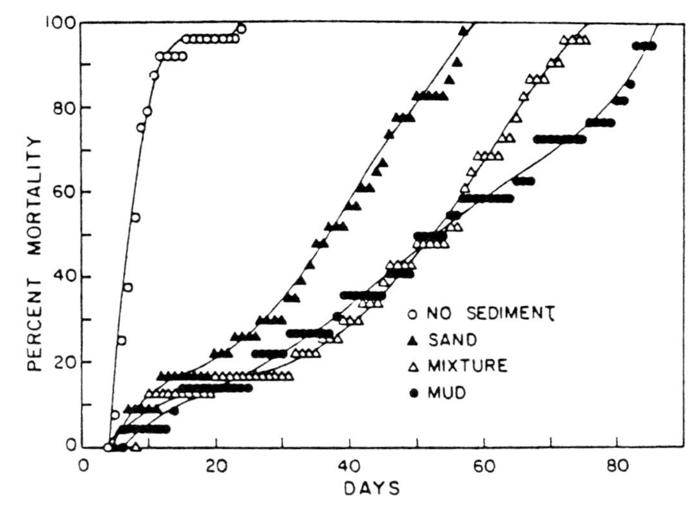


**Fig. 3.** *Neanthes arenaceodentata*. Cumulative percent mortality of adult males exposed to 0.10 mg Cu $l^{-1}$ in seawater in the presence of sand, mud, a mixture of sand and mud, and no sediment. (After Pesch 1979)

coarse sediments are more readily and more rapidly saturated, any adverse effects of waste disposal in areas containing such sediments will show up sooner than in areas with fine muds.

In further experiments accumulation from sediments was tested. Contaminated sediments were moved into clean aquatic environments. In a first experiment (Marquenie et al. 1983), sediments were collected from three sites aligned in a pollution gradient, and *M. balthica* and *S. plana* were exposed to them for 40 days in open containers in an unpolluted estuary. The exposure resulted in a significant increase in contaminant levels, reflecting the gradient situation in the polluted estuary (Fig. 4).

In a second experiment, Marquenie et al. (1985a) tested 24 very different sediments in this way with *M. balthica*. The results showed that exposure to contaminated sediments in natural clean water can lead to accumulation of contaminants as well as mortality.

The last experiment also revealed a very new aspect. The sediments tested contained a wide range of Zn concentrations. In spite of this, the metal was not accumulated by *M. balthica*. It follows that sediment-bound zinc is not bioavailable to this species. In earlier experiments, however, it was shown that this species is capable of accumulating Zn from surface waters. The combined results of these experiments under more or less controlled conditions provide a ready explanation why Luoma and Bryan (1979) failed to relate Zn concentrations in *M. balthica* collected from the field to Zn concentrations in the sediments they were living in.

A bioassay with sediments collected from silts in the Western Scheldt and the port of Rotterdam (Marquenie et al. 1985b) gave indications that cadmium is another metal accumulated from water by *M. balthica*.

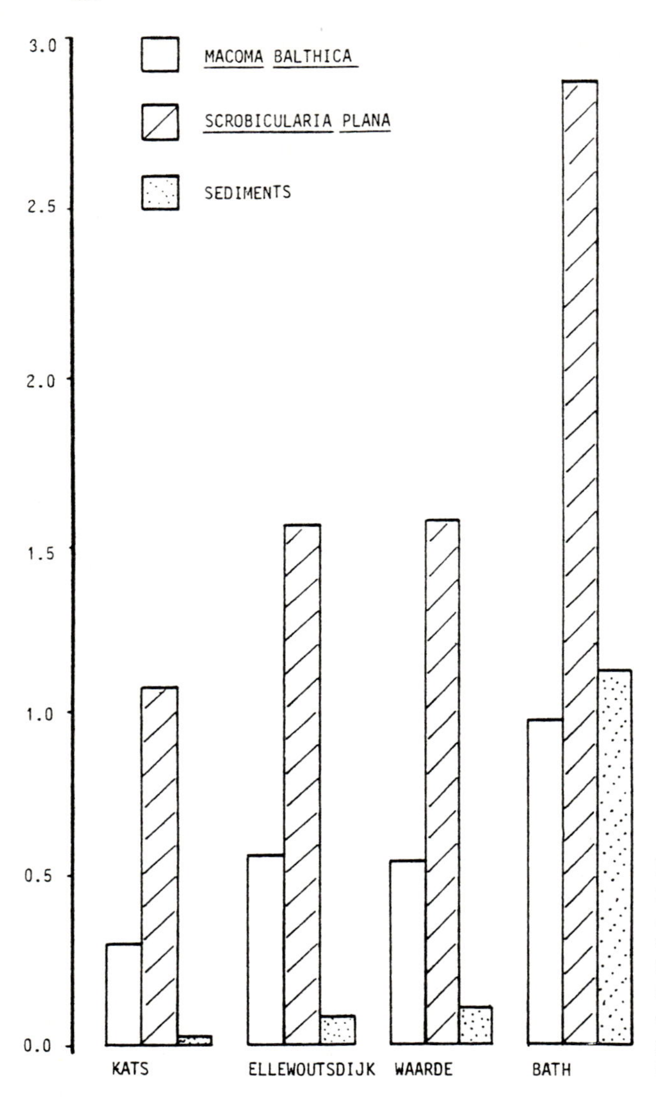


**Fig. 4.** Accumulated levels of cadmium ($\mu g\ g^{-1}$ ash-free dry weight) in *M. balthica* and *S. plana* exposed for 40 days to contaminated sediments (Ellewoutsdijk, Waarde and Bath) in the Eastern Scheldt (Kats). Concentrations in sediments in $0.1 \times \mu g\ g^{-1}$ (dry weight) (After Marquenie et al. 1983)

It was performed in tanks of 2.5 $m^3$, continuously supplied with filtered sea-water at predetermined exchange rates. The tanks were stocked with *Mytilus edulis* (exposed to the surface water only), *Macoma balthica* and *Arenicola marina* (allowed to burrow in the sediment).

It was shown that, after 60 days, two species of molluscs in the tank with sediment from the most eastern silt had accumulated significant amounts of Cd at a water exchange rate of 2% per h. In a tank with the same sediment, but with a ten times higher water exchange rate, accumulation was still noticeable, but strongly decreased in both the filter-feeding and the deposit-feeding bivalves. All these observations indicate that the deposit-feeder *M. balthica* accumulates metals from the water and not from sediment particles. They probably reflect bioavailable metal concentrations in the water at the sediment water boundary layer.

Organic contaminants appear to behave quite differently from metals. Duinker et al. (1983) collected sediments and organisms from the Dutch Wadden Sea and analyzed them for organochlorines, including individual PCB congeners. Because the patterns of pentachloro- and higher chlorinated biphenyls were equal in sediments and in *Macoma balthica* and *Arenicola marina*, they concluded, these congeners not being found in solution, that accumulation occurred from sediment and food only. This conclusion is supported by observations by Marquenie et al. (1985b), who found that an increase in the water exchange rate did not result in a decrease in accumulation of PCB congeners in *M. balthica, M. edulis* or in *A. marina*.

It follows that, when one compares PCB concentrations in sediments with those in organisms, the results may be more meaningful for PCB's than for metals. Controlled bioassays have shown that increasing concentrations of organic matter reduce uptake of PCB's by organisms (Rubinstein et al. 1983). However, this finding has never been confirmed in the field, so the predictive value of this relationship is still in the balance.

It should be remarked at this point that "organic matter" is generally defined as weight loss on ignition at 600°C. It covers a variety of substances differing widely in adsorptive capacity, degradability, etc. For the time being, therefore, it seems to make little sense to try to predict accumulation of contaminants by organisms from contaminant concentrations in sediments corrected for content of compounds of uncertain behavior.

## 5 Ecotoxicological Considerations

The lugworm, *Arenicola marina*, was more or less abandoned as far as contaminant studies are concerned when it was shown that polychaete are poor indicators of metal pollution. Marquenie et al. (1985b) included it in their experiments as a tool for measuring the effect of bioturbation on uptake of contaminants in *M. edulis*. Surprisingly, the worm has found to be the strongest accumulator of PCB's, PAH's and pesticides of all benthic invertebrates tested up to that time. Since it is also a basic food for many birds and fish, it may prove to be of great importance in future. The experiments also revealed an unexpected relation between effects of toxic contaminants and eutrophication. It was found, for instance, that the effect of bioturbation on contaminant uptake was negligible whereas its accelerating effect on leaching of nutrients was tremendous. Since most heavily contaminated sediments also contained large amounts of nutrients, the result was an increased concentration of nutrients in the water. It was found, in addition, that the weight increase of the three species used in the experiment (*A. marina, M. edulis, M. balthica*) was greatest in those tanks which contained the most heavily contaminated sediments. We therefore speculated that, although the contaminants may have had a negative effect on the organisms, this effect may have been obscured by the glut of nutrients and "organic matter", and the algae and bacteria thriving on them. In view of this interaction between potentially harmful contaminants and food substances, the negative effects of the disposal of contaminated sediments on benthic communities manifest themselves mainly in a reduction of species diversity rather than of biomass production. It has also been shown that accumulation of contaminants in such basic food as just described may cause serious harm to their predators. Examples of such ecological disasters are given by several authors. Koeman and Van Genderen (1972) describe mortality of *Sterna sandvicensis* and *Somateria mollissima* in the western Wadden Sea that could be traced to the emission of telodrin in the port of Rotterdam; Risebrough et al. (1970) describe the decline of *Pelecanus occidentalis* due to DDT contamination. Another example is given for *Sterna hirundo* by Nisbet and Reynolds (1983). Mammals are affected as well as birds. The decline of the Dutch seal population in the western Wadden Sea is attributed to their uptake of organochlorines (polychlorinated biphenyls), as reported by Reinders (1980). Strikingly, all these disasters to wildlife occurred in areas of high biological production, apparently without the food organisms themselves being affected. Marquenie et al. (1986) have presented further evidence for the interaction between, on one hand, effects on predators and, on the other, contamination of food organisms with excess nutrients, e.g., nitrates and phosphates and with toxic compounds. They describe a bioassay with two small populations of diving ducks, *Aythya fuligula*, fed freshwater mussels (*Dreissena polymorpha*) for 3 successive years. One group of ducks was fed mussels collected from the polluted Haringvliet Basin, and a reference group mussels from Markermeer. The mussels from the former, though contaminated, were more numerous and in better condition. The authors reported that the contaminants in these mussels were strongly accumulated in the tissues of the ducks and their eggs, with reproduction failure resulting from loss of nest attentiveness, reduced clutch size and weight of eggs, and embryo mortality. The suspected contaminants were organochlorines (PCB's), probably also responsible for the decline of seals in the nearby Wadden Sea, and to a lesser extent DDE and mercury. Furthermore, although the mussels were contaminated with PAH's, heavy metals and a wide range of PCB congeners, only specific PCB congeners and the metals were found to accumulate in the ducks' tissues and only these congeners and mercury were passed on to the eggs. The observations presented led us to believe that the most dangerous situations are those where basic food organisms thrive and at the same time accumulate contaminants above levels critical to their predators. A further implication is that mortality of test organisms is a much less valuable criterion than is often assumed. Although much attention is still being devoted to the effects of heavy metals, almost all ecological disruptions reported so far are related to organic or organometal pollutants.

There is no evidence for any serious effects of the disposal of contaminated sediments in open waters. However, many dumping grounds, particularly those in Europe, are situated in high-energy environments, the result being that contaminants become widely dispersed, ultimately finding their way to sedimentation areas and feeding grounds for birds and seals. The previous section gives several examples of detrimental effects in such areas from the organic contaminants. In earlier sections it was shown that of organisms at the bottom of food chains, when exposed to contaminated sediments, can themselves become severally contaminated, the implication being that uncontrolled disposal is potentially harmful. The harm that can be done cannot be estimated from the concentrations in the sediment, but only from bioassays under realistic conditions. Furthermore, criteria for concentrations of contaminants in sediments cannot be formulated, but should be developed and evaluated for concentrations of contaminants in organisms at the bottom of food chains.

## 6 Concluding Remarks

The first line of research, not dealt with in this chapter, consisted of simple laboratory toxicity tests, which afforded criteria for the formulation of maximum acceptable concentrations (MAC) in many countries. A next line of research, sometimes involving large-scale surveys and sometimes experiments on a semi-field scale, revealed that some communities of organisms were affected by concentrations falling below accepted MAC's, whereas other communities in sometimes natural unpolluted areas were found to be exposed to concentrations in excess of MAC's without suffering harm. This finding stimulated research on what is called, "bioavailability". The problem was studied where it was discovered, namely in the field. These studies were doomed to failure because of conceptual shortcomings. The complexity of the subject necessitated the development of techniques for estimating potentially available contaminants in a direct and realistic way. Application and further development of these techniques (bioassays) gradually revealed why the field approach was bound to fail. However, they also showed that accumulation of contaminants by organisms can often be explained afterwards, but cannot be predicted beforehand, by chemical analysis of sediments.

This means that criteria for the protection of wildlife cannot, for scientific reasons, be based on concentrations in water or in sediments. Such criteria can only be based on concentrations in organisms at the beginning of food chains.

These were shown to lie at a strategic level from which effects on predators can be predicted with more certainty. The tools for assessing potential accumulation have recently become available as bioassays. If properly designed by cooperation between engineers, biologists, and geochemists, the results will reveal the factors responsible for contaminant biomobility. In deciding environmental management strategies, these factor should be taken into account in addition to concentrations of contaminants in the environment. The need for further studies is obvious. Only in this study was it discovered why correlation and regression studies had to fail. With what we now understand, they may in future succeed. Not much is known about long-term processes determining bioavailability. The effects of bioturbation and the composition, characteristics, and fate of organic matter should therefore be studied in more detail. And, last but not least, the interaction between contaminants and nutrients as regards to production and attraction of predators deserves careful attention.

*Acknowledgments.* The authors thank Dr. J.W. Simmers for his contribution to the experimental work, and Mr. G.P.M. Léger, Dr. C.R. Lee and Dr. S.H. Kay for their corrections to the English and other constructive criticism.

## References

Bryan GW (1976) Some aspects of heavy metal tolerance in aquatic organisms. In: Lockwood APM (ed) Effects of pollutants on aquatic organisms. Soc Exp Biol Seminar Ser 2:7–35

Bryan GW, Langston WL, Hummerstone LG (1980) The use of biological indicators of heavy metal contamination in estuaries. Mar Biol Ass UK, Occasional Publ 1:73 pp

Duinker JC, Hillebrand MTJ, Boon JP (1983) Organochlorines in benthic invertebrates and sediments from the Dutch Wadden Sea; identification of individual PCB components. Neth J Sea Res 17 (1):19–38

Jennings CD, Fowler SW (1980) Uptake of $^{55}$Fe from contaminated sediments by the polychaete *Nereis diversicolor.* Mar Biol 56:277–280

Jones AR (1986) The effects of dredging and spoil disposal on macrobenthos, Hawkesbury Estuary, N.S.W. Mar Pollut Bull 17 (1):17–20

Koeman JH, van Genderen H (1972) Tissue levels in animals and effects caused by chlorinated hydrocarbon insecticides, chlorinated biphenyls and mercury in the marine environment along the Netherlands coast. In: Ruivo M (ed) Marine pollution and sea life. FAO (1972) Fishing News (Books), England, pp 428–435

Langston WJ (1980) Arsenic in UK estuarine sediments and its availability to benthic organisms. J Mar Biol Ass UK 60:869–881

Langston WJ (1982) The distribution of mercury in British estuarine sediments and its availability to deposit-feeding bivalves. J Mar Biol Ass UK 62:667–684

Luoma SN, Jenne EA (1976) Estimating bioavailability of sediment-bound trace metals with chemical extractants. In: Hemphil DD (ed) Trace substances in environmental health. Univ of Missouri, Columbia, pp 343–351

Luoma SN, Bryan GW (1978) Factors controlling the availability of sediment-bound lead to the estuarine bivalve *Scrobicularia plana*. J Mar Biol Ass UK 58:793–802

Luoma SN, Bryan GW (1979) Trace metal availability: Modeling chemical and biological interactions of sediment-bound zinc. In: Jenne EA (ed) Chemical modeling in aqueous systems. Am Chem Soc Symp Ser 93:577–609

Marquenie JM, de Kock WChr, Dinneen PM (1983) Bioavailability of heavy metals in sediments. Proc Int Conf Heavy Metals in the Environment. Heidelberg 1983. CEP Consultants, Edinburgh

Marquenie JM, Roele P, Hoornsman G (1986) Onderzoek naar de effecten van contaminanten op duikeenden. Report 86/066, MT-TNO, Delft, 35 pp (In Dutch)

Marquenie JM, Simmers JW, Birnbaum E (1985a) The biological fate of heavy metals after aquatic disposal of dredged materials. Proc Int Conf on Heavy Metals. Athens 1985. CEP Consultants, Edinburgh

Marquenie JM, Simmers JW, Birnbaum E (1985b) An evaluation of dredging in the Western Scheldt (The Netherlands) through bioassays. Report 85/075, MT-TNO, Delft, 56 pp

Nisbet JCT, Reynolds LM (1984) Organochlorine residues in common terns and associated estuarine organisms, Massachusetts USA, 1971–1981. Mar Environ Res 11:33–66

Pesch CE (1979) Influence of three sediment types on copper toxicity to the polychaete *Neanthes arenaceodentata*. Mar Biol 52:237–245

Ray S, McLeese DW, Metcalfe CD (1979) Heavy metals in sediments and in invertebrates from three coastal areas in New Brunswick, Canada, a natural bioassay. ICES CM 1979/E:29:7 pp, Mar Environ Quality Comm

Ray S, McLeese DW (1980) Bioavailability of chlorinated hydrocarbons and heavy metals in sediments to marine invertebrates. 68th Statutory Mtg. Copenhagen, Denmark 6–15 Oct 1980. ICES CM 1980/E:20:13 pp

Reijnders PJH (1980) On the cause of the decrease in the harbor seal population in the Dutch Wadden Sea. Thesis, Wageningen

Risebrough RW, Davis J, Anderson DW (1970) Effects of various chlorinated hydrocarbons. Oregon State Univ Env Health Sci Ser 1:40–53

Rubinstein NI, Lores E, Gregory NR (1983) Accumulation of PCBs, mercury and cadmium by *Nereis virens*, *Mercenaria mercenaria* and *Palaemonetes pugio* from contaminated harbor sediments. Aquat Toxicol 3:249–260

Tent L (1984) Contaminated sediments in the Elbe estuary, ecological and economic problems for the Port of Hamburg. In: Thomas RL (ed) The ecological effects of in-situ sediment contaminants. Proc Workshop Aberystwyth. Wales, August 19–24 1984. US Dept of Commerce

Wildish DJ, Thomas MLH (1985) Effects of dredging and dumping on benthos of Saint John Harbour, Canada. Mar Environ Res 15:45–57

# Oil Exploration and Production and Oil Spills

B. Dicks[1], T. Bakke[2], and I.M.T. Dixon[1]

## 1 Introduction

The recent and considerable oil exploration and production now underway in the waters of the North Sea has drawn attention to potential biological effects. Primary concerns are the maintenance of commercial fisheries, the general health of marine life in the North Sea, and, more locally, whether effects occur in individual oilfields which are sufficiently serious to warrant more stringent controls or costly remedial action. Oil is also shipped throughout the North Sea and refined at many coastal locations, with the accompanying possibilities of spills from these sources as well as from offshore fields. Concerns about impact of spills are usually focussed on shorelines where oil accumulates to cause acute local problems. In this chapter we attempt to summarise the impacts of both sources of oil inputs to the North Sea system and assess their importance in relation to the functioning of the system as a whole.

## 2 Oil Inputs to the North Sea

Recent estimates of oil inputs to the North Sea have been made by Read and Blackman (1980) and Bedborough and Blackman (1986; see Table 1). It can be seen that the proportions from both spills and offshore exploration/production have changed markedly over the period 1980–1986, accepting that there is considerable imprecision in making such calculations. Accidental losses from ships have decreased by more than a factor of 2, whilst oil inputs from offshore activities has increased by a factor of about 10. The former change reflects the general reduction in the crude market over the period and the latter the massive swing from water-based to oil-based muds during drilling, and increasing discharges of produced water (see below). There are important acute and long-term concerns over both sources.

Oil spills can produce locally severe effects and have a high public profile. The North Sea has had no major tanker incidents comparable with the *Torrey Canyon* or *Amoco Cadiz* and, so far, only relatively small "blowouts" from offshore platforms. For example, the Ekofisk Brave blowout in the North Sea spilled some

[1]Oil Pollution Research Unit, Orielton Field Centre, Pembroke, Dyfed. SA71 5EZ, United Kingdom
[2]NIVA, Norwegian Institute for Water Research, P.O. Box 333, Blindern, 0314 Oslo 3, Norway

**Table 1.** Estimated annual inputs of petroleum hydrocarbons to the North Sea (thousand tonnes), from Read and Blackman 1980; Bedborough and Blackman (1986)

| | 1980 | 1986 |
|---|---|---|
| Rivers/land run-off | 50 | 40–80 |
| Accidental shipping losses | 15 | 5–12 |
| Refineries | 11 | 6 |
| Coastal sewage | 8 | 3–14 |
| Dumped sewage sludge | 5 | Not included |
| Dredge spoils | 2.8 | Not included |
| Offshore production | 1.68 | 23 |
| Atmospheric | Not included | 19 |
| Natural seeps | Not included | 0.3–0.8 |
| Oil terminals | 0.61 | 0.8 |
| Other coastal industrial effluents | Not included | 9 |
| | 94.09 | 107–165 |

15,000–22,000 tonnes of crude oil over 7 days (Grahl-Nielsen 1978). None reached shorelines. Although the North Sea spill record is good, a blowout of the scale of Ixtoc 1 (455,000 tonnes spilled over approximately 9 months, Mar. Pollut. Bull. 1981) or a major tanker incident remain a possibility. The coastlines surrounding the North Sea have a high environmental value for ecological, conservation, human, economic and geomorphological reasons and can be regarded overall as highly sensitive to spilled oil.

Exploration and production activities also place stresses upon ecosystems in offshore areas which are largely out of sight of operators, scientists and the public; namely the water column and seabed with their associated biota. In simple terms, North Sea oilfields consist of either individual platforms or a cluster of platforms, and individual fields can be regarded as point sources of chronic contamination and disturbance. Types of input change with time, ranging from drilling discharges through the exploration phase and the construction of production facilities, to long-term and increasing discharges of produced and process water through the life of the field. The major economic importance of North Sea oil means we can expect to have to live with offshore platforms for the next 25–50 years at least, or even longer if discovery of new resources continues and efficiency of oil recovery from the reservoir also carries on increasing. The contribution of oil from these sources requires careful control and monitoring.

## 3 Oil Spill Impact

Thousands of papers on the effects of oil on marine life have been published over the past two decades and several summarise these effects (e.g., RCEP 1981; Oppenheimer 1980; Dicks 1985). Generalisations are difficult but we attempt a brief summary here. Effects depend greatly on the oil itself and upon the type and condition of the receiving environment(s). Each spill is unique as a result of the considerable number of variables which come into play, including:

1. Type and state of oil, especially composition, degree of weathering, mousse formation and biodegradation. The most dramatic impacts result from relatively unweathered materials with high proportions of toxic lighter hydrocarbons, especially aromatics, whilst heavily weathered oils tend to cause damage by smothering. North Sea crudes tend to be light with relatively high proportions of aromatics.

2. The receiving environment. Oil causes most damage in receiving environments where the product of exposure time, degree of retention and toxicity (or smothering effect) is maximised. Sheltered or low energy habitats tend to be most susceptible while exposed shores (e.g. rocky headlands) are least so.

3. The susceptibility of organisms, which ranges from apparently resistant brown seaweeds[3] to sensitive fish eggs and larval stages. Individual habitats may contain a complex range of organisms (existing together but competing for food and space), whose susceptibility to oil pollution may depend on many factors, e.g. stage in life-cycle, mobility, feeding habit and other behavioural modifications and defence mechanisms.

Life forms at the water surface, such as marine birds and mammals (particularly those that spend long periods on the water surface) are at risk, as are the organisms and the larvae or eggs which live just below the surface (neuston and plankton).

In the water column, oil can affect the phytoplankton and the zooplankton. In general, the eggs and planktonic larval stages of pelagic and benthic organisms are particularly sensitive to low oil concentrations. However, significant detrimental impact on plankton has never been demonstrated in the open sea. It has been claimed (McCauliffe 1986) that the dose and duration of oil exposure during a spill will never reach the threshold for effects on open sea plankton. While commercially important fish are a major concern (tainting of their flesh by oil reduces their marketability), the effects of oil on fish populations are much more difficult to quantify, as many species are highly mobile. Furthermore, fish may show avoidance reactions even at very low concentrations of oil in water, and they possess an enzyme system to deal with hydrocarbons in their tissues. Also important in the water column are the micro-organisms (mainly bacteria) which provide pathways for the ultimate removal of oil from marine systems through biodegradation.

On the seabed, benthic micro-organisms also provide routes for the biodegradation of hydrocarbons adsorbed on sediments. Deep waters are unlikely to be significantly affected by oil spills, but in shallow water oil in solution or as dispersed droplets can damage benthos, with effects ranging from detrimental behavioural modifications to mortality. The lack of mobility in many benthic species means that they either endure any oil stress or die. The temporary tainting of many species of shellfish (due to their ability to bio-accumulate hydrocarbons from low ambient oil concentrations and subsequently to depurate these without apparent harm) is also a well-known phenomenon.

[3]Most large brown seaweeds are resistant to single spills, but, like many marine organisms, can be affected by long-term, chronic inputs.

On the shoreline, oil can have a wide range of effects, ranging from subtle changes in individual performance and species composition to extensive mortalities, depending on amounts of oil and the species present. Mortalities of primary grazers on rocky shores can result in temporary massive growths of fast-growing, opportunistic seaweeds, as oil toxicity or smothering declines.

Effects on the biota can persist for long periods of time, especially in sediment systems (e.g. saltmarshes, which provide organic inputs to coastal and estuarine marine food chains). On sedimentary shores oil can penetrate into sediments and biodegradation virtually cease due to oxygen depletion.

Above the high water splash zone, the maritime communities rarely receive oil, but infrequent cases are reported where oil becomes carried by the wind, causing damage to maritime vegetation. Damage to these areas more usually results from clean-up activities.

4. Oil spill clean-up activities. Although mechanical collection of oil may be a biologically preferable option, it can be very difficult to apply unless weather conditions are good and access with bulky equipment is feasible. The application of dispersants may be desirable in certain circumstances to prevent oil arriving ashore in sensitive areas. Apart from the other option of "no immediate response", both response actions can lead to biological damage: dispersants from their toxic effects and mechanical methods from the physical damage caused by repeated trampling or moving heavy equipment and vehicles, especially in the maritime fringe and in saltmarshes.

Ultimately, the impact of a spill is determined by a host of factors, and the best defence is to be prepared with contingency plans which have clearly identified priorities for protection. Table 2 summarises the sensitivity of some common North Sea habitats and resources to oil.

## 4 Offshore Oil Exploration and Production

A summary of discharges and impacts of platforms can be found in Dicks (1982) and Hartley (1982), with more specific details of the impact of drilling activities in Poley and Wilkinson (1983), Davies et al. (1984) and Bakke et al. (1986a,b). The presence of a platform either piled into or resting on the seabed has a number of immediate physical effects, which may be followed later by a variety of effects resulting from discharges during construction, drilling and oil production (see Fig. 1).

The physical presence of structures locally modifies water currents and patterns, and a small area of seabed. It may also act as a large artificial reef which rapidly colonises with a variety of fouling organisms (Wolfson et al. 1979; Society for Underwater Technology 1981), whose larvae settle from the plankton. The form of the growth and its thickness is difficult to predict. Fouling organisms (as well as discharged garbage and sewage) then attract shoals of fish to feed. In the Gulf of Mexico this sort of reef effect has been regarded by local sport fishermen as a distinct environmental improvement. This is only likely to be the case in the North Sea if platforms are developed close inshore, but even then, safety restric-

**Table 2.** Summary of habitat and resource sensitivities to oil for the North Sea

| Habitat/Resource | Sensitivity | Reason(s) |
|---|---|---|
| Saltmarshes | High | Many species susceptible to damage by oil, notably the dominant plant species responsible for marsh building. Often highly productive. Can retain oil for long periods after spills and damage may persist. Often important bird feeding/roosting areas |
| Sheltered inter-tidal flats | High | May have high productivity and support commercial fish/shell-fish or their young which are easily tainted or harmed. Oil can be retained for long periods if it penetrates sediments |
| Sheltered rocky coasts and inlets | High | May have high productivity and retain oil for long periods, with resultant harm to communities. May contain spawn-ing/nursery grounds |
| Sheltered subtidal sediments (< 20 m) | High | Often support seagrass beds spawning grounds nursery areas. May be of high productivity and be easily harmed, and retain oil in sediments for prolonged periods |
| Coastal bird colonies and marine birds | High | Very susceptible to damage by oil. Often have a high interna-tional conservation/wildwide value, and are symbolic of 'con-servation' |
| Marine mammals | High | Very susceptible to damage by oil. Often have a high local or international conservation/wildlife value, and are symbolic of 'conservation' |
| Maritime zone–sand dunes | High | Stabilise long stretches of North Sea coastline and have high wildlife/conservation value. They are easily damaged by man's activities, e.g. during spill clean-up, and erosion may have serious consequences for the hinterland |
| Commercial benthic shell fisheries | High | Breeding stocks are easily tainted and damaged (especially in intertidal zones) by oil |
| Demersal and pelagic fisheries | High | Flesh is easily tainted, which reduces marketability of catch |
| Bathing/recreational beaches | High | Oil has immediate detrimental effects on local economy |
| Exposed tidal sand flats | Medium[a] | Some shell fisheries, but usually poor communities. Oil may penetrate but will usually self-clean |
| Exposed coast subtidal sediments and deep water sediments (>20 m) | Medium | Oil rarely penetrates to these sediments, although they may be rich and productive. If oil should reach the sea bed in any quan-tity effects could be serious |
| Exposed rocky coasts and subtidal rocks | Low[a] | Although biological damage results from oiling, these ha-bitats usually self-clean and recover relatively quickly |
| Exposed gravel or mobile coarse sediments | Low[a] | Usually poor communities and self-clean fairly well during storms |

[a] May assume a higher local priority for various reasons: e.g. some exposed shingle banks are bird roosts, nature reserves or important coastal defences.

Note: Many of the above high priority/high sensitivity resources occur together within estuaries in Europe. In any one area, the relative importance of each resource may vary when deciding priorities for protection.

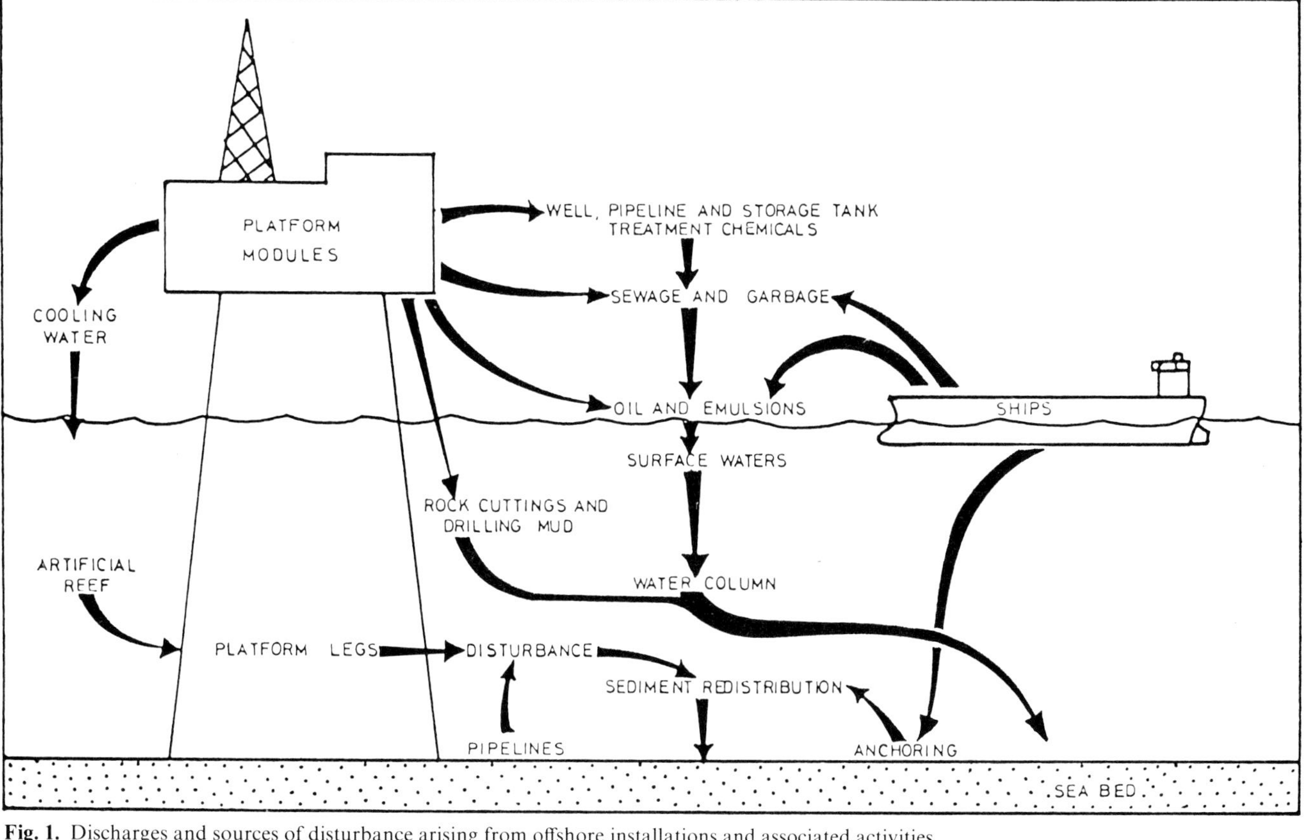


**Fig. 1.** Discharges and sources of disturbance arising from offshore installations and associated activities

tions may preclude all fishing. Localised physical changes, together with the ingress of fouling species and those species feeding on them, serve to modify the environment and alter the feeding relationships of the area.

Most steel structures on platforms require corrosion protection, and in the North Sea offshore platforms are normally protected by a combination of painting with anti-fouling materials and by an electrovoltaic system employing sacrificial electrodes. Anti-fouling paints are not normally very effective for long periods of time, and an electrode system usually forms the main mechanism of protection. These protection systems may have a number of effects on the water column which are difficult to quantify. Anti-fouling paints make small contributions of heavy metals (e.g. copper), and sacrificial electrodes (depending on their type) release a variety of metals by dissolution.

Water discharges from shipping in oilfields and the platforms include rain run-off, cooling water used in various pieces of plant, and water used for operating domestic waste and sewage disposal systems. Water in pipelines or in offshore storage tanks (e.g. those in rig bases or legs) is often treated with chemicals for a variety of reasons, notably corrosion inhibition. Such chemicals are normally discharged to the sea when spent, and the magnitude of the input varies considerably from field to field and with time.

It is likely that all the water-based discharges noted above have only minor or undetectable impacts on systems around the platforms. Furthermore, such effects are largely masked by the much larger impacts which result from drilling activities in particular, and, to a lesser extent, discharged production water.

### 4.1 Drilling Activities

Up to 1985, drilling activities have been one of the major sources of discharges to the offshore environment. A well drilled to a typical North Sea oil formation to ca. 4–5000 m below the seabed produces around 1000 $m^3$ of cuttings (in particulate form) which are routinely discharged to the seabed at the base of the platform. By using deviated drilling, which allows wells to be drilled away from the platform at an angle, up to 50 wells can be drilled from a single platform. The number of wells drilled in the UK sector (oil and gas) has increased from 116 in 1975 to 128 in 1983 and 182 in 1984 (Department of Energy 1985). Oil production from these wells totalled 698 million tonnes in 1984, with recoverable reserves of 1100–4800 million tonnes estimated to remain.

In recent years increasing use has been made during drilling of oil-based muds (OBM's) by the offshore oil industry in the North Sea. These function more efficiently than water-based muds (WBM's) under the rigorous drilling conditions dictated by increasingly efficient exploitation of existing fields, and also of new reserves in deep water or under more complex geological formations. In 1981 approximately 36% of wells drilled on the UK continental shelf were drilled with OBM's leading to an estimated 7000 tonnes of diesel being discharged with drill cuttings. By 1983 approximately 65% of wells were drilled using OBM's, with a loss of 18,100 tonnes of oil to the environment. This corresponded to 90% of the total oil lost by discharge or spillage offshore (data from Davies et al. 1984; Department of Energy 1985).

OBM's were originally formulated on a diesel base, but concern over the probable environmental impact of large quantities of diesel entering the sea led to the development and increasing use of base oils with reduced toxicities. These are more highly refined products with a much lower aromatic hydrocarbon content than diesel. The "low toxicity" oils (low-tox) currently used in the North Sea have a total aromatic content of less than 5%, and a polynuclear aromatic content of less than 0.1%. The corresponding values for diesel are up to 61% total aromatics and at least 20% of these are polynuclear aromatics (Hinds et al. 1983). In comparison to diesel, low-tox base oils appear to be up to 1000 times less toxic than diesel in standard 96 hours $LC_{50}$ tests (Blackman et al. 1982) while in a comparison of whole muds, they vary between 10 and 100 times less toxic than diesel-based mud (Blackman et al. 1983).

Differences of the same order of magnitude were reported by Bakke et al. (1986a) in toxicity tests with used mud/cuttings mixtures on a variety of organisms. They also repeated these tests after the cuttings had been exposed on the seabed for 9 months. For some test organisms, the OBM toxicities were found to have declined by more than 1 order of magnitude, whereas for others the toxicity had increased. Elevated toxicity of used low-tox material was reported by Dow (1984), possibly as a result of other drilling additives.

Field data on seabed sediment hydrocarbons and the biological effects recorded around North Sea OBM operations have been reviewed by Poley and Wilkinson (1983) and Davies et al. (1984). In addition, the results of environmental monitoring around specific North Sea oilfields using OBM's have been reported for Beatrice (Addy et al. 1984) and Statfjord (Matheson et al. 1986). In relation to WBM operations, both the severity and the spatial extent of hydrocarbon contamination and resultant benthic environmental effects are greater following OBM use. However, there is currently insufficient evidence with which to distinguish the chronic or long-term ecological effects (as opposed to the acute toxicological effects) of cuttings contaminated with diesel from those mixed with a low-tox base oil (Davies et al. 1984; Matheson et al. 1986). In fact, the data available indicate that changes to the macrobenthos resulting from the use of both types of OBM share several characteristics. These have been summarised below in relation to distance from the discharge point, as concentrically arranged zones of impact which are usually elongated in the direction of the locally prevailing current regime.

1. Directly beneath the platform, the seabed sediments can consist almost entirely of oily cuttings, and as a result, are hypoxic and support no macrofauna.
2. Beyond the afaunal centre is a zone in which the sediment is still hypoxic with total hydrocarbon concentrations usually exceeding 1000 times the background levels, and in which the macrobenthic communities are impoverished and highly modified. Species richness increases with distance from the discharge point but, within this gradient, there is also, typically, a spatial succession of species populations similar to that described by Pearson and Rosenberg (1978) as a response to point source organic pollution. Most characteristic is an invasion by opportunist species which can cope with the organic loading and disturbance and which generate total individual densities several times higher than background. Survey data indicate that this zone of "major biological impact" is usually confined to within 500 m of the platform.

3. There is then a "transition zone" in macrobenthic diversity and community structure extending to the point beyond which industry-related changes to the fauna become undetectable. Hydrocarbon concentrations range from around 1000 times down to 10 times background levels within this zone which extends to a maximum of about 2000 m from the platform.
4. Elevated hydrocarbon concentrations may be evident at up to twice this distance from the installation without detectable biological effect, particularly in fields with extensive OBM drilling operations.

As Addy et al. (1984) point out, there are a number of likely mechanisms through which the environmental consequences of oily cuttings discharge are mediated. These include direct physical smothering of the seabed and benthic fauna under piles of cuttings; organic enrichment of the sediment by petrogenic hydrocarbons with subsequent oxygen depletion; and the toxic effects of base oils and other mud additives. The way in which the relative importance of each of these factors changes with distance from the platform has never been tested, and has proved impossible to determine from the field data available. It is likely, therefore, that reliable information on how different OBM formulations affect the benthic environment will only be obtained through long-term comparative studies under experimental conditions.

The Department of Agriculture and Fisheries for Scotland (DAFS) has carried out a series of long-term investigations into the chemical, microbiological and meiofaunal aspects of OBM impact in experimental tank systems (e.g. Dow 1984; Leaver et al., in press). These have been complemented by field applications of the same OBM's to seabed plots by the Oil Pollution Research Unit of the Field Studies Council (Dixon 1987). The data from these experiments indicated that faunal disturbance occurred more rapidly following diesel treatment than with low-tox treatment, possibly reflecting the higher acute toxicity of the former. After 1 or 2 months, however, the longer term effects of the low-tox treatment became indistinguishable from those of the diesel OBM. A similar conclusion was reached by Bakke et al. (1986b) in long term field experiments investigating the colonisation of sediments contaminated with diesel or low-tox cuttings by bottom fauna.

What is the long-term fate of these drilling discharges, which are greatly reduced or terminated once drilling is complete? The available experimental evidence suggests that the OBM base oils are degradable and therefore that recovery of communities can take place. So far, we have only very limited field evidence of recovery following completion of drilling (Hannam, pers. comm.) and no long-term data from which rates might be predicted. The drilling discharges are most easily visualised as a mound of material immediately below the platform which may be several metres thick, rapidly thinning to millimetres within tens or hundreds of metres, and only dispersed or current transported material at more than 500 m. Whilst major storms may cause rapid redistribution of the discharges, this is unlikely in deep water. The lack of a normal fauna reworking the sediment also has a stabilising effect on the layers of cuttings. It is probable that the steady action of tide and current will gradually modify the cuttings over long periods of time (5+ years) or that a gradual sedimentation of new, clean material may "dilute" the oil on the cuttings to levels low enough for a normal fauna. In any case, we can expect recovery of biota to be relatively slow. It is likely that the long-term effects of the drilling activities will mask effects of other long-term inputs, like the produced water discharges discussed below.

## 4.2 Produced Water

Whilst drilling discharges constitute a major input of contaminants early in the life of an oilfield, hydrocarbons in water produced from the reservoir with the oil become important during later production. Produced water arises from salt water underlying oil in the reservoir, which may be added to by injection of treated sea water from the platform. This is carried out in order to keep up the pressure within the reservoir and enhance oil recovery. Little is known of the potential for reappearance of the chemicals used to treat injection water (corrosion inhibitors, biocides etc.) in produced water. Volumes of produced water vary greatly from field to field, depending upon the type of reservoir, the location of producing wells and the speed with which oil is produced. In general, the amount of water breaking through into the production wells is small in the early stages, but increases as the oil in the reservoir is depleted. There have been exceptions to this pattern, with significant water breakthrough in the early stages. Produced water, which is often of higher salinity than sea water (2–4 times), is normally treated to 50 ppm oil content or less, and discharged, along with similarly treated displacement water from oil storage tanks. Emulsion-breaking chemicals may also be found in the discharge as a result of de-watering of the crude oil on the platform.

These sources can contribute to sediment contamination by hydrocarbons (Addy et al. 1978) and some of the biological effects recorded by these authors may be ascribed to this cause. Read and Blackman (1980) estimated that by 1985 the discharge of produced water in the UK sector would be three to four times greater than in 1980, and whilst new fields are constantly being developed this trend of steadily increasing inputs is likely to continue.

## 4.3 Detection of Effects in Oilfields

At the present time, most long-term studies in offshore oilfields are conducted on behalf of or financed by oil companies in order to meet quite specific aims; namely, to determine (1) the extent and form of biological changes caused by their operations; (2) how the situation changes with time; and, ultimately, (3) when changes become sufficiently serious to be regarded as damage which requires corrective action to be taken.

Although prime concerns may be damage to economic resources and to the health of offshore ecosystems as a whole, a complete assessment is currently impossible to achieve. Factors such as the complexity of structure in the function of marine ecosystems, their often considerable natural variability, a lack of suitable sampling and analytical techniques, as well as limitations of time and funding, contribute to the difficulties and force the selection of suitable monitors.

Although it is evident that the immediate impacts of acute or chronic liquid discharges from platforms are likely to be on the water column inhabitants in the receiving water body, it is particularly difficult to assess such impacts. The background of natural seasonal variation in these systems, the mobility of water masses, the patchiness in spatial and temporal distribution of organisms and difficulties with quantitative sampling make the identification or isolation of pollution effects from natural variation very difficult. Of particular concern are the eggs and larvae of commercial fish and shellfish. Experimental evidence demonstrates the possibility of effects at low concentrations of some oil fractions, but the significance of such effects against a high and variable natural mortality is hard to assess, the more so since a significant impact would not be manifest until several years later when the year class so affected would be expected to join the harvestable stock. We are thus at present unable to predict or measure directly the impact of oilfields on these 'front-line' receivers. Like the plankton, fish and birds are poor monitors of oil pollution as natural communities are hard to quantify and sample reliably. Some are undoubtedly at risk and these may be of considerable ecological or economic importance.

The sublittoral benthic macrofauna of soft sediments have proven to be the most useful indicators of pollution impact or of general biological health in many areas (for example Pearson 1975; Addy et al. 1978; Gray and Mirza 1979; Sharp et al. 1979; Hartley and Dicks 1987), mainly because they are stationary and hence integrate pollution impact over time. This also enables repeated sampling of the same community over time (monitoring). The macrobenthos may be expected to reflect changes in other marine communities, both indirectly as a result of the interactions between various sections of the offshore environment (although many are poorly known) and directly by pollution effects on the planktonic or pelagic larval stages of benthic organisms. Benthic monitoring is likely to remain the mainstay of assessment of oilfield impact for the foreseeable future.

## 5 Conclusion

How serious is the problem of oil spills in the North Sea? To date the North Sea record is exemplary, with only a single blowout since offshore production began (Ekofisk, 1977 ca. 15–22,000 tonnes) and a number of smaller shipping spills (e.g. examples which came to the public eye include *Eleni V*, 1978; *Sivand*, 1984). This record stems partly from good offshore operating practice and decreasing shipping activity, but undoubtedly also in part from good fortune. Central governments, industry and local coastal authorities are reasonably well prepared to deal with even a large spill (up to several thousand tonnes) in terms of equipment stockpiles. No-one is equipped to deal rapidly and efficiently with a supertanker spill, a major and prolonged blowout like Ixtoc 1 in the Gulf of Mexico or Nowruz in the Arabian Gulf, or with smaller spills under adverse weather conditions, for economic, logistical and probability reasons. In North Sea oilfields, complacency would be foolhardy, for however small the probability, the risk of a *major* spill (several hundred thousand tonnes) exists. The best defence is to be prepared, and to this end further attention should be given to the environmental aspects of spill response and contingency planning. Many local plans as well as broader guidelines already exist (CONCAWE 1981; Dicks 1985) but much could be achieved by a sensible programme of resource evaluation and oil spill sensitivity mapping for North Sea coasts and waters. Such information should be presented not as huge and incomprehensible resource atlases, but as very simple maps and guidelines usable by the people who deal with spills.

How important are platform inputs in the North Sea? Using the relatively insensitive monitoring techniques we have available, it appears that biological impacts tend to be restricted in most cases to within 1 km of a platform, although in some cases, detectable effects have extended further. On this basis, the present ca. 200 installations in the North Sea represent an area of impact of about 0.1% of the total North Sea area. Read and Blackman (1980) identified total oil inputs from offshore operations as a small and insignificant input in relation to those arising from land-based sources (less than 2%). Bedborough and Blackman (1986) show current inputs to be much increased at between 14 and 21% of the total hydrocarbon input to the North Sea. These inputs are made directly into the central North Sea waters, and are not mediated by dilution through nearshore waters, as are most land-based inputs. In recognition of the potential for contamination and damage, it is fair to say that most governments and oil companies have taken a responsible approach to assessing impacts and monitoring change. However, there are some 140 permanent oil and gas structures in the U.K. sector, and a further 50 in the Norwegian sector, and exploration continues in both southern and northern North Sea. The abundance of platforms, combined with our ignorance of long-term effects and even the basic functioning of marine systems, suggests that it would be foolish indeed to become complacent. Continued biological and chemical monitoring in oilfields, combined with appropriate field and experimental research to improve monitoring procedures and our understanding of oil impact mechanisms, should preclude any danger of impacts going undetected or becoming unacceptable.

The main areas of ignorance lie in the long-term impact of (and potential for recovery of the environment from) drilling inputs and produced water discharges. Recent field studies (Matheson et al. 1986) have shown elevated levels of OBM hydrocarbons in areas where drilling has not been initiated, suggesting a general increase in hydrocarbon contamination. The long-term effect of such subtle increase on benthic function and structure is not known, and can hardly be elucidated by field studies for logistic reasons. In this context soft sediment mesocosm experimental systems, such as those currently being developed at Solbergstrand, Eastern Norway, may have an important role to play, as they have the potential for controlled experimental work using whole sediment communities. They may be able to produce data from which environmental predictions can be made to complement that gained from long-term monitoring. In addition to pollution-related studies, the underlying need for research into ecosystem function, natural variation and taxonomy must be fulfilled, for such information is vital to the accurate interpretation of monitoring results and the attribution of effect to cause.

## References

Addy JM, Levell D, Hartley JP (1978) Biological monitoring of sediments in the Ekofisk oilfield. In: Proc Conference on the Assessment of Ecological Impacts of Oil Spills. Arlington, Va 514–539. American Institute of Biological Sciences

Addy JM, Hartley JP, Tibbets PJC (1984) Ecological effect of low toxicity oil-based mud drilling in the Beatrice oilfield. Mar Pollut Bull 15(12):429–436

Bakke T, Blackman RA, Hovode H, Kjorsvik E, Norland S, Ormerod K, Ostgaard K (1986a) Drill cuttings on the sea bed. Toxicity testing of cuttings before and after exposure on the sea floor for 9 months. In: Proc Symp Oil Based Drilling Fluids, Cleaning and Environmental Effects of Oil Contaminated Drill Cuttings. Royal Garden Hotel, Trondheim, Norway, 24–26 February 1986, pp 79–84

Bakke T, Green NW, Naes K, Pedersen A (1986b) Drill cuttings on the sea bed, Phase 3. Field experiment on benthic community response and chemical changes to thin (0.5 mm) layers of cuttings. In: Proc Symp Oil Based Drilling Fluids, Cleaning and Environmental Effects of Oil Contaminated Drill Cuttings. Royal Garden Hotel, Trondheim, Norway, 24–26 February 1986, pp 33–42

Bedborough D, Blackman RAA (1986) A survey of inputs to the North Sea resulting from oil and gas development. Phil Trans Roy Soc Series B. London, 316:495–509

Blackman RAA, Fileman TW, Law RJ (1982) Oil-Based Drill Muds in the North Sea – The use of Alternative Base Oils. ICES CM 1982/E:13 Marine Environmental Quality Committee, 8 pp + tables and figures

Blackman RAA, Fileman TW, Law RJ (1983) The toxicity of alternative base-oils and drill-muds for use in the North Sea. Internat Council for the Explor of the Sea (ICES) CM 1983/E:11. Marine Environmental Quality Committee, 7 pp + tables and figures

CONCAWE (1981) A Field Guide to Coastal Oil Spill Control and Clean-up Techniques. Report No 9/81. CONCAWE, Den Haag, 112 pp

Davies JM, Addy JM, Blackman RA et al. (1984) Environmental effects of the use of oil-based drilling muds in the North Sea. Mar. Pollut Bull 45(10):363–370

Department of Energy (1985). Development of the oil and gas resources of the United Kingdom 1985. HMSO: London, 84 pp + chart

Dicks B (1982) Monitoring the biological effects of North Sea platforms. Mar Pollut Bull 13:221–227

Dicks B (1985) Strategies for the assessment of the biological impacts of large coastal oil spills: European coasts. Report No. 5/85, CONCAWE, Den Haag

Dixon IMT (1987) Experimental application of oil based muds and cuttings to seabed sediments. In: Fare and Effects of Oil in Marine Ecosystems. Kuiper and van den Brink (eds) Nijhoff, Dordrechr. pp 133–150

Dow FK (1984) Studies on the Environmental Effects of Production Water and Drill Cuttings from North Sea Offshore Oil Installations. PhD Thesis, University of Aberdeen, 243 pp + figures, tables and appendices

Grahl-Nielsen O (1978) The Ekofisk Bravo blow-out: petroleum hydrocarbons in the sea. In: Proc Conference on the Assessment of Ecological Impacts of Oil Spills. Arlington, Va. 476–487. American Institute of Biological Sciences

Gray JS, Mirza FB (1979) A possible method for the detection of pollution induced disturbance on marine benthic communities. Mar Pollut Bull 10:142–146

Hartley JP (1982) Methods for monitoring offshore macrobenthos. Mar Pollut Bull 13:150–154

Hartley JP, Dicks B (1987) Macrofauna of Subtidal sediments using remote sampling. In: Baker JM and Wolff WJ (eds) Biological Surveys of Estuaries and Coasts. EBSA, Cambridge Univ Press, pp 106–130

Hinds AA, Smith SPT, Morton EK (1983) A comparison of the performances, cost and environmental effects of diesel-based and low-toxicity oil mud systems. In: Proc Conf Offshore Europe '83. September 6–9, Aberdeen. SPE 11891, 19 pp

Leaver MJ, Murison D, Davies JM, Raffaelli D (1986) Experimental studies of the effects of drilling discharges. Phil Trans R Soc Ser B., 316

Marine Pollution Bulletin (1981) Ixtoc 1 aftermath report. Mar Pollut Bull 12:143

Matheson I, Kingston PF, Johnston CS, Gibson MJ (1986) Statfjord Field environmental study. In: Proc Symp Oil Based Drilling Fluids, Cleaning and Environmental Effects of Oil Contaminated Drill Cuttings. Royal Garden Hotel, Trondheim, Norway, 24–26 February 1986, pp 1–16

McCauliffe CD (1986) Organism exposure to volatile hydrocarbons from untreated and chemically dispersed crude oils in the field and laboratory. In: Proc 9th Arctic Marine Oilspill Programme Technical Seminar. Edmonton, Canada, 10th – 12th June 1986, pp 497–526

Oppenheimer CA (1980) Oil Ecology. Marine Environmental Pollution. I. Hydrocarbons.Elsevier North Holland, Amsterdam, pp 21–35

Pearson TH (1975) The benthic ecology of Loch Linhe and Loch Eil, a sea loch system on the west coast of Scotland. IV. Changes in the benthic fauna attributable to organic enrichment. J Exp Mar Biol Ecol 20:1–41

Pearson TH, Rosenberg R (1978) Macrobenthic succession in relation to organic enrichment and pollution of the marine environment. Oceanogr Mar Biol Ann Rev 16:229–311

Poley JP, Wilkinson TG (1983) Environmental impact of oil-based mud cuttings discharges – A North Sea perspective. In: Proc Conf IADC/SPE 1983 Drilling Conference. New Orleans, Louisiana, 20–23 February 1983, pp 335–342

Read AD, Blackman RAA (1980) Oily water discharges from North Sea installations: A perspective. Mar Pollut Bull 11:44–47

Royal Commission on Environmental Pollution (1981) 8th Report. HMSO, 307 pp

Sharpe JM, Appan SG, Bender ME, Linton TL, Reish D, Ward, CH (1979) Natural variability of biological community structure as a quantitative basis for ecological impact assessment. In: Conf Proc Ecological Damage Assessment. Society for Petroleum Industry Biologists, pp 257–284

Society for Underwater Technology (1981) Marine Fouling of Offshore Structures. Conf Proc, Vol 1, 19–20th May, 1981

Wolfson A, Blaricom G van, Lewbel GS (1979) The marine life of an offshore oil platform. Mar Ecol Prog Ser 1:81–89

# Fishery Effects

H.J.L. HEESSEN[1]

## 1 Introduction

The commercial fish stocks in the North Sea are all heavily exploited and, although they form renewable resources, an essential feature is that they are finite. Because of the concern for overfishing and while fish are not usually restricted to the waters of one particular nation, there is a need for an international fishery policy, with the European Community as the first interested party for most of the North Sea. The biological advice for fishery management is provided by the International Council for the Exploration of the Sea (ICES) which was founded already in 1902. The most important task for ICES is to monitor the fish stocks in the north east Atlantic and to give advice for fishery management, not the management itself. Besides, ICES is a forum for marine research in general. Fishery research in the North Sea has a fairly long history and, in this respect, the North Sea is one of the best-investigated areas in the world.

Fisheries are directed at different species within a year or when they concentrate on the same species throughout the year, different grounds are exploited. In this way fishermen try to spread their own risks, and they have to, while money on the table is their primary concern. Quite a number of different fishing methods exist. Some of them are highly selective and catches consist almost exclusively of one single species, others result in the catch of a mixture of species of which two or three dominate. Traditionally most important in the North Sea is the human consumption fishery, but since the 1960's an industrial fishery developed, the catch of which is used for reduction into fish meal and fish oil.

In this book on pollution of the North Sea, a chapter on the effects of the fishery may seem a bit out of scope. It is by no means the intention to discuss here the possible influence of pollution on fish stocks, but to see what the impact of the fishery is as a human intervention in the North Sea ecosystem.

The annual removal of a vast amount of fish is known to have an effect on the exploited stocks and thus on the ecosystem as a whole. To give a first impression of the importance of the fishery, the changes in the total yield during this century will be discussed. Then three species are chosen to look at in greater detail. Fishery research traditionally focuses on single species, but it is well known that different fish species influence one another, for example as predator and prey or as competitors for the same food resource. This interference has recently attracted

[1]Netherlands Institute for Fishery Investigations, P.O. Box 68, 1970 AB Ymuiden, The Netherlands

special attention and the first results of these investigations will be mentioned. As certain fish methods have a greater influence than just catching fish, because they also influence the bottom fauna, some comments on this influence will be given. At the end of the chapter some concluding remarks are given.

## 2 Total North Sea Yield

Changes in the total North Sea yield and simultaneous changes in the fleets are described by Daan (1986). The total fish catch in the North Sea (Fig. 1) increased from 1 million tonnes at the beginning of the century to 2 million tonnes in 1956, with marked interruptions caused by the two world wars. In the late 1950's catches

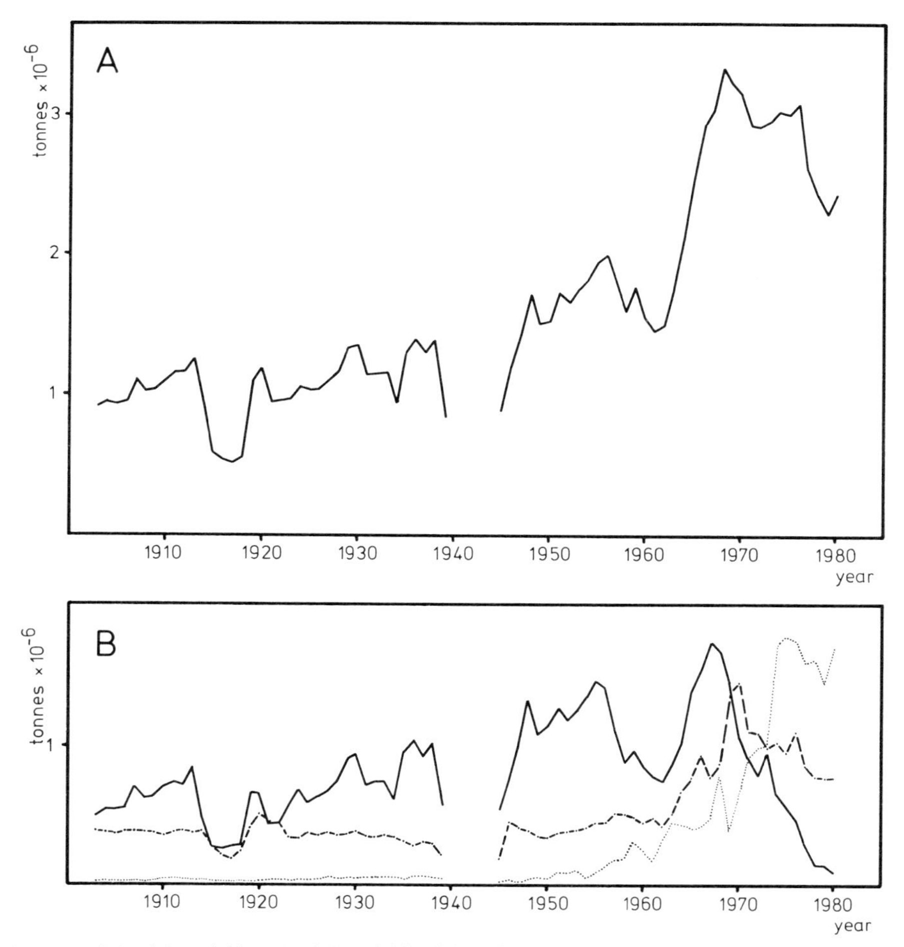


**Fig. 1. A** Total North Sea yield. **B** North Sea yield split into three components (*solid line* pelagic species; *broken line* demersal species; *dotted line* industrial species)

declined to less than 1.5 million tonnes, but a pronounced increase followed in the early 1960's. For a decade the catch fluctuated rather steadily around 3 million tonnes, but in recent years the catches decreased to 2.5 million tonnes. The total biomass of fish in the North Sea has been estimated at about 10 million tonnes in both 1977 and 1978 (Yang 1982a), and thus in recent years the fishery removes annually about 25% of the total fish biomass.

Compared with the human consumption and industrial fishery the fishery for crustaceans (mainly shrimps) and molluscs is negligible.

The landings of fish can be split in demersal and pelagic species used for human consumption, and in fish which is used for reduction. For human consumption the most important demersal species are cod, haddock, whiting, and plaice, and for the pelagic species these are herring and mackerel. The landings from the industrial fisheries mainly consist of Norway pout (a small gadoid), sprat and sandeels. Till the late 1950's, the catch of demersal species remained remarkably stable (Fig. 1) and the annual fluctuations in the total catch largely reflect the variations in the pelagic catch. The change in the total catch in the early 1960's, however, coincided with a marked change in the contribution of the different components. Initially the catch of all three categories increased, but for the pelagic species the increase lasted only 2 or 3 years and a sharp decline followed. By 1980 the pelagic species contributed only a fraction of the long-term average, whereas now, due to the recovery of the herring stock, the contribution of the pelagic species increases again.

The demersal species peaked some years later. Although also followed by a decline, also in recent years the catch is still higher than ever recorded before 1965. Finally, the industrial species appear to have reached a ceiling by the end of the 1970's and their catch has, as yet, not regressed. Holden (1978) discusses in greater detail the changes in the catches of more than 30 species for the period 1909–1973.

Undoubtedly, over this century the fleets fishing in the North Sea have undergone marked changes. For instance, the development of more powerful engines allowed the introduction of the herring trawl fishery after the second world war. The passive driftnet fishery went and the active search for fish schools by means of hydroacoustic equipment became a major component of the fishing effort. This development made species with schooling behavior, such as herring and mackerel, extra vulnerable. In the 1960's, with the introduction of the power block in the purse seine fishery, the catch potential of a vessel even became almost solely dependent on the ability to locate schools and on the loading capacity of the vessel. This also changed the nature of the fishery, because the huge catch rates of the purse seine fleet did not allow processing of the fish for human consumption, and a considerable part of the total herring and mackerel catches were turned into fish meal.

In the demersal fisheries, changes in efficiency are less pronounced. Whereas the Danish seine survived in traditional fisheries, long lining disappeared and steam trawling was gradually replaced by motor trawling. Some new techniques, such as pair trawling and gill netting, have been introduced. In the flatfish sector the very efficient beamtrawl has become the important gear.

Particularly noteworthy is the development of the industrial fishery in the 1960's and 1970's, which includes both purse seiners and relatively small trawlers employing small meshed nets.

Despite all discussions about overfishing, it is clear from what is stated above that the North Sea is by no means emptied by the fishery. The increase in total yield cannot be explained from an increasing eutrophication (Ursin and Andersen 1978), but has to be the result of the fishery itself and possibly of natural causes as well. Qualitatively there have been quite a number of changes in effort and in the balance between different sectors of the international fishery as a whole, which undoubtedly have had their effect on the changes in total catch composition. Detection of changes that may have occurred within the populations of single species or in the species composition cannot be determined from gross catches alone: it requires additional information on the structure of individual stocks.

## 3 Effects of the Fishery on Individual Stocks

One of the primary effects of exploitation on a fish stock is a shift in the age composition: the older individuals are removed and the younger age-groups simply do not get the chance to grow old. Another phenomenon is that fishing a previously unexploited stock may lead to faster growth and earlier breeding amongst members of the surviving population. For the survivors the per capita food supply increases, which can result in faster growth and earlier maturity (density-dependent effects). But fishing could also be a major selective force acting on the genetic composition of fish stocks. Fish has an intrinsic variability; for example, growth rate and age of first maturity can be genetically determined. Fishing therefore may represent a massive and novel selection on the population (May 1984). Fishing reduces the population size and ultimately, heavy exploitation can result in overfishing, as could be observed, for example, in the North Sea herring stock.

Environmental variables can also be responsible for changes in fish stocks. A dominant characteristic for many stocks is the variability at short-term frequencies. Especially variations in reproduction success and survival of the juvenile stages can be very striking, and these subsequently can result in large annual fluctuations in recruitment. But also long-term changes, which are possibly induced by changes in the environment, are observed. Data for Atlanto-Scandian herring suggest the existence of regular fluctuations in stock levels occurring approximately every 100 years. Within these long cycles large year-to-year fluctuations in recruitment may be observed (Devold 1963).

The total annual catch of a certain species, and in this respect not only the actual landings but also the undersized individuals which are discarded at sea should be included, can be considered as a function of population size and fishing effort. But do the changes in the annual catch reflect the real changes in population size? To derive the necessary information, fishery biologists set up an extensive market sampling program which on a routine basis determines the age composition and the length and weight composition of the landings of the most important species. In addition to the data from the market sampling, different kinds of surveys give information on the size of recruiting year classes, the size of the total stock or the spawning stock. The age-structured data from the market sampling, together with data from surveys, are used to estimate the changes in different population parameters. Important parameters are the size of the spawning stock, the fishery mortality, which is a direct measure of the fishing effort to which the population is subjected, and the size of the year classes that recruit to the fishery.

An extensive review on the changes in North Sea fish stocks and their causes was given during a symposium held in 1975 (Hempel 1978). Here only three examples will be discussed: herring, cod and plaice.

*Herring* (Fig. 2). The herring has in the past provided the major component of the North Sea pelagic yield. The catch fluctuated between 500 and 800 thousand tonnes during most of the century until in the mid sixties the catch peaked over 1 million tonnes. This was followed by a dramatic decline until the closure of the fishery in 1977. After 5 years, the fishery was opened again. Estimates of the spawning stock biomass reveal that the stock had been declining from the second world war onwards, with two revivals of short duration. Apparently, the biomass had been decimated already by the end of the 1960's. Nevertheless, the total catch remained high for another few years, and in fact the decision to close the fishery was taken at a time when there was hardly any herring left. The main concern about the herring stock has been the possibility of recruitment overfishing, i.e., reducing the spawning stock to a level that is too low to produce strong year classes. Rebuilding of the stock was slower than was expected, due to a sequence of extremely small year classes. This phenomenon, however, was probably not due to the minimized spawning stock, but to a disruption of the transport of herring larvae in the mid 1970's, from their spawning areas in the western North Sea to the nurseries in the eastern North Sea, caused by a change in the North Sea circulation (Corten 1986). Since 1980, the conditions for herring recruitment appear to have improved and the total stock biomass has increased from a minimum of 0.2 million tonnes in 1977 to 2.5 million tonnes in 1984. The picture presented here is necessarily a simplification. The stock has at least three major components, which behaved differently (see also Burd 1978). However, the effect of the fishery on the total stock size was evident. Most of the individuals of the present herring stock are not more than 3 to 4 years old, whereas in an unexploited population fish up to 10 years old are quite normal. In herring an increase in growth rate has been observed (e.g., Saville 1978). This increase is thought by some scientists to be induced by the fishery (Burd 1984), but others are inclined to believe that environmental factors are responsible (Saville 1978). The increase in growth rate is associated with an increase in fecundity at age as well. The fishery has reduced the ecological diversity of the stock. In earlier years fishermen in the southern North Sea and eastern Channel exploited several spawning populations, each of which had its own specific spawning time and spawning place: several populations spawned on specific grounds that extended from the Seine Bay up to North Hinder. From the

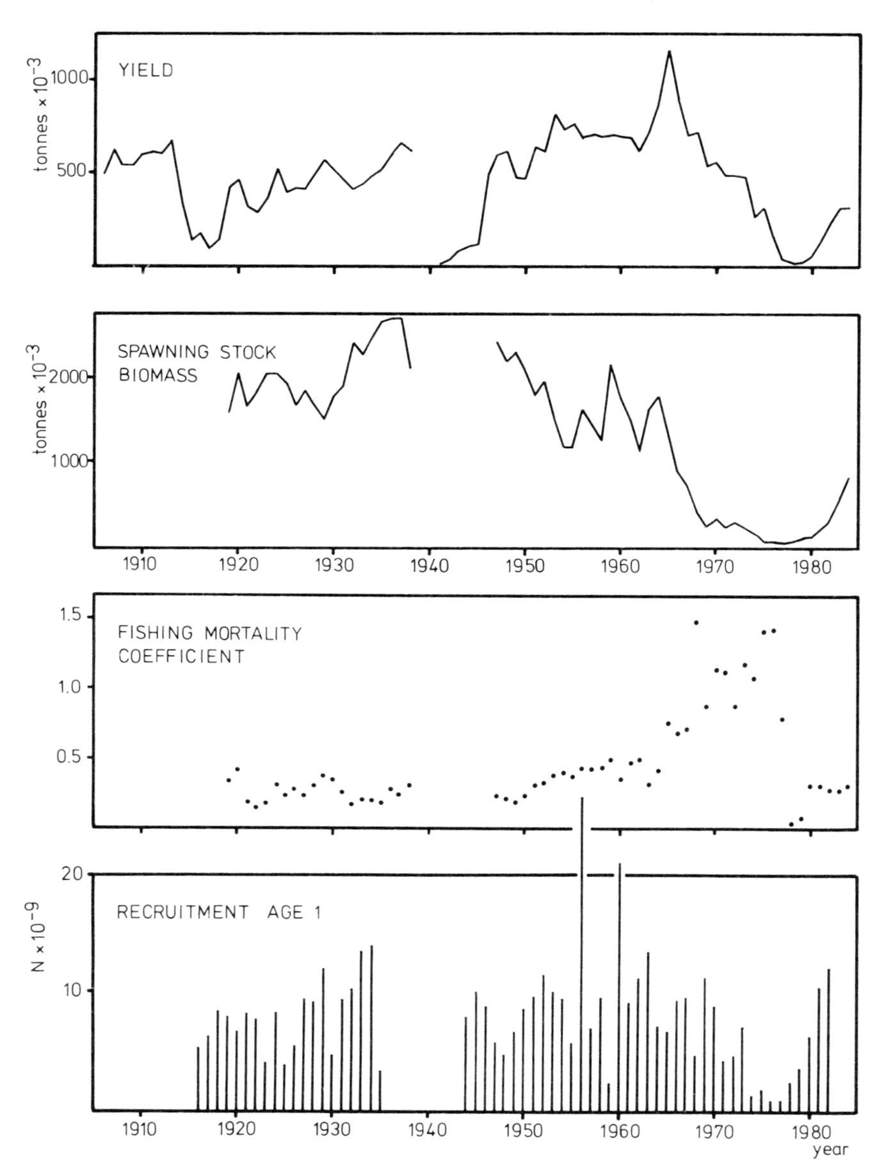


**Fig. 2.** Population parameters North Sea herring

positions of samples of spawning herring from the Dutch fishery it can be concluded that in fact only one of these populations has recovered, the one that spawns off the French coast between Dieppe and Boulogne. At present the fishery concentrates on this single stock and the fishing season for spawning herring, which used to last for 3 months, has now been reduced to not more than a few weeks (A. Corten, pers. commun.).

*Cod* (Fig. 3). The development in the cod stock has been entirely different. Again the landings remained more or less constant over most of the century, but in the mid 1960's, the catch suddenly doubled and remained at a high level afterwards. The increase in the catch was associated with an increased level of recruitment and comparable growth of the spawning stock biomass, but in recent years biomass has fallen back to the original level and has now reached the minimum value for the last 20 years. The estimated fishery mortalities do not indicate any significant increase

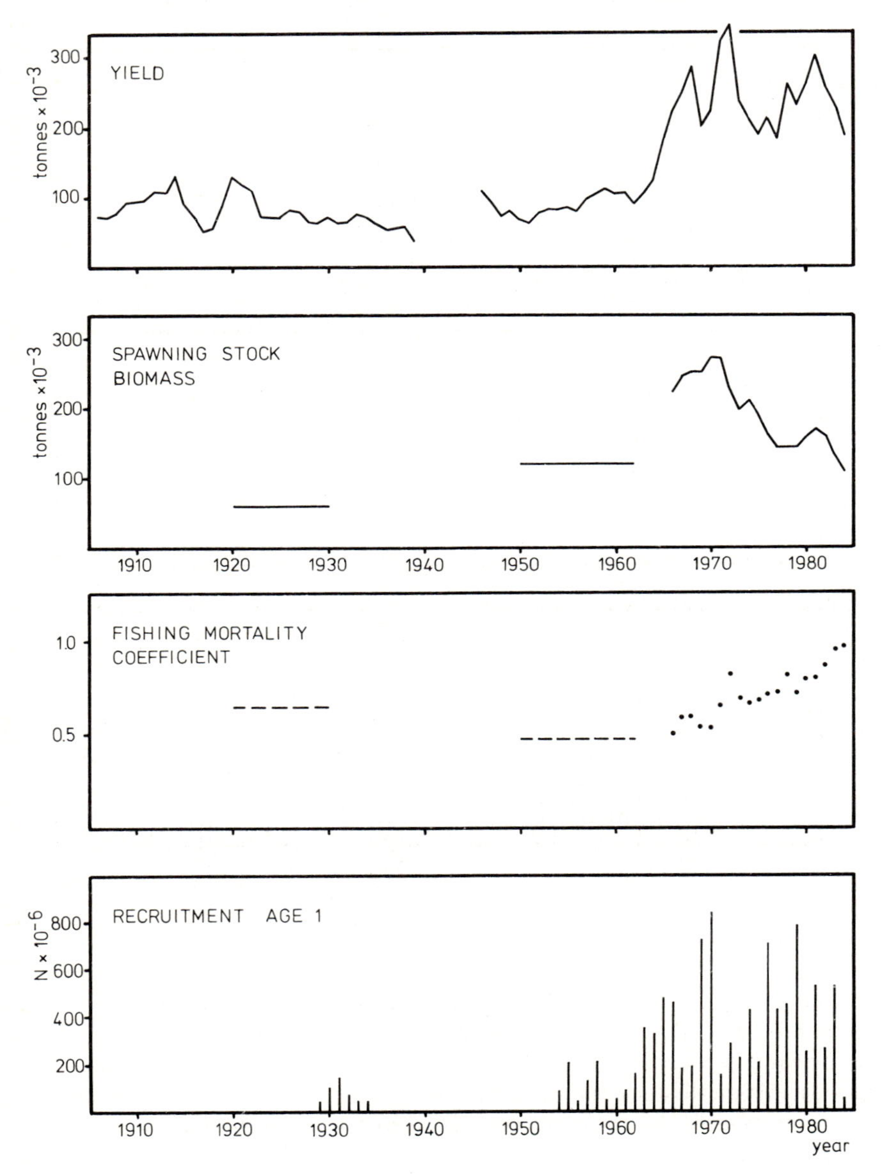


**Fig. 3.** Population parameters North Sea cod

at the time when catch and biomass increased, but the increased exploitation observed in the seventies may have been responsible for the recent decrease in biomass. The present level of fishing mortality is so high that the fishery takes about 60% of all cod of 3 years and older each year. Since a few years the total catch is slightly decreasing again. It is generally accepted that the increase in the cod stock is the result of an increase in the size of the recruiting year classes. Even the poor year classes in recent years appear to be more numerous than before. Thus, the rapid increase in biomass and catch does not seem to be related to a marked change in the fishery and the cause must be found in the sequence of a large number of strong year classes during the last two decades. Apparently, the carrying capacity of the system for the juvenile life stages has changed, but how and why this happened remains unsolved (Daan 1986). Changes in growth rate and age at first maturity are not very clear. Although the spawning stock biomass has decreased, there were, so far, no signs of recruitment overfishing, but it should be noted that year class 1984 seems to be the smallest one ever recorded. The age composition of the stock is such that young fish, up to 3 years old, dominate. The landings consist mainly of fish of 2 and 3 years old, which is in fact too young when their growth potential is concerned and when it is realized that cod mature not before they are at least 3 years old. Even of 4-year-old cod only some 60% become mature. For this species a higher yield could probably be achieved by changing the exploitation pattern. Both an increase in mesh size and a reduction in fishing effort could positively contribute to improve the present situation of growth overfishing in this stock.

*Plaice* (Fig. 4). During the second World War, the plaice stock was rebuilt in the absence of major fisheries. Afterwards the catch remained at a higher level than in the interwar period, and during the last two decades the catch increased even further. Initially, fishing mortality remained at a lower level, and the increased catch and biomass have been explained in terms of an improved exploitation pattern. A most notable feature of the post-1945 period has been the increased frequency of good year classes, which raised the mean level of recruitment (Bannister 1978). The strong 1963 year class has contributed further to the increase in catch and biomass, but since the late 1960's, fishing mortality is increasing and the spawning stock biomass decreasing again. Bannister (1978) found complex changes in growth rate after the second World War, which could not be easily understood. The mean size of the youngest age groups increased, whereas that of older age groups decreased. Rijnsdrop et al. (1983) showed that changes in growth coincided with changes in fecundity and age at first maturity. They concluded that female plaice in recent years spend more energy in reproduction in comparison with the energy spend in somatic growth. Also in plaice the problem of growth overfishing exists. The exploitation pattern could be optimized through an increase in mesh size or better protection of the nursery grounds.

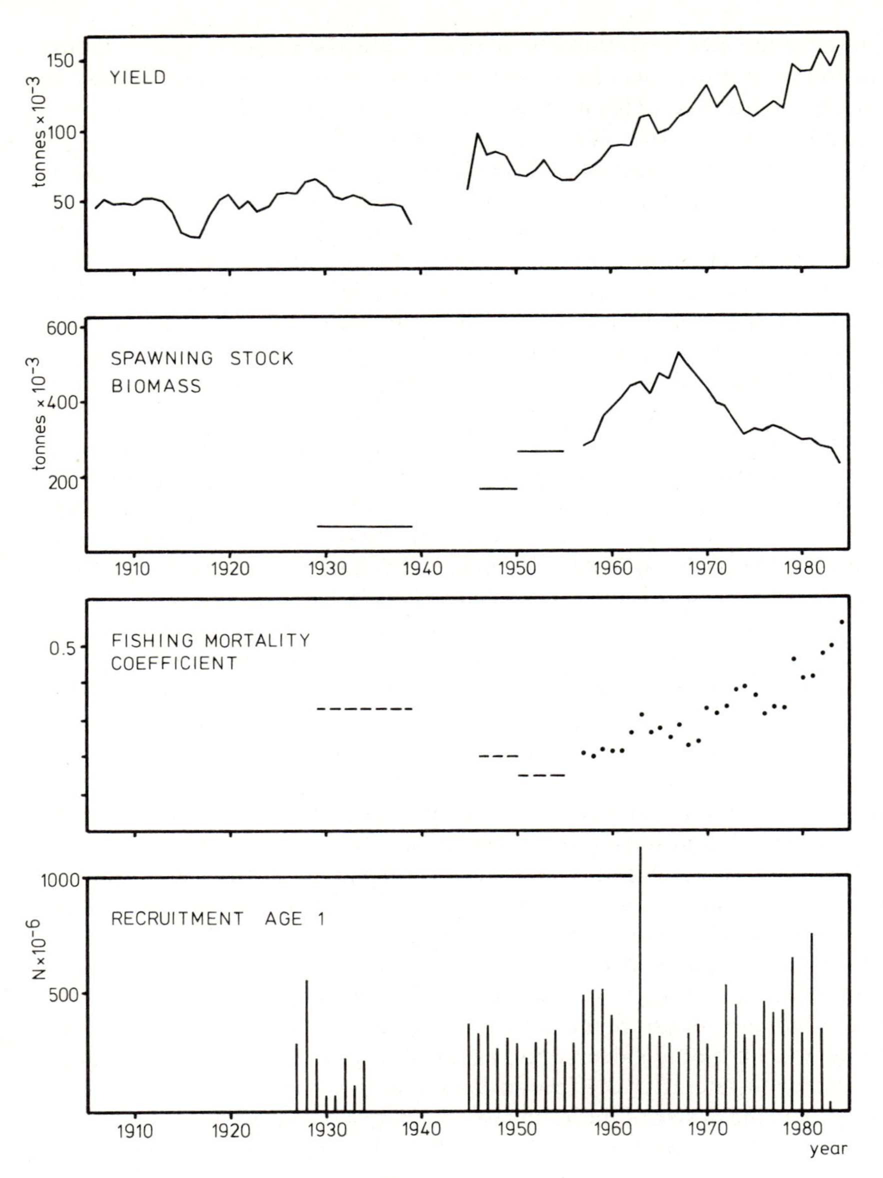


**Fig. 4.** Population parameters North Sea plaice

## 4 Effects on Community Level

As is shown in the examples given above, the stocks of individual species show fairly strong fluctuations which cannot be observed in the total North Sea yield. In the North Sea there has been a very marked shift in species composition. While the pelagic stocks declined, an outburst was observed in gadoid species (Cushing 1980), but the functional relation between these two events is not clear (Cushing 1984). The plankton-eating herring seems to be replaced by other plankton-feeding fishes: sandeels and Norway pout (Andersen and Ursin 1977; Yang 1982b).

Classical fisheries models, such as developed by Beverton and Holt (1957), have typically focused on individual species. In these models the natural mortality of fish has been regarded as constant. With his study of the food consumption of North Sea cod, Daan (1973) drew attention to the enormous influence predators may have on other fish stocks. A few years later, a complex dynamic simulation model for the North Sea was developed by Andersen and Ursin (1977), which showed that there are large gaps in our knowledge with respect to interspecific predation rates. In order to make up for these deficiencies, the theoretical approach had a follow-up in the form of the International Stomach Sampling Project carried out in 1981 under the auspices of ICES. In the course of that year and spread over the whole North Sea, more than 40,000 stomach contents were collected from five fish species that were assumed to be the main fish predators: cod, haddock, whiting, saithe, and mackerel. The first analyses of the results of this project (Anonymous 1986) reveal that the total amount of commercial fish species eaten yearly by these five predatory fishes approximately equals the total fish landings from the North Sea. Available estimates of natural mortality on the youngest age groups of most species, as routinely applied for fish stock assessment, appeared to be far too low. From the results of this project, it is possible to quantify a number of interspecific relations, and to incorporate in a multispecies assessment (Pope 1979; Helgason and Gislason 1979; Sparre 1980), but mutispecies effects are possibly self-cancelling and they may have a zero net effect. So, for short-term predictions, single species models remain valid for many, but perhaps not all species. For management on the longer term, multispecies models may be used in the near future.

It may be clear from what is said so far that most research has been focused on commercial fish species and that relatively little is known about the noncommercial species. One of the few sources of information on other fish species is a rather long series of data from Scottish research vessel surveys (Richards et al. 1978). Such surveys are of special interest in providing unbiased catch rates, while they are independent of the commercial fisheries. The Scottish survey data show changes in abundance throughout the period 1922–1971 in all fish species, irrespective of their state of exploitation. Abundance of most species was constant or declining in the period 1922–39 and during the first years after the war, abundance was as low as just before. This contrasts with the situation recorded for the commercial fisheries, e.g., for plaice and cod. Whereas in the years 1950–71 abundance was steady or increasing, the last decade showed particularly large fluctuations in abundance. In the near future ICES will establish a database with data from the ICES coordinated International Young Fish Survey, a bottom trawl survey which covers the whole North Sea, Skagerrak, and Kattegat and is carried out yearly in February. This data base will contain data from 1970 onwards for all commercial and noncommercial species that are caught. Another interesting data series are the bycatch-data in the German shrimp fisheries in the Wadden Sea in the period 1954–1981 (Tiews 1983). Distinct trends in abundance were observed in ten species. Seven inhabitants of the Wadden Sea have decreased in the investigated period: shore crab, seasnails, butterfish, gobies, eel, solenette, and gurnards. Three species with a distribution that is not restricted to the Wadden Sea: dab, sprat, and cod, have distinctly increased. The abundance of 15 other species showed no specific trend. Some of the changes may reflect interspecific relationships as indicated earlier, but Tiews does not exclude the possibility that the observed decline of some of the species has to be related to pollution input of anthropogenic origin.

## 5 Effects on Benthic Fauna

The effects bottom trawling may have on the benthic fauna have recently been reviewed by de Groot (1984). The impact on the benthic fauna varies with type of gear. A groundtrawl with large bobbins on its groundrope, as used for catching roundfish, will have hardly any effect. Heavy beamtrawls, however, used for fishing sole and plaice, have up to 12 tickler chains which penetrate in the upper bottom layer to activate the fish. This gear undoubtedly damages benthic animals.

Protests against bottom trawling date back to the period of its introduction, which was the 13th century for northwest Europe. The complaints concerned the mesh size as well as the deleterious effects of the use on young fish, fry, and benthic life as a source of food for larger fish. Graham (1955) investigated the effect of the use of tickler chains on the food species for plaice in the southern North Sea. He concluded that the damage to these species cannot be serious. However, since then the power of the engines and the weight of the gears increased substantially.

In the early 1970's ICES stimulated new investigations. From this work it may be concluded that the effect of the passage of a trawl varies greatly with the nature of the sea bed, which may vary from silt to rocky. Under normal working conditions, trawls influence only the top layer of the sea bed: penetration will be up to 30 mm on muddy ground and 10 mm on sandy ground. In tidal waters the sea bed surface is by no means permanent, but is in a constant state of change and surface sediments are constantly being redeposited. In such areas, the physical effect of trawling on the sea bed is probably only of minor importance.

The catch of benthos by a beamtrawl with tickler chains is about ten times as large as that of a groundtrawl without ticklers. Some groups of animals, e.g., hydrozoans and echinoderms may suffer heavy damage, others, such as gastropods and hermit crabs, escape relatively easily. It would seem not unreasonable that there are long-term effects to fishing, which might become apparent as shifts in macrobenthos composition. However, data series are scarce and only from the German Wadden Sea is there an example of the effect of human activities, the mussel culture and the shrimp fishery (Riesen and Reise 1982; Reise 1982). From these data it may be concluded that the trend will presumably be a shift to a relative increase in polychaetes and a relative decrease in molluscs and crustaceans. At present there are no indications that this will lead to a food shortage for fish stocks in the North Sea: there are no signs of reduced growth. Rather the reverse is true, and several species appear to have undergone accelerated growth. It has been suggested that due to the action of trawls, large quantities of benthic animals become available as a food source for fishes. An example that this may occur is given by Arntz and Weber (1970), who found that cod and dab in Kiel Bay (Baltic) were eating large amounts of the bivalve *Ciprina islandica*, a species with solid shells that could only be crushed by the ottertrawl doors. There is also the well-known practice among fishermen to keep fishing along a constant path for even days, during which they see the catches increase. It is asumed that the fish are attracted by odors from the crushed benthic animals.

Although the short-term effect of trawling on the benthic fauna is detectable, no decrease in productivity has yet been observed and so far trawling has not been shown to be detrimental to the North Sea as an ecosystem. The outcome of earlier investigations done so far has recently been questioned (Rauck 1985) as they were restricted to only one single passage of a trawl. With the present effort, however, some grounds are fished three to five times a year, or even more frequent and, furthermore the engine power has since the 1970's increased even more.

## 6 Conclusions

The North Sea ecosystem is very complex, and within this system innumerable interactions exist. The fishery somehow influences the ecosystem, but is itself influenced by technical developments, socio-economic circumstances and natural variations in the fish stocks. Against the very complex background of all possible interactions, it often proves extremely difficult to conclude what exactly is the cause of certain changes, and thus that it is also difficult to distinguish between natural variation and the changes induced or maintained by human activities.

*Acknowledgments.* I should like to thank Dr. D.J. Garrod and Dr. N. Daan for their critical comments on the manuscript.

## References

Andersen KP, Ursin E (1977) A multispecies extension to the Beverton and Holt theory of fishing with accounts of phosphorous circulation and primary production. Medd Dan Fisk Havunders 7:319–435
Anonymous (1986) Report of the ad hoc multispecies assessment working group. ICES CM 1986/Assess:9 (mimeo)
Arntz WE, Weber W (1970) Cyprina islandica L. (Mollusca, Bivalvia) als Nahrung von Dorsch und Kliesche in der Kieler Bucht. Ber Dtsch Wiss Komm Meeresforsch 21:193–209
Bannister RCA (1978) Changes in plaice stocks and plaice fisheries in the North Sea. Rapp P V Reun Cons Int Explor Mer 172:86–101
Beverton RJH, Holt SJ (1957) On the dynamics of exploited fish populations. Fish Invest Lond (2)19:533
Burd AC (1978) Long-term changes in North Sea herring stocks. Rapp P V Reun Cons Int Explor Mer 172:137–153
Burd AC (1984) Density-dependent growth in North Sea herring. ICES C M 1984/H:4 (mimeo)
Corten A (1986) On the causes of the recruitment failure of herring in the central and northern North Sea in the years 1972–1978. J Cons Int Explor Mer 42:281–294
Cushing DH (1980) The decline of the herring stocks and the gadoid outburst. J Cons Int Explor Mer 39:70–81
Cushing DH (1984) The gadoid outburst in the North Sea. J Cons Int Explor Mer 41:159–166

Daan N (1973) A quantitative analysis of the food intake of North Sea cod, Gadus morhua. Neth J Sea Res 6(4):479–517

Daan N (1986) Results of recent time-series observations for monitoring trends in large marine ecosystems with a focus on the North Sea. In: Sherman K, Alexander LM (eds) Variability and management of large marine ecosystems. AAAS selected symposia series 99: 319 pp

Devold F (1963) The life history of the Atlanto-Scandian herring. Rapp P V Reun Cons Int Explor Mer 154:98–108

Graham M (1955) Effect of trawling on animals of the sea bed. Deep Sea Res 3:1–6

Groot SJ de (1984) The impact of bottom trawling on benthic fauna of the North Sea. Ocean Manage 9:177–190

Helgason T, Gislason H (1979) VPA-analysis with species interaction due to predation. ICES C M 1979/G:52 (mimeo)

Hempel G (ed) (1978) North Sea fish stocks – Recent changes and their causes. Rapp P V Reun Cons Int Explor Mer 172

Holden M (1978) Long term changes in landings of fish from the North Sea. Rapp P V Reun Cons Int Explor Mer 172:11–26

May RM (ed) (1984) Exploitation of marine communities, Report of the Dahlem Workshop on Exploitation of marine Communities, Berlin 1984, April 1–6. Springer, Berlin Heidelberg New York, 366 pp

Pope JG (1979) A modified cohort analysis in which constant natural mortality is replaced by estimates of predation levels. ICES C M 1979/H:16 (mimeo)

Rauck G (1985) Wie schädlich ist die Seezungenbaumkurre für Bodentiere? Inf Fishwirtsch 4:165–168

Reise K (1982) Long-term changes in the macrobenthic invertebrate fauna of the Wadden Sea: are polychaetes about to take over? Neth J Sea Res 16:29–36

Richards J, Armstrong DW, Hislop JRG, Jermyn AS, Nicholson MD (1978) Trends in Scottish research-vessel catches of various fish species in the North Sea, 1922–1971. Rapp P V Reun Cons Int Explor Mer 172:211–224

Riesen W, Reise K (1982) Macrobenthos of the subtidal Wadden Sea: revisited after 55 years. Helgol Meeresunters 35:409–423

Rijnsdorp AD, Lent F van, Groeneveld K (1983) Fecundity and the energetics of reproduction and growth of North Sea plaice (Pleuronectes platessa) ICES C M 1983/G:31 (mimeo)

Saville A (1978) The growth of herring in the northwestern North Sea. Rapp P V Reun Cons Int Explor Mer 172:164–171

Sparre P (1980) A goal function of fisheries (legion analysis). ICES C M 1980/G:40 (mimeo)

Tiews K (1983) Über die Veränderungen im Auftreten von Fischen und Krebsen im Beifang der deutschen Garnelenfischerei während der Jahre 1954–1981 – Ein Beitrag zur Ökologie des Deutschen Wattenmeeres und zum biologischen Monitoring von Ökosystemen im Meer. Arch Fishereiwiss 34 (Beih 1): 1–156

Ursin E, Andersen KP (1978) A model of the biological effects of eutrophication in the North Sea. Rapp P V Reun Cons Int Explor Mer 172:366–377

Yang, Jiming (1982a) An estimate of the fish biomass in the North Sea. J Cons Explor Mer 40:161–172

Yang, Jiming (1982b) A tentative analysis of the trophic levels of North Sea fish. Mar Ecol Prog Ser 7:247–252

# Ecological Impacts During the Completion of the Eastern Scheldt Project

C.J. VAN WESTEN[1] and J. LEENTVAAR[2]

## 1 Introduction

The storm flood of February 1953 caused large areas of land to be inundated. Nearly 2000 people and innumerable cattle were killed. This disaster motivated a speed-up of the planned reduction of the length of the Dutch coastline. The "Delta Plan", in force as law since 1958, recommended that five of the seven estuaries in the southwest Netherlands should be closed off by dams between the islands, with additional barriers being built on the landward side where necessary, and that the sea defences along the Dutch coast should be strengthened. The smaller areas would be closed off first, so that the larger enterprises could benefit from the experience gained from constructing the smaller dams.

So far the execution of the Delta Plan has resulted in the creation of two freshwater lakes and two saltwater lakes. The damming of the last estuary, the Eastern Scheldt (Fig. 1), was planned for 1978. However, by the mid 1970's, opinions and priorities had changed to such an extent that the execution of the plan was reconsidered. This resulted in new plans in which the Eastern Scheldt would not be closed completely. Instead, the size of the basin would be reduced by means of two compartment dams (Philipsdam and Oesterdam), reducing its intertidal area. The remaining tidal basin would be safeguarded against too high water levels by the construction of a storm-surge barrier, which would be closed when there was a risk of flooding.

Once the construction of the storm-surge barrier and compartment dams in the Eastern Scheldt has been completed, they will provide adequate flood protection for the adjoining land areas, while the tidal saltwater environment will be preserved in the main body of the estuary. After the decision in 1976 to construct the storm-surge barrier and the compartment dams several detailed plans have been subject to policy analyses. For example, based on thorough investigations, it has been decided to close the compartment dams with sand; for budgetary reasons the completion of the Philipsdam has been delayed by about 1 year. The needs of the saltwater ecosystem are also being borne carefully in mind while the construction works are carried out. The integration of ecological and fishery aspects in the design and construction stage has contributed positively to the conservation of the ecosystem. During the final construction stage, tidal movement in the estuary as a whole will be affected, particularly because the storm-surge barrier will be used to control it while work is being done on the dams and on the barrier itself.

[1]Tidal Waters Division, Water Systems Delta Area, Grenadierweg 31, 4338 PG Middelburg, The Netherlands
[2]Present address: Inland Waters Division, Maerlant 6, 8200 AA Lelystad, The Netherlands

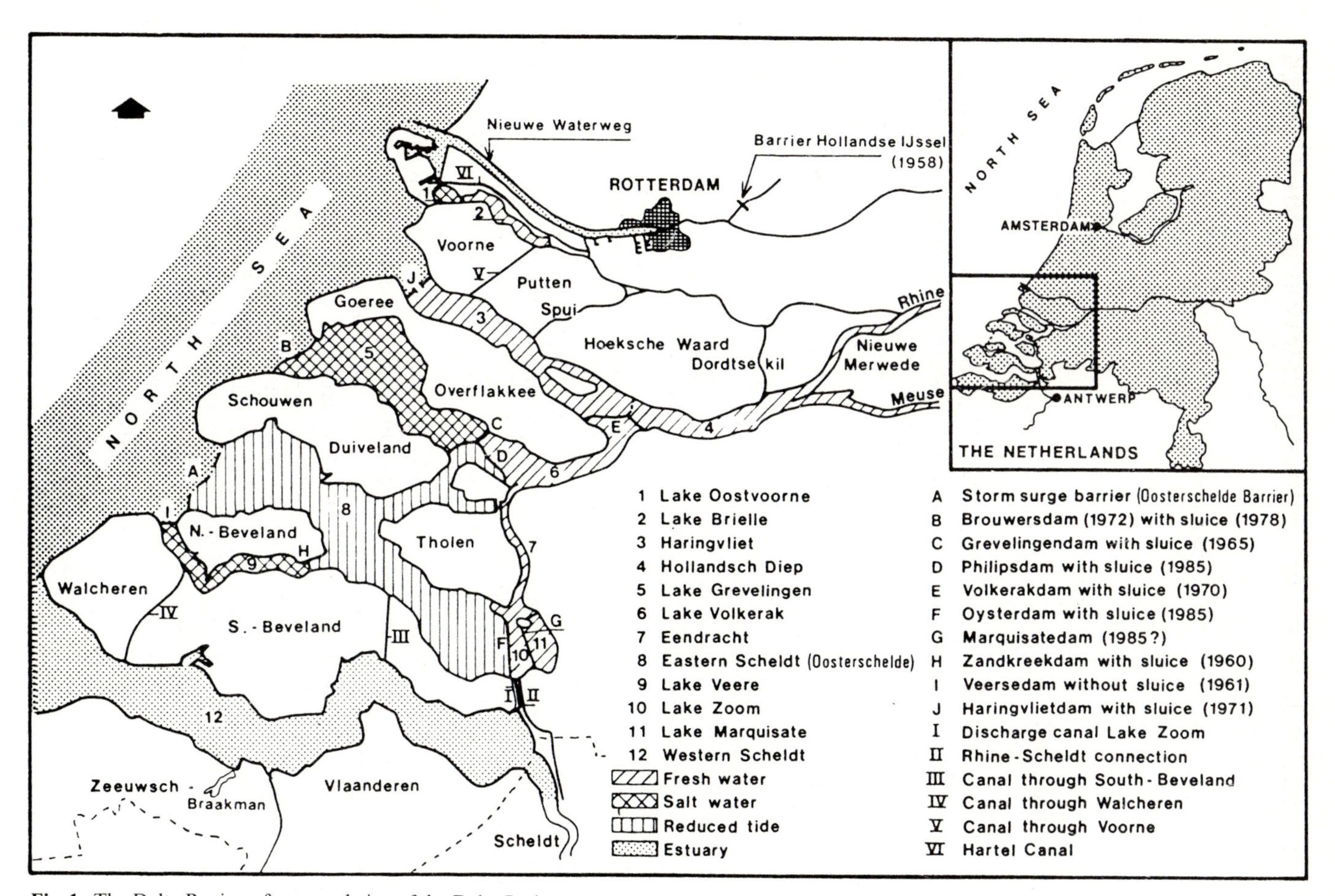


**Fig. 1.** The Delta Region after completion of the Delta Project

A research project has started to investigate the effects of the reduction of the tidal movements; the second aim of the research project is to indicate means to minimize these effects. The aim of this paper is to predict the effects of the plans for the final stage, and possible variants thereof, insofar as it is possible to make such predictions in the light of current knowledge.

## 2 The Project

In the mouth of the Eastern Scheldt, an open storm-surge barrier is under construction, which will consist partly of two dam sections and three barrier sections in the Hammen (675 m width), Schaar (720 m width) and Roompot (1440 m width). When the barrier is open, the tides will pass through the 62 openings between the 65 piers. Due to the construction of the barrier the open cross-section will be reduced from 80,000 $m^2$ to about 16,000 $m^2$. In the eastern part of the Eastern Scheldt, two compartment dams with shipping locks are under construction: the Oesterdam and the Philipsdam. These dams will cut off the eastern part of the present system from tidal influence. The area will be flushed with fresh riverwater, leading to the creation of a freshwater lake.

The main reasons for the compartimentation are in short:

1. To retain a sufficient tidal range in the remaining estuary in spite of the reduction of the cross-section,
2. to counter salination of the Hollandsch Diep and Haringvliet,
3. to create a tidal-free canal between Antwerp and Rotterdam. As a result of the reduced open cross-section and the reduced volume of the estuary by the compartment dams the tidal amplitude will decrease from the present average 3.50 m to the expected value of 3.05 m.

## 3 The Estuary: Present and Future Situation

In the Eastern Scheldt, many changes will occur during and after the completion of the storm-surge barrier and the compartment dams (Knoester et al. 1984) (Fig. 2). In order to evaluate the effects on the estuary while the works are under construction, it is necessary to describe the present situation together with a prognosis of the future situation once the project has been completed.

### 3.1 Present Situation

The tides in the estuary are adequately described in terms of their horizontal and vertical components. The mean tidal range varies from 2.8 m at the entrange to 3.7 m at the most inland site. The maximum flow velocity during flood and ebb is about 1.5 m $s^{-1}$. In each tidal cycle about 1250 million $m^3$ of water flows into and out of

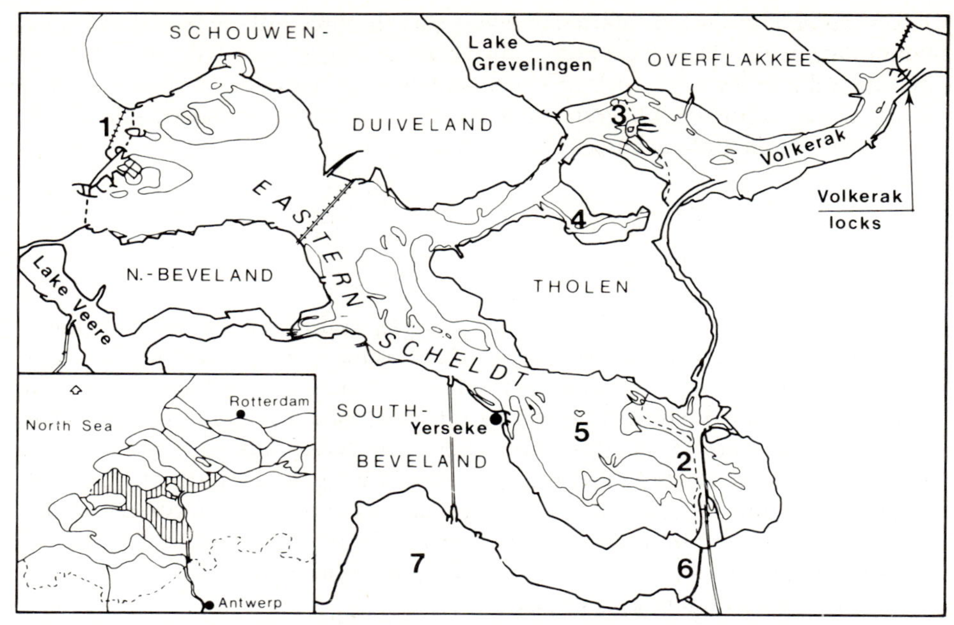


**Fig. 2.** Eastern Scheldt *1* Storm-surge barrier; *2* Oesterdam; *3* Philipsdam; *4* Krabbenkreek; *5* Kom; 6 Bath discharge canal; *7* Western Scheldt

the Eastern Scheldt, with the average rate of flow being approximately 56000 $m^3 s^{-1}$. The area receives on average 55 $m^3 s^{-1}$ of water from all other sources (the Volkerak locks; inflowing small rivers; and excess water from Lake Veere and Lake Grevelingen, and from polders).

Water quality in the Eastern Scheldt is primarily influenced by the sea, but there are local effects of the larger freshwater sources, such as the Volkerak locks and the rivers in the northern arm of the basin. Salinity levels in the Eastern Scheldt are generally within the range 15.5–17.5 g Cl $l^{-1}$, with levels in the northern arm being somewhat lower than elsewhere.

The morphology of the estuary is mainly characterized by tidal channels, intertidal areas and salt marshes. The maximum depth of the tidal channels is about 50 m. Water, intertidal areas and salt marshes together provide the habitat for a rich variety of flora and fauna. The principle features of the biota in the Eastern Scheldt are its productivity and its great diversity of species. The water, intertidal area, and salt marshes form the habitat for a rich flora and fauna. A large variety of organisms are also to be found on and near the hard substratum. The hard peat layers and the materials introduced by man, e.g., for stone dyke facings, offer many places where organisms can live. The Eastern Scheldt is of great importance for bird species, especially waders and ducks. The Delta Region is on one of the migration routes of the northern European bird population.

The Eastern Scheldt plays an important role as a nursery for shellfish. Mussel cultivation makes a major claim on the estuary bottom. Near Yerseke a large area of shallow water is used for the rewatering and storage of mussels. This is because of the presence of a hard peat substratum in the area, combined with certain hydraulic conditions. Cockles, shrimps, eel, and crab are also harvested in the Eastern Scheldt.

The Eastern Scheldt is crossed by a number of important inland waterway routes from Rotterdam to Antwerp. In and around the estuary water sports, angling, shore recreation, and diving are practised.

### 3.2 Future Situation: Major Changes

The government decided that a minimum average tidal range of 2.70 m was necessary at Yerseke to preserve an adequate ecosystem after completion of the storm-surge barrier and compartment dams. Yerseke was chosen as the standard because of the shellfish beds in that area. The average salinity in the Kom is 15.5 g $Cl^- l^{-1}$: reduction below this value to a minimum of 13.5 g $Cl^- l^{-1}$ was deemed admissable for a short period. The salinity in the Krabbenkreek was to be kept above 13.0 g $Cl^- l^{-1}$ to safeguard the salt marsh vegetation. After completion of the storm-surge barrier and compartment dams, it was expected that the average tide would be reduced by 14% and the current velocities by about 30%.

The chloride level is an important parameter for the ecosystem of the Eastern Scheldt due to the relation between salinity, diversity, and productivity. Calculations of salinity changes show that if the tidal range is reduced from 3.50 m to 3.05 m in the final situation, together with reduced freshwater discharge due to the construction of the compartment dams, the salinity in the estuary will increase by 0.5 g $Cl^- l^{-1}$ in winter and by 1.0 g $Cl^- l^{-1}$ in summer (Fig. 3).

The construction of the compartment dams and the tidal reduction will result in the area of salt marshes and the intertidal area being reduced (Fig. 4).

## 4 The Use of the Storm-Surge Barrier

In order to enable the construction of the storm-surge barrier (Fig. 5) and compartment dams in the Eastern Scheldt to be completed, it will be necessary to close the barrier wholly or partially during certain periods. The final construction stage may be divided into three parts:

*Period A* begins in March 1986, when the tidal range at Yerseke will be reduced to less than the ultimate tidal range after all work has been completed (3.05 m). This is because of the placing of the sill beams (Fig. 6) in position, an operation which will be completed by mid-1986, when the tidal range will fall to approximately 2.60 m.

*Period B* will run from June to November 1986, during which time the tidal range at Yerseke will be around 2.30 m, since a number of gates in the barrier will be closed almost constantly. This is to reduce turbulence in the water while the upper beams are mounted and deposits of heavy material are made around the barrier. For a period of up to a fortnight in July 1986, while the upper beams are being mounted in the middle of the Roompot, this channel will have to be

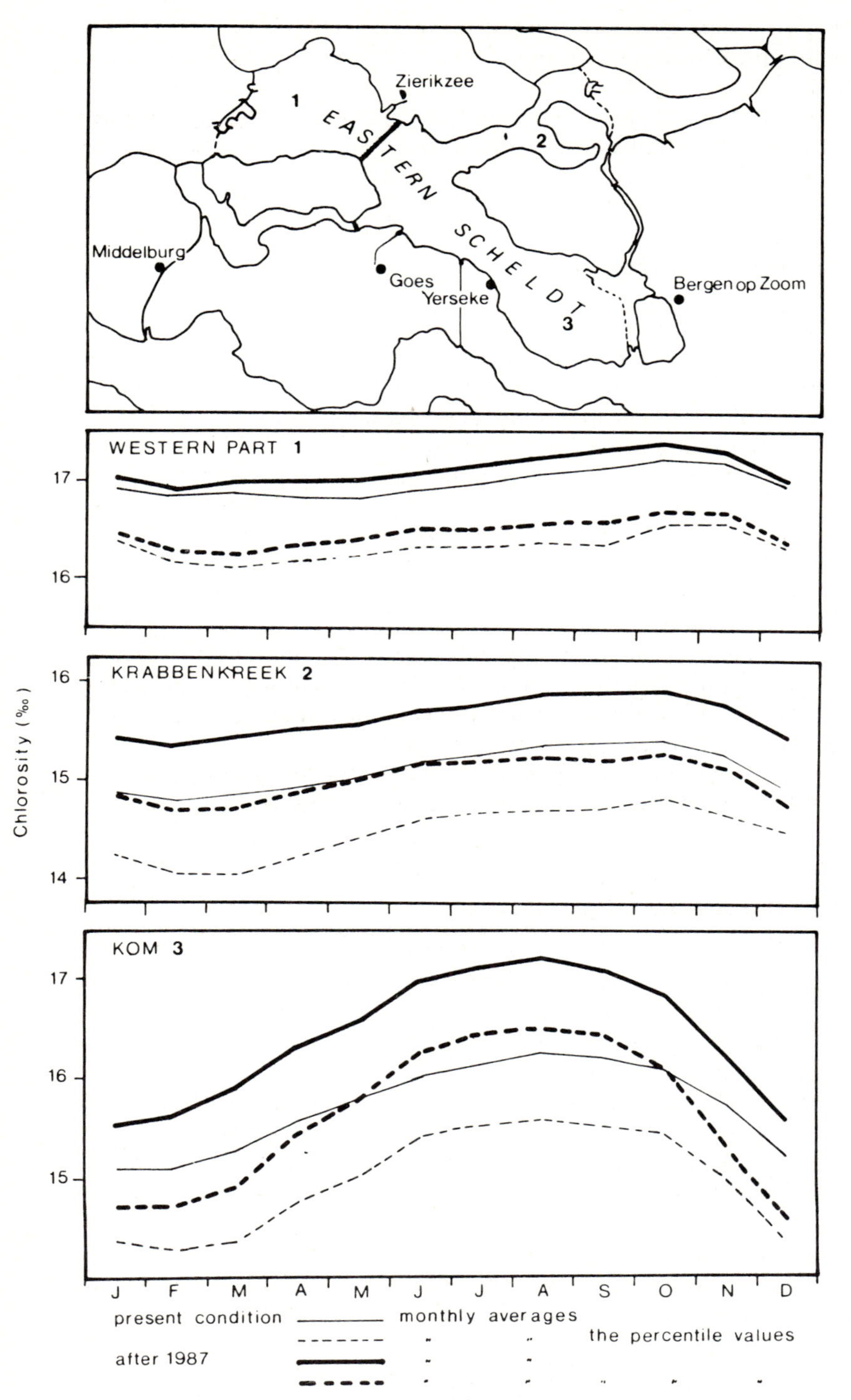


**Fig. 3.** Monthly average chloride content and tenth percentile values at various sites in the Eastern Scheldt under present and future conditions

| MHT1 MHT2 MLT1 MLT2 salt marsh new intertital area above high tide shallow water | Above High Water Level<br>salt marshes | Intertidal Area<br>mudflats sandshoals | Under Mean Low Water Level<br>water |
|---|---|---|---|
| Present Situation Tidal range 3,40 Yerseke | 16,2 | 151,8 | 274,0 |
| After compartimentation Tidal range 3,40 | 6,0 | 114,0 | 230,0 |
| After reduction of tidal range to 2,70 | 6,0 | 103,2 | 240,8 |

**Fig. 4.** Changes in surface (km²) of different tidal areas due to compartimentation and tidal reduction to at least 2.70 m

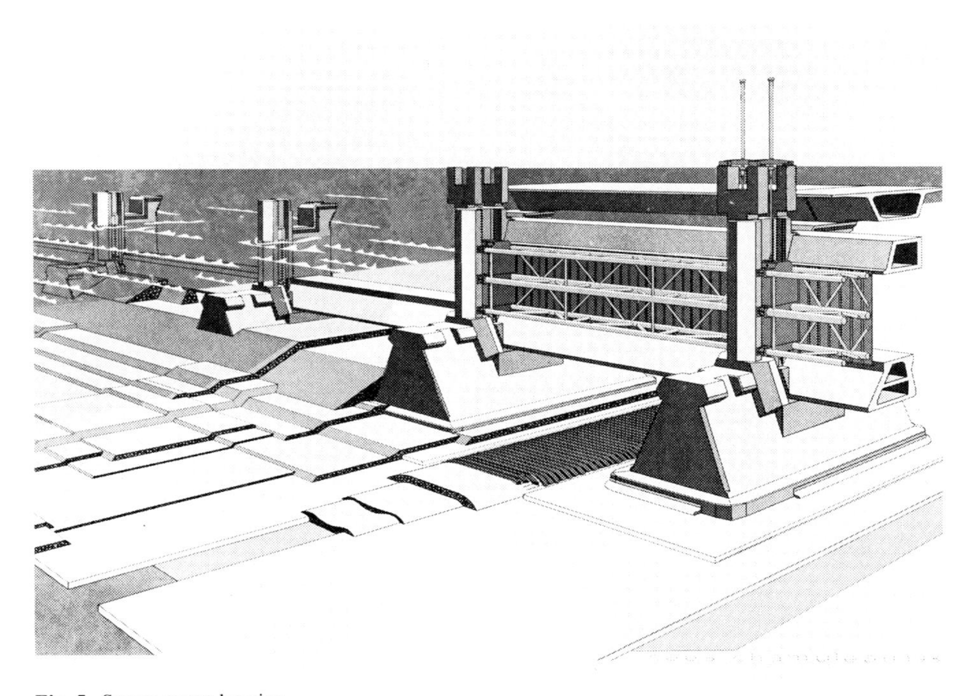

**Fig. 5.** Storm-surge barrier

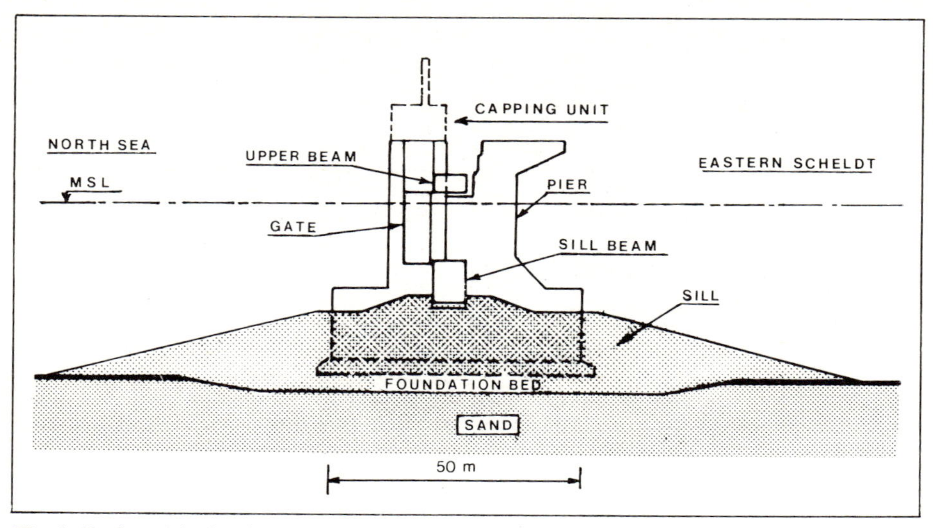


**Fig. 6.** Design of the barrier

completely closed for 12 h each night (tidal range: approximately 1.20 m) and opened during the day and at the weekend (tidal range: approximately 2.60 m). In October 1986, when the Oesterdam is being completed, the storm-surge barrier will be closed completely for a maximum of 2 days, and the tidal range will also be reduced for several days beforehand.

*Period C* runs from December 1986 to April 1987. With the Oesterdam completed and the Philipsdam not yet to become operational, the tidal range will be approximately 2.70 m. In order to enable the Philipsdam to be completed, the frequency and range of the tides will be reduced for several weeks. Thereafter, the storm-surge barrier will be closed completely for up to 2 days.

## 5 Effects

The effects of the completion of the storm-surge barrier are compared with the standards that were set up for salinity and tidal difference in the final situation and also with the predicted situation of the ecosystem in this final situation when the storm-surge barrier and compartment dams have been completed.

A summary of the possible effect has been given by Leentvaar and Nijboer (1986); these effects (Fig. 7) will be described in so far they are essential. The criterion for hazardous effects on the ecosystem is defined that such effects may be produced when the effects are slight or last for such a short interval that they remain within the natural fluctuations. This applies for species that will be present in the final situation. Irreversible adverse effects on the ecosystem should be avoided.

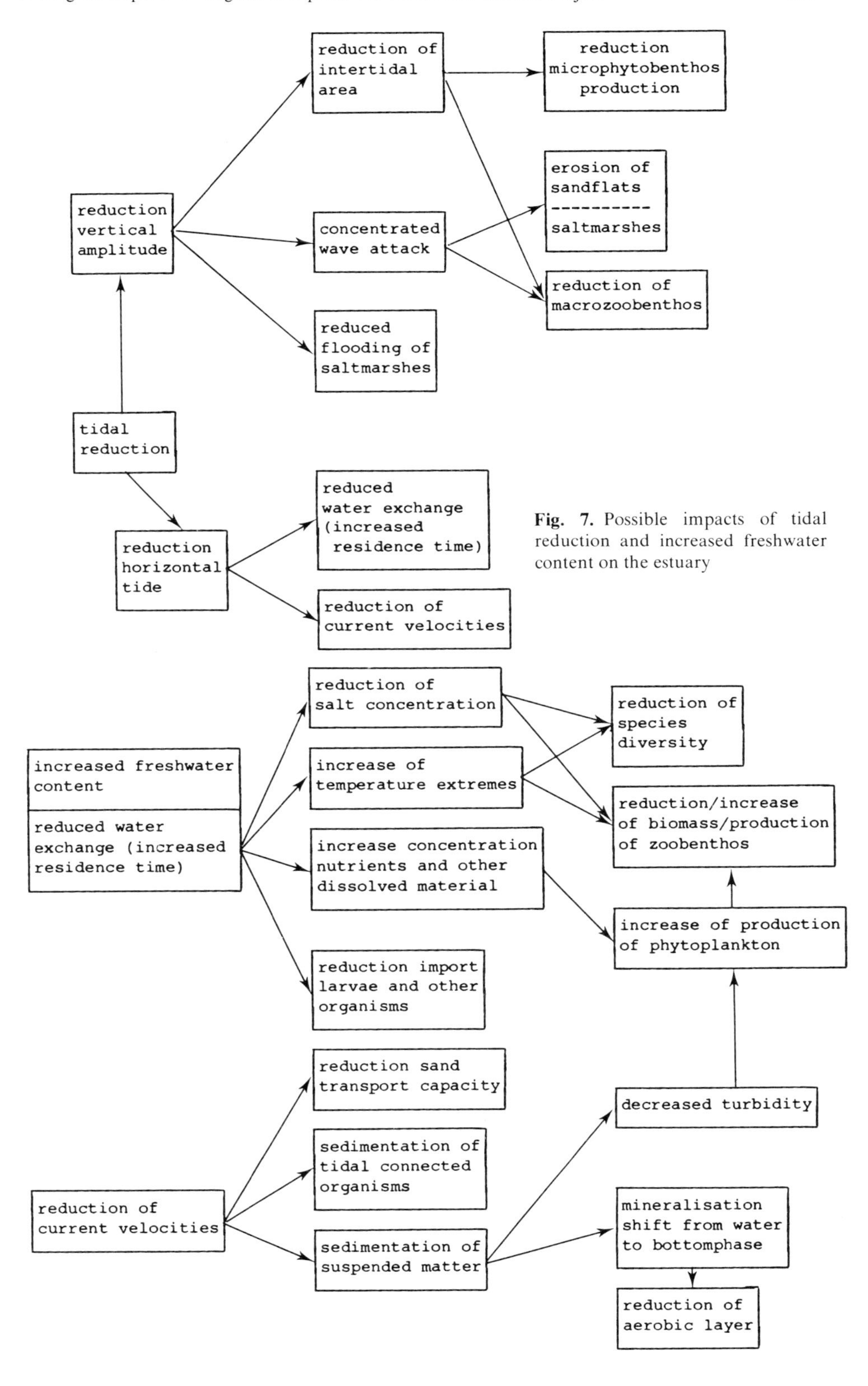


**Fig. 7.** Possible impacts of tidal reduction and increased freshwater content on the estuary

## 5.1 Hydraulic and Physical Effects

Tidal range will be reduced in periods A, B and C by means of the storm-surge barrier. As a result, the intertidal zone will be 7%, 16%, and 6% smaller in area respectively than it will eventually be after all work has been completed. Currents in the estuary will also be reduced, especially during period B, when the section of the barrier in the Roompot will be alternately opened and closed, and while the finishing touches are being put to the compartment dams.

The salt content in the estuary will be reduced during the transitional phase because the flow of freshwater from inland will continue as normal while the influx and outflow of saltwater from the North Sea are reduced. Salt content in the central and more inward sections of the estuary will fall to a level between 0.5 and 1.0 g $l^{-1}$ $Cl^-$ lower than it is now. During the period as a whole, salt contents reduced by 2 to 3 g $l^{-1}$ $Cl^-$ will occur in the Krabbenkreek. Various compensatory measures can be taken which will almost entirely balance out these reductions in salt content.

According to the current state of knowledge, a satisfactory mixed water system can be preserved in the Eastern Scheldt so long as the tidal range does not fall below 2.30 m and the normal tidal frequency is maintained. There will be a danger of stratification during the alternate opening and closing of the barrier in the Roompot and during the final stages of completion of the dams. Calculations suggest that when the barrier is fully or partially closed, the water could become separated into layers of differing salt content within a matter of a few days, with least salt remaining in the top layer. This could cut off the oxygen supply to the deeper strata.

A number of measures are necessary in order to minimize the danger of excessively low salt content and stratification occurring. In particular, research has being conducted into whether the flow of freshwater to the Volkerak via the Volkerak locks and the rivers of Brabant can be reduced or diverted altogether. An other possibility is to heighten the salt content of the Zijpe-Krammer area (12–14 g $Cl^-$ $l^{-1}$) by letting in water from Lake Grevelingen (16–17 g $Cl^-$ $l^{-1}$) via the sluice in the Grevelingen dam. The possibilities and consequences of diverting water through the Bath discharge canal before and during the completion of the Philipsdam are also being investigated. Concerning the realization of these measures, consultations have being made with the several directorates. At this moment the plans are to reduce the flow of freshwater via the Volkerak locks and to raise the salt content by the inlet of water via the sluice in the Grevelingen dam.

During the transitional phase the suspended solids content of the water will be reduced because of the deceleration of currents, and the water will thus become clearer. On the other hand, local and temporary increases in suspended solids could occur due to increased erosion of holes in the bed of the estuary near the barrier and to increased erosion of the intertidal zone and during the deposition of sand in connection with the completion of the compartment dams.

## 5.2 Effects on the Environment

*Salt Marshes.* No essential changes will occur in the composition and zoning of the salt marsh vegetation because the periods of tidal reduction are too short to change the flooding frequency of the salt marshes. The salt marshes may settle somewhat more than usual because of drying out.

*Intertidal Zone.* A decrease in the tidal range affects the frequency of inundation of the lower intertidal area. Vegetation and benthic fauna (Hummel et al. 1985) will die in the parts of the intertidal zone which are left permanently dry. These are the parts higher than mean sea level (M.S.L.) + 1 m and are relatively poor in biomass; some flora and fauna between M.S.L. + 0.5 m and M.S.L. + 1 m will also die because of excessive exposure to the air.

The reductions in tide will concentrate the action of the waves on a smaller area, and increased erosion will occur on the margins of tidal flats and sandy shoals, particularly in spring and autumn. Current research on the distribution of species indicates that all species also occur in the subtidal zone. After completion, common species will recover rather quickly. Rare species are expected to be recovered after several years. No loss of diversity or biomass will be found as a consequence of the reduction of the tidal range in the transitional phase.

*Embankment of Dikes.* The flora and stationary fauna above M.S.L. + 1 m will die by drying out. Effects between M.S.L. + 0.5 m and M.S.L. + 1 m could be considerable because of the reduction in flooding frequency. The vegetation in the lowest areas will remain permanently submerged, but this will not lead to any damage. Due to the low current velocities some sponges may be covered by bacteria within 10 days (Leentvaar and Nijboer 1986). With reference to the recoveryspeed of the benthic fauna it can be assumed that 80–90% of the original biomass will be present again in 2–3 years. Common species will recover in months; the recovery of rare species will last several years. No loss of biomass or diversity will be found at the long term.

*Birdlife.* The number of birds in the Eastern Scheldt varies during the year; especially in the months October-December many birds are present (Fig. 8; after Meininger et al. 1984). The reduction in tides will particularly reduce the foraging area and the length of exposure of areas rich in food. This will lead to lower bird numbers and some birds will move to other areas. These effects will mainly arise during period B and while the dams are being completed.

### 5.3 Effects on Fishing

*Mussel Cultivation.* The reductions in tides will not lead to exposure of mature musselbeds or rewatering plots. However, if no compensatory measures were taken, the salt content in the latter areas could fall too low. The reduction in current speeds, notably during period B, will not have any deleterious effect on the carriage of food to or pseudofeces away from musselbeds. When the Roompot is being opened and closed and during the completion of the dams, the water depth over the young musselbeds will be less than it will be after all work has been completed, and they will therefore be less accessible by boat. Difficulties arising from this can be limited through consultation with the fishermen.

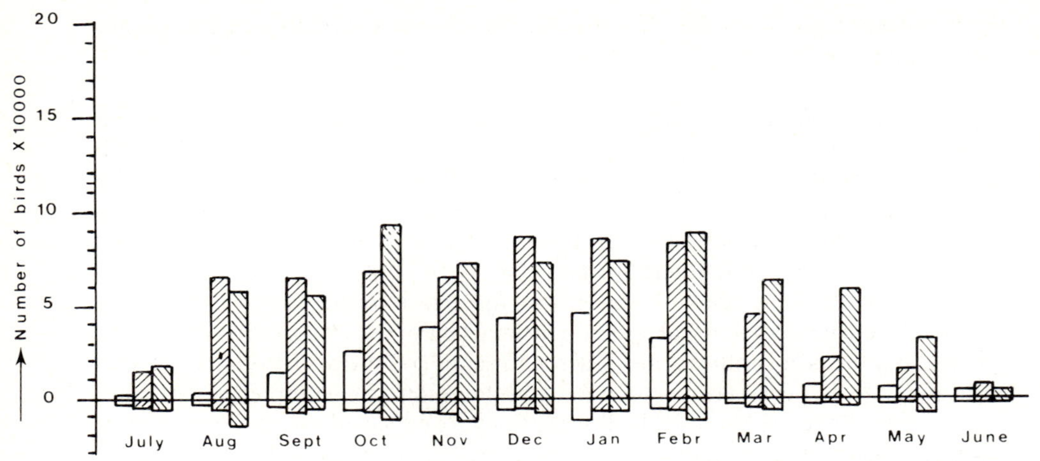


**Fig. 8.** Average number of geese and ducks (□), oyster catchers (▨), and other waders (▧) in the Eastern Scheldt, 1975–1980. The histograms below the X-axis refer to the Krammer-Volkerak area

*Oyster Cultivation.* At this moment plans exist to restart the fresh oyster cultivation after a period in which no cultivation was possible due to the occurrence of an oyster disease. It is advisable to delay any plans for fresh oyster cultivation in the Eastern Scheldt until after April 1987, as in this way it will be possible to guarantee that no harm will come to it.

*Cockle Fishing.* The Eastern Scheldt construction works will have only limited consequences for cockle fishing, since most cockle banks are in the western part of the estuary. There may be some mortality among cockles on the very highest banks if they dry out for long periods.

*Other Fishing.* The reduction in tides will reduce the effectiveness of a type of fishing which is peculiar to the area, in which V-shaped willow containers are placed in shallow areas as the tide is falling. Fishing by these means will be impossible during period B and while the dams are being completed. Fishing using other types of fish trap will suffer little or no interruption, as will fish cultivation and shrimpfishing.

## 5.4 Other Functions of the Estuary

*Shipping.* The consequences for shipping will be very slight. Only access to two smaller harbors will be reduced.

*Drainage.* There will be no negative effects on drainage except possibly from the Lake Veere and rivers of Brabant. This may make it necessary to keep the Lake Veere at winter level (M.S.L.–0.7 m) into the spring of 1987. The summer level of M.S.L. can be reached in June.

*Recreation.* During periods A, B and C, dead flora and fauna may cause unpleasant smells throughout the borders of the Eastern Scheldt.

## 6 Keeping Effects on the Environment and on Fishing to a Minimum

If changes in the plan are being considered, damage to the environment and to fishing can be kept to a minimum by the following means:

1. by making every effort to ensure that work which would entail an extra reduction in tides (over and above the reduction caused by the sill beams) is carried out before all the sill beams are in place;
2. by keeping the period of extra reduction in tides after all the sill beams have been placed in position as short as possible;
3. by taking account of the spring/neap tide cycle when scheduling work and (in the case of the Roompot) possibly also taking account of daily differences in tides;
4. by ensuring that the mean tidal range of 2.30 m at Yerseke is not reduced below this figure.

It is not possible to carry out all these measures simultaneously. While the dams are being completed, the impact can be reduced by:

1. adhering to a sequence in which tidal frequency is reduced, tidal range is reduced and tidal range is further reduced while the mean level is increased;
2. altering normal tidal frequency as little as possible;
3. minimizing the likelihood of critical periods being extended and of the barrier having to be closed for longer than planned as a result.

## Addendum

The storm-surge barrier became operational in October 1986. The Oester- and Philipsdam were completed in October 1986 and April 1987 respectively. During the final construction stage some minor effects on the ecosystem were found. The only exception was a more severe erosion of the salt marshes. In the present situation the kidal range is 3.25 m. The salt content is according to the predictions.

## References

Hummel H, Meyboom A, Wolf L de (1985) De effecten van het gebruik van de stormvloedkering op de bodemdieren van de Oosterschelde. DIHO, Yerseke

Knoester M, Visser J, Bannink BA, Colijn CJ, Broeders WPA (1984) The Eastern Scheldt project. Water Sci Technol 16:51–77

Leentvaar J, Nijboer SM (1986) Ecological impacts of the construction of dams in an estuary. Water Sci Technol 18:181–191

Meininger PL, Baptist HJM, Slob GS (1984) Vogeltellingen in het Deltagebied 1976/76–1979/80. Deltadienst, Middelburg

# Part IV
# Biological Effects and Monitoring

# Accumulation by Fish

U. HARMS[1] and M.A.T. KERKHOFF[2]

## 1 Introduction

Among potentially harmful contaminants certain trace metals (mercury, cadmium, lead, copper, zinc), organochlorines (PCB's, DDT, HCB, HCHs, toxaphene) and hydrocarbons (polycyclic aromatic hydrocarbons) have received particular interest in ecotoxicology as well as priority and significance in the context of the existing conventions for the prevention of marine pollution. When considering the occurrence of "trace metals" in the marine environment, there is principally no zero level. Residues in uncontaminated samples will reflect local geochemical characteristics, the so-called natural background level, which is usually subject to considerable variability. Also for various organic compounds natural background levels exist, but for the xenobiotic organochlorines their presence already means a contamination.Determinations of potentially harmful contaminants in fish have to be considered as a complement to levels in water and sediment, as they provide information on the bioavailable part of the compounds in the different compartments of the environment. Moreover, for the assessment of pollution-induced effects on aquatic biota, it is possible to relate levels of suspected harmful contaminants to health conditions.

## 2 Bioaccumulation Mechanisms

### 2.1 Inorganic Compounds

For the intake by fish, trace metals must exist in a biologically available form, which usually means that they must occur in a dissolved species in the water phase. Elements in particulate form, both suspended in the water phase and in the sediments, are seldom available directly to aquatic organisms. A predominant entrance level to the food web is represented by phytoplankton. For higher organisms like fish, however, an immediate uptake through the water phase via gills, or after contact with outer membranes constitutes an at least efficient route of incorporation (Pentreath 1973; Pentreath 1976a,b,c,d). The precise relationship between the actual trace metal concentrations in fish and the metal burden of the

[1]Bundesforschungsanstalt für Fischerei, Labor für Radioökologie der Gewässer, Wüstland 2, D-2000 Hamburg 55, FRG

[2]Tidal Waters Division, Public Works Department, Van Alkemadelaan 400, 2597 AT 's-Gravenhage, The Netherlands

relevant habitat is complicated, since several factors have to be taken into consideration.

The organism's physiology (sexual cycle) and changing of ambient water quality parameters (temperature, salinity, oxygen content) can markedly alter the uptake rate and metabolic turnover, and thus the possible concentrations in fish. The dependence on rates of feeding, and on diet, may in addition result in variations of metal concentrations in tissues of fish investigated. Also, many authors have reported age-dependent variations in trace metals in fish (Phillips 1980, and references cited there) which in return may be a function of any one of age-dependent biological variables. The rates of uptake and elimination of trace metals by fish is not only affected by their abundance but also by their chemical forms. For example, microbial remobilization in settlement material and sediment increases the biological availability of mercury through transformation of inorganic mercury compounds to methylmercury (Jensen and Jernelöv 1969; Wood et al. 1968). Methylmercury is taken up faster than inorganic mercury and is more slowly excreted (Windom and Kendall 1979). Recently, it could be shown by Grimås et al. (1985) that the accumulation of zinc, copper, lead, and cadmium by cod liver was significantly influenced by the fat content of the tissue analyzed. These observations indicated, as the authors concluded, that metals in liver were connected to specific sites and molecules in both the protein and hydrocarbon fraction, which varied considerably in dependence on the lipid fraction. Irrespective of the still outstanding proof that the above-cited findings are also relevant for other fish species, the study of Grimås et al. (1985) make clear that alterations in the organism's metabolism can lead to substantial fluctuations in the measured contamination level, which make more difficult the detection of changes in ambient water quality. The fact that most harmful trace metals are found at lowest concentrations in skeletal muscle of marine and freshwater fish may be indicative of certain regulatory, metabolic exclusion mechanisms which prevent incorporation or at least enrichment of such metals in muscle tissue (Chow et al. 1974; Phillips 1980). The obvious exception seems to be mercury, which commonly occurs in muscle tissue of fish at concentrations approaching or even exceeding those in the other organs.

## 2.2 Organic Compounds

In an aquatic ecosystem organic compounds move from compartment to another via several transition processes (Fig. 1). When a fish is exposed to organic compounds in the water phase, these compounds will pass the membranes in the gills and the skin. In this way a bioconcentration/elimination equilibrium exists between the lipid tissues in the fish and the surrounding water. Since 1974 this direct exchange has been considered the main route of uptake of poorly water-soluble lipophilic compounds by fish (Neely et al. 1974). The contents in the fish are determined by the lipophilicity of the compounds, their concentrations in the water, and the lipid contents in the fish tissues. The degree of lipophilicity of a compound is often indicated by the bioconcentration factor (BCF), which is the ratio between the concentration in the fish ($C_f$) and that in the water ($C_w$) at steady state, where uptake and elimination rates are equal. The bioconcentration factors

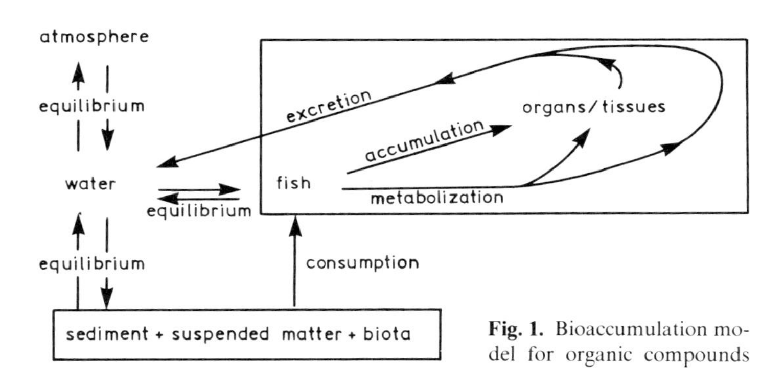


**Fig. 1.** Bioaccumulation model for organic compounds

expressed on a lipid basis resemble the partition coefficients between a lipid solvent such as n-octanol and water ($P_{ow}$):

$$C_f/C_w = BCF \approx P_{ow}$$

The partition coefficients are often used to predict the bioaccumulation in fish. The higher $P_{ow}$ or BCF is, the higher will be the residue in fish: p,p′-DDT with a log $P_{ow}$ of 6.36 is more accumulated than dichlorobenzene with a log $P_{ow}$ of 1.8 (Zaroogian et al. 1985; Davies and Dobbs 1984). As organic compounds are transferred from the sediment to the water and the air and in the opposite directions (Fig. 1), their concentrations in the water more or less reflect the concentrations in sediment, suspended matter, and atmosphere (Larsson 1985; Karickhoff and Morris 1985). High ultimate concentrations in the water will result in proportionally high residues in the fish, except if the molecules have too large dimensions for a good membrane passage. Compounds like hexabromobenzene are too large for this penetration process and will hardly be observed in fishes. Although the bioconcentration model gives a good first impression of the accumulation possibilities, uptake of compounds through consumption of particles and biota out of water and sediment may also contribute to accumulation. Laboratory experiments have shown dietary PCB contributions to total body burdens of 42% for eel (Larsson 1984) and of 53% for spot (Rubinstein et al. 1984).

# 3 Data Evaluation

## 3.1 Biologically Induced Variance

If in environmental programs fish are used as bioindicator, full account must be taken of several factors which influence the data. The above processes together, dominated by some of them, determine the body burden in the fish. However, in addition, residues will be influenced by biological factors such as size, age, sex, sexual cycle, migration, feeding habits, and body composition (lipid content).

Differences in the growth and habits of individuals in a single fish population cause a considerable range in contaminant levels (30–50% variability), more likely demonstrating skewness than Gaussian normality (Kruse and Krüger 1984; Gordon et al. 1980). Beside this natural within-sample variability, spatial or temporal variabilities on scales smaller than those of interest may occur. Much of the biologically induced variance and the sample handling and sample preparation variability can be limited by following particular procedures for sample collection and preparation prior to analysis (ICES 1984; Phillips and Segar 1986). If these procedures are not followed, data obtained cannot be considered as realistic measure of contaminant bioavailability.

### 3.2 Analytical Quality

Often the inherent within-sample variability is additionally superimposed by analytical biases. Laboratories performing analysis of micropollutants should have a high degree of analytical expertise and should use quality assurance procedures with requirements on, e.g., precision, accuracy, blanks, detection limits, and verification. In this context, criticism arises with respect to the effectiveness of existing monitoring programs, primarily because the ability of analysts (or laboratories) to produce accurate and reproducible results, and to achieve an acceptable level of uniformity, is partly limited.

During the last decade the International Council for the Exploration of the Sea (ICES), Copenhagen has organized several intercomparison exercises for the determination of trace metals, organochlorines, and hydrocarbons in biological tissues. Intercomparison exercises provide analysts with the opportunity to compare and check the reliability of their methods with those of other labs, and if needed to improve their analytical procedures. The criteria for producing "satisfactory" data are not precisely defined, but rather depend on the judgement of those who use the results. However, several recent intercomparison exercises have set a certain stage, which allows a more realistic definition of what constitutes acceptable performance of measurement of contaminants in biota.

On the basis of the findings of the 5th ICES trace metal intercomparison exercise (Holden and Topping 1981), it seemed reasonable to conclude that the majority of participating laboratories were able to report comparable and accurate data for copper, zinc and mercury at concentration levels commonly observed in tissues of fish and shellfish. For mercury, even at a level which was considered as low or background level (i.e., below 0.1 mg Hg $kg^{-1}$), the results were very encouraging. In contrast, the comparability of results for cadmium and lead was relatively poor, as a consequence of which a further (the 6th) ICES trace metal intercomparison exercise was organized. The results of this exercise led to the conclusion (Topping 1982) that all participants could produce comparable data for cadmium at a concentration of ca. 1 mg $kg^{-1}$ dry weight ($\cong$ 0.2 mg $kg^{-1}$ wet weight). The results of lead suggested that the majority of participants could produce accurate and comparable data at a concentration of ca. 2 mg $kg^{-1}$ dry weight ($\cong$ 0.4 mg $kg^{-1}$ wet weight) in biological tissue, but that only a minority of participants might be capable of producing accurate and comparable data at a concentration of ca. 0.5 mg $kg^{-1}$ dry weight (≅ 0.1 mg $kg^{-1}$ wet weight) in biological material. Also, the outcome of the first phase of the 7th ICES trace metal intercomparison exercise was discouraging, as considerable discrepancies still occurred for the analysis of lead and no improvement over the 6th intercomparison exercise was observed (Berman 1984). Thus, the present situation implies that data on lead in biological tissue must be regarded with great caution.

After five ICES organochlorine intercalibrations, the levels of agreement among the participating laboratories were still not good enough to allow them all to take part in multilaboratory monitoring programs. The problems exist especially when a variety of analytical methods are used, and Musial and Uthe (1983) identified systematic interlaboratory errors as the most likely source of variance. Fortunately in the last few years improvements in PCB analysis have been made by using capillary gas chromatography and by determining individual PCB congeners instead of total-PCB. In a collaborative interlaboratory project organized by the Community Bureau of Reference (BCR) in Brussels – in cooperation with the State Institute for Quality Control of Agricultural Products (RIKILT) in Wageningen (The Netherlands) – the parts of the analytical procedures have been studied in succession. The preliminary studies gave particular attention to the detection by optimization of gas chromatographic settings, to the identification and to the quantification of seven selected chlorobiphenyls (Tuinstra et al. 1985).

Application of the knowledge obtained in a fish oil interlaboratory study with different clean-up procedures resulted in coefficients of variation ranging from 10 to 21% at levels between 0.2 and 1 mg $kg^{-1}$ per congener and ranging from 15 to 50% at 0.05–0.2 mg $kg^{-1}$ (Tuinstra and Roos 1985). As more (interfering) organochlorines were present (e.g., toxaphene) the worse the results became; it also became clear that a good gas chromatographic separation is a necessity. As a consequence of the progress in PCB quantification, improved results for other well-separated organochlorine compounds may be expected, if indeed the laboratories have developed an optimization procedure according to the BCR protocol. However, at present many of the organochlorine residues in biological tissue must be regarded with great caution in view of the analytical uncertainties still existing in several laboratories.

While the situation for organochlorine analyses has changed in the right direction, the results for the hydrocarbons are still so poor that the currently produced data have to be considered as unreliable. Intercomparison studies on the analysis of individual aliphatic and aromatic hydrocarbons in marine biological tissues show considerable variations in reported concentrations. Sometimes the results differ more than 1 order of magnitude (Law and Portmann 1982; Uthe et al. 1985). Before organizing any cooperative research program to these compounds in marine biota considerable improvements in the analytical procedures have to be made.

## 4 Synopsis on Prevailing Levels of Potentially Harmful Contaminants in North Sea Fish

Repeated and regular measurements of selected contaminants within coordinated international surveys and monitoring programs very probably provide the best means to produce a detailed picture of the state of the marine environment under study and to identify areas with particularly high contamination. Activities within the framework of the International Council for the Exploration of the Sea (ICES) and the Joint Monitoring Programme (JMP) of the Oslo and Paris Commissions are examples of this. In practice, the analyses have concentrated on a few fish species such as cod (*Gadus morhua*), plaice (*Pleuronectes platessa*), flounder (*Platichthys flesus*), hake (*Merluccius merluccius*) and herring (*Clupea harengus*), all of which are regarded as representative for selected sampling areas.

In the light of experience from the aforementioned international activities, it can be concluded that the elements copper and zinc represent no serious problems in the context of environmental safety. Both elements are regarded as essential with a wide safety margin both with respect to human health and marine ecosystems. The concentrations in fish muscle tissue for zinc and copper were variously reported as below 10 and 1 mg $kg^{-1}$ wet weight respectively, and there was little variation in the concentrations found, regardless of the source of fish. The corresponding liver data ranged from about 10 to 40 mg $kg^{-1}$ (wet weight) for zinc and from about 2 to 15 mg $kg^{-1}$ (wet weight) for copper. From the Six-Year Review of ICES Coordinated Monitoring Programs (ICES 1984), from the results of the running Joint Monitoring Program (JMG 1985) and from other relevant publications (Stoeppler and Nürnberg 1979; Kruse and Krüger 1984) it may be concluded that typically the concentrations of cadmium and lead in liver of fish from North Sea offshore fishing grounds are clearly below 1 mg $kg^{-1}$ (wet weight) with a considerable variation in the concentrations irrespective of the species of fish. From the results of the aforementioned intercalibration exercises, it became clear that previously reported rather high lead and sometimes cadmium values in muscle tissue of marine fish were obviously influenced by contamination during sampling and the analytical procedure used, and were therefore of doubtful validity. Further, considerable methodological uncertainties must be taken into consideration, which form the background for a variety of sources of systematic errors (Harms et al. 1982). A recent methodological approach presented by Harms (1985), based on improved analytical sensitivity and effective contamination control, may be regarded as a realistic contribution to harmonize analytical results with the actual lead contents in marine fish samples. This analytical procedure applied to some specimens of cod and plaice collected in offshore waters of the German Bight revealed that extremely low lead concentrations in the range of 0.0005 to 0.004 mg $kg^{-1}$ (wet weight) prevailed in the muscle tissue of the fish analyzed. It may be inferred that the "true" Pb values in fish muscle tissue have the same order of magnitude as those for cadmium (i.e., in the range of 0.00 X mg $kg^{-1}$), and that Cd and Pb values in the liver seem to be reflected only weakly (if at all) in the muscle. Numerous studies on mercury in fish have provided basic informations, which allow the conclusion that for most species the "natural" level seems to be in the region of 0.05 (or less) to 0.1 mg kg$^{-1}$ on a wet weight basis. This statement can also be made for fish collected in the North Sea, with the exception of those areas which are likely to be affected by industrial inputs (compare the section on geographical variation of trace metals in fish).

In some of the larger and fast-growing oceanic fish species, such as tuna and swordfish, there is an extreme tendency to accumulate mercury (Westöö and Ohlin 1975), with the effect that considerably higher values than here given are found. Also, some North Sea fish species can show relatively high levels of mercury, possible due to their long life span in the sea (Krüger and Nieper 1978).

To supply a reliable view on the occurrence of organochlorine compounds in North Sea fishes only recently produced data of laboratories, that participated in the BCR capillary gas chromatography interlaboratory exercises have been used. In Table 1 the residues in cod (muscle, liver) plaice (muscle) and herring (muscle, all caught in the southern part of the North Sea, are summarized. The presentation of residue ranges has been chosen because of the undoubtedly present biologically induced variability. Data are given for four selected individual chlorobiphenyls (instead of total-PCB) and five organochlorine pesticides. No particularly high concentrations have been found. The occurrence of higher levels in more lipid-rich tissues is clearly demonstrated in the order: cod liver > herring muscle > cod muscle = plaice muscle. The 2,2',4,4',5,5'-hexachlorobiphenyl (no. 153) contamination is comparable to that of total-p,p'-DDT. As the total-PCB level is about ten times the level of this congener, this means that the PCB's as a group are the most often occurring organochlorine contaminants in North Sea fishes. The hexachlorobiphenyls (e.g., no. 153) are the dominant congeners, followed by the penta- (e.g., no. 101) and hepta-(e.g., no. 180) chlorobiphenyls. The tetrachlorobiphenyls – the 2,2',5,5'-tetra substituted one (no. 52) often occurs in the highest

**Table 1.** Ranges in organochlorine residues ($\mu$g kg$^{-1}$ on wet weight basis) in cod, plaice and herring caught in the southern part of the North Sea

| | | Muscle | | | Liver |
|---|---|---|---|---|---|
| | | Cod | Plaice | Herring | Cod |
| Chlorobiphenyls | no. | | | | |
| 2,2',5,5'-tetra- | (52) | <2 | 2–5 | 5–10 | 100–200 |
| 2,2',4,5,5'-penta- | (101) | <2 | 2–5 | 10–20 | 200–500 |
| 2,2',4,4',5,5'-hexa- | (153) | 2–5 | 5–10 | 10–20 | 500–1000 |
| 2,2',3,4,4'5,5'-hepta- | (180) | <2 | 2–5 | 5–10 | 200–500 |
| HCB | | <2 | 2–5 | 2–5 | 20–50 |
| α-HCH | | <2 | <2 | 5–10 | 20–50 |
| γ-HCH | | 2–5 | 2–5 | 5–10 | 20–50 |
| dieldrin | | 2–5 | 2–5 | 10–50 | 100–200 |
| Σp,p'-DDT | | 2–5 | 5–10 | 20–50 | 500–1000 |
| Lipid percentage (%) | | <2 | <2 | 10–20 | 20–50 |

Source. Data bank: Netherlands Institute for Fishery Investigations IJmuiden, The Netherlands LAC monitoring program: State Institute for Quality Control of Agricultural Products, Wageningen, The Netherlands concentration – are not very important. Despite the more than 10 years' ban on the use of DDT, it still occurs in relatively high levels. The $\alpha$-, $\gamma$-HCH and HCB contents are ten times lower than the total DDT levels and in the southern part of the North Sea dieldrin takes up a medium position. As in fish metabolic degradation is an important and rapid elimination route for hydrocarbons even fishes from sites with low-level chronic contamination do not seem to show hydrocarbon levels, that are higher than those of uncontaminated sites. Only fish caught near industrial areas may emit a characteristic odor and have hydrocarbon levels. The bioconversion of benzo-a-pyrene and other polycyclic aromatic hydrocarbons also explains why they are usually not detected in fish tissues (Neff 1979; Vassilaros et al. 1982).

## 5 Geographic Distribution of Potentially Harmful Contaminants in North Sea Fish

Most investigations carried out so far suggest that the pollution problems of the North Sea probably lie in inshore waters and river estuaries. This fact is due to river inputs and geochemical circumstances, which become significant in areas where mixing between continental runoff and marine water is most pronounced.

In the light of relatively high metal contamination of water, settlement material and sediments in coastal zones, especially in the vicinity of estuaries, the question arises whether these metals are bioavailable, leading possibly to enhanced contamination levels in fish living in these areas. Regrettably, there is lack of evidence that generally fish from coastal areas of the North Sea contain higher concentrations of trace metals than offshore fish, and furthermore, that fish from areas of suspected industrial input contain higher trace metal concentrations than those from other areas. Only for mercury were several areas detected in concentrations clearly above the background level in fish (0.1 mg $kg^{-1}$) that could be attributed to land-based discharges into estuaries or the sea. As reported by the Joint Monitoring Group (JMG 1985), more than half the data for 1983 measured in flatfish (flounder and plaice) were in the medium range (0.1–0.3 mg $kg^{-1}$), with crucial points of contamination in the German Bight, the Netherlands' Wadden Sea, along the Belgian coast, near the Humber estuary, and the Forth estuary. During recent investigations on the occurrence and distribution of contaminants in fish from coastal waters of the North Sea and the Baltic (Luckas and Harms 1987), it was recognized that pollution within the German Bight was apparently influenced to a large extent by the inflow of waters from the major rivers, mainly the Elbe. From an inspection of the data obtained, a decrease in the mercury concentration when going north from the Elbe estuary along the coast of Jutland could be recognized. While a concentration range between 0.1 and 0.3 mg $kg^{-1}$ prevailed in the German Bight, data near or below 0.1 mg $kg^{-1}$ were measured off the Danish coast near Hirtshals (Skagerrak). The concentration gradient measured in the fish samples correlated with the prevailing direction of the outer river plume of the Elbe, which normally tends northward east of the island of Helgoland and can be discriminated from surrounding seawater often as far as the island of Sylt.

This means that once the plume has been mixed with rather unpolluted water from the open sea, there will be a clear tendency of values to lower (natural) levels. These findings also gave evidence for the prevailing unfavorable hydrographic conditions in the German Bight, which are characterized by relatively weak residual currents, as compared to other coastal waters in the North Sea (Eisma 1981; ICES 1983). This has the effect that substances supplied from the Elbe, Weser, and Ems and also from the west tend to be kept and concentrated in the German Bight.

The geographical distribution of organochlorine compounds in North Sea fishes can be illustrated by residue levels in cod and hake livers. Because cod and hake have a limited migration, they can be used as representative bioindicators for the areas in which they have been caught. In Table 2 data of cod and hake liver studies from the Netherlands Institute for Fishery Investigations are given. For most organochlorine compounds, the highest contents occur in the southern North Sea with a decline in northerly direction and comparable levels in the northern North Sea and the English Channel. The PCB's exhibit the most marked differences, with four to five times higher levels in the southern part than in the northern part. For the organochlorine pesticides the ratios between the contaminant levels in the southern and northern North Sea are between 2 and 3. The overall pattern of distribution and concentration levels has not changed obviously during the last five years except for HCB. A remarkable decrease of the HCB input of the river Rhine has resulted in a decrease in the HCB contents in cod livers from the southern North Sea from 0.2 (1981) to 0.05 mg $kg^{-1}$ wet weight (1984). The DDT residues in cod livers from the southern Baltic are ten times higher than those in the southern North Sea, which appears to give a somewhat higher level in the central North Sea. The $\alpha$-, $\gamma$-HCH and HCB contamination levels in the Baltic are comparable to those in the southern North Sea (Falandysz 1984). Cod livers caught off the east coast of Canada have the same HCB, dieldrin, $\alpha$- $\gamma$-HCH, and DDT residues as livers from the northern North Sea (Freeman et al. 1984). On a global scale relatively higher concentrations of organochlorines have been found in the northern than in the southern hemisphere (Subramanian et al. 1983).

**Table 2.** Mean residue levels of organochlorines (mg $kg^{-1}$ on wet weight basis) in cod livers (homogenates of about 25 specimen per year) sampled during the winter in the southern (1979–1984), central (1980–1984) and northern (1981–1984) part of the North Sea, compared with levels in hake livers from the English Channel (1980, 1982)

| | | sNS | cNS | nNS | E.Ch. |
|---|---|---|---|---|---|
| Chlorobiphenyls | no. | | | | |
| 2,2′,5,5′-tetra- | 52 | 0.18 | 0.06 | 0.04 | 0.04 |
| 2,2′,4,5,5′-penta- | 101 | 0.43 | 0.14 | 0.07 | 0.07 |
| 2,2′,4,4′,5,5′-hexa- | 153 | 1.0 | 0.33 | 0.14 | 0.19 |
| 2,2′,3,4,4′,5,5′-hepta- | 180 | 0.37 | 0.10 | 0.06 | 0.09 |
| HCB | | 0.10 | 0.06 | 0.03 | 0.03 |
| $\alpha$-HCH | | 0.06 | 0.07 | 0.04 | 0.04 |
| $\gamma$-HCH | | 0.04 | 0.02 | 0.01 | 0.01 |
| dieldrin | | 0.16 | 0.13 | 0.07 | 0.08 |
| $\Sigma$p,p′-DDT | | 0.70 | 0.82 | 0.45 | 0.20 |

Source. Data bank Netherlands Institute for Fishery Investigations IJmuiden, The Netherlands

## 6 Summary

Residue levels in fish represent the bioavailable part of contaminants in the different compartments of the aquatic environment. Several biotic and abiotic factors induce a considerable variability in residues of individuals of a single fish population. Due to methodological uncertainties and discrepancies, it is still very difficult to compare data from different laboratories. More efforts should be made to promote procedures that provide accurate environmental analytical data and to develop sampling strategies adequate to the objectives of the programs concerned.

At the present stage the material available is barely sufficient to provide an overall picture on actual levels of potentially harmful contaminants in fish from the North Sea.

An evaluation of meaningful data obtained so far allows the conclusion that mercury shows rising concentrations from the central part to the coastal areas of the North Sea. Crucial points of contamination are being reported in the southern part of the North Sea, especially in the vicinity of the outflows of major rivers. For most organochlorine compounds the highest concentrations occur in fish collected in the southern North Sea with a decline of values in northerly direction. PCB's represent the dominant group of organochlorine compounds in North Sea fish. With the exception of HCB, which shows a reasonable decrease during the last 5 years, residue levels and overall pattern of distribution of organochlorine pesticides and polychlorinated biphenyls in fish have shown no pronounced tendency.

## References

Berman SS (1984) ICES Seventh round intercalibration for trace metals in biological tissue (Part 1) ICES Report CM 1984/E:44 MEQC

Chow TJ, Patterson CC, Settle D (1974) Occurrence of lead in tuna. Nature 251:159–161

Davies RP, Dobbs AJ (1984) The prediction of bioconcentration in fish. Water Res 18:1253–1262

Eisma D (1981) Supply and deposition of suspended matter in the North Sea. Spec Publ Int Assoc Sediment 5:415–428

Falandysz J (1984) Organochlorine pesticides and polychlorinated biphenyls in livers of cod from Southern Baltic, 1981. Z Lebensm Unters Forsch 179:311–314

Freeman HC, Uthe JF, Silk PJ (1984) Polychlorinated biphenyls, organochlorine pesticides and chlorobenzenes content of livers from Atlantic cod (*Gadus morhua*) caught off Halifax, Nova Scotia. Environ Monit Asses 4:389–394

Gordon M, Knauer GA, Martin JH (1980) *Mytilus californianus* as a bioindicator of trace metal pollution: variability and statistical considerations. Mar Pollut Bull 11:195–198

Grimas U, Gothberg A, Notter M, Olsson M, Reutergardh L (1985) Fat amount – a factor to consider in monitoring studies of heavy metals in cod liver. Ambio 14:175–178

Harms U (1985) Possibilities of improving the determination of extremely low lead concentrations in marine fish by graphite furnace atomic absorption spectrometry. Fres Z Anal Chem 322:53–56

Harms U, Berman SS, Jensen A, Topping G (1982) Identification of problems in the determination of lead in biological material. ICES Report CM 1982/E:8 MWQC

Holden AV, Topping G (1981) Report on further intercalibration analyses in ICES pollution monitoring and baseline studies. ICES Coop Research Report No 108

ICES (1983) Flushing times of the North Sea. ICES Coop Research Report No 123

ICES (1984) The ICES coordinated monitoring programme for contaminants in fish and shellfish, 1978 and 1979 and 6-year review of ICES coordinated monitoring programmes. ICES Coop Research Report No 126

Jensen S, Jernelöv A (1969) Biological methylation of mercury in aquatic organisms. Nature 223:753–754

JMG (1985) Assessment of the results of the Joint Monitoring Programme for 1983. Document JHM 10/3/1-E, Oslo Commission, Paris Commission (London)

Karickhoff SW, Morris KR (1985) Impact of tubificid oligochaetes on pollutant transport in bottom sediments. Environ Sci Technol 19:51–56

Krüger K, Nieper L (1978) Bestimmung des Quecksilbergehaltes der Seefische auf den Fangplätzen der deutschen Hochsee- und Küstenfischerei. Arch Lebensmittelhyg 29:165–168

Kruse R, Krüger K (1984) Untersuchungen von Nordseefischen auf Gehalte an toxischen Schwermetallen und chlorierten Kohlenwasserstoffen im Hinblick auf lebensmittelrechtliche Bestimmungen. Arch Lebensmittelhyg 35:128–131

Larsson P (1984) Uptake of sediment released PCBs by the eel, *Anguilla anguilla*, in static model systems. Ecol Bull NFR 36:62–67

Larsson P (1985) Contaminated sediments of lakes and oceans act as sources of chlorinated hydrocarbons for release to water and atmosphere. Nature 317:347–349

Law RJ, Portmann JE (1982) Report on the first ICES intercomparison exercise on petroleum hydrocarbon analyses in marine samples. ICES Coop Research Report No 117

Luckas B, Harms U (1987) Characteristic levels of chlorinated hydrocarbons and trace metals in fish from coastal waters of North and Baltic Sea. Int J Environ Anal Chem 29:215–225

Musial CJ, Uthe JF (1983) Interlaboratory results of polychlorinated biphenyl analyses in herring. J Assoc Off Anal Chem 66:22–31

Neely WG, Branson DR, Blau GE (1974) The use of the partition coefficient to measure the bioconcentration potential of organic chemicals in fish. Environ Sci Technol 8:1113–1115

Neff JM (1979) Polycyclic aromatic hydrocarbons in the aquatic environment; Sources, fates and biological effects. Applied Science, Essex, England

Pentreath RJ (1973) The accumulation and retention of $^{65}$Zn and $^{54}$Mn by the plaice, *Pleuronectes platessa L.* J Exp Mar Biol Ecol 12:1–18

Pentreath RJ (1976a) Some further studies on the accumulation and retention of $^{65}$Zn and $^{54}$Mn by the plaice, *Pleuronectes platessa L.* J Exp Mar Biol Ecol 21:179–189

Pentreath RJ (1976b) The accumulation of organic mercury from sea water by the plaice, *Pleuronectes platessa L.* J Exp Mar Biol Ecol 24:121–132

Pentreath RJ (1976c) The accumulation of mercury from food by the plaice, *Pleuronectes platessa L.* J Exp Mar Biol Ecol 25:51–65

Penthreath RJ (1976d) The accumulation of mercury by the thornback ray, *Raja clavata L.* J Exp Mar Biol Ecol 25:131–140

Phillips DJH (1980) Quantitative aquatic biological indicators. Applied Science, London (Pollut Monit Ser)

Phillips DJH, Segar DE (1986) The use of bio-indicators in monitoring conservative contaminants: programme design imperatives. Mar Pollut Bull 17 (1):10–17

Rubinstein NI, Gilliam WT, Gregory NR (1984) Dietary accumulation of PCBs from a contaminated sediment source by a demersal fish (*Leiostomus xanthurus*) Aquat Toxicol 5:331–342

Stoeppler M, Nürnberg HW (1979) Comparative studies on trace metal levels in marine biota. III. Typical levels and accumulation of toxic trace metals in muscle tissue and organs of marine organisms from different European seas. Ecotoxicol Environ Saf 3:335–351

Subramanian BR, Tanabe S, Hidaka H, Tatsukawa R (1983) DDTs and PCB isomers and congeners in Antarctic Fish. Arch Environ Contam Toxiol 12:621–626

Topping G (1982) Report on the 6th ICES trace metal intercomparison exercise for cadmium and lead in biological tissue. ICES Coop Research Report No 111

Tuinstra LGMT, Roos A, Griepink B, Wells D (1985) Interlaboratory studies of the determination of selected chlorobiphenyl congeners with capillary gas chromatography using splitless and on-column injection techniques. J HRCCC 8:475–480

Tuinstra L, Roos A (1985) Determination of selected chlorobiphenyl congeners in fish oil with capillary gas chromatography (BCR Interlaboratory study 4/1985) RIKILT Report 85.76 25–10–1985

Uthe JF, Musial CJ, Sirota GR (1985) Report on the intercomparitive study 03/HT/BT on the determination of polycyclic aromatic hydrocarbons in biological tissue. ICES Report CM 1985/E:46 MEQC

Vassilaros DL, Stoker PW, Booth GM, Lee ML (1982) Capillary gas chromatographic determination of polycyclic aromatic compounds in vertebrate fish tissue. Anal Chem 54:106–112

Westöö G, Ohlin B (1975) Methylmercury in fish and shellfish. Vår Fõda 27:4–30

Wood JM, Kennedy F, Rosen CG (1968) Synthesis of methylmercury compounds by extracts of methanogenic bacterium. Nature 220:173–174

Windom HL, Kendall DR (1979) Accumulation and biotransformation of mercury in coastal and marine biota. In: Nriagu JO (ed) The biogeochemistry of mercury in the marine environment. Elsevier/North-Holland Biomedical, Amsterdam, pp 303–323

Zaroogian GE, Heltshe JF, Johnson M (1985) Estimation of bioconcentration in marine species using structure-activity models. Environ Toxicol Chem 4:3–12

# Accumulation by Birds

W.R.P. Bourne[1] and G. Vauk[2]

## Human Impact upon North Sea Birds

The fertile waters of the North Sea form one of the major resorts for sea, shore and water birds of the world, and are also crossed by many migrating landbirds. A rough estimate of their numbers has been made by Evans (1973), who calculated that they may include over 2.5 million breeding seabirds in summer and over 2 million in the winter, with a little over 1.4 million migrant shorebirds or waders in the autumn and a little under 1.3 million in the winter, and a hundred thousand breeding wildfowl in summer, nearly half a million moulting birds in the autumn, and nearly three quarters of a million in the winter. The passing migrants may run into tens of millions. The total which is most easily checked, for breeding seabirds, subsequently proved to exceed four million (Bourne 1983a), and even this may also have been an underestimate. There are probably also a good many more sea duck, especially if one includes another population of a similar size frequenting the approaches to the Baltic (Atkinson-Willes 1975).

In the past, breeding seabirds and their eggs must have formed an important source of food for men in this area at the time when it was scarcest in the spring, so primitive hunters and fishermen were normally also fowlers, who must have greatly reduced all the more accessible bird populations before they learnt to manage some of them, such as the gannets *Morus bassanus* nesting on the Bass Rock, as renewable resources (Gurney 1913; Nelson 1978). The seabird colony on Flamborough Head on the east coast of Britain continued to be exploited into recent times (Table 1), but by the time that the first surviving manuscripts such as the Old English poem *The Seafarer* describing the scene around the Bass were written in the seventh century (Fisher 1966) the emphasis was already changing, and birds had become a source of aesthetic inspiration instead. The trend has continued ever since, and apart from a limited amount of hunting for sport, notably for wildfowl in Denmark, and doubtfully predation on fisheries (Furness 1982, 1984; Bourne 1983a), birds are now of little economic importance in this area, whereas the hypothetical threats to their welfare (Bourne 1972a; Evans 1984; Barrett and Vader 1984) have become an important political issues.

Concern about the welfare of birds in the North Sea dates from the start of international cooperation in the development of the ornithology of the area. In so

[1]Department of Zoology, Aberdeen University, Tillydrone Avenue, Aberdeen AB9 2TN, Scotland
[2]Inselstation Helgoland des Instituts für Vogelforschung, Vogelwarte Helgoland, Postfach 12 20, D-2192 Helgoland, FRG

**Table 1.** Fluctuations in seabird numbers at Flamborough Head, Yorkshire

Mid 18th century: One of the largest British colonies (Ray 1678).

1769: Innumerable birds seen from the sea on 3 July (Pennant 1771).

Early nineteenth century: birds reported very numerous (Cordeaux, below).

Late 1860's: Numerous reports by John Cordeaux and others that seabirds were declining due to shooting for sport by people arriving on a new railway and for the plume trade (one man said to have received an order for 10,000 kittiwakes in a season), and also egg-collecting (Zoologist 1864:9243–7, 9292–5, 9325–6, 1866: 21–27, 1867–8: 1008–29, 1869: 1512).

Early 1870's: Cordeaux reported an immediate increase following Seabird Protection Act in 1869, to more birds than within the last 10–12 years though less than within the memory of the oldest egg-collector, associated with the appearance of fish-shoals offshore and the return of birds to the colonies from November (Zoologist 1870: 2262, 1871: 2822–8, 1873: 3530).

1880's: Guillemots laying two, sometimes three, eggs in a season when robbed, so that a team of three climbers could collect 200–300 eggs daily for up to 5 weeks (Seebohn 1885). The number of teams was not specified, but if it was comparable to that in the 1900's, this implies that possibly they may have been taking up to 35,000 eggs annually from 12,000 pairs of birds.

1900's: Guillemots returning to colony in fine weather about Christmas and laying three eggs per season, so that four teams each of four men could take 300–400 eggs daily for up to 6 weeks, calculated at the time to amount to 130,000 eggs per season (Wade 1903, Nelson et al. 1907), and later to imply that there were at least 43,000 pairs of birds.

1918: First prospecting fulmars in England, bred from 1922 (Fisher 1952).

1924: First recent prospecting gannets on mainland of Britain, bred from 1937.

Early 1930's: Severe losses of auks from oil pollution March-April 1930, March 1931, and especially May-June 1932 among other dates (Bird Notes).

1940: Great decrease in guillemots, although only one team of climbers still active and kittiwakes starting to increase (Nicholas 1940).

1952: 7150 Guillemots counted 21–23 June (Brownsey and Peakall 1953); decrease attributed to egg-collecting and oil pollution (Chislett 1952).

1954: Egg-collecting terminated.

1964: 12,851 Guillemots counted 5 and 14 June (Williams and Kermode 1968).

1969: 12,570 Guillemots counted 4–8 June (A.J. Williams).

1975: Site now an RSPB Nature Reserve: 12,200 Guillemots (Stowe 1982).

1977: At least 700 guillemots (many probably from elsewhere?) found killed by oil in February (and others probably missed); then 1475 found dead in April; 9224 subsequently counted at the colony (Stowe 1982; Bourne 1982d).

1978: 13,250 guillemots counted during regular surveys (Stowe 1982).

1987: 32,600 guillemots, 836 pairs of fulmars, 780 pairs of cannets (Seabird Colony Register per Dr Clare Lloyd).

far as this can be associated with any single event, it appears to have begun when an English farmer from Lincolnshire, John Cordeaux (Pashby 1985), who had been active in promoting the first British seabird protection legislation following the institution of railway excursions to enable people to shoot the seabirds breeding on Flamborough Head in 1869, happened to visit the island of Helgoland off north-west Germany (then temporarily a British possession) for 5 days in 1874 to meet the German artist Heinrich Gätke and inspect his collection of stuffed birds. They began a lifelong correspondence about bird migration, which soon led to the organization for the British Association for the Advancement of Science of a great international cooperative inquiry into the occurrence of migrating birds at lighthouses all round the North Atlantic, which incidentally also revealed a good deal of information about the movements and mortality of seabirds and its relation to fisheries and the weather (Harvie-Brown et al. 1880–1889).

In 1887 it was felt that the main British Association inquiry was producing diminishing returns, so it was suspended pending further analysis of the results and the development of various special studies. The most important were the organization of bird-marking schemes, starting in Denmark. The British had little success in catching adult birds at first, but meanwhile German ornithologists had turned Helgoland into the first west European bird observatory with a phenomenally successful fixed trap (Vauk 1972), so in the autumn of 1933 a group of leading British ornithologists went to watch it in operation (Alexander 1934). This led them to establish their own network of observatories for the purpose of maintaining similar traps all round the coasts of Britain and Ireland, which incidentally also provided useful bases for research on seabirds. Meanwhile the Dutch had also been organizing direct observations of migrating birds on the island of Texel, which were soon also taken up in Britain and Scandinavia [Ibis 95(2), 1953], and extended all round the North Sea as more portable mist- and rocket-nets became available in the 1950's.

The development of this network of observers led to a growing awareness of the need for international action to deal with bird conservation problems. Initially, attention was concentrated on the control of hunting for sport and the plume trade, and egg-collecting for consumption and trophies. These were gradually brought under control over the course of a generation by legislation and the creation of reserves for breeding birds by a number of different local and national voluntary organizations such as the (British) Royal Society for the Protection of Birds (RSPB), originally formed in 1889 by a small circle of fashionable ladies to oppose the use of feathers for millinery, which has since gradually developed into a major landowner with some 400,000 senior and 200,000 junior members and nearly 500 staff. Eventually, as these local problems were brought under control, some more general ones began to emerge which were less easy to deal with by local action, and an International Council for Bird Preservation (ICBP) was set up to coordinate the approach to them in 1922, starting with oil pollution of the sea.

There are a limited number of early references to the pollution of seabirds by natural oil seeps and the accidental spillage of ship's cargoes, but the problem does not appear to have become serious until the apparently widespread but poorly documented conversion of coal-burning shipping to use fuel oil, with small consideration for the consequences of spills, during World War I. This attracted little attention while hostilities persisted, but afterwards oiled bird bodies were found to have become numerous all round the southern North Sea. The bird protection organizations protested vigorously, and soon secured action to control the problem, which in Britain involved fitting separators to shipping to remove superfluous oil before discharging bilge-water. There was a rapid but temporary improvement, so that it has been necessary to renew the agitation at regular intervals whenever standards began to decline again ever since, notably after World War II, when the ICBP organized a more general Advisory Committee on Pollution of the Sea including many other interests, which now deals with many aspects of the problem.

At first the evidence produced in support of this agitation was entirely anecdotal. Gradually more systematic counts of the dead birds were made along the shore, and when the first Seabird Group was formed in Britain in 1966 it was decided that the RSPB would ask their members to make more methodical beached bird surveys in order to obtain better evidence for the nature of the birds occurring offshore and their mortality. These plans proved extremely opportune when shortly afterwards the western approaches to the English Channel were suddenly subjected to unprecedented oil pollution following the wreck of the 117,000-t oil tanker *Torrey Canyon* off Cornwall in March 1967 (Bourne et al. 1967; Bourne 1968) at a time when exploration for oil was just beginning in the North Sea. The superfluous money subscribed by the British public for the relief of the oiled birds, few of which survived long enough to benefit from it, was therefore partly devoted to a survey of the British breeding seabirds which might be at risk (Cramp et al. 1974), and similar surveys were also carried out in north-west France and Norway (Brien 1970; Brun 1979).

When it was found that similar developments were also taking place elsewhere in Europe, it was next agreed to coordinate one of the reorganized British monthly winter beached bird surveys with an established annual beach survey carried out by youth groups in the Low Countries during a public holiday at the time of the annual peak in seabird mortality in the late winter to form a European beached bird survey (Bourne and Bibby 1975; Stowe 1982). These surveys soon revealed a variety of causes for bird mortality in addition to oil pollution (Bourne 1976a), as follows:

1. There appear to be certain seasons at which seabirds are regularly liable to die in numbers at intervals of years, either because the weather is so bad that they are unable to feed and become so weak so that they are blown ashore, or because there is a shortage of food, or both. This is commonest during the late winter, as, for example, when there was a major wreck of auks after the collapse of the sprat *Clupea sprattus* population in the northeast North Sea early in 1983 (Edwards and McKay 1984; Underwood and Stowe 1984; Blake 1984), and when the young birds leave the nest and the old birds go into moult at the end of the breeding season, as when bad weather appears to have broken up the stratification of the Irish Sea while the moulting auks and their chicks were flightless on the water in the autumn of 1969 (Holdgate 1971; Bourne 1976b, 1982c).

2. In some areas birds may also be poisoned from time to time by microorganisms which may occur either naturally or in association with pollution, as for example where shags *Phalacrocorax aristotelis* were killed by a "red tide" caused by *Gonyaulax tamarensis* in 1968 (Coulson et al. 1968), and gulls were killed by botulism due to *Clostridium botulinum* in 1975 (Lloyd et al. 1976) along the north-east coast of Britain. In the hot May of 1933 the diatom *Coscinodiscus concinnus* also proliferated to such an extent that it killed auks in the same manner as oil pollution (Wolfe Murray 1936; Táning 1952).

3. Birds may also suffer from epidemics of various infectious diseases, some of which may also infect man or his domestic animals. Examples include an epidemic of ornithosis affecting fulmars *Fulmarus glacialis* which also killed a number of people in the Faroes in the 1930's (Rasmussen-Ejde 1938), a paratyphoid B epidemic in the south-east North Sea which may have spread from gulls to people in the 1940's, although it is apparently unusual for gulls to spread Salmonellae in this way (Bourne 1977a; Vauk-Hentzelt 1986), and mortality of fulmars and kittiwakes *Rissa tridactyla*, which may possibly have been due to influenza in the eastern North Sea in the summer of 1959 (Joensen 1961; Bourne in press).

4. In polluted areas birds are also liable to accumulate toxic chemicals (Bourne 1976a; Bourne et al. 1978; Figge et al. 1976; NERC 1983; Falandysz and Szefer 1984; Szefer and Falandysz 1983, 1986; Barrett and Vader 1984; Delbeke and Joiris 1985; Goede 1985; Goede and de Bruin 1985a,b; Goede and de Voogt 1985), and especially organochlorines which are stored in their fat, which is then also deposited in their eggs, so that both the birds and their chicks may become poisoned when they are starved, the fat is consumed, and its contents liberated, as with some of the birds which died during the auk mortality in the Irish Sea mentioned above (Holdgate 1971; Bourne 1976b), and with the eiders *Somateria mollissima* and Sandwich terns *Sterna sandvicensis* among other species poisoned by chlorinated cyclodiene pesticide factory effluent in the Netherlands in the 1960's (Koeman et al. 1967; Koeman 1971, 1972).

5. Birds may also encounter problems with the growing variety of artifacts deposited by man in the oceans (Hartwig et al. 1985; Vauk and Schrey 1987), either by consuming the smaller objects, such as fragments of plastic, which obstruct or cause ulceration of their digestive systems (Bourne 1976a; van Franeker 1985), and the lead weights from fishing-lines, which may poison them (Simpson et al. 1979), or by becoming entangled in the larger objects, notably fragments of nylon fishing-net, which are often also used for nest material, and the perforated sheets of plastic used to link cans of soft drinks (Bourne 1976a, 1977b; Hartwig et al. 1985; Schrey and Vauk 1987).

6. Under certain weather conditions, notably with high atmospheric pressure to the north-east and low pressure to the south-west (Fig. 1), migrating landbirds may be drifted out to sea by offshore winds and become lost in overcast or foggy weather, when they are liable to be attracted by lights at night and killed at lighthouses or the gas flares at offshore petroleum installations, and eventually land exhausted in vast numbers on the coast, the installations and ships, or failing these fall into the sea, owing to entirely natural causes (Bourne 1979, 1982b).

It has been realized that birds are liable to be washed up dying and dead on the shore at intervals for a very long period, but in the past, as for example during the British Association inquiry into migration at the coastal lights in the 1880's, this was usually attributed to storms. The contribution of human activities to such mortality is extremely difficult to assess because it is liable to be confused with other natural and unnatural fluctuations in bird populations of variable duration. Thus, for example, there appears to have been a long-term alternation over many centuries at western British seabird colonies between a predominance of a seabird of

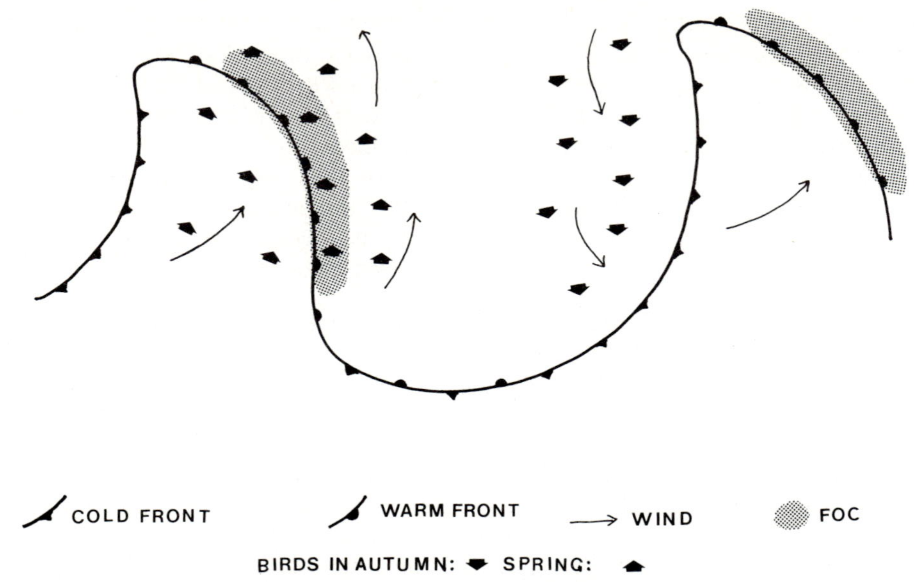


**Fig.1.** The weather conditions under which migrating birds are seen at sea. In general, they tend to set out to the south and west in autumn as the weather clears with the onset of a cold northerly airstream behind a depression moving east across the area, and cross the sea too high to be seen. When they encounter the overcast weather associated with the fronts approaching ahead of the next depression arriving from the west they settle if over land, but if on descending they find themselves over water they may become lost in fog and drift north-west further out to sea with the associated south-east winds until they appear exhausted on offshore islands, ships, or at lighthouses and gas flares at night, where they wait until the weather clears before heading back to the mainland. When they return north and east in the spring they tend to set out in the fine weather with a warm southerly wind ahead of an approaching depression, sometimes flying on through its fronts, when they may become disoriented and drift out to sea again. The whole process may become much more marked if an anticyclone develops over Scandinavia associated with the formation of more fog over the North Sea. (Bourne 1979, 1982a)

subtropical origin, the Manx shearwater *Puffinus puffinus* (known in French as the puffin des anglais), which appears to have been more numerous during warm periods, and another of northern origin, the Atlantic puffin *Fratercula arctica*, which may have been more numerous during cold periods, with the result that even their names have become confused (Bourne and Harris 1979; Bourne 1982c, 1983b; Evans and Nettleship 1985; Fig. 2).

It is possible that similar fluctuations at least partly due to climatic changes (Beverton and Lee 1965) have also occurred at the most southerly major pelagic seabird colony in the North Sea on Flamborough Head (Table 1). Thus it appears to have flourished during a cold period extending through the seventeenth and eighteenth to the early nineteenth centuries, but then went into a sudden decline in the 1860's, attributed at the time to excessive exploitation, although it terminated remarkably suddenly with the appearance of shoaling fish offshore. The birds then flourished again despite ruthless egg-collecting during a further cold period at the end of the last century, but went into another decline attributed to a combination of a lower level of egg-collecting and oil pollution during the warm period in the

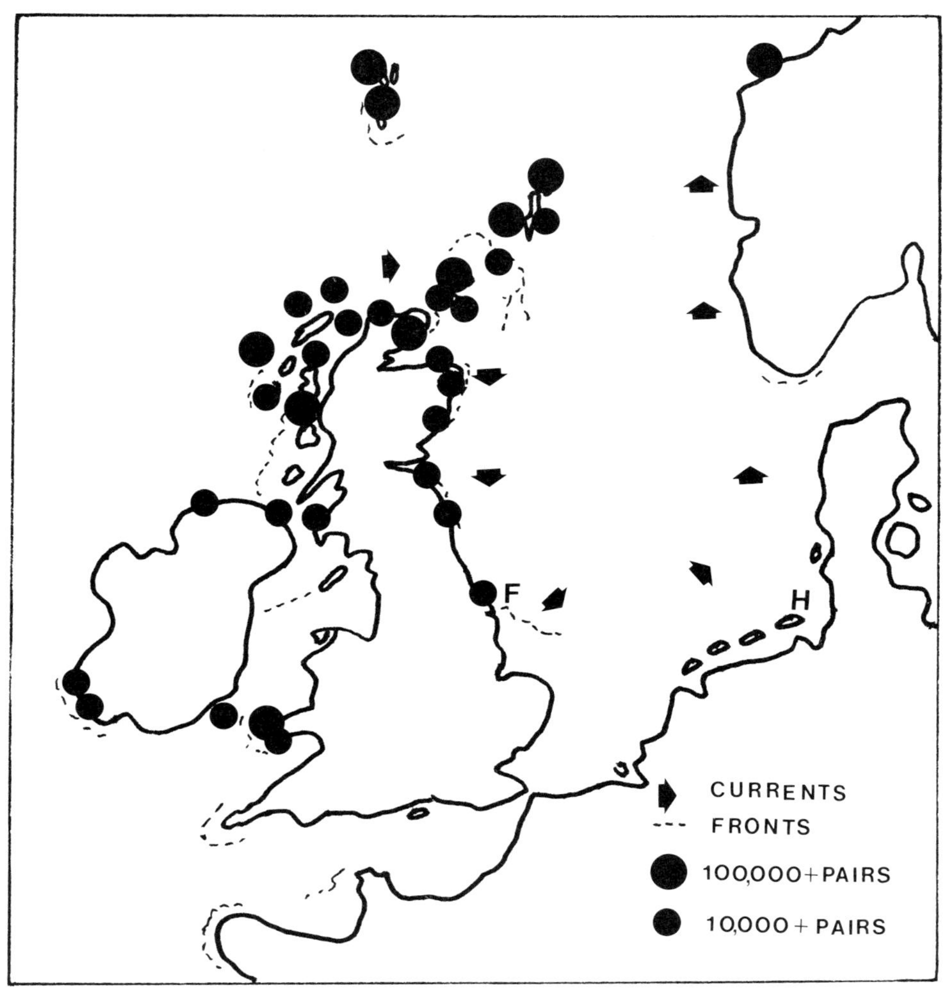


**Fig. 2.** The North Sea, showing the direction of water movement, some of the marine fronts visible on infra-red satellite images (Pingree and Griffiths 1978), and the larger associated pelagic seabird colonies (Cramp et al. 1974; Bourne 1980, 1982a). The low-lying south-eastern coasts of the North Sea also have a large but more dispersed population of eiders, gulls and terns. *F* Flamborough Head; *H* Helgoland

first half of this century. This trend has recently been reversed again following another deterioration in the climate despite the persistence of the oil pollution, and following protection many of the same species have also begun to increase further east on Helgoland as well (Fig. 3).

With the institution of regular beach surveys to assess the impact of oil pollution, it soon became clear that in addition to local pollution incidents and deaths from natural or artificial poisoning of known origin birds were also periodically dying in large numbers from unknown causes. The first of these incidents in the Irish Sea in the autumn of 1969 was at first thought to be associated

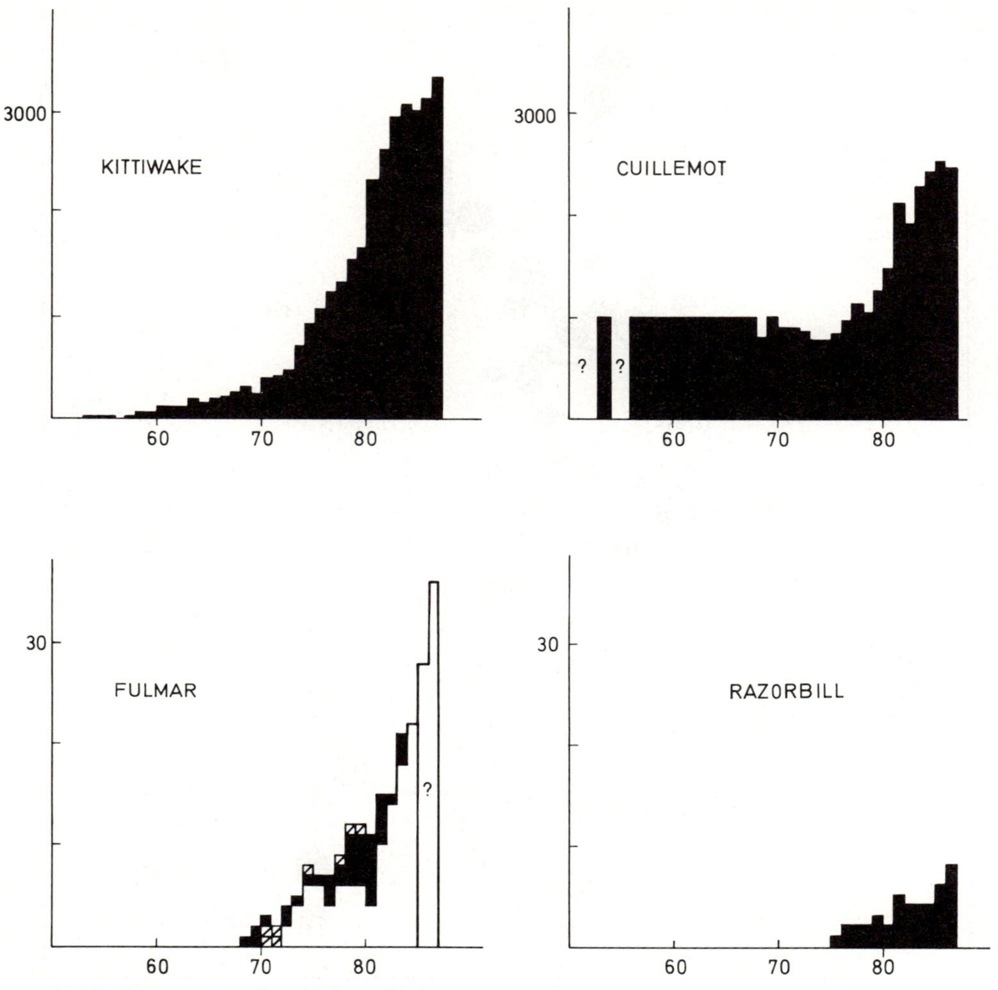


**Fig. 3.** Recent fluctuations in the seabird populations of Helgoland. (G. Vauk, in press)

with polychlorinated biphenyl (PCB) chemical pollution (Holdgate 1971; Bourne 1976b), and the second off north-east Britain early in 1970 with oil pollution of unknown origin (Greenwood et al. 1971). These explanations alone appeared increasingly unsatisfactory when further mortality of starving birds occurred on the east side of the North Sea early in 1981 (Baillie and Mead 1982; Blake 1983) and then the west side early in 1983 (Underwood and Stowe 1984; Hudson and Mead 1984; Hope-Jones et al. 1984; Osborn et al. 1984), and the mortality coincided with a decline in the sprat stock (Edwards and McKay 1984; Blake 1984).

There appears to be a regular mechanism for the development of these incidents. The birds normally congregate throughout the year in areas of water-mixing with large plankton, fish and shellfish populations, with the pelagic species accumulating over the inflow of oceanic water into the north-west North Sea (Bourne in Cramp et al. 1974; Joiris 1978), and the coastal species in the area where sediment is deposited downstream to the south-east (Bourne 1980). Here they become very vulnerable to oil and toxic chemical pollution drifting down upon them with the wind and tide (Bourne 1976a, Figs. 3, 4, 5, 1981), as for example with oil pollution of the Atlantic inflow around north-east Britain (Greenwood et al. 1971; Bourne and Johnston 1971), in the Firth of Forth (Bourne 1972b; Campbell et al. 1978), the south-east English estuaries (Harrison and Buck 1967), the channels between the Frisian Islands (Rittinghaus 1956; Swennen and Spaans 1970; Koeman 1972; Reineking and Vauk 1982; Vauk 1984), and the approaches to the Baltic (Joensen 1972a, b, 1973; Blake 1983). The oil and oiled birds, and also any birds that may be weakened owing to toxic chemical pollution, "red tides", bad weather, or failures of the food-supply, then also drift with the wind and tide, possibly encountering oil in the process if they have not met it before, until they either sink after about 10 days, or wash up on a lee shore which may be far away from the original source of any pollution (Hope-Jones et al. 1970; Bibby 1981).

Since seabirds normally have a delayed maturity and low reproductive rate, it was postulated that any unnatural mortality might have a very serious effect on their numbers (Bourne 1968). In the event, even the most spectacular losses, such as those from pesticides in the Dutch Waddenzee in the 1960's (Koeman 1972), from the possible food failure in the Irish Sea in 1969 (Bibby 1973), from oil pollution at Flamborough Head in 1977 (Stowe 1982), and from a food failure at the Isle of May in 1983 (Harris and Wanless 1984), have had little lasting effect. Thus it now seems doubtful whether pollution in particular has had much permanent influence on bird populations in this area (Table 1, Figs. 3, 5; Bourne 1982c,d,e; Clark 1984), and more likely that in fact the low reproductive rate of seabirds is in fact part of a flexible adaptation for conditions of acute overcrowding and intraspecific competition for food at the breeding colonies (Ashmole 1963) which also permits a rapid response to compensate for any reduction of the population, possibly through the earlier recruitment of young birds (Bourne 1982d,e). Thus recent human developments in the North Sea appear to have had little permanent effect on bird populations (Dunnet 1982, 1987).

In recent years less attention has been paid to landbird migration than in the past [although novel observations are now being made at the British oil installations by a North Sea Bird Club (1979–1985)], and more attention has been devoted to surveys of seabirds both at sea and the breeding-sites, so that the number of individual investigations is becoming difficult to follow, though they include large surveys in both the United Kingdom (NERC 1977; Bourne 1982c; Blake et al. 1984) and Norway (Røv et al. 1984), often with rather inconclusive results. The main current problems appear to be as follows (Bourne 1976a; Evans 1984; Barrett and Vader 1984).

*1. Disturbance at the Breeding Sites.* Seabirds are normally so highly specialized for a marine life that they have become extremely clumsy and vulnerable when they come ashore to breed. The pelagic species which breed on remote cliffs and rocky islands are now usually safer than in the past owing to the relaxation of human exploitation, and are showing an increasing tendency to occupy artificial breeding sites in the south of the area where they are scarce, such as breakwaters and buildings. The more coastal species such as the ducks, waders, gulls and especially the terns, which are adapted to breed on beaches, inland marshes and lagoons, and lower islets have now begun to suffer increasingly from the progressive reclamation of these places or their use by man for recreation, however, to the extent that their continued existence in many areas has often become dependent upon the provision of protected breeding sites (Fig. 5). Gulls, *Larus* sp., and oystercatchers, *Haematopus ostralegus*, have also taken to nesting increasingly upon house roofs in several areas, but the terns, *Sterna* sp., and cormorants, *Phalacrocorax* sp., do not appear to have made much progress with this in Europe yet, though they do so elsewhere, and the terns have responded well to the provision of artificial floating islands free from the risk of inundation by floods or the tide.

*2. Predation.* Increasing human activity has often led to a change in the balance of bird and animal populations and the introduction of additional predators which present a threat to some seabirds. The proliferation of gulls feeding upon agricultural land and garbage has sometimes caused them to usurp all the few breeding-sites available to other species such as the terns and ducks, though these may also sometimes avoid predators by nesting in gull colonies. Human activity has also sometimes led to the proliferation of predators such as cats and rats which prey upon seabird colonies, and an important new predator, the mink, *Mustela vison*, has been introduced into both Britain and Norway, where it may be responsible for a decline of black guillemots *Cepphus grylle* and eiders *Somateria mollissima* (Barrett and Vader 1984). Representations have also had to be made to prevent the establishment of mink farms in both Orkney and Shetland with their vast seabird populations, though there is already one mink farm in Shetland from which animals have escaped in the past, but were recaptured.

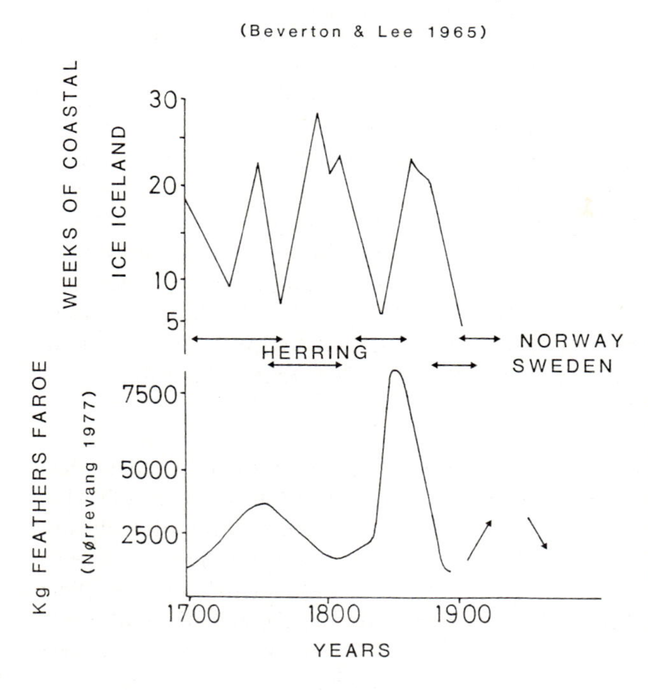


**Fig. 4.** Relationship between the fluctuations in the amount of ice around Iceland, the Herring *Clupea harengus* fisheries in northern and southern Scandinavia, and the amount of auk (probably mainly puffin, *Fratercula arctica*) feathers collected in the Faroes (W.R.P. Bourne, from contribution on seabirds of Europe, ICBP Seabird Workshop, Cambridge, 1982)

*3. Exploitation.* Hunting and egg-collecting must clearly have been important limiting factors for marine bird populations in the past, but have been declining in most countries in recent times, though constant disturbance of the breeding colonies may now have a similar effect. Few eggs now appear to be taken anywhere, possibly because few birds still nest in vulnerable places, and while some seabirds are still shot in Scandinavia, the hunting of the vulnerable larger auks was terminated in Norway in 1979. Although more wildfowl and some waders are still shot in the winter all round the North Sea, they have higher reproductive rates (and most of the marine wildfowl are rather unpalatable ), and do not appear to suffer unduly. Indeed, the numbers of the once scarce brent goose *Branta bernicla* have now increased to such an extent with protection that it has taken to feeding in the gardens of seaside towns and is becoming an agricultural pest in the south of England.

*4. Loss of Food-Supplies.* There has recently been much speculation whether the over-exploitation of fish-stocks may be affecting birds. The continual, long-established fluctuations of some seabirds such as the auks appear to be related to those of some fish such as the clupeids, but it is apparently debatable to what extent these are natural or caused by man. The expansion of demersal fisheries appears to be directly beneficial to seabirds because at least 10% of the weight of the otherwise inaccessible fish caught is lost as vomited stomach contents or discarded as offal at sea, where it is taken by a wide variety of birds, many of which, such as the skuas, gulls, the fulmar, *Fulmarus glacialis*, and gannet, *Sula bassana* (Fisher 1952; Nelson 1978), have been increasing spectacularly since the development of the fisheries, though more offal may be retained for conversion into fishmeal in future. Hypothetically the removal of the larger fish should also leave more smaller organisms for birds, but the use of smaller fish for conversion into fishmeal in the last 20 years may also affect this, though a good deal of fish is still spilt. Most seabirds do not appear to be suffering unduly, and even the puffins *Fratercula arctica* of Røst in Norway which failed to breed for nearly 20 years following the collapse of the local herring *Clupea harengus* stock are now said to be reproducing again.

*5. Nets and Other Floating Objects.* It has been known for over 80 years (McIntosh 1903) that large numbers of diving birds are sometimes lost in fishermen's drift nets, and in recent years the losses are thought to have increased with the introduction of less conspicuous monofilament nylon nets elsewhere, though there appears to be little information for the North Sea, where they are illegal (but by no means unknown) in Scotland. Thousands of birds must also be lost annually in those fixed salmon nets set close to seabird colonies along the east coast of Scotland (Melville 1973), where I have seen up to ten dead auks in the same net, yet the colonies continue to thrive. This problem grades into that caused by other floating objects since fishermen commonly dispose of fragments of indestructible netting at sea or use it to make lobster-pots from which it may eventually float free to join a wide variety of other discarded objects with which seabirds such as gannets frequently become entangled (Schrey and Vauk 1987). Some also collect these materials to make their nests, leading to further losses of their chicks, and indeed plastic materials form the majority of some gannet's nests at Flamborough Head. Smaller objects are also eaten freely by a variety of species (Bourne 1976a); the fulmars washed up in the Netherlands now normally contain an average of a dozen pieces of plastic (van Franeker 1985), and puffins have also picked up fragments of rubber (Parslow and Jefferies 1972). While entanglement in or consumption of such objects often leads to an extremely slow, cruel death, observations along the shore suggest that it only causes a small part of seabird mortality.

6. *Oil and toxic chemical pollution.* Oil pollution commonly causes seabirds to die in an even more conspicuous, unpleasant manner on a much larger scale, and there has been a public outcry at regular intervals ever since the introduction of oil as fuel for ships at the beginning of the century whenever standards of care in its use began to decline, leading to an improvement in the situation. Fortunately one of these periods occurred as the result of a series of tanker disasters just before the start of offshore oil production in the North Sea, so that it has been carried out with care, and conditions have as a result, if anything, improved. The examination of cleaner birds which apparently died mainly as the result of a failure of the food supply in the Irish Sea in the autumn of 1969 revealed that a few were also contaminated with a variety of toxic chemicals, which appears to be a general phenomenon (NERC 1983), but they appear to be a minority which have probably been feeding in polluted places. Following the control of the pesticide effluent which caused a decline of waterbirds in the Netherlands in the 1960's (Koeman 1972; Fig. 5) there appears to be little evidence for further problems of this type.

In conclusion, most seabird populations frequenting the North Sea appear to be in a healthy state at the present time, and the worst problem appears to be the availability of sufficient undisturbed breeding sites, though clearly there is also a continuing need for perpetual vigilance over the control of all types of pollution.

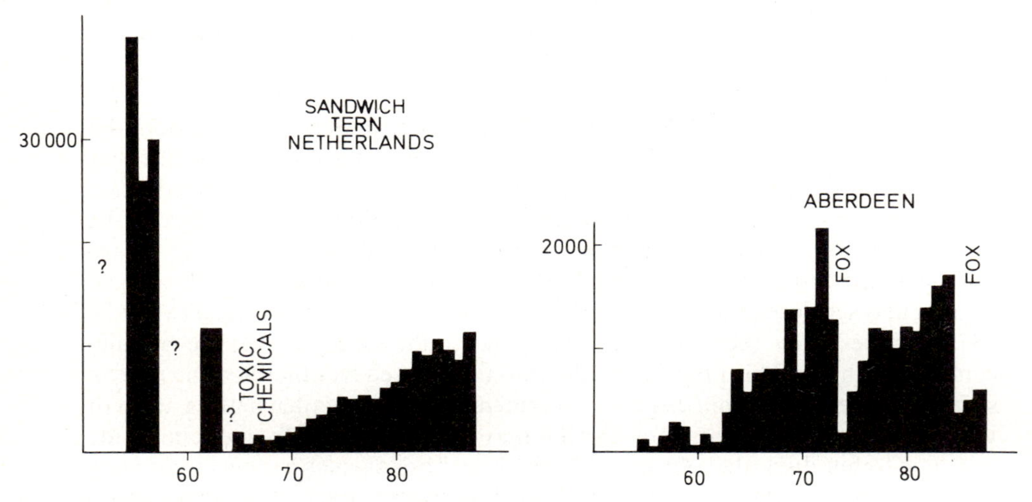


**Fig. 5.** Recent fluctuations in the sandwich tern populations of the Netherlands (Koeman et al. 1967 and J. Rooth, in press) and the Sands of Forvie National Nature Reserve near Aberdeen, north-east Scotland (A.J.M. Smith pers. commun.)

The group which appears to have the worst problems is the terns, which nest in very vulnerable sites along the coast, and indeed the largest colony of sandwich terns in Scotland on the Sands of Forvie National Nature Reserve has been intermittently abandoned in recent years owing to a failure to control foxes and visitors. It seems possible that the Caspian tern *Sterna caspia* may have been lost from the North Sea as a result of such treatment in the past, and might recolonize it from the Baltic if it were given more encouragement. Among other species, the roseate tern *Sterna dougallii* is now also reduced to small numbers in north-east Britain, possibly as a result of a combination of climatic deterioration and destruction by man in its winter quarters, and the gull-billed tern *Sterna nilotica* to even smaller numbers in Denmark, possibly as a result of changes in agricultural methods, though it is also a species of warmer climates elsewhere, and both deserve full protection.

In order to keep track of the future welfare of these birds there is a continuing need for further monitoring of their breeding distribution and success, of the location of vulnerable concentrations at sea, and of the appearance of unusual numbers of dead birds on the shore. While it is difficult to carry out most of these procedures with much accuracy for more reasons than it is easy to discuss in a short space here (NERC 1977), it is not really necessary to know the birds' numbers precisely, merely whether they are still present and appear to be successful, so that it would be a waste of effort to attempt much precision. There is, however, a continuing need for more detailed investigation of some potential threats whose impact is more difficult to assess, including the incidence of fluctuations in climate and fisheries, disease and toxic chemical pollution (Table 2), and there is a need for a coordinated international effort to obtain more comparable information about these.

**Table 2.** Toxic chemicals in seabird eggs

| Species | | Year | No | PCB | pp' DDE | HG |
|---|---|---|---|---|---|---|
| Shag *P. aristotelis* | UK | 67–71 | 9 | 4.56 | 2.13 | |
| | Norway | 83 | 20 | 0.58 | 0.31 | 0.15 |
| Herring gull *L. argentatus* | UK | 68–73 | 11 | 2.17 | 0.6 | 0.17 |
| | Norway | 72 | 91 | 8.49 | 1.57 | |
| | | 83 | 43 | 3.67 | 0.85 | 0.09 |
| Kittiwake *Rissa tridactyla* | UK | 67–73 | 35 | 5.67 | 0.45 | 0.34 |
| | Norway | 72 | 43 | 2.87 | 0.37 | |
| | | 83 | 31 | 1.60 | 0.42 | |
| Guillemot *Uria aalge* | UK | 68–73 | 42 | 4.30 | 1.18 | 0.32 |
| | Norway | 72 | 41 | 2.19 | 1.18 | |
| | | 83 | 34 | 0.63 | 0.63 | 0.11 |
| Razorbill *Alca torda* | UK | 67–73 | 9 | 10.3 | 1.31 | 0.78 |
| | Norway | 72 | 18 | 5.39 | 1.2 | |
| | | 83 | 9 | 2.22 | 0.83 | 0.17 |
| Puffin *Fratercula arctica* | UK | 73 | 21 | 5.00 | 0.79 | 0.57 |
| | Norway | 83 | 51 | 1.23 | 0.58 | 0.17 |

Estimations in parts per million, wet weight, from Barrett et al. 1985 and the raw data incorporated in NERC 1983.

*Acknowledgements.* Assistance has been received from R.T. Barrett with the assessment of the situation in Norway, Vogelwarte Helgoland and Vescin Ferdsand zurn Schitzeh Seevigel with north-west Germany, J. Rooth and A.A. Goede with the Netherlands, C. Joiris with Belgium, and P. Yésou with France. Investigations in Britain were initially carried out by the Seabird Group, and subsequently supported by a grant by the National Environment Research Council to Professor G.M. Dunnet at the Department of Zoology, Aberdeen University, who have continued to provide various assistance.

## References

Alexander WB (1934) The Helgoland Bird Observatory. Br Birds 27:284–289
Ashmole NP (1963) The regulation of numbers of tropical oceanic birds. Ibis 103b:458–473
Atkinson-Willes GL (1975) Effectifs et distribution des canards marins dans le nord-ouest de l'Europe, Janvier 1967–1973. Aves 12:254–284
Baillie SR, Mead CJ (1982) The effect of severe oil pollution during the winter of 1980–81 on British and Irish auks. Ringing & Migr 4:33–44
Barrett RT, Vader W (1984) The status and conservation of breeding seabirds in Norway. ICBP Tech Publ 2::323–333
Barrett RT, Skaare JU, Norheim G, Vader W, Frøslie A (1985) Persistent organochlorines and mercury in eggs of Norwegian seabirds 1983. Environ Pollut A39:79–93
Beverton RJ, Lee AJ (1965) Hydrographic fluctuations in the North Atlantic Ocean and some biological consequences. In: Johnson CJ, Smith GP (eds) The biological significance of climatic changes in Britain. Institute of Biology and Academic Press, London, pp 79–107
Bibby CJ (1973) The annual seabird sample census. Seabird 3:12–15
Bibby CJ (1981) An experiment on the recovery of dead birds from the North Sea. Ornis Scand 12:261–265
Blake BF (1983) A comparative study of the diet of auks killed during an oil incident in the Skagarrak in January 1981. J Zool (Lond) 201:1–12
Blake BF (1984) Diet and fish stock availability as a possible factor in the mass death of auks in the North Sea. J Exp Mar Biol Ecol 76:89–103
Blake BF, Tasker ML, Hope Jones P, Dixon TJ, Mitchell R, Langslow DR (1984) Seabird distribution in the North Sea. Nature Conservancy Council, Huntingdon
Bourne WRP (1968) Oil pollution and bird populations. Field Stud Suppl 2:100–218
Bourne WRP (1972a) General threats to seabirds. Int Congr Bird Pres Bull 11:200–218
Bourne WRP (1972b) Ducks die in the Forth. Mar Pollut Bull 3:53
Bourne WRP (1976a) Seabirds and pollution. In: Johnston R (ed) Marine Pollution. Academic Press, London, pp 403–502
Bourne WRP (1976b) The mass mortality of Common Murres in the Irish Sea in 1969. J Wildl Manage 40:789–792
Bourne WRP (1977a) Seabirds and Salmonellae. Mar Pollut Bull 8:194–195
Bourne WRP (1977b) Nylon netting as a hazard to birds. Mar Pollut Bull 8:75–76
Bourne WRP (1979) Birds and gas flares. Mar Pollut Bull 10:124–125
Bourne WRP (1980) The habitats, distribution and numbers of northern seabirds. Trans Linn Soc N Y 9:1–14
Bourne WRP (1981) Winter colony attendance by auks and the danger of oil pollution. Scott Birds 11:254–257
Bourne WRP (1982a) Birds at North Sea oil and gas installations. Mar Pollut Bull 13:5–6
Bourne WRP (1982b) Threats to seabirds. Birds (Lond) 9:63
Bourne WRP (1982c) The distribution of Scottish seabirds vulnerable to oil pollution. Mar Pollut Bull 13:270–273
Bourne WRP (1982d) Recovery of Guillemot colonies. Mar Pollut Bull 13:435–436
Bourne WRP (1982e) Oil pollution and seabird populations. Philos Trans R Soc Lond B Biol Sci 297:428
Bourne WRP (1983a) Birds, fish and offal in the North Sea. Mar Pollut Bull 14:294–296
Bourne WRP (1983b) Seabird problems. In: Hickling R (ed) Enjoying Ornithology. Poyser, Calton, pp 226–231

Bourne WRP (in press) The place of an outbreak of influenza in chickens in Scotland in 1959 in studies of the role of birds in the long-distance dispersal of disease. ICBP Tech Publ
Bourne WRP, Bibby CJ (1975) Temperature and the seasonal and geographical occurrence of oiled birds on west European beaches. Mar Pollut Bull 6:77–80
Bourne WRP, Harris MP (1979) Birds of the Hebrides: seabirds. Proc R Soc Edinb Sect B (Biol Sci) 77:445–475
Bourne WRP, Johnston L (1971) The threat of oil pollution to north Scottish seabird colonies. Mar Pollut Bull 2:117–119
Bourne WRP, Parrack JD, Potts GR (1967) Birds killed in the Torrey Canyon disaster. Nature 215:1123–1125
Bourne WRP, Bogan JA, Wanless S (1978) Pollution and the Sulidae. In: Nelson JB (ed) The Sulidae Gannets and Boobies Oxford University Press, Oxford, pp 977–983
Brien Y (1970) Avifaune de Bretagne. Société pour létude et la Protection de la Nature en Bretagne
Brownsey BW, Peakall DB (1953) Breeding seabirds of Flamborough Head. Naturalist (Leeds) 1953:149–150
Brun E (1979) Present status and trends in populations of seabirds in Norway. US Fish Wildl Serv Wildl Res Rep 11:289–301
Campbell LH, Standring KT, Cadbury CJ (1978) Firth of Forth oil pollution incident, February 1978. Mar Pollut Bull 9:335–339
Chislett R (1952) Birds of Yorkshire.
Clark RB (1984) Impact of oil pollution on seabirds. Environ Pollut A 33:1–22
Coulson JC, Potts GR, Deans IR, Fraser SM (1968) Exceptional mortality of Shags and other seabirds caused by paralytic shellfish poisoning. Br Birds 61:381–404
Cramp S, Bourne WRP, Saunders D (1974) The Seabirds of Britain and Ireland. Collins, London, 287 pp
Delbeke R, Joiris C (1985) Ecotoxicology of organochlorine residues in marine ecosystems. Proc Progress in Belgian Oceanographic Research, Brussels, March 1985, pp 358–367
Dunnet GM (1982) Oil pollution and seabird populations. Philos Trans R Soc Lond B Biol Sci 297:413–427
Dunnet GM (1987) Seabirds and North Sea oil. Philos Trans R Soc Lond B 316:513–524
Edwards JI, McKay DW (1984) Sprat acoustic surveys. Scott Fish Bull 48:36–40
Evans PGH (1984) Status and conservation of seabirds in northwest Europe (excluding Norway and the USSR). ICBP Tech Publ 2:293–321
Evans PGH, Nettleship D (1985) Conservation of the Atlantic Alcidae. In: Nettleship DN, Birkhead TR (eds) The Atlantic Alcidae. Academic Press, London, pp 427–488
Evans PR (1973) Avian resources of the North Sea. In: Goldberg ED (ed) North Sea Science. Cambridge, Mass, pp 400–412
Falandysz J, Szefer P (1984) Chlorinated hydrocarbons in fish-eating birds wintering in the Gdansk Bay, 1981–82 and 1982–83. Mar Pollut Bull 15:298–301
Figge K, Hoerschelmann H, Polzhofer K (1976) Organochlorpestizide und polychlorierte Biphenyle in Vögeln aus den Gebieten südliches Südamerika, Falklandinseln und Norddeutschland. Hosp-Hyg, Gesundheitswesen Desinfektion 68:354–360, 367–374, 393–400, 403–410
Fisher J (1952) The Fulmar. Collins, London
Fisher J (1966) The Shell Bird Book. Ebury, Michael Joseph, London
Franeker JA van (1985) Plastic ingestion in the North Atlantic Fulmar. Mar Pollut Bull 16:367–369
Furness RW (1982) Competition between fisheries and seabird communities. Adv Mar Biol 20:225–307
Furness RW (1984) Seabird-fisheries relationships in the northeast Atlantic and North Sea. In: Nettleship DN, Sanger GA, Springer PF (eds) Marine birds: their feeding ecology and commercial fisheries relationships. Canadian Wildlife Service, Ottawa, pp 162–169
Goede AA (1985) Mercury, Selenium, Arsenic and Zinc in waders from the Dutch Wadden Sea. Environ Pollut A37:287–309
Goede AA, Bruin M de (1985a) Arsenic in the Dunlin (Calidris alpina) from the Dutch Waddenzee. Bull Environ Contam Toxicol 34:617–622
Goede AA, Bruin M de (1985b) Selenium in a shore bird, the Dunlin from the Dutch Waddenzee. Mar Pollut Bull 16:115–117
Goede AA, Voogt P de (1985) Lead and Cadmium in waders from the Dutch Wadden Sea. Environ Pollut A37:311–322

Greenwood JJD, Donally RJ, Feare CJ, Gordon NJ, Waterston G (1971) A massive wreck of oiled birds, northeast Britain, Winter 1970. Scott Birds 6:235–250
Gurney JH (1913) The Gannet: A bird with a history. London
Harris MP, Wanless S (1984) The effect of the wreck of seabirds in February 1983 on auk populations on the Isle of May (Fife). Bird Study 31:103–110
Harrison JG, Buck WFA (1967) Peril in perspective: an account of oil pollution in the Medway estuary. Kent Bird Rep Suppl 16:24
Hartwig E, Reineking B, Schrey E, Vauk-Hentzelt E (1985) Auswirkungen der Nordseevermüllung auf Seevögel, Robben und Fische. Seevögel Sonderband 6:42–47
Harvie-Brown JA, Cordeaux J et al. (eds) (1880–89) Annual reports on the migration of birds 1879–1887. Zoologist (3)4:161–204 and then separately, British Association for the Advancement of Science, London, Edinburgh
Holdgate MW (1971) The Seabird wreck of 1969 in the Irish Sea. Natural Environment Research Council, (abridged version published as Natural Environment Research Council Publ C 4).
Hope-Jones P, Howells G, Rees EIS, Wilson J (1970) Effect of "Hamilton Trader" oil on birds in the Irish Sea in May 1969. Br Birds 63:97–110
Hope-Jones P, Barrett CF, Mudge GP, Harris MP (1984) Physical condition of auks beached in eastern Britain during the wreck of February 1983. Bird Study 31:95–98
Hudson R, Mead CJ (1984) Origins and ages of auks wrecked in eastern Britain during the wreck of February 1983. Bird Study 31:95–98
Joensen AH (1961) (Disaster among fulmars (Fulmarus glacialis (L.)) and kittiwakes (Rissa tridactyla (L.)) in Danish waters. Dan Ornithol Foren Tidsskr 55:212–218 (Danish with English summary)
Joensen AH (1972a) Oil pollution and seabirds in Denmark 1935–1968. Dan Rev Game Biol 6(8):1–24
Joensen AH (1972b) Studies on oil pollution and seabirds in Denmark 1968–71. Dan Rev Game Biol 6(9):1–32
Joensen AH (1973) Danish seabird disasters in 1972. Mar Pollut Bull 4:117–118
Joiris C (1978) Seabirds recorded in the northern North Sea in July: the ecological implications of their distribution. Gerfaut 68:419–440
Koeman JH (1971) Het voorkomen en de toxicologische betekenis van enkele chloorkoolwaterstoffen aan de Nederlandse kust in de periode van 1965–1970. Ph D thesis, Utrecht University
Koeman JH (ed) (1972) Side-effects of persistent chemicals on birds and mammals in the Netherlands. TNO-Nieuws 27:527–632
Koeman JH, Oskamp AAG, Veen J, Brouwer E, Rooth J, Zwart P, Broek E vd, Genderen H van (1967) Insecticides as a factor in the mortality of the Sandwich Tern (Sterna sandvicensis): A preliminary communication. Meded Rijksfac Landbouwwet Gent 32:841–854
Lloyd CS, Thomas GJ, MacDonald JW, Borland D, Standring KT, Smart JL (1976) Wild bird mortality caused by botulism in Britain, 1975. Biol Conserv 10:119–129
McIntosh WC (1903) The effects of marine piscatorial birds on the food fishes. Ann Mag Nat Hist (7)11:551–553
Melville DS (1973) Birds in salmon nets. Seabird Rep 3:47–50
Nelson JB (1978) The Sulidae- Gannets and Boobies. Oxford University Press, Oxford
Nelson TH, Eagle Clarke W, Boyes F (1907) Birds of Yorkshire 2 vols. Brown, London Hull York
NERC (1977) Ecological research on seabirds Nat Environ Res Counc Publ Ser C 18
NERC (1983) Contaminants in marine top predators. Nat Environ Res Counc Publ Ser C 23
Nicholas WW (1940) Birds of Bempton cliffs. Naturalist (Leeds) 1940:217–219
Norrevang A (1977) Fuglefangsten pa faroerne. Rhodos Publ
North Sea Bird Club (1979-1985). Annual Reports. Department of Zoology, Aberdeen University
Osborn FD, Young WJ, Gore DJ (1984) Pollutants in auks from the 1983 North Sea bird wreck. Bird Study 31:99–102
Parslow JLF, Jefferies DJ (1972). Elastic thread pollution of Puffins. Mar Pollut Bull 3:43–45
Pashby BS (1985) John Cordeaux-Ornithologist. Spurn Bird Observatory, 86 pp
Pingree RD, Griffiths DK (1978) Tidal fronts on the shelf seas around the British Isles. J. Geophys. Res 83:4615–4622
Pennant T (1771) A tour of Scotland. London
Rasmussen-Ejde RK (1938) Veber eine durch Stürmvögel übertragbare Lungenerkrankun auf den Färöern. Zentralbl Bakteriol 143:89–93
Ray J (1678) The ornithology of Francis Willughby. London

Reineking B, Vauk G (1982) Seevögel-Opfer der ölpest. Niedevelbe-Verlag Otterndorf NE
Rittinghaus H (1956) Etwas über die 'indirekte' Verbreitung der ölpest in einem Seevogelschutzgebeit. Ornithol Mitt 8:43–46
Røv N, Thomassen J, Anker-Nilssen T, Barrett R, Folkestad AO, Runde O (1984) Sjøfuglprosjektet 1979–1984. Viltrapport 35
Schrey E, Vauk G (1987) Records of entangled Gannets (Sula bassana) at Helgoland, German Bight. Mar Pollut Bull 18:350–352
Seebohm H (1885) A history of British Birds, vol 3. London
Simpson VR, Hunt AE, French MC (1979) Chronic lead poisoning in a herd of mute swans. Environ Pollut 18:187–202
Stowe TJ (1982) Beached bird surveys and surveillance of cliff-breeding seabirds. Royal Society for the Protection of Birds, Sandy, 207 pp
Swennen C, Spaans AL (1970) (Seabird mortality by oil in the Wadden Sea area in February 1969). Het Vogeljaar 18:239–245 (Dutch, English summary)
Szefer P, Falandysz J (1983) Uranium and Thorium content of Long-tailed Ducks (Clangula hyemalis L.). Sci Total Environ 29:277–280
Szefer P, Falandysz J (1986) Trace metals in the bones of Scaup Ducks (Aythya marila L.) wintering in the Gdansk Bay, Baltic Sea. Sci Total Environ 53:193–199
Tåning AV (1952) Oljedoden. Sver Nat 5
Underwood LA, Stowe TJ (1984) Massive wreck of seabirds in eastern Britain, 1983. Bird Study 31:79–88
Vauk G (1972) Die Vögel Helgolands. Parey, Hamburg, Berlin
Vauk G (1984) Oil pollution dangers on the German coast. Mar Pollut Bull 15:89–93
Vauk-Hentzelt E (1986) Krankeiten bei wildlebenden Möwen (Larus spec.) aus dem Bereich der Insel Helgoland. Verhandlungsber 28 Intern Symp Erkrankung der Zootiere, Rostock, pp 129–134
Vauk G, Schrey E (1987) Litter pollution from ships in the German Bight. Mar Pollut Bull 18:316–319
Wade EW (1903) Birds of Bempton Cliffs. Trans Hull Sci Field Nat Cl 3(1):1–26
Williams AJ, Kermode D (1968) A census of the seabird colony at Flamborough Head, June 1964. Seabird Bull 6:15–21
Wolfe Murray D (1936) In Savage RE, Wimpenny RS (1936) Phytoplankton and the Herring. Fish Invest (2)15:2

# Accumulation and Body Distribution of Xenobiotics in Marine Mammals

P.J.H. Reijnders[1]

## 1 Introduction

There are several types of pollution that could affect marine mammals, but it is evident that the title restricts this subject to chemical pollution excluding, for example, noise pollution as a potential threat (Peterson 1981).

Through their global dispersion, environmental contaminants and trace elements occur in marine mammal tissues throughout the entire marine ecosystem. High concentrations of certain pollutants, particularly lipophilic organochlorine compounds, are found in marine mammals. Reviews and qualitative comparisons of residue levels in various species from different geographical regions have been made (Holden 1978; Wagemann and Muir 1981). These procedures have been demonstrated to be inadequate for a comparative assessment of the degree of hazard that chemical compounds pose to marine mammals. This is concluded from recent discussions on the reliability of sampling procedures, analytical techniques and the representativeness of certain tissues due to incomplete knowledge of kinetics after exposure (e.g., Duinker et al. 1980, 1983, Aguilar 1985; Reijnders 1986a, 1987).

Only when all sampling and analytical techniques are carried out in one study (e.g. O'Shea et al. 1980; Tanabe et al. 1983) can the outcome be used as a relative measure for the identification of the pollutant burden in marine mammals from different areas. Therefore this chapter does not provide figures on so-called reference concentrations of different contaminants acquired from not strictly comparable sources, but rather surveys what is known about the different phases in the accumulation processes in marine mammals in order to arrive at a useful analysis of future research needs. This will enable identification of a conceptual strategy for intercomparison of data on pollutants in marine mammals and also enhances the extrapolation of information on the effects of some pollutants obtained from case studies.

## 2 Compounds Identified

The provision of a comprehensive list of all identified contaminants occurring in marine mammals is beyond the scope of this chapter. Moreover, from the outset it will be incomplete, as nearly every trial to search for a specific hitherto unknown chemical will be successful. To improve the manageability of the matter, four major

[1]Research Institute for Nature Management, P.O. Box 59, 1790 AB Den Burg (Texel), The Netherlands

groups of compounds are distinguished; heavy metals (including bromine, iodine); chlorinated hydrocarbons (e.g., polychlorinated biphenyls PCB's, DDT); petroleum hydrocarbons; and radioactive isotopes.

Zitko (1975) presented some information on the presence in aquatic fauna of toxic compounds in petroleum hydrocarbons: the polynuclear aromatic hydrocarbons. However, there is very limited information on possible implications of marine mammals and oil pollution in the North Sea area (Davis and Anderson, 1976; Duguy and Babin, 1975). Therefore, this group of compounds will not be referred to further.

Analogously, this holds for radioisotopes, as literature on this subject, specifically for the North Sea area, is very scarce. To the author's knowledge, only some data exist on their occurrence in harbor seals (Drescher unpubl. data).

Within this chapter no attention is paid to the question whether a specific compound could be of natural origin. Some metallic elements occur naturally in marine sediments and seapage of oil into the sea is natural phenomenon (Gaskin 1982).

## 3 Kinetics

### 3.1 Uptake

The uptake of contaminants by marine mammals occurs primarily via ingested prey. For a long time this evoked the idea that bioaccumulation of chemical compounds resulted from amplification through the food chains and reflected the pattern of geographical pollution. However, with growing data bases on different residue levels both in marine mammals and their food species, it became apparent that the expected one- to ten-fold magnification in every subsequent trophic level is not a general rule (Drescher 1979; Reijnders 1980; ten Berge and Hillebrand 1974; Tanabe et al. 1984; Hidaka et al. 1984). Bioconcentration factors (ratio of concentration of a residual chemical in the organism and in its food) ranged from less than 1 to more than $10^7$. An obvious conclusion is that apparently many contaminants exhibit bioaccumulative properties, but despite extensive studies the complex mechanisms behind the bioaccumulation processes in natural marine ecosystem are still poorly understood. In the next section generic aspects of biological and chemical factors that can partly elucidate the process of deposition and metabolization will be discussed.

One of the factors that can have a large influence on body burdens of contaminants in marine mammals is the mobility of the mammals. A second factor is their general, rather opportunistic feeding behavior. This implies that seasonal differences exist in the uptake of contaminants. Whether this will be reflected in tissue burdens depends on the biological half-life of the chemical in question. For mercury in marine mammals, for instance, the biological half-life time is up to 700 days (Gaskin 1982), which renders a fluctuating exchange rate due to seasonal changes in feeding habits unlikely.

For lipophilic substances as organochlorines this could certainly play a role, however. It has been found that concentration of PCB's in blubber of harbour seals changes inversely to blubber thickness, which then undergoes seasonal changes (Drescher et al. 1977). Especially baleen whales, that carry out seasonal migration from highly productive waters in summer to "poor" tropical zones in winter months, show a tenfold decrease in average daily food intake (Tomilin 1957; Best 1967). This can result in a total decrease in body weight of 25 to 50%, mostly caused by fat mobilization in blubber (Lockyer and Brown 1981). Apart from other mechanisms, as will be discussed later, these changes in fattening condition will influence the concentration of a given chemical in blubber tissue considerably.

### 3.2 Body Distribution

*Heavy Metals.* It is well known that specific chemicals accumulate in specific tissue. For example, nearly 90% of the total body burden for lead will be found in bones (Marcus 1983). The highest levels for cadmium are generally found in the kidney (McKie et al. 1980; Falconer et al. 1980; Honda and Tatsukawa 1983).

The distribution of copper, zinc, iron and manganese is fairly equal over most tissues, which is not surprising as they play a vital role in several enzymatic reactions. A number of these essential components are mostly found in the form of metalloproteins like, for example, hemoglobins. Most of these elements are present in small quantities, but are known to become acutely toxic when excess concentrations occur (Bryan 1976).

In general, the highest levels of heavy metals in marine mammals are almost invariably found in liver tissue, the next highest being kidney tissue, followed by muscle tissue. The mechanisms for the accumulation in specific organs are not well understood. Some postulated mechanisms are different distributions of protein carriers for, e.g., heavy metals – resulting in differential transport to particular organs and inability of metal-protein complexes to pass certain cellular barriers. However, these explanations should be used with caution. As has been demonstrated for mercury found in brain tissue of seals, concentration levels were always inferior to concentrations in liver, kidney, and muscle. This was supposed to be caused by the so-called blood-brain barrier that mercury had to traverse to enter the brain (Heppleston and French 1973). As more data became available, it became apparent that the hematoencephalic barrier is not effective in impeding passage of mercury (Reijnders 1980).

*Chlorinated Hydrocarbons.* The solubility features of chlorinated hydrocarbons implicate a residue concentration in lipid-rich tissue. The highest levels of organochlorines are found in blubber tissue (e.g., Drescher et al. 1977; Reijnders 1980; Wagemann and Muir 1981; Gaskin 1982). When the organochlorine residue levels in various tissues are expressed on a lipid basis instead of a wet weight basis, the large variation disappears even in different organs from the same animals. This indicates that organochlorines are distributed over the different organs relative to their fat content. However, this procedure can also lead to artifacts, as is indicated by Aguilar (1985). The equilibrium partitioning during the deposition process seems to be largely determined by the lipid composition of the tissue concerned (Tanabe et al. 1981a; Schneider 1982; Aguilar 1985). The distribution of lipophilic xenobiotics in *Stenella coeruleoalba* appeared to be even more uniform when the levels of contaminants were expressed per unit of weight of triglicerides (Aguilar 1985; Kawai and Fukushima 1981). Another complicating factor in validating organochlorine levels in blubber is its inhomogeneous tissue composition. Especially in large whales, the lipid composition differs between inner and more superficial layers. Due to this stratification, the distribution of organochlorines over different areas of the blubber in one animal will be affected (Aguilar 1985). Comparison of contaminant levels in blubber between animals from one species or between different species therefore becomes complicated.

Organochlorine residue levels in brain tissue in marine mammals did not follow the general pattern in other tissues of tissue: blubber (expressed on lipid base) close to 1.

The assumed hematoencephalic barrier (Frank et al. 1973) appeared to be very permeable for those liposolubles (Walker 1975). A possible explanation could be that phospholipids as basic constituents of brain lipids, exhibit a greater polarity than the remaining lipids resulting in a lower retention capacity of brain tissue compared with other organs (Reijnders 1980). This is supported by results of Tanabe et al. (1981b) that the brain of striped dolphins contained higher proportions of lower chlorinated PCB's than other organs did. This was attributed to the greater affinity of phospholipids for these PCB's exhibiting greater polarity (Tulp and Hutzinger 1978).

In general it can be stated that in marine mammals, chlorinated hydrocarbons are distributed proportionally to the trigliceride content of the different body tissues (including brain).

### 3.3 Metabolism

*Heavy Metals.* Many heavy metals with the exception of the elemental components occur in marine mammal tissues as metallothionein-complexes, metallothionein is a specific protein synthesized predominantly in the liver (Pitrowski et al. 1974, 1977) Protein-bound heavy metals are supposed to be more soluble and are most likely eliminated from the body while in transition. The biological half-life of mercury in the blood is considerably less than in tissues (Nordberg et al. 1970). Excretion routes for mercury and cadmium are predominantly through feces and urine (Nomiyama and Foulkes 1977; Nordberg et al. 1970). A remarkable phenomenon described by Koeman et al. (1973) is the molecular 1:1 ratio of mercury and selenium found in all marine mammals they analyzed. Mercury and selenium appear to be bound to proteins by sulfydryl linkages. The possible detoxifying role of selenium with respect to mercury and cadmium compounds has been postulated by Parizek et al. (1969). However, it is still unclear whether selenium is really involved in the demethylating process and in binding of mercury (Roberts et al. 1976). Demethylation of the more toxic organic mercury by marine mammals enhances excretion. Most of the mercury in the food of marine mammals is present as methylmercury (up to 90%), whereas especially in older animals only 10–15% of the total mercury consists of methylmercury (Roberts et al. 1976; Kari and Kauranen 1978). Several mechanisms for the demethylation process have

already been suggested earlier in this section. However, so far the exact nature of the transformation process – enzymatical, bacteriological or chemical – has not been clarified (e.g., van de Ven et al 1979).

*Chlorinated Hydrocarbons.* With respect to the metabolism of chlorinated hydrocarbons, three aspects must be considered: the processes as such, the physicochemical properties of chemical compounds, and the physiological state of the animals concerned. In marine mammals the metabolization of most chlorinated hydrocarbons probably occurs via aryl hydrocarbon hydroxylase activity analogously to metabolization in most animal species (Street and Chadwick 1975; Jansson et al. 1975). Phenolic metabolites of PCB's and DDE are found in feces as well as in bile of grey seals (Jansson et al. 1975), suggesting an important excretion route via liver and kidney. Another group of metabolites, the methylsulfones, was found by Jensen and Jansson (1976).

Physicochemical properties of compounds influence metabolization (Sugiura et al. 1978; Tulp and Hutzinger 1978; Hidaka et al. 1984). Unfavorable stereochemistry resulting from different patterns of chlorine substitution in PCB's have been found to lead to decreased metabolization (Hutzinger et al. 1978; Mizutani et al. 1980; Tanabe et al. 1981a; Shaw and Connell 1982). As part of a recent experiment on the effects of PCB's on reproduction in harbour seals (Reijnders 1986b), kinetics of individual PCB congeners have been studied. It was found that PCB's could be divided into persistent congeners and congeners with lowered concentrations based on their molecular structural features. Congeners showing lowered concentrations possessed vicinal H-atoms at either a meta-para position or at an ortho-meta position of one aromatic ring, with chlorine substitution being absent at both ortho positions of that ring (Boon et al. 1987). Changes in the physiological state of marine mammals can have considerable effect on the metabolization process. For instance, fluctuations in nutritional state viz. blubber layer thickness, will be reflected in contaminant burdens (e.g, Drescher et al. 1977). This could be either a consequence of circumannual cycles in blubber layer thickness due to changes in feeding intensity (Drescher et al. 1977; Aguilar 1985) or of changes in diet (Tanabe et al. 1984). Changes in diet are expected to be responsible for the concentration differences of lipophilic xenobiotics in blubber between baleen and toothed whales (Aguilar 1985).

The sex of marine mammals can clearly influence the tissue pollutant burden. In most areas with low to moderate levels of pollution, the tissue burden for organochlorines is usually higher in males than in females (Heppleston 1973; Addison and Smith 1974; Gaskin et al. 1976; Donkin et al. 1981). As the composition of the diet for both sexes in these studies is the same, differences in tissue burden must be a consequence of different metabolization patterns. Transfer of compounds from female lipid tissue to fetus via placental transport and to offspring via lactation is demonstrated in many marine mammals (Gaskin et al. 1976; Addisson and Brodie 1977; Duinker and Hillebrand 1979; Reijnders 1980).

In striped dolphins, the transfer rate of some organochlorines through parturition ranged from 4.0 to 9.4% (Tanabe et al. 1982) and through lactation even from 80 to 90% (Tanabe et al. 1980).

## Summary

Marine mammals can considerably accumulate contaminants from their diet. For this reason they can be considered as integrators of environmental pollution. Physicochemical properties of the compounds concerned, as well as the nutritional and physiological state of the animals, determine the accumulation rate. Studies on kinetics of uptake and elimination of pollutants in marine mammals have just started. Therefore mapping of marine pollution, as well as monitoring with marine mammals, is not feasible at present.

Residual concentrations of contaminants in several tissues are important for toxological evaluation. However, the study of the occurrence and effects of pollutants in marine mammals should be part of an integrative ecosystem approach whereby dynamics and fates of pollutants through food chain members are studied.

To that end, baseline research on occurrence of contaminants, with standardized sampling and analytical techniques have to be carried out. Next to that, detailed studies on the process of uptake, body distribution, and metabolization are necessary to comprehend the bioaccumulation phenomenon. Estimating whole body burdens of residual chemicals seems to be a useful concept.

*Acknowledgments.* The comments of Alex Aguilar, Herman Eijsackers, and Wim Wolff on an earlier version of this manuscript, are greatly appreciated.

## References

Addison RF, Brodie PF (1977) Organochlorine residues in maternal blubber, milk and pup blubber from grey seals (*Halichoerus grypus*) from Sable Island, Nora Scotia. J Fish Res Board Can 34:937–941

Addison RF, Smith TG (1974) Organochlorines residue levels in arctic ring seals: Variation with age and sex. Oikos 25:335–337

Aguilar A (1985) Compartmentation and reliability of sampling procedures in organochlorine pollution surveys of ceteceans. Res Rev 95:91–114

Best PB (1967) Distribution and feeding habits of baleen whales of the Province. Rep S A Div Sea Fish Invest Rep 57

Boon JP, Reijnders PJH, Dols J, Wensvoort P, Hillebrand MTJ (1987) The kinetics of individual polychlorinated biphenyl (PCB-) congeners in female harbour seals (*Phoca vitulina*) with evidence for structure related metabolism. Aqu Toxic (in press)

Bryan GW (1976) Heavy metal contamination in the sea. In: Johnston R (ed) Marine pollution. Academic Press, London New York, pp 185–302

Davis JE, Anderson SA (1976) Effects of oil pollution on breeding grey seals. Mar Pollut Bull 7:115–118

Donkin P, Mann SV, Hamilton EI (1981) Poluychlorinated biphenyl, DDT and dieldrin residues in grey seal (*Halichoerus grypus*) males, females and mother-foetus pairs sampled at the Farne Islands, during breeding season. Sci Tot Environ 19:121–142

Drescher HE (1979) Biologie, Okologie and Schutz der Seehunde im schleswigholsteinischen Wattenmeer. Wiss Schr I, Landesjagdverband Schleswig-Holstein, Meldorf, FRG

Drescher HE, Harms U, Huschenbeth E (1977) Organochlorines and heavy metals in the harbour seal (*Phoca vitulina*) from the German North Sea coast. Mar Biol 41:99–106

Duguy R, Babin PH (1975) Intoxication aigue par les hydrocarbures observées chez un phoque veau-marin (*Phoca Vitulina*). Int Counc Expl Sea, Mar Mamm Ctee, CM/N:5

Duinker J, Hillebrand MTJ (1979) Mobilization of organochlorines from female lipid tissue and transplacental transfer to fetus in a harbour porpoise (*Phocoena phocoena*) in a contaminated area. Bull Environ Contamin Toxicol 23:728–732

Duinker J, Hillebrand MTJ, Palmork KH, Wilhelmsen S (1980) An evaluation of existing methods for quantitation of polychlorinated biphenyls in environmental samples and suggestions for an improved method based on measurements of individual components. Bull Environ Contamin Toxicol 25:956–964

Duinker J, Hillebrand MTJ, Boon JP (1983) Organochlorines in benthic invertebrates and sediments from the Dutch Wadden Sea: identification of individual PCB components. Neth J Sea Res 17:19–38

Falconer CR, Davies IM, Topping G (1980) Selected trace metals in porpoises (*Phocoena phocoena*) from the northeast coast of Scotland. 1980/E:42. Inst Counc Expl Sea, Mar Mamm Ctee, E:42

Frank R, Ronald K, Braun HE (1973) Organochlorine residues in harp seals (*Pagophilus groenlandicus*) caught in eastern Canadian Waters. J Fish Res Board Can 30:1053–1063

Gaskin DE (1982) The ecology of whales and dolphines. Heinemann, London

Gaskin DE, Holdrinet M, Frank R (1976) DDT residues in blubber of harbour porpoise, *Phocoena phocoena*, from eastern Canadian waters during the five year period 1969–1973. ACMRR MM SC 96, pp 1–11

Heppleston PB (1973) Organochlorines in British grey seals. Mar Pollut Bull 4:44–45

Heppleston PB, French MC (1973) Mercury and other metals in British seals, Nature London 216:1274–1276

Heppleston PB, French MC (1978) Mercury and other metals in British seals. Nature London 243:302–304

Hidaka H, Tanabe S, Kawano M, Tatsukawa R (1984) Fate of DDTs, PCBs and chlordane compounds in the Arctic marine ecosystem. Mem Nat Inst Polar Res, Spec Iss 32, pp 151–161

Holden AV (1978) Pollutants and seals. Mamm Rev 8:53–66

Honda K, Tatsukawa R (1983) Distribution of cadmium and zinc in tissues and organs, and their age-related changes in striped dolphins, *Stenella coeruleoalba*. Arch Environ Contamin Toxicol 12:543–550

Hutzinger O, Lelyand van IH, Zoeteman BCJ (1978) Aquatic pollutants: Transformation and biological effects. Pergamon, London New York

Jansson B, Jensen S, Olsson M, Renberg L, Sundström G, Vaz R (1975) Identification by GC-MS of phenolic metabolites of PCBs and pp′-DDE isolated from Baltic guillemot and seal. Ambio 4:93–97

Jensen S, Jansson B (1976) Anthropogenic substances in seals from the Baltic: Methylsulfone metabolites of polychlorinated biphenyls and DDE. Ambio 5:257–260

Kari T, Kauranen P (1978) Mercury and selenium contents of seals from fresh and brackish waters in Finland. Bull Environ Contamin Toxicol 19:273–280

Kawai S, Fukushima M (1981) Relation between lipid composition and the concentration of organochlorine compounds in the various organs of striped dolphins *Stenella coeruleoalba*. In: Fujiyama T (ed) Studies on the levels of organochlorine compounds and heavy metals in the marine organism. Univ Ryukyus

Koeman JH, Peeters WHM, Koudstaal-Hol CHM, Tjioe PS, de Goey JJM (1973) Mercury selenium correlation in marine mammals, Nature London 245:385–386

Lockyer CH, Brown SG (1981) The migration whales. In: Aidley H (ed) Animal migration. Soc Exp Biol Semin Ser 13. Cambridge Univ Press, p 105

Marcus AH (1983) Compartmental models for trace metals in mammals. Sci Tot Environ 28:307–311

McKie JC, Davies IM, Topping G (1980) Heavy metals in grey seals (*Halichoerus grypus*) from the east coast of Scotland. Int Counc Expl Sea, Mar Mamm Ctee, E:41

Mizutani T, Hidaka K, Ohe T, Matsumoto M, Yamamoto K, Tajima G (1980) Comparative study on accumulation and elimination of hexachlorobiphenyls and decachlorobiphenyls in mice. Bull Environ Contamin Toxicol 25:181–187

Nomiyama K, Foulkes EC (1977) Reabsorption of filtered cadmiummetallothionein in the rabbit kidney. Proc Soc Exp Biol Med 156:97–99

Norberg GF, Berlin MH, Grant CA (1970) Methyl mercury in the monkey: autoradiographical distribution and neurotoxicity: Proc 16th Int Congr Occup Health, Tokyo

O'Shea TM, Brownell RL, Clark DR, Walker WA, Gay ML, Lamont TG (1980) Organochlorine pollutants in small cretaceans from the Pacific and south Atlantic Oceans. Pest Mon J 14:35–46

Parizek J, Benes O, Ostadalova, Babicky A, Benes J, Pitha J (1969) The effect of selenium on the toxicity and metabolism of cadmium and some other metals. In: Barltrop D, Burland WL (eds) Mineral metabolism in Paediatics. Blackwell, Oxford, pp 117
Peterson NM (ed) (1981) The question of sound from icebreaker operations. Proc Worksh, Toronto 23–24 Feb, 1981. Artic Pilot Proj, Calgary
Pitrowski JK, Trojanowska B, Sapota A (1974) Binding of cadmium and mercury by metallothionein in the kidneys and liver of zoats following repeated administration. Arch Toxicol 32:351–360
Pitrowski JK, Bem EM, Werner A (1977) Cadmium and mercury binding to metallothionein as influenced by selenium. Biochem Pharmacol 26:2191–2192
Reijnders PJH (1980) Organochlorine and heavy metal residues in harbour seals from the Wadden Sea and their possible effects on reproduction. Neth Sea Res 14:30–65
Reijnders PJH (1986a) Perspectives for studies of pollution in ceteceans. Mar Pollut Bull 17:31–36
Reijnders PJH (1986b) Reproductive failure in common seals feeding on fish from polluted coastal waters. Nature London 324:456–457
Reijnders PJH (1988) Ecotoxiological perspectives in Marine Mammalogy: Research principles and goals for a conservation policy. Mar Mamm Sci 4:91–102
Roberts TM, Heppleston PB, Roberts RD (1976) Distribution of heavy metals in tissues of the common seal. Mar Pollut Bull 7:194–196
Schneider R (1982) Polychlorinated biphenyls (PCBs) in cod tissues from the western Baltic: significance of equilibrium partitioning and lipid composition in the bioaccumulation of lipophylic pollutants in gill-breathing animals. Meeresforschung 29:69–72
Shaw GR, Connell DW (1982) Factors influencing concentrations of polychlorinated biphenyls in organisms from estuarine ecosystem. Aust J Mar Fresh Res 33:1057–1070
Sugiura K, Ito N, Matsumoto N, Nikora Y, Murata K, Tsukakoshi Y, Goto N (1978) Accumulation of polychlorinated biphenyls and polybrominated biphenyls in fish: limitation of correlation between partition coefficients and accumulation factors. Chemosphere 9:731–736
Street JC, Chadwick RW (1975) Ascorbic acid requirements and metabolism in relation to organochlorine pesticines. Ann NY Acad Sci 258:132–143
Tanabe S, Tanaka H, Maruyama K (1980) Bioaccumulation of *Stenella coeruleoalba:* Elimination of chlorinated hydrocarbons from mother striped dolphins through parturition and lactation. In: Fujiyama T (ed) Studies on the levels of organochlorine compounds and heavy metals in the marine organisms. Univ Ryukyus, Okinawa, pp 115–121
Tanabe S, Nakagawa Y, Tatsukawa R (1981a) Absorption efficiency and biological half-life of individual chlorobiphenyls in rats treated with kanechlor products. Agric Biol Chem 45:717–726
Tanabe S, Tatsukawa R, Tanaka, H, Maruyama K, Miyazaki N, Fujiyama T (1981b) Distribution and total burden of chlorinated hydrocarbons in bodies of striped dolphins (*Stenella coeruleoalba*) Agric Biol Chem 45:2569–2578
Tanabe S, Tatsukawa R, Maruyama K, Miyazaki N (1982) Transplacental transfer of PCBs and chlorinated hydrocarbon pesticides from the pregnant striped dolphin (*Stenella coeruleoalba*) to her fetus. Agric Biol Chem 46:1249–1254
Tanabe S, Mori T, Tatsukawa R (1983) Global pollution of marine mammals by PCBs, DDT and HCHs (BHCs). Chemosphere 12:1269–1275
Tanabe S, Mori T, Tatsukawa (1984) Bioaccumulation of DDTs and PCBs in the southern mink whale (*Balaenoptera acutorostrata*). Mem Natl Inst Pol Res Spec Iss 32:140–150
Tomilin AG (ed) (1957) Mammals of the USSR and adjacent countries, vol 9, cetecea. Trs Progr Sci Transl Ser, 1967 ed Heptner VG, Jeruzalem, pp 717
ten Berge WF, Hillebrand MTJ (1974) Organochlorine compounds in several marine organisms from the North Sea and the Dutch Wadden Sea. Neth J Sea Res 8:361–368
Tulp M TH M, Hutzinger O (1978) Some thoughts on aqueous solubilities and partition coefficients of PCBs, and the mathematical correlation between bioaccumulation and physio-chemical properties. Chemosphere 20:849–860
Ven WSM van de, Koeman JH, Svenson A (1979) Mercury and silenium in wild and experimental seals. Chemosphere 8:539–555
Wageman R, Muir DCG (1981) Assessment of heavy metals and organochlorine concentrations in marine mammals of northern waters. Int Counc Expl Sea, Mar Mamm Ctee N:9
Walker CH (1975) Variations in the intake and elimination of pollutants. In: F Moriarty (ed): Organochlorine insecticides: Persistent organic pollutants. Academic Press, London New York, pp 73–130
Zitko V (1975) Aromatic hydrocarbons in aquatic fauna. Bull Environ Contamin Toxicol 14:621–631

# Effects on Invertebrates

V.A. COOPER[1]

## 1 Introduction

The effects of pollution on marine invertebrates are many and varied (Hart and Fuller 1979), but in practical terms the study of the effects of inputs to the North Sea have been restricted to ecological effects and uptake of persistent contaminants such as heavy metals and chlorinated hydrocarbons. In recent years, more attention has been paid to the measurement of sublethal effects on physiology and biochemistry (Bayne et al. 1985) because it is felt that these give a more precise indication of pollution effects before they are gross enough to kill certain species. Various techniques are currently being evaluated in the blue mussel (*Mytilus edulis*) around inputs to the North Sea.

## 2 Ecological Effects

Most surveys on ecological effects are carried out on benthic communities (i.e., animals living on or within the seabed). Benthic invertebrates have relatively sessile lifestyles and cannot escape effects of pollution by swimming to another area. This makes them vulnerable to pollution effects and also relatively easy to study. However, their natural environment is a hostile one – storms and changes in tidal currents can completely disrupt their habitat – and their life history is "opportunistic" in order to compensate for that. If an area of the seabed becomes habitable or rich in foodstuff for a short while, they can invade it, reproduce, and then be washed away or die off within a few weeks. This makes effects of pollution very difficult to interpret. The question must always be asked, "is this change in the benthic community the result of natural or anthropogenic causes?". Any change attributable to pollution is likely to be very localized and may not always be adverse. The oligochaete worms, in particular, thrive on the high levels of organic matter from domestic sewage and sludge, and may well form the basis of a very healthy community in the region around an outfall where the organics are at an optimum level.

A healthy, unstressed ecosystem has a large number of 'primary consumers'. These animals graze on algae, or feed on dead or decayed matter (detritus) in one

[1]Water Research Centre, Henley Road, Medmenham, P.O. Box 16, Marlow, Buckinghamshire SL7 2HD, England

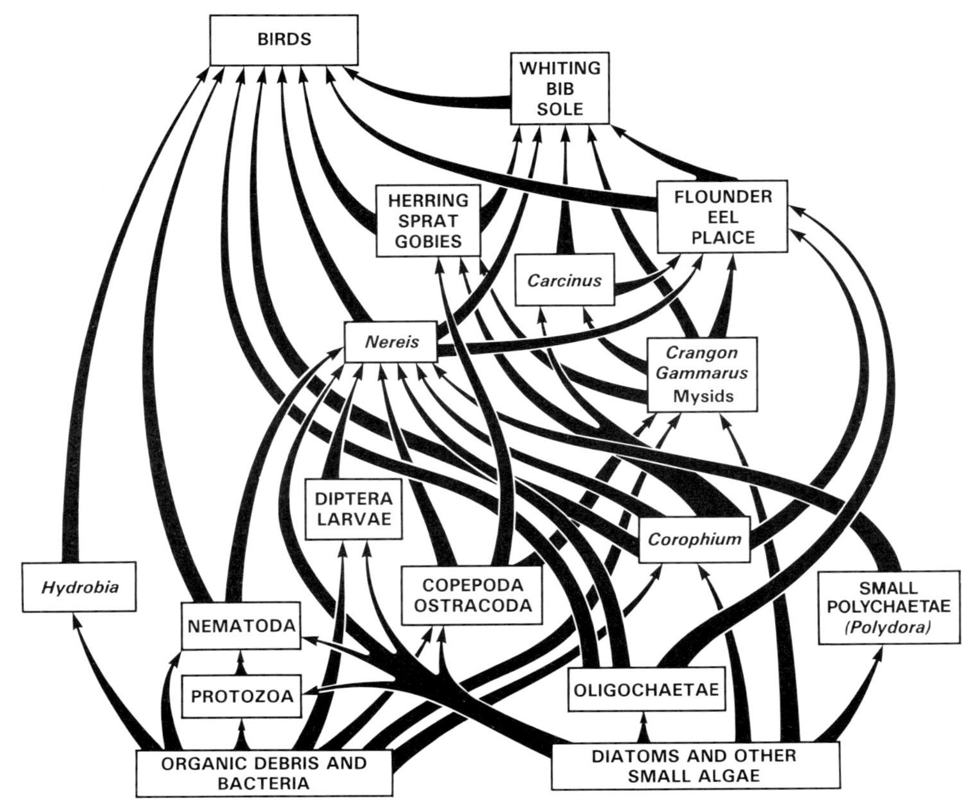


**Fig. 1.** Food web occurring in the Brackish Thames after 1976–1977. Before then oligochaete worms occupied a central position, providing the major diet item for both waterfowl and fishes. (Andrews 1984)

of several ways: shredding dead plant matter; filtering fine pieces of detritus drifting in the current or ingesting the sediment and digesting out any organics in it. These animals are then eaten by larger predators, such as ragworms, starfish or crabs. This forms the basis of a healthy "food web" as shown in Fig. 1. If the ecosystem has only one or two species of primary consumer, then it is unhealthy for two reasons: firstly the limited number of feeding types cannot maintain a balance in the organic matter present in the sediment; secondly the system has no capacity to buffer any further stresses and may be completely wiped out by a minor deterioration in conditions.

The distribution of benthic invertebrate species in the North Sea has been attributed to various physical factors including thermal stability (Glemarec 1973); water masses and their associated plankton (Ursin 1960); depth, wave action, temperature and water mass (Høpner Petersen 1977). Dyer et al. (1983) divided the North Sea into north and south regions (Fig. 2), based on the groups of benthic invertebrates. Within these regions, he identified three subgroups in the south and four in the north. He suggests that the relative importance of any single factor is likely to vary from area to area, but that it is reasonable to assume that factors

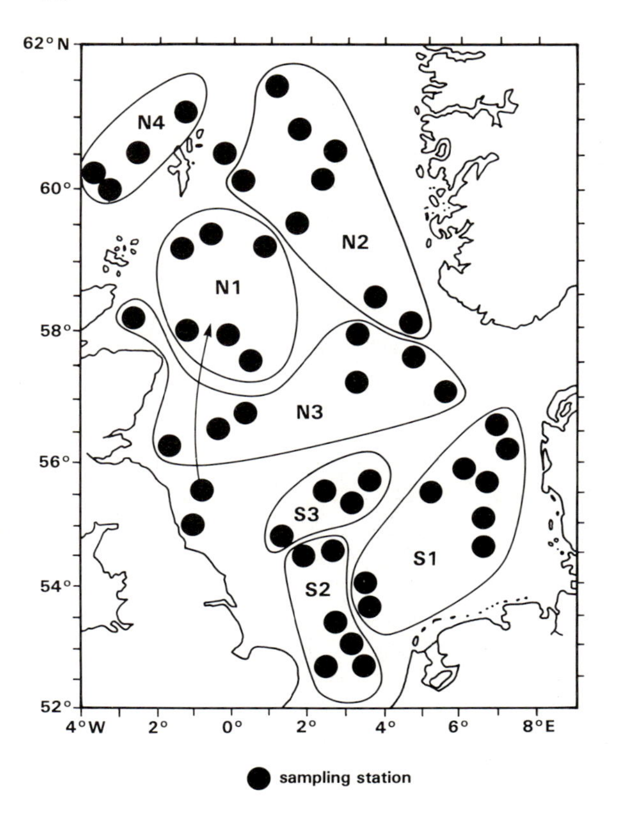


**Fig. 2.** The major groupings of benthic invertebrates in the North Sea. (Dyer et al. 1983)

suggested by Glemarec (1973), Ursin (1960) and Høpner Petersen (1977), plus the type of substratum, all exert some influence on the composition of the benthic invertebrate populations. Certainly the major rivers running into the North Sea do not appear to have widespread effects on the benthic invertebrates. Indeed, the available information suggests that any effects are so localized as to be within a few hundred yards of a discharge, or within an estuary or enclosed bay. Effects of oilspills, too, may be localized in the area where an oil slick washes onto the shore, as in the case of the sinking of the *Amoco Cadiz*, which affected French shorelines, or the oil spill in the Humber estuary in 1983.

Probably the best example of effects of pollution on benthic invertebrates in the North Sea is the change in populations in the Thames estuary since the major clean up in the 1960's and 1970's (Andrews 1984). Before 1964, the benthic community of the Thames estuary was dominated by large numbers of one species of oligochaete worm. Other species were present from time to time, but none could tolerate the conditions for long enough to become established. After 1976, when both domestic and industrial pollution was greatly reduced, a more balanced food web developed (Fig. 1). The system was no longer reliant on the presence of one species and therefore had the capacity to buffer stress. The sludge from the sewage works is now transported into the outer estuary and dispersed in the sea. This has been shown to have no such gross effects on the invertebrate life of the seabed in that area (Talbot et al. 1982).

Other major inputs to the North Sea have been surveyed outside the enclosed estuary areas and hence only minor changes have been reported. The mouth of the Rhine appears to exert some influence upon the type of community found in the Southern Bight (Govaere et al. 1980), but less than would be expected considering the significance of the input to the North Sea from the Rhine (Hill et al. 1984). Similarly, only minor changes have been observed in the Danish Wadden Sea (Madsen 1984), the British coast (Talbot et al. 1982) and the German Bight (Rachor and Gerlach 1978), although Rachor (1980) proposed that the inner German Bight should be considered as a sensitive area and be protected from avoidable additional stress.

## 3 Uptake of Persistent Contaminants

There are many contaminants discharged into the sea which cannot be broken down into harmless constituents. These persistent contaminants include metals (either in atomic form, or as part of a molecule, such as methyl mercury) and organic compounds such as chlorinated or brominated hydrocarbons. Death eventually occurs when the concentration of the compound reaches intolerable levels in the organism or, in the case of pesticides for example, when mobilization of fats releases any stored pesticide and exposes more delicate organs to a sudden peak in concentration. Tolerance limits and ability to regulate persistent contaminants vary according to species. Some, such as the barnacle (*Ballanus balanoides*), do not appear to regulate metals at all, but can tolerate body burdens far in excess of those which would be lethal to other species (Rainbow and White, pers. commun.).

Simplistically speaking, then, sessile marine invertebrates, such as the mussel (*Mytilus edulis*), which do not regulate persistent compounds, integrate the fluctuating concentrations of these materials to which they have been exposed throughout their lives. This has been used by several researchers to compare degrees of pollution in different areas of the North Sea, such as in the Mussel Watch programs. General sampling of the North Sea has been reported by ICES (1974) and more detailed studies have also been carried out round the coast of Scotland (Davies 1980) and Belgium (Meeus-Verdinne et al. 1983), for example.

A large baseline survey was co-ordinated out by ICES in 1972 (ICES 1974) looking at certain heavy metals and persistent organics in fish and shellfish. Subsequent surveys, however, have been rather less comprehensive (ICES 1984). The 1972 survey was successful in identifying areas of poor water quality for subsequent surveillance. For instance, high mercury concentrations were observed in mussels from the Netherlands and the Thames estuary and in one sample from Northern Norway. Concentrations were also slightly raised in the Tyne and Tees estuaries. Cadmium and zinc were also elevated in mussels from the outer Thames estuary. It is interesting that in the 1978 survey (ICES 1984), following the clean-up of the Thames, mercury and zinc in mussels were reduced to levels in keeping with other North Sea areas and cadmium was reduced to half the 1972 values.

Organochlorine and PCB residues were also determined during the same survey, in mussels from all North Sea coastlines except Belgium, France, and Scotland. Dieldrin occurred in higher concentrations in mussels from Germany and the Netherlands than those from England. Minimum DDT concentrations were observed in mussels from the Swedish coast, but the maximum concentration was also observed in Swedish mussels, from the heavily polluted Seläter area. Other areas of high DDT coincided with areas of high industrial or population density, such as the Netherlands, off the Rhine and Schelde estuaries. PCB concentrations were also highest in the Netherlands, any other elevations being very localized.

In more recent years, the ICES Co-ordinated Monitoring Programme has been rather less co-ordinated. Fewer countries seem to be taking part and there is little scope for comparison either spatially or temporally. In 1978 (ICES 1984), mussels were taken only from the English coastline for metal determination and from France for pesticide determination. Improvement in the water quality of the Thames estuary is reflected in the levels of metals found in the mussels, and pesticide concentrations in mussels from the French coast remained similar to previous unelevated levels.

Since over a decade has passed since the ICES baseline survey, it is high time that all the countries with North Sea coastlines joined forces for another such study.

## 4 Physiology and Biochemistry

Recent trends in monitoring effects of pollution have been towards the study of sublethal effects. Although a biological community may have a healthy structure, such as that shown in Fig. 1, organisms may be under stress and thus limited in their ability to tolerate further changes. Bayne et al. (1985) suggest that "sublethal effects on the individual that might be expected to have damaging consequences for the population include depressed rates of growth, reduced fecundity, reduction in egg viability and negative effects on competitive ability". When measuring stress in organisms, it is important to remember that all these effects are inter-related, and the study of a single response may be misleading. For instance, production of metalloproteins to detoxify metals results in increased protein turnover. This requires energies which may otherwise have been available for growth and so the organisms's scope for growth may be reduced.

A suite of responses is more likely to be related to the condition of the population or community than is a single response. Such a suite of physiological and biochemical responses are currently being used to assess the polluting effects of various inputs to the North Sea. The test organism is the mussel (*Mytilus edulis*) which, as seen from the ICES co-ordinated monitoring program, is common in North Sea coastal areas and its sessile lifestyle makes it an ideal test organism.

The mussel cages are deployed around the inputs to include sites suffering various degrees of contamination and at least one control site. Because most of the

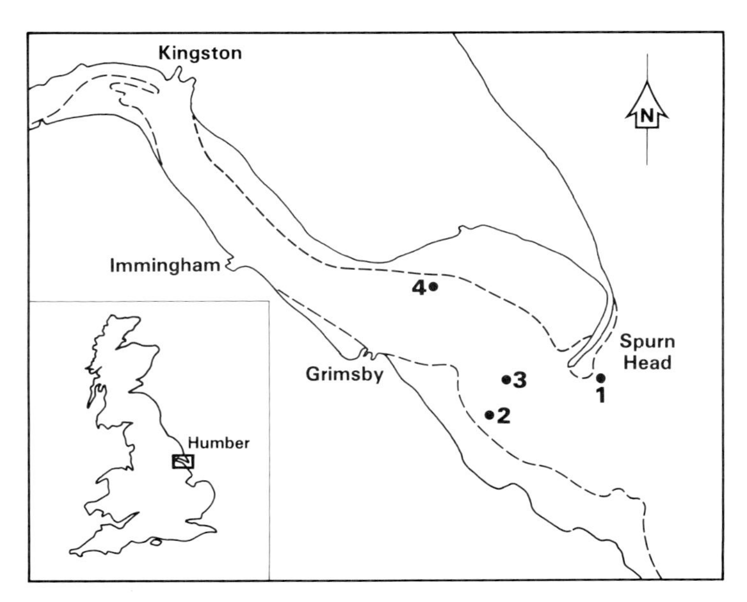


**Fig. 3.** The Humber estuary study area, and locations of mussel deployment sites

major estuaries also lead to ports, the cages obviously have to be sited off the main shipping channels. Figure 3 shows the sites selected on the Humber estuary.

Lack and Johnson (1985) observed some stress in the scope for growth and lysosomal stability of mussels placed around a sludge disposal ground in the English Channel off Plymouth. There was also some evidence of elevation of organic detoxication enzymes and induction of metalloprotein production. The levels of metal-binding proteins present in the mussels were similar to the uptake of zinc by mussels in cages near the seabed (Johnson and Lack 1985), where the effects of the sludge are greatest.

Other field trials of the technique have been carried out in the vicinity of an oil terminal in Sullom Voe, Shetland. Stress indices in native mussels indicated that the populations were in a relatively healthy physiological condition (Widdows et al. 1981). However, the periwinkle (*Littorina littorea*) from certain areas of Sullom Voe showed destabilization of the lysosomal membrane and stimulation of a microsomal detoxication system (Moore et al. 1982). This was interpreted as a response to the presence of oil-derived polynuclear aromatic hydrocarbons.

Trickle-spawning by exposed mussels, which induces stress-like effects on physiological and cytochemical processes, may introduce an undesirable degree of variability into the results. Recent improvements have been made in the technique by selection of mussels from populations with limited spawning periods and the use of histological techniques to assess their sexual maturity. Further development of techniques is in progress and the way ahead for the future is in relating these effects on individual animals to potential population damage. Bayne (1985) sums up the future by saying that there "is an area of responsibility here for the biologists, viz to provide the methodology for an integration across all levels of biological hierarchy and so to progress towards convincing and quantitative assessments of pollution impact in the marine environment".

## References

Andrews MJ (1984) Thames estuary: Pollution and Recovery. In: Sheehan PJ, Miller DR, Butler GC, Bourdeau P (eds) Effects of pollution at the ecosystem level. SCOPE 22, Wiley & Sons, pp 195–227

Bayne BL (1985) Cellular and Physiological Measures of Pollution Effect. Mar Pollut Bull 16 (4):127–129

Bayne BL, Moore MN, Widdows J, Livingstone DR, Salkeld PN (1979) Measurement of the responses of individuals to stress and pollution. Phil Trans Roy Soc Lond B 286:563

Bayne BL, Brown DA, Burns K, Dixon DR, Ivanovici A, Livingstone DR, Lowe DM, Moore MN, Stebbing ARD, Widdows J (1985) The effects of stress and pollution on marine animals. Praeger, New York, 384 pp

Davies IM (1980) Evaluation of a 'Mussel Watch' project for heavy metals in Scottish coastal waters. Mar Biol 57:87–93

Dyer MF, Fry WG, Fry PD, Cranmer GJ (1983) Benthic Regions within the North Sea. J Mar Biol Ass UK 63:683–693

Glemarec M (1973) The benthic communities of the European north Atlantic shelf. Oceanogr Mar Biol Annu Rev 11:263–289

Govaere JCR, Van Damme D, Heip C, de Coninck LAP (1980) Benthic communities in the Southern Bight of the North Sea and their use in ecological monitoring. Helgol Meerunters 33:507–521

Hart CW Jr, Fuller SLH (eds) (1979) Pollution ecology of estuarine invertebrates. Academic Press, London

Hill JM, Mance G, O'Donnell AR (1984) The quantities of some heavy metals entering the North Sea. Water Res Cent Tech Rep TR 205:

Høpner Petersen G (1977) The density, biomass and origin of the bivalves of the central North Sea. Medd Dan Fisk Havunders 7:221–273

ICES (1974) Report of the Working Group for the International Study of the Pollution of the North Sea and its effects on Living Resources and their Exploitation. ICES Coop Res Rep No 39

ICES (1984) The ICES Co-ordinated Monitoring Programme for contaminants in fish and shellfish, 1978 and 1978 and the Six-Year review of the ICES Co-ordinated Monitoring Programme. ICES Coop Res Rep No 126

Johnson D, Lack TJ (1985) Some responses of transplanted *Mytilus edulis* to metal-enriched sediments and sewage sludge. Mar Environ Res, 17:277–280

Lack TJ, Johnson D (1985) Assessment of the biological effects of sewage sludge at a licensed site of Plymouth. Mar Pollut Bull 16(4):147–152

Madsen PB (1984) The dynamics of the dominating macrozoobenthos in the Danish Wadden Sea, 1980–83. Rep Mar Pollut Lab, Charlottenlund, Denmark

Meeus-Verdinne K, van Cauter R, De Borger R (1983) Trace metal content in Belgian coastal mussels. Mar Pollut Bull 14(5):198–200

Moore MN (1980) Cytochemical determination of cellular responses to environmental stressors in marine organisms. Rapp P V Réun Cons Int Explor Mer 170:7–15

Moore MN (1985) Cellular responses to pollutants. Mar Pollut Bull 16(4):134–139

Moore MN, Pipe RK, Farrar SV (1982) Lysosomal and microsomal responses to environmental factors in *Littonna littorea* from Sullom Voe. Mar Pollut Bull 13(10):340–345

Rachor E (1980) The inner German Bight – an ecologically sensitive area as indicated by the bottom fauna. Helgol Meerunters 33:522–580

Rachor E, Gerlach SA (1978) Changes in the macrobenthos in a sublittoral sand area of the German Bight, 1967 to 1975. Rapp P V Réun Cons Int Explor Mer 172:418–431

Rainbow P, White S (Personal communication)

Talbot JW, Harvey BR, Eagle RA, Rolfe MS (1982) The field assessment of effects of dumping wastes at sea: 9. Dispersal and effects on benthos of sewage sludge dumped in the Thames estuary. MAFF Fish Res Tech Rep No 63
Ursin E (1960) A quantitative investigation of the echinoderm fauna of the central North Sea. Medd Dan Fisk Havunders 2:(24) 204
Viarengo A (1985) Biochemical effects of trace metals. Mar Pollut Bull 16(4):153–158
Widdows J (1985) Physiological responses to pollution. Mar Pollut Bull 16(4):129–134
Widdows J, Bayne BR, Donklin P, Livingstone DR, Lowe DM, Moore MN, Salkeld PN (1981) Measurement of the responses of mussels to environmental stress and pollution in Sullom Voe: a baseline study. Proc Roy Soc Edinburgh 80 B:323–338

# Effects of Pollutants on Fish

D. BUCKE[1] and B. WATERMANN[2]

## 1 Introduction

The impact of pollution on fish has received considerable attention over the past few years, and this is partly due to the fact that there appears to be circumstantial evidence that certain diseases in marine fish may be associated with anthropogenic factors (Sindermann et al. 1980; Sindermann 1983). Whereas there are unequivocal results from laboratory experiments that certain individual pollutants can induce diseases in fish by virtue of their carcinogenic and toxic potential direct to cells and tissues (Meyers and Hendricks 1982), the fact remains that outbreaks of diseases in free-living marine fish are the result of multifactoral events involving a variety of natural and anthropogenic factors (Sindermann 1984). Despite the upsurge in investigations which, in the main, have involved limited epidemiological studies, only relatively few scientists have tried to qualify their findings with controlled laboratory studies (Couch and Harshbarger 1985). This presentation sets out to describe some of the more important diseases of North Sea fish and their possible association with pollution.

## 2 Historical Data

Reports on fish diseases in the North Sea date back at least 100 years. Certainly, because of rapid urbanization and industrialization in European countries at that time, it was known that many of the major rivers discharging into the North Sea were heavily polluted (Tetlow 1972). The early reports (McIntosh 1884, 1885; Sandeman 1892) described characteristics of diseases which were essentially the same conditions as those seen to-day. However, assessing information in the older literature is made difficult because records usually refer only to the odd gross clinical anomalies, that is, those most likely to be referred for diagnosis because of their unusual clinical appearance, which has made them a curiosity to fishermen and other lay persons in the first place. An interesting example is a disease of flounders (*Platichthys flesus*) inhabiting a river estuary in eastern England (Lowe

[1]Ministry of Agriculture, Fisheries and Food, Directorate of Fisheries Research, Fish Diseases Laboratory, The Nothe, Weymouth, Dorset DT4 8UB, England
[2]Institut für Zoologie und Museum, Universität Hamburg, Martin-Luther King Platz 36, D-2000 Hamburg 13, FRG

1974). Lowe noticed that many of the flounders exhibited a type of epithelioma or "large fungus growths cropping out over the whole body, the granulations were large and roe-like, and under the microscope consisted of large, nucleated cells". Considering the report was made over 100 years ago, it compared admirably with descriptions of Lymphocystis disease to-day. Further diseases and anomalies in North Sea fish were recorded in the first half of the century (Johnstone 1905, 1920, 1922, 1923, 1924, 1926ab, 1927; Williams 1929, 1931; Schäperclaus 1927, 1934). Amongst these reports there was one important reference to pollution and disease (Johnstone 1922) when, in 1921, North Sea fishermen were catching quantities of cod (*Gadus morhua*), haddock (*Melanogrammus aeglefinus*) and plaice (*Pleuronectes platessa*) exhibiting large, shallow ulcers. Many of the fish were also emaciated or exhibited skeletal deformities. Even the fish processors that year complained of the poor quality of North Sea herrings (*Clupea harengus*). The popular opinion then was that the diseases might be associated with dumping of surplus World War I explosives and mines. As these munitions decomposed, it was speculated that "filthy deleterious chemical compounds could be released, killing off the plankton which subsequently led to starvation, debilitation and diseases in fish stocks" (Johnstone 1922). This type of pollution was also suspected to be associated with mortalities amongst stocks of European flat oysters (*Ostrea edulis*) during the years 1920 and 1921 (Orton 1924).

## 3 Recent Studies

In the 1960's and early 1970's, particularly in the southern North Sea and its estuaries, investigations showed that there was evidence of fish diseases (Koops and Mann 1969; Mann 1970; Wunder 1971; van Banning 1971). Similarly, diseases were recorded from fish in the Irish Sea (Perkins et al. 1972; Shelton and Wilson 1973). These reports and others of diseases in other coastal and offshore areas led to recommendations by ICES to member countries to monitor the marine environment for evidence that fish could be used as a biological model to indicate adverse effects in the marine fauna (Sindermann et al. 1980).

Following a presentation by Dethlefsen and Watermann (1980), there was concern among some scientists and the media that there might be an exacerbation of diseases of fish in the German Bight because of the dumping of industrial wastes, particularly those from the titanium dioxide industry. This led to a series of fish disease surveys (Möller 1979, 1981; Dethlefsen 1980, 1985; Watermann and Dethlefsen 1982; Watermann et al. 1982; Wolthaus 1984; Watermann and Dethlefsen 1985).

During the years of the above and other investigations, an area in the centre of the German Bight was frequently reported to have elevated disease rates in the dab (*Limanda limanda*) (Dethlefsen 1984a; Wolthaus 1984; Mellergaard and Nielsen 1984a, 1985). In addition, higher than average disease rates of dab were also recorded off the Humber estuary and Dogger Bank areas in the southern North Sea (Möller 1979; Dethlefsen 1984a). Statistical data has proved that some epidermal diseases were increased in prevalence according to the age and length of the fish

(Möller 1981; Dethlefsen 1984b; Wolthaus 1984; Mellergaard and Nielsen 1984ab, 1985). Möller (1981) also found a positive correlation between condition factor and disease prevalence, but none of the other cited authors could confirm these findings. Wolthaus (1984) suspected that migration and spawning activities led to seasonal fluctuations in disease prevalence. Although these factors may account for seasonal peaks according to Dethlefsen (1985), they are unlikely to influence the regional differences which were recorded at different times in the year. There is only limited information on the influence of fishing activities, such as bottom trawling, exacerbating disease outbreaks. This information includes a suggestion of an increase in ulcers in Baltic Sea fish (Miljö-Projektor 1981), a link with netting injuries and epidermal diseases of dab (Dethlefsen 1984b) and suspected associations between fin-rot and fishery activities on flatfishes in the Thames estuary (Bucke et al. 1983a).

Simultaneously with the studies cited above, other investigations into the prevalence of diseases and their relation to pollution in the North Sea included reports from the U.K. (Wootten et al. 1982; Bucke et al. 1983b), Ireland (McArdle et al. 1982), Norway (Egidius et al. 1981), the Netherlands (van Banning et al. 1984). These investigations were regarded by Dethlefsen (1985) to be preliminary because they lacked the representative data to compensate for the natural variations which have been mentioned above. It is therefore important to examine both the current status of fish diseases in the North Sea and have an appreciation of the causal agents of such diseases.

## 4 Selected Diseases

*1. Lymphocystis Disease* (Fig. 1) commonly occurs in many fish species throughout the world, but in the North Sea is mostly found in flounder, then dab and plaice. The disease is recognised by the presence of single nodules or multiples of pearly-white clusters of nodules on the external surface of the fish, especially at the edges of the fins. These nodules are hypertrophied epithelial cells stimulated to this greatly increased size by their response to a virus infection. The disease is infectious, but species specific and, in the marine environment, commonly occurs in flatfish over 15 cm length. The condition occurs more frequently in inshore and shallow seas than in deeper water areas. Bucke et al. (1983a) investigated the disease in flounder from three areas: Rye Bay (off the Kent coast), Thames Estuary and off Southwold, Suffolk, where the prevalence was found to be 3.1%, 1.6% and 8.6%, respectively. Much higher prevalences have been recorded in this species, e.g. 18.7% in Dutch coastal waters (Vethaak 1985). Lymphocystis has been recorded at prevalence levels up to 12.1% in dab populations from Danish coastal waters (Mellergaard and Nielsen 1985) and for plaice 4.6% infection was recorded from the northern Irish Sea (Perkins et al. 1972). Up to 12% prevalence levels in dab have been recorded in the German Bight, and on the Dogger Bank up to 20% levels (Dethlefsen 1984b). Experience has shown that prevalence of the disease is often increased where fishing intensity is greatest, probably relating to net damage injuring the integument on escaped fish allowing the entry of foreign bodies. It has

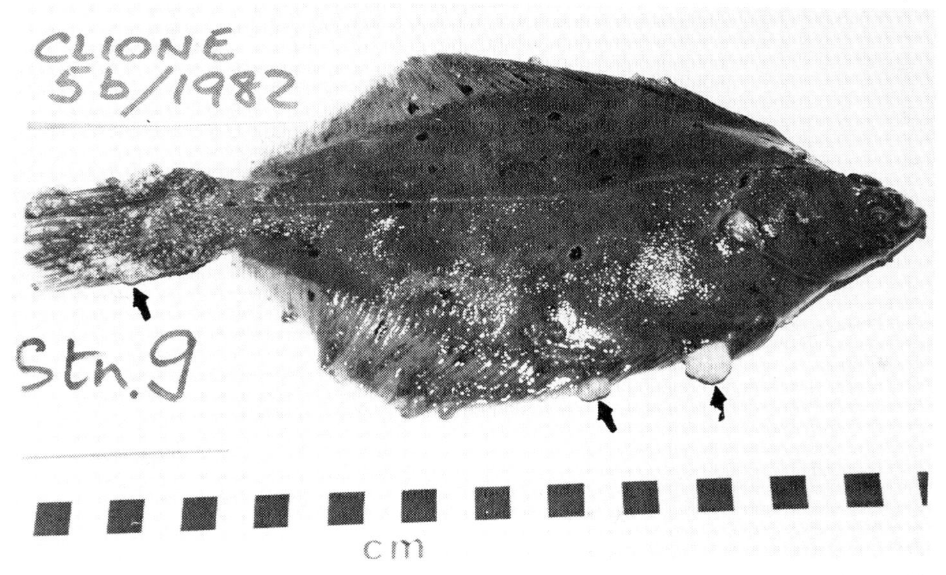


**Fig. 1.** Lymphocystis disease in a plaice (*Pleuronectes platessa*). Nodules of hypertrophied cells (*arrowed*)

been suggested that there are higher prevalences of this disease in polluted waters as compared to less polluted waters (Dethlefsen 1984b; Vethaak 1985). Möller (1984) found elevated prevalence rates of disease in the Elbe estuary, where the body burden of heavy metals was highest, but he suggested the findings were due to the low nutritional status of the fish and salinity fluctuations rather than pollution, despite the fact that his samples were taken in one of the most polluted rivers in Europe.

2. *Epidermal Hyperplasia* (Fig. 2). This disease is recorded frequently in dab and occasionally in whiting (*Merlangius merlangus*) and other gadoids (Watermann and Dethlefsen 1985). Its presence is noted by slightly raised white areas of epidermal cells in patches over the skin surface of these fish. In certain estuaries, especially the Elbe, epidermal hyperplasia is also recorded in smelt (*Osmerus eperlanus*) (Anders and Möller 1985). Epidermal papilloma is possibly the next progressive stage after hyperplasia, and is frequently noted in dab. Papillomas are larger raised nodules, consisting of epidermal cells with supporting and circulatory structures. This condition in the dab should not be confused with the well-known papillomatosis or cauliflower disease of eel (*Anguilla anguilla*), which is frequently found in eels from rivers leading into the German Bight (Peters 1977). Epidermal hyperplasias and papillomas may have a viral aetiology because virus-like particles have been observed in association with some of these conditions (Yamamoto et al. 1983), but virus isolation and transmission experiments in dab have so far failed leaving the aetiology not proven (Bloch et al. 1986). Dethlefsen and Watermann (1980) found higher prevalence rates of epidermal hyperplasia in dab from the

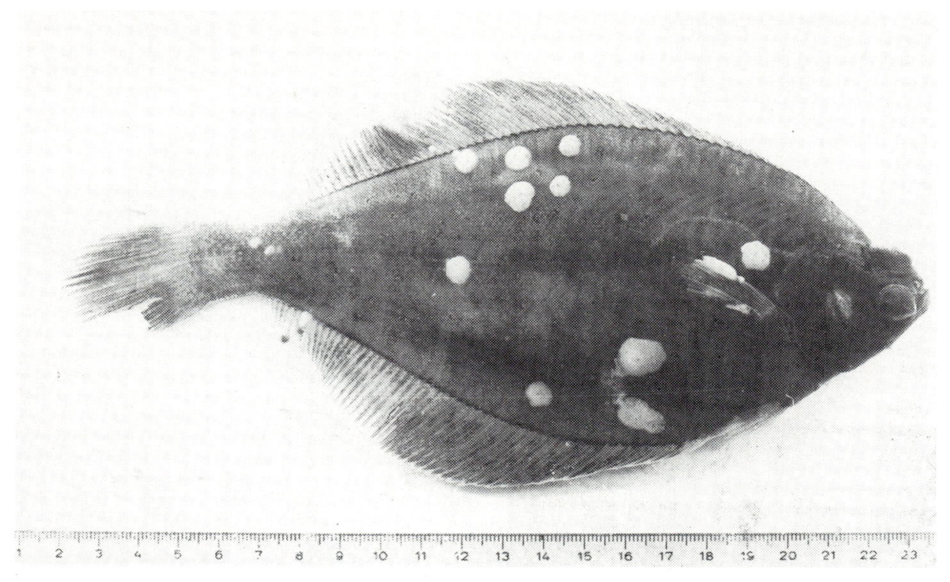

**Fig. 2.** Epidermal hyperplasias in a dab (*Limanda limanda*)

dumping area for $TiO_2$ wastes in the German Bight and even higher prevalences of this disease in dab caught off the Humber estuary and on the Dogger Bank than from other regions of the North Sea (Dethlefsen 1984a). He suspected that these prevalences may be related to heavy metals discharged from the British coast, because in later studies a relationship between high body burdens of chromium and epidermal hyperplasia in dab was suggested (Dethlefsen 1985).

*3. Epidermal Ulcers* (Fig. 3). Skin ulcers have been observed in many fish species, especially flounder, dab, plaice, cod and eel (Dethlefsen 1985). The ulcers vary in size from superficial, petechial lesions to large infiltrating areas involving skeletal muscle and even osseous tissue. Epizootics may occur in estuarine fish such as flounder or eel (Möller 1984). Epidermal ulcers, like lymphocystis, are recorded at higher levels where fishing intensity is greatest and the size range of affected fish is similar (Möller 1981). Micro-organisms such as vibrios or viruses may be the causal agents of skin ulcers (Larsen 1983). Larsen considered that skin ulcers in cod off the Danish coast were dependent on an infectious agent such as bacteria or a virus, but he found circumstantial evidence that increase in prevalence was influenced by high water temperature and discharged wastes containing carbohydrates from sugar and cellulose plants. Investigations into ulcerative conditions occurring in eels along the Dutch and Belgian coasts did not reveal any correlation between the condition and body burdens of heavy metals or organochlorines, but at the population level the prevalence was higher in the contaminated area than in the less polluted areas (van Banning et al. 1984).

**Fig. 3.** Epidermal ulceration in a flounder (*Platichthys flesus*) Note the different stages of the lesions

4. *Fin-Rot.* This condition has frequently been mentioned as a disease specifically associated with infectious agents exacerbated by environmental variations, the abundance of fish and fishing intensity. The complexity of the problem was demonstrated by studies on the high prevalence rates of fin-rot in flatfish species from degraded areas of the New York Bight, where the pathogenesis of the disease remains unknown (Murchelano and Ziskowski 1976). The disease is present in essentially estuarine areas of the North Sea, and despite its unknown aetiology a devised schematic system for its diagnosis was used to record prevalence rates of as high as 13.5% in flounders from the Thames Estuary (Bucke et al. 1983a). However, its use as a diagnostic tool for monitoring fish diseases associated with pollution has been rejected because of differential diagnosis caused by, for example, mechanical damage (Anon 1984).

5. *Pseudobranchial "Tumours" and "X-Cell" Lesions.* The pseudobranchial "tumour" occurs in the pseudobranch gland which is situated at the back of the pharynx, slightly anterior to the gill tissues. The "tumours" occur mainly in cod, although in the North Sea they have been occasionally reported in other gadoids (Watermann and Dethlefsen 1982). The lesions are creamy-white in colour, sometimes situated bilaterally, and can grow to a large size, even forcing open the operculae. Watermann et al. (1982) found a regional coincidence of elevated prevalences in an area of the German Bight which is used for dumping wastes from

$TiO_2$ production (2.1%), but elsewhere in the North Sea prevalences have been far less (0.1%). The condition is predominantly found in fish 25–30 cm length, and the reason why more affected fish were found in the German Bight could have been that there were higher population densities of this sized cod present. Egidius et al. (1981) recorded a 1.7% prevalence level of pseudobranch "tumours" in cod from the Barents Sea, an area not particularly known for its pollution status. Recently, gill pathologies associated with swellings and pallor were observed in North Sea dabs (McVicar et al. 1987). Microscopical examination of the abnormal areas revealed that similar X-cells have been identified in skin lesions of pleuronectids (Wellings et al. 1976) and in pseudobranchial tumours in cod from the North Atlantic (Morrison et al. 1982). The structural uniformity of the cells suggests they have a common origin, and they may not be host cells but protistan cells, multiplying in groups to form Xenomas, and not neoplastic tumours as was originally thought (Dawe 1981). Dethlefsen and Knust (1986) showed that in affected dabs the gonadosomatic index was lower than that of healthy specimens. They recorded the X-cell condition in dab at maximum prevalence rates of 4% on the Dogger Bank. However, Diamant and McVicar (1986) have recorded prevalence levels varying from 0.5% to 60% in dab populations off North East Scotland. This is a condition which warrants further investigation, because little is known about the fate of affected fish.

*6. Skeletal Deformities* (Figs. 4, 5). Malformations such as spinal curvature, compression of vertebrae or "pug-nose" have been recorded in various fish species in the North Sea (Dethlefsen 1980). The cause of these conditions is multifactoral and probably has occurred at the developmental stage of the fish's life (Wedemeyer and Goodyear 1984). Van de Kamp (1977) investigated vertebral deformities in herring from British waters and found varying prevalences with highest levels in the southern North Sea (< 1.0%). His conclusion was that the deformities were associated with pollutants in addition to natural causes. Dethlefsen (1980) and Möller (1981) both recorded prevalences of skeletal deformities in cod less than 1% in the German Bight and elsewhere in the North Sea.

*7. Tumours and Internal Pathologies.* Apart from the epidermal papillomas previously described, tumours (syn. neoplasias), especially internal tumours, are rarely investigated in North Sea fish (Bucke and Feist 1986), whereas studies conducted in polluted waters off North America have demonstrated higher prevalences of liver tumours and other hepatic pathologies in fish from those areas, compared with clean areas (McCain et al. 1977). Bucke et al. (1984), in a preliminary histological study, described a number of morphometric changes in livers and spleens from the North Sea dab. Although the numbers of samples were inadequate for complete statistical analysis, there appeared to be higher prevalence rates for melanin deposits, hepatic granulomas and basophilic hepatic nodules (Fig. 6) from the Dogger Bank than elsewhere in the southern North Sea. Vethaak (1986) reported that old dabs sampled from localized areas in Dutch coastal waters revealed up to 40% prevalence levels of liver tumours. However, there was no substantive evidence that these tumours were neoplastic and cancerous.

**Fig. 4.** Vertebral distortion (scoliosis) at the caudal region in a dab (*Limanda limanda*)

**Fig. 5.** Vertebral compression in a haddock (*upper*). Note the difference in size compared with a normal fish (*lower*)

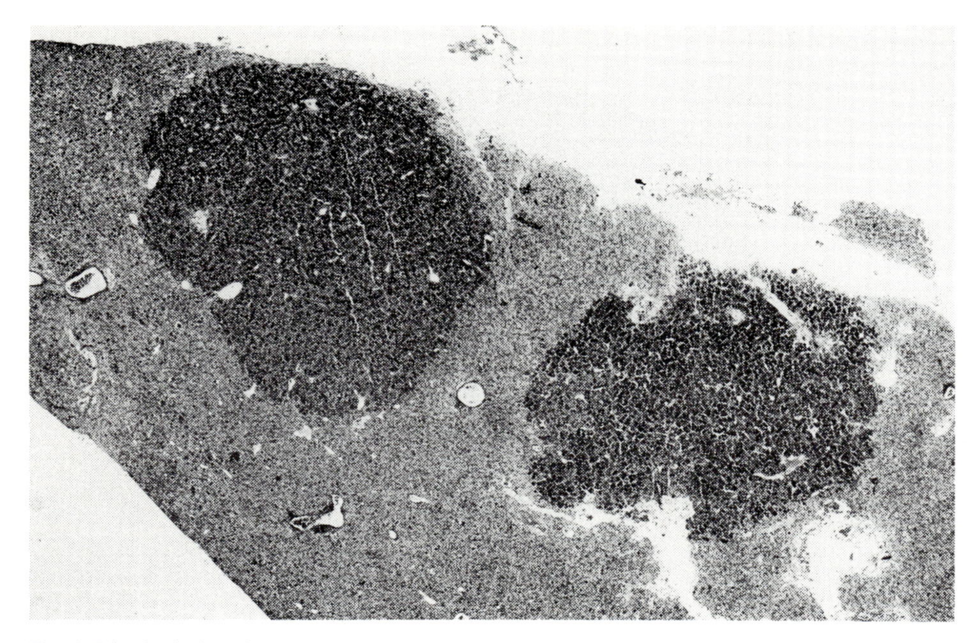

**Fig. 6.** Histological section showing two hyperchromic nodules in a dab (*Limanda limanda*) liver

## 5 Conclusions

Historically, almost all of the currently recorded types of disease have been known for about 100 years, but it is not possible to answer whether the prevalence rates have increased over the years, or even if the older reports did not have a pollution connection associated with the rapid urbanization and industrialization of countries bordering the North Sea.

Currently, indications show that there may be associations with pollutants and diseases of dab in the centre of the German Bight and possibly the Dogger Bank (Dethlefsen 1984a). High levels of disease have been recorded in flatfish and eels in coastal waters and estuaries, as well as internal anomalies in certain estuarine fish species (Möller 1984; Kranz and Peters 1984).

As stated at the beginning of the paper, there have been laboratory and field studies which show that certain pollutants may induce diseases in fish, but so far it has not been possible to establish a clear causal relationship which would help define the impact of pollution on the health of fish stocks in the North Sea. The principal reasons for this uncertainty has to be seen in the fact that we are dealing with highly complex ecosystem interactions where natural and anthropogenic impacts may never be separated unequivocally. The main question, therefore, is whether the pathologist can identify the cause of an outbreak of disease or of an increased prevalence of disease in a natural fish population. Current and future investigations will need to include population dynamics and behaviour, including spawning, migration and nutrition, the influence of length, age and sex, the effects of temperature, salinity and oxygen, fishery activities and the measurement of pollutants in sediments, water, food and biota.

## 6 Summary of Research Needs

1. Fish disease studies should be continued and intensified in areas of the North Sea where so far only sporadic or poorly designed studies have been made.
2. New approaches to present more of the data in objective rather than subjective form.
3. The results of biological data should be correlated as far as possible with chemical and physical data collected by other groups.
4. The sequential pathology of diseases of fish should be studied within the laboratory.

## References


Anders K, Möller H (1985) Spawning papillomatosis of smelt, *Osmerus eperlanus* L. from the Elbe Estuary. J Fish Dis 9:233–235

Anonymous (1984) Seagoing Workshop on Methodology of Fish Disease Surveys. ICES CM 1984/F:17:16

Bloch B, Mellergaard S, Nielsen E (1986) Adenovirus-like particles associated with epithelial hyperplasias in dab (*Limanda limanda*). J Fish Dis 9:281–285

Bucke D, Feist SW (1986) Examples of neoplasia or neoplastic-like conditions in marine fish. In: Vivares CP, Bonami J-R, Jaspers E (eds) Pathology in Marine Aquaculture. Eur Aquac Soc Spec Publ No 9, Bredene, Belgium, pp 311–323

Bucke D, Norton M, Rolfe MS (1983a) The field assessment of effects of dumping wastes at sea. II. Epidermal lesions and abnormalities of fish in the outer Thames Estuary. MAFF Fish Res Tech Rep 72:16

Bucke D, Feist S, Rolfe M (1983b) Fish disease studies in Liverpool Bay and the North East Irish Sea. ICES CM 1983/E:5:9

Bucke D, Watermann B, Feist S (1984) Histological variations of hepato-splenic organs from the North Sea dab, *Limanda limanda* L. J Fish Dis 7:255–268

Couch JA, Harshbarger JC (1985) Effects of carcinogenic agents on aquatic animals: an environmental and experimental overview. J Environ Sci Health Part C Environ 3:63–105

Dawe CJ (1981) Polyoma tumours in mice and X-cell tumours in fish viewed through telescope and microscope. In: Dawe CJ et al. (eds) Phyletic Approaches to Cancer. Jpn Sci Soc, Tokyo, pp 19–49

Dethlefsen V (1980) Observations on fish diseases in the German Bight and their possible relation to pollution. Rapp P V Réun Cons Int Explor Mer 179:110–117

Dethlefsen V (1984a) Diseases in North Sea fishes. Helgol Meeresunters 37:353–374

Dethlefsen V (1984b) Untersuchungen zur Erkennung sublethaler Schadstoffeffekte an Fischen in der Deutschen Bucht: Krankheiten, Biochemie, Physiologie, Rückstände. BMFT-FB-M 84-007, p 427

Dethlefsen V (1985) Review on the effects of dumping of wastes from titanium dioxide production in the German Bight. ICES CM 1985/E:2:42

Dethlefsen V, Knust R (1986) X-cells in gills of North Sea dab (*Limanda limanda* L.) epizootiology and impact on condition ICES CM 1986/E:22:21

Dethlefsen V, Watermann B (1980) Epidermal papilloma of North Sea dab, *Limanda limanda*: histology, epidemiology and relation to dumping of wastes from $TiO_2$ industry. ICES Spec. Meeting on Diseases of commercial Important Marine Fish and Shellfish 8:30
Diamant A, McVicar AH (1986) Internal and external X-cell lesions in common dab (*Limanda limanda* L.), X-cell disease and its implications to flatfish and cod mariculture. Azevado C (ed) Proc 2nd Int Coll Path Mar Aq (Portugal) Univ of Oporto, Portugal, pp 57–58
Egidius EC, Johanessen JV, Lange E (1981) Pseudobranchial tumours in Atlantic cod, *Gadus morhua* L., from the Barents Sea. J. Fish Dis 4:527–532
Johnstone J (1905) Internal parasites and diseased conditions of fishes. Trans Liv Biol Soc 19:278–300
Johnstone J (1920) On certain parasites and abnormal conditions of fishes. Trans Liv Biol Soc 34:120–129
Johnstone J (1922) Diseases and parasites of fishes. Trans Liv Biol Soc 36:286–301
Johnstone J (1923) On some malignant tumours in fishes. Trans Liv Biol Soc 37:145–147
Johnstone J (1924) Disease conditions in fishes. Trans Liv Biol Soc 38:183–213
Johnstone J (1926a) Diseases of fishes. Trans Liv Biol Soc 40:66–71
Johnstone J (1926b) Malignant and other tumours in marine fishes. Trans Liv Biol Soc 40:75–98
Johnstone J (1927) Diseased conditions in fishes. Trans Liv Biol Soc 41:162–157
Koops H, Mann H (1969) Die Blumenkohlkrankheit der Aale. Vorkommen und Verbreitung der Krankheit. Ach FischWiss 20:5–15
Kranz H, Peters N (1984) Pathological conditions in the liver of ruffe *Gymnocephalus cernua* (L.) from the Elbe estuary. J Fish Dis 8:13–24
Larsen NJ (1983) The ulcus syndrome in cod (*Gadus morhua*): a review. Rapp P V Réun Cons Int Explor Mer 182:58–64
Lowe J (1974) Fauna and flora of Norfolk. IV Fishes. Trans Norfolk Norwich Nat Soc 21–56
Mann H (1970) Über den Befall der Plattfische der Nordsee mit Lymphocystis. Ber Dtsch Wiss Komm Meeresforsch 21:219–233
McArdle J, Dunne T, Parker M, Martyn C, Rafferty D (1982) A survey of diseases of marine flatfish from the east coast of Ireland in 1981. ICES CM 1982/E:47:10
McCain BB, Pierce KV, Wellings SR, Miller BS (1977) Hepatomas in marine fish from an urban estuary. Bull Environ Contam Toxicol 18:1–2
McIntosh WC (1884) Multiple tumours in plaice and common flounder. Rep Fish Bd Scotl 3:66–67
McIntosh WC (1885) Further remarks on the multiple tumours of common flounders. Rep Fish Bd Scotl 4:214–215
McVicar AH, Bucke D, Watermann B, Dethlefsen V (1987) Gill X-cell lesions of dab (*Limanda limanda* L.) in the North Sea. Dis Aquat Organisms 2:197–204
Mellergaard S, Nielsen E (1984a) Preliminary investigations on the eastern North Sea and the Skagerrak dab, *Limanda limanda*, populations and their diseases. ICES CM 1984/E:28:23
Mellergaard S, Nielsen E (1984b) Preliminary investigations on the eastern North Sea and the Skagerrak plaice, *Pleuronectes platessa*, populations and their diseases. ICES CM 1984/E:29:28
Mellergaard S, Nielsen E (1985) Fish diseases in the eastern North Sea dab, *Limanda limanda*, population with special reference to the epidemiology of epidermal hyperplasias/papillomas. ICES CM 1985/E:14:15
Meyers JD, Hendricks TR (1982) A summary of tissue lesions in aquatic animals induced by controlled exposure to environmental contamination. Mar Fish Rev 44(12):1–17
Miljö-Projekter (1981) Fiskepatologiske or mikrobiologiske undersogelser i kystnaert, marint miljö. Miljöministeriet miljöstyrelsen, Kobenhavn, 180 pp
Möller H (1979) Review of the geographical distribution of fish diseases in the North East Atlantic. ICES CM 1979/E:13:22
Möller H (1981) Fish diseases in German and Danish waters in summer 1980. Meeresforsch 29:1–16
Möller H (1984)Dynamics of fish diseases in the lower Elbe River. Helgol Meeresunters 37:389–413
Möller H (1985) A critical review on the role of pollution as a cause of fish diseases. In: Ellis AE (ed) Fish and Shellfish Pathology. Academic Press, London, pp 169–182
Morrison CM, Shum G, Appy RG, Odense P, Annand C (1982) Histology and prevalence of X-cell lesions in Atlantic cod (*Gadus morhua*). Can J Fish Aquat Sci 39:1519–1530
Murchelano RA, Ziskowski J (1976) Fin rot disease studies in the New York Bight. Am Soc Limnol Oceanogr Spec Symp 2:329–336

Orton JH (1924) An account of investigations into the cause or causes of the unusual mortality among oysters in English oyster beds during 1920 and 1921. Part 1. Rep MAFF Fish Invest II, VI(3), HMSO Lond, UK, 198 pp
Perkins EJ, Gilchrist JRS, Abbott OJ (1972) Incidence of epidermal lesions in fish of the North East Irish Sea area, 1971. Nature (Lond) 238:101–103
Peters G (1977) The papillomatosis of the European eel, *Anguilla anguilla* L.: analysis of seasonal fluctuations in the tumour incidence. Arch FischWiss 27:251–263
Sandeman G (1892) On the multiple tumours in plaice and flounders. 11th Annu Rep Fish Board Scotland, pp 391–392
Schäperclaus W (1927) Die Rotseuche des Aales im Bezirk von Rugen und Stralsund. Z Fisch 25:99–128
Schäperclaus W (1934) Untersuchungen über die Aalseuchen in Deutschen Binnen-und Küstengewässern 1930–1933. Z Fisch 32:191–217
Shelton RGJ, Wilson KW (1973) On the occurrence of lymphocystis, with notes on other pathological conditions, in the flatfish stocks of the North East Irish Sea. Aquaculture 2:395–410
Sindermann CJ (1983) An examination of some relationships between pollution and disease. Rapp P V Réun Cons Int Explor Mer 182:37–43
Sindermann CJ (1984) Fish and environmental impacts. Arch Fisch Wiss 35:125–160
Sindermann CJ, Bang FE, Christensen NO, Dethlefsen V, Harshbarger JC, Mitchell JR, Mulcahy MF (1980) The role and value of pathobiology in pollution effects monitoring programs. Rapp P V Réun Cons Int Explor Mer 179:135–151
Tetlow JA (1972) Pollution of the water. Med Sci Law 94–103
Van Banning P (1971) Wratzkiekte bij platvis. Visserij 24:336–343
Van Banning P, De Clerk D, Guns M, Stokman G, Vandamme K, Vyncke W (1984) Visaandoeningen en de Mogelijkeld van Relatie met Waterveruiling. Rijksinstituut voor visserijonderzoek, Ijmuiden, CA 84–05, 24 pp
Van de Kamp G (1977) Vertebral deformities of herring around the British Isles and their usefulness for a pollution monitoring programme. ICES CM 1977/E:5:9
Vethaak AD (1985) Prevalence of fish diseases with reference to pollution of Dutch coastal waters. Neth Inst Fish Invest, Ijmuiden, The Netherlands, 70 pp
Vethaak D (1986) Fish diseases, signal for a diseased environment? 2nd Int North Sea Seminar, Rotterdam, pp 20
Watermann B, Dethlefsen V (1982) Histology of pseudobranchial tumours in Atlantic cod, *Gadus morhua*, from the North Sea and the Baltic Sea. Helgol Meeresunters 35:231–242
Watermann B, Dethlefsen V (1985) Epidermal hyperplasia and dermal degenerative changes as cell damage effects in Gadoid skin. Arch Fisch Wiss 35:205–221
Watermann B, Dethlefsen V, Hoppenheit M (1982) Epidemiology of pseudobranchial tumours in Atlantic cod, *Gadus morhua*, from the North Sea and the Baltic Sea. Helgol Meeresunters 35:425–437
Wedemeyer GA, Goodyear CP (1984) Diseases caused by environmental stressors. In: Kinne O (ed) Diseases of Marine Animals. Biolog Anstalt Helgol, FRG 4(1):424–434
Wellings SR, McCain BB, Miller BS (1976) Epidermal papillomas in Pleuronectidae of Puget Sound, Washington. Prog Exp Tumor Res 20:55–74
Williams G (1929) Tumourous growths in fishes. Trans Liv Biol Soc 43:120–148
Williams G (1931) On various fish tumours. Trans Liv Biol Soc 45:98–109
Wolthaus BG (1984) Seasonal changes in frequency of diseases in dab, *Limanda limanda*, from the southern North Sea. Helgol Meeresunters 37:375–387
Wootten R, McVicar AH, Smith JW (1982) Some disease conditions of fish in Scottish waters. ICES CM 1982/E:46:8
Wunder W (1971) Mißbildungen beim Kabeljau (Gadus morhua) verursacht durch Wirbelsäulenverkürzung. Helgol Meeresunters 22:201–212
Yamamoto T, Kelly RK, Nielsen O (1983) Epidermal hyperplasias of northern pike (*Esox lucius*) associated with herpesvirus and C-type particles. Arch Virol 79:255–272

# Ecotoxicology: Biological Effects Measurements on Molluscs and Their Use in Impact Assessment

D.R. LIVINGSTONE, M.N. MOORE, and J. WIDDOWS[1]

## 1 Introduction

Techniques that measure the biological effects of pollutants are critical to any programme of environmental impact assessment (Cairns and Van der Schale 1980; Rosenberg et al. 1981; Bayne 1985). This has been recognised by the IOC Global Investigation of Pollution in the Marine Environment (GIPME) (1976) which requires their application in the third stage of their proposed four-stage systematic approach to the determination of the extent of marine pollution. This stage, called Pollution Assessment, involves the conversion of chemical base-line and mass-balance information into knowledge of biological impact and precedes and renders effective the final stage, the consideration of the need for Regulatory Action.

Many measurements or indices of biological effect have been proposed over the years and their effectiveness and potential reviewed (McIntyre and Pearce 1980; Bayne et al. 1985; Waldichuk 1985). The aim of this chapter is to state the usefulness of molluscs as sentinel organisms and to focus on a number of measurements that either are thoroughly tested and being applied by monitoring bodies, such as scope for growth and lysosomal labilization period (Lack and Johnson 1985; Martin 1985), or offer potential for inclusion in a future programme of impact assessment, such as aspects of the cytochrome P-450 monooxygenase system as a specific indicator of effect by organic xenobiotics (Livingstone 1985; Moore 1985). The principles of biological effects measurements are described and the value of the integrated approach in which responses at the different levels of biological organization can be functionally linked is emphasized. The biological and toxicological bases of the measurements are given in outline only, having been described in detail in several recent reviews, and the application of the measurements illustrated by several field studies. Other uses of the measurements, such as prediction of pollutant impact or application in environmental quality models, are considered and, finally, future research needs are identified.

[1]Natural Environment Research Council, Plymouth Marine Laboratory, Prospect Place, West Hoe, Plymouth, PL1 3DH, United Kingdom

## 2 Principles and Problems of Biological Effects Measurements

A biological effect measurement should fulfil two main criteria: firstly the biological change should result from, or be a response to, an alteration in the environmental factor (pollutant) and, secondly, the biological change must have a detrimental effect on some aspect of animal fitness such as growth, reproduction or survival (Bayne et al. 1976, 1985). Response time should be short, of the order of hours to weeks, as the measurements are used as early-warning systems, and ideally, although not necessarily, the biological response should occur throughout the range from optimal to lethal environmental conditions.

The impact of pollutants on an organism, or the biological response, can occur at different level of functional complexity, from molecular through subcellular and cellular to whole animal. Because each higher level of organization represents an integration of a greater number of processes at the lower level, this results in a suite of biological effects measurements with different features of sensitivity, specificity, quantitativeness and obvious ecological relevance (Widdows 1983; Bayne et al. 1985; Moore et al. 1987). Specificity, in terms of identifying an offending pollutant in a complex environmental situation, can be achieved only at the molecular level whereas, for example, physiological scope for growth represents a non-specific (general) response to the sum of environmental stimuli, providing a measurement of the overall impact of environmental change. Quantitativeness, in the sense of the relationship between biological response and tissue pollutant body burden, has been demonstrated over a wide range of the latter for scope for growth (Moore et al. 1988) but, not surprisingly given considerations of metabolic control at the molecular level (e.g. Newsholme and Start 1973), can vary at the molecular and subcellular level, e.g. the lysosomal labilization period (ß-N-acetyl-hexosaminidase) of the digestive gland of the mussel *Mytilus edulis* is only partially linear with respect to the phenanthrene and anthracene tissue concentrations (Moore and Farrar 1985). The existence of such a quantitativeness is a practical advantage but not an operational necessity as the function of the measurement is to quantify biological effect and not tissue body burden. Ecological relevance is obvious, for example, for scope for growth, but requires a mechanistic theory for molecular events. The consequence of the latter is that a certain amount of fundamental research may be required for interpretation of a particular parameter because, for example, for many biochemical processes fundamental differences exist between the phyla (Livingstone 1985).

The inclusion of a particular biological effects measurement in a programme of environmental monitoring depends upon these feature of specificity, sensitivity etc. and also on the more practical considerations of background variability (signal:noise ratio) and the ease and cost of measurement. These aspects have been discussed extensively elsewhere for molluscs for a number of biochemical, cellular and physiological processes (Bayne et al. 1981, 1985; Livingstone 1984, 1985; Moore 1985; Widdows 1985). Seasonality may be a problem as, in bivalve molluscs, many biochemical and physiological processes show marked but regular patterns of seasonal change, often linked to the annual cycles of food storage and utilization and reproduction. Knowledge of reproductive state and the annual cycle of a particular parameter is, therefore, advantageous, or may be necessary, for interpretation of observed differences between mussel populations in the field. Various approaches are possible where measurements are required on intact living organism: for example, physiological scope for growth has been determined for *M. edulis* on native animals in the field, on transplanted animals in the field and on transplanted animals returned to standard laboratory conditions, and the various benefits of each approach compared and assessed (Widdows 1985).

## 3 The Use of Molluscs in Impact Assessment and the Value of the Integrated Approach

The fundamental concepts embodied in the derivation of biological effects measurements can be applied to any organism. However, molluscs, and in particular mussels, have assumed a global importance in environmental monitoring programmes because they possess a number of attributes: a wide geographic distribution, dominant member of coastal and estuarine communities, accumulate contaminants in their tissues and are responsible to many environmental pollutants but do not show a prolonged handling stress. In addition, a comprehensive knowledge exists on many aspects of the basic biology of the mussel and its responses to intrinsic factors (size, age, tissue specificity, reproductive and nutritional state) and extrinsic factors (temperature, salinity, ration, dissolved oxygen, suspended particulates, season and pollutants), information that may be vital in the planning of monitoring programmes and in the measurement and interpretation of results.

An impressive advantage of the combined or integrated molecular, subcellular, cellular and physiological approach to biological effects monitoring, realized by its application to a single sentinel organism, is the functional inter-relatedness of the different processes (Bayne 1985; Moore et al. 1987, 1988). To illustrate; the lysosomal system of *M. edulis* is a major site for the accumulation and sequestration (detoxication) of metal and organic contaminants (Moore 1985; Viarengo 1985). Sequestration or saturation of the system may lead to alterations in lysosomal structure and function (Sect. 4.2) resulting in increased protein turnover and a reduction in physiological scope for growth and animal fitness (Sect. 4.4). Many other ramifications are possible with the lysosomes and other organelles being primary sites for the formation of highly reactive metabolites (Sects. 4.1 and 4.2) which by various interactions will lead to progressive cellular malfunction and destruction (Sect. 4.3). It is the ability to link the various measurements in a functional and mechanistic manner that considerably strengthens the rationale for their use in impact assessment. In addition, it is only by establishing such links, involving fundamental research, that causality can be demonstrated between initial impact of a particular pollutant and the final consequence of a reduction in animal fitness (see also Sect. 8).

# 4 Measurement and Indices of Biological Effect

## 4.1 Molecular: Cytochrome P-450 Monooxygenase System and Metallothioneins

The cytochrome P-450 monooxygenase or mixed function oxidase (MFO) system is a universally distributed, multi-component, membrane-bound (in the endoplasmic reticulum (microsomes) and other organelles of the cell) enzyme system involved in the detoxication of foreign organic compounds or xenobiotics. It is inducible and its functioning can have deleterious consequences through the conversion of some contaminants to highly reactive and potentially mutagenic and carcinogenic derivatives. Its application as a specific indicator of impact by organic pollutants (Lee et al. 1980; Bayne et al. 1985) is considerably advanced in fish (Kurelec et al. 1977; Stegeman et al. 1986). Although there is considerable evidence for the existence of an MFO system in bivalve and gastropod molluscs, primarily in the digestive gland, virtually nothing is known of the nature of its functioning in vivo (Livingstone 1985; Moore et al. 1988). The levels or activities of its components, particularly cytochrome P-450 and cytochrome P-450 reductase, increase in *M. edulis*, the edible periwinkle *Littorina littorea* and other molluscs following exposure to various organic xenobiotics, but there is very little evidence for a change in overall monooxygenase activity (Livingstone 1985; Livingstone et al. 1985; Moore 1985; Moore et al. 1987, 1988). A better understanding of the molluscan MFO system is, therefore, required before a convincing link-up with animal fitness can be made, although the universality of this enzyme system and the results of exposure studies (Sect. 5) argue for such a connection and for its potential in environmental monitoring. The involvement of the molluscan MFO system in toxication processes has been questioned based on its inability to convert benzo[a]pyrene to mutagenic metabolites in the Ames *Salmonella typhimurium* test (Kurelec 1985; Kurelec et al. 1985). However, this is too simplistic an argument as benzo[a]pyrene is only one of a myriad of xenobiotics that could be converted to toxic metabolites and several mechanisms exist by which the MFO system or its components can give rise to toxic molecular species e.g. the conversion of oxygen to toxic radicals by redox cycling (Kappus and Sies 1981); in addition, non-conversion to mutagens does not mean non-metabolism of a chemical.

The role of metallothioneins in the uptake, homeostasis and toxic effects of trace metals in molluscs and other organisms has recently been reviewed (Viarengo 1985). They are a class of soluble, low molecular weight (about 6800–7000) proteins with a high binding affinity for particular metal cations. Increased synthesis of metallothioneins can occur following exposure of an organism to metals, and cellular toxicity is generally believed to occur only if the capacity of the system is exceeded by the levels of the metals present or being taken up. Induction of metallothioneins, as a specific indicator of biological impact by metals (Lee et al. 1980), has been demonstrated in several tissues of *M. edulis* and *Mytilus galloprovincialis*, both in the laboratory and the field, following exposure to metals (Bayne et al. 1985).

## 4.2 Subcellular: Lysosomal Structure and Function

Lysosomes are organelles involved in the compartmentalization and accumulation of a wide variety of organic chemicals and metals, and are present in high concentration in many cell types of marine molluscs (Moore 1980). They are, therefore, an ideal starting point for investigations of generalized cellular injury, and several aspects of structure and function have been studied, in the laboratory and the field, providing information on mechanisms of cellular toxicity and derivation of biological effects measurements for application in impact assessment (Bayne et al. 1985; Moore 1985; Moore et al. 1987, 1988). Cytochemical techniques such as the measurement of lysosomal membrane stability (based on hydrolase latency) have been particularly successful, and consistent in their applications, and a quantitative relationship has been observed between lysosomal membrane destabilization and physiological scope for growth (Bayne et al. 1979, 1982). Other useful measurements include total, free and latent lysosomal hydrolase activities, lysosomal volume and surface densities and ultrastructural evidence of lysosomal membrane breaks (Moore 1985; Moore et al. 1987, 1988).

## 4.3 Cellular: Digestive Potential and Reproductive State

Although abnormal cellular conditions, such as proliferative neoplastic cells, have been reported in mussels in the field, their aetiology is unknown and therefore their application in impact assessment limited. In contrast, contaminant-related pathological changes in cell structure have been demonstrated in *M. edulis* and other molluscs for both digestive and reproductive processes (Moore 1985; Moore et al. 1987, 1988). Exposure of mussels to hydrocarbons results in atrophy of the epithelium of the digestive tubules and this epithelial "thinning" is quantified by image analysis of histological sections (Lowe et al. 1981); the same response is observed in the periwinkle *L. littorea* and in both mollusc species there is evidence that this may be generalized response to both toxic xenobiotics and physical stressors (Pipe and Moore 1985; Moore et al. 1988). Similarly, exposure of *M. edulis* to hydrocarbons results in a reduction in the volume of storage cells in the mantle, a reduction in volume of ripe gametes and increased degeneration or atresia of oocytes (Livingstone et al. 1985; Lowe and Pipe 1985). Such changes indicate a direct impairment of reproductive processes and a reduction of reproductive capability. In addition to providing quantitative parameters for biological effects monitoring, studies of cellular processes provide a mechanistic link-up with whole animal events (scope for growth, fecundity) and subcellular ones (alterations in lysosomal function), strengthening the rationale for the integrated approach to impact assessment (Moore et al. 1987, 1988).

### 4.4 Organism: Scope for Growth and Oxygen/Nitrogen Ratio

The rate of growth is a fundamental component of physiological fitness, and therefore represents an important index of environmental (pollutant) effect. Determination of the energy available for growth (scope for growth), based on the physiological analysis of the energy budget (Bayne et al. 1985; Widdows 1983, 1985), provides an immediate assessment of the energy status of the animal as well as insight into the individual components (respiration, filtration rate etc.) which effect the changes in growth rate. The concept of scope for growth has been widely used to assess the sublethal biological effects of pollution in marine invertebrates, particularly mussels, and in bivalves good agreement is seen between this indirect estimate of growth and more direct determinations based on detailed population size-class analysis (Widdows 1985). Examples of its application with *M. edulis* are many and include, in addition to those illustrated in section 5, studies in San Francisco Bay (Martin 1985), Narragansett Bay (Widdows et al. 1981) and at a sewage sludge disposal site off Plymouth, England (Lack and Johnson 1985). Growth efficiency is an additional index that can be calculated from the physiological components of the scope for growth equation and is a measure of the efficiency with which food is converted into body tissue (Widdows 1985). In the Narragansett Bay study, reduced net growth efficiency was correlated with increased tissue contaminant levels in *M. edulis* (Widdows et al. 1981). Components of the energy budget can also be used to calculate the O:N ratio (ratio of oxygen consumed and nitrogen excreted) which provides an index of the relative utilization of protein to carbohydrate and lipid in energy metabolism (Bayne et al. 1985). The catabolism of proteins increases and the O:N ratio decreases in *M. edulis* with increased stress and pollution (Widdows 1985).

### 4.5 Genetic Toxicology: Chromosomal Abberations and Damage

Understanding the biological consequences of injury to the genome by environmental chemicals and the mechanisms by which it is effected by molecular events is of prime importance. Similarly, measurements of genetic biological effect would have immediate application in monitoring and impact assessment. Chromosomal damage (increased aneuploidy) has been demonstrated in embryos obtained from *M. edulis* exposed to high levels of pollution in a shipping dock (Dixon 1982). A very sensitive technique which offers potential for the future is sister chromatid exchange (based on switching of labelled arm segments within chromosomes) which has been successfully used in laboratory experiments with *M. edulis* (Dixon and Clarke 1982).

### 4.6 Population and Community Effects

These aspects, particularly in relation to sentinel molluscs, have recently been reviewed (Moore et al. 1987, 1988). Decreased fitness of the individual, due to pollutant effects on cellular and physiological processes, is indicated to result in decreased population performance in the form of reduced recruitment and production and increased mortality. Effects at the community level also occur and a parallel relationship can be discerned between pollutant levels and the nature of the changes in biological effects measurements taken at the individual level (scope for growth) and at the community level (species diversity). This parallel relationship is argued to support the notion that adverse effects measured at the cellular and individual levels ultimately manifest themselves at the population and community levels of organization.

## 5 Examples of Integrated Use of Biological Effects Measurements in Mesocosms and Field Exposures of Molluscs to Hydrocarbons

### 5.1 Solbergstrand Experimental Facility, Norway: Diesel Oil Exposure

Mussels (*M. edulis*) and winkles (*L. littorea*) were exposed long-term to low and high environmentally realistic diesel oil concentrations under conditions closely resembling the field in the Solbergstrand mesocosm facility (outdoor flow-through tanks, natural seawater and food levels, wave and tidal simulation). Hydrocarbons were taken up into the tissues and moreso at the higher oil level (Tables 1 and 2). At the molecular level, cytochromes P-450 and $b_5$ of the MFO system increased and the former was accompanied by a blue shift in the wavelength maximum indicating the synthesis of specific isoenzymes (Table 1). Increases in cytochrome P-450 reductase, another MFO component, were indicated both by biochemical (NADPH-CYTCRED) and cytochemical (NADPH-NTR) assays of this enzyme (Tables 1 and 2). At the subcellular level, lysosomes were destabilized (Tables 1 and

**Table 1.** Tissue aromatic hydrocarbons, reproductive state and responses of the digestive gland lysosomal system and microsomal MFO system of *M. edulis* exposed to 29 and 123 ppb diesel oil for 4 months. (After Livingstone et al. 1985)

| Parameter | Control | 29 ppb | 123 ppb |
|---|---|---|---|
| PNAH[a] | 2.2 ± 0.3 (2) | 18.1 ± 2.8 (2) | 32.8 ± 0.1 (2) |
| Gametes[b] | 0.43 ± 0.09 | 0.17 ± 0.04** | 0.11 ± 0.03** |
| Storage cells[b] | 0.54 ± 0.09 | 0.39 ± 0.06 | 0.41 ± 0.09 |
| Lysosomal stability[c] | 25(25,25) | 5(5,5)** | 2(2,2)** |
| P-450[d] | 46.6 ±14.3 | 93.0 ±14.3* | 89.6 ±21.9 (2) |
| $b_5$[d] | 25.8 ± 2.3 | 40.2 ± 5.8* | – |
| P-450 λ max nm | 449 ± 1 | 445 ± 1** | 445 ± 0 (2) |
| NADPH-CYTCRED[e] | 10.3 ± 1.2 | 13.5 ± 0.9** | 14.7 ± 1.2 (2) |
| NADPH-NTR[f] | 22.2 ± 2.0 | 30.6 ± 1.2** | 40.8 ± 8.1** |

Means ± SEM (n = 5 or 10) or ± range (n = 2). *P≤ 0.1, **P≤ 0.05 comparing 29 or 123 ppb with control.
[a] sum of 2- and 3-ring polynuclear aromatic hydrocarbons ($\mu g\ g^{-1}$ wet total wt.).
[b] Weight in grams in mantle.
[c] Labilization period of β-hexosaminidase (min) (lowest and highest values in parentheses).
[d] Cytochromes P-450 and $b_5$ in pmol $mg^{-1}$ microsomal protein.
[e] NADPH-cytochrome c reductase activity in nmol $min^{-1}\ mg^{-1}$ microsomal protein.
[f] NADPH-neotetrazolium reductase activity as integrated extinction $X10^2$.

**Table 2.** Tissue aromatic hydrocarbons and responses of the digestive gland lysosomal system and microsomal MFO system of *L. littorea* exposed to 29 and 123 ppb for 16 months. (Livingstone et al. 1985)

| Parameter | Control | 29 ppb | 123 ppb |
|---|---|---|---|
| PNAH | 0.3 ±0.1 (2) | 6.6 ±1.3 (2) | 14.0 ±0.8 (2) |
| Lysosomal stability | 24(20,25) | 2.6(2.5)** | 2(2,2)** |
| NADPH-CYTCRED | 7.5 ±0.4 | 11.3 ±0.7* | 21.8 ±1.9** |
| NADPH-NTR | 7.0 ±1.0 | 20.4 ±1.3** | 27.6 ±2.2** |

Means ± SEM (n = 5) or ± range (n = 2) *P≤ 0.1, **P≤ 0.05
Units etc. as for Table 1. except labilization period of $\beta$-glucuronidase measured.

2) and, at the cellular level, reproductive state was affected with a decrease in amount of gametes being seen (Table 1). At the whole-animal level, scope for growth was reduced in mussels and more so at the high oil exposure; following the cessation of oil-dosing, physiological performance recovered and the resulting increase in scope for growth, faster in the high than the low oil condition, paralleled the depuration of hydrocarbons (Fig. 1).

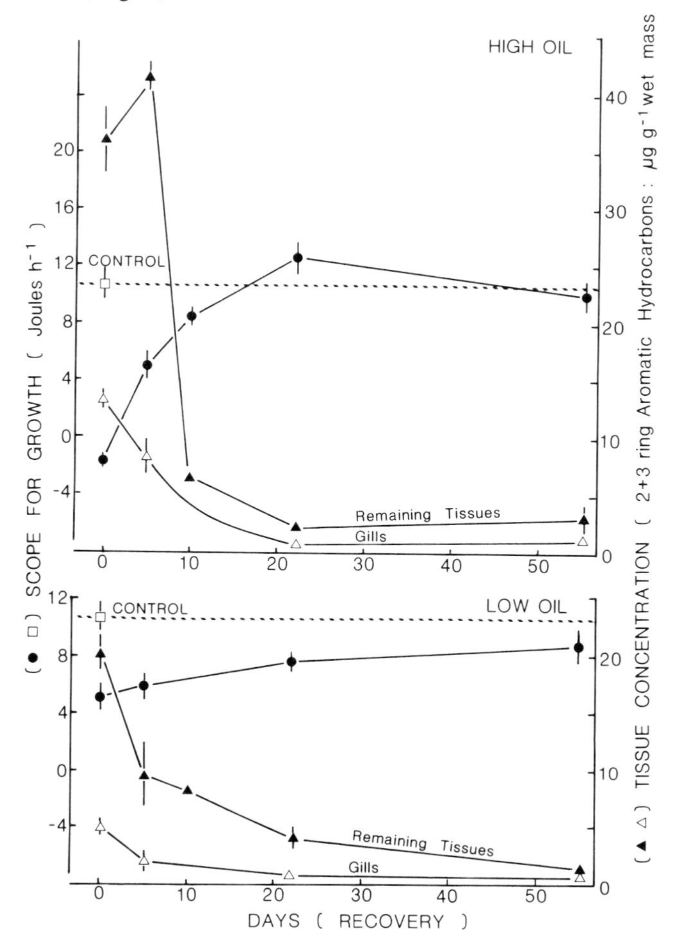


**Fig. 1.** Scope for growth (mean ± SEM; n = 15; standard animal of 0.5 g dry weight) and depuration of 2- and 3-ring aromatic hydrocarbons from gills and remaining tissues of *M. edulis* during recovery from 8 months exposure to low (≈ 30 ppb) and high (≈ 130 ppb) diesel oil. (Data from Widdows et al. 1985)

## 5.2 The Shetland Islands: the Sullom Voe Oil Terminal

The Sullom Voe Oil Terminal was built to receive oil piped ashore from the North Sea oil fields and became fully operational in 1979. No complex refining is carried out at the terminal and the main effluent discharged into the aquatic environment is ballast water removed from the incoming tankers and treated in settlement tanks and lagoons before being released into the sea through a long outfall pipe. The major sources of hydrocarbon contamination at the site are probably a result of shipping and small spillages that occur during the operation of the terminal.

Effects measurements on winkles (*L. littorea*) in the vicinity of the Oil Terminal have indicated localized deleterious responses at the subcellular and whole animal levels of biological organisation. These responses include reduced lysosomal membrane stability, elevated NADPH-NTR, which is believed to be an enzymic activity associated with cytochrome P-450 reductase, and reduced physiological scope for growth (Table 3). In the periwinkle there is evidence that these effects are related to the presence of polycyclic aromatic hydrocarbons in the tissues (Table 3); this is in agreement with previous findings (Moore et al. 1982; Moore 1985; Widdows 1985; Livingstone et al. 1985). The possibility remains, however, that these effects may be in part induced by other factors such as dispersants occasionally used in the treatment of small oil spillages which are probably the major source of polycyclic aromatic hydrocarbons in the periwinkles.

**Table 3.** Application of cytochemical and whole animal effects measurements to the edible periwinkle (*L. littorea*) from the vicinity of the Sullum Voe Oil Terminal. (Data from Widdows et al. 1984)

| Parameter | Sample sites | | | | |
|---|---|---|---|---|---|
| | Gluss Voe (Clean ref.) | Ronas Voe | Mavis Grind (Contaminated ref.) | Scatsa Voe | Tanker Jetty 4 |
| PNAH | 0.08 ± 0.02 | 0.25 ± 0.06 | 0.47 ± 0.03 | 0.71 ±0.07 | 12.69 ± 0.95 |
| Lysosomal stability | 24(20,25) | 21(20,25) | 3.8(2,5)** | 13(10,15)** | 4.4(2,5)** |
| NADPH-NTR | 18.2 ±11.6 | 34.6 ±15.0 | 36.6 ±13.6* | 28.4 ±7.2 | 61.4 ±11.0 |
| Scope for growth[a] | 3.7 | – | 1.31 | – | -2.32 |

Means ± SD (n = 5) or ± range (n = 2) (PNAH). *P ≤ 0.05.
**P ≤ 0.01 (Mann-Whitney U-test comparing with Gluss Voe).
Units etc. as for Table 1.
[a] Joules $g^{-1}$ $h^{-1}$ (pooled animals).

## 5.3 Humberside and the Humber estuary, England: the 'Sivand' Oil Spill

The Sivand oil spill occurred in Immingham Dock in 1983 and the oil spread down both the southern coast and onto the northern headland of the Humber estuary. Cockles, *Cerastoderma edule*, were collected from four sites on the southern coast – Horseshoe Point, Humberstone Fitties, Cleethorpes and Grimsby – in increasing proximity to the oil spill with the latter two also located in urban or industrial areas, and from Spurn Bight on the oil impacted northern headland (for

**Table 4.** Tissue aromatic hydrocarbons, and digestive gland lysosomal membrane stability and microsomal MFO component enzyme activities of the common cockle, *Cerastoderma edule* L., from the Humber Estuary, England, sampled within 8 days of the Sivand oil spill. (Moore et al. 1987)

| | Clean ← | | | → | Oil impact |
|---|---|---|---|---|---|
| Parameter | Horseshoe Point | Humberstone Fitties | Cleethorpes | Grimsby | Spurn Bight |
| PNAH | 2.4 ±0 (2) | 20.1 ±3.0 (2) | 27.7 ±2.0 (2) | 35.5 ± 1.4 (2) | 14.5 ± 1.2 (2) |
| Lysosomal stability | 20(20,25) | 23(20,25) | 15(15,15)** | 10(10,10)** | 17(10,20) |
| NADPH-CYTCRED | 6.7 ±2.2 (2) | 5.3 ±1.3 (3) | 10.5 ±1.2 (3) | 9.1 ± 1.7 (3) | 10.6 ± 2.1 (3) |
| NADPH-NTR | 12.2 ±4.5 | 10.2 ±4.5 | 26.6 ±5.4** | 30.8 ±16.1** | 20.5 ±14.3 |

Means ± SEM (n = 5 unless indicated otherwise) or ± range (n = 2).
**P ≤ 0.05 comparing sites with Horseshoe Point.
Units etc. as for Table 1 except labilization period of β-glucuronidase measured.

further details see Moore et al. 1987). Aromatic hydrocarbons were high in the tissues of cockles from all sites except Horseshoe Point, the site furthest from the oil spill (Table 4). At the molecular level, NADPH-NTR and NADPH-CYTCRED activities (cytochrome P-450 reductase) were elevated in Cleethorpes and Grimsby compared with Humberstone Fitties and Horseshoe Point (using pooled data for the pairs of sites NADPH-CYTCRED was significantly higher at $P \leq 0.05$). At the subcellular level, lysosomal stability was reduced at Cleethorpes and Grimsby. Similar results for Spurn Head were either obtained (NADPH-CYTCRED was elevated compared with Horseshoe Point/Humberstone Fitties; $P \leq 0.1$) or indicated (lysosomal stability and NADPH-NTR). Somewhat complicated differences, indicative of a response, were also seen for cytochrome P-450 (Moore et al. 1987). The lack of change at Humberstone Fitties, where tissue hydrocarbon levels were high, was interpreted as that oil must have reached the site but been insufficient, or the cockle samples taken too early (within 8 days of initial spill), to have affected the organism.

## 6 Applicability of Biological Effects Measurements to Other Phyla

Although the use of sentinel organisms in environmental monitoring has met widespread approval and application (Bayne 1985), there are clearly instances when such measurements will be required on other or any organism. For example, mussels were absent from the area of the Sivand oil spill and cockles were used; input would be required for all the dominant species for complete environmental quality modelling of a localized ecosystem or for making a comprehensive mechanistic link-up between events at the individual level and at the community level. Conceptually the sublethal biological effects approach can be applied to any organism; the difficulties or differences, practical or fundamental, only arise for particular measurements and particular organisms (Bayne et al. 1985). Scope for growth, for example, has had a wide application (Widdows 1985), although the mechanics of obtaining the components of the energy budget may vary. In contrast, the use of the O:N ratio is more limited as, for example, when applied to an organism with a more protein-based metabolism such as carnivorous invertebrates (Stickle et al. 1984). Molluscs are particularly rich in lysosomes, but the fundamental aspects of lysosomal functioning apply to any organism (Pitt 1975). At the molecular level, some aspects such as DNA functioning and damage are likely to be universal, but others, such as pathways and mechanisms of metabolism, may vary; for example, the phenobarbital-type induction responses found in mammals are absent in fish (Stegeman 1981).

## 7 Use of Biological Effects Measurements in the Prediction of Pollutant Impact and the Assessment of Assimilative Capacity

With increased usage of biological effects measurements in the field and collation of data, the effectiveness and predictive application of the approach will be enhanced. Figure 2 provides a synthesis of scope for growth data derived from two studies on *M. edulis* and shows a significant negative correlation ($r^2 = -0.95$)

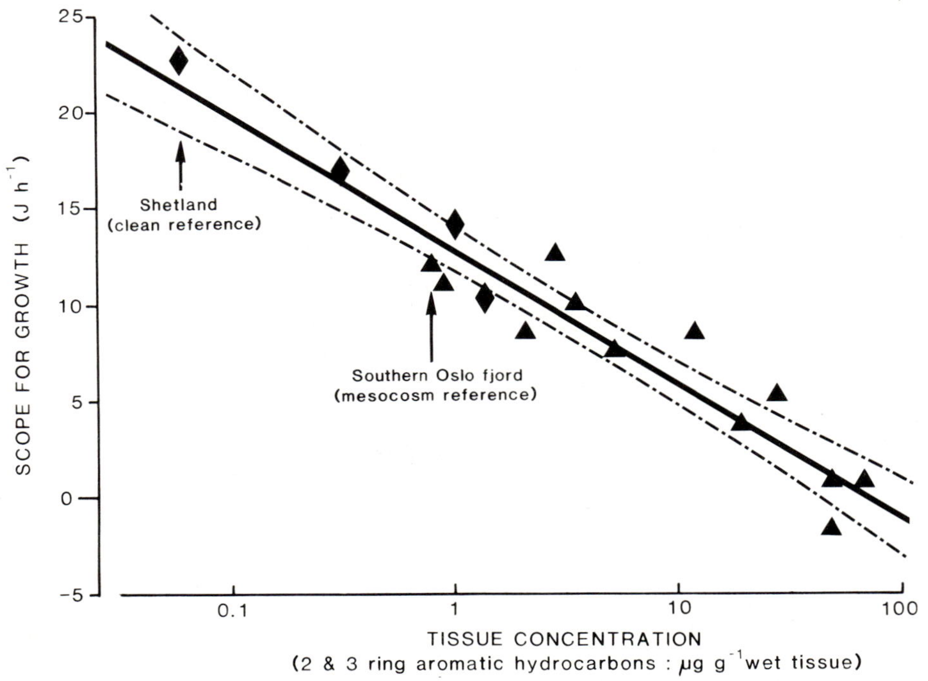


**Fig. 2.** Relationship between scope for growth and whole tissue concentration of 2- and 3-ring aromatic hydrocarbons in *M. edulis* (mean ± 95% confidence limits). ▲ Data from Solbergstrand mesocosm experiment, Oslo fjord, Norway, ♦ Data from Sullom voe, Shetland Islands. (Moore et al. 1987)

between scope for growth and $\log_{10}$ of the tissue concentration of 2- and 3-ring aromatic hydrocarbons (the major component of the accumulated hydrocarbons). The relationship demonstrates that hydrocarbons affect scope for growth over a wide range of tissue concentrations without an apparent threshold concentration of effect. It illustrates, for example, the degree of contamination and reduced physiological performance of "control" mussels from Oslo fjord compared with those from the Shetland Islands. Given information on tissue body burdens, or on water column levels of pollutants and bioconcentration factors, the relationship and others like it can therefore be used to predict the performance of an organism in such an environment (Moore et al. 1987). At a higher level of sophistication such "concentration-response" relationships or biological processes can be integrated with physical and chemical processes into an environmental quality model which can be used both for the prediction of biological effects of potential pollutant levels and in the estimation of the "assimilative capacity" of receiving waters for wastes (Harris et al. 1983; Bayne and Widdows 1985). Assimilative capacity (or rate) may be defined as the quantity of effluent/waste that can be discharged into a body of water per unit time, without producing deleterious or irreversible biological effects (Widdows 1985). These environmental quality models are seen as playing an important role in environmental management, where it is necessary to predict change rather than simply document damage.

## 8 Summary and Future Research Needs

The impact of pollutants on an organism is realized as perturbations at different levels of functional complexity, from molecular through to whole-animal. A number of biological effects measurements are described based on the use of the mussel and other molluscs as sentinel organisms, but with fundamental relevance and possible application to any organism. These measurements include the microsomal cytochrome P-450-dependent monooxygenase system involved in the metabolism of organic xenobiotics, functional and structural responses of lysosomes, quantitative structural alterations in the cells of the digestive and reproductive systems and effects on physiological scope for growth. The various measurements can be linked in a functional manner and the ability to do this is argued to considerably strengthen the rationale for their use in impact assessment. An important attribute of sublethal biological responses is that they are amenable to both laboratory and field measurement, unlike traditional toxicity testing based on $LD_{50}$ and ecological surveys based on community structure. They can be used both to actually monitor effects in the real environment for impact assessment and in environmental quality models for prediction.

Future research must obviously focus on present gaps in knowledge and understanding. The cytochemically measured NADPH-NTR activity has been extensively studied in molluscs and shown to be consistently responsive to organic xenobiotics (Moore 1985). The parallel changes in the cytochromes P-450 and $b_5$ and NADPH-CYTCRED activity support the contention that NADPH-NTR is a measure of part of the MFO system in molluscs. Research is now required to elaborate the in vivo significance of such changes, both with respect to metabolism and toxicity. An increased fundamental understanding and mechanistic integration is required at all levels of biological organization in order to indicate causality, which in combination with empirical correlation will be the most likely means to establish responsibility in an environmental situation. Generalized theories for the various biological processes will be required for all major groups of organisms for comprehensive modelling and prediction and so that biological effects measurements can be effected in any environmental situation.

## References

Bayne BL (1985) Cellular and physiological measurements of pollution effect. Mar Pollut Bull 16:127–129

Bayne BL, Widdows J (1985) Strategies for measuring the biological effects of pollution: Cellular and organism levels. In: SCOPE Symposium on the Chemical Changes in Coastal Regions, Hawaii

Bayne BL, Livingstone DR, Moore MN, Widdows J (1976) A cytochemical and a biochemical index of stress in *Mytilus edulis* L. Mar Pollut Bull 12:221–224

Bayne BL, Moore MN, Widdows J, Livingstone DR, Salkeld P (1979) Measurement of the responses of individuals to environmental stress and pollution: studies with bivalve molluscs. Phil Trans R Soc Lond B 286:563–581

Bayne BL, Clarke KR, Moore MN (1981) Some practical considerations in the measurement of pollution effects on bivalve molluscs, and some possible ecological consequences. Aquat Toxicol 1:159–174

Bayne BL, Widdows J, Moore MN, Salkeld PN, Worrall CM, Donkin P (1982) Some ecological consequences of the physiological and biochemical effects of petroleum compounds on marine molluscs. Phil Trans R Soc Lond B 297:219–239

Bayne BL, Brown DA, Burns K, Dixon DR, Ivanovici A, Livingstone DR, Lowe DM, Moore MN, Stebbing ARD, Widdows J (1985) The effects of stress and pollution on marine animals. Praeger, New York

Cairns Jr J, Schale WH van der (1980) Biological monitoring. Part I – Early warning systems. Water Res 14:1179–1196

Dixon DR (1982) Aneuploidy in mussel embryos *Mytilus edulis* L. originating from a polluted dock. Mar Biol Lett 3:155–161

Dixon DR, Clarke KR (1982) Sister chromatid exchange, a sensitive method for detecting damage caused by exposure to environmental mutagens in the chromosomes of adult *Mytilus edulis*. Mar Biol Lett 3:163–172

Harris JRW, Bale AJ, Bayne BL, Mantoura RFC, Morris AW, Nelson LA, Radford PJ, Uncles RJ, Weston SA, Widdows J (1983) A preliminary model of the dispersal and biological effect of toxins in the Tamar estuary, England. Ecol Model 22:253–284

IOC (1976) A comprehensive plan for the global investigation of pollution in the marine environment and baseline study guidelines. UNESCO, IOC Tech Ser No 14

Kappus H, Sies H (1981) Toxic drug effects associated with oxygen metabolism: Redox cycling and lipid peroxidation. Experientia 37:1233–1241

Kurelec B (1985) Exclusive activation of aromatic amines in the marine mussel *Mytilus edulis* by FAD-containing monooxygenase. Biochem Biophys Res Commun 127:773–778

Kurelec B, Britvic S, Rijavec M, Muller WEG, Zahn RK (1977) Benzo[a]pyrene monooxygenase induction in marine fish – molecular response to oil pollution. Mar Biol 44:211–216

Kurelec B, Britvic S, Zahn RK (1985) The activation of aromatic amines in some marine invertebrates. Mar Environ Res 17:141–144

Lack TJ, Johnson D (1985) Assessment of the biological effects of sewage sludge at a licensed site off Plymouth. Mar Pollut Bull 16:147–152

Lee RF, Davies JM, Freeman HC, Ivanovici A, Moore MN, Stegeman J, Uthe JF (1980) Biochemical techniques for monitoring biological effects of pollution in the sea. In: McIntyre AD, Pearce JB (eds) Biological effects of marine pollution and the problems of monitoring. Rapp P V Réun Cons Int Explor Mer 19:48–55

Livingstone DR (1984) Biochemical differences in field populations of the common mussel *Mytilus edulis* L. exposed to hydrocarbons: some considerations of biochemical monitoring. In: Bolis L, Zadunaisky J, Gilles R (eds) Toxins, drugs and pollution in marine animals. Springer, Berlin Heidelberg New York Tokyo, p 161

Livingstone DR (1985) Responses of the detoxication/toxication enzyme system of molluscs to organic pollutants and xenobiotics. Mar Pollut Bull 16:158–164

Livingstone DR, Moore MN, Lowe DM, Nasci C, Farrar SV (1985) Responses of the cytochrome P-450 monooxygenase system to diesel oil in the common mussel, *Mytilus edulis* L., and the periwinkle, *Littorina littorea* L. Aquat Toxicol 7:79–91

Lowe DM, Pipe RK (1985) Cellular responses in the mussel *Mytilus edulis* following exposure to diesel oil emulsions: reproductive and nutrient storage cells. Mar Environ Res 17:234–237

Lowe DM, Moore MN, Clarke KR (1981) Effects of oil on digestive cells in mussels: quantitative alterations in cellular and lysosomal structure. Aquat Toxicol 1:213–226

Martin M (1985) State mussel watch: toxics surveillance in California. Mar Pollut Bull 16:140–146

McIntyre AD, Pearce JB (eds) (1980) Biological effects of marine pollution and the problems of monitoring. Rapp P V Réun Cons Int Explor Mer 170:346

Moore MN (1980) Cytochemical determination of cellular responses to environmental stressors in marine organisms. In: McIntyre AD, Pearce JB (eds) Biological effects of marine pollution and the problems of monitoring. Rapp P V Réun Cons Int Explor Mer 179:7–15

Moore MN (1985) Cellular responses to pollutants. Mar Pollut Bull 16:134–139

Moore MN, Farrar SV (1985) Effects of polynuclear aromatic hydrocarbons on lysosomal membranes in molluscs. Mar Environ Res 17:222–225

Moore MN, Pipe RK, Farrar SV (1982) Lysosomal and microsomal responses to environmental factors in *Littorina littorea* from Sullom Voe. Mar Pollut Bull 13:340–345

Moore MN, Livingstone DR, Widdows J, Lowe DM, Pipe RK (1987) Molecular, cellular and physiological effects of oil-derived hydrocarbons in molluscs and their use in impact assessment. Phil Trans R Soc Lond B 316:603–623

Moore MN, Livingstone DR, Widdows J (1988) Hydrocarbons in marine molluscs: biological effects and ecological consequences. In: Varanasi U (ed) Metabolism of polynuclear aromatic hydrocarbons by aquatic organisms. CRC, Bocan Raton (Florida) (in press)

Newsholme EA, Start C (1973) Regulation in metabolism. Wiley, London

Pipe RK, Moore MN, (1985) Ultrastructural changes in the lysosomal-vacuolar system in digestive cells of *Mytilus edulis* as a response to increased salinity. Mar Biol 87:157–163

Pitt D (1975) Lysosomes and cell function. Longman, London

Rosenberg DM, Resh VHB, Balling SS, Barnby MA, Collins JN, Durbin DV, Flynn TS, Hart DD, Lamberti GA, McElravy EP, Wood JR (1981) Recent trends in environmental impact assessment. Can J Fish Aquat Sci 38:591–624

Stegeman JJ (1981) Polynuclear aromatic hydrocarbons and their metabolism in the marine environment. In: Gelboin HV, T'So POP (eds) Polycyclic hydrocarbons and cancer. Academic Press, New York, pp 1–60

Stegeman JJ, Kloepper-Sams PJ, Farrington JW (1986) Monooxygenase induction and chlorobiphenyls in the deep-sea fish *Coryphaenoides armatus* Science 231:1287–1289

Stickle WB, Rice SD, Moles A (1984) Bioenergetics and survival of the marine snail *Thais lima* during long-term oil exposure. Mar Biol 80:281–289

Viarengo A (1985) Biochemical effects of trace metals. Mar Pollut Bull 16:153–158

Waldichuck (1985) Methods for measuring the effects of chemicals on aquatic animals as indicators of ecological damage. In: Vouk VB, Bulter GC, Hoel DG, Peakall DB (eds) Methods for estimating risk of chemical injury and non-human biota and ecosystems. SCOPE 26:493–535

Widdows J (1983) Field measurement of the biological impact of pollutants. In: Su J-C, Hung T-C (eds) SCOPE/ICSU Acad Sin, Taipei, Republic of China, pp 111–129

Widdows J (1985) Physiological responses to pollution. Mar Pollut Bull 16:129–124

Widdows J, Phelps DK, Galloway W (1981) Measurement of physiological conditions of mussels transplanted along a pollution gradient in Narragansett Bay. Mar Environ Res 4:181–194

Widdows J, Cleary JJ, Dixon DR, Donkin P, Livingstone DR, Lowe DM, Moore MN, Pipe RK, Salkeld PN, Worrall CM (1984) Sublethal biological effects monitoring in the region of Sullom Voe, Shetland, July 1983. Aberdeen: Shetland Oil Terminal Environmental Advisory Group, 34 pp

Widdows J, Donkin P, Evans SV (1985) Recovery of *Mytilus edulis* L. from chronic oil exposure. Mar Environ Res 17:250–253

# Between Test-Tubes and North Sea: Mesocosms

J. KUIPER[1]* and J.C. GAMBLE[2]

## 1 Introduction

In the marine environment, and the North Sea is no exception, it is often difficult to prove that concentrations of chemicals, elevated above background levels (pollution), cause (reversible of irreversible) effects on the ecosystem. Although the North Sea is one of the most intensively studied seas, our knowledge of its ecology is far from complete (WKP 1986). Measured field data often show large variations, and it is difficult to judge whether extremes in polluted regions fall inside or outside natural ranges. Therefore a widening of our general ecological knowledge is a prerequisite if we are to judge the effects of human interference with the sea. Ecology is a basic science for ecotoxicology. In ecotoxicology the toxic effects of chemical and physical agents on living organisms, especially on populations and communities within ecosystems, are studied together with the transfer pathways of those agents and their interactions with the environment (definition SCOPE, Butler 1978). In this definition the concepts of population and ecosystem take a central position. The key issue is not the protection of individuals of a species, but the conservation of populations and ecosystems. Nevertheless, as in human toxicology, which tries to protect the human individual against undesired effects, most ecotoxicological studies are conducted with individuals of one species. Many authors described the reasons why results of such studies cannot usually be extrapolated to actual field conditions (NAS 1981; Kuiper 1984). Most ecotoxicological tests lack "pollutant and ecological realism" (Blanck et al. 1978). Characteristics of the chemical in the natural environment are not incorporated in the test system. This lack of pollutant realism is related to the chemical form, including speciation and degradation intermediates, concentrations, exposure time to the chemical, and to interactions with other chemicals and physicochemical factors. The influence of these factors on possible effects can be large (e.g., NAS 1981; Kuiper 1984). Lack of ecological realism is often apparent from the choice of test species (sometimes going as far as using tropical freshwater species for predictions in the temperate North Sea!), but in particular from the single-species character of most tests. In the field chemicals act on populations of different species that interact with each other and their surroundings. Interactions in and among species are, for example, apparent in competition and predation. In some eco-

[1]Laboratory of Applied Marine Research, Division of Technology for Society, TNO, Den Helder, The Netherlands
[2]Marine Laboratory, Aberdeen, United Kingdom
*Present address: Ecomare Wadden and North Sea Center, Ruyslaan 92, 1796 AZ De Koog, The Netherlands

toxicological experiments, competition was introduced by dosing chemicals to algal cultures of two or more species (Fisher et al. 1974; Lundy et al. 1984). Heavy metals and PCB's influence the species composition of these mixed cultures in concentrations that do not effect monocultures of the same species. Changes in species composition are often found as a first effect in pollution research with isolated natural phytoplankton communities (O'Connors et al. 1978). Predation can also be influenced by pollutants. It was shown that exposure to chemicals in concentrations that did not influence monocultures of prey increased their mortality rate, when also exposed to predators, via a change in behaviour caused by the chemical (Ward et al. 1976; Pearson et al. 1981). Via bioaccumulation and biomagnification the predator can be affected. In the field, decreasing populations of top predators are often a first sign of ecosystems malfunctioning. The decrease in numbers of the sandwich tern in the Wadden Sea (Koeman and Van Genderen 1972) and the declining populations of the harbor seal in the same area (Reijnders 1980, 1981, 1983) are well-known examples of this phenomenon.

It appears that ecosystems are subject to laws typical for that level of organization. They are cybernetic systems (Jordan 1981; Mann 1982), whose reaction to certain agents cannot be predicted from data valid for a lower level of organization without a detailed knowledge of the laws ruling at the higher level. As long as ecology as a basic science for ecotoxicology is insufficiently developed, experimental methods using more or less complete ecosystems (model ecosystems, mesocosms) are necessary to study the fate and effects of pollutants in the marine environment. This chapter, the title of which is a tribute to Strickland's (1967) *Between beakers and bays*, is devoted to the use of mesocosms in relation to the pollution of the North Sea. Scientists and the authorities responsible for the management of marine resources can benefit in two ways from mesocosms. First, they form an experimental tool that helps to bridge the gap between the artificiality of laboratory test tubes and the inherent variation in measurement data from the (North) Sea. This bridge is necessary as long as our knowledge of ecosystems is insufficient to build confident, predictive mathematical models. Secondly, formalized knowledge of perturbed and unperturbed ecosystems in the form of mathematical models can be validated by mesocosms studies much better than by field studies in terms of costs and data variance (signal-to-noise ratio). General reviews of the use of model ecosystems in environmental research have been given by Giesy (1980); NAS (1981); Grice and Reeve (1982); Kuiper (1984). This chapter presents some examples of experiments with mesocosms that illustrate the types of problems relevant to the North Sea that can be tackled, and the directions further research is likely to take.

## 2 Water Column Ecosystems

### 2.1 Technical Aspects

Model ecosystems of different design and dimension have been used to simulate the structure and functioning of the plankton community in the water column. In the late 1930's, the first land-based containers were built (Pettersson 1939; see

Banse 1982 for historical overview) in which plankton ecology could be studied. In the 1960's flexible plastic materials became available from which translucent enclosures could be made, housing natural communities in situ under near natural conditions. This approach has also been used by many investigators around the North Sea (France: Lacaze 1974; U.K.: Davies and Gamble 1979; F.R.G.: Brockmann et al. 1974, 1983; Norway: Dahl et al. 1983; The Netherlands: Kuiper 1977, 1981a; Ireland: Patching 1981). Figure 1 shows some of the designs used.

## 2.2 General Ecological Research

Plankton model ecosystems have been used for studying many aspects of plankton ecology, one of their major advantages over field research being that the same experimentally manipulated water mass can be sampled repeatedly. Table 1 lists some mesocosm experiments, together with the main aspects studied. The ob-

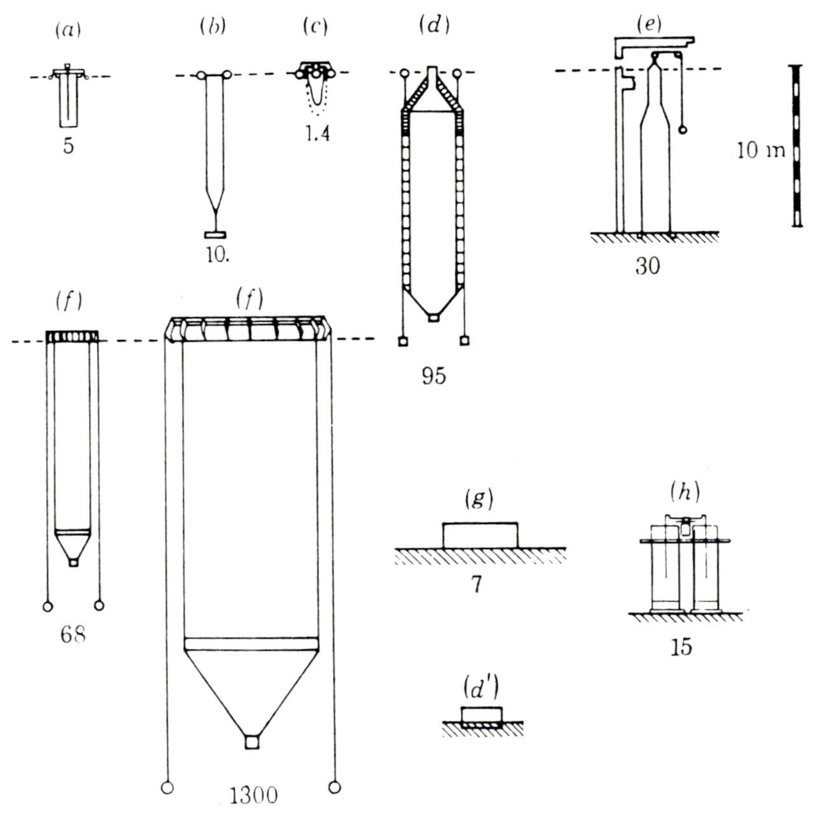


**Fig. 1a-h.** Marine enclosures design after Zeitzschel (1978); the capacity is indicated in $m^3$. **a** Brockmann et al. (1974), Helgoland; **b** Berland et al. (1975), Marseille; **c** Kuiper (1977), Den Helder, Netherlands; **d** Gamble et al. (1977), Loch Ewe, Scotland; **e** von Bodungen et al. (1976), Kiel; **f** Menzel and Case (1977), CEPEX, British Columbia; **g** Saward et al. (1975), Loch Ewe; **h** Pilson et al. (1977) MERL, Narragansett Bay, Rhode Island. (Davies and Gamble 1979)

**Table 1.** Some examples of marine mesocosm studies concerning basic problems in plankton ecology

| Objective study | Enclosure/Tank | Reference |
|---|---|---|
| Phytoplankton photosynthesis | E | Strickland and Terhune 1961 |
| Effects storm event | T | Oviatt et al. 1981 |
| Effects flagellates or diatoms on secondary production | E | Grice et al. 1980 |
| Effects tertiary producers | E | Sullivan and Reeve 1982 |
| Cycling organic nitrogen | E | Hollibaugh et al. 1980 |
| Effects glucose on relations bacteria phytoplankton | E | Parsons et al. 1980 |
| Effects nutrients on primary, secondary and tertiary productions | E | Parsons et al. 1977a,b |
| Phytoplankton succession in nutrient poor seawater | E | Brockmann et al. 1977 |
| Phytoplankton spring bloom | E | Brockmann et al. 1983 |
| Effects nutrients and light on phytoplankton succession | E | Parsons et al. 1978 |
| Production extracellular materials plankton | E | Brockmann et al. 1979 |
| Effects turbulence on plankton | E | Sonntag and Parsons 1979 |

jectives vary widely and cover the complete field of plankton ecology. Different studies resulted in quantitative and qualitative data on relations between organisms at the ecosystem level, showing that laws and rules typical for that level are functioning. It has been shown, for example, that amount and species composition of the phytoplankton, being limited by nutrients, limited the production of the copepods and other secondary producers (Parsons et al. 1977a). Apart from this qualitatively expected result, which shows how the basic food supply (primary production) shapes the next step in the food chain, it has also been shown that amount and type of secondary producers are of prime importance in structuring the phytoplankton community through selective grazing of larger species (Ryther and Sanders 1980).

The practical importance of this increase in basic ecological knowledge is that quantitative data on ecological relations are necessary to construct (validated) mathematical models of ecosystems. These mathematical models (together, of course, with experimental studies) will play an increasing role in the management of ecosystems, including management of pollution.

## 2.3 Applied Ecological and Ecotoxicological Research

When it had been shown that identically treated enclosed plankton communities develop similarly (Takahashi et al. 1975; Kuiper 1977), this method could be used to study the effects of experimental manipulations in comparison with untreated controls. A large proportion of the mesocosm literature is devoted to applied research into fisheries-related and environmental problems. Apart from typical aquaculture studies, enclosures and large outdoor basins have been used in studies on larval development, predation potential, food selection, species interactions, energy budgets etc. For a review see Øiestad (1982). In Scotland Gamble et al. (1985) reared herring larvae in enclosures, showing that growth was dependent on food availability and larval stocking density but also that the larvae were very susceptible to predation. Reviews on the use of mesocosms in the study of the fate and effects of pollutants on natural plankton communities have been given by Menzel and Steele (1978), Davies and Gamble (1979), Kuiper (1982, 1984). In most experiments a pollutant was added in a single dose at the start of the experiment and the development of the enclosed ecosystem was compared with that of untreated controls. Such a single dose is similar to actual pollution events in the sea, when the source of pollution is a river, or similar pollution point source, e.g., a pipeline, dumping ship, or platform (Menzel and Case 1977). After addition of the pollutant, the development of the phyto- and zooplankton, the bacteria and a set of physicochemical factors were monitored, including the fate of the added pollutant.

As an example, Fig. 2 shows the inhibition of primary production per unit of phytoplankton biomass (measured as chlorophyll) after addition of 5 $\mu$g Hg $l^{-1}$ in an experiment (during POSER) in the Rosfjord, Norway (Kuiper et al. 1983). The development of mercury concentration is also shown. Inhibition of the growth rate of one or more compartments of the enclosed ecosystem often manifests itself as a first effect of pollution. Figure 2 also shows that after a period of time similar production rates are measured in contaminated systems and controls. The resumption of phytoplankton activity may be due to the mercury concentration falling below a critical level, or to conversion of mercury into a different form (speciation), or to adaptation of the community. These results reveal a major advantage of this method: the biogeochemical fate of a chemical (e.g., speciation of heavy metals) can be studied under near-natural conditions *simultaneously* with the biological phenomena in the system including possible ecological effects. Differences in effects of chemicals between ecosystems and laboratory tests can be

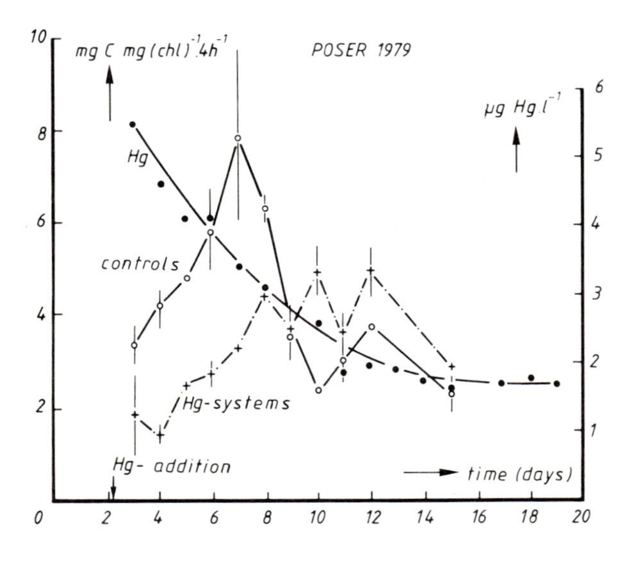


**Fig. 2.** Relative primary productivity in the controls and in MOP's after addition of 5 $\mu$g Hg $l^{-1}$ during POSER. Mercury concentrations in the water are also indicated

caused by differences in geochemical and biological fate (complex formation of metals, biodegradation and photooxidation of organic chemicals, adsorption to suspended particles or sediments, volatization etc.). For example, after addition of mercury (II) chloride to plankton enclosures it was found that, apart from formation of methylmercury, transformation of ionic mercury to the volatile metallic mercury occurred. This disappeared from the system by evaporation (Kuiper 1981a).

For organic pollutants, photooxidation and biodegradation are important processes in the sea. Kuiper and Hanstveit (1987) report on a project in which laboratory die-away biodegradation tests with North Sea water were performed simultaneously with model plankton studies in enclosures. Water from the enclosures was often used in the die-away tests. Many organics were tested, such as substituted phenols, dichloroaniline (DCA), PCB's and a detergent (tetrapropylenebenzenesulphonate, TPBS). For some of these chemicals qualitative degradation results in the laboratory test and in the enclosure agreed reasonably well. This was the case particularly for easily degradable compounds such as 4-chlorophenol, and for very persistent compounds such as PCB's and DCA. Sometimes differences in degradation rates were found (e.g. with phenol) which were probably due to nutrient limitation in the plankton enclosures. Sometimes large differences were found betwen lab and semi-field degradation tests. An example was 4-nitrophenol. This chemical was not degraded in three different enclosure studies lasting 37–49 days. In laboratory experiments, using water from the enclosures, complete degradation was found in 11 to 35 days (Kuiper and Hanstveit 1987). In laboratory degradation tests (in accordance with OECD guidelines), complete degradation was found in a few days. With TPBS similar results were obtained. These differences in degradation rates were probably due to lack of light in the laboratory test and other factors working at the ecosystem level such as competition between different species of bacteria for substrate and inorganic nutrients, competition between bacteria and phytoplankton for inorganic nutrients, etc. The model plankton ecosystem experiments have shown that results from laboratory tests must be interpreted with great care, even if chemicals are found to be "readily degradable" by OECD tests. The importance of a simultaneous study of fate and effects of a pollutant in one test system also becomes clear when the pollutant is converted into more toxic metabolites. In current laboratory tests this phenomenon is not detected, because degradation and toxicity experiments are performed separately. In mesocosms it is easily detected. Figure 3 shows the development of the primary production (measured at a depth of 0.5 m in situ from 10–14 h) in an experiment in which on day 5 1.0 mg $l^{-1}$ 4-chlorophenol (4-CP) was added. After day 20, 4-CP concentrations in the water decreased, and after day 23, concentrations were below the detection limit (0.05 mg $l^{-1}$). Remarkably, after 20 days inhibition of the phytoplankton was at its maximum, as was also apparent from chlorophyll concentrations which decreased and remained low in the enclosure that had received 1 mg 4-CP $l^{-1}$. The inhibition was probably due to intermediates formed during degradation of 4-CP (Kuiper and Hanstveit 1984b). Apart from indirect effects caused by changes of the form of the pollutant, indirect effects can also arise from biological interactions in the ecosystem. Such interactions work on the same trophic level (e.g., competition between different

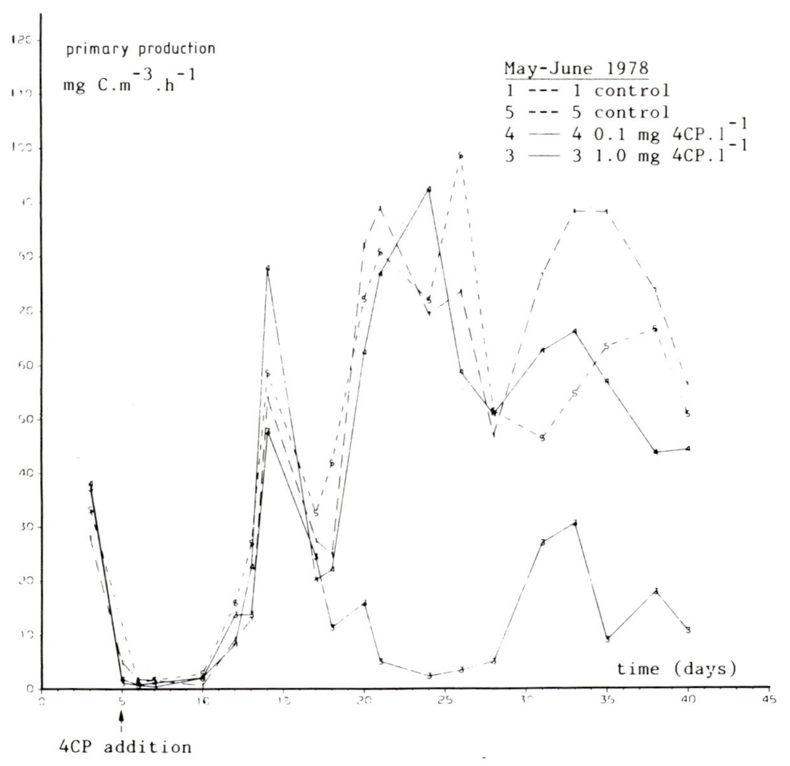


**Fig. 3.** Primary productivity in controls and in MOP's contaminated with 0.1 or 1.0 mg 4 CP $l^{-1}$

species of bacteria or phytoplankton), as well as between trophic apparently small changes, such as the extinction or inhibition of one species, can have large influences on the further development of the total ecosystem via these interactions. Kuiper (1981b) reported a model plankton experiment in which addition of cadmium in concentrations of 1–50 $\mu$g Cd $l^{-1}$ resulted in lower densities of the jelly, *Pleurobrachia pileus,* with respect to the controls. *P. pileus* is a predator of copepods, and due to the differences in population densities of the jellies, the predation pressure on the herbivorous zooplankton differed considerably among the enclosures. This resulted in lower densities of copepods in the controls in comparison with cadmium-contaminated enclosures. High concentrations of cadmium inhibited the copepod development directly, the result being that at one time the controls and the enclosures with the highest cadmium concentration had the same densities of copepods. Differences in species composition of the plankton between controls and contaminated systems were often found. In experiments in which we added a single dose of 0.5 or 5.0 $\mu$g PCB's $l^{-1}$ (Aroclor 1254), higher numbers of $\mu$-flagellates were found in contaminated enclosures than in the controls in the days following the additions, probably as a result of the differences in sensitivity of different species for the PCB's.

Sometimes differences in species composition occurred as a result of interactions in the ecosystem. In nearly all experiments in which addition of contaminants resulted in lower densities of herbivorous zooplankton (copepods), such as after the addition of mercury, 4-CP, PCB's, Nigerian or Forties crude oil, the phytoplankton species composition differed always in such a way that larger species were relatively more important in the contaminated system than in the controls. Figure 4 shows, as an example, the particle size distribution of the phytoplankton on day 40 of an experiment in which DCA was added. Selective grazing of copepods on larger phytoplankton species probably causes these differences in species composition (cf Ryther and Sanders 1980; Steele and Gamble 1982). The studies discussed so far concern general ecotoxicological problems involving the fate and effects of pollutants in plankton ecosystems. The experiments have shown that the fate of pollutants can at best be predicted only qualitatively from the results of laboratory tests, and that contaminants had effects on the plankton at concentrations which were often lower than no-effect levels derived from laboratory studies. Therefore this kind of information is of general interest in setting standards and norms in the marine environment. The mesocosm

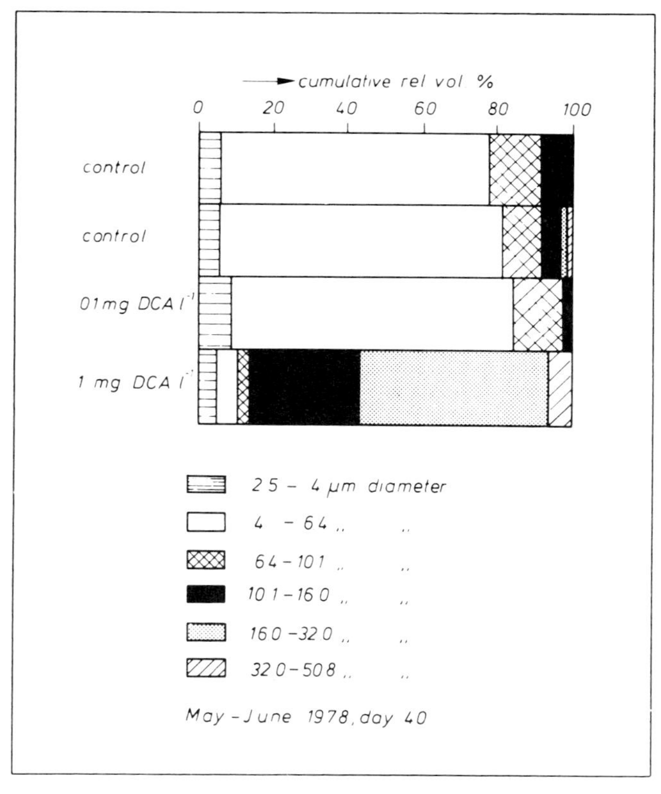


**Fig. 4.** Particle size distribution of the phytoplankton on day 40 of an experiment in which on day 5 single doses of 0.1 or 1.0 mg DCA $l^{-1}$ were added

methodology has also been used to tackle specific pollution problems in the marine environment, including the North Sea. The studies by Horstmann (1972) on the influence of waste water on plankton in the Kiel Bight are an early example of this approach. Here some attention will be given to problems related to the offshore industry in the North Sea. Davies et al. (1981) report on a study in which the fate and effects of an oily effluent from a North Sea production platform (Auk) were tested in MOP's. They showed strong inhibition of the development of the zooplankton at oil concentrations of 5–15 $\mu$g oil $l^{-1}$, likely to occur at a distance of 500–1000 m from the platform. These results help the authorities responsible for the protection of the marine environment to demand extra measures to decrease the oil concentrations in the effluent water (see also Davies et al. 1980). Another example of a study devoted to a particular problem is the MOP's study of Kuiper et al. (1986) in which the ecological effects of a simulated rupture of the pipeline transporting oil from the F3-block in the North Sea were studied. Figure 5 shows the numbers of calanoid copepods in contaminated MOP's and controls. The study showed that F3-oil is a relatively toxic oil in comparison with other oils. Several studies with MOP's were aimed at quantifying the effects of oil combat measures, like spraying with dispersants (Tjessem et al. 1984; Kuiper et al. 1984c). Figure 6 shows, as an example, the development of the zooplankton in MOP's treated with oil or with oil plus dispersant in comparison with the controls. Results indicated that effects of dispersant-treated oil on the plankton are much stronger than those of untreated oil, mainly as a result of the very high oil concentrations in the water column after dispersant use. The general conclusion from this and other studies was that unless large possibilities for dilution are available, or severe effects on other

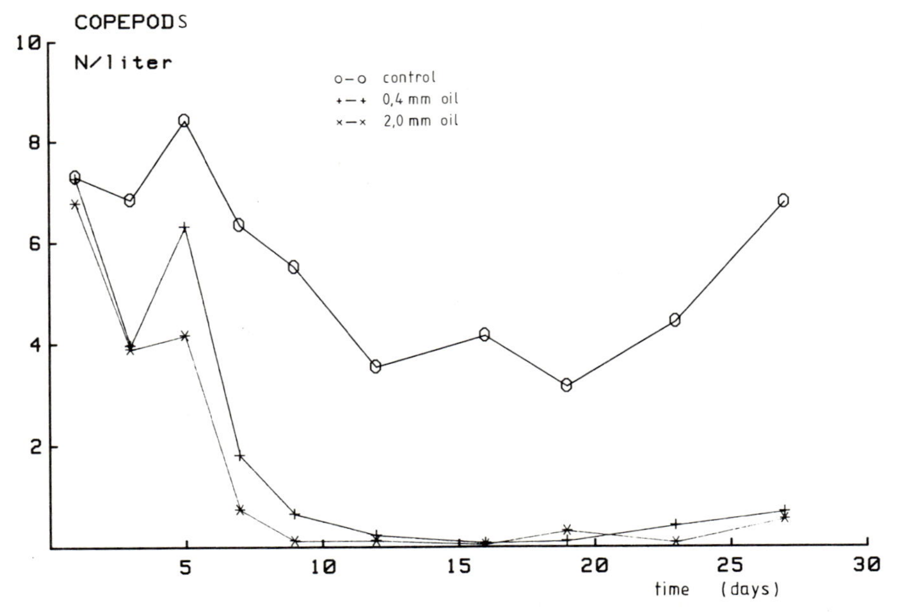


**Fig. 5.** Numbers of calanoid copepods during an experiment in which MOP's were contaminated with a single dose of an oil layer (0.4 or 2.0 mm initial thickness)

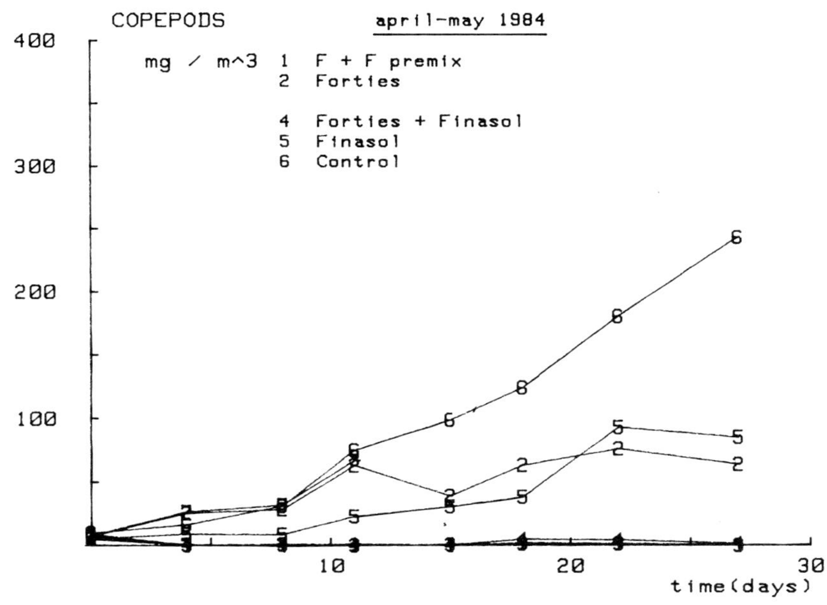


**Fig. 6.** Development of copepod biomass (mg ash-free dry weight) in control and MOP's contaminated with oil (0.5 mm), dispersant, or oil and dispersant (two applications)

important compartments of the ecosystem (damage to birds), or economical damage (beaches with tourists) are expected, the use of a dispersant is not to be recommended on an oil spill in the open sea. The MOP's method can also be used in a variety of other environmental problems to study the effects of dumping of chemicals wastes, harbor sludges, speciation and bioavailability of pollutants, eutrophication, sanitation measures, etc.

## 3 Water-Sediment Ecosystems

De Wilde and Kuipers (1977) were among the first to develop a marine model ecosystem containing water and sediment for basic ecological studies on tidal mudflat ecosystems. In later years their experience was used to construct Model Tidal Flat ecosystems (MOTIF's) to study the fate and effects of oil and oil combat measures in the Wadden Sea (OPEX Oil Pollution Experiment, Kuiper et al. 1984a,b). MOTIF's consist of concrete basins with a mudflat with an area of 18 $m^2$ in which a double tide is simulated. It is a flow-through system directly supplied with water from the Wadden Sea. Macrofauna species are introduced at the start of an experiment in natural densities. A MOTIF experiment usually lasts 1 year during which development of the biota (macrofauna, meiofauna, phytobenthos, bacteria, phyto and zooplankton) is monitored, together with a set of physico-chemical factors, including oil concentrations in water, sediment and biota. After showing in the first 2 years research that the variation between identically treated

MOTIF's was acceptably small and that the MOTIF's structure and function developed very similarly to that of the tidal flat ecosystem in the Wadden Sea, a 3-year project started in 1984 aimed at the study of fate and effects of oil spills and combat measures against these spills in the Wadden Sea. From 1984–1986 the studies were focused on the evaluation of the use of dispersants. To this end, small and large spills of the North Sea crude Forties were treated with the dispersant Finasol OSR5 and its effects (short- and long-term) were compared with Forties or Finasol. The addition of oil and oil dispersants led to an increase in the mortality of several macrofauna species such as *Macoma balthica, Cerastoderma edule* and *Mytilus edulis*. As an example the cumulative mortality of *C. edule* is shown in Fig. 7, measured by counting dead cockles at the sediment surface. The mortality due to the oil treated with dispersant is somewhat stronger than that caused by the oil alone. Addition of dispersant only to the MOTIF's had moderately acute effects, as is also reflected by the cockle mortality in Fig. 7. Interesting is the increased cockle mortality in the oil-treated MOTIF's at the end of the experiment. This high mortality followed frost periods in January and February 1985. Even though the pollution with oil had taken place months before, the stress still persisting, obviously rendered the cockles more sensitive to the cold than the control populations, which did not show this increased mortality after the frost. The main conclusion as to the use of dispersants was that they did not abate the effects of the oil spill in the MOTIF's. In particular, the short-term effects of treated oil were very strong, due to the high oil concentrations in the water column resulting from dispersant application. Based on criteria measured in the OPEX project, application of dispersants in the Wadden Sea is not justified. If factors outside the experimental set-up (e.g. damage to bird populations) are included in the decision-making process, dispersants may be used on special occasions. A second

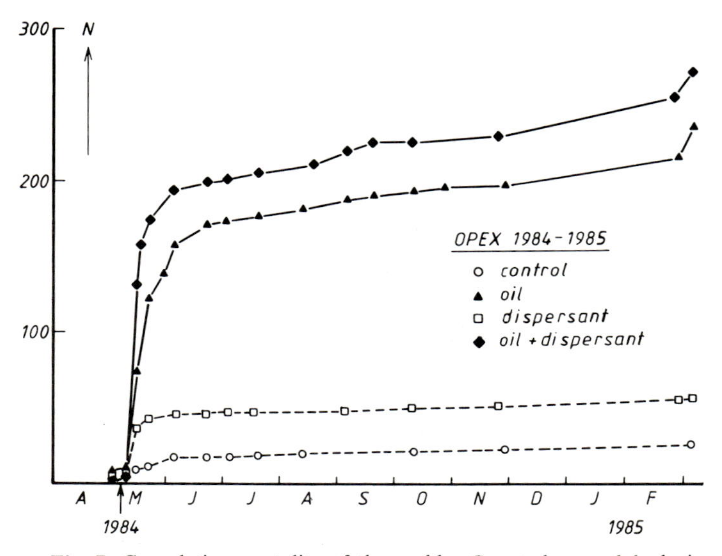


**Fig. 7.** Cumulative mortality of the cockle, *Cerastoderma edule* during OPEX 1984–1985. *Arrow* indicates the addition of the oil and/or dispersant

conclusion, which is of general interest to ecotoxicology, is that sublethal stress by pollutants (e.g., accumulated oil in organisms) on top of sublethal stress by natural factors (extreme temperatures) can be lethal. Figure 8 shows the numbers of *Corophium volutator*, a macrofauna species not experimentally brought into the MOTIF's, but grown from eggs and/or larvae which had entered the MOTIF's with the suppletion water. Data from both experiments are shown. During the first experiment (large spill, 1984–1985) the small numbers present suffered high mortality directly after the addition of the oil, and populations did not recover

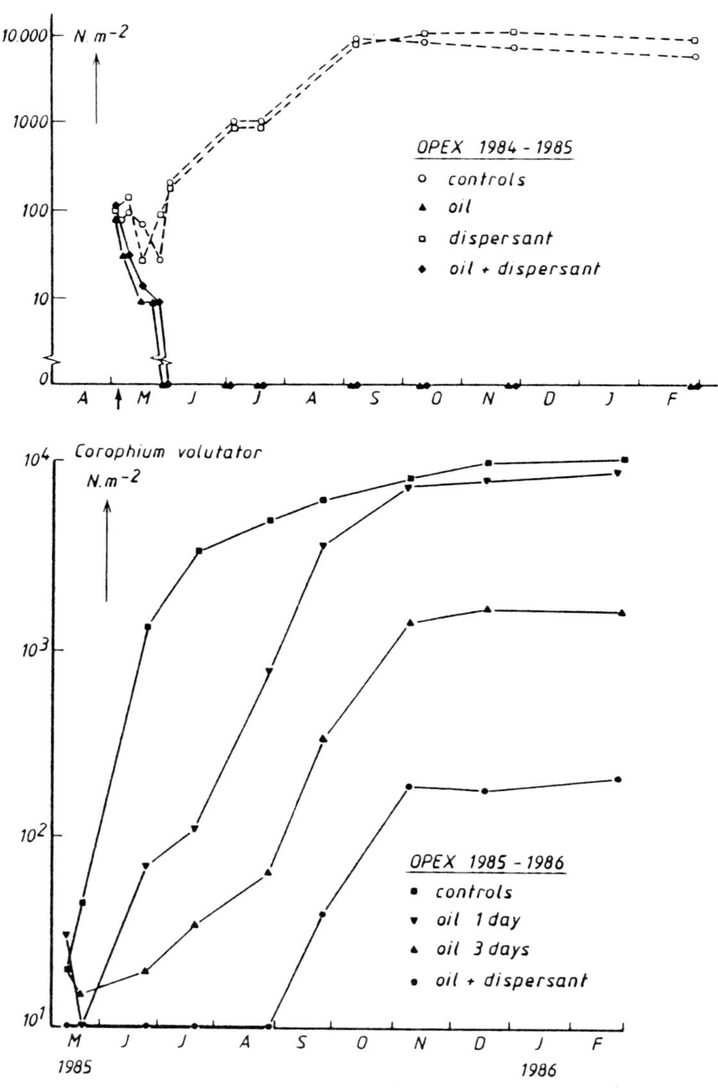


**Fig. 8.** Development of numbers of *Corophium volutator* during two experiments in model tidal flat ecosystems. (OPEX 1984–1986)

during the experiment. In the controls, and also in the MOTIF's treated with dispersant only, high populations built up during the experiment to a density quite natural for the Wadden Sea. During the second experiment (small spill, 1985–1986) populations also collapsed after the oil additions, but they recovered in all systems, first in the MOTIF's only treated with oil for 1 day, then in the MOTIF's treated with oil for 3 days and finally, many months later, in those treated with oil and dispersant. Apart from the importance of studying long-term effects of pollutants, these data also show that model ecosystem studies can provide useful knowledge on indicator species. *C. volutator* is apparently a species which can be used to monitor effects of oil pollution in the field (cf. v. Bernem 1982). Mesocosms were also used by others around the North Sea to study pollution problems (Farke and Günther 1984; Bakke et al. 1982). Gillan et al. (1986) reported on a joint project between Norway and the U.K. involving two different types of benthic mesocosms for studying the degradation rates of oil-based drilling muds (ODM). A second objective was to determine whether addition of nutrients or an adapted bacterial flora stimulated the biodegradation of oil compounds from ODM containing diesel oil or so-called low-tox oil. On the coast of Loch Ewe, Scotland, the benthos was simulated in large flow-through polyester tanks, on the bottom of Loch Ewe large cylinders with lids were pressed into the sediments and contaminated with ODM. Results were compared with those from monitoring studies around oilrigs in the field. It could be shown that part of the oil adhering to drill cuttings was removed by physical processes and biodegradation, both from the land-based tanks and the sea-bed cylinders. Diesel was not degraded faster or slower than low-tox oil and neither addition of nutrients nor of adapted bacteria increased the degradation rates of oil from the tanks. Degradation rates were comparable to those observed around the platforms in the North Sea.

Results showed that these mesocosms can be applied with great success for studying specific environmental problems in the North Sea. In the US the MERL mesocosms, simulating the benthic and pelagic ecosystem of Narragansett Bay, have been used intensively during the last years for environmental and basic ecological and geochemical studies (Grice and Reeve 1982; Oviatt et al. 1984).

## 4 Conclusions and Consequences for Future North Sea Research and Management

In general, mesocosms can play a key role in increasing our basic knowledge of the biological, physical and geochemical aspects of marine ecosystems. Much of the variation inherent in data-sets from field research is lost in mesocosms, without loss of the extrapolation possibilities to the natural system. Collection of data in relevant model ecosystems is very important for the construction *and* validation of descriptive and predictive mathematical models of marine ecosystems. Validated models will gain increasing importance in the management of intensively used seas like the North Sea. Apart from the important contributions to our basic knowledge of marine ecosystems, mesocosms play a major role in different fields of applied science. For risk analysis of new chemicals, laboratory tests (often with single species) have received most attention. There is, however, an increasing amount of evidence, showing that single-species tests alone are not an appropriate basis for estimating environmental risks. Also the OECD (1981) recognized that: "where appreciable environmental concentrations of chemicals are likely to be involved and/or some indication of possible environmental hazard exists, it may be necessary to assess the effects in experimental systems more closely approaching something like natural conditions, especially with regard to interspecific relations and the functioning of multispecies systems". Mesocosms can play an important role by providing a means of bridging the gap between the laboratory and the field. By inclusion of this type of research in testing schemes, mesocosms provide important information for setting standards in relation to pollution of the sea (general standards).

Risk analysis should be a continuous process in which predictions of possible ecological effects in ecosystems are tested by monitoring the environment (Baker 1980). It is also in this field of study that mesocosms can provide valuable contributions in that on the one hand indicator processes and indicator organisms to be monitored can be forwarded, and on the other hand hypotheses from monitoring studies, which remain hypotheses due to the large measurement variations in the field, can be tested under near-natural conditions. As a fourth area of application, mesocosms could be used to study existing environmental problems such as the dumping of drilling muds, waste of titanium dioxide manufacturing and other industrial wastes, sanitation sludges and dredged materials from harbors etc. Mesocosms can also be used to test measures for combating existing or future pollution (dispersants of oil, addition of nutrients to ODM etc.). The main advantage in all these applications is that the biogeochemical fate of contaminants can be studied experimentally, simultaneously with their ecological effects in one experimental system under near-natural conditions. Which mesocosm is best suited to provide answers to certain questions depends on the aims of the investigations and on the questions asked. The "best" model ecosystem does not exist, and this is one reason why there is no point in standardization. A second reason why standardization is undesirable is that "the" marine ecosystem does not exist either. The variation among various ecosystems is very large, resulting in large differences in the fate and effects of pollutants in them.

## References


Baker JM (1980) Ecological impact assessment. Mar Environ Res 3:245–248

Bakke T, Dale T, Thingstad TF (1982) Structural and metabolic responses of a subtidal sediment community to water extracts of oil. Neth J Sea Res 16:524–537

Banse K (1982) Experimental marine ecosystem enclosures in a historical perspective. In: Grice GD, Reeve MR (eds) Marine Mesocosms. Springer, Berlin Heidelberg, New York, pp 11–24

Berland BR, Bonin DJ, Maestrini SY (1975) Isolement in situ d'eau de mer naturelle dans les enceintes de grands volume. Application à l'étude d'une eutrophisation; intérêt et prospective. CNEXO, Rapp Sci Techn 21:2–18

Bernem KH van (1982) Effect of experimental crude oil contamination on abundance, mortality, and resettlement of representative mud flat organisms in the mesohaline area of the Elbe estuary. Neth J Sea Res 16:538–546

Blanck HGD, Dave G, Gustafsson K (1978) An annotated literature survey of methods for determination of effects and fate of pollutants in aquatic environments. Rep Natl Swed Environ Prot Bd 398 p

Bodungen B von, Bröckel K von, Smetacet J, Zeitschel B (1976) The plankton tower. I. A structure to study water/sediment interactions in enclosed water columns. Mar Biol 34:369–372

Brockmann UH, Eberlein K, Junge HD, Trageser H, Trahms KT (1974) Einfache Folientanks zur Planktonuntersuchung in situ. Mar Biol 24:163–166

Brockmann UH, Eberlein K, Hosumbek P, Trageser H, Maier-Reimer E, Schöne HK, Junge HD (1977) The development of a natural plankton population in an outdoor tank with nutrient-poor seawater. I. Phytoplankton succession. Mar Biol 43:1–17

Brockmann UH, Eberlein K, Junge HD, Maier-Reimer E, Siebers D (1979) The development of a natural plankton population in an outdoor tank with nutrient-poor seawater. II. Changes in dissolved carbohydrates and amino acids. Mar Ecol Prog Ser 1:283–291

Brockmann UH, Dahl E, Kuiper J, Kattne G (1983) The concept of POSER (Plankton Observation with Simultaneous Enclosures in Rosfjorden). Mar Ecol Prog Ser 14:1–8

Butler GC (ed) (1978) Principles of ecotoxicology. SCOPE Rep No 12. Wiley, New York, 350 p

Dahl E, Laake M, Tjessem K, Eberlein K, Bøhle B (1983) Effects of Ekofish crude oil on an enclosed planktonic ecosystem. Mar Ecol Prog Ser 14:81–91

Davies JM, Gamble JC (1979) Experiments with large enclosed ecosystems. Philos Trans R Soc Lond Biol Sci 286:523–544

Davies JM, Baird IE, Mossic LC, Hay SJ, Ward AP (1980) Some effects of oil-derived hydrocarbons on a pelagic food web from observations in an enclosed ecosystem and a consideration of their implications for monitoring. Rapp PV Réun Cons Int Explor Mer 179:201–211

Davies JM, Hardy R, MacIntyre AD (1981) Environmental effects of North Sea oil operations. Mar. Pollut Bull 12:412–416

Farke H, Günther CP (1984) Effects of oil and a dispersant on intertidal macrofauna in field experiments with Bremerhaven caissons and in the laboratory, pp 219–235 In: Persoone G, Jaspers E, Claus C (eds) Ecotoxicologial testing for the marine environment, vol 2. State Univ Ghent and Inst Mar Sci Res, Belgium, 584 p

Fisher NS, Carpenter EJ, Remsen CC, Wurster CF (1974) Effects of PCB on interspecific competition in natural and gnotobiotic phytoplankton communities in continuous and batch cultures. Microb Ecol 1:39–50

Gamble JC, Davies JM, Steele JM (1977) Loch Ewe bag experiment, 1974. Bull Mar Sci 27:146–175

Gamble JC, MacLachlas P, Seaton DD (1985) Comparative growth and development of autumn- and spring-spawned Atlantic herring larvae reared in large enclosed ecosystems. Mar Ecol Prog Ser 26:19–33

Giesy JP (ed) (1980) Microcosms in ecological research. Tech Inf Center, US Dep Energy Springfield, 1110 p

Grice GD, Reeve MR (eds) (1982) Marine mesocosms. Springer, Berlin Heidelberg New York, 430 p

Grice GD, Harris RP, Reeve MR, Heinbokel JF, Davis CO (1980) Large-scale enclosed water column ecosystems. An overview of foodweb I, the findal CEPEX experiment. J Mar Biol Assoc UK 60:401–414

Gillan AH, O'Carroll K, Wardell JN (1986) Biodegradation of oil adhering to drill cuttings. In: Proceedings conference on oil-based drilling fluids. Norwegian Petrol Society, Trondheim, February 1986, pp 123–136

Hollibaugh JT, Carruthers AB, Fuhrman JA, Azan F (1980) Cycling of organic nitrogen in marine plankton communities studied in enclosed water columns. Mar Biol 59:15–21

Horstman U (1972) Über den Einfluss von häuslichem Abwasser auf das Plankton in der Kieler Bucht. Keil Meeresforsch 28:178–198

Jordan CF (1981) Do ecosystems exist? Am Nat 118:284–287

Koeman JH, Genderen H van (1972) Tissue levels in animals and effects caused by chlorinated hydrocarbon insecticides, chlorinated biphenyls and mercury in the marine environment along the Netherlands coast, pp 428–435. In: Ruivo M (ed) Marine pollution and sea life. FOA Fish News Books, England, 625 p

Kuiper J (1977) Development of North Sea coastal plankton communities in separate plastic bags under identical conditions. Mar Biol 44:97–107

Kuiper J (1981a) Fate and effects of mercury in marine plankton communities in experimental enclosures. Ecotoxicol Environ Safe 5:106–134

Kuiper J (1981b) Fate and effects of cadmium in marine plankton communities in experimental enclosure. Mar Ecol Prog Ser 6:161–174

Kuiper J (1982) The use of enclosed plankton communities in aquatic ecotoxicology. Thesis Wageningen, 256 p

Kuiper J (1984) Marine ecotoxicological tests: multispecies and model ecosystem experiments. In: Persoone G, Jasperse E, Claus C (eds) Ecotoxicological testing for the marine environment, vol. 1. State Univ Ghent and Inst Mar Sci Res, Belgium, pp 527–588

Kuiper J, Hanstveit AO (1987) Biodegradation rates of xenobiotic compounds in plankton communities. In: Capuzzo JM, Kester DR (eds) Oceanic processes in marine pollution. Vol. 1. Biological processes and wastes in the ocean. Krieger publ comp, Malabar, Florida pp 79–88

Kuiper J, Hanstveit AO (1984a) Fate and effects of 3,4-dichloroaniline (DCA) in marine plankton communities in experimental enclosures. Ecotoxicol Environ Safe 8:34–54

Kuiper J, Hanstveit AO (1984b) Fate and effects of 4-chlorophenol (4-CP) and 2,4-dichlorophenol (DCP) in marine plankton communities in experimental enclosures. Ecotoxicol Environ Safe 8:15–33

Kuiper J, Brockmann UH, Groenewoud H van het, Hoornsman G, Roele P (1983) Effects of mercury on enclosed plankton communities in the Rosfjord during POSER. Mar Ecol Prog Ser 14:93–105

Kuiper J, Hoornsman G, Groenewoud H van het (1984c) The fate and effects of dispersants and dispersant-treated crude oil in marine model ecosystems with different mixing regimes. Report MT-TNO R 84/167, Delft 79 pp

Kuiper J, Wilde P de, Wolff W (1984a) Effects of an oil spill in outdoor model tidal flat ecosystems. Mar Pollut Bull 15:102–106

Kuiper J, Wilde P de, Wolff W (1984b) Oil pollution experiment (OPEX). I. Fate and effects of an oil mousse in a model ecosystems representing a Wadden Sea tidal mudflat, pp 331–359. In: Persoone G, Jaspers E, Claus C (eds) Ecotoxicological testing for the marine environment, vol. 2. State Univ Ghent and Inst Mar Sci Res, Belgium, 584 p

Kuiper J, Groenewoud H van het, Admiraal N, Hoornsman G, Meer M van de, Schulting F, Verkoelen E (1986) Fate and effects of F3-oil in an enclosed Wadden Sea plankton community. Report MT-TNO P86/002, 23 p

Lacaze JC (1974) Ecotoxicology of crude oils and the use of experimental ecosystems. Mar Pollut Bull 5:153–156

Lundy P, Wurster CF, Rowland RG (1984) A two-species marine algal bioassay for detecting aquatic toxicity of chemical pollutants. Water Res 18:187–194

Mann KH (1982) Ecology of coastal waters: a systems approach. Blackwell Scient, Oxford, 322 p

Menzel DW, Case J (1977) Concept and design: controlled ecosystem pollution experiment. Bull Mar Sci 27:1–7

Menzel DW, Steele JH (1978) The application of plastic enclosures to the study of pelagic marine biota. Rapp P V Réun Const Int Explor Mer 173:7–12

NAS (1981) Testing for effects of chemicals on ecosystems. Report of Committee to review methods for ecotoxicology. Natl Acad Press, Washington DC, 103 p

O'Connors HB, Powers CD, Biggs DC, Rowland RG (1978) Polychlorinated bipheyls may alter marine pathways by reducing phytoplankton size and production. Science 201:737–739

OECD (1981) Guidelines for testing chemicals. Organisation for Economic Cooperation and Development, Paris, 800 pp

Øiestad V (1982) Application of enclosures to studies on the early life history of fishes. In: Grice ED, Reeve MR (eds) Marine mesocosms. Springer, Berlin Heidelberg New York, pp 49–62

Oviatt CA, Hust CD, Vargo GA, Kopchynski KW (1981) Simulation of a storm event in marine microcosms. J Mar Res 39:605–626

Parsons TR, Thomas WH, Seibert D, Beers JR, Gillespie P, Bawden C (1977a) The effect of nutrient enrichment on the plankton community in enclosed water columns. Int Rev Gesamten Hydrobiol 62:565–572

Parsons TR, Bröckel K von, Koeller P, Reeve MR, Holm-Hansen O (1977b) The distribution of organic carbon in a marine planktonic food web following nutrient enrichment. J Exp Mar Biol Ecol 26:235–247

Parsons TR, Albright LJ, Whitney F, Wong S, Williams PJLE (1980) The effect of glucose on the productivity of seawater: an experimental approach using controlled aquatic ecosystems. Mar Environ Res 4:229–242

Parsons TR, Harrison PJ, Waters R (1978) An experimental simulation of changes in diatom and flagellate blooms. J Exp Mar Biol Ecol 32:285–294

Patching JW (1981) A study of biodegradation of oil in coastal waters and the effects of oil on microbial activity in the water column. In: EEC, second environmental programme 1976–1980. Reports on the research programme under the second phase 1979–1980. EEC, Brussels, pp 728–732

Pearson WH, Woodruff DL, Sugarman PC, Olla BL (1981) Effects of oiled sediment on predation on the littleneck clam, *Prototheca staminea*, by the Dunganese crab, *Cancer magister*. Estuarine Coastal Shelf Sci 13:445–454

Pettersson H (1939) The plankton shaft. In: Pettersson H, Gross F, Koczy F (eds) Large-scale plankton cultures. Grötesong's K Vetensk, Vitterh Samh Handl Ser 5 B6(13):1–24

Pilson MEQ, Vargo GA, Gearing P, Gearing JN (1977) The marine ecosystems research laboratory: a facility for the investigations of effects and fates of pollutants. Proc 2nd Nat Conf Interagency Energy Environ, R and D Program. Washington DC

Pilson MEQ, Nixon SW (1980) Marine microcosms in ecological research, pp 724–741. In: Giesy GP (ed) Microcosms in ecological research. Tech Inf Center, US Dep Energy, Springfield, 1110 p

Reijnders PJH (1980) On the causes of the decrease in the harbour seal (*Phoca vitulina*) population in the Dutch Wadden Sea. Thesis, Univ Wageningen

Reijnders PJH (1981) Management and conservation in the harbour seal (*Phoca vitulina*) population in the international Wadden Sea area. Biol Conserv 19:213–221

Reijnders PJH (1983) Man-induced environmental factors in relation to fertility changes in Pinnipeds. ICES, Mar Mammals Comm, Ref Mar Environ Qual Comm, CM 1983/N:11 10 p

Ryther JH, Sanders JG (1980) Experimental evidence of zooplankton control of species composition and size distribution of marine phytoplankton. Mar Ecol Prog Ser 3:279–283

Saward D, Stirling A, Topping G (1975) Experimental studies on the effects of copper on a marine food chain. Mar Biol 29:351–361

Sonntag NC, Parsons TR (1979) Mixing an enclosed, 1300 $m^3$ water column: effects on the planktonic food wes. J Plankton Res 1:85–102

Steele JH, Gamble JC (1982) Predator control in enclosures. In: Grice GD, Reeve MR (eds) Marine mesocosms. Springer, Berlin Heidelberg New York, pp 227–238

Strickland JDH (1967) Between beakers and bays. New Sci 2:276–278

Strickland JDH, Terhune LDB (1961) The study of in-situ marine photosynthesis using a large plastic bag. Limnol Oceanogr 6:93–96

Sullivan BK, Reeve MR (1982) Comparison of estimates of the predatory impact of ctenophores by two independent techniques. Mar Biol 68:61–65

Takahashi M, Thomas WH, Siebert DLR, Beers J, Koeller P en, Parsons TR (1975) The replication of biological events in enclosed water columns. Arch Hydrobiol 76:5–23

Tjessem K, Pedersen D, Aaberg A (1984) On environmental fate of a dispersed Ekofisk crude oil in sea-immersed plastic columns. Water Res 18:1129–1136

Ward DV, Howes BL, Ludwig DF (1976) Interactive effects of predation pressure and insecticide (Temefos) toxicity on populations of the marsh fiddler crab *Uca pugnax*. Mar Biol 35:119–126

Wilde PAWJ de, Kuipers BR (1977) A large indoor tidal mud-flat ecosystem. Helgol Wiss Meeresunters 30:334–342

WKP (1986) Water Quality Management Plan North Sea. Ministry of Water works and transport. Directorate North Sea, The Hague

Zeitzschel B (1978) Controlled environment experiments in pollution studies. Proc Oceanol 1978, Tech Sess B (Biol Mar Tech), pp 21–32

# The Role of Biological Monitoring

A.R.D. Stebbing and J.R.W. Harris[1]

## 1 Introduction

If we initiate pollution monitoring programmes, we must not only consider the monitored ecosystem to be worth conserving, but also that by maintaining deliberate awareness of its health we could anticipate and prevent its decline. We cannot expect to prevent organisms from coming into contact with toxic agents, but can try to ensure that the level of toxic stress is within their capacity to cope.

Despite the numerous techniques that are now available for detecting and measuring the biological effects of contaminants (McIntyre and Pearce 1980; Persoone et al. 1984), the emphasis in monitoring programmes has been on chemistry; but it is the biological effect of wastes discharged into the North Sea that is of primary concern, so biological indices of toxic effects should have a central position in any monitoring programme (McIntyre 1984). Purely chemical monitoring presupposes the contaminants likely to be toxic are known; a deficiency clearly demonstrated by the recent problem caused by tributyl tin, whose culpability for deleterious effects in coastal waters has taken over 15 years to establish.

The delayed application of biological techniques in monitoring has meant that chemical and biological approaches have developed independently, and monitoring programmes rarely integrate the two. If the objective is pollution control, it is the interface – the effect of water chemistry upon the biota – that is of primary concern. GESAMP (1980) have addressed the question of formulating an integrated strategy, and more recently an ICES Study Group has developed it further, exploring its possible application to hypothetical problems in the North Sea (ICES 1985).

The purpose of this chapter is to discuss present thinking on monitoring strategies, to suggest an approach to the biological monitoring of the North Sea, to consider the extent to which the existing approach to monitoring already conforms to such a strategy and finally to discuss the ways in which future monitoring programmes might be improved. Our recommended objective is primarily to allow regulation of effluent waste disposal within an environmental capacity, below the maximum sustainable discharge rate. While we acknowledge their importance, it is not our intention to consider aesthetic and conservation aspects of marine pollution, as they inevitably involve subjective judgement of values, proper to a more political arena.

[1]Plymouth Marine Laboratory, Prospect Place, West Hoe, Plymouth, PL1 3DH, Great Britain

## 2 An Appropriate Level of Control

The concept of "environmental capacity" or "assimilative capacity" (Preston 1979; Goldberg 1979, 1981) has been the subject of much attention (Cairns 1977a, 1981; Su and Hung 1982; Pravdic 1985). In principle it is possible to define the capacity of coastal waters to assimilate wastes and effluents without unacceptable harm; however, the balance between the conflicting requirements of environmental protection and waste disposal depends crucially on a subjective assessment of acceptability. This subjectivity is inherent in the social action needed to protect the environment, but it inhibits the definition of scientific goals, particularly in an international context.

Part of an ecosystem's ability to assimilate contaminants without undue damage rests with the attributes of its component organisms. For instance, one important contributor to environmental capacity of turbid estuaries is the biogenic suspended particulate load, which adsorbs contaminants, influencing their subsequent movement and bioavailability (Harris et al. 1984).Clearly contaminants bound to particulates will eventually become deposited in sediments. Binding and complexation of metals by dissolved organic components is a well-understood mechanism of metal detoxification (Lewis et al. 1972; Sunda and Guillard 1976).

Another example involves a single species, a member of the benthic infauna of the Irish Sea, *Maxmulleria lankesteri*. This echiuroid burrows deeply into the sediments, and in doing so transfers plutonium from the surface of the seabed to a depth where it is less likely to be remobilised, or taken up by other organisms (Kershaw et al. 1983, 1984). If this activity is important in reducing the bioavailability of plutonium, water quality should be maintained to protect this species in particular.

By definition, if contaminant concentrations rise too high, deleterious biological effects result. It can be seen that these may themselves reduce environmental capacity, producing a positive feedback which accelerates the process of environmental degradation. To avoid this, it is important to recognise and quantify the contribution of the biota to environmental capacity and set criteria for water quality which protect this contribution. This constitutes a criterion which is intrinsic to the process of waste disposal and quite independent of any clash with fishery, conservational or aesthetic considerations.

It must be clear that are we to utilise the capacity of the North Sea to accept wastes, we must at the very least avoid deleterious effects on those biota which play a part in the degradation and recycling processes, by keeping discharges below a sustainable rate. Such "maximum sustainable discharge rate" provides an objective upper limit to environmental capacity and hence a suitable scientific goal and framework for social decision-making. Once criteria are accepted, it is obviously preferable to calculate the potential effect of any discharge in advance of permission for its release being granted, so that the environment's capacity is not determined by exceeding it.

As many variables determine the maximum sustainable discharge rate, numerical simulation provides an appropriate tool. A certain predictive capability is already within the scope of environmental quality models, which attempt to simulate toxicant inputs, dispersion, chemical speciation and uptake and impact on the biota (Harris et al. 1984; Van Pagee et al. 1985; Klomp et al. 1985; Novak 1985; Taylor 1987).

Fate and distribution models are widely used in pollution control to predict the dispersal of wastes from outfalls, or of specific contaminants into larger water masses from numerous input points (Radford et al. 1981), but the biological component is needed to complete the chain from input to impact. Current biological submodels are relatively unsophisticated, predicting little that is not obvious from the relationship between predicted levels of contaminants and their toxicity. In time they will doubtless be capable of simulating physiological effects of toxic substances and the capacities of organisms to counteract these effects, with more realism.

## 3 An Evolving Strategy for Monitoring

At present, acceptable rates of discharge of a new effluent are typically predicted from laboratory toxicity tests with various organisms and a knowledge of the hydrography of the receiving waters (Lloyd 1984). Procedures of this kind can only provide approximate estimates of likely impact and must rely on a safety margin of dilution that may vary, depending on the perceived values of effluent discharge and the biological resources at risk.

Recognising the uncertainty involved in both the determination and the observance of acceptable discharge levels, a monitoring programme must not only provide an estimation of conditions in the receiving waters, but also a means of verifying that the initial predictions were sound (Portmann and Lloyd 1986). Such monitoring of low level and chronic inputs needs to be applied in the context of an overall strategy that links the various steps between the environment and those who have control over discharges. Such a strategy has been evolving from meetings of GESAMP (1980) and the Study Group of ICES (1985) on the use of biological effects techniques, and has been used elsewhere (Bayne et al. 1985) to provide a framework in which to consider monitoring techniques. However, emphasis is given in this strategy to monitoring as a discontinuous process directed to dealing with water quality problems, failing to address monitoring sensu stricto as a continuous or periodic measurement (Collocott 1971). Thus we feel that greater emphasis be given to "environmental" monitoring (Holdgate 1979) designed to maintain awareness of the continued health of coastal waters, rather than directed to specific polluting events, or discharges. Few monitoring programmes at present are sufficiently "open-ended" in space and time to serve this purpose.

Once problems have been detected by routine monitoring or by some other means, the Identification (Ia) phase involves initial attempts at defining the geographic and temporal scale of the problem. Quantification (Ib) involves determining the extent of the impact on the indigenous biota and deploying various biological techniques to quantify the effect on biological water quality. The next phase, Causation (II), requires that the link be established between the toxic contaminants and the biological effects for which they are responsible, so that Evaluation (III) of the case for Control Action (IV) can be made.

The hierarchical organisation of biological systems affects the choice of monitoring index, and the way in which its measurements are interpreted. Toxic agents are a natural constituent of the marine environment; organisms have evolved the means of both producing (allelochemicals and antibiotics) and of metabolising or sequestering them (MFO systems and metallothioneins).

For this reason indices of toxic effect at lower levels of organisation tend to provide more rapid and more sensitive indices of stress; first because processes at lower levels are faster and second because – in an organisational sense – such indices are closer to the point of toxic impact on biochemical processes. Thus, indices such as the activity of the cytochrome P-450 mono-oxygenase system and lysosomal latency (Moore et al. 1986) provide "early warning" of toxic stress by their sensitivity. Only at higher concentrations or over longer periods will such toxicants have deleterious effects upon organisms, then populations, and finally ecosystems. Effects of toxicants must pass from lower to higher levels of this organisation and if their impact is dissipated at each level, the organisation provides a sequence of buffers to destabilisation, conferring what Bronowski (1970) termed "stratified stability". Consequently low levels of toxin would be expected to have little effect at the organismal level until the capacity of the organism's various mechanisms to detoxify, to sequester and to counteract are exceeded, so that indices based on whole organism responses may be expected to be less sensitive.

The sensitive tests involving the responses of gametes, embryos and larvae may appear an exception to this generalisation, but in effect such life stages of metazoans exist transiently at lower levels of organisation than their parents, and so their greater sensitivity is to be expected. Adaptive responses to stress also provide sensitive indices in that on exposure to toxicants the response increases, while the action of homeostatic control mechanisms tends to minimise and obscure perturbation of the major physiological processes by low levels of toxic stress.

Biological techniques which may be deployed in this true monitoring phase, and in the subsequent quantification of a problem and the establishment of causal links fall into three categories:

1. Surveys and analysis of indigenous communities – pelagic, planktonic and benthic. One major problem in detecting pollution effects at the community level is that the pollution "signal" may be obscured by the "noise" of natural variability. Nevertheless, sensitive methods of analysing community data have been developed and used to detect significant pollution-related change in, for example, benthic communities where it was not as apparent before (Gray 1979; Warwick 1986; Field et al. 1982). Seasonal and spatial variation in the plankton is so great that it appears impracticable to use this group in such a way. Nevertheless, long-term trends in plankton abundance have been tentatively attributed to pollution effects (Glover et al. 1971) of global scale, although long-term climatic patterns are more likely to have accounted for the decline in zooplankton abundance over the 30 years to 1980 in the North Sea, particularly in view of the reversal of this trend since 1980 (Colebrook et al. 1984).

2. Bioassays of water samples taken at various times or depths and covering the area in question. Laboratory-based bioassays of environmental water samples provide a rapid and relatively inexpensive means of estimating water quality. They have the obvious advantage of permitting water quality to be assessed wherever and whenever it is possible to take a water sample, providing a degree of spatial and temporal definition not attainable in any other way, and making them the ideal reconnaisance technique with which to identify "hot spots", or to monitor large areas. Available techniques of this kind that could be incorporated into monitoring programmes were reviewed by Stebbing et al. (1980). In Europe the bioassay approach has been applied primarily to specific problems, such as the effect of algal blooms on water quality (Thain and Watts 1984), the dumping of acidic effluent (Byrne et al. 1985), assessing the impact of an experimental oil slick with and without the use of detergent (Cormack 1982), and on dumping grounds (Lloyd and Thain 1982; Klockner et al. 1985) or as part of research programmes (Bayne et al. 1985; Stebbing 1979; Stebbing et al. 1983). One of the chief constraints of bioassays that depend on early developmental stages is that their availability is likely to be seasonal. Cellular systems which can be grown for indefinite periods, as clones in laboratory culture would seem preferable from this point of view. The nearest current approach to this is the use of microalgae (Jensen 1984).

3. Deployment of caged or transplanted organisms, such as mussels or fish, along the gradient of contamination. Various physiological, biochemical, genetic or other indices of stress can then be applied to subsamples of these populations at intervals.

There are now a number of examples where caged or transplanted organisms have been deployed along known pollution gradients to provide a quantitative measure of biological water quality. Unlike the bioassay techniques, such systems provide an integration of pollution over periods of weeks or months, but they tend to be more expensive. Examples include the use of microalgae in dialysis bags (Jensen et al. 1972; Eide and Jensen 1979) and mussels (Widdows et al. 1980–81; Lack and Johnson 1985). Despite the greater costs at present involved, the integrative nature of these techniques combined with the ability to indicate subcritical doses of contaminants, makes them ideal for the sentinel role (Phase Ia in Fig. 1). Since responses are in general quantitative, they will also contribute to the quantification of any problem as it arises (Phase Ib of Fig. 1).

As mentioned earlier, biological techniques have so far failed to fulfill earlier expectations that many held for them (McIntyre and Pearce 1980), but to depend on chemical monitoring alone presupposes that all the contaminants likely to be of biological significance are known. This is a dangerous assumption when many new, toxic organic compounds find their way into the marine environment each year. Organisms provide an integration in the most appropriate terms of the combined effects of all the contaminants present that contribute to the toxic load, automatically including synergistic or antagonistic effects.

Of course, once significant responses have been detected, analysis of body burdens in transplanted organisms, or of water samples taken for bioassay, is the first step towards establishing causal links. The synthesis of the biological and chemical data in a way that permits effects to be linked to the contaminants responsible, so that causality can be established constitutes the second phase (Fig. 1). Techniques specifically designed to link biological effects to their chemical causes have been proposed (Stebbing 1979; Stebbing et al. 1980), but not yet developed.

Specific biochemical indices, perhaps arising from work with cytochrome P-450 or metallothioneins, remain attractive possibilities, but none are yet operational. In the absence of appropriate techniques for establishing causality experimentally or with specific indices, less rigorous criteria are required. To be demonstrably responsible for an observed effect, a contaminant must be present either in the water or the tissues at concentrations that exceed the threshold for the observed effect in laboratory tests and its distributions in water samples, and body burdens should correlate in space and time with biological effects data.

It has been suggested that relationships between tissue concentrations and their biological effects established in laboratory experiments could provide a means for predicting deleterious effects using body burden data from organisms collected in the environment (Stebbing et al. 1980). Obviously, this approach could

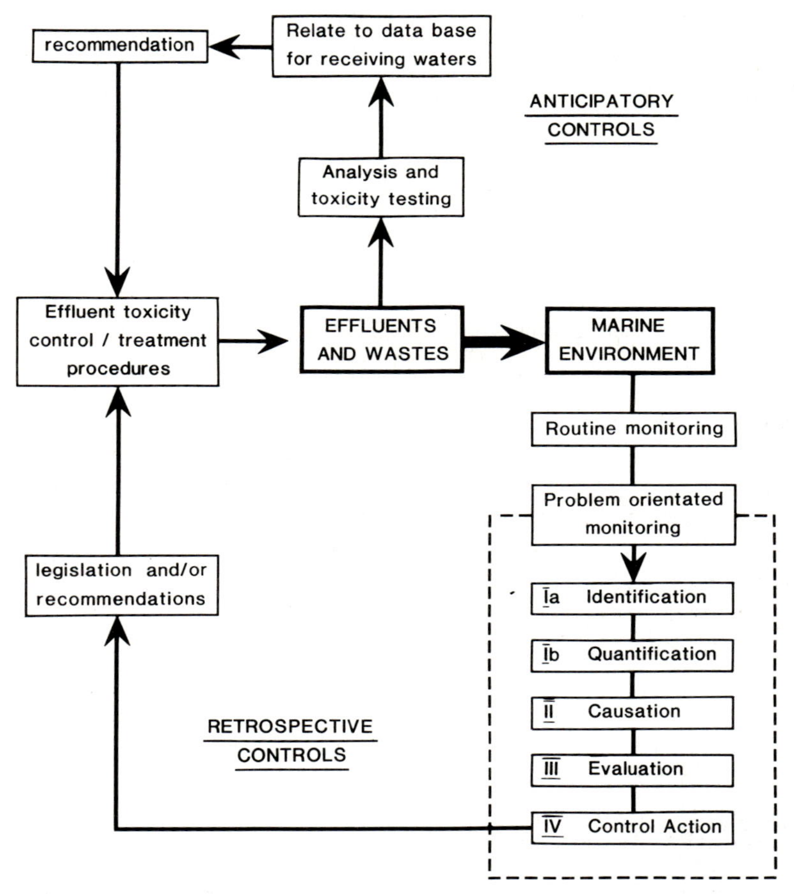


**Fig. 1.** Diagram indicating the flow of information in pollution control systems. For further details see text

be used with either indigenous or transplanted populations. Such relationships have been established between scope for growth and tissue concentrations in *Mytilus*. Furthermore, as there is a tight linear relationship between levels of hydrocarbons in water and those in *Mytilus* tissues (Widdows et al. 1987), it is possible in principle to predict the water concentrations likely to result in a significant depression of scope for growth (Donkin and Widdows 1986). Using the data from the North Sea from Massie et al. (1985), the expected depression in "scope for growth" from levels of hydrocarbons in seawater may be predicted. This approach is a cost-effective method of deriving biological effects data from water and tissue concentrations of hydrocarbons in caged mussel populations deployed around oil rigs in the North Sea (Wilkinson 1982; Somerville and Wilkinson 1987). Now that this principle has been established, it should find application in analogous situations.

It is obviously necessary, to evaluate (Fig. 1, Phase III) not only the consequences of depressed water quality on valued biological resources in the receiving waters, but also what control action is worthwhile. However, the scientific priority in pollution control is to provide a scale of impact measurement and the means to identify the contaminants responsible.

In considering the design of any monitoring programme for the North Sea, it is obvious that some areas are more susceptible than others; not simply because they are closer to points of input, but because of the hydrographic characteristics of the region. One role of simulation models is to predict the times and places where factors combine to reduce or overload environmental capacity, so that the monitoring techniques of phase Ia (Fig. 1) may be concentrated in these regions. For example, strongly stratified waters at a time of high natural primary productivity are susceptible to anoxia and mortality of benthic organisms in response to an influx of effluents with high BOD. Another example of particularly susceptible areas is the centre of gyres which tend to accumulate particulate material (Pingree 1978). These will concentrate contaminants adsorbed to particulates, and there is evidence of depressed water quality at their centres (Stebbing et al. 1987). Such locations are predictable using hydrographic models (Pingree 1978; Uncles 1982), so they can be used to identify where biological monitoring might be most effectively deployed. On a larger scale, hydrographic models of the North Sea which can be used to predict the likely distribution of contaminants with reasonable assurance are already available (Taylor 1987). These models may be expected to improve rapidly in sophistication, to allow the unambiguous identification of sensitive areas on a regional scale. Already areas such as the German Bight appear critical.

## 4 Examples of Biological Monitoring in the North Sea

Fishery statistics are the longest biological time series which relate to the North Sea, and which, while more restricted in intent, monitor an aspect of its ecological well-being (Jones 1982; MAFF 1986). Still associated with the wish to conserve fish stocks, but less subject to the vagaries of the fishing industry, the Continuous Plankton Recorder provides data which may truly be regarded as monitoring changes in the sea's pelagic community. This constitutes one of the longest continuous series of data on a biological community, and reveals previously mentioned trends in plankton abundance since 1945, although this trend has apparently reversed in recent years (Colebrook et al. 1984; Colebrook 1986).

Directed towards other ends, neither of these long-term, large-scale monitoring efforts lends itself to the demonstration of contaminant-induced changes. Monitoring the UK which is directed specifically at detecting the biological effects of pollution has been targeted to the dump sites and the exploitation of oil in the North Sea. In 1982/83 MAFF spent £0.43 million on monitoring the biological effects of dumping wastes at sea (Royal Commission 1984). Most effort has been in surveys of benthic communities, complemented by appropriate chemical, hydrographic and sedimentological data, however fish and shellfish bodyburdens of contaminants and fish disease incidence have been monitored on occasion (Murrey and Norton 1982; Bucke et al. 1983). Bioassay techniques using oyster larvae have been deployed in Liverpool Bay (Lloyd and Thain 1982), and there is obvious scope for their greater exploitation (Klockner et al. 1985). Similarly, community effects (Dicks 1982; Hartley 1982; Davies et al. 1984), monitoring organisms (Somerville et al. 1987) and bioassay (R.R. Stephenson, pers. commun.) have all been proposed and used to assess the impact of drilling for oil on the local biota.

Although a significant depression in environmental quality can be demonstrated by the use of sufficiently sensitive techniques in an area subject to long-term oil contamination (Widdows et al. 1987), and there is a risk of major catastrophies such as those of the *Torrey Canyon* and *Amoco Cadiz*, there is growing agreement that long-term effects of oil exploitation are unlikely (Royal Commission 1981, 1984; Kornberg 1982). Similarly, although dumping practices may exceed "the dispersive capacity" of a dump site, as at the Barrow Deep in the Thames Estuary (Norton et al. 1981), and produce severe ecological damage, effects are generally local. A more significant and insidious threat to the biota of the North Sea is the domestic, agricultural and industrial effluent which enters it from the surrounding land masses, both directly via rivers and their estuaries, and from the atmosphere. The quantities of metals, for instance, that enter the North Sea in this way is an order of magnitude greater than that which enters by dumping (Royal Commission 1984) yet the monitoring effort directed to detecting their effects is derisory.

## 5 Recommendations

1. Environmental capacity should be defined in terms that can be objectively expressed in order to maximise the productivity of international scientific co-operation on its determination.
2. Many organisms contribute to environmental capacity and criteria for water and sediment quality must protect these organisms, if this capacity is to be sustained.
3. It follows from 1 and 2 that the upper limit of environmental capacity be expressed as the maximum sustainable discharge rate.

4. Simulation modelling should be used to estimate this maximum sustainable discharge rate, so that environmental capacity may be determined within this, before consents to discharge are granted.
5. Geographic areas of particular susceptibility should be identified by inter alia numerical simulation techniques.
6. Much greater biological effects monitoring effort should be directed towards detecting the effects of riverine and atmospheric inputs, particularly in these susceptible regions.
7. Biological and chemical monitoring programmes must be integrated in a way that reflects the requirements to establish causal relationships to effect control.
8. A monitoring programme is needed that is not "targeted to" specific inputs, which requires the use of sensitive biological techniques. Only then may we hope to anticipate and prevent irreversible effects on the biota and to maintain the health of the marine environment, so that it can be fully utilised.

*Acknowledgements.* This chapter has benefited from the membership of one of us (ARDS) on the ICES Study Group on Biological Effects Techniques, but the views expressed are our own. We thank Prof I Dundas, who read and improved this chapter. We also appreciate helpful discussions of these matters with our colleagues at PML.

## References

Bayne BL, Brown DA, Burns K, Dixon DR, Ivanovici A, Livingstone DR, Lowe DM, Moore MN, Stebbing ARD, Widdows J (1985) The effects of stress and pollution on marine animals. Praeger, New York

Bronowski J (1970) New concepts in the evolution of complexity. Synthese 21:228–246

Bucke D, Norton MG, Rolfe MS (1983) The field assessment of effects of dumping wastes at sea: 11 Epidermal lesions and abnormalities of fish in the outer Thames Estuary. Fish Res Tech Rep 72:1–16

Byrne CD, Thain JE, Law RJ, Fileman TW (1985) Measurement of the dispersion of liquid industrial waste discharged from ships. 2 studies using U/V flourescence and bioassay techniques. ICES CM 1985/E 25:1–7

Cairns J (1977a) Quantification of biological integrity. In: Ballentine RK, Guarraia LJ (eds) The integrity of water. US Govt, Washington, pp 171–187

Cairns J (1977b) Aquatic ecosystem assimilative capacity Fisheries 2: 5–7 and 24

Cairns J (1981) Discussion of: A critique of assimilative capacity. J Wat Pollut Contr Fed 53:1653–1655

Colebrook JM (1986) Environmental influences on long-term variability in marine plankton. Hydrobiologia 142:309–325

Colebrook JM, Robinson GA, Hunt HG, Roskell J, John AWG, Bottrell HH, Lindley JA, Collins NR, Halliday NC (1984) Continuous plankton records: a possible reversal in the downward trend in the abundance of the plankton of the North Sea and north-east Atlantic. J Cons Int Explor Mer 41:304–306

Collocott TC (ed) (1971) Dictionary of Science and Technology. Chambers, Edinburgh

Cormack D (1982) The use of aircraft for dispersant treatment of oil slicks at sea. Mar Pollut Control Unit, London

Davies JM, Addy JM, Blackman TA, Blanchard JR, Ferbrache JE, Moore DC, Somerville HJ, Whitehead A, Wilkinson T (1984) Environmental effects of the use of oil-based drilling muds in the North Sea. Mar Poll Bull 15:363–370

Dicks B (1982) Monitoring the effects of North Sea platforms. Mar Poll Bull 13:221–227

Dicks B, Hartley JP (1982) The effects of repeated small oil spillages and chronic discharges. Phil Trans R Soc Lond B 297:285–307

Donkin P, Widdows J (1986) Scope for growth as a measurement of environmental pollution and its interpretation using structure-activity relationships. Chem Ind 3 Nov:732–737

Eide I, Jensen A (1979) Application of in situ cage cultures of phytoplankton for monitoring heavy metal pollution in two Norwegian fjords. J Exp Mar Biol Ecol 37:271–286

Field JG, Clarke KR, Warwick RM (1982) A practical strategy for analysing multispecies distribution patterns. Mar Ecol Prog Ser 8:37–52

GESAMP (1980) Monitoring biological variables related to pollution. Rep No 12, UNESCO, Paris

Glover RS, Robinson GA, Colebrook JM (1971) Plankton in the North Atlantic – an example of the problems of analysing variability in the environment. In: Ruivo M (ed) Marine pollution and sea life. Fishing News, London, pp 439–445

Goldberg ED (1981) The oceans as waste space: the argument. Oceanus 24:1–9

Goldberg ED (ed) (1979) Assimilative capacity of US coastal waters for pollutants. NOAA, Washington

Gray JS (1979) Pollution-induced changes in populations. Phil Trans R Soc Lond B 286:545–561

Harris JRW, Bale AJ, Bayne BL, Mantoura RFC, Morris AW, Nelson A, Radford PJ, Uncles RJ, Weston SA, Widdows J (1984) A preliminary model of the dispersal and biological effect of toxins in the Tamar Estuary, England. Ecol Modelling 22:253–284

Hartley JP (1982) Methods for monitoring offshore macrobenthos. Mar Poll Bull 13:150–154

Holdgate MW (1979) A perspective of environmental pollution. CUP, Cambridge

ICES (1985) Report of the study group on biological effects techniques. ICES CM 1985/E:48 1–46

Jensen A (1984) Marine ecotoxicological tests with phytoplankton. In: Persoone G, Jaspers E, Claus C (eds) Ecotoxicological testing in the marine environment Vol 1. State Univ Ghent Inst Mar Sci Res, Belgium, pp 195–213

Jensen A, Rystad B, Skoglund L (1972) The use of dialysis culture in phytoplankton studies. J Exp Mar Biol Ecol 8:241–248

Jones R (1982) Population fluctuations and recruitment in marine populations. Phil Trans R Soc Lond B 297:353–368

Kershaw PJ, Swift DJ, Pentreath RJ, Lovett MB (1983) Plutonium redistribution by biological activity in Irish Sea sediments. Nature (Lond) 306:22–29

Kershaw PJ, Swift DJ, Pentreath RJ, Lovett MB (1984) The incorporation of plutonium, americium and curium into the Irish Sea seabed by biological activity. Sci Tot Environ 40:61–81

Klockner K, Rosenthal H, Willfuhr J (1985) Invertebrate bioassays with North Sea water samples. 1. Structural effects on embryos and larvae of serpulids, oysters and sea urchins. Helgol Meeresunters 39:1–19

Klomp R, van Pagee JA, Glas PCG (1985) An integrated approach to analyse the North Sea ecosystem behaviour in relation to waste disposal. Waterloopkdig Lab Sep No 85/05, The Netherlands

Kornberg H (1982) Oil pollution of the sea: an assessment. Phil Trans R Soc Lond B 297:429–432

Lack TJ, Johnson D (1985) Assessment of the biological effects of sewage sludge at a licensed site off Plymouth. Mar Poll Bull 16:147–152

Lewis AG, Whitfield PH, Ramnarine A (1972) Some particulate and soluble agents affecting the relationship between metal toxicity and organism survival in he calanoid copepod *Euchaeta japonica*. Mar Biol 17:215–221

Lloyd R (1984) Marine ecotoxicological testing in Great Britain. In: Persoone G, Jaspers E, Claus C (eds) Ecotoxicological testing in the marine environment. Vol 1. State Univ Ghent Inst Mar Sci Res, Belgium, pp 39–55

Lloyd R, Thain JE (1982) Use of oyster embryo bioassay to monitor coastal water quality. ICES WGMPNA 5–1/2: 1–6

MAFF (1986) Fishing prospects. MAFF, Lowestoft

Massie LC, Ward AP, Davies JM, Mackie PR (1985) The effects of oil exploration and production in the northern North Sea: Part 1 – the levels of hydrocarbons in water and sediments in selected areas, 1978–1981. Mar Environ Res 15:165–213

McIntyre AD (1984) What happened to biological effects monitoring? Mar Poll Bull 15:391–392

McIntyre AD, Pearce JB (eds) (1980) Biological effects of marine pollution and the problems of monitoring. Rapp P V Réun Cons Int Explor Mer 179:1–346

Moore MN, Lowe DM, Livingstone DR, Dixon DR (1986) Molecular and cellular indices of pollutant effects and their use in environmental impact assessment. Wat Sci Tech 18:223–232

Murray AJ, Norton MG (1982) The field assessment of effects of dumping wastes at sea: 10 Analysis of chemical residues in fish and shellfish from selected coastal regions around England and Wales. Fish Res Tech Rep 69:1–42

Norton MG, Eagle RA, Nunny RS, Rolfe MS, Hardiman PA, Hampson BL (1981) The field assessment of effects of dumping wastes at sea: 8 Sewage sludge dumping in the outer Thames Estuary. Fish Res Tech Rep 62:1–62

Novak R (1985) Pollution clean-up techniques examined with computer simulation. Simulation 45:306–307

Pagee JA van, Gerritsen H, de Ruijter WPM (1985) Transport and water quality modelling in the southern North Sea in relation to coastal pollution research and control. Waterloopkdig Lab, The Netherlands

Persoone G, Jaspers E, Claus C (eds) (1984) Ecological testing for the marine environment. 2 Vols. State Univ Ghent Inst Mar Sci Res, Bredene

Pingree RD (1978) The formation of the Shambles and other banks by tidal stirring of the seas. J Mar Biol Ass UK 58:211–226

Portmann JE, Lloyd D (1986) Safe use of the assimilative capacity of the marine environment for waste disposal – is it feasible? Wat Sci Tech 18:233–244

Pravdic V (1985) Environmental capacity – is a new scientific concept acceptable as a strategy to combat marine pollution? Mar Poll Bull 16:295–296

Preston A (1979) Standards and environmental criteria: the practical application of the results of laboratory experiments and field trials to pollution control. Phil Trans R Soc Lond B 286:611–624

Radford PJ, Uncles RJ, Morris AW (1981) Simulating the impact of technological change on dissolved cadmium distribution in the Severn Estuary. Water Res 15:1045–1052

Royal Commission (1981) Oil pollution of the sea. Eighth report of the Royal Commission on Environmental Pollution. HMSO, Lond

Royal Commission (1984) Tackling pollution – experience and prospects. Tenth report of the Royal Commission on Environmental Pollution. HMSO, Lond

Sommerville HJ, Bennett D, Davenport JN, Holt MS, Lynes A, Mahieu A, McCourt B, Parker JG, Stephenson RR, Watkinson RJ and Wilkinson TG (1987) The environmental effect of produced water from North Sea oil operations. Mar Poll Bull 18:549–588

Stebbing ARD (1979) An experimental approach to the determinants of biological water quality. Phil Trans R Soc Lond B 286:465–481

Stebbing ARD, Akensson B, Calabrese A, Gentile JH, Jensen A, Lloyd R (1980) The role of bioassays in marine pollution monitoring. Rapp P V Réun Cons Int Explor Mer 179:322–332

Stebbing ARD, Cleary JJ, Brinsley M, Goodchild C (1983) Responses of a hydroid to surface water samples from the River Tamar and Plymouth Sound in relation to metal concentrations. J Mar Biol Ass UK 63:695–711

Stebbing ARD, Cleary JJ, Brown L, Rhead M (1987) The problem of relating toxic effects to their chemical causes in waters receiving wastes and effluents. In: Hood DW, Schoener A, Park PK (eds) Scientific monitoring strategies for ocean water disposal Vol 4. Krieger, Melbourne

Su J-C, Hung T-C (eds) (1982) Assimilative capacity of the oceans for man's wastes: proceedings of symposium 26–30 April 1982. SCOPE/ICSU Academia Sin, Taipei, China

Sunda WG, Guillard RRL (1976) The relationship between cupric ion activity and toxicity of copper to phytoplankton. J Mar Res 34: 511–529

Taylor AH (1987) Modelling contaminants in the North Sea. Sci Tot Environ 63:45–67

Thain JE, Watts J (1984) The use of bioassay to measure changes in water quality associated with a bloom of *Gyrodinium aureolum* (Halbert). ICES CM 1984/D:3:1–14

Uncles RJ (1982) Computed and observed residual currents in the Bristol Channel. Oceanol Acta 5:11–20

Warwick RM (1986) A new method for detecting pollution effects on marine macrobenthic communities. Mar Biol 92:557–562

Widdows J, Phelps DK, Galloway W (1980–81) Measurement of physiological condition of mussels transplanted along a pollution gradient in Narraganset Bay. Mar Environ Res 4:181–194

Widdows J, Donkin P, Salkeld P, Evans SV (1987) Measurement of scope for growth and tissue hydrocarbon concentrations of mussels (*Mytilus edulis*) at sites in the vicinity of the Sullom Voe oil terminal: – a case study. In: Kuiper J, Van den Brink WJ (eds) Fate and effects of oil in marine ecosystems. Nijhoff, Dordrecht

Wilkinson T (1982) An environmental programme for offshore oil operations. Chem Ind 20 Feb: 115–123

# Model-Monitoring Relationships

P.J. RADFORD[1]

## 1 Introduction

The concept of monitoring has no meaning except in the context of testing measurements against some model of system performance. The most simple model might be that the system is not expected to change with time. Such an hypothesis is inadequate for an ecosystem which experiences large asymetric seasonal cycles dependent upon forcing functions such as rainfall, solar radiation and wind-induced mixing, each delivered with a large stochastic component. It may well be true that such a system is operating in some kind of dynamic equilibrium, but to measure and monitor significant changes would be virtually impossible, due to natural variability which results in statistical uncertainty. When considering a large extended ecosystem such as the North Sea, the difficulty is compounded by problems of the geographical heterogeneity and the local patchiness of almost all system variables. A good model will explain a large proportion of the observed variance and average over much of the random noise found in the basic data.

## 2 Local Patchiness

Biological variables such as phytoplankton and zooplankton biomass are notorious for the degree of local patchiness they exhibit. The causes may be hydrodynamical, linked to small-scale eddy currents, or biological, related to the shoaling of predators or the feeding behaviour of copepods, but are of little consequence to long-term monitoring. When sampling, it is therefore important to average across local heterogeneity to obtain a meaningful statistic. This may be achieved by taking continuous samples which bridge, or composite samples which mechanically average over these structures. The Continuous Plankton Recorder survey (Colebrook 1960; Robinson and Hiby 1978) has sampled the zooplankton of the North Sea for some 50 years and currently covers a total of 24,000 miles a year along the routes shown in Fig. 1. Local heterogeneity is avoided by analysing samples which represent a 10 nautical (n.) mile track and by averaging results over large geographical areas. The size of these areas has been determined with due consideration of the negative binomial distribution of samples and the precision

[1]IMER, Prospect Place, The Hoe, Plymouth PL1 3DH, United Kingdom

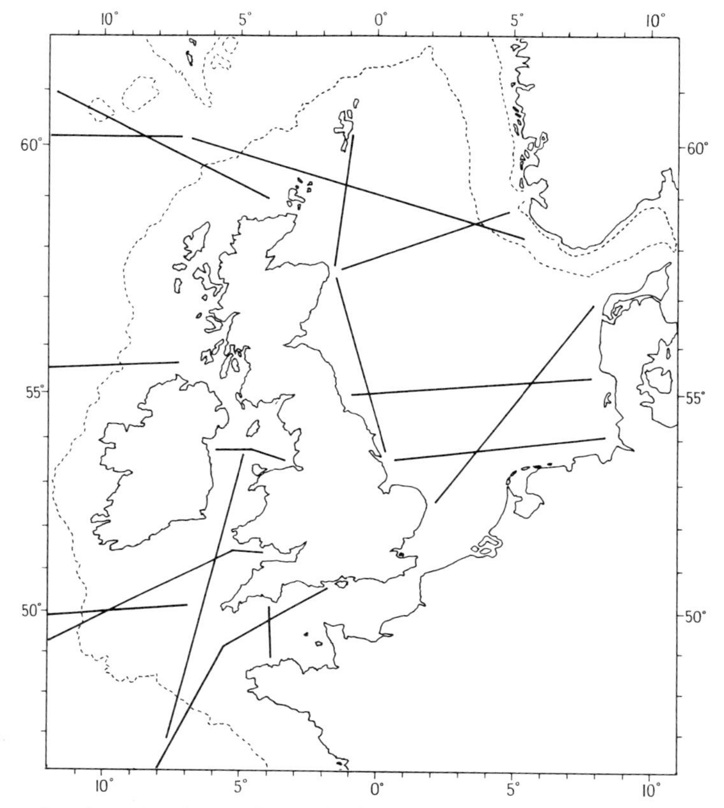


**Fig. 1.** Chart of North Sea and environs showing regular monitoring routes taken by the Continuous Plankton Recorder Survey. (Colebrook 1960)

required for different purposes. Monthly mean values are averaged over $325{,}000^2$ n. miles but meaningful annual averages may be expressed for regions of $3{,}600^2$ n. miles. These empirical results have been found to give adequate signal/noise ratios for detecting significant trends and cycles (Colebrook 1985). If remote sensing techniques are used, the choice of an appropriate grid size is often much more obvious from a close inspection of the complete spacial distribution.

## 3 Tidal Oscillations

Tidal ranges over the North Sea vary from zero at the three amphidromic points to values in excess of 5 m near the coastlines and at the mouths of the larger estuaries. The tidal excursions associated with the higher tidal ranges can be considerable ($>$ 10 km) and these displacements can give added noise to survey data if not taken into account. In the Severn Estuary, this was achieved by the construction of a simple tidal correction model based upon published data of tidal currents (Radford 1983).

Such data are available for the North Sea (US Naval Oceanographic Office 1965) and could be used as the basis for correcting sampling positions to a specific state of tide (e.g. local HW, LW or MW) or to a specific point in time (e.g. HW at Dover). Alternatively, a theoretical hydrodynamic model could serve the same purpose. The disadvantage of making tidal corrections is that samples taken on a regular monitoring grid or sampling route became notionally displaced and chaotic. This is not too devastating, because the next stage, of averaging over regions, is often best done by contouring methods, which in general do not require a regular sampling grid as input data. An example of contoured data set computed from tidally corrected data is given in Fig. 2. The advantages of fitting a contoured surface to data is that it allows for sensible non-linear interpolation between sample points and for a logical method of averaging over any subregion by integration of the fitted surface.

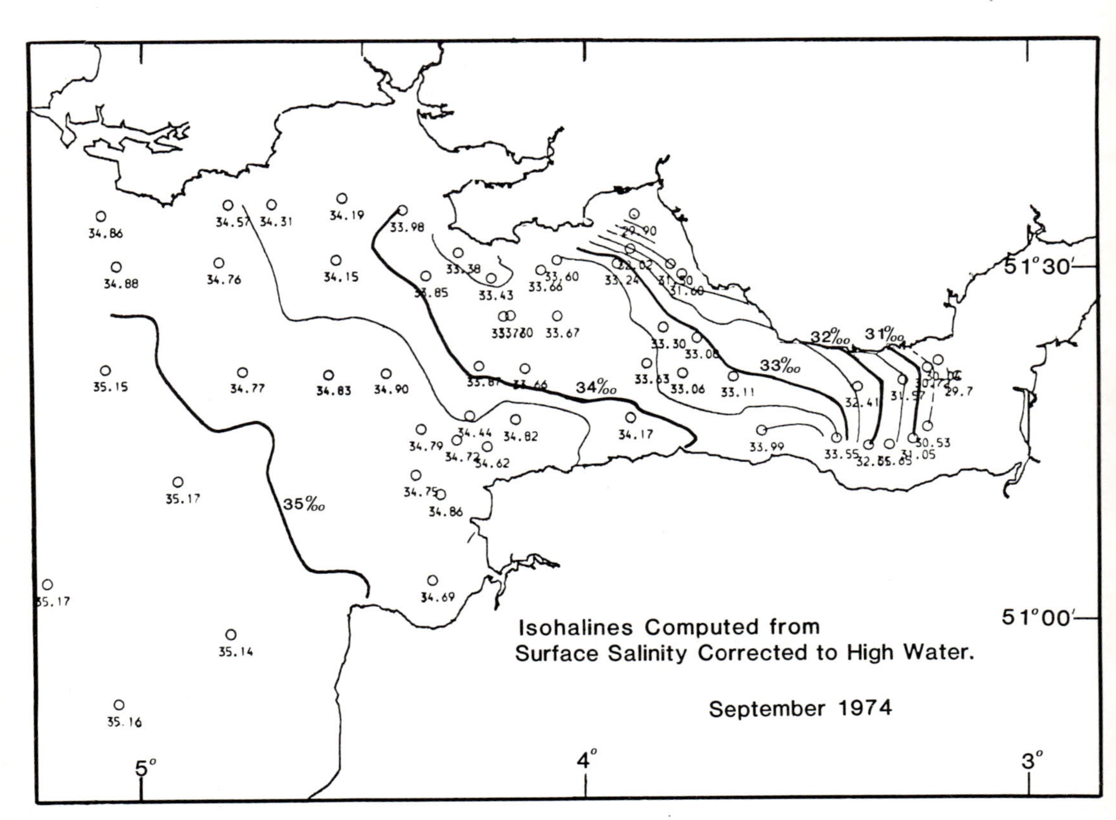


**Fig. 2.** Contour chart of the Bristol Channel fitted to salinity data (‰) sampled on a regular grid but corrected to their positions at the time of high water, Swansea

## 4 River Run-Off

Salinity data averaged over large geographical areas still exhibit large asymetric seasonal fluctuations. For example the tidally averaged salinities of two different regions of the Bristol Channel (Fig. 3) have very different long-term mean values and both show large annual oscillations which make the detection of small but persistent changes very hard to detect. Radford and West (1986) have calculated that for a monitoring scheme, required to detect a 10% change in mean annual salinity, some 46 samples per year would be needed. However, much of the observed variance can be accounted for as the effects of river run-off, a phenomenon that is well understood and regularly measured for most important rivers. A model which explains a large proportion of the observed variance becomes the preferred hypothesis against which monitoring measurements may be tested. In the case of the Severn data, some 80% of the variance was explained by the simple dispersion/advection model which predicted the continuous lines of Fig. 3. There already exists a similar model for the southern North Sea, which explains 65% of the variance of annual mean salinities (Taylor et al. 1983) which is capable of further development. It could undoubtedly be applied with greater precision to smaller regions and over shorter time intervals if linked to an appropriate hydrodynamic model. The ability to predict salinity is of paramount importance because it is a natural tracer of the movement of water masses, and hence an indicator of the advection and dispersion of all dissolved components in both fresh and salt water. To predict salinity with good precision is an essential precurser to the prediction of all dissolved and planktonic components of the water column. For example, in the Severn (Radford and West 1986), the hydrodynamic model used to predict salinity was subsequently adapted to facilitate monitoring of the fate of dissolved and particulate cadmium.

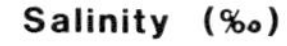


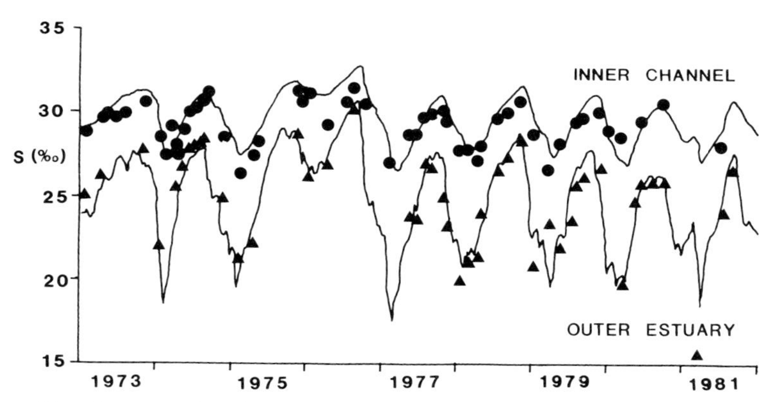


**Fig. 3.** Three years of baseline survey data followed by 6 years of monitoring data of salinity (‰) for two regions of the Bristol channel and Severn Estuary. The *points* indicate data and the *continuous lines* indicate model predictions using GEMBASE in each case

## 5 Temperature

Temperature is a further exogenous variable which induces asymetric seasonal changes in both chemical and biological processes. Although specific water masses often exhibit rather predictable, sinusoidal seasonal variation the tidal and residual mixing of adjacent bodies of water can combine in rather variable ways to produce less uniform distributions. Fortunately these processes are capable of being modelled (Taylor and Stephens 1983), so providing a basis for comparison with future monitoring data. Also the gross effect of temperature on chemical/biological variables will depend upon their local concentrations and/or stage of development. For example the rate of nitrification of ammoniacal nitrogen may be assumed to confirm to first order kinetics, the rate constant being 0.1 per day at 20°C, but changing by 1.7% per °C deviation from 20°C (Gilligan 1972). Similarly, the rate of mortality of faecal coliforms (Ecoli) is strongly dependent upon the temperature and salinity of receiving waters.

## 6 Ecosystem Models

The biological components of ecosystems are strongly influenced by river run-off, temperature, tidal mixing and solar radiation because these exogenous variables interact to provide the moving platform which is their environment. All must be modelled if realistic simulations of the ecosystem are to be made. It is beyond the scope of this chapter to explain how this might be achieved, but examples are given by Radford (1983) and Harris et al. (1984) and others. Here we are concerned with the relationship between model predictions and monitoring data. Irrespective of the structure of a model, it is useful if it can explain a large proportion of the variance observed in field measurements. The General Ecosystem Model of the Bristol Channel and Severn Estuary (GEMBASE) (Radford and Joint 1980) was calibrated against results of a 3-year baseline cruise programme and validated against a 5-year monitoring scheme. Precise details of the design of these two sampling programmes are given by Morris (1983) and the methods of averaging data over sub-regions are explained by Radford (1983). As might be expected, different components of the system are predicted with different precision, but for a number of the key components, sufficient variance is explained by the model to detect real system changes against the background of a highly variable environment.

Salinity results from GEMBASE have been presented in Fig. 3, and as has already been stated, this model explains 80% of the variance of these data. It is therefore much more efficient to monitor salinity changes against the model rather than a simple "no system change" hypothesis. The test of system change then becomes the detection of an unexpectedly large deviation of the monitoring data (a point) from the model (the line). These deviations are plotted for the Outer Estuary Region in Fig. 4b. A control chart technique suggested by Radford and West (1986) is then used to look for abnormal deviations of the model from the

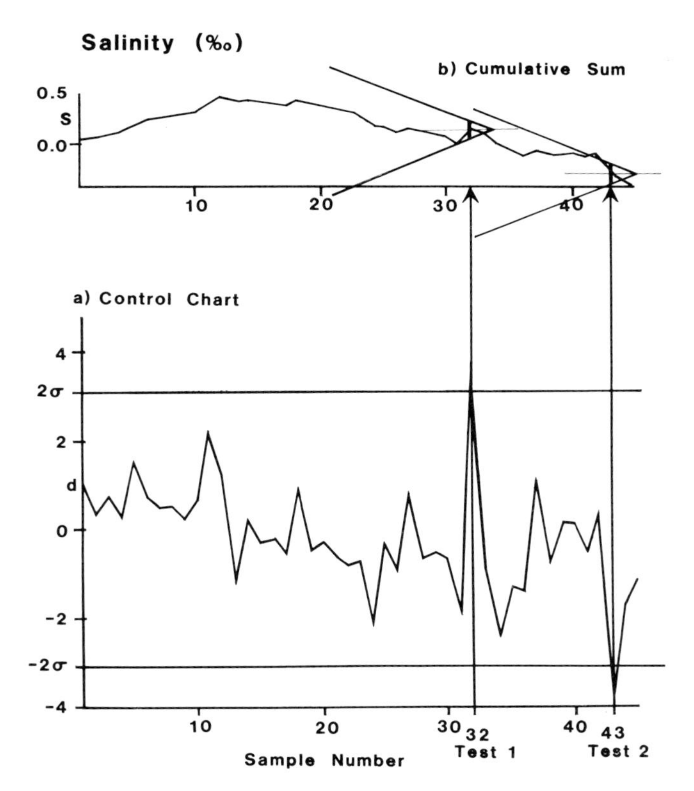


**Fig. 4.** **a** Control chart of the differences between observed and simulated salinities plotted against sample number for the Severn Outer Estuary Region. **b** Cumulative sum of the differences plotted in **a**. The "V" masks indicate significant divergence between model and measurement

monitoring data. Since the deviations only cross the 2$\sigma$ limits for two odd samples out of 45 (i.e. samples No. 32 and No. 43), there is no reason to suppose that the model is not a good representation of the data. A more sensitive test for detecting consistent over- or underestimates of the data by the model is given in the "V" tests (Radford and West 1986) shown on the Cumulative Sum Chart of Fig. 4b. As the "V" is moved from sample to sample (1–31), its arms do not intersect the cumulative sum line. This again indicates that there is no reason to conclude that the model is not an adequate representation of the system as measured by the monitoring data. The slight, almost tangental intersections which occur for samples 32 and 42 again provide insufficient evidence to reject either the model or the data. In contrast the control charts for the Inner Channel Region shown in Fig. 5 indicate significant inconsistencies in these data. The deviations of samples 25, 26 and 28 in the Control Chart indicate the possibility that some data points might be unexpectedly low, but the V tests of Fig. 5b indicate consistent overestimate by the model from the 25th to the 49th (via 35th). In this case, the fault could be traced to the model, which could be corrected for future use. In a similar way, unexpected but consistent changes in the system may be detected.

Phytoplankton results from GEMBASE are simulated with less precision than salinities. Figure 6 presents the logarithm of phytoplankton biomass (gC $m^{-3}$) as measured on the baseline and monitoring cruises and as simulated for the Outer

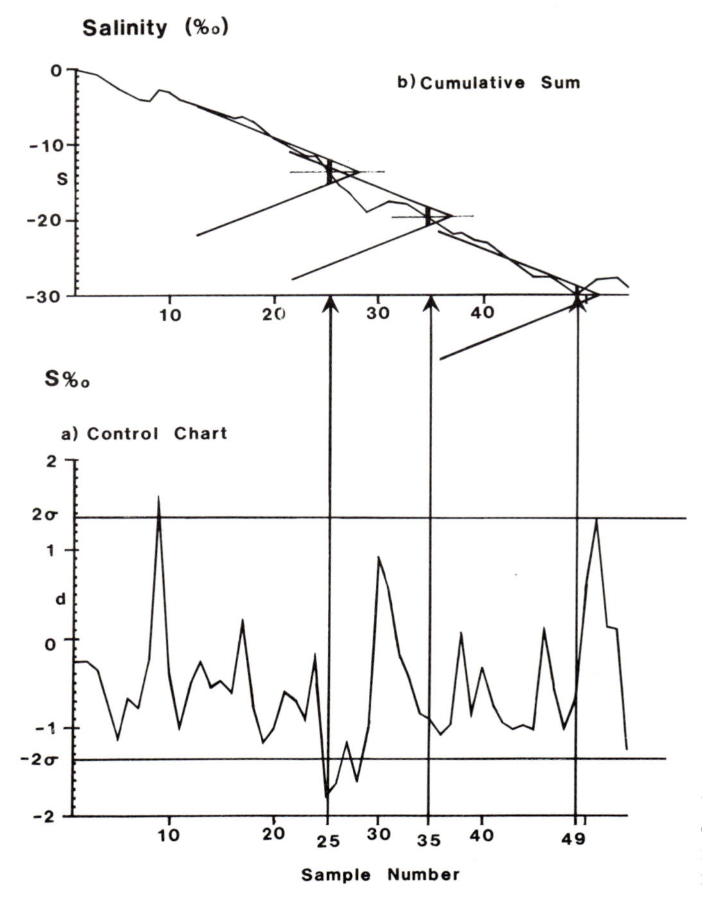


**Fig. 5. a** Control chart of the difference between observed and simulated salinities plotted against sample number for the Severn Inner Channel Region. **b** Cumulative sum of the differences plotted in **a**. The "V" masks indicate a consistent overestimate of the measurements by the model from samples No. 25 to 49

Channel South region only. The linear correlation between observed and modelled phytoplankton for this region is the highest of all regions studied and the regression explains about 60% of the variance. A study of the relevant control charts (Fig. 7) will demonstrate the consistency of the model and data, given this precision of prediction. From Fig. 7b it can be seen that the deviation between model and data remains within $\pm 2\sigma$ limits except for sample No 34. Similarly the "V" test of Fig. 7b encloses the cumulative sum line up until sample No 34, where the upper area of the "V" intersects it. Both of these tests suggest that the model and data are in satisfactory agreement, except perhaps for sample No. 34, which shows a greater than 3 deviation from zero. No satisfactory explanation could be found by a study of the data from that cruise to suggest the cause of this low phytoplankton value. Under a properly constituted monitoring programme, such an anomaly would be investigated immediately it occurred and if necessary a repeat cruise would be arranged to check for a persistent cause.

Some components of the ecosystem, such as carnivorous and omnivorous zooplankton, could not be modelled with adequate precision to make it worthwhile

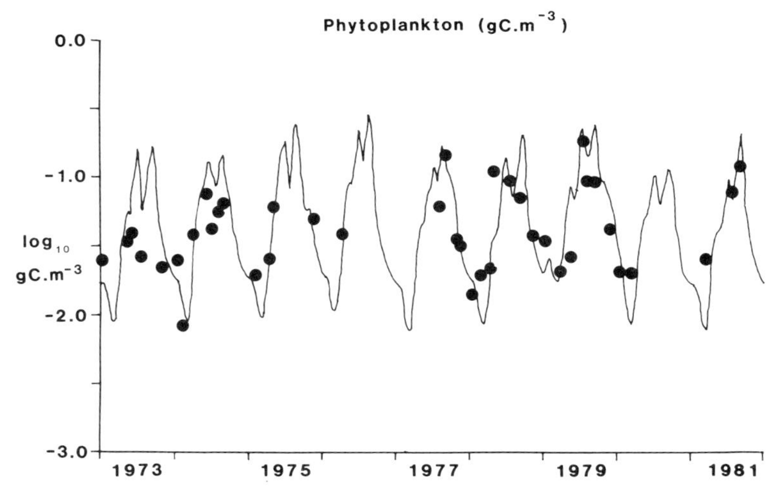


**Fig. 6.** Three years of baseline survey measurements followed by subsequent monitoring data of the logarithm of phytoplankton concentrations (g m$^{-3}$) for the Outer and Southern Region of the Bristol Channel. The *points* indicate observed data and the *continuous line* indicates model prediction using GEMBASE

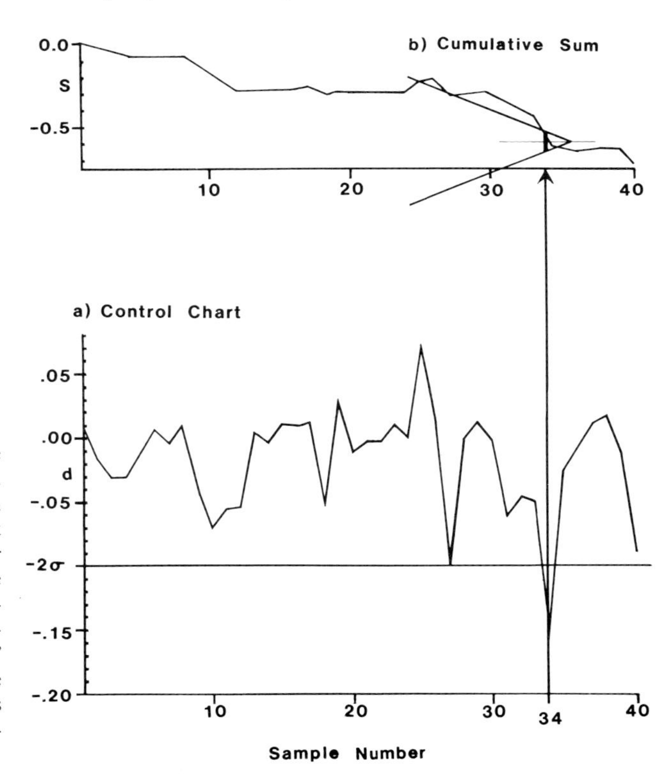


**Fig. 7.** **a** Control chart of the differences between observed and simulated phytoplankton concentration plotted against sample number for the Outer and Southern Region of the Bristol Channel. **b** The cumulative sum of the differences plotted in **a**. The "V" mask indicates that only one monitoring measurement was significantly lower than predicted by the model

to use model predictions as a basis for judging monitoring data. If regressions explain less than 50% of the variance, it is better to consider the deviations of the data from the historic mean concentrations rather than deviations from the model. Even in these situations, control chart techniques are useful in detecting extreme events or consistent trends in the data. As ecosystem models improve, it will become increasingly sensible to use their predictions as a standard against which to interpret the results of monitoring measurements.

## 7 Conclusions and Research Needs

It would be inefficient to embark upon any monitoring programme of the North Sea without postulating a model which could predict expected system performance. Such a model should be capable of explaining the natural seasonal variability of the system in terms of its exogenous, driving variables such as the hydrodynamics, (advection and dispersion), river run-off, temperature and solar radiation. The model and monitoring data should be averaged over relatively large water masses to negate the effect of local heterogeneity and allowance should be made for the variability induced by tidal oscillations. Simulation models are valuable tools for exploring the inherent variability of dissolved and planktonic components of the water column and substrate. When model results are combined with monitoring measurements in Control Charts (Shewhart 1931) and cumulative sum techniques (Woodward and Goldsmith 1964), it is possible to test for anomalies, and on the basis of this knowledge detect significant changes in the system or errors in the model with a degree of objectivity. These methods will efficiently detect large pollution events, but will also reveal more subtle but consistent changes in levels of pollution with minimal sampling. The frequency of monitoring measurements is not critical, provided sampling is often enough to catch large isolated pollution events yet not so frequent that information is duplicated because of the inherent retention time of the water mass.

*Acknowledgements.* The General Ecosystem Model of the Bristol Channel and Severn Estuary (GEMBASE) is a product of the former Estuarine and Near Shore System Ecology Group of the Institute for Marine Environmental Research, a component of the UK Natural Environment Research Council. This work received partial financial support from the UK Department of the Environment under contract numbers DGR 480/48 and PECD7/7/077-139/82.

## References

Colebrook JM (1960) Continuous plankton records: methods of analysis. 1950-1959. Bull Mar Ecol 5:51-64

Colebrook JM (1985) Sea surface temperature and zooplankton North Sea 1948 to 1983. J Cons Int Explor Mer 42:179-185

Gilligan RM (1972) Forecasting the effects of polluting discharges on estuaries. Part III. Chem Ind (Lond) 23:950-958

Harris JRW, Bale AJ, Bayne BL, Mantoura RFC, Morris AW, Nelson LA, Radford PJ, Uncles RJ, Weston SA, Widdows J (1984) A preliminary model of the disposal and biological effect of toxins in the Tamar Estuary England. Ecol Model 22:253–284

Morris AW (1983) Practical procedures for estuarine studies. In: Morris AW (ed) Nat Environ Res Counc, pp 262

Radford PJ (1983) Systems modelling of estuaries. In: Morris AW (ed) Practical Procedures for Estuarine Studies. Nat Environ Res Counc, pp 239–262

Radford PJ, Joint IR (1980) The application of an ecosystem model to the British Channel and Severn Estuary. Wat Pollut Control 244–250

Radford PJ, West J (1986) Models to minimise monitoring. Water Res 20, 8:1059–1066

Robinson GA, Hiby AR (1978) The continuous plankton recorder survey. In: Sowria A (ed) Phytoplankton. UNESCO, pp 59–63

Shewhart WA (1931) The economic control of the quality of manufactured product. Macmillan, New York

Taylor AH, Reid PC, Marsh TJ, Stephens JA, Jonas TD (1983) Year to year changes in the salinity of the southern North Sea 1948–1973: A budget. In: Sundermann J, Lenz W (eds) North Sea dynamics. Springer, Berlin Heidelberg New York, pp 200–219

Taylor AH, Stephens JA (1983) Seasonal and year-to-year changes in the temperatures of the English Channel and the southern North Sea 1961–1976: A budget. Oceanologica Acta 6:1 63–72

US Naval Oceanographic Officer (1965) Oceanographic Atlas of the North Atlantic Ocean. Sect I. Tides and Currents 700–75

Woodward RH, Goldsmith PL (1964) Cumulative sum techniques. Monograph No 3 mathematical and statistical techniques for industry. Oliver and Boyd, Edinburgh, pp 66

# Subject Index